# Manual of Techniques in Invertebrate Pathology

## Second Edition

# Manual of Techniques in Invertebrate Pathology

Second Edition

Edited by

LAWRENCE A. LACEY

IP Consulting International
Yakima, Washington, USA

ACADEMIC PRESS
An imprint of Elsevier Science
Amsterdam • Boston • Heidelberg • London • New York
Oxford • Paris • San Diego • San Francisco • Sydney • Tokyo

Academic Press is an imprint of Elsevier
32 Jamestown Road, London NW1 7BY, UK
225 Wyman Street, Waltham, MA 02451, USA
525 B Street, Suite 1800, San Diego, CA 92101-4495, USA

Second edition



**Cover image credits:** Upper left corner: Japanese beetle, *Popillia japonica*, larva with milky disease caused by the bacterium, *Paenibacillus popilliae.* (Photo by Michael Klein.) Upper left corner: Pre-adult *Romanomermis culicivorax* emerging from larval *Culex quinquefasciatus.* (Photo courtesy of Tokuo Fukuda.) Upper center micrograph: Infective juvenile of *Steinernema carpocapsae.* (Micrograph by S. Patricia Stock.) Bottom right: Electron micrograph of *Lymantria dispar* nucleopolyhedrovirus occlusion bodies. (Micrograph by Regina Kleespies.) Bottom left: Light micrograph of larval fat body of *Spodoptera littoralis* infected by nucleopolyhedrovirus. (Micrograph by A.M. Huger.) Bottom center: Forest locust, *Schistocerca* sp., infected with *Cordyceps* sp. (Photo courtesy of Harry Evans.)



**British Library Cataloguing-in-Publication Data**
A catalogue record for this book is available from the British Library

**Library of Congress Cataloging-in-Publication Data**
A catalog record for this book is available from the Library of Congress

ISBN : 978-0-12-386899-2

For information on all Academic Press publications
visit our website at elsevierdirect.com

Typeset by TNQ Books and Journals Pvt Ltd.
www.tnq.co.in

Printed and bound by CPI Group (UK) Ltd, Croydon, CR0 4YY

Transferred to digital print 2012

# Dedication

The second edition of the *Manual of Techniques in Invertebrate Pathology* is dedicated to Donald W. Roberts in recognition of the enormous impact he has had on insect pathology through his research as well as the wide variety of mycological studies, from basic to applied, carried out by his students, post-doctoral fellows, and visiting scientists in his laboratory over the past 50 years. Of his many achievements, Don considers his most significant contribution to be the mentoring of students and post doctoral scientists from several countries, many of which have gone on to successful scientific careers.

His several scientific achievements have been documented in 269 publications from 1966 through to the present. Most significant is his career-long research on *Metarhizium anisopliae*: its toxins, microbial control efficacy, formulation, and environmental stress factors affecting survival and entomopathogenic activity. After earning his PhD under Prof. Ed Steinhaus, Dr. Roberts was awarded an NSF Postdoctoral Fellowship (1964–1965) to work at the Swiss Federal Institute of Technology in Zurich, Switzerland with Prof. George Benz. Following his studies there, Don worked for the Boyce Thompson Institute for Plant Research (BTI) from 1965 until his retirement in 1997. His research at BTI included basic and applied studies on several fungi for control of crop pests and mosquitoes. In addition to several research projects in the United States during his tenure at BTI, Don conducted collaborative research with numerous scientists in a multitude of countries on every continent except Antarctica. He has worked for extended periods of time on projects in India, the Philippines, and Brazil. Upon leaving BTI, Don began his "retirement" as a Research Professor at the Department of Biology, Utah State University. His research, mentoring, and international activities have continued ever since.

Don is an active member of several scientific societies including the Entomological Society of America (ESA), Sociedade de Entomológica do Brasil (SEB) and the Society for Invertebrate Pathology (SIP). He is a Founding Member of SIP and has remained one of the Society's most dedicated members since its inception in 1968. He has served SIP as Treasurer, Vice-President and President from 1986 to 1990. His many awards include the SIP Founders'

Lecture speaker and honoree; Honorary Member of SEB, their highest honor; and the ESA L.O. Howard Distinguished Achievement Award.

To present Dr. Roberts' career achievements in complete detail would require the addition of another chapter to this book. For more detail, the reader is referred to the SIP Founders' Lecture presented by Raymond St Leger in 2010: (Society for Invertebrate Pathology 2009 Founders' Lecture, Donald W. Roberts—50 Years of leadership in insect pathology. *J. Invertebr. Pathol.* **105,** 211–219).

# Contributors

**James J. Becnel** *Center for Medical, Agricultural, and Veterinary Entomology, United States Department of Agriculture, Agricultural Research Service, 1600 S.W. 23rd Drive, Gainesville, FL 32608, USA*

**Karolin E. Eberle** *Institute for Biological Control, Julius Kühn-Institut, Federal Research Centre for Cultivated Plants, Heinrichstr. 243, 64287 Darmstadt, Germany*

**Jørgen Eilenberg** *University of Copenhagen, Department of Agriculture and Ecology, Thorvaldsensvej 40, 1871 Frederiksberg C., Denmark*

**Juerg Enkerli** *Agroscope Reckenholz-Tanikon, Reckenholzstrasse 191, Switzerland*

**T.W. Fisher** *University of Arizona, School of Plant Sciences, Tucson, AZ 85721, USA*

**S.F. Garczynski** *USDA-ARS, Yakima Agricultural Research Laboratory, Wapato, WA 98951, USA*

**Travis R. Glare** *Bio-Protection Research Centre, P.O. Box 84, Lincoln University, Lincoln 7647, New Zealand*

**Mark S. Goettel** *Agriculture and Agri-Food Canada Research Centre, Lethbridge, Alberta, Canada T1J 4B1*

**Heidi Goodrich-Blair** *Department of Bacteriology, University of Wisconsin, Microbial Sciences Building, Room 4550, 1550 Linden Drive, Madison, WI 53706, USA*

**Ann E. Hajek** *Cornell University, Department of Entomology, Ithaca, NY 14853–2601, USA*

**Richard A. Humber** *USDA-ARS Biological Integrated Pest Management Research Unit, Robert W. Holley Center for Agriculture and Health, 538 Tower Road, Ithaca, NY 14853-2901, USA*

**G. Douglas Inglis** *Agriculture and Agri-Food Canada Research Centre, Lethbridge, Alberta, Canada T1J 4B1*

**Mark A. Jackson** *United States Department of Agriculture, Agriculture Research Service, Crop Bioprotection Research Unit, National Center for Agricultural Utilization Research (NCAUR), 1815 N. University Street, Peoria, IL 61604, USA*

**Trevor A. Jackson** *AgResearch, Biocontrol, and Biosecurity, Lincoln Research Centre, Lincoln, Canterbury, Private Bag 4749, New Zealand*

**Stefan T. Jaronski** *United States Department of Agriculture, Agriculture Research Service, Pest Management Research Unit, Northern Plains, Agricultural Research Laboratory (NPARL), 1500 N. Central Avenue, Sidney, MT 59270, USA*

**Johannes A. Jehle** *Institute for Biological Control, Julius Kühn-Institut, Federal Research Centre for Cultivated Plants, Heinrichstr. 243, 64287 Darmstadt, Germany*

**Regina G. Kleespies** *Institute for Biological Control, Julius Kühn-Institut, Federal Research Centre for Cultivated Plants, Heinrichstr. 243, 64287 Darmstadt, Germany*

**Michael G. Klein** *Department of Entomology, Ohio State University, OARDC, Wooster, OH 44691, USA*

**Albrecht M. Koppenhöfer** *Department of Entomology, Rutgers University, New Brunswick, NJ 08901, USA*

**Lawrence A. Lacey** *IP Consulting International, Yakima, WA 98908, USA*

**Maureen O'Callaghan** *AgResearch, Private Bag 4749, Christchurch 8140, New Zealand*

**Bernard Papierok** *Institut Pasteur, Collection des Champignons, 25—28, rue du Dr Roux, 75724 Paris Cedex 15, France*

**Joel P. Siegel** *USDA/ARS, Commodity Protection and Quality Unit, San Joaquin Valley Agricultural Sciences Center, Parlier, CA 93648, USA*

**Leellen F. Solter** *Illinois Natural History Survey, Prairie Research Institute, University of Illinois, 1816 S. Oak Street, Champaign, IL 61820, USA*

**S. Patricia Stock** *Department of Entomology, University of Arizona, Forbes Building, Room 410, 1140 E. South Campus Drive, Tucson, AZ 85721-0036, USA*

**Jiri Vávra** *Institute of Parasitology, Academy of Sciences of the Czech Republic and Department of Parasitology, Faculty of Science, University of South Bohemia, Ceske Budejovice, Czech Republic, and Department of Parasitology, Faculty of Science, Charles University, Prague, Czech Republic*

**Jörg T. Wennmann** *Institute for Biological Control, Julius Kühn-Institut, Federal Research Centre for Cultivated Plants, Heinrichstr. 243, 64287 Darmstadt, Germany*

**Michael J. Wilson** *AgResearch, Ruakura Research Centre, East Street, Private Bag 3123, Hamilton 3240, New Zealand*

# Preface to first edition

The beginnings of practical insect pathology can be traced into antiquity to work with beneficial insects. As a discipline, however, it is a fairly young branch of science. There are several accounts of published research in insect pathology in the early literature dating from Bassi's incrimination of *Beauveria bassiana* as a pathogen of the silkworm in 1835, but the field came into its own in the 1940s and 1950s with the development of formal course work in insect pathology and the publication of *Principles of Insect Pathology* (Steinhaus, 1949). Over the past two decades, interest in the use of alternative insecticides due to environmental and human health concerns has stimulated increased efforts in the development of microbial control agents as components of integrated pest management systems.

The literature in book form is a veritable cornucopia of analytical information covering both basic and applied aspects of invertebrate pathology. Atlases and manuals are available that enable the identification of a wide range of insect pathogens. Some of these have also included a limited number of techniques to be used predominantly for preparing entomopathogens for isolation and identification. A large number of techniques for the isolation, identification, production and evaluation of insect pathogens are scattered throughout the literature. However, a single comprehensive manual of techniques in insect pathology has heretofore not been available. When confronted with the need to work on a new pathogen group, those of us trained in a particular area of invertebrate pathology must scour the literature in search of instructions for working with the new organisms.

In this *Manual* an international group of experts have brought together a broad array of techniques for the identification, isolation, propagation/cultivation, bioassay and storage of the major groups of entomopathogens. This *Manual* was designed to provide general and specific background to experienced insect pathologists, biologists and entomologists who are beginning work with pathogen groups that are new to them. It will also be useful as a laboratory manual for courses in insect pathology and biological control and related areas of study. It is our hope that this *Manual* will also provide practical information to other researchers, students, biotechnology personnel, entomologists working in integrated pest management, and government regulators concerned with the more technical side of regulatory issues. Chapters on safety testing of entomopathogens in mammals and complementary techniques for the preparation of entomopathogens and diseased specimens for more detailed study using microscopy and molecular techniques, broaden the subject matter of the *Manual* beyond classical insect pathology. To provide in-depth background to the user, this *Manual* will be an ideal complement to the book, *Insect Pathology* recently published by Tanada and Kaya (1993).

The style of presentation differs somewhat between pathogen groups due to the variety of backgrounds of the authors and diversity of subject matter and also inherent differences between the individual pathogen groups. We have concentrated on the 'how to' aspects of the techniques, but have also tried to provide the reader with an appreciation for why they are used as well as to provide a spectrum of supplemental literature and recipes for media, fixatives and stains. Owing to the extensive possibilities resulting from the diversity of pathogens and their hosts and a finite page limit for the *Manual*, we have had to be somewhat selective in the number of techniques that were covered and the amount of literature that could be referenced.

Lawrence A. Lacey
May 1996
Montpellier, France

Steinhaus, E. (1949) *Principles of Insect Pathology*, McGraw Hill, New York, 757 pp.
Tanada, Y. & H. K. Kaya (1993) *Insect Pathology*, Academic Press, New York, 666 pp.

# Preface to second edition

The field of insect pathology has grown steadily since its inception as a formal academic discipline in the mid-1940s (Steinhaus, 1949). Concomitantly, the applied side of insect pathology has also grown considerably resulting in the commercial development of bacterial, viral, fungal and nematode entomopathogens as microbial control agents. Their use by organic and conventional growers for the microbial control of arthropod pests of food and fiber crops; by forestry personnel for protection of forests; as well as by health-related and abatement agencies for control of insects of medical and veterinary importance, has also increased. They have been employed in all of the major types of biological control: classical, conservation and augmentative (Kaya & Vega, 2012).

Since the publication of the first edition of the *Manual of Techniques in Insect Pathology* (Lacey, 1997), the methods available for the study and implementation of entomopathogens have increased substantially especially with regard to the use of molecular tools. The second edition comprises 15 chapters covering methods for the study of virtually every major group of entomopathogen, as well as methods for discovery, diagnosis and initial identification of entomopathogens, the use of complementary methods for microscopy and an overview of safety testing. The *Manual* is a collaborative work written by 29 scientists originating from 10 countries that are experts in the fields of invertebrate pathology including, but not limited to virology, microbiology, mycology, nematology, biological control and integrated pest management.

The coverage of classical techniques in the *Manual* includes overviews of each group of entomopathogen and the methodology used for their identification, isolation, propagation, quantification, bioassay and preservation. These have been updated to include developments in classic methodology during the past 15 years. Molecular techniques are presented for identification, determination of phylogeny, mode of action and virulence factors, factors responsible for resistance to entomopathogens, recombinant technology and several other aspects of the science of invertebrate pathology. These extend basic techniques with specific procedures proven to be successful for research on entomopathogens.

The intended audience of the *Manual* includes invertebrate pathologists, students, technicians, those working in the biopesticide industry, others conducting research in invertebrate pathology, and those researching and implementing environmentally friendly integrated management of invertebrate pests of food, fiber and ornamental crops, structural pests, forests and vectors of disease producing organisms in humans and domestic animals. In addition to recently developed methodology, the second edition of the *Manual* is also intended to function as a complementary resource for books that comprehensively cover insect pathology (Vega & Kaya, 2012), field study and application of entomopathogens (Lacey & Kaya, 2007) and other texts on related subjects (Grewal *et al.*, 2005; Stock *et al.*, 2009).

I am especially grateful to the authors for their expertise, time, and effort in contributing to the second edition of the *Manual*. We the authors also appreciate the huge number of individuals who contributed literature, unpublished data and observations, graphics, scientific advice and reviews of the chapters. I am indebted to Cynthia Lacey for support and advice. I thank the Elsevier editorial staff, especially Pat Gonzalez, Kristi Gomez, and Caroline Johnson for their management and editorial skills and attention to detail for the production of the second edition of the *Manual.*

Lawrence A. Lacey
October 2011

Grewal, P.S., Ehlers, R. U & Shapiro-Ilan, D. (2005) *Nematodes as Biological Control Agents*. CABI Publishing, Wallingford.

Kaya, H. K. & Vega, F. E. (2012) Scope and basic principles of insect pathology. In *Insect Pathology* (eds F. E. Vega & H. K. Kaya), pp. 1–12. Academic Press, San Diego.

Lacey, L. A. (1997) *Manual of Techniques in Insect Pathology.* Academic Press, London.

Lacey, L. A. & Kaya, H. K. (eds) (2007) *Field Manual of Techniques in Invertebrate Pathology.* Springer, Dordrecht.

Steinhaus, E. A. (1949) *Principles of Insect Pathology.* McGraw Hill, New York.

Stock, S. P. Vandenberg, J., Boemare, N. & Glazer, I. (eds) (2009) *Insect Pathogens: Molecular Approaches and Techniques.* CABI Publishing, Wallingford.

Vega, F. E. & Kaya, H.K. (eds) (2012) *Insect Pathology 2nd ed.* Academic Press, San Diego.

# Contents

CHAPTER I

# Initial handling and diagnosis of diseased invertebrates

LAWRENCE A. LACEY* & LEELLEN F. SOLTER[†]

*IP Consulting International, Yakima, WA 98908, USA
[†]Illinois Natural History Survey, Prairie Research Institute, University of Illinois, 1816 S. Oak Street, Champaign, IL 61820, USA

## 1. INTRODUCTION

Over the past 70 years the discovery and development of new strains and species of microbial agents for the control of invertebrate pests has been largely driven by the increasing need for effective and long-term methods to protect food and fiber resources. The increase in the human population and the concomitant increase in agricultural production and pressure on natural landscapes require that these efforts continue in the 21st century in order to intensify production while safeguarding the environment (Altieri, 1999; Lacey *et al.*, 2001). Some discoveries of microbial control agents have been serendipitous, but, for the most part, successful implementation of microbial control has been the result of systematic surveys and exhaustive investigations on pathogen biology, host–pathogen interactions, and application techniques. This book is intended to be a research tool to advance the discovery and management of invertebrate pathogens that may prove useful for microbial control efforts or must be mitigated in beneficial insects or mass-rearing situations. This chapter will provide general guidelines for the recognition, handling, and initial diagnosis of diseased insects and the identification of major entomopathogen groups. Some of the key terms in insect *pathology* that are used in this chapter are italicized in the text and are defined in Martignoni *et al.* (1984) and in a publically available glossary (Onstad *et al.*, 2006). For an introduction to the principles of the field of insect pathology, the reader is referred to Steinhaus (1949), Tanada & Kaya (1993), and Kaya & Vega (2012).

Insects are associated with a broad diversity of microorganisms in a variety of *symbiotic* relationships including: *commensalism, mutualism,* and *parasitism.* Internal mutualistic organisms are critical to the survival of the *host*, such as *symbionts,* which are found in *mycetocytes* and *mycetomes* of many invertebrate species. Although mutualistic organisms such as the protists associated with termites may be abundant in their insect hosts, they are not by definition pathogenic to the host insect. Entomopathogens, on the other hand, result in a variety of conditions in host insects that are subtly to distinctly unfavorable to the host. *Disease* in insects results from the direct effects of infection and/or from *toxemia*.

An astronomical number of entomopathogens cause *disease* and the number of insect hosts in which to find them is equally enormous. Distinguishing one disease from another based on the *signs* and *symptoms* of the disease, its *etiology*, *pathogenesis*, and other characteristics, is the process of *diagnosis*. Steinhaus (1963a) described the diagnosis of insect diseases as one of the most important and complex branches of pathology. The elements and background of diagnosis of insect disease are presented in detail by Steinhaus & Marsh (1962) and Steinhaus (1963a). Essentially, diagnosis is divided into two main categories: the gathering of facts concerning insect disease and their analysis.

Disease and death in insects is not always an indication of infection with entomopathogens. Information on noninfectious diseases in insects due to non-microbial causes (poisoning, mechanical and physical injuries, and

MANUAL OF TECHNIQUES IN INVERTEBRATE PATHOLOGY
ISBN 9780123868992

nutritional and metabolic diseases) is presented by Steinhaus (1949), several authors in Steinhaus (1963b), and Tanada & Kaya (1993). This chapter focuses on infectious diseases of invertebrates that are caused by pathogens and presents general guidelines for their recognition and initial diagnosis. Greater diagnostic detail using a variety of microscopic techniques as well as microbiological, biochemical, and molecular techniques for the identification of specific pathogens are provided in subsequent chapters.

## 2. COLLECTION OF DISEASED INVERTEBRATES AND ENTOMOPATHOGENS

Both living and dead invertebrates that are patently infected with entomopathogens can be found in virtually every setting inhabited by invertebrates including natural terrestrial and aquatic ecosystems, agroecosystems, and in laboratory and commercial insect colonies. Often, the researcher or technician most familiar with healthy insects first notices the presence of diseased insects. Thus, a visual search of habitats of interest or of insect colonies in the laboratory for insects with atypical appearance is a common means of collecting diseased specimens. While many pathogens are usually present at low levels in insect field populations as *enzootic diseases,* they are most easily discovered during *epizootics* when there is an unusual abundance of diseased insects.

### A. Recognition of diseased insects, mites and mollusks in the field—gross pathology

The recognition of diseased insects in the field or subsequently in the laboratory initially relies on *gross pathology* and presence of *patent* infections. Insects that are *patently infected* with entomopathogens often manifest characteristic *symptoms* and *signs* (*syndrome*) of disease, e.g., striking color changes, luxuriant growth of the pathogen on the outside of the specimen, signs of the *pathogen* or *etiological agent* inside the host that are visible through the cuticle, *dysentery*, peculiar behavior including lack of feeding or unusual position on host plants, tremors, *mummification*, fragility or hardening of the integument, noticeable difference in size, aborted molt or pupation, and other signs and symptoms. In some cases, symptoms of disease may be very subtle or not initially apparent, for example, sublethal effects such as *parasitic castration*, reduced longevity, etc. Some pathogens may be *occult* (see *occult virus*).

Some of the most common aspects of gross pathology are:

#### *1. Color changes*

Color changes not found in healthy members of an insect population are often one of the first symptoms of diseased insects to be noticed. This phenomenon is observed in both living and dead diseased insects. Color changes in living insects due to entomopathogens are usually associated with insects that have semi-transparent to transparent integuments, such as white grubs (Figure 5.1 in Chapter V, and cover) and larval forms of weevils, hymenopteran larvae, and many groups of aquatic insects, most notably the Diptera. The shift in color from that of the normal variation observed in the insect may be subtle to drastically different. Blue iridescence in beetle grubs, mosquito larvae, and certain lepidopteran larvae, for example, is not associated with the color of the healthy larvae and indicates an iridescent virus infection. The iridescence rapidly disappears with the death of the host.

A broad array of color changes are seen in dead insects, ranging from white, gray, red, orange, yellow, blue, and green to brown and black. *Melanized* areas of the insect cuticle (usually in the form of black spots) are often due to immune responses of the host as a result of invading nematodes, fungi, baculoviruses, and microsporidia. Melanization can also result from exposure to scarifying agents such as corn starch.

#### *2. Physical signs of the entomopathogen*

Often the causal agent of the disease can be observed in direct association with the infected insect or cadaver. Fungi, for example, frequently produce luxuriant and colorful growth over the surface of the insect while caterpillars infected with many baculoviruses exhibit characteristic lysis. Infective and reproductive forms of nematodes can be observed in the hemocoel of living and dead hosts (see cover). Virus infections and several species of fungi may be observed through the cuticle of living insects. For example, European corn borer larvae infected with *Beauveria bassiana*, may turn pale pink in color prior to death. In insects with transparent cuticles, certain protist-caused infections may result in

**Figure 1.1** Equipment and materials used for collection and initial processing of aquatic arthropods. A. Paraphernalia for collection and transport of specimens in habitat water; large debris are removed from samples at the collection site. B. Sieving field-collected water and specimens for additional removal of debris and separation of different sized arthropods. C. A one sieve subsample to facilitate closer examination of specimens. D. Funnel for separation of small specimens. E. Observation of mosquito larvae in a spot plate using a dissecting microscope. (Photographs provided by James Becnel.) Please see the color plate section at the back of the book.

*hypertrophied* tissues that are abnormally opaque and/or colored and are observable in insects with transparent cuticles; mosquitoes infected with microsporidia are readily distinguished by the white appearance of fat body tissues observed through the cuticle. The location of infected tissues may also be characteristic for certain entomopathogens and may be visible through the cuticle. Cytoplasmic polyhedroviruses (CPV) infection in black fly larvae is found characteristically in the gastric caeca and the posterior portion of the midgut while nucleopolyhedroviruses (NPV) of mosquito larvae are found throughout the entire midgut.

### *3. Aberrant behavior*

Irritability, unusual dispersal or aggregation, and cessation of feeding in normally voracious insects are behaviors that may indicate presence of an infectious disease in an insect population or colony. Insects infected with certain fungi or viruses often climb to high points on host plants and become attached to the plant just prior to death, a behavior often called 'summit disease'.

### *4. Changes in form and texture*

A variety of physical changes occur in diseased insects, most often after death. The cadavers may become mummified, soft or liquefied, firm, 'cheese-like' or 'bread-like' internally, or leathery and dry. Anatomical abnormalities such as prolapsed rectum, a characteristic sign of certain viral infections and other morphological aberrations, are observed in living, *moribund*, and dead insects.

### *5. Odor*

In certain cases, cadavers may become odiferous. Odors are usually associated with insects when body tissues are liquefied. For example, European foulbrood infection in honey bee larvae is associated with a sour, rotten-meat odor.

## B. Collection and initial handling of specimens

Hand picking of specimens allows selectivity and conservation of space, if that is an important consideration. Collection of large numbers of living insects using standard insect collecting techniques with subsequent screening for diseased individuals in the field or laboratory is another strategy. The same methods of collection used for survey of healthy insect populations are utilized (traps, sweeping, hand picking, aquatic nets and dippers). Figure 1.1 depicts the type of equipment for collection and initial processing of mosquito larvae and other aquatic invertebrates. Various methods and equipment for use in collection and initial processing of terrestrial forest invertebrates are shown in Figure 1.2.

There are two major considerations when collecting invertebrates to detect pathogens: avoiding exposure to UV radiation and excessive heat in the field and in vehicles, and avoiding death and decomposition of the host during transport. To a large extent, any temperature or other maintenance requirements of the pathogen are dependent on the conditions necessary to keep the host alive during the collection period and transport to the laboratory. Insects that are not cannibalistic can be held *en masse* in cages or boxes filled with vegetation or other material to prevent jostling and transported in a cooler/ice chest. It is important to avoid condensation and ice melt in coolers; small insects can easily drown in very small amounts of condensate and vials will leak if submerged in water. Commercial ice packs may be preferable to bags of ice. For microsporidia and viruses, portable liquid nitrogen dewars can be used to freeze and store collected insects in the field.

Dead or moribund insects, which may have succumbed to a pathogen, should be collected separately into sealable bags (for a very short period of time to avoid undue saprobic fungal and bacterial growth) held in a separate cooler so that the pathogens are not transmitted to living larvae. Other than quick (chilled) transport of dead insects or purified pathogens, sealed containers should not be used.

In the field, diseased terrestrial insects should be removed from the substrate upon which they are found with fine forceps and placed individually, if possible, in clean dry containers. To avoid damaging cadavers that are tightly attached to the substrate, the portion of the host plant upon which they are fixed should be collected with the insect attached. Minute insects, such as scales and whiteflies, can also be collected in this manner. Containers for collecting live hosts may include any type of cage that prevents escape, including plastic vials with pinholes in the lids or with cotton plugs for air circulation and prevention of condensation. The addition of a drying agent, such as silica gel, to the container used for temporary storage will slow or prevent germination of entomopathogenic fungi and bacteria, and help eliminate the growth of saprobic fungi on specimens (Figure 1.3). Aquatic invertebrates and specimens containing nematodes and certain protistan *parasites* should not be allowed to dry. Fragile insect bodies, especially those from aquatic habitats that can be dried, should be placed on filter paper or a microscope slide and allowed to dry or placed in a drop of 4% formalin in a vial (Weiser & Briggs, 1971). Once insects are returned to the laboratory, they may be refrigerated, frozen or maintained in a dried or aqueous state at room temperature depending on biological requirements. Researchers should be familiar with the appropriate storage methods for the host and pathogen taxa under study. These are presented for virus, bacteria, fungi, microsporidia, and nematodes in subsequent chapters.

*Patently infected* living insects should be kept in the medium in which they are found (i.e. on foliage, in soil or water) and transported to the laboratory as soon as possible. Aquatic insects may also be transported on damp aquatic plants (see Chapters IV and XI). Cool temperatures will help reduce *stress* on the organisms and retard unwanted microbial growth until they can be examined. When an insect is suspected of being infected with an entomopathogen, it should be examined as soon as possible after collection. The invasion of dead specimens by fast growing saprobic organisms may complicate the determination of the true *etiological agent.*

Healthy insects should be also collected for comparative purposes and held separately under conditions that enable good survival. Where diseased individuals are rare or infections are unapparent, large numbers of apparently healthy individuals can also be used for rearing under conditions that are less favorable for survival. Rearing in stressful conditions may accelerate the *incubation* period and the development of pathogens that are present at low levels, occult, or in

**Figure 1.2** Research methods for forest invertebrate pathology. A. Trees are banded with burlap to collect foliage feeding Lepidoptera and other tree dwelling arthropods. B. Beat sheets are useful for collecting foliage-feeding insects. C. Tree branches are infested with infected and non-infected foliage-feeding hosts and enclosed in fabric bags to study disease transmission. D. Microsporidia-infected gypsy moth larva reared for release in a classical biological control program. E. Silk glands of a lepidopteran larvae infected with a microsporidium. F. Isolation of pathogen infective units in a Ludox® HS-40 gradient. (Photographs by LFS and Gernot Hoch.) Please see the color plate section at the back of the book.

an *eclipse period* at the time of collection. Stressing insects by crowding, starving, or other conditions may result in an overt infection through the *induction* of an occult virus or the appearance of other diseases that might not be apparent in the field (Steinhaus, 1958). However, Weiser & Briggs (1971) caution that crowding and/or inclusion of excess food may cause asphyxia or stimulate the development of facultative or otherwise saprobic bacteria. In addition, some species of insects become aggressive or even cannibalistic when crowded. Large numbers of apparently healthy larvae also may be mass processed by trituration followed by differential centrifugation to concentrate and detect low prevalence pathogens either microscopically or using molecular methods.

## C. Extracting entomopathogens from natural habitats

It is possible to collect entomopathogens without ever needing to find a diseased natural host. The use of baiting with *surrogate* host insects is used commonly for the isolation of entomopathogenic fungi and nematodes. Details on techniques using the wax moth, *Galleria mellonella*, and other insects as bait are presented in

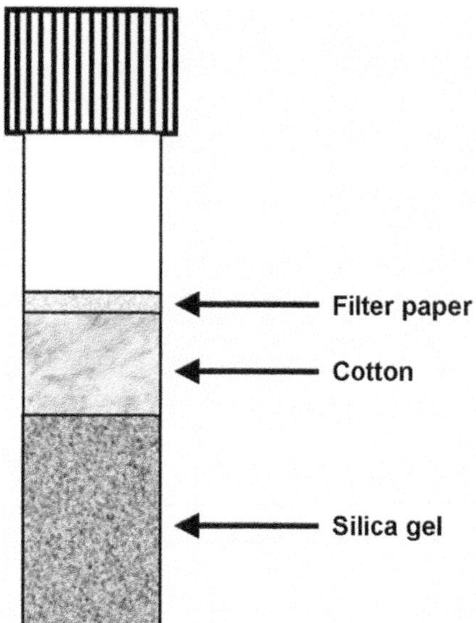


**Figure 1.3** Vial for dry preservation of small insects infected by fungi.

Chapters VII and XII. Selective media and procedures for the isolation of fungi from insect habitats are presented in Chapter VII, and Avery & Undeen (1987) detailed methods for centrifugation of water from mosquito habitats for the isolation of microsporidia and other pathogens.

### D. Recording field data

At the time of collection, detailed information should be recorded regarding:

- the insect (species, stage, gross pathology, aberrant behavior, location in the environment);
- host plant or habitat;
- *prevalence* of disease (did an epizootic occur or were diseased insects less numerous?);
- location, elevation, and climatic conditions (GPS position data is now expected);
- date.

Additional information pertaining to specific pathogen groups is presented in subsequent chapters.

## 3. EXAMINATION IN THE LABORATORY

Diagnosis based on gross pathology alone can be quite misleading if not followed up with microscopic examination. Several classes of signs and symptoms are common to different groups of entomopathogens (color, odor, behavior) or may result from other causes such as chemical insecticides, poor food quality, etc. Details of general laboratory conditions and good laboratory practices (cleanliness, sterile instruments, etc.) that are suitable for examination of diseased insects are presented by Wittig (1963), Weiser & Briggs (1971), and Thomas (1974).

### A. Logging diseased specimens

A system for complete record keeping was devised by Steinhaus & Marsh (1962; also published in Steinhaus, 1963a, and Thomas, 1974) to cover the spectrum of information related to the accession of diseased insects, their examination, and eventually for evaluation and recording of diagnostic conclusions. Along with field data and other collection information (date, collector, number of specimens, etc.), an accession number is assigned to the specimens. The accession number is the means by which the specimens are tracked through the various examinations, disease diagnosis, culturing and bioassays.

### B. Preparation of specimens for initial laboratory examination and/or isolation of suspected pathogens

To enable detailed examination or isolation of the causal agent, the specimen or specimens will usually require further preparation. The following are general procedures for entomopathogens.

#### *1. Surface sterilization*

When a non-contaminated sample of diseased tissues is needed for inoculating media or injecting into healthy insects, surface sterilization of diseased insects will usually be necessary. Figure 1.4 shows a typical set up for surface sterilization of insects. A suggested sequence follows:

1. Place insect in 70% alcohol for a few seconds to facilitate wetting of the specimen.
2. Rinse briefly in distilled water.
3. Place in dilute sodium hypochlorite (NaClO) for 1 min (commercially available household bleach, typically 3–5% NaClO, is diluted with water to the

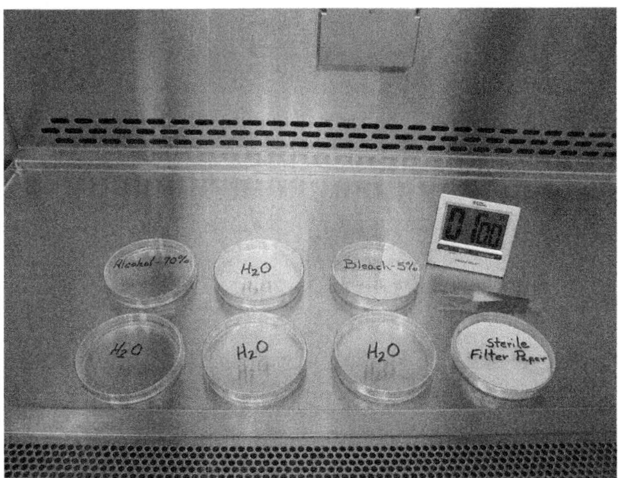


**Figure 1.4** Set up used for surface sterilization of specimens.

appropriate concentration, between 0.5–1% NaClO depending on the size and condition of the insect).

4. Rinse briefly in two to three changes of sterile water.
5. Blot dry with sterile filter paper.

Variations of the above method include more or less time in alcohol, greater or lower concentration of NaClO, or fewer rinses in water. Due to the hydrophobic nature of insect cuticle, the addition of a surfactant such as Tween 80 to the NaClO may increase effectiveness. Alternative substances including Hyamine and Zephiran chloride are also used for surface sterilization (Martignoni & Milstead, 1960; Wittig, 1963).

Surface sterilization of very small insects will kill the entomopathogen inside the host (e.g., whiteflies infected with entomopathogenic fungi). In this case, the removal of spores from the tips of *conidiophores* (see Glossary in Chapter VI) is accomplished by touching the most distal spores to a minute amount of sterile media on the tip of a sterile minuten pin (mounted on a match stick) and then inoculating an appropriate medium in a Petri dish.

### *2. Dissection*

To examine individual organs and tissues, careful dissection of diseased specimens is necessary. Dissecting instruments such as fine-tipped forceps, micro-scalpels, iris scissors, stainless steel minuten pins mounted on match sticks, and the like, are useful for this type of precision dissection. It may be desirable to dissect the insect in a drop of saline solution. Quarter-strength Ringer's solution provides a medium that is more osmotically compatible with tissues and pathogens than water. Recipes for Ringer's solution and other dissecting fluids are presented in Chapter XV.

### *3. Preparation of slides*

#### *a. Unstained slides*

Whole small insects, organs, blood, and other tissues can be mounted on slides in quarter-strength Ringer's solution, water, or other medium for observation using phase contrast microscopy. Drops of regurgitate or diarrhea should also be placed on a slide and covered with a coverslip for observation. When examining wet-mount preparations of various tissues, organs, or other body components, care should be taken to make the preparation as thin as is practical (a one-cell layer thickness is optimum) to assist visualization under phase microscopy. To prevent rapid drying of these preparations, the edges of the coverslip can be sealed with melted paraffin, petroleum jelly, or mineral oil using a small camel-hair brush or cotton-tipped applicator stick.

#### *b. Stained slides*

Tissues and exudates can also be stained with a variety of materials that facilitate coloration of insect cells and entomopathogens. In addition to wet-mount materials, cellular debris and other materials can be spread thinly on the slide using a pair of fine-tipped forceps and allowed to air dry, followed by fixation if required and staining with a differential stain. Stains and procedures for their use for specific pathogen groups are presented in Chapters II, III, VI, XI, XII, and XV.

### *4. Preparation of tissues for histological sections and subsequent molecular studies*

A more detailed examination often will be required to elucidate the *histopathology* and pathogenesis of disease in order to make an accurate diagnosis. The non-occluded viruses, in particular, require examination using transmission electron microscopy (TEM) or polymerase chain reaction (PCR) and DNA sequencing to make identifications. Fixatives and procedures for preparing tissues for light and electron microscopy are presented in Chapter XV. Procedures used in molecular studies of entomopathogenic virus, bacteria, fungi, and nematodes are presented in Chapters II, III, V, VII, IX, XI, XII, and XIII.

*5. Preparation of inoculum for transmission studies*

Satisfying Koch's postulates to confirm the presence of an infectious agent is the final step in the diagnosis process (see Section 3E of this chapter). The preparation of inoculum for such tests may require isolation and culture of the putative pathogen or, when this is not possible for obligate pathogens, purification of inoculum from field-collected or laboratory-infected hosts. In addition, some obligate parasites require production in an *intermediate host.* Procedures for isolation, cultivation/propagation, and determination of the *pathogenicity* and *virulence* of particular microorganisms or nematodes is covered in Chapters II, IV, V, VII, IX, XI, XII, and XIII. General procedures for sterile techniques are covered by Thomas (1974), and in a variety of microbiology manuals.

## C. Microscopic examination

*1. External and internal examination using the dissecting microscope*

Preliminary examination of diseased and healthy specimens with the dissecting microscope (magnification 6–50×) should be conducted in conjunction with dissection. Prior to making an incision in the insect, observe and record any abnormal behavior, signs of regurgitation, dysentery, external lesions or growths, abnormal morphology or coloration, and presence of structures seen through the cuticle that are not present in healthy hosts. Upon incising the host, note any changes in color, size (*atrophy* or *hypertrophy*), or structure (*hypoplasia* or *hyperplasia*) of organs and tissues.

*2. Comparison of healthy and diseased tissues using light microscopy*

A prerequisite for recognizing diseased organs and tissues is to become familiar with corresponding tissues in healthy insects. Diseased tissues may be differently colored, *atrophied* or even missing (*aplasia*), hypertrophied, or undergoing an increase in cells (*hyperplasia*). The alimentary tract (especially the midgut), fat body, Malpighian tubules and blood of diseased insects often exhibit patent signs of disease and are good starting points for comparison with healthy insects. Slides that were made with diseased and healthy tissues and observed using light microscopy (preferably phase contrast microscopy) are compared and contrasted. Note the color, external and internal evidence of etiological agents, or abnormalities in organs.

Phase-contrast microscopy not only provides an excellent means of observing non-stained specimens, but also produces a characteristic *refringence* of certain pathogens such as microsporidian spores. When observed through polarizing filters, uric acid crystals commonly present in the Malpighian tubules of insects are *birefringent* while superficially similar viral *polyhedra* are not birefringent. Fat cells may sometimes be confused with similarly shaped polyhedra. After staining with Sudan III, fat cells become red whereas polyhedra do not take up the stain (see Chapter XV).

*3. Localization of infection*

Specific organs or tissues that are infected (tissue tropism) may be characteristic of a particular disease. For example CPVs are usually restricted to certain portions of the midgut of host insects. The location of pathogens within cells can also be used to distinguish the type of entomopathogen; as their names imply, CPV is found in the cell cytoplasm and NPV infects the cell nuclei.

*4. Size and shape of the causal agent*

The size ranges and shapes of the various pathogen groups are presented in Section 4B below, and in greater detail in the following chapters. In addition to measuring and recording the size and shape of suspected pathogen stages, record the presence of *occlusion bodies* (OBs) within cells and associated crystals and other structures.

## D. Identification of major entomopathogen groups

*1. Key to the major entomopathogen groups*

**1a.** Distinct external growth on insect. 2
**1b.** No external growth on insect. 3
**2a.** Mass of wormlike (non-segmented) organisms over surface of insect or inside its body. Color of cadaver may be different from natural coloration. Usually red or cream to light-brown in color. Many with a distinct unshed second stage integument. NEMATODES

**2b.** Thread-like growth of organisms on surface of insect, often becoming powdery (e.g., white, green, red, etc.) and sometimes limited to intersegmental areas, or growth macroscopic and club-like (see cover). FUNGI

**3a.** Insect usually normal in appearance but may be stunted or slightly malformed; upon dissection body may contain one or more multicellular organisms with many segments and a distinct to indistinct head region with mandibles, possess a tracheal system marked by various types and arrangements of spiracles; in some cases body of organism may protrude through the host's integument or even may be feeding externally on host tissues. PARASITOIDS

**3b.** Insect may be normal or abnormal in appearance but upon dissection does not contain metazoan parasites, may contain wormlike, non-segmented organisms or small, unicellular particles represented by motile or non-motile rod-shaped to spherical organisms or various life-cycle stages of other microorganisms including spores, cysts, OBs or hypha-like structures. 4

**4a.** Aquatic or other insect with essentially transparent cuticle (integument). 5

**4b.** Terrestrial insects or those with basically non-transparent cuticle. 12

**5a.** Insect iridescent or specific tissues (especially fat body) iridescent. 6

**5b.** Insect non-iridescent. 7

**6a.** Insect (scarabaeid grubs) white–bluish to blue–grayish in coloration, containing crystals and minute bacterial-like forms (often pleomorphic in shape) that are only visible by light microscopy. BACTERIA/RICKETTSIAE

**6b.** Insect orange to green to blue in coloration, infectious agent not visible with light microscopy. VIRUSES

**7a.** Hemolymph (blood) as seen through cuticle milky in coloration, rod-shaped, motile cells often with refringent spore giving a footprint appearance under phase microcopy (see Figure 5.1 in Chapter V, and cover). BACTERIA

**7b.** Hemolymph clear, essentially normal in appearance. 8

**8a.** Wormlike, or rapidly motile via cilia, visible through cuticle at low microscope magnification. 9

**8b.** Nematodes or ciliated organisms absent. 10

**9a.** Organisms wormlike and elongate, nonsegmented (see Figures 12.6 and 12.17 in Chapter XII, and cover). NEMATODES

**9b.** Organisms pyriform and motile by cilia (Chapter XI). PROTISTS

**10a.** Intestine (gut) abnormally opaque white, particles polyhedral in shape, visible with phase microscopy in cytoplasm of gut cells (Figure 2.2B in Chapter II). VIRUSES

**10b.** Intestine essentially normal in appearance. 11

**11a.** Abnormal whitish masses in hemocoel associated with various tissues (e.g., fat body) or in hemolymph itself, masses composed of round or ovoid to pyriform spores refringent under phase microscopy (Chapter XI). PROTISTS

**11b.** Hemocoel filled with hyaline hypha-like bodies or thick-walled structures, colorless to variously colored, with smooth to sculptured walls. FUNGI

**12a.** Insect body usually hardened, mummified and cheesy in consistency, filled with hyaline hyphae, hypha-like bodies or spherical to ovoid resting spores (Chapter VI). FUNGI

**12b.** Insect not hardened or cheesy, may be stunted or with malformed body parts. 13

**13a.** Insect flaccid and usually discolored, integument may be fragile. 15

**13b.** Insect may be stunted or malformed, integument usually normal in appearance. 14

**14a.** Various body tissues and cells containing refringent, nonmotile spores or cysts (oval, elongate oval, pyriform or spherical in shape) best visualized by phase contrast microscopy; in some cases the alimentary tract may contain motile, flagellated organisms or relatively large, slow moving, septate organisms often occurring in pairs or chains. PROTISTS

**14b.** Various body tissues and cells containing minute, non-motile, bacterial- to pleomorphic-like cells just visible by light microscopy, cells usually exhibit Brownian movement and are highly refringent, often occurring in pairs or chainlike structures, crystals may or may not be present. BACTERIA/RICKETTSIAE

**15a.** Wormlike, nonsegmented organisms in body tissues which may be liquified and creamy, grayish to reddish in coloration. NEMATODES

**15b.** Nematodes not present in body tissues. 16

**16a.** Insect often with putrid odor, usually brown, black or reddish in coloration, body tissues may be liquified with rod-shaped, motile organisms that may contain refringent spores evident under phase microscopy. BACTERIA (exclusive of Rickettsiae)

**16b.** Insect flaccid and discolored, integument may be very fragile, liquefied body tissues filled with refringent spherical to polyhedral-shaped OBs which may also occur in the cytoplasm or nuclei of intact cells (see Figures 2.1–2.5 in Chapter II), occlusions usually dissolved by a weak basic solution (NaOH or KOH; dead caterpillars may be attached to host plant in inverted v-shaped manner hanging by abdominal prolegs. VIRUSES

A number of other keys are available for the identification of the major pathogen groups (Weiser & Briggs, 1971; Poinar & Thomas, 1984); including keys in Portuguese (Alves, 1998) and Italian (Deseö-Kovács & Rovesti, 1992).

### *2. General characteristics of insect disease caused by the major groups of entomopathogens*

Brief descriptions of signs and symptoms of insect disease caused by the major groups of entomopathogens and additional information are provided below to aid in initial diagnosis. Detailed descriptions of each group are provided in subsequent chapters and by Tanada & Kaya (1993) and in several chapters in Vega & Kaya (2012).

#### *a. Viruses*

Viruses are reported from virtually every insect order and are the smallest of the entomopathogens. Virions of the non-occluded forms range in size from 0.01 to 0.3 μm while the polyhedra and other OBs that occlude the *virions* of the *occluded* viruses range from 1.0 to 15 μm in size. Some of the more virulent viruses produce widespread epizootics resulting in dramatic collapses in host populations. Most of the non-occluded (virions not occluded in a protein matrix) or non-aggregated viruses are not visible under light microscopy. The occluded viruses: granuloviruses (GV), NPV, entomopoxviruses (EPV) and CPV, are the most commonly observed due to the incorporation of the virus particles into a protein matrix which is large enough to be visible under light microscopy. The protein matrix of the OBs of NPV, GV, and CPV dissolve in basic solutions such as 1 N potassium hydroxide (KOH) and sodium hydroxide (NaOH), enabling their distinction from other crystalline structures such as uric acid crystals. While the infected insect is alive, OBs of NPVs or granules of GVs are sometimes shed in drops of diarrhea or regurgitate.

Color changes due to virus infections are observed in both dead and living insects. The blue coloration associated with iridescent viruses may be pale to a deep blue–purple with a distinct iridescence. Less commonly, orange iridescent virus may also be observed in mosquito larvae. Chalky white zones in the midgut and fat body are observed with some viruses. In aquatic Diptera, NPV in mosquitoes and CPV in mosquitoes, black flies and others are distinctly observed in the midgut region as chalky white areas. Color changes in Lepidoptera due to patent infections of NPV and GV may result in a change from normal color to white, gray, or light brown. The normal color of the integument may fade somewhat and the translucent areas of the integument (e.g., the prolegs, ventrum, and cervix) become milky in color due to the presence of OBs in the hemolymph. Insects infected with these viruses often die attached to substrates by the prolegs. At this point the insect is flaccid, the integument is easily ruptured and the insects appear to disintegrate.

#### *b. Bacteria including Rickettsiae*

Bacteria found in insects include forms that may or may not form spores. Entomopathogenic bacteria and related organisms come in a range of shapes (rods, cocci, spiral, and pleomorphic) and sizes (0.5–50 μm), occur singly or in chains, are gram-negative or gram-positive and are aerobic or anaerobic. Unfortunately, dead insects make excellent media for a broad diversity of saprobic species. Even insects that have been killed as a result of some of the other entomopathogens may be invaded by non-pathogenic bacteria. However, these invasions usually result in a population of mixed species. When only one or a predominant bacterial species is found, it is an indication that the insect possibly died of bacterial septicemia. Insects that have been recently invaded by nematodes with bacterial symbionts may also be filled with a single bacterial species.

Some bacterial entomopathogens are not initially lethal to their insect hosts and signs and symptoms may be observed in living insects (e.g., *Paenibacillus popilliae*, one of the bacteria that causes milky disease in scarabs; see Figure 5.1 in Chapter V, and cover) and *Serratia entomophila*, the species that causes amber disease in scarabs (see Figures 5.2 and 3 in Chapter V). Both of these are covered in detail in Chapter V. Changes in color due to some bacteremias of insects are quite distinct. Infected larvae may appear white, red, amber, black or brown. Recently killed insects may be odiferous, flaccid, and fragile. Cadavers that have aged somewhat usually shrivel and dry into a hard scale.

The most commonly used microbial control agent, *Bacillus thuringiensis*, may produce a range of symptoms in insects depending upon the subspecies or strain of the bacterium and the target insects. Because the predominant mode of action is midgut toxicity, insects may be killed due to toxemia with or without subsequent reproduction of the bacterium in the hemocoel. Prior to death many species stop feeding and may wander from their original feeding site or even from the host plant (Chapter III).

Although they are less commonly observed than many of the viral infections in insects, rickettsial infections may occasionally be obvious in certain insect populations. Rickettsiae apparently have a broad insect host range. They are small (0.2–0.6 μm), rod shaped, gram-negative organisms that look like bacteria and behave like viruses (i.e. they are obligate intracellular pathogens). Species in the genus *Wolbachia* produce unapparent infections in insects and are seldom harmful to their hosts. Species in the genus *Rickettsiella* are pathogenic for insects and are reported from Coleoptera, Diptera, Lepidoptera, and Orthoptera. *Rickettsiella* spp. that infect larvae of several scarab species produce a bluish cast to infected fat body. Krieg (1963) describes the color as a bluish-green iridescence similar to blue iridescent virus, but not as intense. Also known as blue disease, these rickettsioses produce somewhat chronic infections (Krieg, 1963). Large birefringent crystals are an accompanying characteristic of many of the infections caused by *Rickettsiella* spp.

Some important human pathogens in the genus *Rickettsia* (causal agents for typhus, Rocky Mountain spotted fever and others) are transmitted by arthropod intermediate hosts and are pathogenic for the arthropod vectors as well as humans. *Rickettsiella melolonthae* has also been reported as being pathogenic for mammals (Krieg, 1963). Due to some doubt regarding their specificity and potential danger for humans and the need for *in vivo* production, *Rickettsiella* pathogens of insects have not been developed as microbial control agents and will not be treated in greater detail in this *Manual*. For more information on techniques for their isolation, identification, cultivation, bioassay, and storage refer to Krieg (1963) and Poinar & Thomas (1984).

### *c. Protists*

Unlike most other types of entomopathogens, protists usually produce chronic infections manifested by such signs and symptoms as irregular growth, sluggishness, loss of appetite, malformed pupae or adults, or adults with reduced vigor, fecundity, and longevity. Such characteristics are seldom *pathognomonic* in nature, although black, pepper-like spots on the integument of silk worm larvae are essentially diagnostic for the microsporidian disease known as 'pebrine'. Whitish masses of microsporidian spores may be visible in the fat body tissues of insects with transparent cuticles (e.g., mosquitoes and other aquatic insects) (see Chapter XI). However, the majority of hosts infected with protists must be dissected and the tissues examined for the presence of vegetative forms and cysts or spores. Spores typically range in size from 2 to 20 μm. While the reproductive forms (spores or cysts) are readily recognized when examined in wet-mount preparations by phase microscopy, it is usually necessary to stain wet-mounted, smears of various tissues to examine the vegetative stages of development. The life-cycles of microsporidia and neogregarines are often extremely complex, sometimes involving an intermediate host, and specific identifications can only be made with the assistance of a specialist.

Protists, especially the microsporidia, are relatively host specific and usually can be found within specific host species. However, under laboratory conditions, many species can be cross-transmitted to a wide range of hosts that may be helpful in carrying out infectivity tests involved in Koch's postulates. Almost all entomopathogenic protists are obligate parasites and can not be grown on artificial media. Some exceptions are presented in Chapter XI. Detailed investigations of most species require examination of infected hosts by TEM and/or DNA sequencing as presented in Chapter XV.

### *d. Fungi*

Some of the most spectacular infections in insects are produced by fungal entomopathogens and many result in colorful and/or striking outgrowths of the fungus as shown on the cover of this *Manual*. Some fungal species that forcibly discharge spores from the host or grow on to the substrate from the host, may produce a distinct halo around the infected insect. The entomopathogenic fungi are a broad and diverse group taxonomically and biologically and infect virtually every insect order. Due to the mode of entry through the host cuticle by most species of entomopathogenic fungi, with the exception of microsporidia, they are the only entomopathogens found in sucking insects (Hemiptera). [Molecular evidence suggests that microsporidia are related to fungi but the exact nature of this relationship is debated (see Chapter XI).] Infectious propagules come in a broad array of shapes and sizes (5 μm to several mm) and may be motile, projected from host cadavers, wind-borne, dispersed by water or by the insect hosts themselves (Chapters VI, VII, and IX).

Most of the entomopathogenic fungi kill their hosts relatively soon after infection occurs. Following death, infectious spores are usually produced on the surface of the insect. Larger, thick-walled resting spores of many fungal species can also be found inside the host. Insect cadavers are often mummified due to mycoses and may persist in the environment for several weeks, enabling isolation of the pathogen long after death of the host. Developmental and reproductive stages of entomopathogenic fungi can also be found in living hosts, most commonly in larvae of aquatic Diptera.

*e. Nematodes*

Except in the early stages of infection, signs of nematode infection are readily apparent to the observer; if conspicuous, one or several nematodes may be seen through the cuticle. With many of the more commonly observed nematode species, the host may be alive up to the moment the nematode emerges (see cover) or is killed shortly after infective forms invade the host. Host insects that are attacked by heterorhabditids or steinernematids often change color. Usually, it is red–orange in heterorhabditids and honey- or cream-colored in steinernematids. Change of color is due to the presence of nematode bacterial symbionts. Infective forms of these nematodes usually have a distinct second cuticle. In addition to causing distinctive coloration, many species of the symbionts of heterorhabditid nematodes, *Photorhabdus* spp., are luminescent in the dark. The nematodes found in insects may range from less than 1 mm to several cm in length.

Identification of most species of entomopathogenic nematodes requires adult and the third stage juvenile (i.e. infective juveniles). Stages leaving the host insect are usually not adults. Dissection of the insect when infective juveniles begin leaving will provide the adult stage (see Chapter XII).

Several atlases of insect diseases provide color and black and white photographs of diseased insects that can aid in the recognition of diseased insects in the field. These include: Weiser (1977), Poinar & Thomas (1978, 1984), Samson *et al.* (1988), and Adams & Bonami (1991).

## E. Conclusive diagnosis—satisfying Koch's postulates

The positive identification of a suspected pathogen from a diseased insect does not always incriminate the organism as the causal agent of the disease. Careful analysis of the facts gathered in the field, from laboratory examinations, study of the progress of the disease and other etiological information outlined by Steinhaus (1963a) is necessary when the diagnosis is critical or other information on hand is not conclusive. Satisfying Koch's postulates is the most definitive way to make a conclusive diagnosis. The following steps are taken to confirm that the isolated microorganism is the causal agent of the disease (modified from Steinhaus, 1963a; Agrios, 2004; Kaya & Vega, 2012):

1. The pathogen must be isolated from all of the diseased insects examined, and the signs and/or symptoms of the disease recorded.
2. The pathogen must be grown in *axenic culture* on a nutrient medium (for facultative pathogens) or in a susceptible insect (obligate pathogens), and it must be identified and/or characterized.
3. The pathogen must be inoculated on/into healthy insects of the same or a related species to the original, and signs and symptoms of disease must be the same.
4. The pathogen must be isolated in *axenic culture* or in a susceptible insect once again and its characteristics must be exactly like those observed in Step 2.

When it is possible to culture a suspected pathogen on artificial media and produce infectious propagules, satisfying Koch's postulates is a relatively straightforward process. However, many organisms are obligate pathogens (all viruses and Rickettsiae, many pathogenic protists, fungi and nematodes, and some bacteria). In these cases, the infectious agent is produced *in vivo* and purified using methods presented in Chapters II, V, VII, IX, XI, and XII. It should be noted that some organisms that were previously regarded as impossible to produce on artificial media have since been successfully cultured on complex media that satisfy specific nutritional requirements that enable production of infectious propagules (some Entomophthorales, *Lagenidium giganteum*). Obligate parasites that require intermediate hosts (some protists and fungi) may be even more problematic (see Chapter XI), especially if the requirement is suspected and the intermediate host is not yet known. Additionally, stress inherent to laboratory rearing may exaggerate virulence and exacerbate susceptibility to marginal or facultatively pathogenic organisms.

With pathogens that are obligate parasites or are submicroscopic in size, a variety of techniques must be used in carrying out Koch's postulates. The use of the electron microscope is essential in detecting and characterizing such intracellular entomopathogens such as viruses and Rickettsiae. It is also more difficult to obtain pure cultures of such obligate entomopathogens and various techniques such as density gradient or rate-zonal centrifugation must be utilized. Tissue cultures can also be used to obtain pure cultures of viruses or protists but care must be taken to avoid contamination. In addition, identification may involve the use of various molecular techniques such as those presented in Chapters II, III, V, VII, IX, and XI–XIII. A comparison of molecular Koch's postulates (Falkow, 2004) with that of Koch's original three postulates is presented by Kaya & Vega (2012). The molecular methods are used to associate a specific gene or genes with factors responsible for pathogenicity of a species or strain of pathogenic microorganism. Deletion of the gene(s) results in a loss of pathogenicity and their reintroduction should restore pathogenicity.

## 4. SAFETY CONSIDERATIONS

Although most pathogens found in insects are selective for insects, care should be taken when handling these organisms until identifications are made and their safety determined. With the exception of the Rickettsiae, the safety to vertebrates and non-target organisms of each of the entomopathogen groups covered in this *Manual* are discussed in detail by several authors in Laird *et al.* (1990) and Hokkanen & Hajek (2003).

## ACKNOWLEDGMENTS

We thank Heather Headrick for preparation of some of the figures; Dr James Becnel for providing aquatic insect collection photographs; Richard Humber, Patricia Stock, Steven Garczynski, Kelli Hoover, Regina Kleespies, Harry Kaya, Steven Arthurs, and Cynthia Lacey for their constructive comments.

## REFERENCES

Adams, J. R., & Bonami, J. R. (1991). *Atlas of Invertebrate Viruses*. Boca Raton, FL: CRC Press, Inc., pp. 684.

Agrios, G. N. (2004). *Plant Pathology* (5th ed.). San Diego, CA: Elsevier, pp. 952.

Atieri, M. A. (1999). The ecological role of biodiversity in agroecosystems. *Agric. Ecosys. Environ., 74*, 19–31.

Alves, S. B. (Ed.). (1998). *Controle Microbiano de Insetos* (2nd ed.). Piracicaba, Brazil: Fundação de Estudos Agrárias Luiz de Queiroz, pp. 1163.

Avery, S. W., & Undeen, A. H. (1987). The isolation of microsporidia and other pathogens from concentrated ditch water. *J. Amer. Mosq. Control Assoc., 3*, 54–58.

Deseö-Kovács, K. V., & Rovesti, L. (1992). *Lotta Microbiologica Control i Fitofagi Teoria e Pratica*. Bologna, Italy: Edagricole-Edizioni Agricole., pp. 296.

Falkow, S. (2004). Molecular Koch's postulates applied to bacterial pathogenicity — a personal recollection 15 years later. *Nat. Rev. Microbiol., 2*, 67–72.

Hokkanen, H. M. T., & Hajek, A. E. (Eds.). (2003). *Environmental Impacts of Microbial Insecticides: Need and Methods for Risk Assessment*. Dordrecht: Kluwer Academic Publishers.

Kaya, H. K., & Vega, F. E. (2012). Scope and basic principles of insect pathology. In F. E. Vega, & H. K. Kaya (Eds.), *Insect Pathology* (pp. 1–12). San Diego, CA: Academic Press.

Krieg, A. (1963). Rickettsiae and rickettsioses. In E. A. Steinhaus (Ed.), *Insect Pathology, An Advanced Treatise, Vol. 1* (pp. 577–617). New York: Academic Press.

Lacey, L. A., Frutos, R., Kaya, H. K., & Vail, P. (2001). Insect pathogens as biological control agents: Do they have a future? *Biol. Control, 21*, 230–248.

Laird, M., Lacey, L. A., & Davidson, E. W. (Eds.). (1990). *Safety of Microbial Insecticides*. Boca Raton, FL: CRC Press, pp. 259.

Martignoni, M. E., Krieg, A., Rossmore, H. W., & Vago, C. (1984). *Terms Used in Invertebrate Pathology in Five Languages: English, French, German, Italian, Spanish*. Portland, OR: Publ. PNW-169, U. S. Dept. Agric., Forest Serv., pp. 195.

Martignoni, M. E., & Milstead, J. E. (1960). Quaternary ammonium compounds for the surface sterilization of insects. *J. Insect Pathol, 2*, 124–133.

Onstad, D. W., Fuxa, J. R., Humber, R. A., Oestergaard, J., Shapiro-Ilan, D. I., Gouli, V. V., Anderson, R. S., Andreadis, T. G., & Lacey, L. A. (2006). *An abridged glossary of terms used in invertebrate pathology* (3rd ed.). http://www.sipweb.org/glossary/.

Poinar, G. O., Jr., & Thomas, G. M. (1978). *Diagnostic Manual for the Identification of Insect Pathogens*. New York, NY: Plenum Press, pp. 218.

Poinar, G. O., Jr., & Thomas, G. M. (1984). *Laboratory Guide to Insect Pathogens and Parasites*. New York, NY: Plenum Press, pp. 392.

Samson, R. A., Evans, H. C., & Latgé, J.-P. (1988). *Atlas of Entomopathogenic Fungi*. Berlin, Germany: Springer-Verlag, pp. 187.

Steinhaus, E. A. (1949). *Principles of Insect Pathology*. New York, NY: McGraw Hill.

Steinhaus, E. A. (1958). Crowding as a possible stress factor in insect disease. *Ecology, 39*, 503–514.

Steinhaus, E. A. (1963a). Background for the diagnosis of insect diseases. In E. A. Steinhaus (Ed.), *Insect Pathology, An Advanced Treatise, Vol. 2* (pp. 549–589). New York, NY: Academic Press.

Steinhaus, E. A. (Ed.). (1963b), *Insect Pathology, An Advanced Treatise, Vol. 1*. New York, NY: Academic Press, pp. 661.

Steinhaus, E. A., & Marsh, G. A. (1962). Report of diagnoses of diseased insects 1951–1961. *Hilgardia, 33*, 349–490.

Tanada, Y., & Kaya, H. K. (1993). *Insect Pathology*. San Diego, CA: Academic Press, pp. 666.

Thomas, G. M. (1974). Diagnostic techniques. In G. E. Cantwell (Ed.), *Insect Diseases, Vol. 1* (pp. 1–48). New York, NY: Marcel Dekker.

Vega, F. E., & Kaya, H. K. (2012). *Insect Pathology* (2nd ed.). San Diego, CA: Academic Press.

Weiser, J. (1977). *An Atlas of Insect Diseases*. Prague, Czech Republic: Academia, pp. 240.

Weiser, J., & Briggs, J. D. (1971). Identification of pathogens. In H. Burges, & N. W. Hussey (Eds.), *Microbial Control of Insects and Mites* (pp. 13–66). New York, NY: Academic Press.

Wittig, G. (1963). Techniques in insect pathology. In E. A. Steinhaus (Ed.), *Insect Pathology, An Advanced Treatise, Vol. 2* (pp. 591–636). New York, NY: Academic Press.

CHAPTER II

# Basic techniques in insect virology

KAROLIN E. EBERLE, JÖRG T. WENNMANN, REGINA G. KLEESPIES & JOHANNES A. JEHLE

Institute for Biological Control, Julius Kühn-Institut, Federal Research Centre for Cultivated Plants, Heinrichstr. 243, 64287 Darmstadt, Germany

## 1. INTRODUCTION

The importance of insect viruses for the well-being of humans can not be overestimated. As naturally occurring pathogens of many insect species, they contribute to the natural regulation of insect populations. Many of these insects are vectors of the causal agents of human diseases or pests of agricultural and horticultural crops and defoliators of forest trees. Thus, the interest in insect viruses has been fostered for the last 50 years by the recognition of the possibility to use them as control agents of many pest insect species (Hunter-Fujita *et al.*, 1998; Cory & Evans, 2007). Most developments in insect virology followed the concepts and discoveries in other fields of virology, especially bacteriophages, which provided a simple model organism for virus—host interaction. However, some methods developed with insect viruses, such as the establishment of the baculovirus expression vector system and its exploitation to produce heterologous proteins in insect cell culture, have paved the way for manifold progresses in numerous scientific areas, including vertebrate virology and microbiology as well as human and veterinarian medicine (Smith *et al.*, 1983).

Actually, the early beginning of insect virology was somewhat delayed by the fact that the infectious virus particles of many insect viruses, including baculoviruses, entomopoxvirses, and cypoviruses are embedded in large occlusion bodies (OBs), a unique feature of several taxa of insect viruses. These occluded viruses were filterable and did not fit into the early definition of viruses as non-filterable, sub-microscopic infectious agents (Glaser & Chapmann, 1913; Glaser, 1918). In the following decades the focus of insect virus was on disease symptoms, the etiology of disease and the characteristics of cytopathological effects, all examined using light microscopy. Later, these studies were more and more complemented by transmission electron microscopy studies, the elucidation of physico-chemical characters by using density gradient centrifugation, the determination of the chemical nature of the virus genomes (RNA or DNA) as well as studies of the relationship of virus components by using immunological approaches, such as radioimmunoassays (RIA) and enzyme-linked immunosorbent assays (ELISA) (McCarthy & Gettig, 1986).

Two techniques have basically revolutionized our understanding of virus function and virus diversity: cell culture techniques and molecular methods. Numerous standard volumes covering these methods are available (Vlak *et al.*, 1996; O'Reilly *et al.*, 1994; Sambrook & Russell, 2001). Replicating viruses in cell culture allowed the propagation of viruses independent from host insects and dissection of the virus replication cycle on cellular and molecular levels. Since the establishment of the first insect cell line (Grace, 1962) several hundred cell lines, many of them permissive to insect viruses, have been described from more than 100 insect species. By applying molecular methods, including gene cloning, reverse genetics, DNA/DNA and DNA/RNA hybridization, and polymerase chain reaction (PCR) based approaches, the function of thousands of single virus genes has been characterized during the last two decades. However, most of what we

MANUAL OF TECHNIQUES IN INVERTEBRATE PATHOLOGY
ISBN 9780123868992

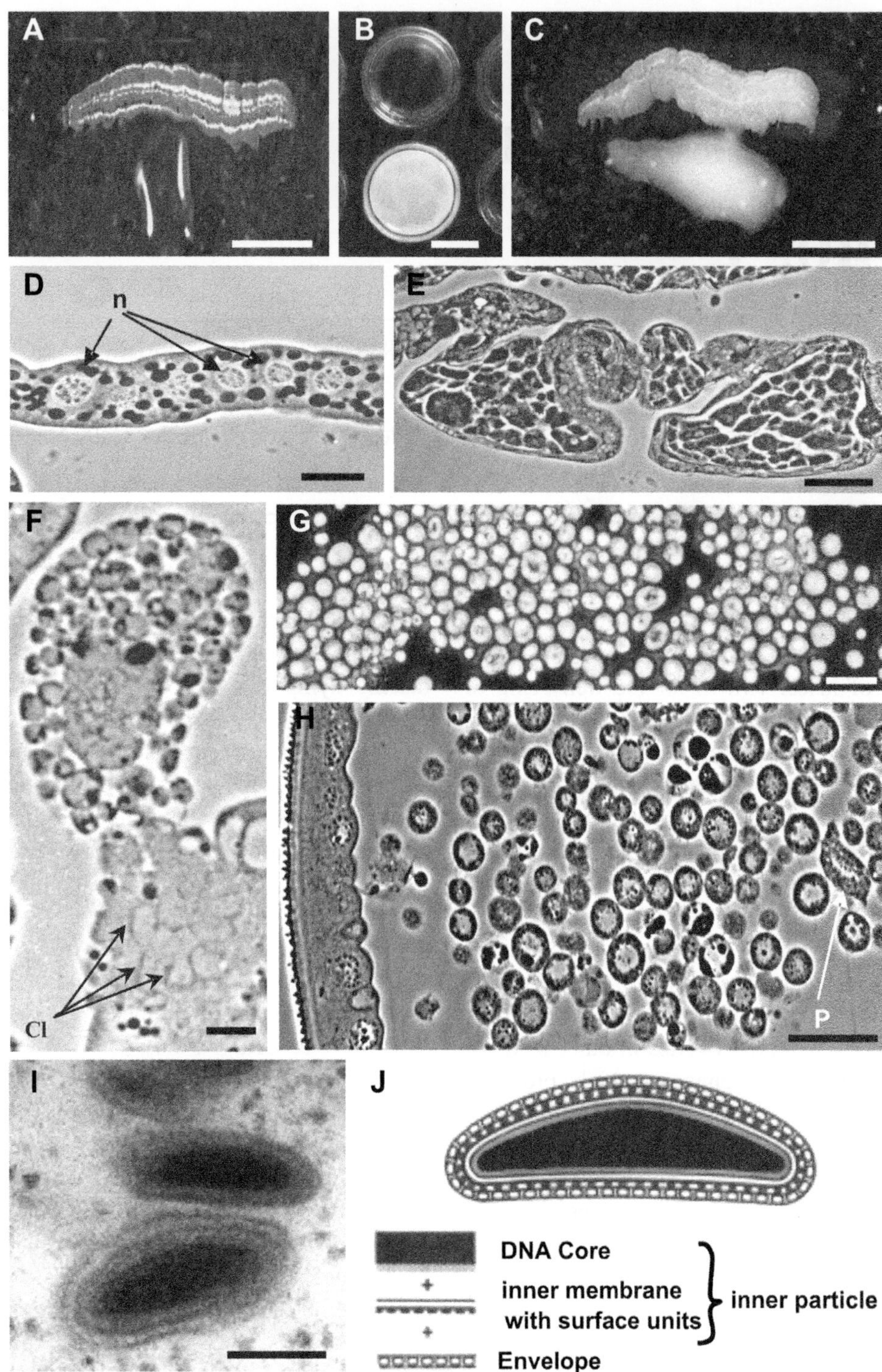


**Figure 2.1** Typical pathology associated with ascovirus infections observed in the cabbage looper, *Trichoplusia ni*. A. Healthy larva. The clear pool of liquid below the larva is normal hemolymph. B. Comparison of the appearance of healthy hemolymph (upper well) with hemolymph from an ascovirus-infected larva (lower well) at 7 days post-infection. C. ascovirus-infected *T. ni* larva showing typical milky-white appearance of infected hemolymph. White opaque hemolymph in lepidopteran larvae is generally diagnostic of ascovirus infection. The dense opacity is due to the accumulation of virion-containing vesicles in the hemolymph (scale bars in A–C = 1 cm). (D and E, respectively) Sections through healthy

know about viral gene functions was determined in an individual approach focusing on a single gene in a more or less artificial *in vitro* system of a cell culture environment. For an improved understanding of the biology of insect viruses it will be essential to transform this knowledge and the established concepts into the *in vivo* environment of the infected host insect. The development of the most recent omics technologies provides the chance to obtain a more comprehensive insight into the function of the total virus genome and its interaction with the host. Microarray technologies, and high throughput DNA and RNA sequencing will contribute not only to a better knowledge of virus diversity but they will also help to reveal the interaction with and responses of the infected host. The following chapter will provide an update of the methods described by Evans & Shapiro (1997). Nevertheless, it is far from a complete description of the multitude of techniques used in insect virology. Our aim is to focus on recently refined technologies for insect virus identification and diagnostics as well as on basic molecular methods necessary to close the gap of knowledge between *in vitro* and *in vivo* functions of insect viruses.

## 2. BRIEF OVERVIEW OF THE DIFFERENT GROUPS OF INSECT VIRUSES

This brief overview provides a basic introduction to the physico-chemical, molecular, and biological key characteristics of most of the known insect viruses. Typical external and internal symptoms for identification of the principal virus groups are also given. The descriptions are based on the virus online database of the International Committee on Taxonomy of Viruses (ICTVdB), the Eighth Report of the International Committee on Taxonomy of Viruses (Fauquet *et al.*, 2005), and on Asgari & Johnson (2010).

### A. DNA Viruses

#### *1. Ascoviruses*

Members of the family of *Ascoviridae* are double-stranded DNA (dsDNA) viruses that infect lepidopteran insects and cause the unique pathology of forming virion-containing vesicles in the hemolymph of infected hosts. The presence of the vesicles gives the hemolymph a milky white appearance, which is a major characteristic of the disease (Figure 2.1A–E). Millions of these virion-containing vesicles (Figure 2.1F–H) begin to disperse from infected tissues 48–72 h after infection into the hemolymph. The circulation of virions and vesicles in the hemolymph facilitates mechanical transmission by parasitic wasps during oviposition. Although ascoviruses (AVs) appear to be very common, only seven species are currently recognized, all from Noctuidae, with the type species *Spodoptera frugiperda ascovirus 1*.

**Structure:** Virions are bacilliform, ovoid, and allantoid. They consist of an envelope, a core, and an internal lipid membrane associated with the inner particle (Figure 2.1I, J). The virus capsid measures 130 nm in diameter, and 200–240 nm in length.

**Genome:** AVs have circular double-stranded DNA genomes ranging from 119 to 186 kbp.

**External symptoms:** Infected larvae exhibit a retarded development with difficulties in completing a larval molt (Table 2.1).

**Internal symptoms:** AV-infected larvae have a milky white hemolymph (Table 2.1, Figure 2.1B, C).

#### *2. Baculoviruses*

The baculoviruses are a family (*Baculoviridae*) of large rod-shaped viruses that are classified into four genera: *Alpha-*, *Beta-*, *Gamma-*, and *Deltabaculovirus*. Based on the morphology of their occlusion bodies (OBs), nucleopolyhedroviruses (NPVs) and granuloviruses (GVs) are distinguished (Figure 2.2A–F). The OBs of NPVs

and infected fat body tissue of a larva. Note the extensive hypertrophy of the ascovirus-infected cells in E. (n, nuclei; scale bars in D and E = 10 μm). F. Section through a lobe of fat body in which a greatly hypertrophied cell is cleaving into viral vesicles (arrows), (Cl = cleavage planes throughout a cell, scale bar = 5 μm). G and H. respectively, wet mount of ascovirus-infected hemolymph shown in B and C, and the appearance of blood and spherical viral vesicles as observed in plastic sections, (P = plasmatocyte, a type of insect blood cell, scale bar of G = 5 μm, of H = 20 μm). I. Transmission electron micrograph of two ascovirus virions of *T. ni* (scale bar = 100 nm). J. Schematic interpretation of ascovirus virion structure based on the appearance of virions in ultrathin sections and negatively stained preparations. (Courtesy of B. A. Federici, University of California, Riverside, USA.) Please see the color plate section at the back of the book.

**Table 2.1** Recorded symptoms and host ranges of the principal insect viruses.

| *Insect virus* | *Main symptoms* | *Recorded host orders* | *Usual host stage* |
|---|---|---|---|
| *DNA Viruses* | | | |
| Ascoviruses (Figure 2.1A–J) | **External:** Extended development with difficulties in completing a larval moult.<br>**Internal:** Milky white hemolymph (Figure 2.1C). | Lepidoptera (Noctuidae only) | Larvae |
| Baculoviruses (Figure 2.2A–F) | **External:** Discoloration (brown and yellow), stress with regurgitation and refusal to eat, getting lethargic with slow-moving or no movement at all. After death, the outer skin ruptures easily, releasing the liquefied body content.<br>**Internal:** White gut, but most organs are affected, too, typical fragile hypodermis. | Diptera, Hymenoptera, Lepidoptera | Larvae, sometimes pupae or adults |
| Densoviruses (Figure 2.3A, B) | **External:** Lepidoptera, infections start with anorexia and lethargy followed by flaccidity and inhibition of moulting or metamorphosis. Larvae become whitish and progressively paralysed.<br>Cockroaches and the house cricket, typical symptoms of hind legs paralysis and uncoordinated movements are displayed.<br>**Internal:** Wide tissue tropism; hypertrophy of nuclei is the salient histopathological feature of all DNV-infected cells (Figure 2.3A, B). | Diptera, Blattoidea, Lepidoptera, Odonata, Orthoptera | Larvae, pupae, adults |
| Entomopoxviruses (Figure 2.3C, D) | **External:** White or light blue body, extremely extended longevity and a decrease in feeding.<br>**Internal:** Primary site of replication is the fat body, but other tissues can be infected, too. Hemolymph appears white, filled with occlusion bodies (Figure 2.3C) and the fat body disintegrates. | Coleoptera, Diptera, Hymenoptera, Lepidoptera, Orthoptera | Larvae, pupae, adults |
| Hytrosaviruses (Figure 2.4A–C) | **External:** Inhibition of reproduction and/or disrupting mating behavior.<br>**Internal:** Replication in nuclei of salivary gland cells in adult flies, inducing gland enlargement. They cause overt salivary gland hypertrophy symptoms, testicular degenerations and ovarian abnormalities (Figure 2.4A, C). | Diptera | Adults |
| Iridoviruses (Figure 2.3E–G) | **External:** Blue, green, or purple, in some cases also an orange to brown iridescence. Reductions in fecundity and longevity and a decrease in feeding.<br>**Internal:** Replication in the majority of host tissues, especially the fat body (Figure 2.3E, G), hemocytes, and epidermis. Paracrystalline arrangements of virus particles in infected tissues (Figure 2.3G). | Coleoptera, Diptera, Homoptera, Hymenoptera, Lepidoptera, Orthoptera, and terrestrial isopods (Crustacea) | Larvae |

**Table 2.1** Recorded symptoms and host ranges of the principal insect viruses—Cont'd

| *Insect virus* | *Main symptoms* | *Recorded host orders* | *Usual host stage* |
|---|---|---|---|
| *DNA Viruses—Cont'd* | | | |
| Nudiviruses (Figure 2.5A–D) | **External:** Larvae are getting lethargic with slow-moving or no movement at all. Infected grubs of *Oryctes rhinoceros* develop diarrhea, and their body gradually becomes turgid and assumes a glassy, beige-waxen or pearly appearance.<br>**Internal:** Tissues appear greatly hypertrophied. Nuclei at the advanced stage of infection have a dense dark ring-zone adjacent to the nuclear membrane and a lighter center (Figure 2.5D). | Lepidoptera, Trichoptera, Diptera, Siphonaptera, Hymenoptera, Neuroptera, Coleoptera, Homoptera, Thysanura, Orthoptera, Acarina, Aranaeina, (Crustacea) | Larvae, adults |
| Polydnaviruses (Figure 2.5E, F) | **External:** There are no obvious pathological effects on the parasitoid hosts.<br>**Internal:** Replication is restricted to specialized cells in the ovaries of females. | Parasitic Hymenoptera | Adults |
| *RNA Viruses* | | | |
| Dicistroviruses (Figure 2.6A) | **External:** Reduced longevity and fecundity. Symptoms range from flaccidity of the body to paralysis.<br>**Internal:** Tissue paralysis of the gut, and often also fat body, trachea, muscle, brain, and nerves can also be infected. | Diptera, Hemiptera, Hymenoptera, Lepidoptera, Orthoptera | Larvae, adults |
| Iflaviruses (Figure 2.6D, E) | **External:** Iflaviruses may cause diarrhea and developmental malformations.<br>**Internal:** Primary sites of replication are variable: *Infectious flacherie virus* first infects goblet cells of the midgut, while *Deformed wing virus* spreads in the whole body. | Lepidoptera, Hymenoptera, Hemiptera, bee parasitic mites (Acarina) | Larvae, adults |
| Nodaviruses (Figure 2.6B, C) | **External:** Mostly, infections are inapparent but can result in a slower development or a slight reduction in the total population due to a reduction of egg viability. In *Heteronychus arator* larvae may become flaccid and lose pigmentation of the hypodermis.<br>**Internal:** Virus replication takes place exclusively in the cytoplasm of the host cells (Figure 2.6C), and the formation of paracrystalline arrays is typical. Primary infection sites are muscle cells, hemocytes and hypodermis. | Diptera, Coleoptera, Lepidoptera | Larvae, adults |
| Tetraviruses | **External:** Reduction in larval and pupal size. Underdevelopment and malformed wings of adults can be found.<br>**Internal:** The virus multiplies primarily in the cells of the midgut. The integrity of the internal tissues is largely unaffected. | Lepidoptera | Larvae |
| Cypoviruses (Figure 2.6F–H) | **External:** Cypovirus (CPV) infections lead to an extended development and reduced feeding. Longevity and reproductive performance of infected adults is lowered.<br>**Internal:** CPV infection is restricted to the gut which may become white or yellow. | Lepidoptera Diptera, Hymenoptera, Coleoptera | Larvae, pupae, adults |

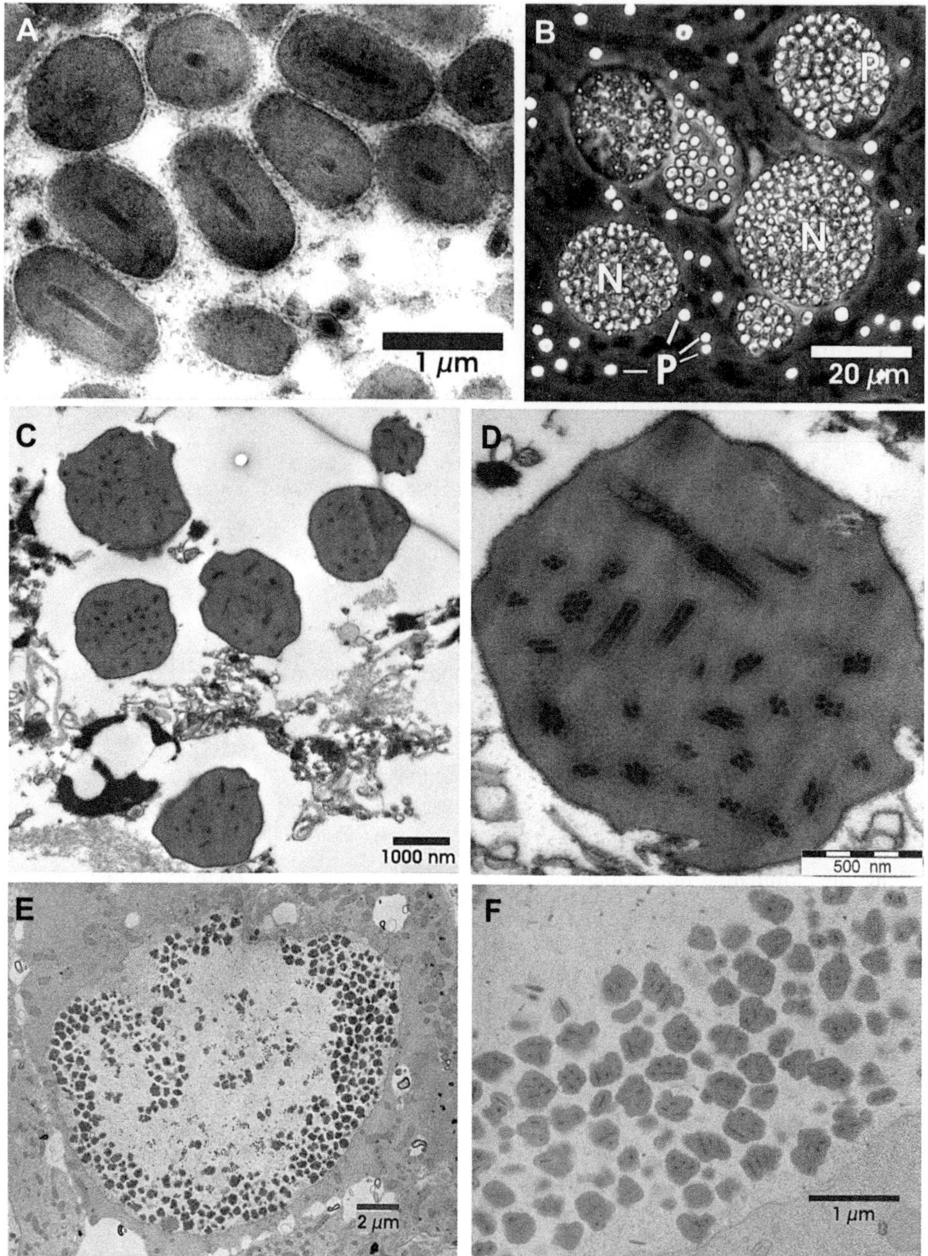


**Figure 2.2** A. Transmission electron micrograph of a thin section of larval fat body of *Cydia pomonella* infected by *Cydia pomonella* granulovirus. Note the capsular virus occlusion bodies (OBs), each containing one enveloped virus rod. [Courtesy of A. M. Huger, JKI Darmstadt, Germany.] B. Phase contrast light micrograph of larval fat body of *Spodoptera littoralis* infected by NPV. Hypertrophied nuclei (N) are filled with virus polyhedra (P) [Courtesy of A. M. Huger, JKI Darmstadt, Germany.] C, D. Transmission electron micrographs of thin sections of *Lymantria dispar* multiple NPV (LdMNPV) in a nucleus of the fat body. E, F. Transmission electron micrographs of a midgut cell from larval stages of *Culex quinquefasciatus* infected with *Culex nigripalpus* NPV (CuniNPV); E. nucleus of a midgut cell containing occlusion bodies of CuniNPV; F. occlusion bodies of CuniNPV containing multiple singly enveloped rod shaped virions. [E and F, courtesy of J. J. Becnel, USDA, ARS, CMAVE, Gainesville, Florida, USA.]

typically enclose several to many virions per OB. The virions contain either multiple (MNPV) or single (SNPV) nucleocapsids (Figure 2.2C, D and E, F, respectively). In contrast, GVs enclose only one virion per OB (Figure 2.2A). *Alphabaculovirus* comprise lepidopteran specific NPVs, *Betabaculovirus* lepidopteran specific GVs, *Gammabaculovirus* hymenopteran specific NPVs, and *Deltabaculovirus* dipteran specific NPVs, respectively.

**Structure:** The OBs of GVs and NPVs are composed of a crystalline protein matrix, mainly consisting of a single protein, the so-called polyhedrin/granulin. The OBs of GVs measure 120–300 nm × 300–500 nm, those of NPVs have a diameter of 0.5–5 (15) μm, respectively. Virion dimensions are in the size range of 30–60 nm × 260–360 nm for GVs and 40–140 nm × 250–400 nm for NPVs. The rod-shaped nucleocapsids are about 30–60 nm × 250–300 nm. The budded virus (BV) phenotype, produced during virus replication, has no diagnostic value.

**Genome:** The genome consists of a circular dsDNA molecule of 80–180 kbp.

**External symptoms:** Larvae infected by *Alpha*- and *Betabaculoviruses* display a discoloration (white, brown or yellow), with regurgitation and refusal to eat. Infected larvae become lethargic with slow or no movement. After death, the integument ruptures easily, releasing the liquefied body contents. Infection by *Gamma*- and *Deltabaculoviruses* is restricted to the midgut (Table 2.1).

**Internal symptoms:** The gut of larvae infected with *Gamma*- and *Deltabaculoviruses* becomes white, whereas in those infected with *Alpha*- and *Betabaculoviruses*, several tissues are infected such as fat body, tracheal matrix cells, and hypodermal cells (Table 2.1).

### *3. Densoviruses*

The family *Parvoviridae* includes the insect specific genus *Densovirus* (DNV) (subfamily *Densovirinae*). DNVs are single-stranded DNA (ssDNA) viruses, which have been isolated from different insect orders, including Lepidoptera, Hemiptera, Orthoptera, Diptera, and Dictyoptera. The type species is *Juonia coenia densovirus*. Other members of this genus are *Aedes albopictus densovirus* and *Galleria mellonella densovirus*.

**Structure:** The virions consist of non-enveloped isometric capsids of either 18–22 nm or 20–26 nm in diameter. The capsids have an icosahedral symmetry and consist of 60 capsomers, with a structure described as "a quadrilateral 'kite-shaped' wedge"; the surface is said to have a rough appearance with small projections.

**Genome:** The non-segmented genomes contain a single linear molecule of ssDNA, 4–6 kb in length. It could either be negative sense or positive sense.

**External symptoms:** In most Lepidoptera, symptoms associated with DNV infections start with anorexia and lethargy followed by flaccidity and inhibition of molting or metamorphosis. Larvae become whitish and progressively paralyzed. Typical symptoms of infected cockroaches and house crickets are paralysis of hind legs and uncoordinated movements (Table 2.1).

**Internal symptoms:** Most DNVs present a wide tissue tropism (Table 2.1). The hypertrophy of nuclei is the salient histopathological feature of all DNV-infected cells (Figure 2.3A, B).

### *4. Entomopoxviruses*

The family *Poxviridae* comprises two subfamilies, the vertebrate-specific *Chordopoxvirinae* and the insect-specific *Entomopoxvirinae*. Three genera belong to the *Entomopoxvirinae*: (1) *Alphaentomopoxvirus* with the type species *Melolontha melolontha entomopoxvirus*. Viruses of this genus infect beetles (Coleoptera). (2) *Betaentomopoxvirus* with the type species *Amsacta moorei entomopoxvirus*. Members of this genus infect Lepidoptera (Figure 2.3C, D). (3) *Gammaentomopoxvirus* with the type species *Chironomus luridus entomopoxvirus*. These viruses infect flies and mosquitoes.

**Structure:** Virions of *Alpha*- and *Betaentomopoxviruses* are of ovoid shape, and of 450 × 250 nm and 250 × 50 nm in size, respectively (Figure 2.3C, D). *Gammaentomopoxviruses* are characteristically brick shaped (320 × 230 × 110 nm). Entomopoxviruses are occluded within the paracrystalline protein spheroidin.

**Genome:** The genome of the *Entomopoxvirinae* is a single linear molecule of dsDNA of 200–390 kbp. The genomes are highly A+T-rich and contain inverted terminal repeats at the ends.

**External symptoms:** Entomopoxvirus infections are associated with color changes including white or light-blue body, a decrease in feeding, and, the most striking characteristic, the extremely extended longevity of infected insects (Table 2.1).

**Internal symptoms:** The primary site of replication is the cytoplasm of fat body cells, but other tissues can also be

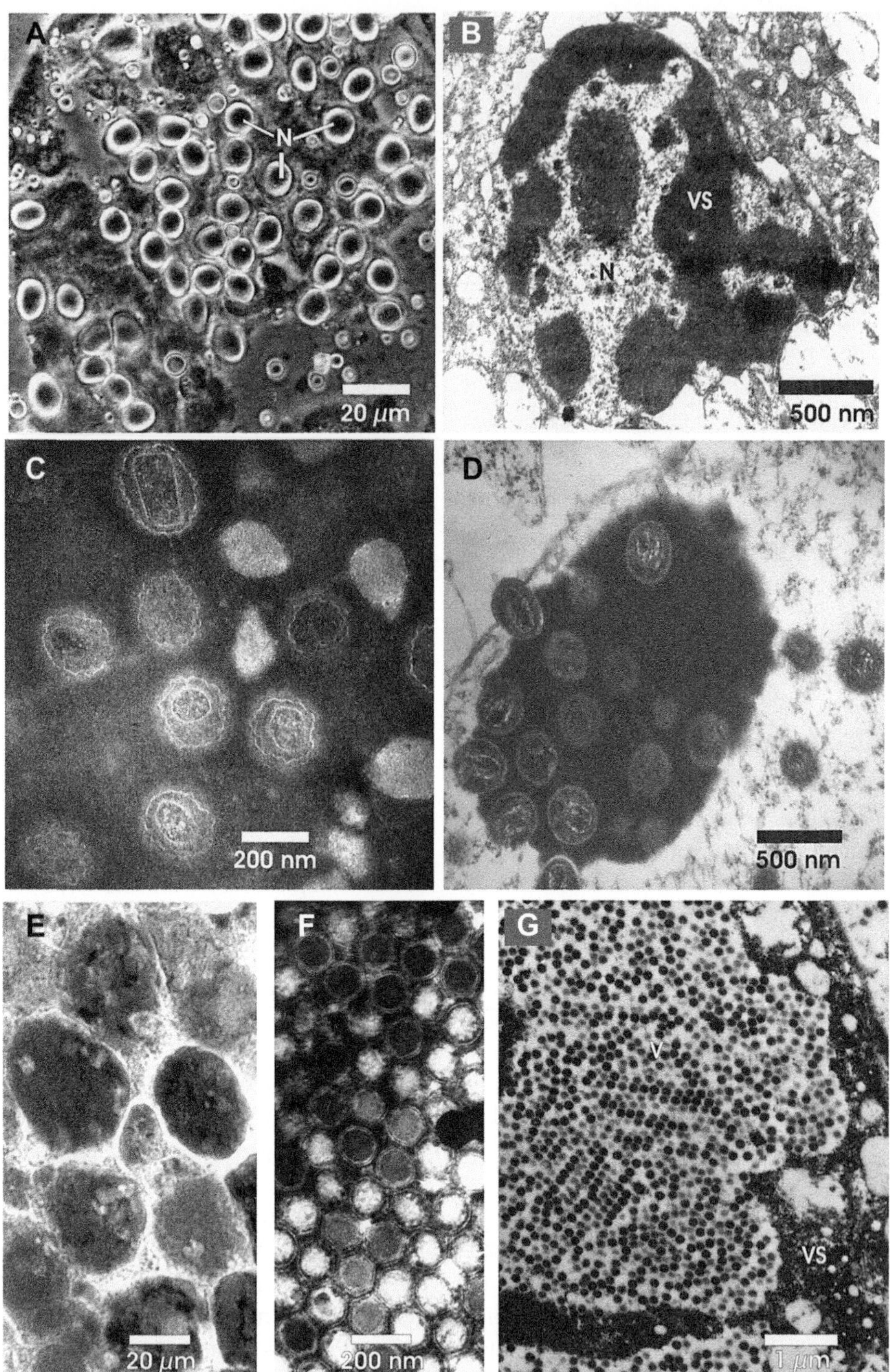


**Figure 2.3** A. Phase contrast light micrograph of larval fat body of *Tenebrio molitor* infected by Densovirus causing hypertrophied dense nuclei (N). [Courtesy of A. M. Huger, JKI Darmstadt, Germany]. B. Transmission electron micrograph of a Densovirus infected nucleus (N) of *Acheta domesticus* (VS, virogenic stroma). C, D. Transmission electron micrographs of *Choristoneura biennis* entomopoxvirus. C. Negative contrast stain with sodium phosphotungstate; D. section of an infected cell. [C and D, courtesy of B. Arif, Laboratory for Molecular Virology, GLFC, Sault Ste. Marie (Ontario), Canada.] E–G. Cricket Iridescent Virus and pertinent cytopathological changes of infected *Acheta domesticus*. E. Light

infected. The hemolymph appears white, filled with OBs and the fat body disintegrates (Table 2.1, Figure 2.3D).

*5. Hytrosaviruses*

The Salivary Gland Hypertrophy Viruses (SGHVs) (proposed family *Hytrosaviridae*) comprise unclassified members of entomopathogenic dsDNA viruses which cause symptoms of salivary gland hypertrophy in their hosts (Figure 2.4A). They have been reported from three genera of Diptera: the tsetse fly *Glossina pallidipes* (GpSGHV), the housefly *Musca domestica* (MdSGHV) (Figure 2.4A–C), and the narcissus bulb fly *Merodon equestris* (MeSGHV).

**Structure:** Virions are non-occluded, and enveloped, (Figure 2.4B–C). The size of the nucleocapsids varies among the different SGHVs; those of MdSGHV and MeSGHV are rod-shaped and measure about 500–600 nm in length by 50–60 nm in diameter, whereas the nucleocapsids of GpSGHV are filamentous and significantly longer and measure 800–1200 nm in length by 50–60 nm in diameter. The enveloped virions, measuring 70–80 nm in diameter, consist of an inner membrane that encloses the nucleocapsid and an outer membrane separated from the inner membrane by a narrow space (Figure 2.4C).

**Genome:** SGHVs possess a circular dsDNA genome ranging from 120 to 190 kbp. The G+C ratios range from 28% to 44%.

**External symptoms:** Viral infection inhibits reproduction by suppressing vitellogenesis, causing testicular aberrations, and/or disrupting mating behavior (Table 2.1).

**Internal symptoms:** SGHVs replicate in the nuclei of salivary gland cells in adult flies, inducing gland enlargement with little obvious external disease symptoms (Figure 2.4A, C). They cause overt salivary gland hypertrophy symptoms, testicular degenerations, and ovarian abnormalities (Table 2.1).

*6. Iridoviruses*

Invertebrate Iridescent Viruses (IIVs) (family *Iridoviridae)* are known to infect a number of agricultural pests, medically important insect vectors, and terrestrial isopods that live in damp or aquatic habitats. The major characteristic of this family is the presence of iridescent blue, green, orange, or purple coloration in heavily infected individuals.

**Structure:** The icosahedral non-enveloped virions are 130–180 nm in size (Figure 2.3F). They can be arranged in paracrystalline arrays (Figure 2.3G), thus producing the characteristic iridescence of this family.

**Genome:** IIVs have linear circularly permuted and terminally redundant dsDNA genomes of 150–280 kbp.

**External symptoms:** Heavily infected individuals display a blue, green, or purple, in some cases also an orange to brown, iridescence. Reductions in fecundity and longevity and a decrease in feeding have been observed (Table 2.1).

**Internal symptoms:** IIVs replicate extensively in the majority of host tissues, especially the fat body (Figure 2.3E, G), hemocytes, and epidermis. Paracrystalline arrangements of virus particles can be observed in infected tissues by electron microscopy (Figure 2.3G, Table 2.1).

*7. Nudiviruses*

This virus group has previously been considered as 'Non-occluded baculoviruses', because of the lack of an occlusion body and similar virus–host interactions. Now, the name 'Nudiviruses' (Latin *nudi* = bare, naked, uncovered) is proposed. Nudiviruses form a highly diverse group of rod-shaped and enveloped viruses with dsDNA. The *Oryctes rhinoceros* nudivirus (Figure 2.5A–D) was discovered in 1963 in Malaysia and has been successfully used to control the most important pest of the coconut palm, the coconut beetle, *O. rhinoceros*. A variety of nudiviruses and nudivirus-like viruses have been reported from various host species, including *Gryllus bimaculatus, Helothis zea*, and others.

**Structure:** The size of the rod-shaped virus particles range between 370 × 130 nm and 120 × 50 nm (Figure 2.5A–D).

**Genome:** Nudiviruses contain a circular dsDNA genome of 97–228 kbp. The G+C ratios range from 28% to 42%.

micrograph of virus-infected hypertrophied fat body cells; Hematoxylin Heidenhain stain. F. Icosahedral Iridovirus particles of *A. domesticus* from fat body cells; negative contrast stain with sodium phosphotungstate. G. Ultrathin section of an infected fat body cell of *Gryllus campestris*. Note the accumulation of icosahedral-shaped virions (V) in the cytoplasm, partly in paracrystalline aggregation, and the virogenic stroma (VS).

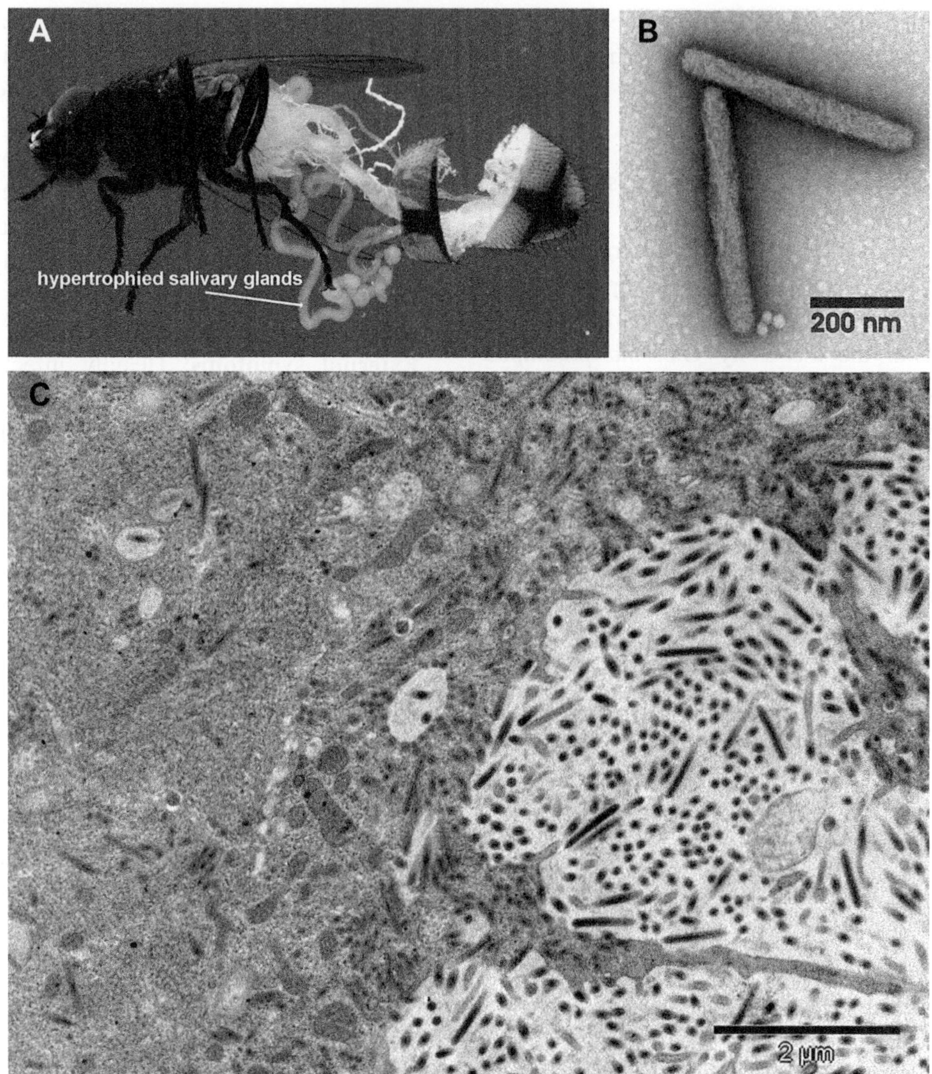


**Figure 2.4** Symptoms and ultrastructure of the housefly *Musca domestica* salivary gland hypertrophy virus (MdSGHV). A. Hypertrophied salivary glands of *M. domestica*. B. Purified MdSGHV; negative contrast. C. Transmission electron micrograph of MdSGHV showing the lumen/cell region of the salivary gland displaying the numerous enveloped MdSGHV virions. [Courtesy of D. Boucias, University of Florida, Gainesville, USA.]

**External symptoms:** Infected larvae become lethargic with slow or no movement. Infected grubs of *O. rhinoceros* develop diarrhea, and their body gradually becomes turgid and has a glassy, beige-waxen or pearly appearance (Table 2.1).

**Internal symptoms:** Infected tissues appear greatly hypertrophied. Nuclei at the advanced stage of infection have a dense dark ring zone adjacent to the nuclear membrane and a lighter center (Figure 2.5D, Table 2.1).

### *8. Polydnaviruses*

The family *Polydnaviridae* contains insect Polydnaviruses (PDVs) of two genera: *Bracovirus* and *Ichnovirus*. The bracoviruses can be found in braconid wasps and ichnoviruses in ichneumonid wasps (Hymenoptera). As the name implies, a major characteristic of these viruses is the presence of polydispersed superhelical dsDNA molecules in virus-like particles. PDVs exist in two forms. In wasps, they persist and are transmitted to offspring as stably integrated proviruses. Virus-like

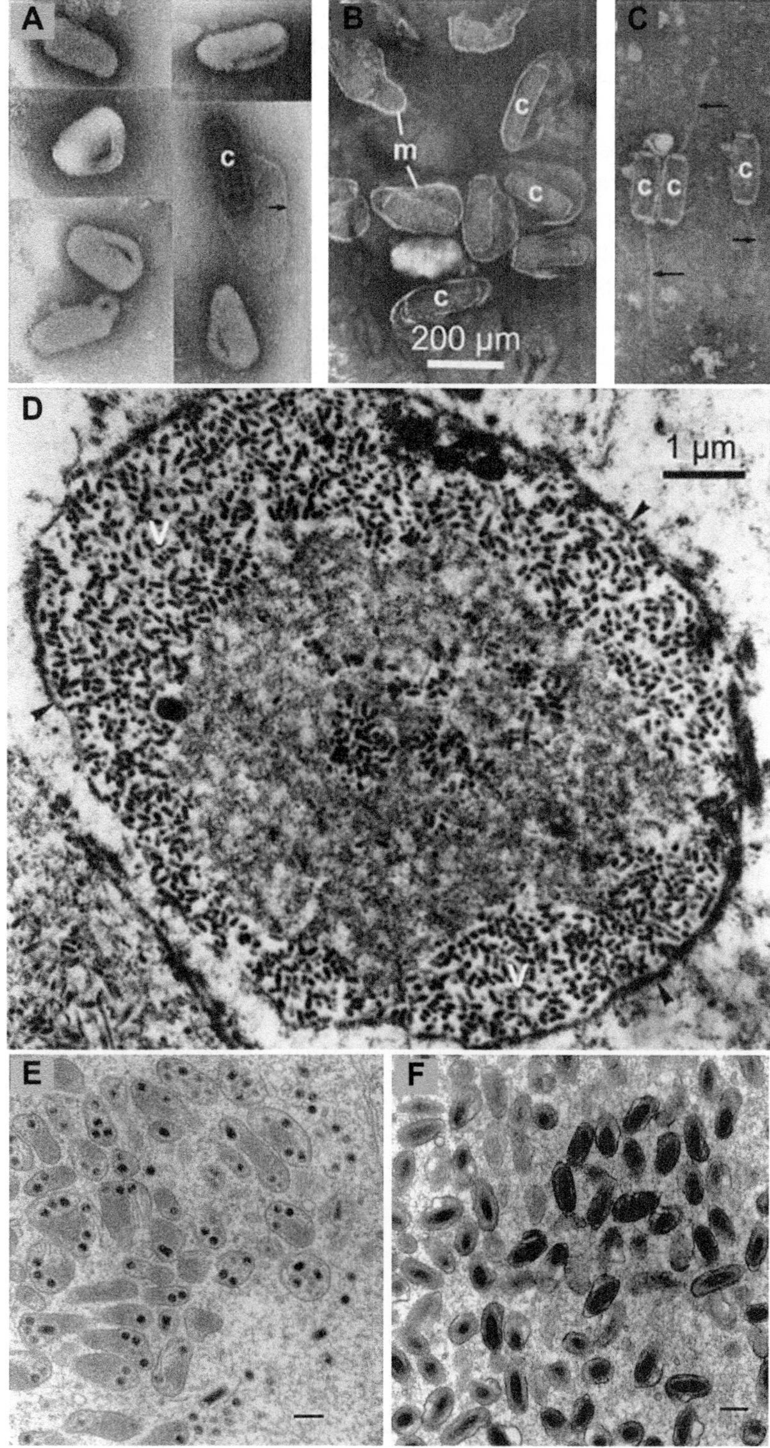


**Figure 2.5** A–D. Electron micrographs with structural details of *Oryctes* virus rods, negatively stained with phosphotungstic acid. A. Virions unpenetrated by stain, often being artificially mug-shaped; middle right: the virus membrane (arrow) is shed off from the capsid (c). B. Virions with longer penetration by stain, thus displaying the capsids (c) and the surrounding viral membrane (m). C. Three capsids (c) showing the typical thread-like appendix (arrows). D. Electron micrograph of a thin section through a nucleus from the myriads of cells in the midgut lumen of a virus vectoring rhinoceros

particles are produced in specialized cells of the ovaries of females, the calyx cells. Here, multiple amplified copies of cellular-like genes are encapsidated into virus-like particles, which become injected together with the wasp egg into the parasitoid's host insect. The encapsidated virus-like particles do not replicate in the host but they express gene products that suppress the immune response of the host and thereby facilitate the successful development of the parasitoids's offspring. The viral genes encoding the structural proteins of the virus-like particle are not encapsidated. Thus, PDVs and hymenopteran wasps have a symbiotic relationship. The extracellular virus-like particles supports the survival and development of the parasitoid in the parasitized host larvae, thereby contributing to the success of vertical transmission of the proviral genes of the PDVs in the parasitoid. Thus, there is mutualistic relationship between PDVs and parasitoid.

**Structure:** Bracoviruses and ichnoviruses differ in their morphology. Bracovirus virions consist of a single unit membrane enveloping one or more cylindrical nucleocapsids that possess tails (Figure 2.5E). Capsids, excluding tails, are uniform in width (30–50 nm) but vary greatly in length (30–150 nm). In contrast, ichnoviruses have two unit membranes that envelope a single fusiform nucleocapsid with a short, tail-like appendage (Figure 2.5F). Capsids (without tails) are approximately 85 × 330 nm.

**Genome:** The virus genome is integrated as a provirus into the parasitoid's genome. The encapsidated DNA is composed of multiple segments of superhelical dsDNA. The DNA is packed in capsid proteins and a single-layer (bracoviruses) or double-layer (ichnoviruses) envelope. It is suggested that bracoviruses and ichnoviruses evolved independently, as little or no sequence homology exists between the two genera. Phylogenetic analyses suggest that bracoviruses and nudiviruses have a common origin.

**External symptoms:** No pathological effects have been observed with the parasitoid hosts (Table 2.1).

**Internal symptoms:** Production of virus-like particles is restricted to specialized cells in the ovaries, the calyx cells (Table 2.1).

## B. RNA Viruses

### *1. Dicistroviruses*

The family *Dicistroviridae* is a member of the order *Picornavirales* and contains more than a dozen virus species that infect a wide range of arthropods, especially of the insect orders Diptera, Hemiptera, Hymenoptera, Lepidoptera, and Orthoptera. The type species of the genus *Cripavirus* is *Cricket paralysis virus* that causes paralysis of gut, fat body, trachea, muscle, and brain.

**Structure:** The non-enveloped virions have pseudo-T = 3 icosahedral symmetry and are 30 nm in diameter (Figure 2.6A).

**Genome:** The genome of the *Dicistroviridae* is a positive sense ssRNA molecule of 9–11 kb. In contrast to other members of the *Picornavirales* (*Iflaviridae*, *Picornaviridae*, and *Sequiviridae*), dicistroviruses have a bicistronic genome which codes the non-structural proteins at the 5′-end and the structural proteins at the 3′-end, respectively.

**External symptoms:** Infected insects exhibit reduced longevity and fecundity. Symptoms range from flaccidity of the body to paralysis, but commonly, infection is chronic and may become acute in response to stress induced by overcrowding (Table 2.1).

**Internal symptoms:** Most often the primary site of infection of dicistroviruses is the gut displaying tissue paralysis; but fat body, trachea, muscle, neurons, and nerves can also be infected (Table 2.1).

### *2. Iflaviruses*

Iflaviruses were previously considered as *Dicistroviridae* but due to their genome organization they have been recently reclassified into a separate family, *Iflaviridae*. The type species of the family is *Infectious flacherie virus*

beetle adult. The rod-shaped virions (V) are accumulating in the marginal 'ring zone' of the nucleus, while virus assembly is progressing in the nuclear center. Arrowheads show nuclear membrane. [Reprinted from Huger (2005) with permission from Elsevier]. E, F. Virion morphology of Bracoviruses and Ichnoviruses. Electron micrographs of the E. *Protapanteles paleacritae* bracovirus and the F. *Hyposoter exiguae* ichnovirus illustrate the morphological differences in members of the two Polydnavirus genera. Scale bars in E and F = 500 nm. [Reprinted from Fath-Goodin & Webb (2008) with permission from Elsevier.]

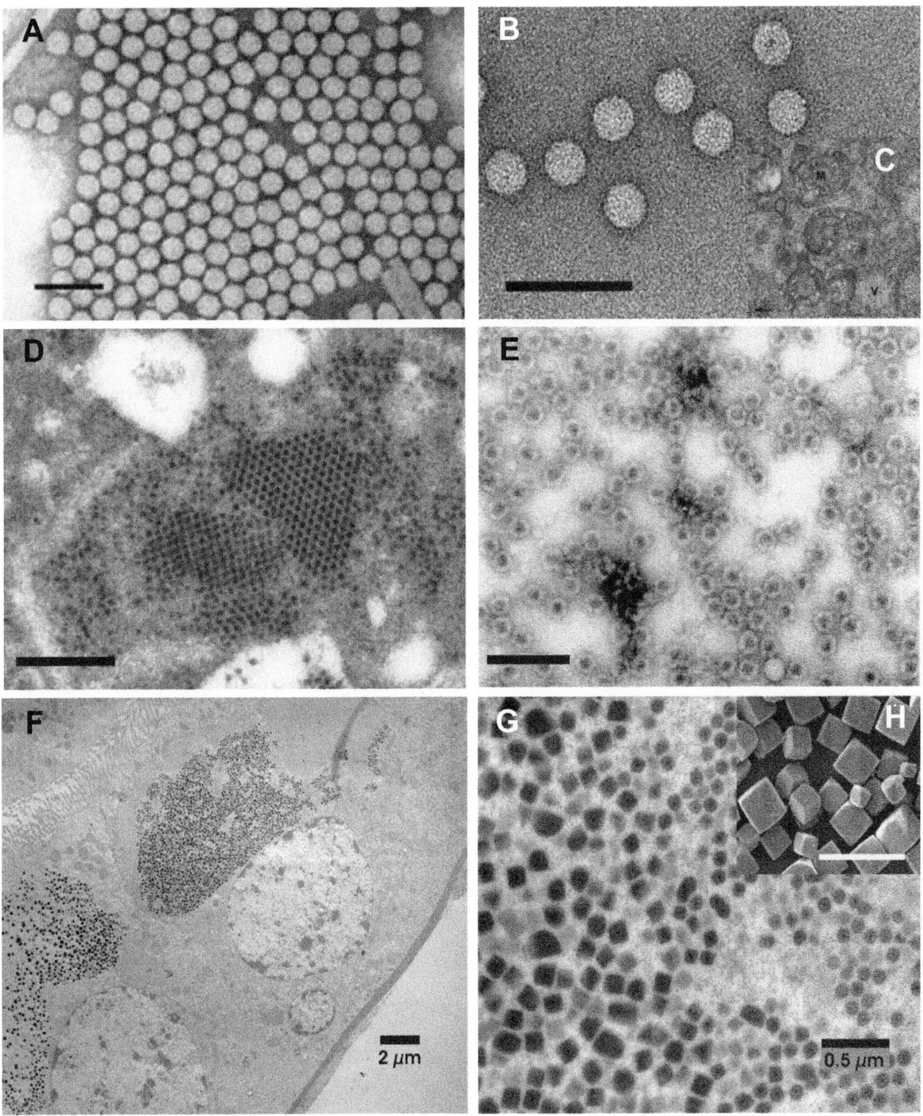


**Figure 2.6** A–G. Transmission electron micrographs. A. *Cricket paralysis virus* (CrPV) virions. Scale bar = 100 nm. [Reprinted from Fauquet *et al.* (2005) with permission from Elsevier.] B, C. *Alphanodavirus flock house virus* (FHV). B. Virion particles, negatively stained with 2% uranyl acetate. Scale bar = 100 nm. C. Sections through FHV-infected *Drosophila melanogaster* cells (10 h post-infection). 'M' designates mitochondria, which are extensively deformed as a result of FHV replication on the outer mitochondrial membrane, while 'V' points to small aggregates of virions in the cytoplasm. Scale bar = 200 nm. [Courtesy of P. A. Venter, A. Schneemann and M. R. Wood, Scripps Research Institute (La Jolla campus), USA.] D, E. Virus particles of the *Varroa destructor virus-1* (VDV-1). D. VDV-1 particles in mite tissue. E. Purified VDV-1 particles. Scale bars represent 300 nm D. and 100 nm E. [Courtesy of M. van Oers and D. Peters, Wageningen University, The Netherlands.] F, G. Midgut cells from the larval stages of *Uranotaenia lowii* infected with *Uranotaenia sapphirina cypovirus* (UrsaCPV). F. Cytoplasm of a midgut cell containing occlusion bodies of UrsaCPV. G. Occlusion bodies of UrsaCPV containing icosahedral virions, usually one per occlusion body. [Courtesy of J. J. Becnel, USDA, ARS, CMAVE, Gainesville, Florida, USA.] H. Scanning electron micrograph of occlusion bodies of *Bombyx mori cypovirus* (BmCPV). Scale bar = 10 μm. [Courtesy of P. Metcalf, University of Auckland, New Zealand.]

of silkworms. The proposed family belongs to the order *Picornavirales* and comprises insect viruses infecting Lepidoptera, Hymenoptera, Hemiptera, and bee parasitic mites (Acarina) (Figure 2.6D, E).

**Structure:** Iflaviruses form non-enveloped, icosahedral particles of approximately 30 nm in diameter (Figure 2.6D, E).

**Genome:** Iflaviruses have a positive sense ssRNA genome of 8.8–10.3 kb.

**External symptoms:** The viruses may cause diarrhea and developmental malformations. Sacbrood disease is an example of a fatal Iflavirus infection, causing a failure of honey bee larvae to pupate. Deformed wing virus (DWV) causes wing deformities, shortened abdomens, and discoloring of adult bees (Table 2.1).

**Internal symptoms:** The primary sites of replication are variable. Infectious flacherie virus, for example, first infects goblet cells of the midgut, while DWV spreads within the whole body. There is insufficient detail in electron microscope examination to determine the family characteristics (Table 2.1).

### 3. *Nodaviruses*

The nodaviruses represent a family of non-enveloped, icosahedral viruses. Members of the genus *Alphanodavirus* are insect-pathogenic. The type species is *Nodamura Virus* (NoV) isolated from mosquitoes. Additional well known members of the *Nodaviridae* family are *Black beetle virus* (BBV) from *Heteronychus arator* (Coleoptera: Scarabaeidae) and *Flock house virus* (FHV) from the scarab beetle *Costelytra zealandica,* both from New Zealand (Figure 2.6B, C).

**Structure:** Nodaviruses are non-enveloped icosahedral viruses with virions of 29–32 nm in diameter.

**Genome:** The nodaviral genome is bipartite and consists of two segments of positive sense ssRNAs, designated RNA1 (3.0–3.2 kb) and RNA2 (1.3–1.4 kb), respectively.

**External symptoms:** Most nodaviruses and, in particular, NoV and FHV produce unapparent or persistent infections, resulting in a slower development of the insect or a slight reduction in the total population due to a reduction of egg viability. However, infected larvae of *H. arator* may become flaccid and lose pigmentation of the hypodermis (Table 2.1).

**Internal symptoms:** Nodaviruses are morphologically identical to other members of the order *Picornavirales* of insects and are, thus, difficult to distinguish by electron microscopy. Virus replication takes place exclusively in the cytoplasm of the host cells (Figure 2.6C), and the formation of paracrystalline arrays is typical. Primary infection sites are muscle cells, hemocytes and hypodermis (Table 2.1).

### 4. *Tetraviruses*

Members of family *Tetraviridae* exclusively infect larvae of Lepidoptera, many of which are important agricultural pests. Little is known about the biology of tetraviruses and the interaction between virus and host. Most tetraviruses were isolated from saturniid, limacodid, and noctuid moths. The type species is the *Nudaurelia-β virus*, which was isolated from *Nudaurelia cytherea capensis* (Saturniidae).

**Structure:** Virus capsids are non-enveloped icosahedra with a T = 4 symmetry and a diameter of 35–38 nm.

**Genome:** The genome consists of one or two positive sense ssRNA molecules.

**External symptoms:** Tetravirus infections of neonate larvae produce a chronic disease, where the only gross symptom is a marked reduction in larval and pupal size. Virus administered to late-stage larvae appear to have no effect but can persist as inapparent infections. Underdevelopment and malformed wings of adults can be found (Table 2.1).

**Internal symptoms:** The virus multiplies primarily in midgut cells. The integrity of the internal tissues is largely unaffected (Table 2.1).

### 5. *Cypoviruses*

Cypoviruses (CPVs) are insect specific members of the family *Reoviridae* and are characterized by producing OBs which embed large numbers of CPV virions (Figure 2.6F, G). CPVs have an RNA genome and replicate in the cytoplasm of the infected cells, whereas NPVs have a dsDNA genome and replicate in the nucleus. CPVs have been isolated from more than 250 insect species.

**Structure:** CPVs are characterized by the presence of capsids made up of concentric protein layers with icosahedric symmetry (Figure 2.6G). The virions are approximately 60 nm in diameter with 12 prominent turrets which extend approximately 4 nm outwards from the surface at the icosahedral five-fold symmetry axes.

**Genome:** The genome is composed of 10–11 segments of dsRNA.

**External symptoms:** CPV infections lead to a retarded development and reduced feeding. Longevity and breeding performance of infected adults is lowered (Table 2.1).

**Internal symptoms:** CPV infection is restricted to the gut which may become white or yellow (Table 2.1).

## 3. IDENTIFICATION

A detailed description of preparation of diseased insects for light and electron microscopy is given by Becnel in Chapter XV.

### A. Light microscopy

A very important tool for initial differentiation between virus groups is the use of light microscopy with phase-contrast and bright-field microscopes. Comprehensive overviews of preparation procedures and diagnostic techniques were described by Adams & Bonami (1991), Evans & Shapiro (1997), and in Chapter XV.

#### *1. Preparation of microscope slides*

Preparation procedures adapted from Evans & Shapiro (1997) are presented below:

1. Prepare smears from tissues of all organs from the whole insect body.
2. Thinly spread the tissues, ideally as a monolayer of cells, enabling nuclear and cytoplasmic details to be viewed.
3. The instruments for preparations of smears essentially consist of forceps to handle and tease apart the specimens and mounted needles to aid spreading of body contents.
4. If the specimen is already somewhat desiccated or is very small, it may be necessary to use a saline solution (Chapter XV).
5. The use of a permanent marking system is needed such as a diamond marker or, for frosted slides, a resistant marker to indicate any of the reagents employed in the staining procedure.
6. Prevent contamination between smears by decontamination of all instruments by rinsing in alcohol and wiping or flaming.
7. Air dry preparations before staining.
8. Use of a fixative is dependent on the stain, but is not always necessary.

#### *2. Staining methods for light microscopy*

For virus diagnosis, particularly of those producing OBs, there are three most important stains that are commonly employed: (1) Buffalo Black 12B; (2) Giemsa's stain; and (3) Heidenhain's hematoxylin stain. All three stains are permanent and can be stored in light-protected boxes for many years.

##### *a. Buffalo Black 12B stain (Evans & Shapiro, 1997)*

1. Air dry the preparation to be stained.
2. Heat the Buffalo Black solution to 40–50°C in a staining rack on a hotplate.
3. Immerse the slide into the Buffalo Black solution for 5 min.
4. Wash the slide under running tap water for 10 s.
5. Dry the slide and examine under oil immersion for the presence of OBs.
6. Protein is stained blue–black, the crystalline protein of virus OBs can be distinguished from other cells and cellular components.

##### *b. Giemsa's stain (adapted from Evans & Shapiro, 1997)*

1. Immerse slides with air-dried smears for 2 min in Giemsa's fixative.
2. Rinse slides under running tap water for 10 s.
3. Stain for 45 min in 10% Giemsa stain in 0.01 M phosphate buffer, pH 6.9. Giemsa's azur eosin methylene blue solution (Merck) is known to work well.
4. Rinse under running tap water for 10 s.
5. Air dry the slide and apply a mounting medium (Entellan, Merck, or other suitable mounting medium) and a coverslip to improve longevity of the stain.
6. Examine under oil immersion.
7. Giemsa's stain is a differential stain that allows distinguishing nuclear and cytoplasmic cellular details. Therefore, it can be used for determination of the sites of replication of various virus groups. Using Giemsa's stain, nuclei appear red, cytoplasm blue, while OBs of NPV or CPV remain colorless but with a distinct edge.

*c. Heidenhain's hematoxylin stain (Langenbuch, 1957)*

1. Air dry the preparation to be stained.
2. Immerse for 5 min in xylene to remove fat.
3. Use a decreasing EtOH concentration series (100%, 95%, 70%, 60%) for 3 min each.
4. Fixation according to Bouin Dubosq Brazil (Romeis, 1989) (see Appendix of Chapter XV) for 1 h.
5. Immerse in 80% EtOH for 10 min.
6. Use a decreasing EtOH concentration series (100%, 95%, 70%, 60%, 50%, distilled water) for 3 min each.
7. Immerse in 96% glacial acetic acid for 1 min (this procedure was superseded by treatment with NHCl) (Huger, personal communication).
8. Rinse under distilled water for 10 s, mordant in 2.5% iron alum for 1 h.
9. Rinse under distilled water for 10 s, stain in Heidenhain hematoxylin for 1 h.
10. Rinse in slow-running tap water for 5 min.
11. De-stain in iron alum until nuclei stand out distinctly, observe under a bright-field microscope.
12. Rinse in slow-running tap water for 15 min.
13. Dehydrate in graded EtOH and xylene for 3 min each.
14. Mount in Entellan (Merck) or other suitable mounting medium.
15. Polyhedral OBs are stained deep black with this classic procedure that can be used for tissue smears or sectioned material.
16. A special staining technique to differentiate between OBs of NPV and CPV was described by Evans & Shapiro (1997) based on the method of Wigley (1980a). This is useful, if it is suspected that laboratory cultures may be contaminated with CPV. A glass slide with a smear of virus-infected tissues is stained in three zones for a differential staining with (1) Giemsa, (2) Picric acid Giemsa, and (3) Buffalo Black to get different colors of the OBs.

## B. Electron microscopy

A very useful tool for diagnosis is scanning (SEM) and particularly transmission electron microscopy (TEM). Viruses, especially GVs and small non-occluded viruses cannot be distinguished by light microscopy, but are easily by SEM or TEM. The direct layering of virions on grids and the use of ultrathin sections for examination under the TEM enables the investigation of virion structure and provides considerable insights into the virus taxa.

To detect viruses with TEM, *negative staining* is a rapid and robust method (see Chapter XV). Purified or semi-purified material can be placed directly onto formvar or other grids and layered with aqueous phosphotungstic acid to create a darker background around the virus particle. By a short treatment of OBs with 0.1 N NaOH or KOH resulting in a partly solvation of the OBs, virus particles can also be observed by negative staining.

For detailed ultrastructural studies, special tissue preparation, fixation, dehydration, staining, embedding in resin, sectioning, and placing on a grid for examination are necessary. Thin sections of body tissues or purified viral preparations can reveal most diagnostic features necessary for identification. Sections of a thickness of 60 to 80 nm, prepared with an ultramicrotome, show fine details of cellular organization [see Adams & Bonami (1991), Evans & Shapiro (1997), and Chapter XV].

## C. DNA restriction endonuclease analysis

Due to the lower precision of discriminative morphological and immunological characters of viruses of the same family, analytical methods based on genotypic traits became the most powerful routine in virus identification on the species and strain level. For invertebrate viruses with a dsDNA genome, restriction endonuclease (REN) digestion combined with agarose gel electrophoresis is the method of choice for a fast and reliable identification and discrimination of virus isolates. RENs are enzymes, prevalent in eubacteria and archaea, which cut dsDNA at specifically recognized nucleotide sequences, the so-called restriction site. Among the three different classes of RENs, Type II RENs that recognize palindromic sequences of six base pairs (bp) are the most important ones for analyzing virus DNAs. These enzymes produce blunt ends, $3'$ or $5'$ overhangs, a feature which is not important for the REN analysis but eventually for downstream cloning experiments. Since their first discovery in 1970, thousands of RENs have been discovered and several hundred are cloned and commercially available (Roberts *et al.*, 2007). The enzymes are normally delivered with a suitable buffer system and further instructions for choosing the optimal reaction temperature, which may range between 25 and 60°C.

The frequency of a given REN site in a given genome depends on the length of the REN site and the nucleotide composition. A 4-bp REN site is statistically 16-times more frequent on a DNA molecule and hence generates many more and much smaller fragments than a 6-bp REN site or even a rare cutter with an 8-bp recognition site. RENs with AT-rich recognition sequences cut AT-rich DNAs more often than GC-rich DNAs. Hexanucleotide RENs, e.g., *Eco*RI (G\AATTC), statistically cut every 4000 nucleotides and typically generate five to 30 fragments of a genome of 100–200 kbp.

REN analysis of virus genomic DNA has become an extremely important method in the pre-PCR and pre-sequencing era, as it was the only powerful method for species and strain identification and the basis for the construction of physical maps, which define the relative location and ordering of REN sites to each other in a genome. Nowadays, PCR amplification followed by sequencing (Section 3D) and genome sequencing (Section 3F) complement REN analyses and provide excellent information at a low price. REN mapping has been fully replaced by genome sequencing.

The second step in a REN analysis is the separation of the REN fragments through an agarose gel in an electric field. Due to the uniform negative charge of the DNA's phosphate backbone, DNA molecules move in an electric field from the cathode to the anode. The electrophoretic mobility is a function of the fragment length: short fragments move faster than longer ones due to the sieving effect of the agarose matrix. The DNA fragments are separated in horizontal gels containing 0.5–1.0% agarose, which give the best separation for fragments of 100–25,000 bp. Smaller fragments are typically separated in vertical polyacrylamide gels, larger fragments need to be applied to pulse field electrophoresis (Sambrook & Russell, 2001). The rate of migration of linear DNA fragments is further determined by the agarose concentration, the strength of the electric field, and the buffer system. Best electrophoretic results for REN digests of large DNA viruses are obtained with large horizontal gels of at least 20–25 cm length and 20 cm width. The electrophoresis needs to be conducted at a low voltage (20–40 V) for 10–14 h, which can be easily performed overnight. The buffer of choice for separating large DNA fragments is a 1×TAE buffer consisting of Tris-acetate buffer supplemented with EDTA. The separation of fragments larger than 5000 bp is significantly better in TAE buffer than in TBE buffer (Tris-borate, EDTA), which is superior for separating small fragments (< 5000 bp). The disadvantage of using TAE buffer is its low buffer capacity and the rapid exhaustion of the buffer, resulting in an acidification of the anodal buffer compartment. Therefore, this buffer can be used only once; exhaustion can be avoided by exchanging or recirculating the buffer during the run.

Before loading the DNA digest into the gel slot, it needs to be mixed with loading buffer consisting of a loading substance (e.g., sucrose, glycerol, or Ficoll) and a dye (e.g., bromophenolblue and/or xylene cyan FF). The loading substance facilitates the sedimentation of the sample in the gel slot, whereas the negatively charged dyes co-migrate with the DNA and indicate the rough position of a linear dsDNA fragment of 300 bp (bromophenol blue) or of 5000 bp (xylene cyan FF). As the latter may interfere with the detection of 4000–6000-bp fragments, loading buffers with only bromophenol blue are recommended. The DNA fragments are separated along with a dsDNA size standard containing fragments of defined size (e.g., lambda DNA digests, 1-kb ladders, etc.). After the gel run is completed the gel is submerged and gently agitated in a bath with fluorescent intercalating dyes, such as ethidium bromide (EtBr, 1 μg/ml) or SybrGold. Note that EtBr is mutagenic, carcinogenic, and teratogenic and needs to be handled with care and disposed of according to local safety requirements. SybrGold is considered less harmful and has a > 1000-times greater fluorescent enhancement than EtBr. However, it is significantly more expensive.

Stained gels are typically excitated at 302 nm and photographed using a charged couple device (CCD) camera or image system with appropriate filters.

DNA REN digestion and agarose gel electrophoresis are described in detail in many method books on molecular biology (Sambrook & Russell, 2001). A standard protocol based on experience with restriction analysis of baculovirus DNA is given below.

Material

- Purified virus DNA at a concentration of ~ 100 ng/μl.
- Restriction endonuclease and REN buffer (commercially available).
- 1 x TAE buffer (40 mM Tris, 20 mM acetic acid, 1 mM EDTA pH 8.0).
- 1 x TBE buffer (90 mM Tris, 90 mM boric acid, 1 mM EDTA, pH 8.0).

- 5 × Loading buffer [0.25% bromophenol blue, 15% Ficoll (Type 400) Pharmacia].
- EtBromide stock solution (10 mg/ml in $H_2O$).

*1. Restriction endonuclease digestion*

1. Pipette 0.8−1 μg DNA (dissolved in less than 10 μl TE buffer), 2 μl 10 × REN buffer (supplied with enzyme), 1 μl REN (5−20 units) into a reaction tube and fill up to 20 μl with $ddH_20$. Mix the sample by snipping the reaction tube (no vortex to avoid DNA shearing) and spin the solution down for a few seconds.
2. Incubate the mixture at the appropriate temperature for 2 h.
3. Take an aliquot of 2 μl of the digest for electrophoresis on a minigel and freeze the remaining 18-μl digest at −20°C.

*a. Minigel electrophoresis*

1. This is performed for an initial examination of REN digests for the amount and the digestibility of the DNA. Mix the 2-μl aliquot with 0.4 μl 5 × Loading buffer and run the sample on a 0.7% agarose minigel prepared with TBE buffer. Minigels can be directly stained by adding EtBr at a final concentration of 0.5 μg/ml to the agarose solution prior to cooling.
2. Run the minigel at 40 V for 1−3 h.
3. If only a smear of DNA or a mixture of weak and strong bands is visible (most likely an incomplete digest), the incubation time of the frozen aliquot needs to be extended for another two or more hours at the appropriate digestion temperature. Adding fresh REN may increase the efficacy of digestion. Depending on the quality of the DNA preparation it might be necessary to incubate the DNA overnight.

*b. Horizontal gel electrophoresis*

1. When the digests for all chosen enzymes are completed, then add 1/5 volume of 5 × Loading buffer to the DNA digests and apply the samples into the slots of a large 0.7−1% agarose gel buffered with TAE. Add size standards to the most right and most left lanes of the gel.
2. Run the gel overnight at low voltage, e.g., 20 V.
3. When the electrophoresis is completed, immerse the gel for 1 h in a solution of 1 μg/ml EtBr while gently shaking. De-staining in water for 30 min may help to enhance the signal-to-noise ratio. This time the gel should be run in the absence of EtBr for the following reasons: (1) EtBr reduces the electrophoretic mobility of a linear dsDNA molecule by about 15% resulting in less sharp bands. (2) As EtBr is positively charged it migrates in the gel from the anode to the cathode resulting in a gradient of dye. For the detection of submolar bands or double bands and their quantitative interpretation, however, an even distribution of EtBr in the gel is essential.
4. Visualize the fragments in the gel using a gel documentation system with an appropriate orange filter.

Beyond the identification of a given virus, REN analysis is still a very important source of information, e.g., for the identification of genome mutations and/or for the estimation of the homogeneity of a virus sample allowing to determine whether a virus preparation is a mixture of different genotypes. Virus genomes may undergo point mutations or insertions/deletions and both can be detected by REN analysis. Point mutations within a restriction site result in its loss and thus in a fusion of neighboring fragments. Such a mutation, however, can only be detected when using the REN enzyme specific for the mutated restriction site, any other digest is not affected by it. In contrast, an insertion/deletion at a specific genomic location is theoretically visible in any digest, if the length difference between the original and the mutated fragment is big enough to be resolved on the agarose gel.

REN analysis is also powerful in identifying mixtures of genotypes in a virus sample. When the DNA of a homogenous (single genotype) DNA preparation is digested with REN, each REN fragment is generated at an equimolar ratio (Figure 2.7). When separated on an agarose gel and stained with EtBr or other intercalating dyes, the intensity of each band is proportional to the amount of DNA in each band and due to the equimolarity of fragments proportional to the fragment size. This means that the intensity of bands continuously increases with the size of a fragment. If two different REN fragments have the same size, they coincide in the same band and such a band would have a double fluorescent intensity. If the intensity of a band does not fit into the gradient of fluorescence observable from the smaller to the larger fragment it can be assumed to be submolar, indicating a second genotype in the virus preparation. Thus REN analyses allow determination of the homogeneity of

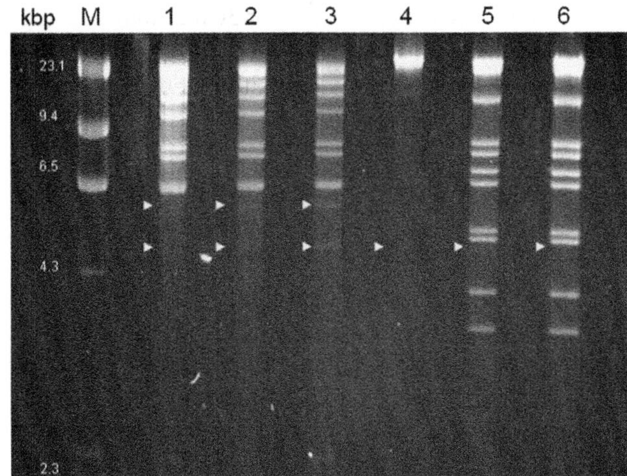


**Figure 2.7** Agarose gel with submolar bands in *Pst*I (lanes 1–3) and *Xho*I (lanes 4–6) digests of DNA of *Cydia pomonella* granulovirus (isolate I12) caused by (i) star activity or (ii) incomplete digestion. The DNA samples were digested with 5 U (lanes 1 and 4) 10 U (lanes 2 and 5) and 20 U (lanes 3 and 6) of *Pst*I and *Xho*I, respectively. If a digest shows star activity (see *Pst*I digests), submolar bands (arrows) increase with the duration of digestion and/or the concentration of restriction endonuclease. In an incomplete digestion (see *Xho*I digests) the intensity of submolar bands (arrows) decrease with the duration of digestion and/or concentration of restriction endonuclease. Lane M = size marker lambda DNA/*Hin*dIII in kbp.

a virus preparation. This technique can also be applied to quantify genotypes in a mixture of two or more viruses (Arends *et al.*, 2005)

Interpretation of REN patterns needs some experience as the results can be obscured by incomplete digests, best visible in the presence of numerous submolar bands, and by star activity of some RENs caused by non-optimal reaction conditions, such as high ionic strength, inappropriate pH, temperature, or high concentration of glycerol. Star activity results in a cleavage at a non-canonical REN site and thus in a more or less pronounced loss of specificity of the given REN (Figure 2.7).

If the REN patterns of two viruses are very similar or identical it can be concluded that they are narrowly related or identical. However, if the REN patterns are different from any other virus, nothing can be said about the relationship. Even closely related viruses encoding highly similar proteins cannot be identified if there is a change in the AT content that is restricted to the third (whobble) position of the codon (Lange & Jehle, 2003). To make a statement about the relationship of a new virus to known viruses, concepts of molecular evolution need to be applied. However, these are infeasible for REN analyses.

## D. PCR, partial sequencing, and phylogeny

To overcome the limitation of REN analysis for virus identification, the amplification of conserved gene fragments by PCR, followed by sequencing and the phylogenetic analysis of these fragments have proven to be extremely efficient and useful. Methods have been developed for baculoviruses (Lange *et al.*, 2004; Herniou *et al.*, 2004; Jehle *et al.*, 2006) and may be adopted in principle for members of other virus families. All baculovirus genomes sequenced so far share 30 conserved genes. Some of these core genes contain highly conserved sequence motifs, which has allowed the design of degenerate oligonucleotides specific for *Alpha*- and *Betabaculoviruseses*, which represent the two largest baculovirus genera. Degenerate oligonucleotides contain some ambiguous positions in their sequence, especially at the 3′-end, which allow binding to highly conserved DNA motifs with a few sequence variations. Fragments of three highly conserved genes, the *polyhedrin/granulin (polh/gran),* the *late expression factor 8 (lef-8),* and the *late expression factor 9 (lef-9)* were targeted for primer design, after it had been demonstrated that the concatenated sequences/combined analysis of these fragments produced similar phylogenetic tree topologies to the complete set of the core genes, thus reproducing a robust phylogeny of the virus and enabling predictions about the relationship of an uncharacterized virus to known viruses (Lange *et al.*, 2004). At their 5′-end, the oligonucleotides were fused to universal sequencing primer sequences allowing direct sequencing of the PCR products.

### *1. DNA extraction from diseased insects*

1. Grind infected tissue or larvae using liquid nitrogen or a disposable mini pistil.
2. Extract DNA from homogenized sample using standard techniques, DNeasy Tissue Kit (Qiagen, Hilden, Germany) according to the instructions of the manufacturer.
3. Elute DNA in a volume of 50 µl elution buffer or $ddH_2O$.

*2. Perform PCR on prepared DNA samples*

1. Add the following components to a 50-µl reaction vial:
   5 µl of 10 × reaction buffer (normally provided by Taq polymerase manufacturer)
   100 µM each of dATP, dCTP, dGTPm and dTTP
   52 mM $MgCl_2$
   1 unit Taq DNA polymerase
   0.2 µM of each primer
   1–100 ng of prepared template DNA
   Add water to 50-µl reaction volume
2. Perform PCR reaction according to the following conditions:
   Amplification reactions with *polh/gran* specific primers: Initial denaturing step of 95°C for 3 min followed by 36–40 cycles of 95°C for 30 s, 50–53°C for 1 min, 72°C for 1 min, and a final extension step at 72°C for 10 min.
   Amplification reactions with *lef-8* specific primers: Initial denaturing step of 95°C for 4 min, 30–35 cycles of 95°C for 2 min, 38–48°C for 1 min, 72°C for 1 min, and a final extension step 72°C for 5 min.
   Amplification reactions with *lef-9* specific primers: Initial denaturing step of 95°C for 4 min, followed by 35 cycles of 72°C for 1 min, 95°C for 2 min, 45–54°C for 1 min, and a final extension step 72°C for 2 min.
3. Run the fragments on a 1% agarose gel (see Section 3C) and purify the PCR products by using the GFX PCR DNA and Gel Band Purification kit (Amersham) or any other appropriate method.

*3. Sequence analyis of PCR fragments*

1. Supply the purified PCR fragments to DNA sequencing by using the sequencing primers specific for the 5′-ends of the oligonucleotide primers used in step 1 and sequence both strands of the DNA.
2. Merge sequences of both strands to a single sequence and perform BLAST search to GenBank (NCBI).
3. Verify that the partial *polh/gran*, *lef-8* and *lef-9* gene sequences consistently group within the *Alpha- or Betabaculoviruses.* If identical hits are found in GenBank the virus of interest is identified.
4. In order to determine the phylogenetic relationship to other known baculoviruses perform phylogenetic analyses using MEGA, Phylip or any other appropriate software.

A sequence based approach for classification was suggested by Jehle *et al.* (2006): Consider the isolate of interest as belonging to the same baculovirus species as another isolate, if their Kimura 2-Parameter (K-2-P) distance between single and/or concatenated *polh*, *lef-8*, and *lef-9* nucleotide sequences is smaller than 0.015. If the K-2-P distance between single and/or concatenated sequences is larger than 0.050, then consider two viruses as different virus species. For K-2-P distances between 0.015 and 0.050, no sequence-based prediction about their classification can be given and complementary biological information is needed.

## E. Identification of RNA viruses

Detection of RNA virus infection in an insect population requires more attention than detection of DNA viruses. RNA viruses often cause chronic infections which are more difficult to detect in the host population. Infection can be restricted to only one tissue and pathogenicity may vary between virus isolates of the same family. Despite the fact that some viruses have descriptive names like the honeybee infecting Deformed wing virus (DWV) or the Acute bee paralysis virus (ABPV), there is often no reliable relationship between symptoms and virus which would be suitable as a diagnostic feature (Erlandson, 2008). In contrast to insect DNA viruses, the virions of most RNA viruses are icosaeders smaller than 40 nm and are not occluded.

The usual way of RNA virus identification from a sample starts with the observation of viral particles by electron microscopy (EM). In contrast to molecular or serological methods, electron microscopy does not require specific probes or reagents which would require further information about the unknown sample, in order to choose the correct chemicals and protocols. By using EM, it is possible to identify most viruses to the family level, a pre-requisite for a more detailed characterization by biochemical methods (Goldsmith & Miller, 2009). Numerous publications of micrographs and descriptions of virion size and shape for virus identification are available (Fauquet *et al.*, 2005; Hunter-Fujita *et al.*, 1998). Physical properties like stability in buffer or sedimentation are then further features for differentiation between various viruses of one family.

For identification on the nucleic acid level, a total RNA isolation of infected insects is performed

(Section 4D). Viral RNA isolated from an insect can be either dsRNA or ssRNA. The dsRNA and ssRNA can be separated on hydroxyapatit using different temperature and salt conditions, as was shown for the ssRNA tobacco mosaic virus (TMV) and the dsRNA CPV (Kalmakoff & Payne, 1973). Precise information about the viral genome requires analysis on the molecular level. Before the advent of molecular techniques, the identification of RNA viruses relied on serological techniques and the production of antisera. These methods have some disadvantages like low specificity and sensitivity and do not detect latent infections (Teixeira *et al.*, 2008).

Today, molecular methods such as reverse transcriptase polymerase chain reaction (RT-PCR) allow a sensitive and specific identification of an RNA virus in all insect life stages or tissues, even at low concentrations. Furthermore, multiple virus infections can be detected in one reaction (multiplex RT-PCR). The RT-PCR reaction is based on two main steps: first, RNA is transcribed by the enzyme reverse transcriptase into complementary DNA (cDNA); second, the resulting cDNA is the template for DNA amplification resulting in a dsDNA molecule. Two types of primers can be applied for RT-PCR detection of RNA viruses: (1) Random oligomers are oligonucleotides with a statistical base composition. These primers can hybridize at different positions in the RNA and start a cDNA synthesis over the complete RNA length. (2) Sequence-specific primers presume the knowledge of either virus specific sequences, or sequences, which are conserved in the virus family or sequences of a closely related virus.

*Standard reverse transcription protocol using SuperScript II reverse transcriptase*

1. Mix the following reagents in a 0.5-ml tube on ice:
   1 μl primer (50–250 ng random primer or 2 pmol sequence specific primer)
   *x* μl RNA (1 μg)
   *x* μl $H_2O$ to a total volume of 12 μl.
2. Incubate the solution for 5 min at 65°C for denaturation, put it then directly on ice.
3. Add the following solutions:
   4 μl 5 × first strand buffer
   2 μl DTT (0.1 mM)
   1 μl dNTP mix (10 mM)
4. Incubate for 2 min at 42°C.
5. Add 1 μl of SuperScript II.
6. First strand cDNA synthesis is performed in cycler using the following conditions:
   42°C for 50 min
   70°C for 15 min
7. Store cDNA at −20°C.

Two principal methods can be distinguished in RT-PCR protocols. During *two step*-RT-PCR, reverse transcription and cDNA synthesis is performed first and a sample of this RT reaction is then used as a template in the following PCR. This allows the adjusting of optimal conditions for both enzymatic reactions. In a *one step*-RT-PCR, reverse transcription and PCR are performed in the same tube, reducing the risk of contamination. Here, sequence-specific primers are a pre-requisite, as one of the primers is used for the first strand cDNA synthesis.

Various RNA virus genomes have been partially or completely sequenced. Thus, numerous virus-specific primers and RT-PCR protocols are already described in the literature or can be designed according to published sequences (Table 2.2).

## F. Genome sequencing

In the 1970s, Frederick Sanger presented a method for nucleotide sequence determination, which is also known as the chain-termination method or Sanger method (Sanger *et al.*, 1977). In the following decades, this method was continuously improved and it is still one of the most frequently used techniques for DNA sequencing. Although many procedures have been automated, time- and labor-consuming sample preparation and nucleotide fragment separation are still necessary. However, the Sanger method is still the most appropriate technique to sequence small DNA molecules, like plasmids, PCR fragments, or small genomic regions. To obtain a higher sequencing throughput, next-generation sequencing (NGS) techniques were invented and have revolutionized the genomic era. The NGS techniques include the 454 pyrosequencing technology (Life Science/Roche), Illumina sequencing technology, and SOLiD sequencing (Applied Biosystems). These three NGS techniques parallelize processes and are characterized by producing thousands or millions of sequencing reads. Thus complete genomes can be *de novo* sequenced with a single run. This revolutionary progress in DNA sequencing also provides new opportunities in insect virus genome analyses. But it also entails extensive data

**Table 2.2** Sequenced RNA virus genomes of different virus families and PCR protocols for their detection and identification.

| *RNA virus (*family*)* | *Genome structure/ GenBank accession no.* | *Host* | *PCR primer/protocol* |
|---|---|---|---|
| Acute bee paralysis virus (ABPV) (*Dicistroviridae*) | (+) ssRNA, NC_002548 | *Apis mellifera scutellata* (Hymenoptera: Apidae) | (Teixeira *et al.*, 2008) (Benjeddou *et al.*, 2001) |
| Black queen cell virus (BQCV) (*Dicistroviridae*) | (+) ssRNA, NC_003784 | *Apis mellifera scutellata* (Hymenoptera: Apidae) | (Teixeira *et al.*, 2008) (Benjeddou *et al.*, 2001) |
| *Bombyx mori* CPV (*Reoviridae*) | dsRNA (segmented), AY388398.1 (VP1 gene) | *Bombyx mori* (Lepidoptera: *Bombycidae*) | (Hagiwara *et al.*, 1998) |
| Cricket paralysis virus (*Dicistroviridae*) | (+) ssRNA, NC_003924 | *Teleogryllus* spp. (Orthoptera: Gryllidae) | (Johnson & Christian, 1996) |
| Deformed wing virus (DWV) (*Iflaviridae*) | (+) ssRNA, NC_004830 | *Apis mellifera* spp. (Hymenoptera: Apidae) | (Yue & Genersch, 2005; Teixeira *et al.*, 2008) |
| *Drosophila C* virus (*Dicistroviridae*) | (+) ssRNA, NC_001834 | *Drosophila* spp. (Diptera: Drosophilidae) | (Kapun *et al.*, 2010) |
| *Drosophila* X virus (*Birnaviridae*) | dsRNA, NC_004177 (segment A), NC_004169 (segment B) | *Drosophila melanogaster* (Diptera: Drosophilidae) | (Zambon *et al.*, 2005) |
| Flock House virus (*Nodaviridae*) | ssRNA, NC_004146 (RNA1), NC_004144 (RNA2) | *Costelytra zealandica* (Coleoptera: Scarabaeidae) | (Perera *et al.*, 2009) |
| *Nudaurelia β* virus (*Tetraviridae*) | (+) ssRNA, NC_001990 | *Nudaurelia cytherea capensis* (Lepidoptera: Saturniidae) | (Walter *et al.*, 2008) |
| Sigma virus (DmelSV) (*Rhabdoviridae*) | (−) ssRNA, NC_013135 | *Drosophila melanogaster* (Diptera: Drosophilidae) | (Carpenter *et al.*, 2007) |
| *Solenopsis invicta* virus-3 (SINV_3) | (+) ssRNA, FJ528584 | *Solenopsis invicta* (Hymenoptera: Formicidae) | (Valles & Hashimoto, 2009) |

processing. Virus genomes are generally too small to be sequenced alone by NGS platforms. Usually, several sequencing projects need to be handled and processed during a single run and a huge amount of output data has to be processed and interpreted. Data processing can take up to several weeks and has to be considered in whole genome sequencing experiments.

What all sequencing techniques have in common is that the whole genome is randomly sheared into smaller fragments, which are sequenced separately to obtain *reads*. Each read is a short nucleotide sequence of 100–1000 nucleotides (depending on the sequencing technique applied) and the combined reads of a sequencing project contain ideally the whole genome sequence. Because several of genomic molecules are randomly fragmented and sequenced, the sequence reads consist of a large number of overlapping random fragments. These overlapping reads are used for assembling the target genome by computer software.

Sequencing an insect virus genome usually requires 10–30 μg of viral DNA or cDNA. In most cases, the virus has to be propagated first (Section 4A) before genomic DNA/RNA can be isolated and purified (Sections 4C, D). For virus propagation, the appropriate host insect or cell line is required, which can be challenging in some cases. A method based on multiple displacement amplification by using Φ29 DNA polymerase and requiring only nanogram amounts of DNA has been successfully applied for sequencing the genome of the *Oryctes rhinoceros* nudivirus (Wang *et al.*, 2008).

If compared with genomes of cellular organisms, insect virus genomes are small in size. The largest completely sequenced insect virus genomes are found in the family of the *Entomopoxviridae* (> 230,000 bp). In general, genomes of DNA viruses are larger than the genomes of RNA viruses. With the exception of the densoviruses (*Parvoviridae*) the genomes of DNA insect viruses are double-stranded and larger than 80,000 bp (Table 2.3A). Except for cypoviruses, the genomes of RNA insect viruses are mainly single stranded, unsegmented or segmented, and smaller than 10,000 bases (Table 2.3B).

Most insect virus genomes were sequenced by a whole genome shotgun sequencing technique based on the Sanger method. For many years, this technique was the key to deciphering the nucleotide sequences of genomes of cellular organisms and of viruses. Sanger sequencing is costly but accurate; NGS techniques will enable access to more genomic data within a shorter time. NGS techniques will simplify future attempts in terms of sequencing and comparison of several insect virus genomes at once.

### *1. Sanger sequencing*

This technique is based on the chain termination method (Sanger method) and has already been used for sequencing the human genome (Venter *et al.*, 1998). Most of the recent DNA and RNA insect virus genomes were completely sequenced using this technique.

For sample preparation, the genome is randomly sheared by sonification or nebulization or the genome is cut into specific fragments using RENs. After separation of fragments by agarose gel electrophoresis only parts of the fragments with a defined size from 0.5 to 8 kb are extracted and cloned into an appropriate vector. Clones are taken from this library and their inserts are sequenced by the Sanger method. If enough clones were sequenced each part of the genome should be covered and an assembly of all obtained sequences should form the whole genome. Sequencing length is up to 1000 and there are more nucleotides per read. In general, this method works well for small genomes like those of insect viruses. But it has one crucial step: some genomic fragments are difficult to clone into a vector and are consequently missing in the library. Sequencing of more clones of the library or 'primer walking' can fill occurring sequence gaps. This requires additional working steps like sample preparations and primer design, which are time consuming and increase costs.

### *2. 454 Sequencing*

This 'sequencing by synthesis' technique by 454 Life Science (Roche) is based on pyrophosphate detection (Ronaghi *et al.*, 1998). The target sequence is physically sheared and fragments are separately linked to small beads. Each fragment is clonally amplified by emulsion PCR (Margulies *et al.*, 2005) so that each bead is linked with several identical copies of a DNA fragment. The beads are separately loaded into wells of a PicoTiterPlate™ where the actual sequencing reactions take place. Similar to the Sanger method, a ssDNA molecule is used as a template and a complementary strand is synthesized with the aid of an oligonucleotide primer and a DNA polymerase. During the sequencing reaction the four nucleotides (dATP, dGTP, dCTP and dTTP) are successively added to the plate. After each nucleotide addition a washing step resets the system. Whenever a desoxynucleotide triphosphate (dNTP) is complementarily added to the template, a pyrophosphate ($PP_i$) is released which is subsequently detected by a cascade of enzymatic reactions as a light signal. The intensity of a dNTP-specific signal increases with the number of incorporated nucleotides at each sequencing cycle allowing the estimation of the numbers of incorporated dNTPs. Although this works well with single or a few dNTPs of the same type, it creates the problem of resolving homomers of five or more identical nucleotides. Sequencing reactions are performed in each well of the microtiter plate and provide sequence lengths of about 400 nucleotides on average. Because of the length of single *reads* the assembly of genomic DNA is less problematic and, therefore, the 454 technology is well applicable for *de novo* sequencing. To date, only two published insect virus genomes of Salivary Gland Hypertrophy Viruses (Abd-Alla *et al.*, 2008; Garcia-Maruniak *et al.*, 2008) (Table 2.3A) have been sequenced using this technique.

### *3. Illumina Sequencing*

This technique is also based on the principle of 'sequencing by synthesis' and can be regarded as

**Table 2.3A** List of completely sequenced and published genomes of invertebrate DNA viruses. All genomes were published at GenBank of the National Center for Biotechnology Information (NCBI).

| *Family (Subfamily)*<br>*Genus*<br>*Genome type* | *Virus* | *Abbreviation* | *Length (bp)* | *% GC* | *Accession* |
|---|---|---|---|---|---|
| *Ascoviridae* (–) | *Diadromus pulchellus* ascovirus 4a | DpAV-4a | 119343 | 49.7 | NC_011335 |
| ***Ascovirus*** | *Heliothis virescens* ascovirus 3e | HvAV-3e | 186262 | 45.9 | NC_009233 |
| dsDNA, circular | *Spodoptera frugiperda* ascovirus 1a | SfAV-1a | 156922 | 49.3 | NC_008361 |
| | *Trichoplusia ni* ascovirus 2c | TnAV-2a | 174059 | 35.2 | NC_008518 |
| *Baculoviridae* (–) | *Adoxophyes honmai* NPV | AdhoNPV | 113220 | 35.6 | NC_004690 |
| ***Alphabaculovirus*** | *Adoxophyes orana* NPV | AdorNPV | 111724 | 35 | NC_011423 |
| dsDNA, circular | *Agrotis ipsilon* NPV | AgipNPV | 155122 | 48.6 | NC_011345 |
| | *Agrotis segetum* NPV-A | AgseNPV-A | 147544 | 45.7 | NC_007921 |
| | *Antheraea pernyi* NPV | AnpeNPV | 126629 | 53.5 | NC_008035 |
| | *Anticarsia gemmatalis* multiple NPV | AgMNPV | 132239 | 44.5 | NC_008520 |
| | *Autographa californica* multiple NPV | AcMNPV | 133894 | 40.7 | NC_001623 |
| | *Bombyx mandarina* NPV | BomaNPV | 126770 | 40.2 | NC_012672 |
| | *Bombyx mori* NPV | BmNPV | 128413 | 40.4 | NC_001962 |
| | *Choristoneura fumiferana* DEF multiple NPV | CfDefNPV | 131160 | 45.8 | NC_005137 |
| | *Choristoneura fumiferana* NPV | CfNPV | 129593 | 50.1 | NC_004778 |
| | *Chrysodeixis chalcites* NPV | ChchNPV | 149622 | 39 | NC_007151 |
| | *Clanis bilineata* NPV | ClbiNPV | 135454 | 37.7 | NC_008293 |
| | *Ecotropis obliqua* NPV | EcobNPV | 131204 | 37.6 | NC_008586 |
| | *Epiphyas postvittana* NPV | EppoNPV | 118584 | 40.7 | NC_003083 |
| | *Euproctis pseudoconspersa* NPV | EupsNPV | 141291 | 40.3 | NC_012639 |
| | *Helicoverpa armigera* multiple NPV | HearMNPV | 154196 | 40.1 | NC_011615 |
| | *Helicoverpa armigera* NPV NNg1 | HearNPV NNg1 | 132425 | 39.2 | NC_011354 |
| | *Helicoverpa armigera* NPV C1 | HearSNPV-C1 | 130759 | 38.9 | NC_003094 |
| | *Helicoverpa armigera* NPV G4 | HearSNPV-G4 | 131405 | 39 | NC_002654 |
| | *Helicoverpa zea* SNPV | HzSNPV | 130869 | 39.1 | NC_003349 |
| | *Hyphantria cunea* NPV | HycuNPV | 132959 | 45.5 | NC_007767 |
| | *Leucania separata* NPV | LsNPV | 168041 | 48.6 | NC_008348 |
| | *Lymantria dispar* multiple NPV | LdMNPV | 161046 | 57.5 | NC_001973 |
| | *Lymantria xylina* NPV | LyxyNPV | 156344 | 53.5 | NC_013953 |
| | *Mamestra configurata* NPV-A | MacoNPV-A | 155060 | 41.7 | NC_003529 |
| | *Mamestra configurata* NPV-B | MacoNPV-B | 158482 | 40 | NC_004117 |
| | *Maruca vitrata* multiple NPV | MaviNPV | 111953 | 38.6 | NC_008725 |

**Table 2.3A** List of completely sequenced and published genomes of invertebrate DNA viruses. All genomes were published at GenBank of the National Center for Biotechnology Information (NCBI)—Cont'd

| *Family (Subfamily)*<br>*Genus*<br>*Genome type* | *Virus* | *Abbreviation* | *Length (bp)* | *% GC* | *Accession* |
|---|---|---|---|---|---|
| | *Orgyia leucostigma* NPV | OrleSNPV | 156179 | 39.9 | NC_010276 |
| | *Orgyia pseudotsugata* multiple NPV | OpMNPV | 131995 | 55.1 | NC_001875 |
| | *Plutella xylostella* NPV | PlxyNPV | 134417 | 40.7 | NC_008349 |
| | *Rachiplusia ou* multiple NPV | RoMNPV | 131526 | 39.1 | NC_004323 |
| | *Spodoptera exigua* multiple NPV | SeMNPV | 135611 | 43.8 | NC_002169 |
| | *Spodoptera frugiperda* multiple NPV | SfMNPV | 131331 | 40.2 | NC_009011 |
| | *Spodoptera litura* NPV | SpltNPV | 139342 | 42.8 | NC_003102 |
| | *Spodoptera litura* NPV II | SpltNPV II | 148634 | 45 | NC_011616 |
| | *Trichoplusia ni* single NPV | TnSNPV | 134394 | 39 | NC_007383 |
| ***Betabaculovirus*** | *Adoxophyes orana* granulovirus | AdorGV | 99657 | 34.5 | NC_005038 |
| dsDNA, circular | *Agrotis segetum* granulovirus | AgseGV | 131680 | 37.3 | NC_005839 |
| | *Choristoneura occidentalis* granulovirus | ChocGV | 104710 | 32.7 | NC_008168 |
| | *Chlostera anachoreta* granulovirus | ClanGV | 101487 | 44.4 | NC_015398 |
| | *Cryptophlebia leucotreta* granulovirus | CrleGV | 110907 | 32.4 | NC_005068 |
| | *Cydia pomonella* granulovirus | CpGV | 123500 | 45.3 | NC_002816 |
| | *Helicoverpa armigera* granulovirus | HearGV | 169794 | 40.8 | NC_010240 |
| | *Phthorimaea operculella* granulovirus | PhopGV | 119217 | 35.7 | NC_004062 |
| | *Pieris rapae* granulovirus | PrGV | 108592 | 33.2 | NC_013797 |
| | *Plutella xylostella* granulovirus | PlxyGV | 100999 | 40.7 | NC_002593 |
| | *Pseudaletia unipuncta* granulovirus | PsunGV | 176677 | 39.8 | NC_013772 |
| | *Spodoptera litura* granulovirus | SpltGV | 124121 | 38.8 | NC_009503 |
| | *Xestia c-nigrum* granulovirus | XecnGV | 178733 | 40.7 | NC_002331 |
| ***Deltabaculovirus***<br>dsDNA, circular | *Culex nigripalpus* NPV | CuniNPV | 108252 | 50.9 | NC_003084 |
| ***Gammabaculovirus*** | *Neodiprion sertifer* NPV | NeseNPV | 84264 | 33.4 | NC_005905 |
| dsDNA, circular | *Neodiprion lecontii* NPV | NeleNPV | 81755 | 33.3 | NC_005906 |
| | *Neodiprion abietis* NPV | NeabNPV | 86462 | 33.8 | NC_008252 |
| *Hytrosaviridae*<br>*Glossinavirus* | *Glossina pallidipes* salivary gland hypertrophy virus | GpSGHV | 190032 | 28 | NC_010356 |
| *Muscavirus*<br>dsDNA, linear | *Musca domestica* salivary gland hypertrophy virus | MdSGHV | 124279 | 43.5 | NC_010671 |
| *Iridoviridae*<br>***Chloriridovirus*** | *Aedes taeniorhynchus* iridescent virus | AtIV | 191100 | 47.9 | NC_008187 |
| dsDNA, linear | (Invertebrate iridescent virus 3) | | | | |

(*Continued*)

**Table 2.3A** List of completely sequenced and published genomes of invertebrate DNA viruses. All genomes were published at GenBank of the National Center for Biotechnology Information (NCBI)—Cont'd

| *Family (Subfamily)*<br>*Genus*<br>*Genome type* | *Virus* | *Abbreviation* | *Length (bp)* | *% GC* | *Accession* |
|---|---|---|---|---|---|
| ***Iridovirus*** | *Chilo* iridescent virus | CIV | 212482 | 28.6 | NC_003038 |
| dsDNA, linear | (Invertebrate iridescent virus 6) | | | | |
| *Polydnaviridae* (−)<br>***Ichnovirus*** | *Campoletis sonorensis* ichnovirus | CsIV | - | - | NC_007985-008008 |
| dsDNA, proviral | *Glypta fumiferanae* ichnovirus | GfIV | - | - | NC_008837-008941 |
| | *Hyposoter fugitivus* ichnovirus | HfIV | - | - | NC_008946-009003 |
| ***Bracovirus*** | *Cotesia congregata* bracovirus | CcBV | - | - | NC_006633-006662 |
| dsDNA, proviral | *Microplitis demolitor* bracovirus | MdBV | - | - | NC_007028-007041,<br>NC_007044 |
| *Parvoviridae* | *Aedes aegypti* densovirus | AaeDNV | 3776 | 37.2 | NC_012636 |
| (Densovirinae) | *Aedes albopictus* densovirus | AalDNV | 4176 | 38.2 | NC_004285 |
| ***Brevidensovirus*** | Mosquito densovirus BR/07 | BR/07 MDV | 3893 | 37.1 | NC_015115 |
| ***Densovirus*** | *Blattella germanica* densovirus | BgDNV | 5335 | 39.6 | NC_005041 |
| (−)/(+)ssDNA, linear | *Diatraea saccharalis* densovirus | DsDNV | 5941 | 35.5 | NC_001899 |
| | *Galleria mellonella* densovirus | GmDNV | 6039 | 36.4 | NC_004286 |
| | *Junonia coenia* densovirus | JcDNV | 5908 | 37 | NC_004284 |
| | *Mythimna loreyi* densovirus | MlDNV | 6034 | 37.9 | NC_005341 |
| | *Myzus persicae* densovirus | MpDNV | 5499 | 42 | NC_005040 |
| | *Planococcus citri* densovirus | PcDNV | 5380 | 37.6 | NC_004289 |
| ***Iteravirus*** | *Bombyx mori* densovirus 5 | BmDNV | 5078 | 39.3 | NC_004287 |
| (−)/(+)ssDNA, linear | *Casphalia extranea* densovirus | CeDNV | 5002 | 37.8 | NC_004288 |
| | *Dendrolimus punctatus* densovirus | DpDNV | 5039 | 37.6 | NC_006555 |
| ***Pefudensovirus*** | *Periplaneta fuliginosa* densovirus | PfDNV | 5454 | 38.1 | NC_000936 |
| (−/+)ssDNA, linear | *Acheta domesticus* densovirus | AdDNV | 5234 | 39.9 | NC_004290 |
| ***Unclassified*** | *Culex pipiens* densovirus | CpDNV | 5759 | 37.7 | NC_002685 |
| | *Fenneropenaeus chinensis* hepatopancreatic densovirus | FcHPDNV | 6085 | 39.4 | NC_014357 |
| | *Penaeus merguiensis* densovirus | PmergDNV | 6321 | 41.7 | NC_007218 |
| *Poxviridae*<br>(Entomopoxvirinae)<br>***Betaentomopoxvirus*** | *Amsacta moorei*<br>entomopoxvirus 'L'<br>*Melanoplus sanguinipes* | AMEV<br>MSEV | 232392<br>236120 | 17.5<br>18.3 | NC_002520<br>NC_001993 |

**Table 2.3A** List of completely sequenced and published genomes of invertebrate DNA viruses. All genomes were published at GenBank of the National Center for Biotechnology Information (NCBI)—Cont'd

| *Family (Subfamily)* *Genus* *Genome type* | *Virus* | *Abbreviation* | *Length (bp)* | *% GC* | *Accession* |
|---|---|---|---|---|---|
| ***unclassified*** dsDNA, linear | entomopoxvirus | | | | |
| ***Unclassified*** | *Oryctes rhinoceros* nudivirus | OrNV | 127615 | 41.6 | NC_011588 |
| ***(Unclassified)*** | *Heliothis zea* nudivirus1 | HzNV-1 | 228089 | 41.8 | NC_004156 |
| ***Unclassified*** dsDNA, linear | *Gryllus bimaculatus* nudivirus | GbNV | 96944 | 28 | NC_009240 |

a further development of the Sanger method. An Illumina Genome Analyzer manages to conduct up to 120 million single sequencing reactions per run and therefore produces a huge amount of data. However, the read length is considerably shorter (75−100 nucleotides). Such short reads are more complicated to assemble unless a reference sequence is available.

With Illumina technology, the DNA template is immobilized and amplified on the surface of a chip. As in the Sanger method, a DNA strand is synthesized complementary to an unknown template by adding nucleotides. The used nucleotides of the Illumina technology are reversible terminators linked to a specific fluorescent marker (four different colors for the four nucleotide types) (Ju *et al.*, 2006). Whenever a nucleotide is added to an extending DNA strand the fluorescent type is detected. Subsequently, terminator and fluorescent marker are cleaved off and strand synthesis can continue by adding new cleavable fluorescent nucleotide reversible terminators to the sequences. This method was successfully applied to completely sequence the genome of *Clanis bilineata* (Liang *et al.*, 2011).

### *4. SOLiD Sequencing*

The third NGS method, called SOLiD™ System (Applied Biosystems™), is based on sequencing by ligation. The target DNA is sheared, hybridized to 1-μm beats and clonally amplified by emulsion PCR. For sequencing, small di-base probes labeled with a fluorescent dye are hybridized to the template DNA. These oligonucleotides are successively ligated to a complementary strand. After each ligation step the fluorescent marker is cleaved from the probe and detected, which gives information about the target sequence. This method applies a sophisticated color-coding system, by which each DNA fragment is read several times. The short read length and multiple sequencing of a DNA fragment result in the high accuracy of this technique. No published insect virus genome has been sequenced using this method to date.

## 4. VIRUS PROPAGATION

### A. *In vivo* propagation

Despite the fact that large-scale production of virus in living insects is labor intensive, *in vivo* propagation is still a highly feasible method. Also on a smaller laboratory scale, the feeding of virus followed by harvesting of infected insects is the standard for virus stock production. *In vitro* production of virus in cell culture is possible but could be associated with higher costs for cell culture media and sterile disposable material. In addition, the establishment of a specific insect cell line susceptible for the virus is difficult and not always successful. Additionally, virus propagation in cell culture includes the risk of losing genetic features not selected for in the *in vitro* system. For example the baculovirus *per os* infectivity factors (PIFs) are essential for virus infection of living insects but not required in tissue culture. Genetic features not essential during *in vitro* replication might be deleted from the viral genome. Concerning occluded viruses, there is also no selective pressure for producing OBs, so

**Table 2.3B** List of completely sequenced and published genomes of invertebrate RNA viruses. All genomes were published at GenBank of the National Center for Biotechnology Information (NCBI).

| *Family* *Genus* | *Genome type* | *Virus* | *Abbreviation* | *Length (bases)* | *% GC* | *Accession* |
|---|---|---|---|---|---|---|
| *Dicistroviridae* | | | | | | |
| ***Cripavirus*** | +ssRNA, linear | Aphid lethal paralysis virus | ALPV | 9812 | 38.6 | NC_004365 |
| | | Black queen cell virus | BQCV | 8550 | 40.2 | NC_003784 |
| | | Cricket paralysis virus | CrPV | 9185 | 39.3 | NC_003924 |
| | | *Drosophila* C virus | DCV | 9264 | 36.6 | NC_001834 |
| | | Himetobi P virus | HiPV | 9275 | 39.6 | NC_003782 |
| | | *Homalodisca coagulata* virus 1 | HoCV-1 | 9345 | 45.3 | NC_008029 |
| | | *Plautia stali* intestine virus | PSIV | 8797 | 36.4 | NC_003779 |
| | | *Rhopalosiphum padi* virus | RhPV | 10011 | 38.8 | NC_001874 |
| | | *Solenopsis invicta* virus 1 | SInV-1 | 8026 | 38.8 | NC_006559 |
| | | Triatoma virus | TrV | 9010 | 35.9 | NC_003783 |
| ***Unassigned*** | | Acute bee paralysis virus | ABPV | 9491 | 35.4 | NC_002548 |
| | | Kashmir bee virus | KBV | 9524 | 37.7 | NC_004807 |
| | | Taura syndrome virus | TSV | 10205 | 43.2 | NC_003005 |
| ***Unassigned*** | | Israel acute paralysis virus | IAPV | 9499 | 37.9 | NC_009025 |
| | | Mud crab dicistrovirus | MCDV | 10436 | 40.7 | NC_001793 |
| *Iflaviridae* | | | | | | |
| ***Iflavirus*** | +ssRNA, linear | Infectious flacherie virus | IFV | 9650 | 42.8 | NC_003781 |
| | | *Perina nuda virus* | PnPV | 9476 | 44.2 | NC_003113 |
| | | Sacbrood virus | SBV | 8832 | 40.6 | NC_002066 |
| | | *Varroa destructor* virus 1 | VDV-1 | 10112 | 38.6 | NC_006494 |
| ***Unclassified*** | | *Brevicoryne brassicae* picorna-like virus | BrBV | 10180 | 33.6 | NC_009530 |
| | | Deformed wing virus | DWV | 10140 | 38.1 | NC_004830 |
| | | *Ectropis obliqua* picorna-like virus | EoPV | 9394 | 44.5 | NC_005092 |
| | | Kakugo virus | KV | 10152 | 38.3 | NC_005876 |
| | | Slow bee paralysis virus | SBPV | 9505 | 37.5 | NC_014137 |
| *Nodaviridae* | | | | | | |
| ***Alphanodavirus*** | +ssRNA, linear, two segments | Black beetle virus | BBV RNA1 | 3106 | 48.3 | NC_002037 |

**Table 2.3B** List of completely sequenced and published genomes of invertebrate RNA viruses. All genomes were published at GenBank of the National Center for Biotechnology Information (NCBI)—Cont'd

| *Family* *Genus* | *Genome type* | *Virus* | *Abbreviation* | *Length (bases)* | *% GC* | *Accession* |
|---|---|---|---|---|---|---|
| | | | BBV RNA 2 | 1399 | 49.3 | NC_001411 |
| | | Boolarra virus | Bov RNA 1 | 3096 | 49.4 | NC_004145 |
| | | | Bov RNA 2 | 1305 | 47.3 | NC_004142 |
| | | Flock house virus | FHV RNA 1 | 3707 | 48.3 | NC_004146 |
| | | | FHV RNA 2 | 1400 | 49.2 | NC_004144 |
| | | Nodamura virus | NoV RNA 1 | 3204 | 56.7 | NC_002690 |
| | | | NoV RNA 2 | 1336 | 51 | NC_002691 |
| | | Pariacato virus | PaV RNA 1 | 3011 | 51.7 | NC_003691 |
| | | | PaV RNA 2 | 1311 | 52.3 | NC_003692 |
| *Reoviridae* | | | | | | |
| ***Cypovirus*** | dsRNA, linear, 10–12 segements | *Lymantria dispar* cypovirus 1 | LdCPV-1 | - | - | NC_003016-003025 |
| | | *Heliothis armigera* cypovirus 5 | HaCPV-5 | - | - | NC_010661-010670 |
| | | *Lymantria dispar* cypovirus 14 | LdCPV-14 | - | - | NC_003006-003015 |
| | | *Trichoplusia ni* cypovirus 15 | TnCPV-15 | - | - | NC_002557-002567 |
| *Tetraviridae* | | | | | | |
| ***Betatetravirus*** | +ssRNA, linear | *Euprosterna elaeasa* virus | EeV | 5698 | 52.2 | NC_003412 |
| | | *Nudaurelia capensis* β virus | NβV | 6625 | 54.1 | NC_001990 |
| | | Providence virus | PrV | 6155 | 51.8 | NC_014126 |
| ***Omegatetravirus*** | +ssRNA, linear, two segments | *Helicoverpa armigera* stunt virus | HaSV RNA1 | 5312 | 59.7 | NC_001981 |
| | | | HaSV RNA2 | 2478 | 59 | NC_001982 |
| ***Unclassified*** | | *Dendrolimus punctatus* tetravirus | DpTV RNA1 | 5492 | 57.2 | NC_005898 |
| | | | DpTV RNA2 | 2490 | 55.9 | NC_005899 |

that a stock of occluded viruses might degenerate into a virus stock defective in OB production.

Working with host-specific insect viruses implies that each virus has to be propagated in its homologous host or an alternative host that is susceptible to the virus. In any case, it is important to check the identity of the virus obtained after *in vivo* propagation using molecular tools, e.g., REN analysis or PCR. Propagation in an alternative host insect might select for another virus present at a non-visible level within the virus stock or as a latent infection in the host insect. To guarantee a stable quality of virus, a healthy laboratory colony of the host insect needs to be established and reared over the whole year. Insect rearing must be kept strictly separate from all virus working places. The insect population should be monitored regularly for latent virus infection (Jenkins & Grzywacz, 2000). This can be performed by isolation of insect DNA (Section 4C) followed by (real-time) PCR (Section 5B) using virus-specific primers.The principal method for *in vivo* propagation of insect viruses has not changed during the last decades (reviewed by Shapiro, 1986).

### *1. Propagation of* Cydia pomonella granulovirus *(CpGV) in* C. pomonella *larvae*

1. Rear neonate codling moth larvae for 10 days on virus-free artificial diet.
2. After 10 days, larvae reach instar L4–L5.
3. Pipette 1 µl virus suspension containing 1000 OBs on the surface of a small piece of diet (about 2 × 2 × 2 mm).
4. Keep larvae single in a well containing one piece of contaminated diet and incubate them for 24 h at 26°C.
5. Transfer larvae, which have eaten the contaminated diet completely, to virus-free diet.
6. Incubate larvae at 26°C until infection is visible (usually after 5–6 days) and monitor daily.
7. Collect infected larvae before tissue rupture.
8. Collected larvae can be frozen at −20°C until virus purification.

Virus propagation follows the same protocol for small- and large-scale production. Infection of about 200 codling moth larvae following this protocol will usually result in 5 ml purified virus suspension with a concentration of about $10^{11}$ OB/ml. It is recommended to feed more larvae, because not every larva will eat its diet plug completely or show symptoms of virus infection.

To minimize production costs and to gain maximum virus yield, the following points have to be optimized for every insect–virus system:

#### *a. Inoculum dosage*

Optimizing the yield of virus production has been focused on determining the appropriate virus dose for a specific instar, in order to reach a maximum larval weight before the insect dies of virus infection. The amount of virus produced is usually positively correlated to the larval weight (Jones, 2000). The relationship between virus dose and maximum virus yield has to be determined for every system (Shapiro, 1986). Comparison of different inoculation methods of *Hyblaea puera* NPV (HypuNPV) in 5$^{th}$ instar larvae of the teak defoliator *H. puera* revealed that the most efficient method was rearing larvae continuously on contaminated diet instead of feeding a certain dose for a short time (Biji *et al.*, 2006). However, a very high dose of OBs might damage the insect's organs, causing insect death without efficient virus propagation. If the concentration is too low, virus yield will also be suboptimal because the insects might be able to pupate. Generally, a uniform response of 100% infection or mortality is desired (Shapiro, 1986). For an unknown insect–virus system, it is reasonable to determine first the optimal concentration by previously testing a range of concentrations using one larval instar. Estimation of the $LC_{50}$ and $LC_{90}$ values (see Section 6) provides information about the course of infection for the desired virus–host system necessary for choosing the optimal virus concentration.

#### *b. Larval age*

Generally, larvae infected at a too early instar die before they reach their maximum body mass, whereas larvae infected too late are more resistant to virus infection (Briese, 1986), both resulting in reduced virus production. Usually, the optimal time for infection is one to two days after molting into the last larval instar (O'Reilly *et al.*, 1994). To determine the optimal larval age of *Helicoverpa armigera* for inoculation of *H. armigera* NPV (HearNPV), Gupta *et al.* (2007) fed virus-contaminated diet (1 × $10^4$ OB/larvae) to 5-, 7-, 8-, 9-, 10-, and 11-day-old larvae. The optimal virus yield was recorded when 7- and 8-day-old larvae were harvested. With younger instars, the yield per larva was significantly lower than with older instars. In older stages, a decrease in larval harvest was reported; with increasing age, the larvae started to pupate. A maximum virus yield of different HearNPV isolates was obtained when a concentration of $5 \times 10^5$ OB/larvae was administered to early 5$^{th}$ instar larvae (Mehrvar *et al.*, 2007).

For production of *Leucoma salicis* NPV (LesaNPV) in the satin moth *L. salicis*, the 4th instar was found to be the optimal age for virus replication (Ziemnicka, 2008). Propagation of *Spodoptera exigua* multiple NPV (SeMNPV) in *S. exigua* larvae was increased by a factor 2.7–2.9 when producing the virus in a supernumerary host instar. This 6th instar was induced by application of juvenile hormone analogs (Lasa *et al.*, 2007). Propagation of *Neodiprion abietis* NPV (NeabNPV) in four different larval stages of the balsam fir sawfly *N. abietis* revealed that 3rd instar larvae inoculated with $10^5$ OB/ml and 4th and 5th instars inoculated with $10^7$ OB/ml produced the most virus. Taking into account the virus used for inoculation, the longer incubation time to 4th and 5th instars and the resources required to rear the larvae until then, the authors concluded that the infection of 3rd instar larvae is the age of choice for optimized virus inoculation (Li & Skinner, 2005).

*c. Incubation*

Incubation is usually carried out between 20 and 26°C (Shapiro, 1986). Comparison of virus yield of three HearNPV isolates at different incubation temperatures indicated that the maximum yield was obtained at 25°C. Maximum larval mortality occurred at room temperature, followed by 30°C (Mehrvar *et al.*, 2007). Infection of *Heliothis zea* using two concentrations of *H. zea* single NPV (HzSNPV) and incubation at five different temperatures indicated that high temperatures (40°C) could inhibit virus multiplication. Under normal field conditions (13–35°C), virus-induced mortality was not significantly dependent on temperature (Ignoffo, 1966).

*d. Time of harvest*

For *in vivo* production of different HearNPV isolates in *H. armigera*, the maximum OB yield was obtained when harvesting dead larvae (Gupta, 2007; Mehrvar *et al.*, 2007). Time of harvest was found to be crucial for propagation of *Spodoptera littoralis* NPV in *S. littoralis* (Grzywacz *et al.*, 1998). Until 4 days post infection, virus replication was in the primary phase and did not result in production of OBs, which are mostly produced during secondary infection. The main productivity was found between 6 and 8 days after infection. In general, early harvesting of infected larvae might result in a decreased biological activity of the virus stock due to incomplete, non-occluded virus particles. Otherwise, if tissue rupture occurs before harvesting, virus spreads on the artificial diet and might be difficult to collect. A compromise is to collect virus-infected larvae shortly prior to death when they have already stopped feeding and incubate them in tubes until tissue rupture. This results in mature virus particles and the virus suspension can be easily rinsed from the tube for virus purification. Post-harvest incubation has also been described to increase OB yields. Concurrently, this harbors the risk of bacterial contamination (Cherry *et al.*, 1997).

*e. Quality control*

Virus propagation in living insects may lead to product variation in terms of composition and purity. Virus content and activity may be subjected to fluctuations; other microorganisms present in the insects might contaminate the final product. Therefore, stringent quality controls are necessary. The virus stock used for propagation should be a strain characterized previously by DNA restriction analysis to determine whether it is composed of a single isolate or a mixture of genotypes. Purification of the inoculum by gradient centrifugation minimizes the risk of contamination with protozoa or bacterial spores that might interfere with viral replication (Jenkins & Grzywacz, 2000). For every virus batch produced, the yield should be estimated by counting (Section 5A) or other appropriate methods and tested for its biological activity by bioassay (Section 6). Ideally, the strain is then further characterized by REN analysis (Section 3C). If these tasks are performed routinely following every propagation, any changes or problems concerning virus propagation will become directly visible.

### B. Virus isolation from insects

Virus isolation from an infected host is usually the next step following virus propagation. For several downstream applications like DNA/RNA isolation or electron microscopy, a highly purified virus suspension is needed. Also for bioassays, the virus suspension should be free of material or microorganisms that might interfere with the infection process. The details of purification procedure vary according to the desired virus (occluded, non-occluded, NPV, GV), but it is always based on the following steps:

*1. Homogenization and filtration*

Dead insects are homogenized in a low-concentration anionic detergent to facilitate insect tissue rupture and the release of virus particles or OBs. Freezing of the larval material prior to homogenization is helpful to disrupt the cells (Tompkins, 1991). The homogenized suspension still contains residues of the rearing diet, head capsules, and large integument portions and is therefore filtered through cheese-cloth or gauze.

*2. Centrifugation*

The filtered virus suspension still contains fat, bacteria, and other fine, non-viral particles. By several centrifugation steps, the virus is separated from these contaminants and subjected to several washing steps. Centrifugation procedure (time, gradient) varies according to the virus to be purified and equipment and common procedure in the laboratory.

A commonly used method for the purification of OB of GVs and NPVs are given in the following. The details of these methods were worked out for virus-infected 3rd/4th instar larvae and do not require ultra-centrifugation. Purification of viral OBs from formulated products can be performed following the same procedure, introducing the formulated suspension instead of infected larvae.

*3. Purification of OB of GVs*

1. Autoclave glassware-homogeniser, water, pipette-tips, 15- and 50-ml Falcon tubes, gauze, cotton wool or baby diaper.
2. Prepare all solutions in sterile water.
3. Collect virus-infected 3rd/4th instar larvae (about 150) in a 50-ml Falcon tube and incubate on ice for 20–30 min.
4. Add 20 ml 0.5% sodium dodecyl sulfate (SDS) and grind the larvae with a homogenizer, wash pistil and glassware with another 10 ml SDS.
5. Filter the suspension through a sandwich filter (gauze filled with a piece of cotton wool) into a 50-ml Falcon tube.
6. Rinse the filter with 30 ml water and squeeze the liquid out of the filter.
7. Spin down at 18,000 × *g* for 30 min at 12°C.
8. Discard supernatant.
9. Resuspend OB pellet in 20 ml water for washing.
10. Spin down at 18,000 × *g* for 30 min at 12°C.
11. Repeat washing step one more time.
12. Resuspend OB pellet in maximum 16 ml water.
13. Vortex.
14. Sonicate for 5 min.
15. Prepare glycerol layer gradient in a Falcon tube in the following steps, beginning at the bottom: (1) 80% (v/v); (2) 70%; (3) 65%; (4) 60%; (5) 55%; (6) 50%; (7) 30% (top gradient layer).
16. Use 2–4 ml for every layer.
17. Load about 2 ml of virus suspension on top of the gradient.
18. Spin for 45 min at 3200 × *g* at 12°C in a swing out rotor.
19. Remove the supernatant until reaching the 60% glycerol gradient layer into a new 50-ml tube.
20. Add $H_2O$ until 50 ml to dilute the glycerol.
21. Spin for 45 min at 3200 × *g* at 12°C.
22. Resuspend OB pellet in 45 ml $H_2O$.
23. Spin for 45 min at 3200 × *g* at 12°C, repeat once more.
24. Resuspend pellet in maximum 10 ml water.
25. Load the resuspended pellet again on a 30–80% glycerol layer gradient.
26. Spin for 45 min at 3200 × *g* at 12°C in a swing out rotor.
27. Remove the supernatant until reaching the 60% glycerol layer in a new 50-ml tube.
28. Add $H_2O$ to dilute the glycerol.
29. Spin for 45 min at 3200 × *g* at 12°C.
30. Wash pellet twice with water.
31. Resuspend OB pellet in 10–20 ml water.
32. Store at −18°C or colder.

The number of washing steps depends on the amount and condition of the virus suspension. A purified virus suspension is clear (low virus concentration) or whitish (high virus concentration). A dark OB pellet indicates that further rounds of purification over a glycerol gradient are necessary. Gray and pink residues derive from larval debris. Formulated products usually need several rounds over a glycerol gradient, as the formulation can inhibit DNA isolation of viral OBs.

*4. Purification of OBs of NPVs*

Solutions:

TTE buffer

0.1 M TRIS/HCL, pH 7.2

0.1% v/v Thioglycollic acid
0.5% w/v EDTA
Adjust the pH to 7.2

Sucrose solutions (62.5% w/v, 50% w/v, or 30% w/v)
62.5 g, 50 g, or 30 g of sucrose
Add TTE until 100 g
The TTE contains phenyl methyl sulfenil fluoride (0.1% v/v of a freshly prepared 0.1M solution in isopropanol)

TTE/SDS solution (make fresh)
2 volumes of TTE pH 7.2
1 volume of 10% SDS
Adjust pH to 7.2

1. Infect insect larvae (ca. 100) *per os* (the most convenient instar is early L4) with the appropriate NPV by applying OBs to the medium.
2. Incubate the larvae at 26°C until they are dead or moribund (*ca.* 4–5 days).
3. Place dead or moribund larvae at −20°C for 2–3 h to facilitate harvesting of the larvae with forceps.
4. Grind the larvae with the same amount w/v of TTE in a Sorvall mixer. Mix for 5 min at 4°C.
5. Filter the suspension through two layers of cheese cloth to remove large structures such as heads, etc. Layer the filtrate on to 100 ml of 30% w/w sucrose in a 250-ml bucket (GSA-Sorvall) and centrifuge for 20 min at 23,500 × *g* and 6°C.
6. Discard the supernatant, resuspend pellet in 50 ml TTE, centrifuge for 20 min at 10,400 × *g* and 6°C in a GSA-Sorvall and wash the pellet and the bucket with TTE twice.
7. Resuspend pellet in 10 ml TTE/SDS solution, transfer the solution into a 20-ml tube and wash the bucket with 2.5 ml TTE twice. Combine these solutions.
8. Sonicate the combined suspensions for 2 min in a Bransonic bath-sonicator to homogenize the suspension.
9. Place10–15 ml of the sample suspension on to 10 ml of 50% w/w sucrose solution which has first been layered over 10 ml of a 62.5% w/w sucrose solution in a SW28 centrifuge tube.
10. Balance the tubes with TTE and centrifuge at 113,000 × *g* for 60 min and 4°C in a SW28 Beckmann ultracentrifuge.
11. Remove the polyhedra from the 50–62% sucrose interface with a syringe and transfer to a Sorvall SS34 centrifuge tube.
12. Centrifuge for 10 min at 27,100 × *g*. Resuspend the OBs in TTE and vortex, centrifuge for 10 min at 27,100 × *g* and 6°C.
13. Resuspend the pellet in 4 ml TTE and 1 ml TTE/SDS solution. Wash the centrifuge tube with 5 ml TTE twice. Combine these solutions.
14. Sonicate the combined suspensions for 1 min in a Bransonic bath-sonicator to homogenize the solution.
15. Repeat the steps 9–14 once.
16. Resuspend the pellet in 5 ml TTE and wash the centrifuge tube with 5 ml TTE.
17. Keep the OB suspension in dark conditions at −20°C.
18. Yield about $10^9$ OBs/larvae; about $10^{11}$ polyhedra total.

Further detailed protocols for every virus group are summarized in Tompkins (1991). A brief overview of the differences in homogenization and centrifugation steps according to the virus to be purified is given in Table 2.4.

## C. DNA isolation

For the isolation and purification of viral DNA it is essential to release the DNA from the virus capsid and in case of occluded viruses (baculoviruses, entomopoxviruses) also from OBs by dissolving them in sodium carbonate. The release of DNA from virus capsids is achieved by treatment with a detergent (SDS) and proteinase K degradation of capsid proteins followed by phenol/chloroform extraction (Sambrook & Russell, 2001). A basic protocol for DNA isolation and purification from occluded and non-occluded DNA viruses is given below. It is generally used with 10 mg of OBs and can be easily scaled up or down.

### *1. Protocol for DNA isolation and purification from occluded viruses*

1. Liberate virus particles by suspending the OBs in 500 µl $H_2O$ and incubate the suspension with 0.05 volumes 1 M sodium carbonate ($Na_2CO_3$) for 30–60 min at 37°C. The suspension should become clear, if not add more $Na_2CO_3$ and incubate longer.

**Table 2.4** Isolation and purification procedures for invertebrate viruses from infected hosts.

| *Sample* | *Homogenization* | *Centrifugation* | *Literature* |
|---|---|---|---|
| CPV OBs from *C. quinquefasciatus larvae* | Homogenize larvae in deionized water and filter through nylon screen | 16,000 × *g* for 30 min on HS-40 Ludoxcontinous gradient<br>Wash viral band in 0.1 mM NaOH | Green *et al.*, 2006 |
| CPV virions from *B. mori* midguts | Homogenize infected midguts with Teflon-glass homogenizer | At 10,000 × *g* for 10 min<br>Virions remain in supernatant<br>Purification over sucrose gradient | Hayashi *et al.*, 1970 |
| Densonucleosis virus from *A. aegypti* larvae | Homogenize larvae in PBS using a grinder | Centrifuge at 6500 × *g* for 30 min<br>Centrifuge supernatant at 40,000 × *g* for 2 h at 4°C on a glycerol layer<br>Dissolve pellet in PBS<br>Centrifuge at 100,000 × *g* for 4 h at 4°C on a discontinuous CsCl gradient<br>Collect virus band | Sivaram *et al.*, 2009 |
| Dicistrovirus from *A. mellifera* larvae | Homogenize larvae in detergent containing buffer | Two cycles of differential centrifugation:<br>10,000 × *g* for 20 min and 100,000 × *g* for 3 h<br>Suspend pellet in CsCl<br>Centrifuge at 100,000 × *g* for 24 h<br>Collect virus band | Maori *et al.*, 2007 |
| NPV from *A. ipsilon* larvae | Homogenize larvae in 0.5% sodium dodecyl sulfate (SDS) with an Ultra-Turrax and filter homogenate through two layers of cheesecloth and wire mesh, wash with 0.5% SDS | Centrifuge at 750 × *g* for 10 min to pellet polyhedra<br>Wash pellets by resuspending twice in 0.1% SDS<br>Resuspend once in 0.5 M NaCL<br>Pellet by centrifugation after every washing step<br>Resuspend polyhedra in deionized distilled water<br>Solubilize with $Na_2CO_3$ | Harrison, 2009 |
| NPV from *H. armigera* insects | Homogenize insect cadavers in 10 times their volume of 1% SDS and incubate over night at room temperature, filter through four layers of muslin | 5000 × *g* for 15 min<br>Resuspend pellet in distilled water<br>500 × *g* for 10 min<br>Wash pellet three times with distilled water<br>Resuspend pellet | Christian *et al.*, 2001 |

**Table 2.4** Isolation and purification procedures for invertebrate viruses from infected hosts—Cont'd

| *Sample* | *Homogenization* | *Centrifugation* | *Literature* |
|---|---|---|---|
| RNA virus from *H. armigera* larvae | Homogenize larvae in 50 mM Tris-HCL pH 7.4 | Centrifuge at 10,000 × *g* for 30 min<br>Centrifuge supernatant through 10% sucrose at 100,000 × *g* for 3 h<br>Suspend virus pellet overnight in buffer at 4°C<br>Centrifuge virus suspension for 12 h at 200,000 × *g* over 60 and 30% CsCl layer<br>Remove light-scattering virus band<br>Pellet at 100,000 × *g* for 3 h<br>Resuspend pellet | Hanzlik *et al.*, 1993 |
| SINV-3 from *S. invicta* insects | Homogenize insects in NT buffer (10 mM Tris-HCL pH 7.4, 100 mM NaCl), filter through 8 layers of cheesecloth, extract with chloroform | 5000 × *g* for 5 min<br>Transfer layer to discontinuous CsCL gradient 190,000 × *g* for 2 h<br>Remove virus band and centrifuge at 330,000 × *g* for 16 h<br>Collect virus band | Valles & Hashimoto, 2009 |

2. Lower the pH of the suspension to pH 8.0 by titrating with 1 M HCl, check pH with pH paper.

### 2. *Protocol for DNA isolation and purification from non-occluded viruses and virus particles released from OBs*

1. Add DNase-free RNase A (45 μg/ml) to the virus suspension and incubate at 37°C for 10 min.
2. Disrupt virus particles by adding 0.1 volumes of 10% SDS, 0.1 volumes of 0.5 M EDTA and proteinase K to a final concentration of 2 mg/ml; incubate the mixture at 65°C for 1–2 h.
3. Extract twice with Tris/EDTA (TE) saturated phenol/chloroform/isoamyl alcohol (25:24:1), then once with chloroform/isoamyl alcohol (24:1).
4. Perform an EtOH precipitation of the supernatant by adding 0.2 volumes of 3 M sodium acetate (pH 5.2) and 2 volumes of 96% EtOH. Pellet the precipitated DNA by centrifuging at 10,000 × *g*. Wash pellet twice with 70% EtOH. Gently air dry DNA pellet until liquid is just evaporated and dissolve DNA in 100 μl TE (fully dried genomic DNA is hard to solubilize).
5. Alternatively to EtOH precipitation, the virus DNA can be dialyzed against TE buffer: dialyze supernatant of phenol/chloroform extraction for 24 h against T-buffer, change buffer three times.

## D. RNA isolation

### 1. *Working with RNA*

Unlike DNA, RNA is more susceptible to rapid degradation. Due to the 2′-hydroxyl groups adjacent to the phosphodiester linkages, RNA is highly susceptible to hydrolysis. Ribonucleases (RNases) do not require metal ions as co-factors but take advantage of the reactive 2′-hydroxyl group as a nucleophile. RNases are ubiquitous and very stable, therefore, it is necessary to avoid any contamination with RNase, when working with

RNA, Several precautions have to be taken during RNA handling (Nielsen, 2011):

- RNA working space, chemicals, and equipment are separated from other applications.
- Always wear gloves, never touch surfaces that come in contact with RNA.
- Disposable, sterile plasticware is recommended.
- Autoclaved glassware (250°C, more than 4 h) is recommended, as RNases can survive autoclaving.
- Glassware can be treated for RNase inhibition with 0.1% diethyl pyrocarbonate (DEPC) overnight at 37°C and then autoclaved to eliminate DEPC residues.
- Treat water and solutions with 0.1% DEPC and autoclave for removal of DEPC.
- Test water for RNase activity by incubating a RNA sample with water and check on agarose gel for degradation.
- Keep RNA samples on ice to reduce enzyme activity.
- Keep pH neutral or slightly acidic.
- Discard any material that may have come into contact with non RNase-free surfaces.

*2. Standard protocol for RNA isolation*

Methods for RNA isolation are based on cell lysis using chemicals which concurrently destroy RNAses. During the next steps, RNA is separated from the cellular components to obtain purified RNA for downstream experiments (reverse transcription, northern blot, etc.). The different protocols for RNA isolation are all based on the same procedure:

1. Homogenization of cells/tissues
2. Cell lysis
3. Phase separation
4. RNA precipitation
5. Washing steps
6. RNA elution
7. Determination of RNA concentration and integrity

In detail, protocols vary in the disintegration method of cells which depends on the type of cells introduced (prokaryotic, yeast, plant, animal) and in the inactivation of cellular RNases. It is not possible with every protocol to separate DNA completely. Therefore, a DNA hydrolysis by an RNase-free DNAse is necessary for sensitive downstream applications.

Applicable for total RNA isolation from tissues and cells is the so-called single-step method according to Chomczynski & Sacchi (1987) using Trizol® reagent. Trizol® induces cell lysis and is a strong inhibitor of RNase. Additionally, Trizol® reagent contains phenol for dissolving DNA and proteins.

*3. Total RNA isolation from codling moth larvae:*

1. Homogenize 5th instar larvae (50−100 mg) in a 1.5-ml Eppendorf tube in 1 ml Trizol® reagent.
2. Incubate for 5 min at room temperature.
3. Centrifuge at 12,000 × *g* and 4°C for 10 min.
4. Remove fat-containing top layer, transfer RNA containing supernatant to a fresh 1.5-ml tube.
5. Add 0.2 ml chloroform and shake tube vigorously by hand for 15 s.
6. Incubate for 2−3 min at room temperature.
7. Centrifuge at 12,000 × *g* and 4°C for 10 min.
8. Remove supernatant and wash the RNA pellet with 1 ml 75% EtOH by vortexing.
9. Centrifuge at 7500 × *g* and 4°C for 5 min.
10. Remove supernatant, air dry RNA pellet.
11. Dissolve in RNase-free water and store at −80°C.

Currently, various kits for RNA isolation are available from commercial suppliers. After cell lysis, RNA is thereby bound to a silica-based column, washed with buffer and eluted. Depending on the tissue (bacteria, virus, plant, mammalian cells), the desired application (PCR, isolation of RNA and protein, small RNA analysis) and the amount of starting material, various systems are available. The advantage of RNA preparation using commercial kits is that the reagents are standardized and tested by quality control.

*4. Quality control of RNA preparations*

Spectrophotometric measurement is usually used to estimate the amount and purity of nucleic acids. For RNA preparations without protein contamination, the ratio of absorbance at 260 and 280 nm should range between 1.7 and 2. An optical density (OD) of 1 at 260 nm corresponds to an amount of about 40 μg ssRNA. However, only significant protein contamination will be detected, as nucleic acid absorbance at 260 nm is very strong. Additionally, pH and ionic strength of the solution affect the specific absorbance of RNA. Therefore, this method is only useful for purified preparations of RNA (Sambrook & Russell, 2001).

The integrity of the RNA sample can be checked quickly by gel electrophoresis. Electrophoresis of RNA is, like for DNA, carried out in agarose gels separating by molecule size. A rapid method for testing the RNA integrity is to check for the ribosomal RNA (rRNA) molecules in a 2% agarose gel running for 1 h. The rRNA molecules are the most yielding type of RNA. They appear as discrete bands on the gel; 23 S and 16 S for prokaryotes and 18 S and 28 S for eukaryotes. If these bands are visible, the other RNA components can be expected also to be intact. Any contamination of DNA or degradation of RNA is clearly visible on the gel (Wilson & Walker, 2005).

For gene expression studies using sensitive downstream applications like microarray analysis or RT-PCR, it is recommended to use a bioanalyzer platform (Lab-on-a-Chip) instead of the conventional methods. Thereby, RNA samples are separated on a microfabricated chip by electrophoretic separation and detected by laser-induced fluorescence. Improved reproducibility, minimal sample and chemical consumption, comparability between different labs, digital data output, and short analysis times are the advantages of this instrumental analysis of RNA size, quantity, and quality. The industry standard for RNA analysis is the RIN algorithm (RNA integrity number), estimated by bioanalyzer software. This algorithm is applied to different features from the electrophoretic measurements which contribute information about RNA integrity (Auer & Lyianarachchi, 2003; Schroeder *et al.*, 2006).

## 5. VIRUS QUANTIFICATION

For bioassays and bio-molecular analysis it is important to know the concentration of a virus suspension. Several techniques have been investigated for virus quantification, each technique with its advantages and disadvantages. Some of these techniques are listed below, to provide a brief overview about virus enumeration. The technique of choice mainly depends on the virus size and, also, on the experimental objective.

In general, virus particles are too small to be counted under the light microscope (ascoviruses, iridoviruses, densoviruses, iflaviruses, and many others). Smaller viruses have to be observed using SEM or TEM, increasing costs and effort of quantification. Due to their size (> 500 nm), occlusion body forming baculoviruses (NPVs and GVs), CPVs and entomopoxviruses (EPVs) are easily countable by light microscope techniques. In most cases, the quantification of OBs is sufficient, though the issue is complicated by a variable number of virus particles per OB. It has to be taken into account that especially OBs of the genus *Alphabaculovirus* can contain up to hundreds of infectious virus particles. This is not the case in the genus *Betabaculovirus*. Here, one OB contains only one virion and it can be assumed that the number of OBs equals the number of virions.

In nature, OBs achieve horizontal transmission of an insect virus and should be regarded as an aggregation of infectious units. Therefore, quantification of OBs by counting is a reliable procedure to measure the potency of a virus suspension although the exact virus particle concentration remains unknown.

### A. Counting

#### *1. Counting by hemocytometer*

A hemocytometer was invented to count blood cells but in biology it also became a useful tool to count microscopic particles. There are many kinds of different hemocytometers but for OBs, such as GVs or NPVs of the family *Baculoviridae*, a hemocytometer with improved Neubauer ruling is the best choice. Counting polyhedra of different insect viruses under light microscope conditions is a well-established standard method and is also described by Wigley (1980b), Hunter-Fujita *et al.* (1998), and Jones (2000). An improved Neubauer hemocytometer is mainly a thick glass microscope slide, on which one or two fine grids of perpendicular lines are engraved. A grid consists of 25 large squares and each square is subdivided into 16 smaller squares. A coverslip is placed above the grid at a defined distance. The area of a small square (0.0025 mm$^2$) and the distance of the coverslip create a space of a known volume. The space between grid and coverslip can be filled by a sample of virus suspension and polyhedra of a certain volume can be counted microscopically. By extrapolation the concentration of the virus suspension, from which the sample was taken, can be calculated, taking previous dilutions in account.

The virus suspension should be pure or semi-pure to avoid mistaking other particles for OBs (for identification of insect viruses, see Section 3). The large and intense refracting OBs formed by alphabaculoviruses, cypoviruses, and entomopoxviruses are countable microscopically under bright-field illumination, using an Improved Neubauer hemocytometer with a depth of 0.1 mm (Figure 2.8A). Betabaculoviruses are much smaller (0.15–0.5 μm) and more difficult to count. Usually a hemocytometer with a smaller chamber depth, like a 0.02-mm-deep Petroff–Hausser counting chamber, is used. In this case OBs are counted under phase-contrast and dark-field illumination (Figure 2.8B).

Tips:

Make an appropriate dilution series of your sample. Crowded OBs are too difficult to count. Six occlusion bodies per small square are optimal; lower values lead to an increasing statistical variation (Jones, 2000).

Virus suspension should be homogeneous. Mix samples thoroughly to avoid clumping and aggregation of occlusion bodies.

Coverslip and slide should be free of any dust and dirt. If necessary rinse coverslip and slide with 70% EtOH and use a dust- and fiber-free tissue to carefully clean the equipment.

It is not necessary to count the whole counting grid. Five large squares (each 0.04 mm$^2$) should be enough. Multiply the counted number with the dilution factor to obtain the concentration of the virus stock (OBs/ml).

The following steps should be repeated twice or several times.

1. Make a dilution series of your sample and take one dilution for counting.
2. Place and firmly press down the coverslip on top of the chamber. Humidifying the junctions of cover glass and chamber by breathing on them helps to fix the cover glass by adhesion (Newton's rings should be visible).
3. Dispense a drop (5–8 μl) of virus suspension at the edge of the coverslip. Wait until the liquid has been sucked completely into the counting chamber by capillary action. The surface of the grid should be completely covered by the virus suspension.
4. Wait a few minutes to reduce the Brownian motion of the OBs and let them settle.
5. Examine microscopically.
6. Count five large squares of the counting grid. Large squares of an improved Neubauer hemocytometer are bordered by three gridlines. Only the line in the middle borders the square. Count all 16 small squares of each of the five large squares (Figure 2.8A). Only OBs, which touch the top and left gridline, are included in the square. This avoids double counting of OBs.
7. Extrapolate from the counted number of OBs per square to the total concentration of the stock (OBs/ml):

$$\text{Total number of OBs per ml} = \frac{(\text{mean number of OBs per large square} \times \text{dilution factor} \times 10^3)}{(\text{area of one large square} \times \text{depth of chamber})}$$

Area of a small square: 0.0025 mm$^2$

Area of a large square: $16 \times 0.0025$ mm$^2$ = 0.04 mm$^2$

Mean number of OBs per square: Total number of OBs divided by the number of counted large squares.

Depth of counting chamber depends on used hemocytometer: 0.1 mm or 0.02 mm (Petroff–Hausser).

Dilution factor: dilution of counted virus suspension.

Multiply the result by $10^3$ to calculate the total number of OBs per ml.

*2. Electron microscope estimation*

Non-occluded insect viruses are too small to be visualized and quantified by light microscopy techniques. By use of a TEM, any size of virus can be observed. A method for quantification by using a TEM was described by Hunter-Fujita *et al.* (1998). In summary, a purified virus suspension is mixed with a known concentration of polystyrene beads. These beads should be the same size as the virus but also be clearly distinguishable. Droplets of this evenly mixed suspension are applied and fixed on carbon-coated Formvar grids. These grids are examined in the electron microscope and the number of both the beads and the virus particles are counted within a certain

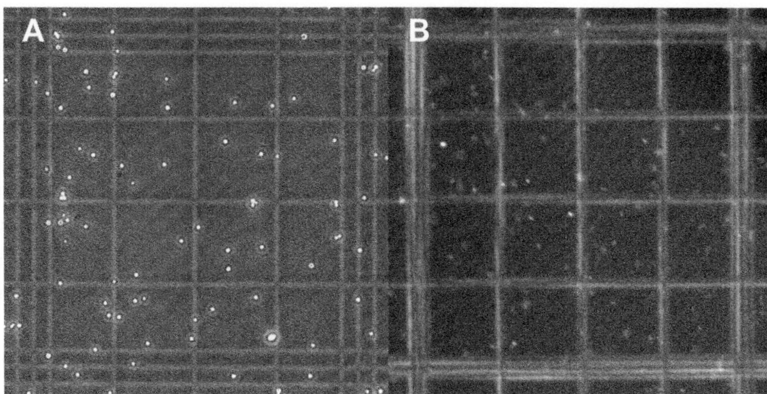


**Figure 2.8** A. Example of a NPV [*Agrotis segetum* NPV Oxford strain (AgseNPV-B)] under the light microscope using an improved Neubauer hemocytometer (depth 0.1 mm). The edge of each small square is 0.05 mm × 40 (for further information see text). B. A granulovirus [*Cydia pomonella* granulovirus (CpGV)] as seen by dark-field illumination under the light microscope. For observation a Petroff–Hausser counting chamber (depth 0.02 mm) is used. The edge of each small square is 0.05 mm × 40.

area. The virus concentration is determined by the ratio of polystyrene beads to virus particles.

*3. Other counting techniques using microscopy*

The dry counting technique (Wigley, 1980b) allows OB enumeration of unpurified virus preparations by light microscopy (Jones, 2000). In summary, this method is based on quantitative detection of OBs on a known area. The virus preparation is applied to a microscope slide with circles of a given area in which all OBs are counted. By calculations, the virus concentration is achieved. Detailed protocols are described in Evans & Shapiro (1997), Hunter-Fujita *et al.* (1998), and Jones (2000).

Elleman *et al.* (1980) described a method for direct visual enumeration of OBs on plant surfaces by fixing them carefully to an adhesive tape (impression film) followed by counting them under a light microscope. OBs have to be stained to avoid confusion with other polyhedral-like particles. Protocols for this method can be found in Evans & Shapiro (1997) and Hunter-Fujita *et al.* (1998).

*4. Indirect counting techniques*

Insect viruses can also be detected and quantified by ELISA. This method was used as a double-antibody sandwich methodology to quantify SpliNPV (Jones, 2000).

## B. Virus quantification using PCR

Quantification of viruses can also be achieved by using quantitative polymerase chain reaction (qPCR) techniques that are mainly based on conventional PCR protocols. This method requires knowledge of at least some part of the genome in order to generate specific primers, which bind uniquely to the sequence of interest. Quantitative thermocyclers allow a fast and accurate quantification of PCR products during their amplification (*real time*) by qPCR based on fluorescent-labeled probes or on fluorescent dyes. Fluorescent-labeled probes are oligonucleotides that specifically bind to the target sequence and allow quantification by fluorescence. Three types of probes with different molecular background are mainly applied: *TagMan*® principle (Livak *et al.*, 1995), *molecular beacons* (Tyagi & Kramer 1996), and *hybridization probes*. They allow an accurate way of PCR product quantification and a simultaneous application to quantify several different PCR products even within the same sample (*multiplex* PCR). However, quantification using fluorescent dyes intercalating with the amplified DNA, such as SYBR® green, is a fast and cost-saving alternative but does not allow multiplexing. Additional information concerning SYBR® green or design, choice, and use of fluorescent-labeled probes is usually supplied by the manufacturer and can also be found in the advanced literature.

But how does real-time quantification of PCR products actually work? Instead of measuring the absolute amount of PCR products, the kinetic of PCR reaction is utilized for quantification. In the beginning of a PCR reaction the amplification is exponential and becomes linear in the later course of reaction until it is exhausted and finally stops (Figure 2.9A). The moment the fluorescent signal is

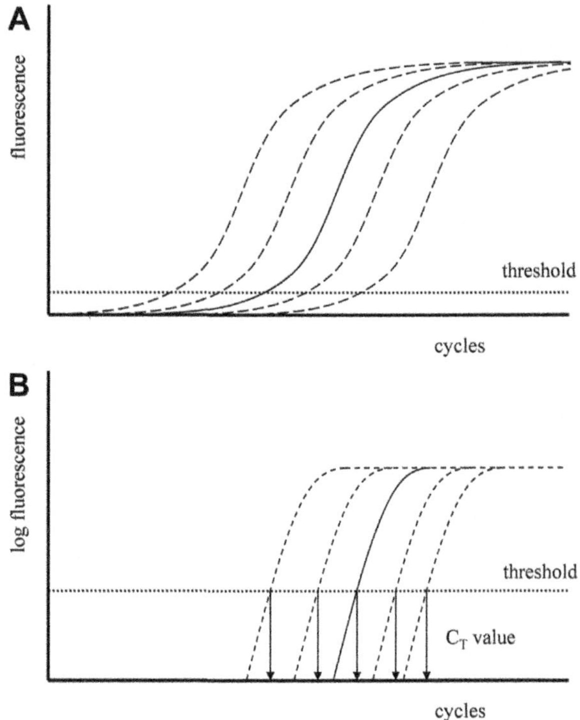


**Figure 2.9** A. Theoretical course and outcome of quantitative PCR analysis. The fluorescent curves of five different PCR samples are shown. The four sigmoid dashed lines represent the standard samples for standard curve calculation. The continuous line represents the actual samples. B. The same fluorescent curves as in part A but shown at a logarithmic scale. If the fluorescent signal of each reaction curve shows exponential growth and the signal is clearly distinguishable from the background signal, the threshold is chosen at this cycle ($C_T$).

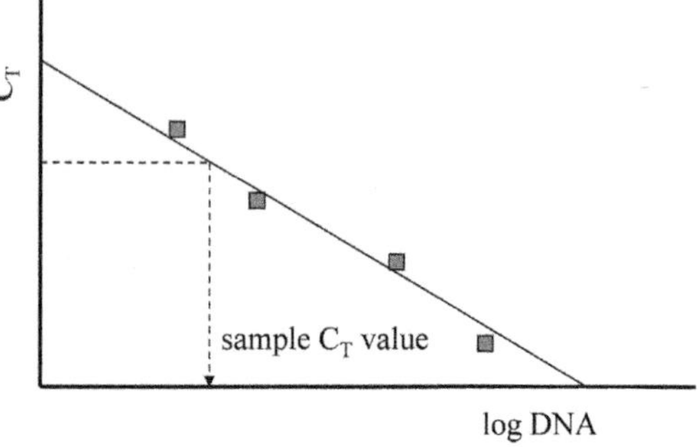


**Figure 2.10** The standard curve is gained by plotting the logarithmic scale of the initial DNA concentration against the $C_T$ value of each reaction and the original DNA concentration of a sample can be calculated.

clearly above the background signal a *cycle treshold* ($C_T$ value) is chosen. At this time, the amplification is exponential and in theory the amount of template could be directly calculated if the amplification ratio is known. But the reaction is influenced by an unknown number of factors and rather a standard is applied for each qPCR reaction. A standard is usually a dilution series of the template in a known concentration. It is not necessarily the virus genome itself, but a plasmid or DNA fragment that contains the target DNA sequence for PCR primer binding. Also, non-target sequence standards or external standards are possible, as long as the PCR conditions and the time of PCR product synthesis are similar.

The standard and the sample are separately amplified under the same reaction conditions and the $C_T$ value is chosen for all samples (Figure 2.9B). If the standard $C_T$ values are plotted against the logarithm of the initial DNA concentration, a standard curve is obtained by regression (Figure 2.10). Finally, the original DNA concentration of the sample is calculated by using the standard curve and the sample $C_T$ value.

Especially small viruses (e.g., densoviruses, iridoviruses, most RNA viruses) that are not countable by light microscopic techniques can be quantified by qPCR. A problem may occur if the virus genome is segmented (nodaviruses, cypoviruses, tetraviruses). Though quantification of viruses using qPCR requires some optimization, quantification of virus genomes by qPCR techniques is a well-established method. Below, examples for qPCR quantification of insect viruses are listed.

### *1. Budded virus quantification*

Quantitative PCR can also be used for quantification of BV of baculoviruses, when cell lines are not available for performing end point dilution assay (Section 7C). BVs are not visible under a light microscope but play an important role during secondary infection and in cell culture techniques (see Section 7). They are usually found in the hemolymph of an infected insect where they spread the infection, but are also obtained from the supernatant of infected cells in cell culture. Mukawa & Goto (2006) and Asser-Kaiser *et al.* (2011) described protocols for quantifying BVs of NPVs and GVs, respectively.

A short protocol to quantify BVs of CpGV from Cp14 cells is given below:

1. Remove supernatant from transfected Cp14 cells; the supernatant contains the BVs.

2. Include the supernatant of untreated cells as a negative control.
3. Isolate DNA of the BV suspension using a commercial kit for DNA isolation from tissue, according to the manufacturer`s protocol (Ron`s Tissue Kit, Bioron; DNeasy kit, Qiagen).
4. Perform the same steps with the untreated control.
5. Prepare a dilution series of CpGV DNA of known concentration as a standard for qPCR.
6. Include CpGV DNA isolated from OBs as a positive control.
7. For every sample, prepare the reaction mix as follows:
   20 μl SYBRGreen
   2 μl primer_forward (10 pm%l/μl)
   2 μl primer_reverse (10 pm%l/μl)
   11 μl dd$H_2O$
   5 μl DNA (10 ng/μl)

The number of DNA copies calculated by the software refers to the 5-μl sample introduced. Calculate the DNA copies in the original suspension used for DNA isolation.

### 2. Quantification of densoviruses

A *TaqMan*®-based method for the quantification of the mosquito densoviruses (MDV) *Aedes aegypti* densovirus (AeDNV), *Aedes* Peruvian densovirus (APeDNV) and *Hemagogus equinus* densovirus (HeDNV) was developed by Ledermann *et al.* (2004). *TagMan*® probe and primers are specific for a conserved region of the genomes and a plasmid (pANS1-GFP) is used as a standard. The plasmid is serially diluted to concentrations of $1 \times 10^7$ to $1 \times 10^1$ genome equivalents (geq) per microliter. A similar approach was used for quantification of MDV (BR/07) by Mosimann *et al.* (2011). Again a plasmid serially 10-fold diluted to concentrations of $10^{10}$–$10^1$ geq/μl was used as a genome equivalent for standard curve construction.

### 3. RT-PCR based quantification of honeybee viruses

A number of RNA viruses are able to infect the honeybee *Apis mellifera*. Also co-existence of different viruses in a single bee mite was observed by Chantawannakul *et al.* (2006) and methods for identification, discrimination, and quantification of these viruses were developed. Prior to quantification by real-time PCR, the viral RNA has to be isolated and transcribed into complementary DNA (cDNA) by reverse transcription. Procedures are established for a couple of honeybee viruses: DWV, Kashmir bee virus, Chronic bee paralysis virus (CBPV), Acute paralysis virus, Sacbrood virus (SBV), Black queen cell virus (BQCV), Apis iridescent virus (AIV) (Chen *et al.*, 2005; Chantawannakul *et al.*, 2006; Blanchard *et al.*, 2007; Olivier *et al.*, 2008).

## 6. QUANTIFICATION OF VIRUS ACTIVITY (VIRULENCE)

The biological properties of a virus are expressed among others in terms of pathogenicity and virulence. Whereas the term pathogenicity is more general, describing the potential ability to produce disease, virulence is a measurable characteristic (Shapiro-Ilan *et al.*, 2005). Virulence is measured in bioassays as dose–response or time–response relationships, expressed in terms of median lethal dose ($LD_{50}$), median lethal concentration ($LC_{50}$) or median lethal or survival time ($LT_{50}$ or $ST_{50}$). The pathogen dose or dosage applied, the inoculation method, and intervals of recording data depend on the virus–host system, on the feeding habits and on the availability of the virus stock. The method used for estimating the time–response relationship is principally the same as that applied for estimating the amount of virus required to kill the insect (Hunter-Fujita *et al.*, 1998). Usually, the following inoculation methods are applied in insect bioassays (Figure 2.11):

**Droplet feeding:** Neonate larvae drink from a droplet of virus suspension colored with food dye. The uptake of virus suspension is then visible by the colored gut. Larvae are transferred to virus-free diet and reared until response (usually death of insect) or pupation. Dosage is expressed as particles/ml.

**Surface contamination:** A specified volume of virus suspension is pipetted on to the surface of artificial diet. Larvae feed for a certain time and are then reared on uncontaminated diet until response or pupation. Dosage is expressed as particles/$mm^2$.

**Diet plug assay:** The virus dose is given on a small plug of artificial diet, suitable for full consumption by each test larva. After ingestion, larvae are reared on virus-free diet until response or pupation. Dosage is expressed as particles/larva.

**Diet incorporation:** Virus is mixed directly into artificial diet at a known concentration. Larvae feed

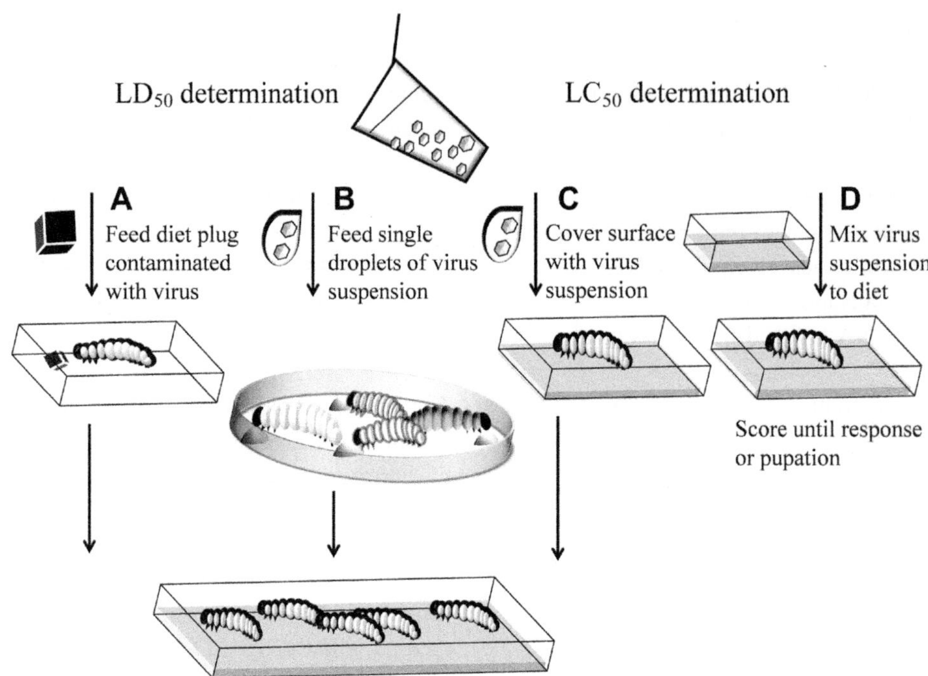


**Figure 2.11** Basic bioassay procedures for $LD_{50}$ and $LC_{50}$ determination. A. For the diet plug method, a known dosage of virus suspension is pipetted on a small piece of diet and fed to one test larvae each until full consumption. Larvae are then reared on virus-free diet. B. Droplet feeding: single droplets of virus suspension containing a known dosage of virus mixed with food dye are fed to single larvae. Larvae which have ingested the droplet are then reared on virus-free diet. C. For surface contamination, virus suspension is added to a known unit of artificial diet to cover the surface. After a short time of feeding, larvae are transferred to virus-free diet. Concentration is given per $mm^2$. D. Diet incorporation: virus suspension is mixed directly to a measured volume of artificial diet. Test larvae feed on the contaminated diet until the end of the bioassay.

on the contaminated diet until response or pupation. Dosage or concentration is expressed as particles/ml.

The dose–response assays should cover a mortality range between 10 and 90%. Therefore, the optimal range of dose/concentration for a certain host stage should be established in advance. First, a large range of doses or dosages should be applied to assess the shape and position of the response curve. Then, the doses or dosages are arranged symmetrically around the assessed $LC_{50}$ or $LD_{50}$ point (Cory & Bishop, 1995). Usually, early larval instars are used because of their higher susceptibility, lower rearing costs and lower variation in weight. $LC_{50}$ values for different stages of *S. litura* were found to increase from $1 \times 10^3$ OBs for 2-day-old larvae to $1.5 \times 10^9$ OBs for 8-day-old larvae (Trang & Chaudhari, 2002). As a general rule, at least 30 larvae should be introduced for every virus dose or dosages, encompassing five doses or dosages (Jones, 2000). Two independent controls should be performed for every bioassay experiment and the whole assay should be repeated independently at least three times. The independent repetition of the bioassay is necessary to randomize effects related to uncontrollable laboratory procedures and conditions and thereby estimating the experimental error (Robertson & Preisler, 1992). To obtain an independent replication, the diet should be prepared freshly for every bioassay experiment. Also the virus dilutions should be prepared freshly for every bioassay. It is important that the concentration of a virus suspension is quantified exactly before it is diluted for bioassay preparation. The concentration should be estimated for every single replicate. To reduce variation among assays, rearing and bioassay conditions (temperature, day length, humidity, food) need to be controlled and kept constant. In case an insect species cannot be reared on artificial diet, leaf assays using the food plant of the insect can be used. The method is principally the same as for surface contamination but the way of presenting the virus-contaminated food has to be optimized for the specific insect (David *et al.*, 1971). Depending on the aim of the experiment, it also might be more useful to feed insects on contaminated leaves or fruits to mimic field conditions.

## A. Determination of $LC_{50}$ values

Where a precise dosage determination of ingested virus is not possible, the $LC_{50}$, the concentration required to kill 50% of the test insects, is determined. $LC_{50}$ values may vary between tests more than $LD_{50}$s, because the amount of virus ingested depends on the insect's behavior (Cory & Bishop, 1995). $LC_{50}$ determination is often performed by surface contamination. This method is preferable for larvae that feed on the surface of the diet and is possible with every larval instar. The inoculum needs to be spread evenly over the surface of the diet.

Diet incorporation is also applicable for $LC_{50}$ estimation. This method is suitable for insects which bore into the diet, such as the codling moth *C. pomonella*. Virus suspension is mixed with artificial diet; before mixing, the diet must be cooled below 40°C to avoid thermal damage of the virus. For diet incorporation, a greater amount of virus suspension is required than for surface contamination. In favor of this method, all of the diet consumed by the test larvae contains virus.

### *1. Determination of $LC_{50}$ for CpGV in neonate larvae of* C. pomonella

1. Perform bioassays in autoclavable 50-well plates containing 45 ml of artificial diet (Ivaldi-Sender, 1974) mixed with 5 ml of OB suspension of different concentrations per plate.
2. For control plates, mix the diet with 5 ml $H_2O$ instead of OB suspension.
3. Prepare diet with half the amount of agar–agar to assure that diet can cool down to 40°C without solidifying.
4. Allow freshly made diet plates to dry for at least one day before setting up larvae.
5. The concentration–mortality relationship is determined at five different concentrations: $3 \times 10^2$, $1 \times 10^3$, $3 \times 10^3$, $1 \times 10^4$, and $3 \times 10^4$ OB/ml.
6. Use at least 30–35 freshly hatched neonate larvae (L1) for each concentration, and double the amount for the untreated controls.
7. In order to exclude those larvae from the experiment that died from handling, determine larval mortality the first day following the experimental setup. Only larvae alive at day 1 are recorded for experimental evaluation.
8. Incubate bioassay at 26°C with a relative humidity of 60% and a light–dark period of 16:8 h.
9. Determine virus mortality after 7 and 14 days.
10. Conduct three independent replicates.

## B. Determination of $LD_{50}$ values

Where the uptake of a precise virus dose is determined, the response is expressed as $LD_{50}$. Determination of $LD_{50}$ is more suitable for comparison of experiments, because it is independent of differences in individual behavior patterns of the insect. It is often determined by the diet plug method, which is suitable for larvae from 3rd instar on (Jones, 2000). Also surface contamination can be applied for $LD_{50}$ estimation, given that a precise amount of contaminated diet is consumed in a short time. Though, within-assay and between-assay variation was considered very high if the acquisition time is too long and when larvae ingest the virus at different times.

Therefore, the droplet feeding method was described as a fast method to apply a uniform volume of virus suspension to each test larva (Hughes & Wood, 1981). The volume ingested by neonate larvae of different lepidopteran larvae was shown to be very constant (Jones, 2000).

### *1. Droplet feeding methods for neonate* C. pomonella *larvae*

1. Mix virus suspension with blue food color (8%) to visualize virus ingestion.
2. Prepare five suspensions with different OB concentrations.
3. For controls, prepare virus-free suspension only containing food color.
4. Pipette droplets of 2 μl virus/control suspension in a circle of about 1.5 cm in diameter on a sheet of parafilm or another hydrophobic surface.

5. Place 20–30 L1 larvae in the center of the circle using a fine brush.
6. After 10–15 min, transfer larvae with blue guts to plates with virus-free artificial diet
7. Incubate at 26°C and a 16:8 h light–dark period.
8. Exclude larvae from the experiment which are dead on the first day after experimental setup.
9. The dose ingested is calculated by the virus concentration and the drinking volume of a neonate larva [$6.8 \pm 2.4$ µl (Chowdhury, 1992)].
10. Record mortality after 7 days.
11. Conduct three independent replicates.

### C. Determination of time–mortality response

Time–mortality response is measured either as the median lethal time ($LT_{50}$) by continually exposing test insects to virus or as median survival time ($ST_{50}$), when the virus inoculum is only ingested at the beginning of the experiment. Hughes & Wood (1981) suggested that time–mortality response is useful for virulence estimation due to its high reproducibility and sensitivity to slight differences. Bioassays of HearNPV against different populations of *H. armigera* showed no significant difference in $LD_{50}$ values, but the $LT_{50}$ values differed significantly by about 10–15 h (Grant & Bouwer, 2009). The droplet feeding method is especially appropriate for $LT_{50}$ estimation, as known doses are acquired in a short time period. Droplet feeding allows a synchronous uptake of the virus suspension, removing the acquisition-time component of $LT_{50}$ values (Hughes & Wood, 1981). Therefore, it is considered the most suitable method for time-effect assays (Jones, 2000). Usually, a single dose is tested for each sample, e.g., a previously determined $LD_{80}$. The bioassay is then monitored daily until the first virus infected cadavers occurs. Then, mortality is recorded at 8-h intervals.

### D. Evaluation of mortality data

The most common method used for analysis of dose/concentration–mortality data is probit analysis (Finney, 1971), determining the $LC_{50}$ and $LD_{50}$ values. The dose/concentration–mortality response curves are sigmoid. Easier for comparison are a log transformation of the doses/concentrations and a probit transformation of the mortalities obtained. This results in a regression line with the formula $y = ax + b$, where $y$ is the expected mortality, $a$ the slope, $x$ the log dose/concentration, and $b$ the intercept. Heterogeneity based on chi-square estimation and fiducial limits are determined. Within the fiducial limits, the true values are likely to lie with a selected range of certainty, usually 95%. The regression lines for different treatments can be compared for their slopes (hypothesis of parallelism). If the slopes of the probit–log dose lines are parallel, two virus suspensions can be compared based on their $LD_{50}$ or $LC_{50}$. If the probit lines are not parallel, the difference between suspensions will vary at different mortality levels. In this case, $LD_{50}$ or $LC_{50}$ *and* slope need to be given when comparing bioassays (Jones, 2000). The same is applied when comparing samples based on their relative potencies, which is the ratio between the $LC_{50}$ of a test suspension and the $LC_{50}$ of a standard. Normally, mortalities observed in the treatments are corrected for control mortality using Abbott`s formula (Abbott, 1925). Common programs for probit analysis are implemented in SAS or Toxrat Solutions software packages.

For time–mortality assays, probit analysis is only applicable, if the observations are independent, i.e. if each observation is made with an independent cohort of test animals. If the observations are repeatedly made with the same test animals (until they die), and hence violate the presupposition of data independence as it is required for probit analysis, then parameter-free survival analysis using the Kaplan–Meier estimator needs to be applied (Kaplan & Meier, 1958).

## 7. CELL CULTURE TECHNIQUES

Cell culture systems allowing the replication of a virus are of crucial importance for the molecular characterization of virus gene function, as they allow easy and synchronized infections of permissive host cells. Insect cells grow in nutrient-rich media, which are commercially available and might be supplemented by fetal bovine serum (FBS). As the cultures are highly nutritious, microbial growth needs to be avoided by applying good microbial practices and by working in a sterile tissue culture hood (laminar flow cabinet). For further details on cell culture techniques of insect cells, the laboratory manual of O`Reilly *et al.* (1994) is recommended.

The following precautions need to be considered:

- Switch on tissue culture hood at least 10 min before starting work.
- Carefully wash hands with soap before working in the flow cabinet.
- Carefully clean the working surface in the hood with 70% EtOH.
- Use only sterile, disposable pipettes, pipette tips, cell culture flasks, dishes, and plates.
- Avoid moving hands in and out of the flow cabinet, if not necessary, as this will disturb the vertical laminar flow and will increase the probability of contamination.
- Never touch pipette tips or the inner surface of cell culture plates, flasks, or lids with your fingers.
- Never touch the outer surface of cell culture plates, flasks, or lids with the pipette tip.
- Always dispose of pipettes, pipette tips, cell culture plates, flasks, and other material, if you are not sure that they are sterile.
- After finishing the experiment, carefully clean the working surface in the hood with 70% EtOH.

## A. Transfection

The term transfection generally describes the introduction of viral DNA into cells for virus replication and production of infectious virus. 'Transfection' was coined from transformation–infection, as the introduction of foreign DNA into cells is called DNA-mediated transformation.

An efficient and widely used method for transfecting insect cells with viral DNA is the lipofectin procedure. Thereby, virus DNA is bound to cationic liposomes. DNA and liposomes form a complex which is transported into the cultured cells. The liposome complex sequesters the DNA completely, delivering viral DNA into the cell. The transfection mix is prepared by diluting the lipofectin reagent and the DNA to be introduced using tissue culture (TC) medium; both solutions are then combined and incubated to form the DNA/lipofectin complex. It is important that these steps are carried out in polystyrene tubes, as the lipofectin reagent binds to polypropylene surfaces. Insect cells should be in the log phase of growth and the confluency for transfection should be about 70%.

### *1. Transfection of* Spodoptera frugiperda *(Sf) cells with baculovirus DNA*

1. Transfer $10^6$ Sf cells in 35-mm TC Petri dishes and let cells sediment for 1 h. (For some cell lines, it can be useful to incubate cells overnight to allow attachment to the plate.)
2. Mix 6 μl lipofectin reagent to 100 μl Sf900 TC medium in a polystyrene tube.
3. Mix 5 μl of DNA with 100 μl Sf900 TC medium in a second polystyrene tube.
4. Mix both solutions gently and incubate for 30 min at room temperature.
5. Add 800 μl Sf900 TC medium to the mix and mix gently (= transfection mixture).
6. Remove TC medium from sedimented cell culture and replace with transfection mixture.
7. Incubate for 3–5 h at 27°C.
8. Remove transfection mixture and replace with 2 ml Sf900 TC medium.
9. Incubate cells for 4 days at 27°C.
10. Remove TC medium and filter it, the supernatant contains BVs.

The optimal concentrations of lipofectin and DNA vary and have to be determined for every virus–cell system. Today, various reagents optimized for the transfection of specific insect cell lines are available (FectoFly™, BaculoFECTIN™, Insect GeneJuice™).

## B. Infection

The viral infection of insect cells is influenced by several culture conditions such as temperature, pH, or nutrient composition of the TC medium. A temperature between 25 and 28°C and a pH of 6.2 are suitable for most lepidopteran cell cultures. For baculoviruses, BV suspension can be produced by transfection of cell cultures with viral DNA (Section 7B). Alternatively, hemolymph of infected insects can be collected and frozen at −80°C until used for infection of cultured cells. Cells should be used in the mid- to late-log phase. Confluency should range between 60 and 80%.

### *1. Method for the infection of monolayer cell cultures*

1. Seed cells to an appropriate density and allow attachment for 1 h.

2. Remove TC medium and add BV virus or hemolymph suspension. In case no specific titer is desired, add enough virus suspension to completely cover the cells.
3. Incubate cells for 1–5 h at 27°C.
4. Tilt flask once per hour to circulate the virus suspension on the cell monolayer.
5. Remove inoculum and rinse cells with medium.
6. Add fresh medium.
7. Incubate for 2–4 days at 27°C.
8. Transfer supernatant to a fresh tube and store at 4°C. The supernatant contains BVs.

### C. End-point dilution assay

As described in Sections 6A and 6B, the virus activity of a given virus can be characterized in the host larvae by determining the $LC_{50}$ or $LD_{50}$. For the titration of a virus sample in cell culture an equivalent method, the end-point dilution assay, can be applied. The concept of this assay is to infect a number of cell cultures with different dilutions of a virus stock and to calculate that dilution, which would infect 50% of the cultures. This value is termed the *50% Tissue Culture Infective Dose* or $TCID_{50}$. A standard procedure working well for AcMNPV is given below. The duration of the experiment as well as the concentration of cells might need to be adjusted for other viruses and cell lines.

*1. Procedure for end-point dilution assay*

1. Dilute Sf cells (or other cell line) in appropriate cell culture medium (e.g., Hinks + 10% FBS or serum-free cell culture medium) to a concentration of $1.5 \times 10^6$ cells/ml.
2. Prepare a dilution (90 µl) of the virus suspension from $10^{-1}$ to $10^{-9}$ in a reaction vial and add to each tube 90 µl Sf cell suspension. Mix well. Use a new pipette tip for each dilution.
3. Add to each well of a microtiter plate (6 × 10 wells) 10 µl of the virus/cell suspension. Fill six wells per dilution (A–F), starting with the lowest concentration. Add to the last six wells uninfected Sf cell suspension as an untreated control.
4. Incubate the plates at 27°C for 5 days in a moist incubator.
5. For viruses containing a LacZ marker gene add 10 µl X-Gal solution to each well using clean tips for each well. Incubate overnight at 27°C in a moist incubator.
6. Count the number of infected wells (polyhedra containing cells, cytopathogenic effects, gfp expression, X-Gal staining) and uninfected wells per dilution and calculate the $TCID_{50}$ as shown in Box 2.1.

**Box 2.1.**

| Microtiter Plate (A B C D E F) | Dilution | Infected Wells | Uninfected Wells | Accumulated Infected | Accumulated Uninfected | % Infected |
|---|---|---|---|---|---|---|
| 1 ●●●●●● | $10^{-1}$ | 6 | 0 | | | |
| 2 ●●●●●● | $10^{-2}$ | 6 | 0 | | | |
| 3 ●●●●●● | $10^{-3}$ | 6 | 0 | | | |
| 4 ●●●●●● | $10^{-4}$ | 6 | 0 | **11** | **0** | **100.0** |
| 5 ○●●○●● | $10^{-5}$ | 4 | 2 | **5** | **2** | **71.4 (= A)** |
| 6 ○●○○○○ | $10^{-6}$ | 1 | 5 | **1** | **7** | **12.5 (= B)** |
| 7 ○○○○○○ | $10^{-7}$ | 0 | 6 | **0** | **13** | **0.0** |
| 8 ○○○○○○ | $10^{-8}$ | 0 | 6 | | | |
| 9 ○○○○○○ | $10^{-9}$ | 0 | 6 | | | |
| 10 ○○○○○○ | Control | *0* | *6* | | | |

*2. $TCID_{50}$ calculation*

Accumulate the number of infected and uninfected wells per dilution, starting with the lowest concentration for infected wells and with the highest concentration for uninfected cells.

Interpolate the dilution that would have given a 50% response by calculating the proportional distance (PD) between the response below and above 50% infections: A = percentage of accumulated portion above 50% and B = percentage of accumulated portion below 50% by applying the following formula:

$$PD = (A - 50)/(A - B)$$

Calculate the $TCID_{50} = 10^{(N-PD)}$, where N = log of the dilution giving a response > 50% (in the example above, N = −5).

Calculate the virus titer as the reciprocal $1/TCID_{50}/5$ µl (volume of virus suspension per well), thus the virus titer is $(1/TCID_{50} \times 200)$/ml.

Express the virus titer as *plaque forming units* (pfu) by applying the relationship pfu/ml = $(1/TCID_{50} \times 200) \times 0.69$/ml.

A detailed mathematical background for the relationship between $TCID_{50}$ and pfu is given in O'Reilly *et al.* (1994).

## D. Injection of budded virus

For experiments such as host range studies or the detection of barriers to viral infection, injection of BV directly into the insect's hemocoel can be useful. Furthermore, recombinant viruses or viral bacmids often do not form OBs because they are occlusion-negative (occ−). After production of BV by transfecting the cell culture, BV can be injected into the insect`s hemocoel. One to ten microliters are injected with a microsyringe into a proleg, the volume thereby depending on the insect's size. The proleg is chosen because the muscle closes the wound and therefore prevents bleeding or contamination.

Protocol for injection of BV into *C. pomonella* larvae:

1. Rinse syringe with 70% EtOH then sterile water and clamp it on to a stand.
2. Anesthetize freshly molted 4th instar larvae with diethyl ether vapor for 2−3 min by adding a few droplets of diethyl ether to a filter paper in a covered beaker.
3. Prior to injection, disinfect larvae ventrally with 0.4% hyamine solution using a sterile cotton bud.
4. Grasp the insect with two forceps and insert the needle longitudinally through the second proleg (Hamilton syringe, 0.21 mm diameter). Insert the needle just underneath the cuticle without piercing the midgut.
5. Inject 1−2 µl BV suspension
6. Let larvae recover for several minutes.
7. Transfer larvae to virus free artificial diet and incubate at 26°C.
8. Record mortality data daily.

The method is time consuming and requires quick and sterile handling of larvae and syringe in order to avoid bacterial infection. BV is sterile filtered through a 45-nm filter before injection. The concentration of the BV suspension is determined previously by qPCR (Section 5B) or titered in cell culture (Section 7C). BV suspensions can be diluted to the desired concentrations using TC (tissue culture) medium. Control experiments injecting TC medium only need to be performed in parallel. Hemocoelic injection of BV has been described by others for 5th instar larvae of *B. mori* (Katsuma *et al.*, 2006), 4th instar of *T. ni* and 5th instar of *S. frugiperda* (Prikhod'ko *et al.*, 1999), 3rd instar larvae of *S. faculata* (Chejanovsky *et al.*, 1995), and 4th instar of *C. pomonella* (Asser-Kaiser *et al.*, 2011). For most insects, the ideal time for injection is one to two days before molting into the final instar (O'Reilly *et al.*, 1992).

## 8. REVERSE GENETICS

Reverse genetics describes an approach for studying the biological function of viral genes in experimental virology. The methods are based on infectious full-length clone methodology, viruses with a genome derived from cloned cDNA (Neumann *et al.*, 2002). Mutations can be introduced directly into the cloned genomes at any location of the viral genome. After silencing a viral gene or inserting deletions or point mutations, the effect on the resulting phenotype is analyzed by transfecting host cells. Virus interactions with host cells, viral life cycle, assembly or the role of viral proteins are questions investigated using this approach, which has been developed further since the early 1980s.

### A. Infectious full-length clones of RNA viruses

Infectious RNA full-length clones consist of the cDNA copy of the viral RNA genome in a bacterial plasmid. In short, viral RNA is transcribed by RT-PCR into the more stable cDNA (Section 3E), which is then cloned into a plasmid. The plasmids containing these genomes can be introduced into cell culture by transfection (Section 7A). RNA transcripts produced then by *in vitro* transcription are replicas of the genomic RNA and are ideally able to induce infection in host cells. Infectious full-length clones are today established for positive and negative sense RNA viruses, including those with segmented and non-segmented genomes.

#### *1. Positive sense RNA viruses*

Positive sense RNA viruses were the first RNA viruses used for reverse genetics systems and are successfully constructed for several virus families (Table 2.5). In positive sense RNA viruses, the genome serves as mRNA and can be directly translated. This means that the release of RNA transcribed from cloned cDNA into host cells initiates viral protein expression and therefore virus replication (Pekosz *et al.*, 1999; Yamshchikov *et al.*, 2001).

In contrast to the first attempts to produce infectious clones in the early 1980s, the construction of a full-length cDNA is currently facilitated by the fact that genome sequences of numerous viruses are determined, allowing the design of sequence specific primers. In addition, DNA polymerases are currently improved for higher fidelity and amplification of long PCR products.Various systems are commercially available for this purpose.

After RNA isolation from a virus stock, the first step in constructing a full-length clone is the production of ss cDNA from viral RNA. This is done by reverse transcription. Primer design allows the inclusion of recognition sites for restriction enzymes useful for cloning or for introducing (RNA polymerase) promoter sequences into the 5′-region of the primer sequence (Benjeddou *et al.*, 2002; Boyer & Haenni, 1994). For reverse transcription into cDNA, diverse commercial kits are available.

The complete genome is then synthesized from the single-stranded template as a dsDNA by PCR. Use of high-fidelity DNA polymerase allows the full-length amplification of viral genomes. By treating the PCR product with the enzyme RNase H, any RNA residues are degraded from the resulting product. dsDNA can now be cloned into a bacterial vector, or directly used for *in vitro* transcription of viral RNA using commercially available *in vitro* transcription reaction mixtures (Benjeddou *et al.*, 2002). Another approach is cloning of DNA into the baculovirus expression system under control of the *polyhedrin* (*polh*) promoter (Boyapalle *et al.*, 2008; Krishna *et al.*, 2003; Pal *et al.*, 2007). Expression can then be performed in Sf21 cells and therefore does not require host-specific susceptible cell lines. In addition, large-scale production of RNA viruses usually expressed at a low level is possible (Figure 2.12).

Directed alterations of the RNA virus genome can be introduced at different steps of the protocol. When performing cDNA synthesis using specific primers, restriction enzyme recognition sites can be included into the primer sequence to allow further cloning of the resulting PCR product. For this approach it is assumed that the sequence is known and that the recognition sites are not part of the viral sequence. For downstream

**Table 2.5** Infectious full-length clones constructed from insect RNA viruses.

| *Virus (family)* | *Genome* | *Literature* |
|---|---|---|
| Black queen cell virus BQCV (*Dicistroviridae*) | (+) RNA | Benjeddou *et al.*, 2002 |
| Pariacoto virus PaV (*Nodaviridae*) | (+) RNA | Johnson *et al.*, 2000 |
| *Nodamura* virus NoV (*Nodaviridae*) | (+) RNA | Johnson *et al.*, 2003 |
| Black beetle virus BBV (*Nodaviridae*) | (+) RNA | Dasmahapatra *et al.*, 1986 |
| *Rhopalsiphum padi* virus RhPV (*Dicistroviridae*) | (+) RNA | Boyapalle *et al.*, 2008 |
| *Helicoverpa armigera* stunt virus HaSV (*Tetraviridae*) | (+) RNA | Gordon *et al.*, 2001 |
| Infectious flacherie virus IFV (*Picornaviridae*) | (+) RNA | Isawa *et al.*, 1998 |

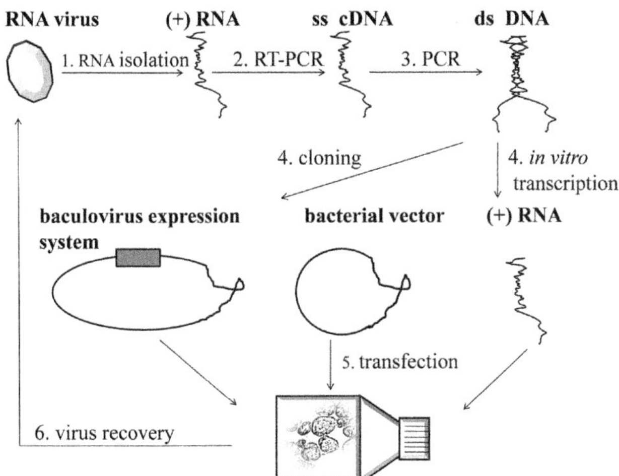


**Figure 2.12** Full length clone construction of positive sense RNA viruses. Following RNA isolation from a virus stock (1), single-stranded positive sense RNA is transcribed to cDNA by the enzyme reverse transcriptase, using either specific or random primers (2). By PCR using specific primers, single-stranded cDNA is converted into a double-stranded DNA molecule (3). The dsDNA can either directly be used for *in vitro* transcription by RNA polymerase into positive sense RNA or cloned into a vector molecule (4). Transfection of susceptible cell lines using these constructs (5) allows expression of the cloned genome and recovery of infectious virus (6).

work like *in vitro* synthesis, promoter sequences like bacteriophage T7 promoter for DNA-dependent RNA polymerase can be included into the 5′-end of the oligonucleotide primer. Directed mutations (insertions, deletions, substitutions) are introduced by homologous recombination into the cloned genome. The mutagenized plasmids are then transfected into host cells and changes in virulence parameters evaluated.

*2. Negative sense RNA viruses*

From negative sense RNA viruses, the first completely cloned cDNA was obtained in 1994 (Neumann *et al.*, 2002). In contrast to positive sense RNA virus genomes, negative sense RNA genomes alone are not infectious. For initiation of viral replication and transcription, co-expression of the virally encoded polymerase complex (RNA dependent RNA polymerase) is required (Boyer & Haenni, 1994). Beyond that, negative sense RNA viruses are generally enveloped. Their ribonucleocapsids are embedded in a lipid bilayer derived from the host cell. Into this envelope, viral glycoproteins involved in receptor-binding and host entry are inserted (Garciasastre & Palese, 1993). Two different experimental solutions have been developed to obtain infectious DNA clones of minus strand non-segmented or segmented genomes.

*a. Non-segmented viruses*

Negative sense ssRNA viruses with non-segmented genomes consist of single RNA molecules. Full-length minus strand genomic RNA is not infectious. Introduction of minus sense RNA into cells that produce viral proteins required for mRNA production does not result in infectious virus. To obtain infectious virus, host cells are transfected with a recombinant vaccine virus that synthesizes T7 RNA polymerase and are transformed with a plasmid containing the plus strand copy of the viral genome and expression plasmids for polymerase, phosphoprotein and nucleocapsid protein, all under the control of T7 RNA polymerase. After transfection with the plasmids, plus sense RNA is copied into minus sense strands, which are templates for mRNA synthesis and genome replication (Neumann *et al.*, 2002; Walpita & Flick, 2005).

*b. Segmented viruses*

Segmented viral genomes consist of several different RNA molecules (Garciasastre & Palese, 1993; Neumann *et al.*, 2002). To obtain full-length clones, cloned viral DNA of each segment is inserted between an RNA polymerase I promoter and termination signal, and an RNA polymerase II promoter and a polyadenylation signal. When plasmids are introduced into cells, minus strand RNA is synthesized from the cellular RNA polymerase I and transcribed by the viral polymerase.

*c. Technical pitfalls*

Cloning strategy, sequences bordering the viral insert and cDNA synthesis strongly influence the biological activity of the full-length clone. Especially for positive sense ssRNA viruses, problems with the stability of virus-encoding plasmids during propagation in *Escherichia coli* are described. *In vivo* transcription of viral cDNA cassettes under control of eukaryotic promoters can result in the production of products toxic for the host (Mishin *et al.*, 2001). To avoid such toxic effects on host bacteria, viral RNA genomes can be divided into several fragments when cloned into a vector.

During extension of large genomes, fidelity of the PCR reaction reduces proportionally. The use of high-fidelity reverse transcriptase and DNA polymerase can lower the rate of mismatches enormously. The presence of nonviral nucleotides at the 5′-end of viral transcripts strongly reduces infectivity. Generally, extensions on the 3′-end are more easily tolerated. In contrast, even extensions of one to two nucleotides on the 5′-end reduce infectivity (Boyer & Haenni, 1994).

## B. Bacmid technology

### *1. Construction of recombinant baculovirus by homologous recombination*

Recombinant baculoviruses were mainly constructed for the expression of heterologous proteins in insect cells. For this reason a foreign open reading frame (ORF) was usually set under the control of the *polyhedrin* (major occlusion body protein) promoter, which is characterized by its high transcription level during the very late stage of infection. The foreign ORF was then expressed in insect cells. Due to eukaryotic post-transcriptional modification the expressed gene was often functionally similar to its counterpart, which is the major advantage of the Baculovirus Expression Vector System (BEVs) compared to bacterial expression systems.

In the past, construction of recombinant baculoviruses was mainly based on homologous recombination, by which the polyhedrin ORF was replaced by a foreign ORF. The foreign ORF had been previously cloned into a transfer vector downstream of a *polyhedrin* promoter and was flanked by viral sequences that target the construct to a particular region in the viral genome (O'Reilly *et al.*, 1992). Here, homologous regions of both viral DNA and plasmid hybridize and crossover recombination events occur. Consequently, the foreign gene replaces the viral genomic sequence. For purification and identification of recombinant baculoviruses, plaque assays (O'Reilly *et al.*, 1994) have to be performed, which can take up to several weeks. Different approaches were taken to increase the recombination frequency and to facilitate the screening of recombinant viruses (O'Reilly *et al.*, 1994; Jarvis, 1997). However, all of them needed plaque screening and purification in insect cell culture.

### *2. The bacmid technology*

Luckow *et al.* (1993) presented a new and highly efficient method for generating recombinant baculoviruses, which was based on the construction of a baculovirus shuttle vector (bacmid). This method took advantage of the circular dsDNA genome of baculoviruses, which can behave like a bacterial plasmid after a couple of modifications. As a shuttle vector, it can replicate as a plasmid in *E.coli* and is still capable of infecting lepidopteran cells and insects as a virus. To achieve this, a DNA cassette, containing a selectable kanamycin resistance marker, a bacterial mini-F replicon, and a bacterial transposon Tn7 attachment site (*att*Tn7) was cloned into a plasmid pVL1393 and then introduced into the genome of AcMNPV by homologous recombination (Luckow *et al.*, 1993). Most elegantly, the *att*Tn7 is positioned within a lacZα gene that allows bacterial colony screening by α-complementation. Transposition of a foreign gene into the *att*Tn7 site within the lacZα locus takes place in *E. coli* and recombinant baculovirus bacmids can be easily screened by blue–white selection after adding X-Gal to the bacterial colonies. The bacmid DNA is propagated in *E. coli* and then used to transfect insect cells. This technique provided the basis for the commercially available AcMNPV Bac-to-Bac® Baculovirus expression System (Invitrogen).

Beyond the use of the bacmid technology for protein expression, it is an ideal tool to study the function of single genes in baculoviruses. So far, bacmids of different baculoviruses, including SeMNPV (Pijlmann *et al.*, 2002), HearNPV (Wang *et al.*, 2003), CpGV (Hilton *et al.*, 2008), and others, have been constructed and subjected to molecular analyses. As the bacmids can be constructed to be occlusion positive and to produce viable OBs, OBs derived from bacmids can be isolated from transfected insect cells and then further be used to infect insects. This allows the dissecting of gene function not only in cell lines but also under *in vivo* conditions in the host insect. The generation of baculovirus bacmids is greatly facilitated by sub-cloning pVL1393 and introducing a unique *Bsu*36I restriction site resulting in the vector BAC-Bsu36I (Figure 2.13) (Pijlman *et al.*, 2002). Others used the bacterial cassette or BAC-Bsu36I for a simplified construction of new bacmids (Table 2.6). Below, their construction strategies are outlined.

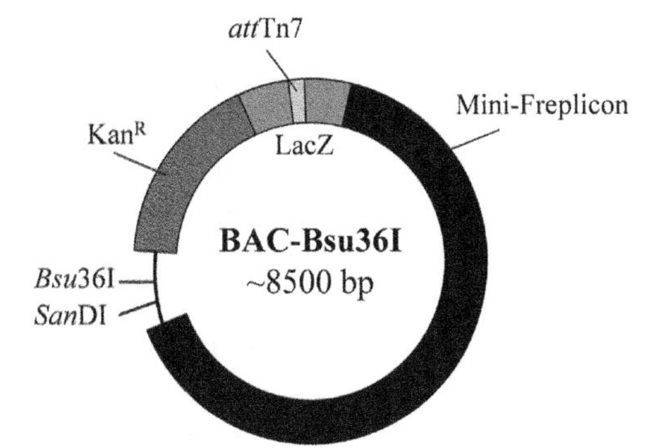


**Figure 2.13** Bacterial cassette for baculovirus shuttle vector construction (bacmid). This cassette contains a mini F-replicon, akanamycin resistance gene and a LacZ, in which an attachment site for the bacterial transposon Tn7 is located (*att*Tn7). This cassette was modified at its ends, self-ligated (Pijlman *et al.*, 2002) and designated BAC-Bsu36I. This vector has two unique restriction sits (*San*DI, *Bsu*36I).

*3. Cloning into a unique restriction site*

Some baculovirus genomes have unique REN sites (see Section 3C) allowing the linearization of the circular genome at this site and the ligation of the bacterial replication cassette into this site (Figure 2.13). As in SeMNPV, the unique restriction site (*San*DI) is located within the coding region of the *polyhedrin* gene and the insertion of BAC-Bsu36I cassette led to a loss of the functional ORF resulting in an occlusion negative phenotype. The occlusion positive phenotype was subsequently restored by transposing the wild-type *polyhedrin* gene into the *att*Tn7 site of the inserted bacterial cassette (Pijlmann *et al.*, 2002). The unique restriction site can also be located within a non-coding area between two ORFs as is the case in CpGV. Here, the bacterial cassette was ligated into the unique *Pac*I site (Hilton *et al.*, 2008). In any case, the bacmid needs be tested and compared with its wild-type baculovirus for its virulence for cell cultures (Section 7) (O'Reilly *et al.*, 1992) or in bioassays (Section 6).

*4. Insertion by homologous recombination*

Wang *et al.* (2003) added the bacterial cassette to the genome of HearNPV by homologous recombination (Figure 2.14). To achieve this, a transfer vector was generated carrying the bacterial cassette that was flanked

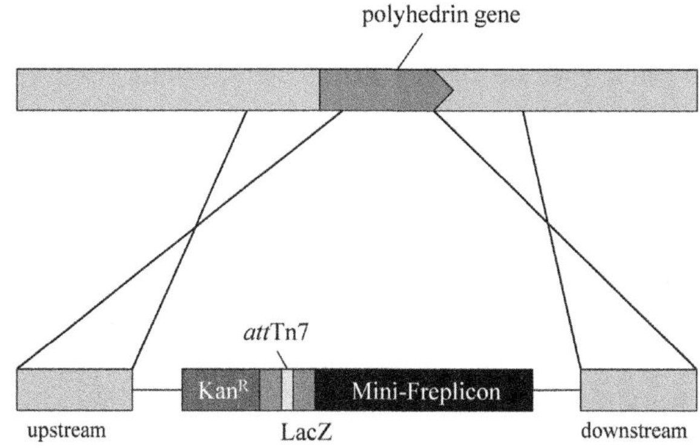


**Figure 2.14** Theory of homologous recombination for generating a bacmid by deletion of the polyhedrin locus. Upstream and downstream fragments of the wild-type polyhedrin locus were ligated to both sites of the linear BAC Bsu36I (see Figure 2.13). The natural polyhedrin sequence is then replaced by homologous recombination in *E.coli*.

**Table 2.6** List of constructed bacmids and their generation strategy.

| *Baculovirus* | *Abbreviation* | *Strategy of bacterial cassette insertion* | *Reference* |
|---|---|---|---|
| *Autographa californica* multiple NPV | AcMNPV | Homologous recombination | Luckow *et al.*, 1993 |
| *Spodoptera exigua* multiple NPV | SeMNPV | Ligation | Pijlman *et al.*, 2002 |
| *Helicoverpa armigera* NPV | HearNPV | Homologous recombination | Wang *et al.*, 2003 |
| *Bombyx mori* NPV | BmNPV | Homologous recombination | Motohashi *et al.*, 2005 |
| *Cydia pomonella* granulovirus | CpGV | Ligation | Hilton *et al.*, 2008 |

by an upstream and downstream segment of the wild-type HearNPV *polyhedrin* gene (Figure 2.14). Insect cells of *Helicoverpa zea* (Hz2e5) were co-transfected with a transfer vector and wild-type HearNPV DNA and the *polyhedrin* gene was deleted by homologous recombination. Here again, the occ$^+$ phenotype needed to be restored by transposing the wild-type *polyhedrin* gene of HearNPV into the *att*Tn7 of the bacterial cassette. A similar approach was also performed for the construction of *Bombyx mori* NPV (BmNPV) bacmid (Motohashi *et al.*, 2005).

## C. siRNA and RNA silencing

Due to the increasing number of sequenced genomes, RNA interference (RNAi) has become an important tool for analyzing loss-of-function of genes even in non-model organisms. Thereby, small interfering RNAs (siRNAs), duplexes of 21 nt RNA molecules, are synthesized chemically or by transcription. They correspond to the gene to be silenced and are introduced into cells by transformation. By causing sequence-specific mRNA degradation, they inhibit the production of specific proteins (Figure 2.15). Application of the RNAi technique is useful for specific analysis of genes without inducing a (possibly lethal) knock out. The successful reduction of gene expression is then analyzed either on the RNA level by quantitative RT-PCR, or by Western blot analysis, provided that a protein-specific antibody is available.

In *Drosophila melanogaster*, the silencing of almost every gene is nowadays possible by the RNAi technique, supporting large-scale screening for insecticide resistance targets (Perry *et al.*, 2011). Beyond the model organism, *D. melanogaster*, RNAi-based experiments have been performed in several insects. The contributions of distinct genes in wing or eye formation, development of legs, or adult morphological diversity have been studied, among others, in the floor beetle *Tribolium castaneum*, the milkweed bug *Oncopeltus fasciatus*, the cricket *Gryllus bimaculatus*, and the wasp *Nasonia vitripennis* (reviewed by Mito *et al.*, 2011).

Design and sequence of siRNA are crucial for a gene silencing approach. Choosing the siRNA target site is the first step in a siRNA experiment. Some rules for siRNA design are as follows (Elbashir *et al.*, 2001; Reynolds *et al.*, 2004; Tuschl, 2002):

- Efficient siRNAs are composed of a 21-nt sense and 21-nt anti-sense siRNA, each providing a 2-nt overhang at the 3′-end.
- The target GC content should range between 30 and 52%.

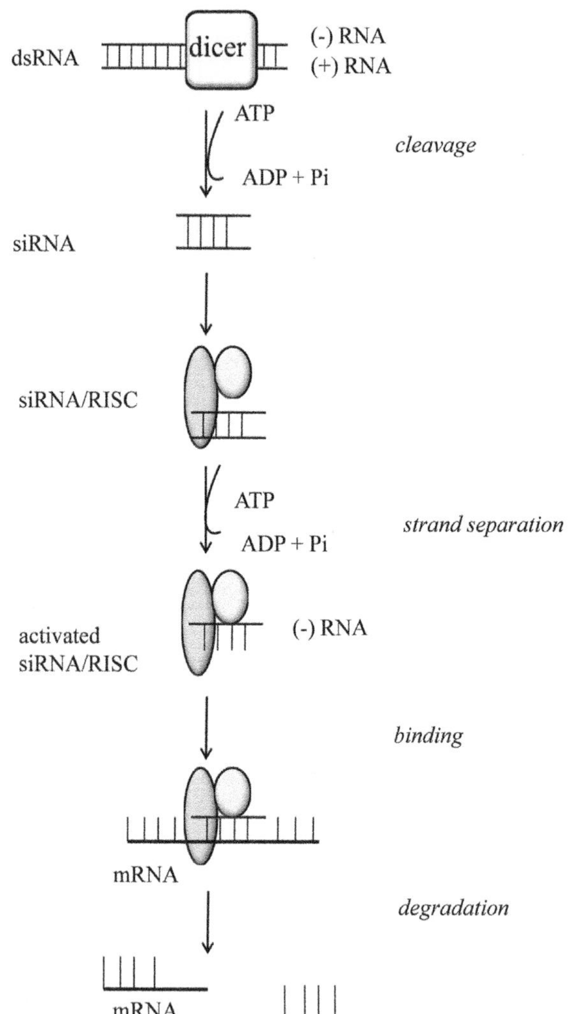


**Figure 2.15** Synthesis and pathway of siRNA. Cytoplasmatic dsRNA activates the ribonuclease protein dicer, which binds to the dsRNA molecule and cleaves it in double stranded fragments of 20–25 bp with unpaired nucleotides at the 3′-end, the siRNAs. The siRNS are associated with the RISC complex (RNA-induced silencing complex) and separated into single strands. One RNA strand is part of the activated RISC complex, the other strand becomes degraded. The activated RISC complex binds to the target mRNA mediated by the siRNA and induces cleavage to avoid translation of the target RNA.

**Table 2.7** Methods available for siRNA synthesis.

| *siRNA preparation* | *Method* | *Reference* |
|---|---|---|
| *In vitro* transcription | *In vitro* transcription of sense and anti-sense DNA coding for siRNA commercial kits available | (Zhu *et al.*, 2005) |
| Expression vector | Synthesis of DNA sense and anti-sense strand coding for siRNA<br>Sense and anti-sense strand are separated by a spacer for hairpin formation<br>Cloning into a commercial expression vector with RNA pol III promoter controlling expression of siRNA-hairpin | (Miyagishi & Taira, 2002) (Sui *et al.*, 2002) |
| PCR expression cassette | PCR production of sense and anti-sense siRNA<br>Flanked by RNA pol III promoter and termination sequence<br>Cloning of siRNA sequence into an expression vector | (Castanotto & Rossi, 2004) (Castanotto *et al.*, 2002) (Brummelkamp *et al.*, 2002) |
| RNAse III cleavage | *In vitro* transcription of long dsRNA<br>Production of siRNA cocktail by digestion with RNAse III enzyme | (Yang *et al.*, 2002) |
| Chemical synthesis | Performed by commercial supplier | |

- The target sequence should be located 50–100 nt downstream of the start codon ATG.
- Search for the 23-nt sequence motif AA($N_{19}$)TT or NA($N_{21}$) or NAR($N_{17}$)YNN, where N is any nucleotide, R a purine (A, G) and Y a pyrimidin base (C, U).

Various siRNA design tools are available online (http://itb.biologie.hu-berlin.de/~nebulus/sirna/siDesign.htm). SiRNA design is also provided by commercial suppliers, such as Ambion, Eurofins MWG Operon, Quagen or Sigma-Aldrich.

Several methods are available for preparing siRNA (Table 2.7). The methods differ in their costs, expenditure of time and stability of expression.

## REFERENCES

Abbott, W. S. (1925). A method of computing the effectiveness of an insecticide. *J. Econ. Entomol.*. (1925), *82*, 265–267.

Abd-Alla, A. M., Cousserans, F., Parker, A. G., Jehle, J. A., Parker, N. J., Vlak, J. M., Robinson, A. S., & Bergoin, M. (2008). Genome analysis of a *Glossina pallidipes* salivary gland hypertrophy virus reveals a novel, large, double-stranded circular DNA virus. *J. Virol.*. (2008), *82*, 4595–4611.

Adams, J. R., & Bonami, J. R. (1991). *Atlas of Invertebrate Viruses*. Boca Raton, FL: CRC Press.

Arends, H. M., Winstanley, D., & Jehle, J. A. (2005). Virulence and competitiveness of *Cydia pomonella* granulovirus mutants: Parameters that do not match. *J. Gen. Virol.*. (2005), *86*, 2731–2738.

Asgari, S., & Johnson, K. N. (Eds.). (2010). *Insect Virology*. (2010). Caister Academic Press.

Asser-Kaiser, S., Radtke, P., El-Salamouny, S., Winstanley, D., & Jehle, J. A. (2011). Baculovirus resistance in codling moth (*Cydia pomonella* L.) caused by early block of virus replication. *Virology*. (2011), *410*, 360–367.

Auer, H., & Lyianarachchi, S. (2003). Chipping away at the chip bias: RNA degradation in microarray analysis. *Nat. Genet.*. (2003), *35*, 292–293.

Ayres, M. D., & Howard, S. C. (1994). The complete DNA sequence of *Autographa californica* nuclear polyhedrosis virus. *Virology*. (1994), *202*, 586–605.

Benjeddou, M., & Leat, N. (2002). Development of infectious transcripts and genome manipulation of Black queen-cell virus of honey bees. *J. Gen. Virol.*. (2002), *83*, 3139–3146.

Benjeddou, M., Leat, N., Allsopp, M., & Davidson, S. (2001). Detection of Acute Bee Paralysis Virus and Black Queen Cell Virus from honeybees by reverse transcriptase PCR. *Appl. Environ. Microbiol.*. (2001), *67*, 2384–2387.

Biji, C. P., Sudheendrakumar, V. V., & Sajeev, T. V. (2006). Influence of virus inoculation method and host larval age on productivity of the NPV of the teak defoliator, *Hyblaea puera (Cramer)*. *J. Virol. Methods*. (2006), *133*, 100–104.

Blanchard, P., Ribiere, M., Celle, O., Lallemand, P., Schurr, F., Olivier, V., Iscache, A. L., & Faucon, J. P. (2007). Evaluation of a real-time two-step RT-PCR assay for quantitation of Chronic bee paralysis virus (CBPV) genome in experimentally-infected bee tissues and in life stages of a symptomatic colony. *J. Virol. Methods*. (2007), *141*, 7–13.

Boyapalle, S., Beckett, R. J., Pal, N., Miller, W. A., & Bonning, B. C. (2008). Infectious genomic RNA of *Rhopalosiphum padi* virus transcribed *in vitro* from a full-length cDNA clone. *Virology*. (2008), *375*, 401–411.

Boyer, J.-C., & Haenni, A.-L. (1994). Infectious transcripts and cDNA clones of RNA viruses. *Virology*. (1994), *198*, 415–426.

Briese, D. T. (1986). Host resistance to microbial control agents. In J. M. Franz (Ed.), *Biological Plant and Health Protection*. (1986) (pp. 233–256). New York, NY: Fisher.

Brummelkamp, T. R., Bernards, R., & Agami, R. (2002). A system for stable expression of short interfering RNAs in mammalian cells. *Science*. (2002), *296*, 550–553.

Carpenter, J. A., Obbard, D. J., Maside, X., & Jiggins, F. M. (2007). The recent spread of a vertically transmitted virus through populations of *Drosophila melanogaster*. *Mol. Ecol.*. (2007), *16*, 3947–3954.

Castanotto, D., Li, H., & Rossi, J. J. (2002). Functional siRNA expression from transfected PCR products. *RNA*. (2002), *8*, 1454–1460.

Castanotto, D., & Rossi, J. J. (2004). Construction and transfection of PCR products expressing siRNAs or shRNAs in mammalian cells. In M. Sioud (Ed.), *Ribozymes and siRNA Protocols*. (2004) (pp. 509–514). Humana Press.

Chantawannakul, P., Ward, L., Boonham, N., & Brown, M. (2006). A scientific note on the detection of honeybee viruses using real-time PCR (TaqMan) in Varroa mites collected from a Thai honeybee (*Apis mellifera*) apiary. *J. Invertebr. Pathol.*. (2006), *91*, 69–73.

Chejanovsky, N., Zilberberg, N., Rivkin, H., Zlotkin, E., & Gurevitz, M. (1995). Functional expression of an alpha anti-insect scorpion neurotoxin in insect cells and lepidopterous larvae. *FEBS Lett.*. (1995), *376*, 181–184.

Cherry, A. J., Parnell, M. A., Grzywacz, D., & Jones, K. A. (1997). The optimization of *in vivo* nuclear polyhedrosis virus production in *Spodoptera exempta* (Walker) and *Spodoptera exigua* (Hübner). *J. Invertebr. Pathol.*. (1997), *70*, 50–58.

Chen, Y. P., Higgins, J. A., & Feldlaufer, M. F. (2005). Quantitative real-time reverse transcription-PCR analysis of deformed wing virus infection in the honeybee (*Apis mellifera* L.). *Appl. Environ. Microbiol.*. (2005), *71*, 436–441.

Chomczynski, P., & Sacchi, N. (1987). Single-step method of RNA isolation by acid guanidinium thiocyanate-phenol-chloroform extraction. *Anal. Biochem.*. (1987), *162*, 156–159.

Chowdhury, S. (1992). *Characterisation of some aspects of the isolates of* Cydia pomonella *virus (CpGV)*. Littlehampton, UK: Department of Microbiology, University of Reading and HRI.

Christian, P. D., Gibb, N., Kasprzak, A. B., & Richards, A. (2001). A rapid method for the identification and differentiation of Helicoverpa NPVes (NPV Baculoviridae) isolated from the environment. *J. Virol. Methods*. (2001), *96*, 51–65.

Cory, J. S., & Bishop, H. L. (1995). Use of baculoviruses as biological insecticides. In C. D. Richardson (Ed.), *Methods in Molecular Biology. Baculovirus Expression Protocols*. (1995) (pp. 277–294). Humana Press.

Cory, J. S., & Evans, H. F. (2007). Viruses. In L. A. Lacey, & H. K. Kaya (Eds.), *Field Manual of Techniques in Invertebrate Pathology*. (2007) (2nd ed.). (pp. 149–174) Dordrecht: Springer.

Dasmahapatra, B., Dasgupta, R., Saunders, K., Selling, B., Gallagher, T., & Kaesberg, P. (1986). Infectious RNA derived by transcription from cloned cDNA copies of the genomic RNA of an insect virus. *Proc. Natl. Acad. Sci. USA*. (1986), *83*, 63–66.

David, W. A. L., Ellaby, S. J., & Gardiner, B. O. C. (1971). Bioassaying an insect virus on leaves: I. The influence of certain factors associated with the virus application. *J. Invertebr. Pathol.*. (1971), *17*, 158–163.

Elbashir, S. M., Martinez, J., Patkaniowska, A., Lendeckel, W., & Tuschl, T. (2001). Functional anatomy of siRNAs for mediating efficient RNAi in *Drosophila melanogaster* embryo lysate. *EMBO J.*. (2001), *20*, 6877–6888.

Elleman, C. J., Entwistle, P. F., & Hoyle, S. R. (1980). Application of the impression film technique to counting inclusion bodies of nuclear polyhedrosis viruses on plant surfaces. *J. Invertebr. Pathol.*. (1980), *3*, 129–132.

Erlandson, M. (2008). Insect pest control by viruses. In B. W. J. Mahy, & M. H. V. v. Regenmortel (Eds.), *Encyclopedia of Virology*. (2008) (pp. 125–133). Oxford, UK: Academic Press.

Evans, H., & Shapiro, M. (1997). Viruses. In L. A. Lacey (Ed.), *Manual of Techniques in Insect Pathology*. (1997) (pp. 17–53). London, UK: Academic Press.

Fath-Goodin, A., & Webb, B. A. (2008). Polydnaviruses: General features. In B. W. J. Mahy, & M. H. V. v. Regenmortel (Eds.), *Encyclopedia of Virology*. (2008) (pp. 256–261). Oxford, UK: Academic Press.

Fauquet, C. M., Mayo, M. A., Maniloff, J., Desselberger, U., & Ball, L. A. (Eds.). (2005). *Virus Taxonomy. 8th Report of the International Committee on Taxonomy of Viruses*. (2005). New York, NY: Academic Press, Elsevier.

Finney, D. J. (1971). *Probit Analysis*. Cambridge, UK: Cambridge University Press.

Garcia-Maruniak, A., Maruniak, J. E., Farmerie, W., & Boucias, D. G. (2008). Sequence analysis of a non-classified, non-occluded DNA virus that causes salivary gland hypertrophy of *Musca domestica*, MdSGHV. *Virology*. (2008), *377*, 184–196.

Garciasastre, A., & Palese, P. (1993). Genetic manipulation of negative-strand RNA virus genomes. *Ann. Rev. Microbiol.* (1993), *47*, 765–790.

Glaser, R. W. (1918). A new disease of gypsy moth caterpillars. *J. Agricult. Res.*. (1918), *13*, 515–522.

Glaser, R. W., & Chapman, J. W. (1913). The wilt disease of gypsy moth caterpillars. *J. Econ. Entomol.*. (1913), *6*, 479–488.

Goldsmith, C. S., & Miller, S. E. (2009). Modern use of electron microscopy for detection of viruses. *Clin. Microbiol. Rev.*. (2009), *22*, 552–563.

Gordon, K. H. J., Williams, M. R., Baker, J. S., Gibson, J. M., Bawden, A. L., Millgate, A. G., Larkin, P. J., & Hanzlik, T. N. (2001). Replication-independent assembly of an insect virus (Tetraviridae) in plant cells. *Virology*. (2001), *288*, 36–50.

Grace, T. D. (1962). Establishment of four strains of cells from insect tissues grown *in vitro*. *Nature*. (1962), *195*, 788–789.

Grant, M., & Bouwer, G. (2009). Variation in the time-mortality responses of laboratory and field populations of *Helicoverpa armigera* infected with NPV. *J. Appl. Entomol.*. (2009), *133*, 647–650.

Green, T. B., Shapiro, A., White, S., Rao, S., Mertens, P. P., Carner, G., & Becnel, J. J. (2006). Molecular and biological characterization of a Cypovirus from the mosquito *Culex restuans*. *J. Invertebr. Pathol.*. (2006), *91*, 27–34.

Grzywacz, D., Jones, K. A., Moawad, G., & Cherry, A. (1998). The *in vivo* production of *Spodoptera littoralis* nuclear polyhedrosis virus. *J. Virol. Methods*. (1998), *71*, 115–122.

Gupta, R. K., Raina, J. C., & Monobrulla, M. D. (2007). Optimization of *in vivo* production of NPV in homologous host larvae of *Helicoverpa armigera*. *J. Entomol.*. (2007), *4*, 279–288.

Hagiwara, K., Tomita, M., Kobayash, J., Miyajima, S., & Yoshimura, T. (1998). Nucleotide sequence of *Bombyx mori* cytoplasmic polyhedrosis virus segment 8. *Biochem. Biophys. Res. Comm.*. (1998), *247*, 549–553.

Hanzlik, T. N., Dorrian, S. J., Gordon, K. H., & Christian, P. D. (1993). A novel small RNA virus isolated from the cotton bollworm, *Helicoverpa armigera*. *J. Gen. Virol.*. (1993), *74*, 1805–1810.

Harrison, R. L. (2009). Genomic sequence analysis of the Illinois strain of the *Agrotis ipsilon* multiple NPV. *Virus Genes*. (2009), *38*, 155–170.

Hayashi, Y., Kawarabata, T., & Bird, F. T. (1970). Isolation of a cystoplasmic-polyhedrosis virus of the silkworm, *Bombyx mori*. *J. Invertebr. Pathol.*. (1970), *16*, 378–384.

Herniou, E. A., Olszewski, J. A., O'Reilly, D. R., & Cory, J. S. (2004). Ancient coevolution of baculoviruses and their insect hosts. *J. Virol.*. (2004), *78*, 3244–3251.

Hilton, S., Kemp, E., Keane, G., & Winstanley, D. (2008). A bacmid approach to the genetic manipulation of granuloviruses. *J. Virol. Methods*. (2008), *152*, 56–62.

Huger, A. M. (2005). The Oryctes virus: Its detection, identification, and implementation in biological control of the coconut palm rhinoceros beetle, *Oryctes rhinoceros* (Coleoptera: Scarabaeidae). *J. Invertebr. Pathol.*. (2005), *89*, 78–84.

Hughes, P. R., & Wood, H. A. (1981). A synchronous peroral technique for the bioassay of insect viruses. *J. Invertebr. Pathol.*. (1981), *37*, 154–159.

Hunter-Fujita, F. R., Entwistle, P. E., Evans, H. F., & Crook, N. E. (1998). *Insect Viruses and Pest Management*. New York: Wiley & Sons, pp. 620.

Isawa, H., Asano, S., Sahara, K., Lizuka, T., & Bando, H. (1998). Analysis of genetic information of an insect picorna-like virus, infectious flacherie virus of silkworm: evidence for evolutionary relationships among insect, mammalian and plant picorna(-like) viruses. *Arch. Virol.*. (1998), *143*, 127–143.

Jehle, J. A., Lange, M., Wang, H., Hu, Z.-H., Wang, Y., & Hauschild, R. (2006). Molecular identification and phylogentic analysis of baculoviruses of Lepidoptera. *Virology*. (2006), *346*, 180–196.

Johnson, K. N., & Christian, P. D. (1996). A molecular taxonomy for cricket paralysis virus including two new isolates from Australian populations of *Drosophila* (Diptera: Drosophilidae). *Arch. Virol.*. (1996), *141*, 1509–1522.

Johnson, K. L., Price, B. D., & Ball, L. A. (2003). Recovery of infectivity from cDNA clones of nodamura virus and identification of small nonstructural proteins. *Virology*. (2003), *305*, 436–451.

Johnson, K. N., Zeddam, J. L., & Ball, L. A. (2000). Characterization and construction of functional cDNA clones of Pariacoto Virus, the first Alphanodavirus isolated outside Australasia. *J. Virol.*. (2000), *74*, 5123–5132.

Ignoffo, C. M. (1966). Effects of temperature on mortality of *Heliothis zea* larvae exposed to sublethal doses of a nuclear-polyhedrosis virus. *J. Invertebr. Pathol.* (1966), *8*, 290–292.

Ivaldi-Sender, C. (1974). Techniques simples pour elevage permanent de la tordeuse orientale, *Grapholita molesta* (lep., Tortricidae), sur milieu artificiel. *Ann. Zool. Ecol. Anim.*. (1974), *6*, 337–343.

Jarvis, D. L. (1997). Baculovirus expression vectors. In L. K. Miller (Ed.), *The Baculoviruses*. (1997) (pp. 389–431). New York, NY: Plenum Press.

Jenkins, N. E., & Grzywacz, D. (2000). Quality control of fungal and viral biocontrol agents—assurance of product performance. *Biocontrol Sci. Technol.*. (2000), *10*, 753–777.

Jones, K. A. (2000). Bioassays of entomopathogenic viruses. In A. Navon, & K. R. S. Ascher (Eds.), *Bioassays of Entomopathogenic Microbes and Nematodes*. (2000) (pp. 95–140). Wallingford, UK: CABI Publishing.

Ju, J., Kim, D. H., Bi, L., Meng, Q., Bai, X., Li, Z., Li, X., Marma, M. S., Shi, S., Wu, J., Edwards, J. R., Romu, A., & Turro, N. J. (2006). Four-color DNA sequencing by synthesis using cleavable fluorescent nucleotide reversible terminators. *Proc. Natl. Acad. Sci. USA*. (2006), *103*, 19635–19640.

Kalmakoff, J., & Payne, C. C. (1973). A simple method for the separation of single-stranded and double-stranded RNA on hydroxyapatite. *Anal. Biochem.*. (1973), *55*, 26–33.

Kaplan, E. L., & Meier, P. (1958). Nonparametric estimation from incomplete observations. *J. Amer. Statist. Assn.*. (1958), *53*, 457–481.

Kapun, M., Nolte, V., Flatt, T., & Schlötterer, C. (2010). Host range and specificity of the *Drosophila* C Virus. *PLoS One*. (2010), *5*, e12421.

Katsuma, S., Horie, S., Daimon, T., Iwanaga, M., & Shimada, T. (2006). *In vivo* and *in vitro* analyses of a *Bombyx mori* NPV mutant lacking functional vfgf. *Virology.* (2006), *355*, 62–70.

Krishna, N. K., Marshall, D., & Schneemann, A. (2003). Analysis of RNA packaging in wild-type and mosaic protein capsids of Flock House Virus using recombinant Baculovirus vectors. *Virology.* (2003), *305*, 10–24.

Lange, M., & Jehle, J. A. (2003). The genome of the *Cryptophlebia leucotreta* granulovirus. *Virology.* (2003), *317*, 220–236.

Lange, M., Wang, H., Zhihong, H., & Jehle, J. A. (2004). Towards a molecular identification and classification system of lepidopteran-specific baculoviruses. *Virology.* (2004), *325*, 36–47.

Langenbuch, R. (1957). Beitrag zur Färbung von Einschlußkörpern (Polyedern) in Blut- und Gewebeausstrichen viruskranker Insekten. *Mikroskopie.* (1957), *12*, 267–268.

Lasa, R., Caballero, P., & Williams, T. (2007). Juvenile hormone analogs greatly increase the production of a NPV. *Biol. Control.* (2007), *41*, 389–396.

Ledermann, J. P., Suchman, E. L., Black, W. C., 4th, & Carlson, J. O. (2004). Infection and pathogenicity of the mosquito densoviruses AeDNV, HeDNV, and APeDNV in *Aedes aegypti* mosquitoes (Diptera: Culicidae). *J. Econ. Entomol..* (2004), *97*, 1828–1835.

Li, S. Y., & Skinner, A. C. (2005). Influence of larval stage and virus inoculum on virus yield in insect host *Neodiprion abietis* (Hymenoptera: Diprionidae). *J. Econ. Entomol..* (2005), *98*, 1876–1879.

Liang, Z., Zhang, X., Yin, X., Cao, S., & Xu, F. (2011). Genomic sequencing and analysis of *Clostera anachoreta* granulovirus. *Arch. Virol..* (2011), *156*, 1185–1198.

Livak, K. J., Flood, S. J., Marmaro, J., Giusti, W., & Deetz, K. (1995). Oligonucleotides with fluorescent dyes at opposite ends provide a quenched probe system useful for detecting PCR product and nucleic acid hybridization. *PCR Methods Appl..* (1995), *4*, 357–362.

Luckow, V. A., Lee, S. C., Barry, G. F., & Olins, P. O. (1993). Efficient generation of infectious recombinant baculoviruses by site-specific transposon-mediated insertion of foreign genes into a baculovirus genome propagated in *Escherichia coli. J. Virol..* (1993), *67*, 4566–4579.

Maori, E., Lavi, S., Mozes-Koch, R., Gantman, Y., Peretz, Y., Edelbaum, O., Tanne, E., & Sela, I. (2007). Isolation and characterization of Israeli acute paralysis virus, a dicistrovirus affecting honeybees in Israel: evidence for diversity due to intra- and inter-species recombination. *J. Gen. Virol..* (2007), *88*, 3428–3438.

Margulies, M., Egholm, M., Altman, W. E., Attiya, S., Bader, J. S., Bemben, L. A., Berka, J., Braverman, M. S., Chen, Y.-J., Chen, Z., Dewell, S. B., Du, L., Fierro, J. M., Gomes, X. V., Godwin, B. C., He, W., Helgesen, S., Ho, C. H., Irzyk, G. P., Jando, S. C., Alenquer, M. L. I., Jarvie, T. P., Jirage, K. B., Kim, J.-B., Knight, J. R., Lanza, J. R., Leamon, J. H., Lefkowitz, S. M., Lei, M., Li, J., Lohman, K. L., Lu, H., Makhijani, V. B., McDade, K. E., McKenna, M. P., Myers, E. W., Nickerson, E., Nobile, J. R., Plant, R., Puc, B. P., Ronan, M. T., Roth, G. T., Sarkis, G. J., Simons, J. F., Simpson, J. W., Srinivasan, M., Tartaro, K. R., Tomasz, A., Vogt, K. A., Volkmer, G. A., Wang, S. H., Wang, Y., Weiner, M. P., Yu, P., Begley, R. F., & Rothberg, J. M. (2005). Genome sequencing in microfabricated high-density picolitre reactors. *Nature.* (2005), *437*, 376–380.

McCarthy, W. J., & Gettig, R. R. (1986). Current developments in baculovirus serology. In R. R. Granados, & B. A. Federici (Eds.), *The Biology of Baculoviruses.* (1986) (pp. 147–158). Boca Raton, FL: CRC Press.

Mehrvar, A., Rabindra, R. J., Veenakumari, K., & Narabenchi, G. B. (2007). Standardization of mass production in three isolates of NPV of *Helicoverpa armigera* (Hubner). *Pak. J. Biol. Sci..* (2007), *10*, 3992–3999.

Mishin, V. P., Cominelli, F., & Yamshchikov, V. F. (2001). A 'minimal' approach in design of flavivirus infectious DNA. *Virus Res..* (2001), *81*, 113–123.

Mito, T., Nakamura, T., Bando, T., Ohuchi, H., & Noji, S. (2011). The advent of RNA interference in Entomology. *Entomol. Sci..* (2011), *14*, 1–8.

Miyagishi, M., & Taira, K. (2002). Development and application of siRNA expression vector. *Nucleic Acids Symposium Series.* (2002), *2*, 113–114.

Mosimann, A. L., Bordignon, J., Mazzarotto, G. C., Motta, M. C., Hoffmann, F., & Santos, C. N. (2011). Genetic and biological characterization of a densovirus isolate that affects dengue virus infection. *Mem. Inst. Oswaldo Cruz.* (2011), *106*, 285–292.

Motohashi, T., Shimojima, T., Fukagawa, T., Maenaka, K., & Park, E. Y. (2005). Efficient large-scale protein production of larvae and pupae of silkworm by *Bombyx mori* nuclear polyhedrosis virus bacmid system. *Biochem. Biophys. Res. Commun..* (2005), *326*, 564–569.

Mukawa, S., & Goto, C. (2006). *In vivo* characterization of a group II NPV isolated from *Mamestra brassicae* (Lepidoptera: Noctuidae) in Japan. *J. Gen. Virol..* (2006), *87*, 1491–1500.

Neumann, G., Whitt, M. A., & Kawaoka, Y. (2002). A decade after the generation of a negative-sense RNA virus from cloned cDNA—what have we learned? *J. Gen. Virol..* (2002), *83*, 2635–2662.

Nielsen, H. (2011). Working with RNA. In H. Nielsen (Ed.), *RNA.* (2011). (pp. 15–28). Humana Press.

O'Reilly, D. R., Miller, L. K., & Luckow, V. A. (1994). *Baculovirus Expression Vectors. A Laboratory Manual.* Oxford, UK: Oxford University Press.

Olivier, V., Blanchard, P., Chaouch, S., Lallemand, P., Schurr, F., Celle, O., Dubois, E., Tordo, N., Thiéry, R., Houlgatte, R., & Ribiére, M. (2008). Molecular characterisation and phylogenetic analysis of Chronic bee paralysis virus, a honey bee virus. *Virus Res..* (2008), *132*, 59–68.

Onstad, D. W., Fuxa, J. R., Humber, R. A., Oestergaard, J., Shapiro-Ilan, D. I., Gouli, V. V., Anderson, R. S., Andreadis, T. G., & Lacey, L. (2006). *An Abridged Glossary of Terms Used in Invertebrate Pathology.* Society for Invertebrate Pathology website. sipweb.org.

Pal, N., Boyapalle, S., Beckett, R., Miller, W. A., & Bonning, B. C. (2007). A baculovirus-expressed dicistrovirus that is infectious to aphids. *J. Virol.*. (2007), *81*, 9339–9345.

Pekosz, A., He, B., & Lamb, R. A. (1999). Reverse genetics of negative-strand RNA viruses: closing the circle. *Proc. Natl. Acad. Sci. USA*. (1999), *96*, 8804–8806.

Perera, N., Aonuma, H., Yoshimura, A., Teramoto, T., Iskei, H., Nelson, B., Igarashi, Il, Yagi, T., Fukumoto, S., & Kanuka, H. (2009). Rapid identification of virus-carrying mosquitoes using reverse transcription-loop-mediated isothermal amplification. *J. Virol. Methods*. (2009), *156*, 32–36.

Perry, T., Batterham, P., & Daborn, P. J. (2011). The biology of insecticidal activity and resistance. *Insect Biochem. Mol. Biol.*. (2011), *41*, 411–422.

Pijlman, G. P., Dortmans, J. C., Vermeesch, A. M., Yang, K., Martens, D. E., Goldbach, R. W., & Vlak, J. M. (2002). Pivotal role of the non-hr origin of DNA replication in the genesis of defective interfering baculoviruses. *J. Virol.*. (2002), *76*, 5605–5611.

Prikhod'ko, E. A., Lu, A., Wilson, J. A., & Miller, L. K. (1999). *In vivo* and *in vitro* analysis of baculovirus ie-2 mutants. *J. Virol.*. (1999), *73*, 2460–2468.

Reynolds, A., Leake, D., Boese, Q., Scaringe, S., Marshall, W. S., & Khvorova, A. (2004). Rational siRNA design for RNA interference. *Nat. Biotechnol.*. (2004), *22*, 326–330.

Roberts, R. J., Vincze, T., Posfai, J., & Macelis, D. (2007). REBASE-enzymes and genes for DNA restriction and modification. *Nucleic Acids Res.*. (2007), *35*(Database issue), D269–70.

Robertson, J. L., & Preisler, H. K. (1992). *Pesticide Bioassays with Arthropods*. Boca Raton: CRC Press.

Romeis, B. (1989). *Mikroskopische Technik*. Wien, Baltimore: Urban and Schwarzenberg, München, pp. 697.

Ronaghi, M., Uhlén, M., & Nyrén, P. (1998). A sequencing method based on real-time pyrophosphate. *Science*. (1998), *281*, 363–365.

Sambrook, J., & Russell, D. W. (2001). *Molecular Cloning. A laboratory manual*. New York, NY: Cold Spring Harbor Laboratory Press.

Sanger, F., Nicklen, S., & Coulson, A. R. (1977). DNA sequencing with chain-terminating inhibitors. *Proc. Natl. Acad. Sci. USA*. (1977), *74*, 5463–5467.

Schroeder, A., Mueller, O., Stocker, S., Salowsky, R., Leiber, M., Gassmann, M., Lightfoot, S., Menzel, W., Granzow, M., & Ragg, T. (2006). The RIN: an RNA integrity number for assigning integrity values to RNA measurements. *BMC Molec. Biol.*. (2006), *7*, 3.

Shapiro, M. (1986). *In vivo* production of baculoviruses. In R. R. Granados, & B. A. Federici (Eds.), *The Biology of Baculoviruses*. (1986) (pp. 31–61). Boca Raton, FL: CRC Press, Inc. II.

Shapiro-Ilan, D. I., Fuxa, J. R., Lacey, L. A., Onstadt, D. W., & Kaya, H. K. (2005). Definitions of pathogenicity and virulence in invertebrate pathology. *J. Invertebr. Pathol.*. (2005), *88*, 1–7.

Sivaram, A., Barde, P. V., Kumar, S. R., Yadv, P., Gokhale, M. D., Basu, A., & Mourya, D. T. (2009). Isolation and characterization of densonucleosis virus from *Aedes aegypti* mosquitoes and its distribution in India. *Intervirology*. (2009), *52*, 1–7.

Smith, G. E., Summers, M. D., & Fraser, M. J. (1983). Production of human beta interferon in insect cells infected with a baculovirus expression vector. *Mole. Cell Biol.*. (1983), *3*, 2156–2165.

Sui, G., Soohoo, C., Affar el, B., Gay, F., Shi, Y., Forrester, W. C., & Shi, Y. (2002). A DNA vector-based RNAi technology to suppress gene expression in mammalian cells. *Proc. Natl. Acad. Sci. USA*. (2002), *99*, 5515–5520.

Teixeira, E. W., Chen, Y., Message, D., Pettis, J., & Evans, J. D. (2008). Virus infections in Brazilian honey bees. *J. Invertebr. Pathol.*. (2008), *99*, 117–119.

Tompkins, G. J. (1991). Purification of invertebrate viruses. In J. R. Adams, & J. R. Bonami (Eds.), *Atlas of Invertebrate Viruses*. (1991) (pp. 31–39). Boca Raton, FL: CRC Press.

Trang, T. T. K., & Chaudhari, S. (2002). Bioassay of nuclear polyhedrosis virus (NPV) and in combination with insecticide on *Spodoptera litura* (Fab). *Omonrice*. (2002), *10*, 45–53.

Tuschl, T. (2002). Expanding small RNA interference. *Nat. Biotech.*. (2002), *20*, 446–448.

Tyagi, S., & Kramer, F. R. (1996). Molecular beacons: probes that fluoresce upon hybridization. *Nat. Biotechnol.*. (1996), *14*, 303–308.

Valles, S. M., & Hashimoto, Y. (2009). Isolation and characterization of *Solenopsis invicta* virus 3, a new positive-strand RNA virus infecting the red imported fire ant, *Solenopsis invicta*. *Virology*. (2009), *388*, 354–361.

Venter, J. C., Adams, M. D., Sutton, G. G., Kerlavage, A. R., Smith, H. O., & Hunkapiller, M. (1998). Shotgun sequencing of the human genome. *Science*. (1998), *280*, 1540–1542.

Vlak, J. M., de Gooijer, C. D., Tramper, J., & Miltenburger, H. G. (Eds.). (1996). *Insect Cell Cultures: Fundamental and Applied Aspects*. (1996). Netherlands: Springer, pp. 324.

Walpita, P., & Flick, R. (2005). Reverse genetics of negative-stranded RNA viruses: A global perspective. *FEMS Microbiol. Lett.*. (2005), *244*, 9–18.

Walter, C. T., Tomasicchio, M., Hodgson, V., Hendry, D. A., Hill, M. P., & Dorrington, R. A. (2008). Characterization of a succession of small insect viruses in a wild South African population of *Nudaurelia cytherea capensis* (Lepidoptera: Saturniidae). *S. Afr. J. Sci.*. (2008), *104*, 147–152.

Wang, H., Deng, F., Pijlman, G. P., Chen, X., Sun, X., Vlak, J. M., & Hu, Z. (2003). Cloning of biologically active genomes from a *Helicoverpa armigera* single-nucleocapsid NPV isolate by using a bacterial artificial chromosome. *Virus Res.*. (2003), *97*, 57–63.

Wang, Y., Kleespies, R. G., Ramle, M. B., & Jehle, J. A. (2008). Nanogram genomic sequencing of a large dsDNA virus, *Oryctes rhinoceros* nudivirus, using multiple displacement amplification. *J. Virolog. Meth.*. (2008), *152*, 106–108.

Wigley, P. J. (1980a). Practical: Diagnosis of virus infections—staining of insect inclusion body viruses. In J. Kalmakoff, & J. F. Longworth (Eds.), *Microbial Control of Insect Pests*. (1980a). *New Zealand Dept. Sci. Indus. Res. Bull.* (1980a), *228*. Wellington, New Zealand.

Wigley, P. J. (1980b). Practical: Counting micro-organisms. In J. Kalmakoff, & J. F. Longworth (Eds.), *Microbial Control of Insect Pests*. (1980b). *New Zealand Dept. Sci. Indus. Res. Bull.* (1980b), *228* (pp. 35), Wellington, New Zealand.

Wilson, K., & Walker, J. (2005). *Principles and Techniques of Biochemistry and Molecular Biology*. New York, NY: Cambridge University Press.

Yamshchikov, V., Mishin, V., & Cominell, F. (2001). A new strategy in design of (+)RNA virus infectious clones enabling their stable propagation in E-coli. *Virology*. (2001), *281*, 272–280.

Yang, D., Buchholz, F., Huang, Z., Goga, A., Chen, C. Y., Brodsky, F. M., & Bishop, J. M. (2002). Short RNA duplexes produced by hydrolysis with *Escherichia coli* RNase III mediate effective RNA interference in mammalian cells. *Proc. Natl. Acad. Sci. USA*. (2002), *99*, 9942–9947.

Yue, C., & Genersch, E. (2005). RT-PCR analysis of Deformed wing virus in honeybees (*Apis mellifera*) and mites (*Varroa destructor*). *J. Gen. Virol.*. (2005), *86*, 3419–3424.

Zambon, R. A., Nandakumar, M., Vakharia, V. N., & Wu, L. P. (2005). The Toll pathway is important for an antiviral response in *Drosophila*. *Proc. Nat. Acad. Sci. USA*. (2005), *102*, 7257–7262.

Ziemnicka, J. (2008). Effects of viral epizootic induction in population of the satin moth *Leucoma salicis* L. (Lepidoptera: Lymantriidae). *J. Plant Prot. Res.*. (2008), *48*, 41–52.

## GLOSSARY

Excerpted from the SIP glossary (Onstad *et al.*, 2006) and from Evans & Shapiro (1997).

**Acute.** Of short duration. Characterized by sharpness or severity. As 'acute disease'.

**Bioassay.** Biological assay. The measurement of the potency of any stimulus, physical, chemical, biological, physiological, or psychological, by means of the response which it produces in living matter.

**Capsid.** The protein coat or shell of a virus particle. The capsid is a 'Surface crystal', built of structure units. The structure units are the smallest functionally equivalent building units of the capsid. The structure unit could be a single polypeptide chain or an aggregate of identical or different polypeptide chains. In a shell with cubic symmetry the structure units can associate in a limited number of ways, forming symmetric clusters. These clusters are the morphological units which may be seen with the electron microscope, and for which the word capsomere has been proposed. See also Nucleocapsid and Virion.

**cDNA.** Complementary DNA, copy DNA. A DNA molecule synthesised by a reverse transcriptase using an mRNA molecule as a template. The cDNA molecule is complementary to the mRNA molecule.

**Core.** Protein structure containing the viral genome which is enclosed by the viral capsid.

**DNA (deoxyribonucleic acid).** Polymer compound of deoxyribonucleotides. The nucleotides consist of an organic base (adenine, thymine, guanine or cytosine), a sugar molecule (desoxyribose) and a phosphoric acid. The genetic information is encoded by the sequence of the different bases.

**Envelope.** An outer lipoprotein bilayer membrane bounding the virion.

**Gene.** DNA segment, which encodes one protein.

**Genome.** The genetic material of an organism. More specifically, a set of chromosomes with the genes they contain. The haploid karyotype.

**Granulin.** The protein of the crystalline body (capsule) surrounding the granulovirus rod. Synonymous with, but preferred to, 'capsule protein', 'matrix protein', 'occlusion body protein', and 'proteinic crystal'.

**Hemocyte.** A colorless blood cell. See Leucocyte for a complete description.

**Hemolymph.** The body fluid of an invertebrate with a hemocoel

**Host.** An invertebrate that harbors or nourishes another organism.

**Hypertrophy.** An increase in size (weight) and functional capacity or an organ or tissue, without an increase in the number of structural units upon which their functions depend. Hypertrophy is usually stimulated by increased functional demands.

***In vitro.*** In the 'test tube', or other artificial environment. Outside a living organism.

***In vivo.*** In the living organism.

**Infection.** The introduction or entry of a pathogenic microorganism into a susceptible host, resulting in the presence of the microorganism within the body of the host, whether or not this causes detectable pathologic effects (or overt disease). In the case of the viruses, an infection has been defined as the introduction into a cell or an organism of an entity able to multiply, able to produce disease, and able to reproduce organized infective entities.

**Inoculum.** The microorganisms used in inoculation.

**Leucocyte (Leukocyte, Hemocyte, and Amebocyte).** A general term for colorless blood cells in non-arthropod invertebrates that do not contain a respiratory pigment. Such cells are typically capable of amoeboid movement and phagocytic activity during some stage(s) of their development. They wander freely through the hemolymph, loose connective tissue and epithelial surfaces, especially in mollusks. Leucocytes are considered to be multifunctional, participating in a variety of activities including wound repair, shell repair, gamete resorption, calcium and other ion transport, glycogen storage and transport, initiation of encapsulation and cellular immune reactions. Leucocytes and hemocytes are currently classified as either granulocytes or hyalinocytes depending on the presence or absence of cytoplasmic granules.

**Median lethal concentration.** A concentration of a pathogen or agent which will produce death in half the test subjects; the experimental method is not sufficiently accurate to determine the precise dose to which the test animals were exposed. Its symbol is $LC_{50}$. (contrast with median lethal dose).

**Median lethal dose.** A more restricted concept of median effective dose. The dose which will produce death in half the test subjects. Its symbol is $LD_{50}$.

**Median lethal time.** In a time-dependent biological assay procedure, this is the period of exposure to a pathogenic (including toxicological) stimulus which will produce death in half the test subjects. The length of exposure is a direct measure of dosage, and an increase in the period of exposure results in an increase in uptake and true dose in the same ratio. Its symbol is $LT_{50}$.

**Median survival time.** The time at which death occurs in half the test subjects after exposure to a pathogenic (including toxicological) stimulus. Its symbol is $ST_{50}$, not a direct measure of dosage, and it is not to be confused with the Median lethal time ($LT_{50}$), which is a direct measure of dosage. Survival analysis (e.g. Kaplan-Meier) procedures have been developed to account for censored data, which are an intrinsic characteristic of survival data.

**Non-occluded.** Said of those viruses in which the virions are not occluded in a dense protein crystal. Preferable to 'non-inclusion'.

**Nucleocapsid.** The structure composed of the capsid with the enclosed viral nucleic acid. Some nucleocapsids are naked, others are enclosed in an envelope (or limiting membrane). See also Virion.

**Occluded.** Viruses in which the virions are occluded in a dense protein crystal, large enough to be visible under a light microscope (e.g. baculoviruses, cypoviruses, entomopoxviruses).

**Occlusion body.** A virus-directed structure that is assembled within the infected cell and contains or occludes infectious virus particles or virions. Occlusion bodies may contain one or many virions depending on the type of virus. In the case of the *Baculoviridae*, occlusion bodies are assembled in the nucleus (NPVs) or the mixed nuclear-cytoplasmic contents after loss of the nuclear membranes. Baculovirus occlusion bodies may contain one or many virions.

**Pathology.** The science that deals with all aspects of disease. The study of the cause, nature, processes, and effects of disease. Any branch of science, or any technique or method or body of facts that contributes to our knowledge of the nature and constitution of disease belongs in the broad realm of pathology. 'Invertebrate pathology' refers to all aspects of disease (including abnormalities) which occur in invertebrate animals. Similarly, 'insect pathology' is that branch of entomology or invertebrate pathology that embraces the general principles of pathology as they may be applied to insects. If biology is defined as that branch of science which deals with the origin, structure, functions, and life history of organisms, then pathology might be defined as 'biology of the abnormal'. For each branch of biology there is a corresponding branch of pathology. Also, in a more limited sense, pathology refers to the structural and functional changes from the normal. Also in a limited sense, general pathology relates to disturbances which are common to various tissues and organs of the body, such as degenerative processes, pigmentations, mineral deposits, circulatory disturbances, specific and nonspecific inflammations, progressive tissue changes such as hyperplasia and hypertrophy, and tumors.

**Polyhedrin.** The protein of the crystalline body (polyhedron) surrounding the polyhedrosis virions. C-polyhedrin and N-polyhedrin designate the polyhedrin of *Cypovirus* and of *NPV*, respectively. Synonymous with, but preferred to, polyhedron protein, matrix protein, inclusion body protein, and proteinic crystal.

**Polyhedron.** The protein of polyhedral occlusion bodies of alpha-, gamma-, deltabaculoviruses and cypoviruses. Note that polyhedrons of alpha- and gammabaculoviruses are homologous to the granulin but not to that of deltabaculoviruses nor cypoviruses.

**Population.** A group of individuals of the same species set in a frame that is limited and defined with regard to both time and space.

**Provirus.** A noninfectious intracellular form of a virus. The genetic material (genome) of a virus—essentially a nucleic acid. It is perpetuated in stable association with the internal structure of the host cell, and for this reason has, so far, not been directly detectable. One of three phases (proviral, vegetative, and infective) in which a virus may exist. (Prophage, which is the nucleic acid of bacteriophage in lysogenic bacteria, is a provirus.).

**Replication.** The unique mechanism by which viruses multiply; the synthesis of new virions inside living host cells. Within a host cell, disappearance of the virion as a structural entity, release of the viral genetic material, integration of such material into the biochemical machinery of the cell, synthesis of viral components, and assembly of complete virions are essential steps of viral replication.

**RNA (ribonucleic acid).** Polynucleotide which is made of a pentose (ribose), a purin (adenine, guanine) or pyrimidin base (cytosine, uracil) and a phosphoric acid. Alternating connections between phosphoric acid and ribose lead to a straight molecule. Within the process of the protein biosynthesis, the transfer RNA (tRNA) functions as donator for amino acids, the messenger RNA (mRNA) transfers the information of the genes and the ribosomal RNA (rRNA) is a component of the ribosomes.

**Single strand.** Term for molecules of nucleic acids, which only consist of one polynucleotide chain.

**Spheroidin.** The virus-coded protein that forms the crystalline protein matrix of spheroids within which the virions of the entomopox viruses are occluded.

**Transmission (of disease).** The conveyance of disease from one individual host to another. The transfer or transport of an infectious agent from reservoir to susceptible host.

**Virion.** The mature virus, the ultimate phase of viral development. The virion is either a naked or an enveloped nucleocapsid. The term 'virus' embraces all phases of the viral development, and it includes the virion.

**Virogenic stroma.** A microscopically differentiable region of viroplasm that develops in virus-infected cells from which virions assemble. In the nucleoplasm of cells infected by nucleopolyhedroviruses or granuloviruses, it is a dark staining network at the edges of which nucleocapsids form. With cypoviruses, it is a dense granular region of the cytoplasm that develops after infection and which gives rise to the icosahedral particles.

**Virus.** Non-cellular entities whose genome is an element of nucleic acid, either RNA or DNA, which replicates inside living cells, and uses intracellular pools of precursor materials and cellular synthetic machinery to direct the synthesis of specialized particles, the virions, which contain the viral genome and transfer it to other cells. Replication and assembly occurs within the cellular cytoplasm or nucleoplasm and are not separated from the host cell contents by a lipoprotein bilayer membrane as with cellular pathogens.

## APPENDIX: RECIPES FOR STAINS

(Adapted from Evans & Shapiro, 1997.)

### Buffalo Black 12B (Naphthalene Black 12B or Amido Schwarz or Acid Black 1) working solution

Mix the solution using the following ingredients and weights/volumes (to produce 100 ml of working solution).

| | |
|---|---|
| Buffalo Black | 1.5 g |
| Glacial acetic acid | 40 ml |
| Distilled water | 60 ml |

### Preparation of working solutions for Giemsa's stain (Romeis, 1989)

Make up 0.01 M phosphate buffer solution:

Solution A: 13.61 g of $KH_2PO_4$ dissolved in 1 l of distilled water

Solution B: 17.8 g of $Na_2HPO_4 \cdot 2H_2O$ dissolved in 1 l of distilled water.

Mix 40 ml of solution A with 60 ml of solution B to get a solution of phosphate buffer with pH 6.9–7.0.

### Giemsa's fixative [Giemsa's azur eosin methylene blue solution (Merck) or other Giemsa's stain products]

| | |
|---|---|
| Giemsa's stain | 1 part |
| 0.1 M phosphate buffer pH 6.9–7.0 | 9 parts |

CHAPTER III

# Isolation, culture, preservation, and identification of entomopathogenic bacteria of the Bacilli

T.W. FISHER* & S.F. GARCZYNSKI[†]

*University of Arizona, School of Plant Sciences, Tucson, AZ 85721, USA
[†]USDA-ARS, Yakima Agricultural Research Laboratory, Wapato, WA 98951, USA

## 1. INTRODUCTION

Bacteria pathogenic to insects can be found in a variety of habitats worldwide, including water, soil, plants, and animals. Entomopathogenic bacteria are found within the kingdom Procaryotae among the Gracilicutes (mainly Gram-negative bacteria), and Firmicutes (Gram-positive bacteria) eubacterial divisions, and mainly occur within the families Bacillaceae, Pseudomonadaceae, Enterobacteriaceae, Streptococcaceae, and Micrococcaceae (Tanada & Kaya, 1993). The most well-known bacteria pathogenic to insects are in the class Bacilli, and some of the more common members of this class are listed in Table 3.1. Since the publication of the first edition of this manual there has been substantial revision to the genus *Bacillus*. Some of the more relevant changes are Family reassignment (see Table 3.1 for current Family assignment) and genus name changes including, *Bacillus sphaericus to Lysinibacillus sphaericus* (Ahmed *et al.*, 2007), *Bacillus laterosporous* to *Brevibacillus laterosporous* (Shida *et al.*, 1996), *Bacillus popilliae* to *Paenibacillus popilliae* (Pettersson *et al.*, 1999) and *Bacillus larvae* to *Paenibacillus larvae* (Heyndrickx *et al.*, 1996). For more details on general bacterial classification, on each genus and on the role of each species in insect infections, one can refer to the following books: Bergey's Manual of Systematic Bacteriology (De Vos *et al.*, 2009), *Bergey's Manual of Determinative Bacteriology* (Holt *et al.*, 2000), *Microbiology* (Wistreich & Lechtman, 1988), *The Prokaryotes Volume 4: Bacteria: Firmicutes, Cyanobacteria* (Stahly *et al.*, 2006) and *Insect Pathology* (Tanada & Kaya, 1993). Descriptive information on bacteria found in soil inhabiting insects, and those associated with nematodes are presented in Chapters V and XII of this manual.

The purpose of this chapter is to provide techniques, both classical and modern, that will enable researchers in any laboratory setting to work with entomopathogenic bacteria. We have updated this current chapter to include the use of 16S rDNA gene sequence determination and comparison, which has led to much of the systematic refinements among the Bacilli. However, these bacteria still have the same biochemical and morphological traits used to determine their identity in the first edition of this manual, therefore, we have retained many of the classical procedures used to isolate, cultivate and identify Bacilli.

## 2. ISOLATION

### A. Prospecting

*Bacillus* spp. are mainly soil bacteria (Nicholson, 2002), but are ubiquitous and can be isolated from a wide variety

MANUAL OF TECHNIQUES IN INVERTEBRATE PATHOLOGY
ISBN 9780123868992

**Table 3.1** Classification of some well-known Bacilli.

| *First edition assignment* | *Current taxonomic assignment*[a] | *Family* |
|---|---|---|
| *Bacillus alvei* | *Paenibacillus alvei*[b] | Paenibacillaceae |
| *Bacillus brevis* | *Brevibacillus brevis*[c] | Paenibacillaceae |
| *Bacillus circulans* | *Bacillus circulans* | Bacillaceae |
| *Bacillus larvae* | *Paenibacillus larvae*[d] | Paenibacillaceae |
| *Bacillus laterosporus* | *Brevibacillus laterosporus*[c] | Paenibacillaceae |
| *Bacillus macerans* | *Paenibacillus macerans*[e] | Paenibacillaceae |
| *Bacillus polymyxa* | *Paenibacillus polymyxa*[f] | Paenibacillaceae |
| *Bacillus pantothenticus* | *Virgibacillus pantothenticus*[g] | Bacillaceae |
| *Bacillus pasteurii* | *Sporosarcina pasteurii*[h] | Planococcaceae |
| *Bacillus sphaericus* | *Lysinibacillus sphaericus*[i] | Bacillaceae |
| *Bacillus anthracis* | *Bacillus anthracis* | Bacillaceae |
| *Bacillus badius* | *Bacillus badius* | Bacillaceae |
| *Bacillus cereus* | *Bacillus cereus* | Bacillaceae |
| *Bacillus coagulans* | *Bacillus coagulans* | Bacillaceae |
| *Bacillus firmus* | *Bacillus firmus* | Bacillaceae |
| *Bacillus lentus* | *Bacillus lentus* | Bacillaceae |
| *Bacillus licheniformis* | *Bacillus licheniformis* | Bacillaceae |
| *Bacillus mycoides* | *Bacillus mycoides* | Bacillaceae |
| *Bacillus pumilus* | *Bacillus pumilus* | Bacillaceae |
| *Bacillus subtilis* | *Bacillus subtilis* | Bacillaceae |
| *Bacillus thuringinensis* | *Bacillus thuringinensis* | Bacillaceae |

[a]Garrity *et al.* (2004).
[b]Ash *et al.* (1993, 1994).
[c]Shida *et al.* (1996).
[d]Ash *et al.* (1993, 1994); Heyndrickx *et al.* (1999).
[e]Ash *et al.* (1993, 1994).
[f]Ash *et al.* (1993, 1994).
[g]Heyndrickx *et al.* (1999).
[h]Yoon *et al.* (2001).
[i]Ahmed *et al.* (2007).

of sources. A general search for *Bacillus* spp. consists of random sampling of soil, leaves, trees, animals, and dead insects (larvae or adults). When searching for a particular species pathogenic towards a specific insect, one must look for diseased insects in their natural habitat. While *L. sphaericus* is ubiquitous in soil, strains toxic to mosquito larvae might best be found in samples of water, mud, and substrates obtained from breeding sites. However, one can also isolate a pathogen from a non-susceptible insect to that pathogen. For example, *L. sphaericus* strain 2362 was isolated from an adult of *Simulium damnosum* in Nigeria while it is strictly pathogenic to Culicidae larvae

(Weiser, 1984). So, there are no hard and fast rules for pathogen prospecting, but thoroughness and perseverance are valuable, along with a bit of luck and good eyesight.

- Collect each sample in a separate container to avoid contamination. Sterile sample bags make good vessels for a variety of solid and semi-solid materials. Sterile screw-capped tubes or specimen cups are good vessels for water or mud samples. Whatever you use, make sure it is sterile so that you only isolate bacteria from your sample.
- It is important to record all information concerning sampling, especially when different people are involved in surveys. In a field notebook, record insect or animal name, physiological state, geographic zone, type of habitat, and date, as well as each important detail of the sampling procedures used. Make sure you coordinate coding labels of collection vessels with the information you record in your field book. Check for information about prior bacterial treatments in sampling areas as spores are very resistant to environmental degradation.

## B. Isolation

After arrival in the laboratory, each sample needs to be processed. Pre-clean the work area by wiping down bench top with 70% ethanol before you start. Make sure you have all the materials you will need for processing prepared and easily accessible. It is important that things are well labeled to avoid mixing up samples. Record all information in your laboratory notebook.

**Note:** When processing samples, keep in mind that there might be some human pathogenic bacteria in each sample.

- For soil samples: each sample is divided into 2–4-g lots and each lot is added to screw-capped tubes containing 10 ml sterile water. Briefly vortex each tube, then proceed with heat treatment and plating described below.
- For insect samples: place the insect into tubes containing 1 ml of sterile water per 0.2–0.4 g of insect. Homogenize the sample (addition of Tween 80 to 0.5% may aid the homogenization), then proceed with heat treatment and plating described below.
- For water samples: bacteria and spores can be concentrated by filtration through a 0.22-μm filter. The filter is then placed in a tube containing 10 ml sterile water. Briefly vortex each tube, then proceed with heat treatment and plating described below.

### *1. Heat treatment and plating*

1. Heat the samples in a water bath at 80°C for 10 min, and then chill tubes rapidly on ice. This step kills most vegetative cells of Bacilli and non-spore forming bacteria, thereby enriching for spores of *Bacillus* species (due to their heat-resistant nature).
2. After allowing the solid content of the tubes to settle, plate 100 μl each of the heated sample and dilutions of the heated sample (usually $10^{-1}$ and $10^{-2}$) on to a Petri dish containing a growth medium. Incubate plates for 24 h at 30°C to allow for bacterial growth. Growth medium can vary and include a sporulation medium (MBS medium; Appendix medium no. 17) or more typical media (UG, Appendix media no. 13, 14 and 15; Nutrient Agar, Appendix media no. 3).
**Note:** These conditions do not allow for selection of all bacteria, just for the most commonly known entomopathogenic Bacilli species. If it is necessary to isolate all bacteria present in a sample, skip step 1 and plate higher dilutions ($10^{-4}$, $10^{-5}$ and $10^{-6}$) in step 2.
3. Examine each plate for bacterial growth. Using a fine sterile loop, transfer each colony to 10 ml growth media in sterile tubes (e.g., UG with glucose, MBS, etc.) and shake at 250 r.p.m. on an orbital shaker for 48 h at 30°C. Check microscopically for purity, and stage of culture.

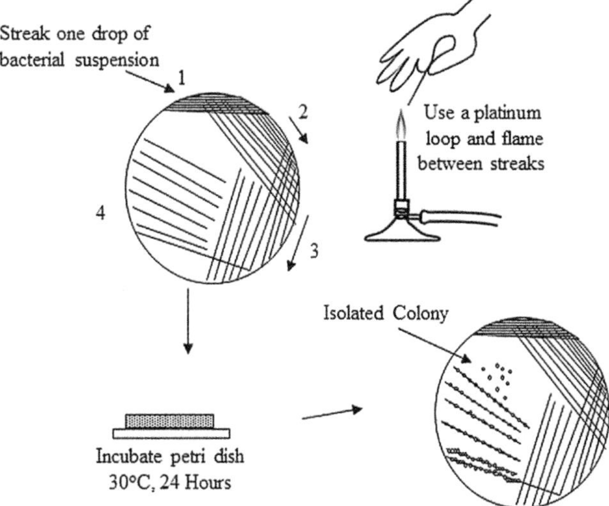


**Figure 3.1** Streaking an agar plate for isolation of single bacterial colonies.

4. Plate a loopful of the suspension to isolate single pure colonies (Figure 3.1). Incubate plates at 30°C for 24–48 h. These single colonies will be used for preparing bacterial suspensions for further cultivation, identification, preservation, and bioassay (see Chapters IV and V).

**Note:** Procedures for isolating other bacteria associated with insects can be found in chapters V and XII and from slugs in chapter XIII.

## 3. CULTIVATION

### A. On artificial media

Factors that influence bacterial growth can include physical requirements (i.e. temperature, pH, oxygen, osmotic pressure) and nutrient requirements (i.e. sources of energy, carbon, nitrogen, minerals, water, growth factors). Growth of *B. thuringiensis* and *L. sphaericus* is typically done at 30°C. *B. thuringiensis* strains can use a carbohydrate source for growth, whereas *L. sphaericus* is unable to metabolize carbohydrates and requires a proteinaceous growth medium. Amino acids are the preferred nitrogen source for *L. sphaericus,* which also requires the vitamins biotin and thiamine to grow and calcium and manganese ions for sporulation (Lacey, 1984; Russell *et al.*, 1989).

While there are many media that can be used to cultivate Bacilli, in this section we give example growth conditions using one medium recipe for the growth of both *L. sphaericus* and *B. thuringiensis*. This medium has provided reliable and reproducible growth, sporulation, and production of parasporal bodies in both cases (see Appendix nos 13 and 14 for recipes of UG medium).

### B. Example of *B. thuringiensis* cultures

#### *1. Preparation of a 10-ml preculture*

From a stock tube (see Section 4) or a colony from a fresh plate, inoculate a tube containing 10 ml UG medium to serve as a preculture. Incubate on a shaker for 48 h at 30°C. After incubation, observe a drop of the culture under a microscope to check for sporulation. After sporulation occurs, the preculture is heat-treated at 80°C for 10 min to kill vegetative cells. Heat treatment allows for a more consistent growth of the new culture to be inoculated with the preculture. This preculture will be used to inoculate the larger cultures below.

#### *2. Preparation of a 200-ml preculture*

Prepare a 1-l Erlenmeyer flask containing up to 200 ml of UG medium (sterilized at 121°C for 15 min). Aseptically add a few drops of preculture (~1 ml per 200 ml of media) which is sufficient for inoculation. Incubate 5–6 h at 30°C in a shaking incubator. This preculture can be used for inoculation of a fermentor (~ 200 ml preculture for 4.5 l of bacterial culture).

#### *3. Bacterial growth for spore/crystal harvesting*

1. Prepare a 1-l Erlenmeyer flask containing 100 ml UG medium. The flask is closed with a cotton plug. After sterilization (121°C for 15 min) add glucose to give a 1% final concentration. Use a stock solution of 10% glucose that has been filter sterilized (alternatively, the glucose solution can be autoclaved at 105°C for 10 min, but be careful as glucose will caramelize at higher temperatures).
2. Inoculate the flask directly with ~ 0.5 ml of a preculture and incubate at 30°C with orbital agitation for 48–72 h until cell lysis is complete. Check culture under a phase-contrast microscope to monitor cell lysis, the sporulation rate and presence of parasporal crystal proteins.
3. Centrifuge the final whole culture 5000 × *g* 15–20 min to pellet spores and crystals. Decant supernatant.
   **Note:** Prior to centrifugation, remove 5 ml of culture for use in determination of cell and spore counts described below.
4. Resuspend the pellet of spores and crystals with 0.5 M NaCl for 15 min to avoid exoprotease activity.
5. Centrifuge the resuspended spore crystal mix 5000 × *g* 15–20 min.
6. Resuspend the pellet in distilled or deionized water. Centrifuge the resuspended spore crystal mix 5000 × *g* 15–20 min. Repeat.
7. Then either:
   a. Resuspend the pellet in a water volume identical to the initial culture (i.e. 200 ml). Aliquot and freeze at −20°C. This material can be used in

bioassays to determine target insects as other characterizations of the strain are being done.

or

**b.** pellet of spores–crystals is kept frozen and can be used to prepare a lactose-acetone powder (Dulmage *et al.*, 1970) or freeze dried for long-term storage (see Section 4).

*4. Cell and spore counting*

The larvicidal activity of *B. thuringiensis* has been determined in many ways, from using crude spore–crystal mixes to preparations of purified activated toxins. No matter which method is used for bioassay, it is important to quantitate the material you are testing. Below are two methods (plate count and direct microscopic count) that can be used to determine the number of cells and spores present in your culture. Counting gives an idea of the growth of a culture and can be used, as can the optical density, to compare various cultures but it is not a valuable tool for evaluating the entomopathogenic potential of a strain as it does not reflect precisely the quantity of parasporal inclusions responsible for toxicity.

*a. Direct microscopic count using a counting chamber*

1. Fill the counting chamber with bacterial or spore culture according to the manufacturer's directions and place under the microscope.
2. Count the number of bacteria or spores according to the manufacturer's instruction sheet and calculate the number of bacteria or spores per ml of original culture from the known volume of the counting chamber and degree of dilution as given in the manufacturer's instructions.

**Note:** Counting bacteria by this method does not distinguish viable and dead cells.

*b. Plate count method*

The number of cells and spores can be evaluated by counting the number of cells (before heat treatment) and number of spores (after heat treatment) present by plating a series of dilutions (0.1 ml per plate) on to solid medium in Petri dishes.

1. Prepare a dilution series of your primary culture or heat-treated spore mix for quantitation (Figure 3.2). Aseptically transfer 1 ml to a tube containing 9 ml of sterile UG medium. Mix well, then transfer 0.1 ml of your dilution to a new tube containing 0.9 ml of sterile UG medium. Repeat a series of 10-fold dilutions from $10^{-1}$ to $10^{-7}$.
2. Plate 0.1 ml of the $10^{-5}$, $10^{-6}$, and $10^{-7}$ dilutions onto Petri dishes containing UG agar in triplicate. Incubate the plates at 30°C for 24 h. Plates that produce 30–150 colonies provide the most reliable counts.

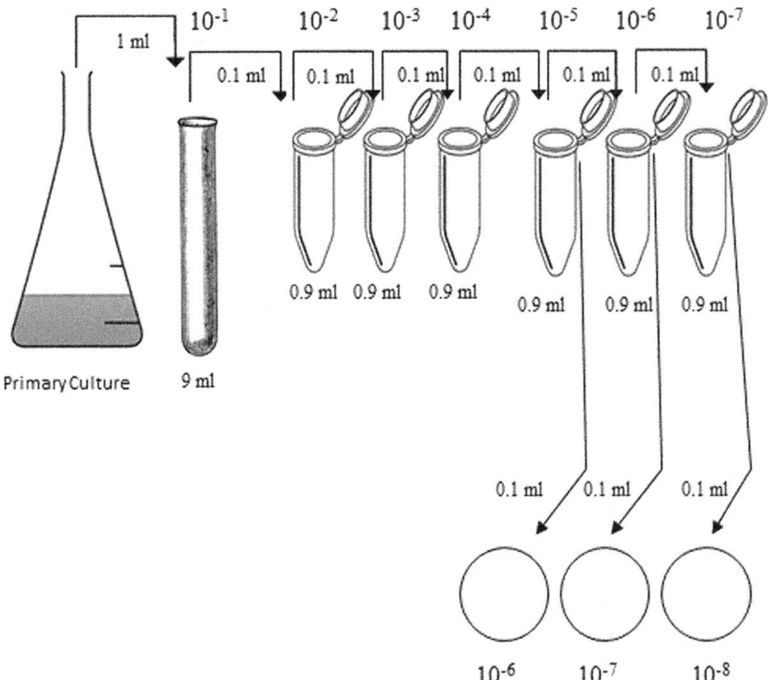


**Figure 3.2** Preparation of a dilution series from a primary culture.

**Note:** Using cultures produced with conditions given in Section 3B above usually produces ca $10^9$–$10^{10}$ cells/ml for *B. thuringiensis* with roughly 100% sporulation, while *L. sphaericus* cultures produce ~ $10^8$–$10^9$ cells/ml with a lower and variable sporulation rate.

## 4. PRESERVATION

Preservation of bacterial strains must maintain both culture viability and cultural characteristics. Numerous problems may appear either during manipulation, or during storage (genetic mutation, plasmid loss, loss of characteristics, selection of resistant populations) until a strain is completely lost. No method is 100% reliable and some are more or less adapted to certain species. Other parameters can influence the choice of a particular technique, such as cost, dispatching, mailing, number of strains to store, management of stocks, etc. Below are some of the more common methods for bacterial storage.

### A. Short-term storage

A simple solution for low-cost, short-term storage of your bacteria is to maintain culture plates of single colony isolates. The advantage is the immediate availability of your bacterial strains, but the inconveniences can be numerous, such as increased chance of contamination and loss of certain characteristics.

After plating your strains on nutrient agar plates for single colony isolation, seal the covers with Parafilm and store the plates at 4°C. Each month, re-inoculate a fresh agar plate as described in Figure 3.1. This technique is compatible with long-term storage and convenient for internal laboratory use.

Maintenance of spore–crystal cultures at 4°C is satisfactory for 1–2 weeks. Most spore–crystal cultures can be stored at −20°C for months before use. It is better to store these cultures in a pelleted form to avoid eventual degradation of protein by excreted proteases.

### B. Long-term storage

Several options exist for long-term storage of Bacilli. Frozen stocks of vegetative cells can be maintained and used as a form of long-term storage for internal laboratory use. For longer term, several methods are presented for storage of spores. Spores can germinate after more than 30 years of storage.

#### *1. Frozen glycerol stocks*

1. Isolate a colony from a nutrient agar plate and inoculate 10 ml of culture medium. Incubate the tube shaking at 250 r.p.m. at 30°C for 24 h.
2. Aseptically transfer 830 µl of the cultured bacteria to cryovials containing 170 µl of sterile glycerol (glycerol acts as a cryoprotectant). Cap the cryovial and vortex to ensure even mixing.
3. Freeze tubes immediately at −80°C.

**Note:** Always freeze several lots as the steps of freezing and thawing reduce bacteria viability.

#### *2. Storage of spores of* Bacillus spp. *on filter paper*

1. Isolate a colony from an agar plate and inoculate 10 ml of UG medium. Incubate the tube with shaking at 30°C for 48 h. Check culture to ensure that sporulation is complete.
2. Heat the culture at 80°C for 12 min to select for spores.
3. Aseptically place a drop of each heated culture onto a piece of sterile filter paper previously placed into a tube that can be heat sealed.
4. Let the filter paper dry in the tube for 2–3 weeks at 37°C.
5. Seal the tube under sterile conditions by melting the glass shut.
6. Store the stock tubes at 4°C until use.
7. To recover filter paper, file the tube to open at the top (be careful not to cut yourself on the glass) and pour the filter paper into the appropriate medium for growth.

These stock-tubes are convenient for sending strains to collaborators.

#### *3. Freeze drying* Bacillus *spore mixtures*

This technique is considered the most efficient for long-term storage and conservation of strain characteristics. Distribution of lyophilized material is practical for supplying strains to other workers as no special storage conditions are required. Freeze drying is particularly useful for *Bacillus* strains that sporulate poorly, for oligosporogenous strains or for strains with spores which survive only for a short time. Apart from collection and storage, freeze drying is also used for bacterial products in order to avoid the problem of hydrophobicity sometimes linked with bacterial powders. The equipment needed is a vacuum pump attached to a quick freeze dryer system, and is frequently found in many laboratory facilities.

**Note:** The freeze dryer operator must be experienced with the equipment. It should not be done for the first time without supervised instruction.

*a. Freeze drying in a manifold lyophilizer*

1. Grow strains in optimum conditions for sporulation to be complete. Cells can be grown on either agar plates or in liquid cultures.
   - **a.** For UG agar plates: spread bacteria to generate a lawn. Incubate plates at 30°C for 48 h or until sporulation is complete (this can be monitored under a microscope). Harvest the spores in sterile physiological saline or water using a cell scraper. Transfer cells in water or saline to a sterile culture tube. Proceed with step 2 below.
   - **b.** For liquid cultures: inoculate 10 ml of UG media with a colony of your bacteria and incubate with shaking at 30°C for 48 h or until sporulation is complete (this can be monitored under a microscope).
2. Centrifuge cultured cells or cells scraped from agar plates 5000 × *g* 15–20 min to pellet spores and crystals. Decant supernatant.
3. Resuspend the pellet of spores and crystals with 0.5 M NaCl for 15 min to avoid exoprotease activity. Centrifuge the resuspended spore–crystal mix 5000 × *g* 15–20 min. Decant supernatant.
4. Resuspend the pellet in sterile distilled or deionized water. Centrifuge the resuspended spore crystal mix 5000 × *g* 15–20 min. Repeat.
5. Resuspend and homogenize the spore pellet in 10 ml sterile physiological saline or water containing 20% horse serum.
6. Place 100 μl into 1–2-ml sterile glass lyophilization ampules with a pipette, avoid dropping any of the suspension on the outer edge of the tube. Cover the tube with sterile cotton or glass wool.
   **Note:** It is important to use glass as water can pass through plastic.
7. Freeze the ampules at −80°C overnight, or immerse the ampules deeply into a dry-ice ethanol bath to freeze the samples quickly (use protective glasses and cryo-gloves).
8. Remove the cotton plugs and attach the tubes to the freeze-dryer as explained in the user instructions.
9. After lyophilization, the tubes are sealed with flame. Lyophilized tubes can be stored at room temperature, 4°C or below −20°C for many years in an active stage.
10. Each lot of freeze-dried spores should be checked periodically. Growth should be tested from one ampule in order to check viability, purity, and characteristics of the strain. To resuspend spores for growth, file one end of the ampule and break open the top (be careful not to cut yourself). Add a few drops of nutrient broth to create a bacterial suspension. Using a sterile loop, plate cells to isolate single colonies.

**Note:** The germination rate of spores of *B. thuringiensis* is enhanced by heat shock (65°C for 30 min; Krieg, 1981).

*b. Freeze drying in a shelf lyophilizer*

1. Prepare spore mixtures as in steps 1–5 above.
2. Place 100 μl into 1–2-ml sterile glass lyophilization vials. Place split stopper on the vials.
3. Freeze the vials at −80°C overnight, or immerse the ampules deeply into a dry-ice ethanol bath to freeze the samples quickly.
4. Place the vials centered on the shelf. Proceed with lyophilization according to machine instructions.
5. When lyophilization is complete, stopper the vials (under vacuum) using the stoppering plate mechanism. Release the vacuum, remove the stoppered vials, and secure the stoppers with foil crimp seals.
6. Store the vials at room temperature, 4°C, or below −20°C for many years in an active stage.
7. Check the freeze-dried spores as in step 10 above.

## 5. IDENTIFICATION

### A. Identification of the major groups of Bacilli

The role of identification keys is to identify strains using a minimum of phenotypic characteristics. For simplification, a key for 22 of the most frequently encountered Bacilli species in nature is presented in Figure 3.3. There

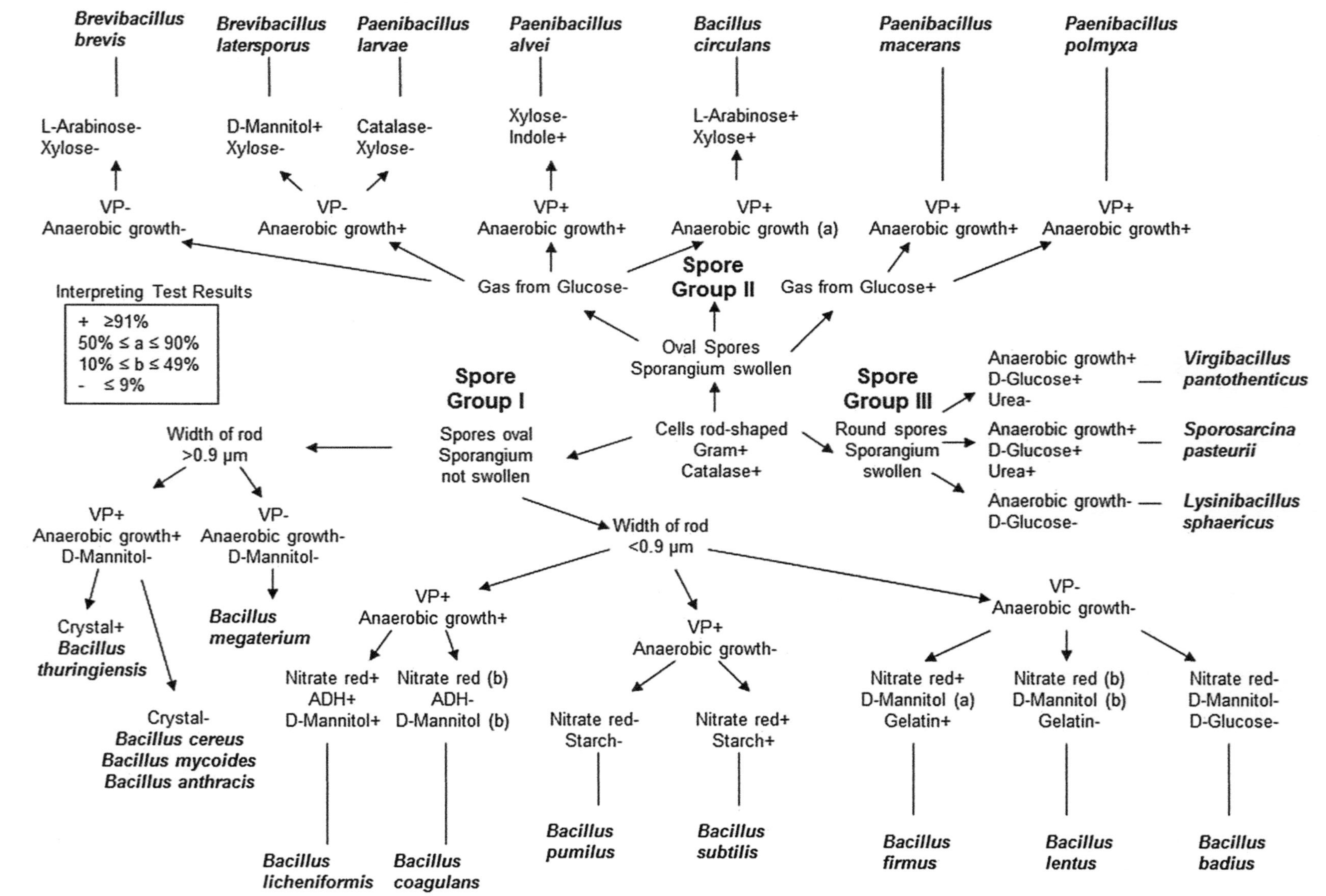


**Figure 3.3** Key for identification of the major groups of Bacilli.

are, however, far more Bacilli species and it may be necessary to consult a more comprehensive text for identifications. Alternatively, using 16S rDNA can help in identification (see Section 5B below). Traditional methods for keying Bacilli to the major species are described below and summarized in Figure 3.3. For a complete description of phenotypic traits and techniques, texts such as *Bergey's Manual of Systematic Bacteriology* (De Vos *et al.*, 2009) should be consulted.

**Note:** Use of aseptic technique and reasonable microbiological caution is necessary when working with environmental samples. Always keep in mind that the sample you are working with might contain human pathogens.

**Note:** To ensure proper identification it is important to follow recipes for culture media precisely, and to pay particular attention to substrates and incubation conditions used for each test. There are many variations of the tests outlined below, and test reagents are available from a variety of commercial sources.

### *1. Microscopic observation of bacterial cells*

For microscopic observations of bacterial morphology, high-quality optical instruments as well as the use of standardized conditions are necessary. A good-quality phase-contrast microscope is essential for examination of spores, as it allows for observation and differentiation of spore refraction from other components in the bacterial cell or medium.

1. After isolation of various colonies from your field sample (see Section 2), use a wire needle (heated until red then cooled) to transfer a small part of one colony, well-isolated on an agar plate, into a sterile tube containing sporulating medium (10 ml sterile liquid medium in a tube; use Appendix medium no. 14 or 17).
2. Grow the culture at 30°C for 16–24 h preferably with agitation (250 r.p.m.).
3. Put a drop of culture between a slide and coverslip. To avoid drying of the suspension, seal the coverslip to the slide with hot paraffin wax as shown in Figure 3.4.
4. Observe the bacterial cells using a light microscope (1000 ×) under oil immersion. Bacilli are rod-shaped cells with rounded extremities. Record the dimensions and cell motility (motility defined as the cell movements in distinct directions). Cell movement can appear to be enhanced around air bubbles stuck between slide and cover slip.

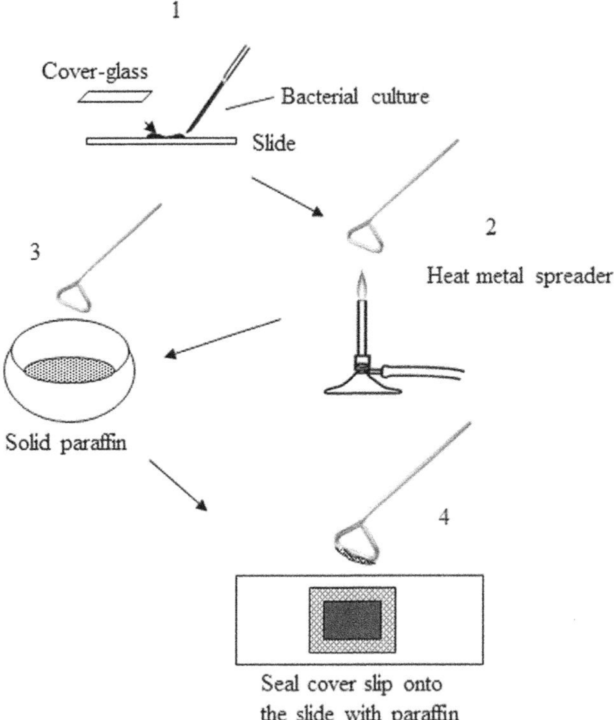


**Figure 3.4** Preparation of a slide for the observation of a bacterial suspension under a microscope.

### *2. Observation of the spores and determination of the Bacilli group*

Gordon *et al.* (1973) arranged *Bacillus* species into three morphological groups based on spore shape and swelling of the sporangium. Even though many of these bacteria are no longer 'Bacillus' species, the original groupings are still useful morphological aspects for bacterial identification. Group I species produce terminal endospores that are oval and do not cause the rod-shaped bacterial cell to swell, and can be divided into two classes: class 1 having a rod width greater than 0.9 μm; and class 2 having a rod width under 0.9 μm. The majority of the Group I bacteria still reside within the genus *Bacillus*. Group II species produce oval endospores that swell the sporangium. Many of the species in this Group II have been reassigned to the family Paenibacillaceae and include bacteria of the genus *Brevibacillus* and *Paenibacillus*. Group III species contain round spores and a swollen sporangium. Included in Group III is *Lysinibacillus sphaericus*. Refer to Figure 3.5 for diagrammatic representations of each of the groups.

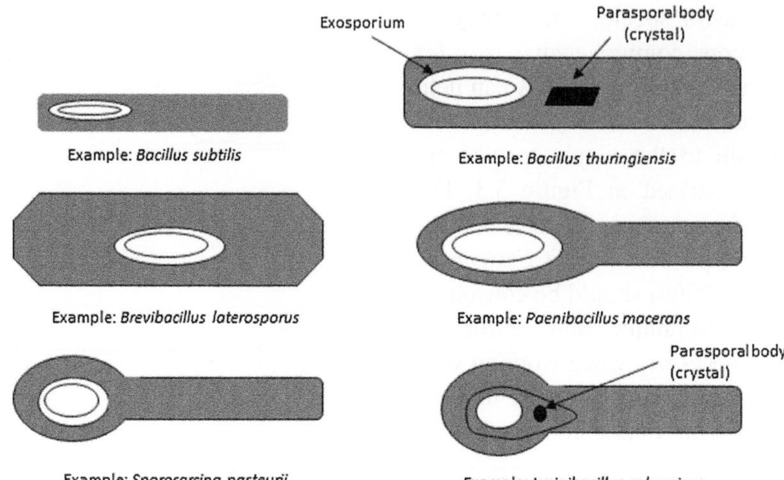


**Figure 3.5** Morphological aspects of Bacilli.

1. After observation of a cell culture during the vegetative growth stage, usually within 24 h of starting the culture, continue to grow the culture at 30°C.
2. After 24–72 h of incubation, one can begin to observe spores. For *B. thuringiensis* and *L. sphaericus*, parasporal bodies also become visible. Usually, under optimal growth conditions, spore refringence appears after 24–48 h incubation at 30°C. Because growth of bacteria can be variable, you may have to incubate your cultures longer to achieve optimal sporulation.
3. After 48–72 h, observation of sporangium lysis, spore liberation into the medium and, for *B. thuringiensis* and *L. sphaericus*, confirmation of the presence of the parasporal inclusion bodies become apparent.

Examine your bacterial culture under a phase-contrast microscope after 48–72 h of growth to observe spores. This step is important for choosing which biochemical tests are applicable to determine the identity of your bacteria. Morphology is sometimes difficult to classify within one of the three spore groups mentioned above. Observation of a swollen sporangium is not always obvious, therefore it may be necessary to compare sporulated cells with those still in the vegetative stage.

**Note:** The shape of the spores and the bacterial cells may be modified if the bacteria are grown under less than optimum growth conditions. The quality of nutrients in the culture medium is important and changes should be made if there are low levels of sporulated cells or slow culture growth is observed.

*3. Gram staining*

Prepare a nutrient agar plate (see Appendix medium no. 3) and streak your plate with your unknown bacteria to isolate single colonies. After incubating the plate for 16–18 h at 30°C, while the bacteria are still in exponential growth phase, pick a colony with a wire loop or needle (heated until red, then cooled), and mix it with a drop of sterile distilled water on a glass microscope slide and spread the bacteria forming a thin smear. Allow the smear to dry and heat-fix the cells by passing slide over a gentle flame. Refer to Figure 3.6 for diagrammatic representation of the procedure below.

**Note:** Thick smears make it difficult to obtain reliable results and are often hard to interpret.

1. Cover the bacterial smear with crystal violet, let stain set for 1 min.
2. Gently rinse slide with water, shake off the excess.
3. Cover the bacterial smear with Gram's iodine, let set for 1 min.
4. Pour off the Gram's iodine.
5. Run 95% ethanol down the slide until the solvent runs clear (about 20 s).
6. Gently rinse slide with water, shake off the excess.
7. Cover the bacterial smear with safranin, let set for 15 s.
8. Gently rinse off the stain with water. Blot dry.
9. Observe under light microscope (1000 × with oil immersion). The Gram-positive bacteria appear dark violet while the Gram-negative bacteria are colored pink. Most Bacilli species are Gram-positive, but some irregular staining might be observed.

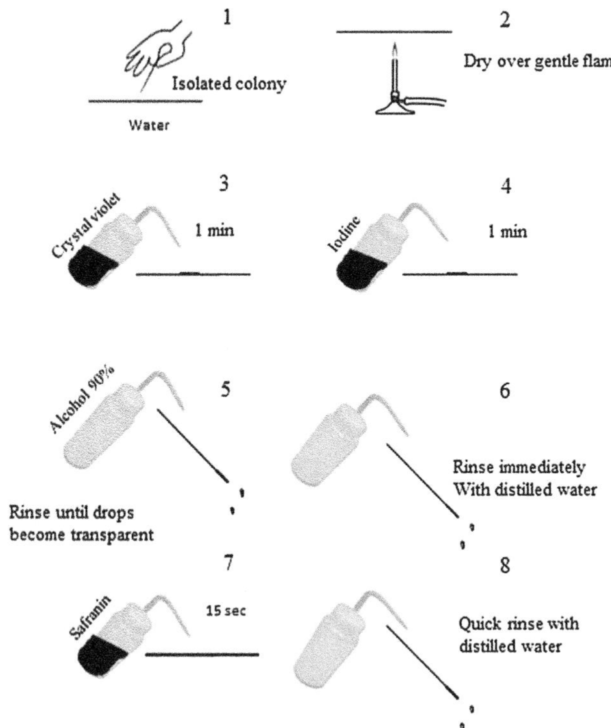


**Figure 3.6** Gram staining procedure.

**Note:** For Gram staining, the culture must be less than 18 h old, because some strains may lose their Gram properties and give false results. Always use controls for Gram staining with two reference strains, one confirmed Gram-positive species such as *Bacillus subtilis* and one confirmed Gram-negative species, such as *Serratia marcesans* or *Escherichia coli.*

*4. Detection of catalase*

1. Add 0.2 ml of 3−6% hydrogen peroxide ($H_2O_2$) into the bottom of a test tube.
2. Using a sterilized loop, pick a single colony from a culture freshly grown on nutrient agar.
3. Smear the bacterial colony on the tube, just above the $H_2O_2$. Cap the tube.
4. Tilt the tube so that the hydrogen peroxide solution covers the bacteria. The appearance of oxygen bubbles indicates the presence of catalase (this reaction is fairly rapid and should occur within 10 s).

**Note:** Do not use colonies grown on a medium containing catalase (e.g., blood agar). A positive example for this test is *E. coli* and a negative example is *Paenibacillus larvae.*

*5. Growth under anaerobic conditions*

1. Melt a semi-solid gelatin medium in a screw cap culture tube (without nitrate; Appendix medium no. 9) in a boiling water bath for 30 min (this eliminates dissolved oxygen).
2. Place the tube containing the melted gelatin medium in a water bath at 50−55°C and inoculate it with a bacterial sample of vegetative cells grown in liquid culture (a melted closed long Pasteur pipette works well). When inoculating the melted gelatin medium, swirl the Pasteur pipette containing inoculum from the bottom of the tube to the surface. Cap the tube and cool rapidly by immersing it into a cold water bath to solidify the agar (Figure 3.7).
3. Incubate at 30°C, for several days if necessary, until colony formation can be observed. Strict anaerobic bacteria as such will grow only on the bottom of the tube. Facultative anaerobic bacteria as such will grow throughout the tube. Strict aerobic bacteria only grow at the top of the tube.

**Note:** *Clostridium* spp. (such as *Clostridium bifermentans* subsp. *Malaysia*) are examples of strict anaerobes, *Pseudomonas* spp, are strict aerobes, and *E. coli* is a facultative anaerobe.

*6. Voges−Proskauer test (Barritt's method)*

During the intermediate steps of glucose metabolism, acetylmethylcarbinol (AMC) is produced by certain

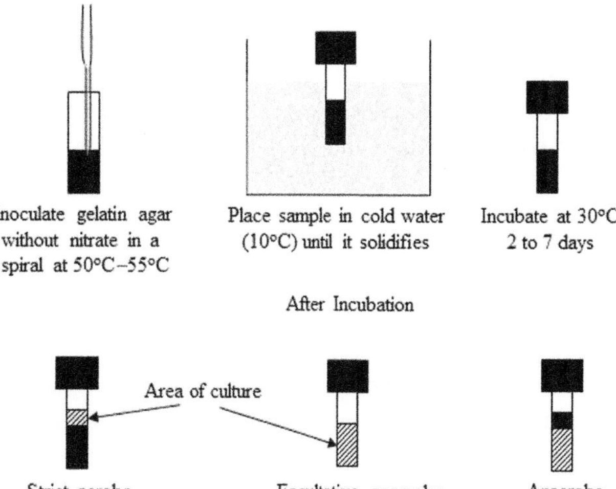


**Figure 3.7** Interpretation of anaerobic growth.

strains of bacteria (from pyruvic acid or during the course of butylen-glycolic fermentation). Detection of this substance is a useful phenotypic test for Bacilli discrimination.

1. Inoculate a tube containing MRVP medium with a single colony (Appendix medium no. 18 or from a commercial source) and incubate at 30°C for 48 h.
2. In a sterile glass tube, mix 2.5 ml culture, 0.6 ml Barritt's reagent A and add 0.2 ml Barritt's reagent B (for VP A and VP B in Appendix reagent no. 21). Shake the tube for 30 s.
3. Place the open tube on a slant to increase the contact with the air. Allow tube to stand for 30 min. Surface changes to a pink or red color indicate a positive test for AMC production.

**Note:** An example of a VP positive bacteria is *Klebsiella* and *E. coli* is VP negative

*7. Nitrate reduction to nitrite (Griess' reaction)*

1. Inoculate a nutrient nitrate medium (Appendix medium no. 2) with a 24-h-old bacterial suspension and incubate at 30°C for 24–48 h.
2. Add a few drops of Reagent $NO_3$ A (see Appendix reagent no. 20).
3. Add a few drops of Reagent $NO_3$ B. Observe the color change.
   The test is positive for nitrites when the mixture turns from yellow to red.
4. If there is no color change, add a pinch of zinc powder. If there are still some nitrates that are not reduced and left in the medium, they will then be transformed into nitrites and the red coloration will appear, implying a negative response to the nitrate-reductase test. The yellow color of the medium proves the complete disappearance of nitrates and reduction beyond the nitrite state, thus the test is positive.

**Note:** A positive example is *E. coli* and a negative example is *Bacillus subtilis*.

*8. Study of carbohydrate metabolism*

To study carbohydrate metabolism, it is necessary to use a medium that has reduced protein, as acid released by sugar fermentation may be masked by liberation of ammonia from proteinaceous material in the medium. A basal salt medium (salts, sugars and yeast) without any peptone will be used for this procedure (see Appendix medium no. 11).

1. Aseptically add 0.5 ml of a 10% solution of the sugar to be tested to a tube containing 10 ml of Basal salt medium to yield a final sugar concentration of 0.5%. **Note:** Prepare 10% solutions of the sugars to be tested and sterilize by filtration through a 0.22-μm filter. Store the sugar solutions at 4°C.
2. Inoculate medium (containing sugar) with a single colony and incubate at 30°C. Incubation can last for 15 days. A positive test for sugar fermentation is indicated by a change from violet to yellow. Make sure the bacteria actually grew in the tube so as not to confuse absence of acidity with lack of growth.

The biochemical differentiation of certain species is based on the difference of metabolism of one or two sugars (see identification flow chart). For more exhaustive testing, 20–50 different sugars can easily be tested.

*9. Test for proteolytic metabolism (proteolysis of gelatin)*

1. Inoculate 10 ml UG broth medium (Appendix medium no. 15) containing 4% (v/v) gelatin nutrient (see Appendix medium no. 8) in tubes. Incubate for at least 48 h at 30°C.
2. To test for proteolysis, place the tube in an ice bath. If protease is produced, the medium will not solidify.

**Note:** Most *Bacillus* and *Pseudomonas* spp. are gelatinase positive while many Enterobacteriacea are gelatinase negative.

*10. Metabolism of amino acids (arginine dihydrolase)*

This test will determine the presence of arginine dihydrolase (ADH). It is particularly useful to distinguish *B. licheniformis* and *B. coagulans*.

1. Inoculate a culture of a medium containing arginine (Appendix medium no. 10), cover the top with ~ 1 cm of sterile mineral oil. As a control, use a tube containing the medium without arginine. Incubate 30°C for 48 h .

In the first step, glucose fermentation decreases the pH of the medium (purple indicator changes to yellow) in the two tubes. In the presence of ADH, arginine is broken up into alkaline products that raise the pH, changing the indicator back to purple.

The ADH test is positive at 48 h if the control tube has changed to yellow and the test tube has reverted to purple.

*11. Test for urease and the production of indole*

Several techniques and different media can be used to detect the presence of urease or indole. For Bacilli, detection of urease can be done with Christensen medium (see Appendix medium no. 12) or urea-indole medium, among others. The Christensen medium contains agar, is less buffered and therefore more sensitive to the presence of urea. The indole medium allows for testing for the presence of indole produced from the bacteria's ability to break down tryptophan.

*a. Urease test*

1. Inoculate the two agar slants, one containing Christensen medium with urea and one without urea, from a single colony. Incubate the slants at 30°C for 6 days.

A positive reaction will turn the slant containing the urea medium red. When compared to the control slant, the medium in the urea slant must be redder than the control tube. If the two tubes are a similar color, the reaction is negative. *Proteus vulgaris* is an example of a urease positive bacterium while *E. coli* is urease negative.

*b. Indole test*

1. Inoculate 4 ml Tryptone broth (see Appendix medium no. 22) with a pure colony.
2. Incubate at 30°C for 48 h.
3. After incubation, add five drops of Kovac's reagent (see Appendix medium no. 19). The appearance of a red ring at the surface indicates the presence of indole.

**Note:** *E. coli* is an example of an indole-positive bacterium and *B. subtilis* is indole negative.

For organization and time-saving purposes, a summary of the biochemical tests is presented in Table 3.2.

*12. Overview of other characteristics and biochemical methods*

*a. Differentiation of* B. thuringiensis *from other members of the* B. cereus *group*

A key feature that distinguishes *B. thuringiensis* from other members of the *B. cereus* group is the production of parasporal crystals during sporulation. Parasporal crystals were determined to be responsible for the insecticidal activity of *B. thuringiensis* (Angus, 1954), and these activities have been recently reviewed (van Frankenhuyzen, 2009). The parasporal crystals can take various shapes (bipyramidal, cuboidal, flat rhomboid, round, amorphous) depending on their protein composition. The *cry* (crystal protein) genes encode the insecticidal proteins, and by 1989, 42 *cry* genes had been sequenced (Höfte & Whiteley, 1989). Initial analyses based on structural similarities and insecticidal activity identified 14 distinct genes falling into two families (*cry* and *cyt*; Höfte & Whiteley, 1989). With the increasing number of *cry* genes being discovered, the classification scheme of Höfte & Whiteley (1989) was insufficient for clear assignments. Crickmore *et al.* (1998) proposed a system that categorizes *cry* genes according to evolutionary divergence. Currently, there are 218 holotype *cry* gene families (*cry1*−*cry68*, with subdivisions) and 11 cytolytic crystal protein gene families (*cyt*1−*cyt*3, with subdivisions) that are included on the *B. thuringiensis* nomenclature website (Crickmore *et al.*, 2011). For further information on the characteristics and mode of action of *B. thuringiensis* toxins refer to one of the many available reviews (e.g., Schnepf *et al.*, 1998; Soberón *et al.*, 2009).

*13. Classification of* B. thuringiensis *and* L. sphaericus *by H-antigen determination*

For over 40 years, serological analysis of flagellar (H) antigens has served as the basis for designating serovars or subspecies of *B. thuringiensis* and *L. sphaericus* (de Barjac & Bonnefoi, 1962). As of 1999, there were 69 *B. thuringiensis* subspecies and 82 serovars contained in the Pasteur Institute collection (Lecadet *et al.*, 1999). If you are interested in further classification of your *B. thuringiensis* and *L. sphaericus* isolates using flagellar typing, refer to procedures for using H-serotyping described by de Barjac (1981), Laurent *et al.* (1996), and Thiery & Franchon (1997).

## B. Identification of Bacilli using molecular based techniques

*1. Identification of Bacilli using 16S rDNA gene sequences*

Traditionally, identification of unknown bacterial species has been achieved through determination of phenotypic traits. Analysis of 16S rDNA gene sequences has been

**Table 3.2** Summary of biochemical tests.

| *Tests* | *Incubation* | *Reagents (delay)* | *Positive reaction* | *Negative reaction* |
|---|---|---|---|---|
| Catalase | | $H_2O_2$(immediate) | Oxygen bubbles | |
| Anaerobiosis | 3–7 days | | Growth in bottom of tube | Growth only on top of the tube |
| VP | 48 h | VP A + VP B (5–30 min) | Pink | Colorless |
| Nitrate | | $NO_3$ A + $NO_3$ B (immediate)<br>Yellow + Zn (5–15min) | Red<br>Yellow | Yellow<br>Red |
| Sugars | 2–15 days | | Yellow | Violet |
| ADH | 48 h | | Violet (control yellow) | Yellow (test and control tube) |
| Urea | 2–6 days | | Red | Orange |
| Indole | 48 h | Kovac (2 min) | Red ring | Yellow ring |

a powerful tool to rapidly determine the identity of isolated bacteria. Below we present procedures useful for identification of your bacterial unknowns using 16S rDNA sequences generated from PCR products.

*a. Harvesting bacteria*

*i. From plated bacterial colonies.*

1. From an agar plate streaked for isolation of single colonies, pick a few colonies and place them in a 1.5-ml microfuge tube.
   **Note:** Use freshly grown bacteria incubated at 30°C for less than 24 h.
2. Grind colonies into a fine powder while frozen with liquid nitrogen using a pre-cooled pestle.
3. Suspend the powder in 100 μl sterile PCR grade water or appropriate buffer for genomic DNA extraction.

Alternatively, selected colonies can be directly suspended and homogenized in water or buffer without freezing (Gols *et al.*, 2007; Machado *et al.*, 2010).

*ii. From bacterial cells grown in liquid culture.*

1. From an overnight culture, transfer 1 ml of cells to a 1.5-ml microfuge tube.
2. Harvest the cells by centrifugation at maximum speed for 2 min in a microfuge. Remove supernatant.
3. The pelleted bacteria can then be freeze-dried and ground to a fine powder. Resuspend the bacterial powder in 100 μl sterile PCR grade water or buffer for genomic DNA extraction. Alternatively, the bacterial pellet can be immediately suspended in water or buffer for genomic DNA extraction.

*b. Extraction of genomic DNA*

Intact, good quality DNA is essential for the identification of bacteria using any molecular biology technique. Numerous methods have been developed for the isolation of genomic DNA. Most methods involve the same key steps: (1) disruption, (2) lysis, (3) removal of proteins and other contaminants, and (4) recovery of DNA (Atashpaz *et al.*, 2010; Wang *et al.*, 2011). However, for saving time and material, crude extraction methods have been developed and utilized. Provided below is a traditional extraction method described by Wilson (2001) and a rapid boil technique (Ji *et al.*, 2004).

**Note:** Versions of traditional protocols provided below are routinely used for the isolation of bacterial genomic DNA but there are many commercially available kits proven to be effective and reliable for reproducible bacterial DNA isolations. These kits are designed to minimize the number of steps as well as reduce or eliminate the use of hazardous chemicals.

*i. Method using cetyltrimethylammonium bromide (CTAB).*

1. Resuspend bacteria from either isolated colonies or liquid culture pellet in 567 μl TE buffer (Appendix media no. 23). Add 3 μl of 10% SDS, 3 μl of 20 mg/ml proteinase K. Mix and incubate for 1 h at 37°C.
   **Optional step:** Prior to addition of SDS and proteinase K, add 5 μl lysozyme (100 mg/ml stock solution) to resuspended bacteria and incubate 5 min at room temperature. The lysozyme step is necessary for hard to lyse Gram-positive bacteria.

2. Add 100 μl of 5 M NaCl. Mix thoroughly.
3. Add 80 μl of CTAB/NaCl solution (Appendix media no. 24). Mix and then incubate 10 min at 65°C.
4. Extract with an equal volume of chloroform:isoamyl alcohol (24:1). Spin 5 min in microcentrifuge.
   **Note:** When using organic solvents, work in a fume hood.
5. Transfer the aqueous phase to a fresh tube. Extract DNA with equal volume of phenol : chloroform : isoamyl alcohol (25 : 24 : 1). Spin 5 min in microcentrifuge.
   **Note:** Use appropriate safety precautions when handling phenol.
6. To precipitate the DNA, transfer aqueous phase to a fresh tube then add 0.6 vol isopropanol (~450 μl for 100% recovery of aqueous phase). Spin tube at top speed in a microfuge for 15–30 min at 4°C.
7. Carefully decant the supernatants being careful not to dislodge the DNA pellet. Add 1 ml 70% ethanol to wash the pellet. Spin tube at top speed in a microfuge for 5–15 min at 4°C.
8. Remove supernatant (being careful not to dislodge the DNA pellet), then briefly dry pellet in in a hood or speed vac.
9. Resuspend DNA pellet in 100 μl TE buffer.

*ii. Rapid boil method.*

1. Selected bacterial colonies or cell pellets in a microfuge tube are suspended in 100 μl of sterile PCR-grade water.
2. The microfuge tubes containing the bacteria are incubated in a boiling water bath for 10 min.
3. Cell debris is removed by centrifugation in a microfuge at 3500 × *g* for 10 min at room temperature.
4. The supernatant contains DNA and can be directly used as the PCR template (5–15 μl/PCR rxn).

### *3. Polymerase chain reaction (PCR) to amplify bacterial 16S rDNA genes*

PCR-based identification of and differentiation between bacterial species has become a standard technique in many laboratories (for examples, see Tailliez *et al.*, 2006; Lacey *et al.*, 2007; Dingman, 2009; Lindh & Lehane, 2011). Amplification and sequence analyses of the universally conserved regions within the bacterial small subunit (16S) ribosomal DNA gene can also be used to determine the presence or absence of bacteria. In a PCR reaction, the oligonucleotide primers are designed against highly conserved DNA sequences of the 16S rDNA gene. Under proper conditions, the oligonucleotide primers will anneal to the bacterial DNA, serving as priming sites for amplification of PCR products containing a copy of the bacterial 16S rDNA gene. Gel electrophoresis can be used to verify the presence of a PCR product. The presence of a band (amplified product) indicates the presence of bacteria. In the following section we will provide a complete PCR protocol for the detection and analysis of entomopathogenic bacteria.

*a. Oligonucleotide primers*

Oligonucleotide primer sets and PCR conditions suitable for the amplification of the 16SrDNA gene from most Eubacteria were described by Weisburg *et al.* (1991). You will need to order oligonucleotide primers with nucleotide sequences that are identical or nearly identical to those corresponding to the 16SrDNA genes from many Bacilli. These can include:

16S FWD: 5′-AGAGTTTGATCCTGGCTCAG-3′
and
16S REV: 5′-ACGGCTACCTTGTTACGACTT-3′

These oligonucleotide sequences are still widely used today (Lane, 1991; Gols *et al.*, 2007; Lacey *et al.*, 2007; Leclerque & Kleespies, 2008).

*b. PCR reagents and reaction set up*

Standard PCR reactions are usually performed in 50-μl volumes containing 1X PCR buffer, 20–100 ng of bacterial genomic DNA, 0.5 μM of each primer, 200 μM of dNTPs, and 1 U Taq DNA polymerase (Kramer & Coen, 2006). Because you are basing your identification on the nucleotide sequence of your unknown, the use of high fidelity or proofreading Taq DNA polymerase is recommended.

31 μl PCR grade water
5 μl 10 × PCR Buffer
10 μl Genomic DNA
1 μl 10 μM 16S FWD primer
1 μl 10 μM 16S REV primer
1 μl 50 × PCR-grade nucleotide mix (10 mM each dNTP)
1 μl High-fidelity Taq polymerase mixture

Combine all components in a 0.2-ml PCR tube and proceed with amplification given below.

*c. Amplification conditions*

PCR can be carried out in a thermal cycler using the following program:

Step 1. Initial denaturation: 94°C for 3 min
Step 2. Denaturation: 94°C for 30 s
Step 3. Primer annealing: 55°C for 30 s
Step 4. Extension: 72°C for 3 min
Repeat steps 2–4 for 28–30 cycles.
Step 5. Final extension: 72°C for 5 min

**Note:** The expected length of the amplified PCR product is ~1.5 kbp and should be confirmed by gel electrophoresis on a 1% agarose gel.

**Note:** No single set of conditions can be applied to all PCR amplifications as thermal cyclers, tubes and reagents vary by manufacturer. Individual reaction component concentrations as well as time and temperature parameters may need to be adjusted within recommended ranges for efficient amplification of specific targets.

*4. Preparation of PCR products for direct sequencing*

Once it has been confirmed that your PCR product is of the correct size and appears as a single band by gel electrophoresis, its DNA sequence can be determined. Direct sequencing is an available option when a single band is visible on the agarose gel. For direct-sequencing of PCR-amplified DNA, the primers used will be the original oligonucleotides used for PCR amplification. A detailed description of sequencing reaction conditions with 16S FWD and 16S REV primers is provided by Anderson & Haygood (2007). Alternately, many universities have facilities and services that perform DNA sequencing. Also, there are many companies which also provide DNA sequencing services.

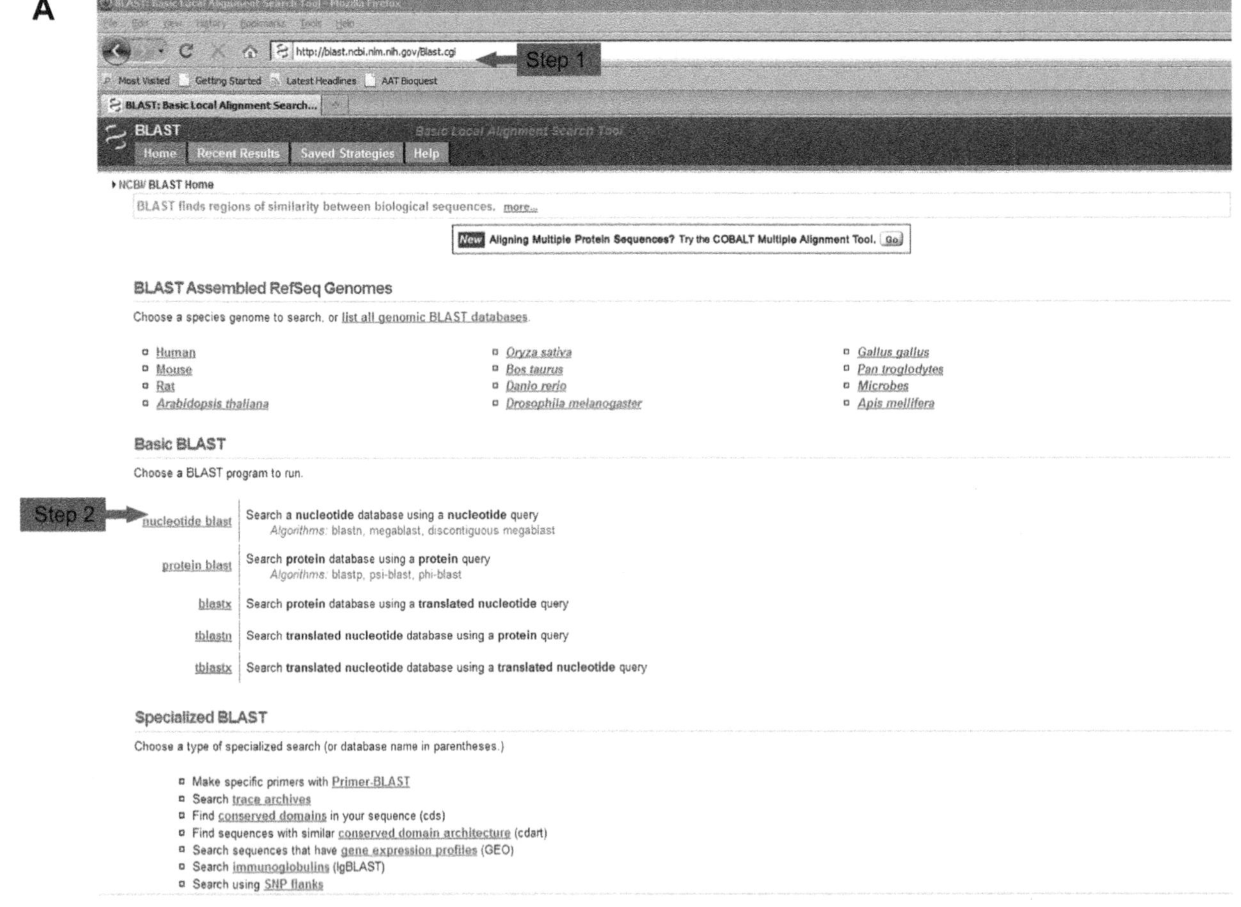


**Figure 3.8** Steps for performing 16S rDNA analysis using NCBI BLAST. (A) Steps 1 and 2. (B) Steps 3–7. (C) Steps 8 and 9.

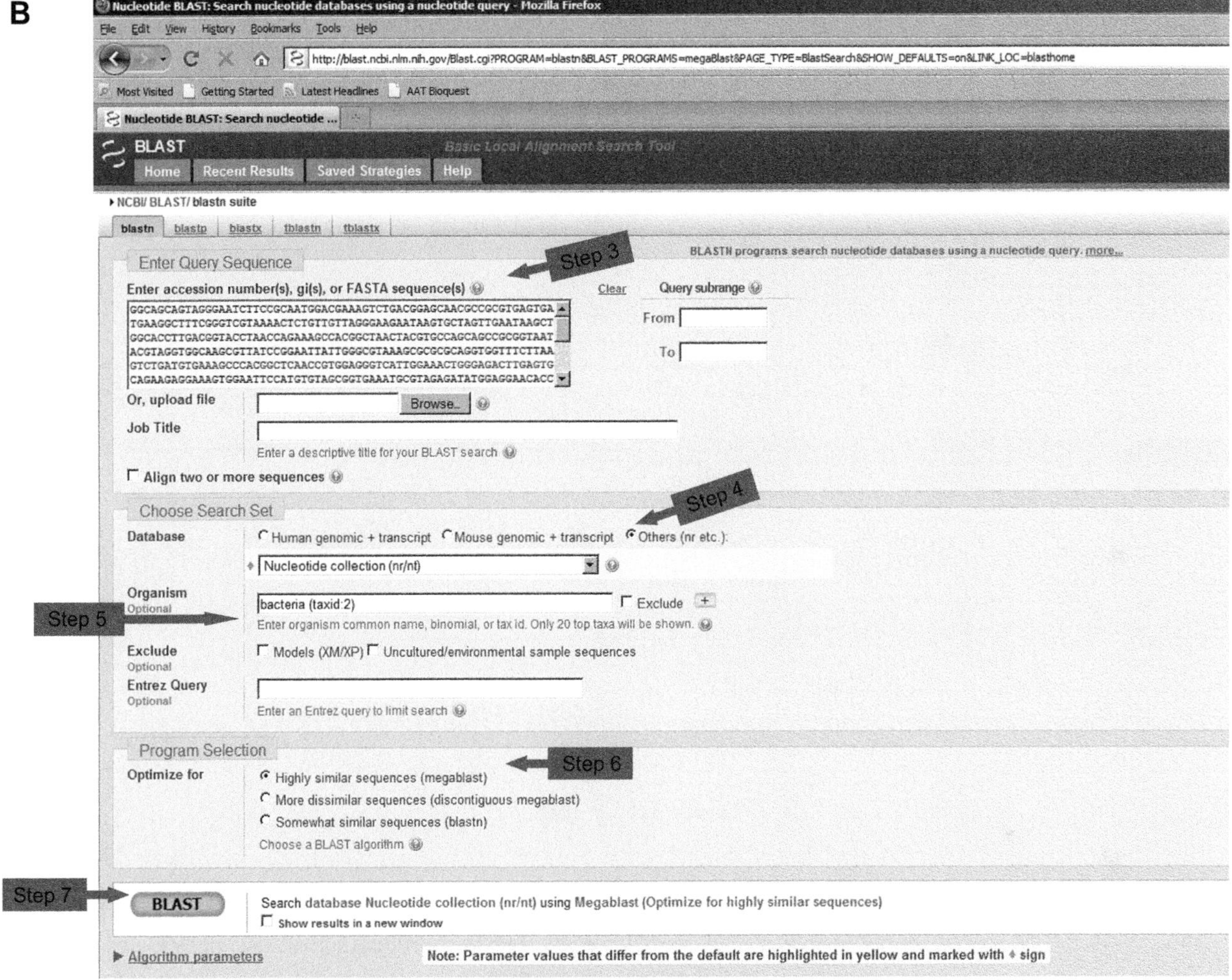


**Figure 3.8** (*continued*).

*a. Template preparation*

The quality of the PCR product often determines the quality of the resulting sequence. The principal elements in PCR reactions that contaminate and interfere with cycle sequencing reactions are dNTPs and oligonucleotide primers. Column purification (Gols *et al.*, 2007) and enzymatic precipitation (Kim & Blackshaw, 2001) are some of the methods commonly used to purify PCR products. Several commercial kits are available to clean up PCR reactions. Below, a basic ethanol precipitation procedure is given (Demkin *et al.*, 2000).

*i. Ethanol precipitation of PCR products.*

1. With the remaining 45 μl of PCR reaction (5 μl is used for gel electrophoresis) prepare a 1.5-ml microcentrifuge tube containing the following:
   5 μl of 3 M sodium acetate, pH 4.6
   100 μl of 95% ethanol
2. Pipet the entire contents of each PCR reaction (45 μl) into a tube containing the sodium acetate/ethanol mixture.
3. Vortex the tubes and incubate at −20°C for 30 min to precipitate the PCR product.
4. Spin the tubes in a microfuge at 4°C for 20 min at maximum speed.
5. Carefully aspirate the supernatant being careful not to dislodge the pellet.
6. Rinse the pellet by adding 300 μl of 70% ethanol to the tube. Vortex briefly to mix.
7. Spin the tubes in a microfuge at 4°C for 5 min at maximum speed. Again, carefully aspirate or decant the supernatant and discard.

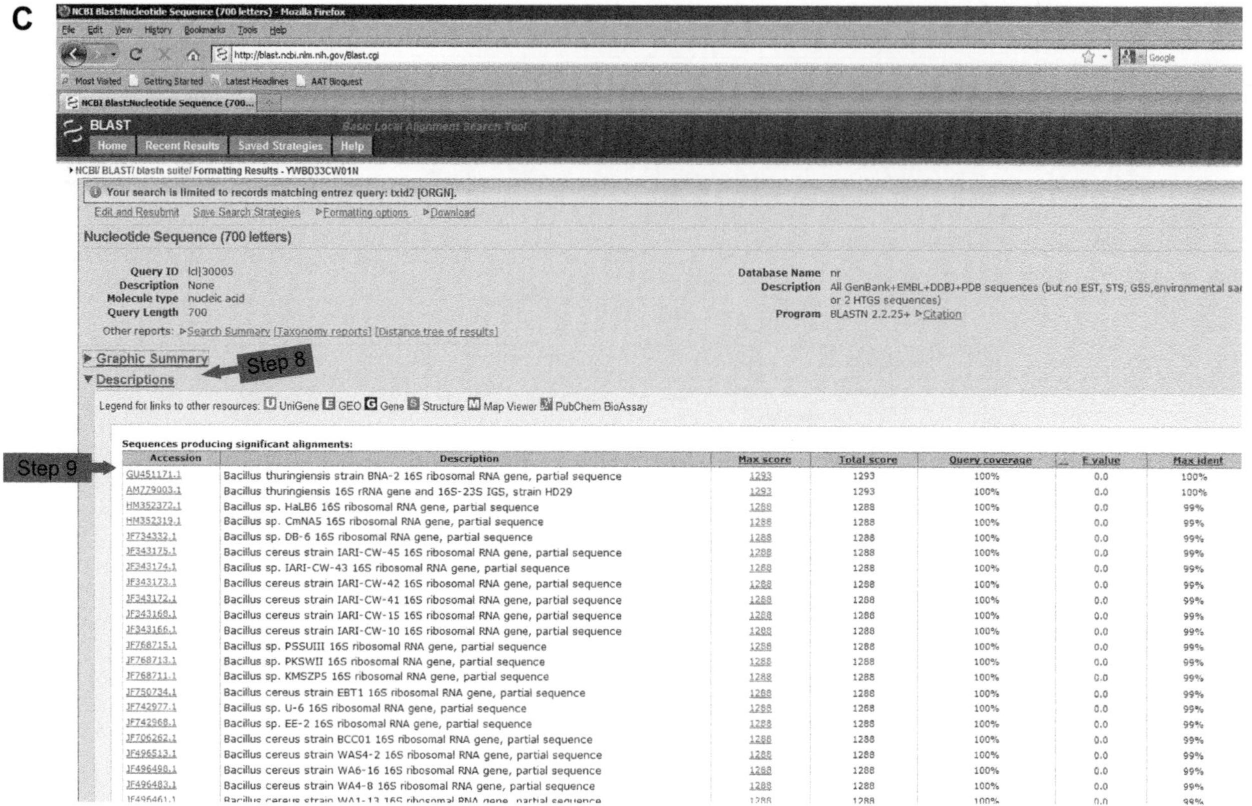


**Figure 3.8** *(continued).*

**Note:** The pellet does not adhere to the wall of the tube very well so be careful not to lose it.

8. Dry the pellet in a vacuum centrifuge for 3 min or until dry. Do not over dry. Resuspend in 50 µl of sterile water.

### *5. Analysis of 16s sequences for bacterial identification*

#### *a. BLAST analysis using National Center for Biotechnology Information website*

The Basic Local Alignment Search Tool (BLAST) algorithm and the computer program that implements it were developed by Altschul *et al.* (1990). BLAST is used for comparing primary biological sequence information such as those generated from your 16S ribosomal DNA amplicons. A BLAST search enables a researcher to compare an unknown query sequence to those deposited in a database of sequences maintained by the National Center for Biotechnology Information (NCBI). Different types of BLAST are available according to the nature of the query sequences and the search patterns desired. The different types of BLAST are available on NCBI website (http://blast.ncbi.nlm.nih.gov/Blast.cgi). Below are steps to perform a nucleotide BLAST. These steps are presented diagrammatically in Figure 3.8.

1. In your internet browser, go to the NCBI BLAST website (http://blast.ncbi.nlm.nih.gov/Blast.cgi).
2. Select nucleotide blast.
3. Locate the section 'Enter Query Sequence'. Paste a copy of your nucleotide sequence in the box under 'Enter accession number, gi, or FASTA sequence'.
4. Locate the section 'Choose Search Set'. Click on 'Others (nr etc)' in the Database Selection area. 'Nucleotide collection (nr/nt)' will automatically appear in the box below.
5. Locate the section 'Organism' and type in bacteria (taxid:2).
6. In the 'Program Selection' section, make sure that the 'Optimize for Highly similar sequences (megablast)' button is highlighted.
7. Press the BLAST button at the bottom of the page for results.

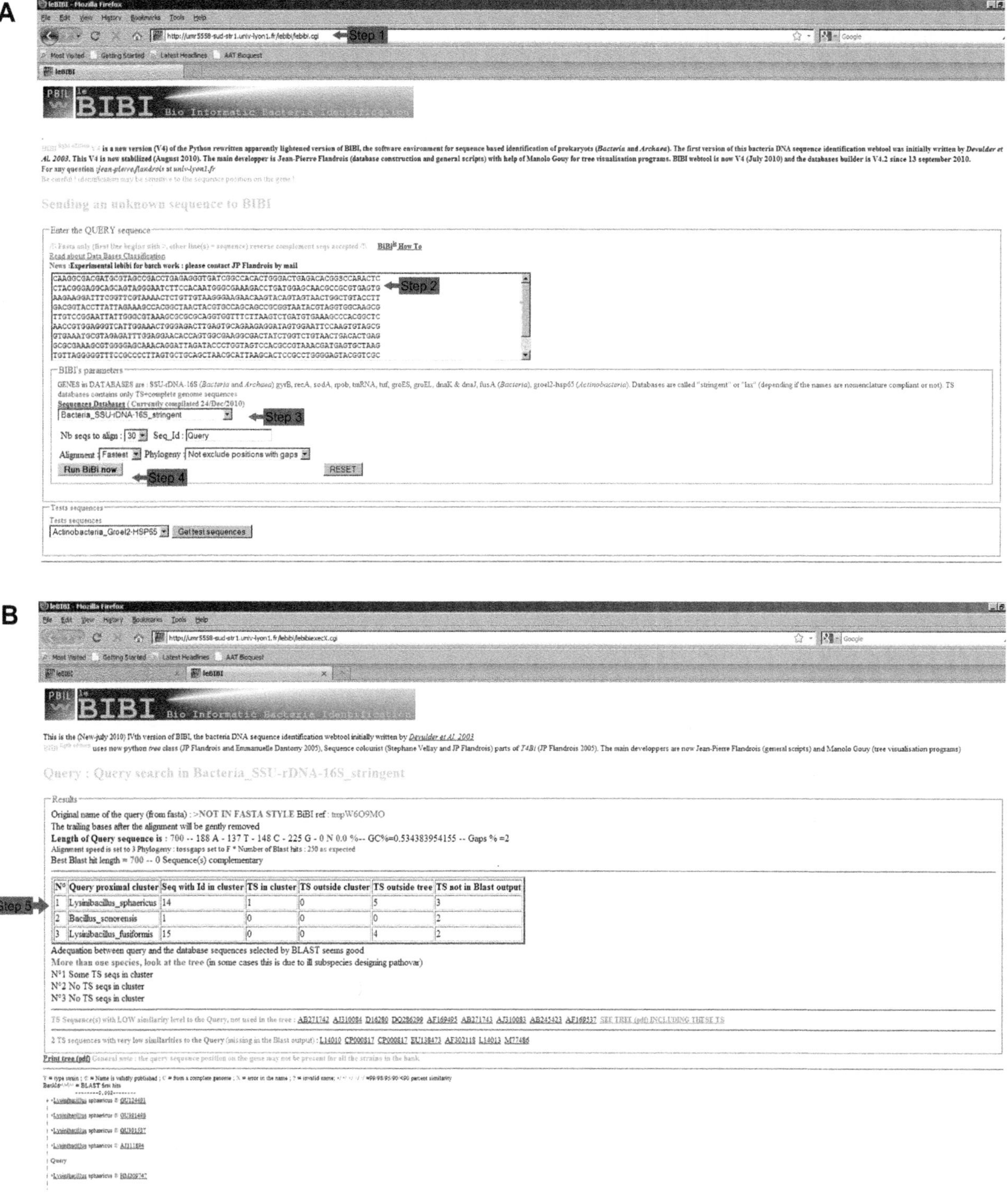


**Figure 3.9** Steps for performing 16S rDNA analysis using le BIBI. (A) Steps 1–5. (B) Step 5.

8. On the results page, locate the 'Descriptions' section.
9. The database sequences which produce significant alignments with the query sequence are listed. The best or highest scoring results are listed first.

*b. Identification of unknowns using the Bio Informatics Bacteria Identification website*

The Bio Informatics Bacteria Identification (BIBI) website is a specific bioinformatics tool dedicated to bacterial identification. The program implements a chaining of two well-known tools: BLAST and CLUSTAL W. This integration, and employment of different sequence databases specifically adapted for bacteria simplifies sequence analysis within a bacterial identification framework (Devulder *et al.*, 2003). Below are steps to perform sequence-based identification using BIBI. These steps are presented diagrammatically in Figure 3.9.

1. In your internet browser, go to http://umr5558-sud-str1.univ-lyon1.fr/lebibi/lebibi.cgi (the link to the BIBI website).
2. Locate the section 'Enter Query Sequence'. Paste your nucleotide sequence in the box under 'Enter accession number, gi, or FASTA sequence'.
3. In the 'BIBI's parameters" section select 'Bacteria_SSU-rDNA-16S_stringent'.
4. Select 'Run BIBI now' for results.
5. BIBI will show the best BLAST hit (significant match) for the query sequence.

## 6. CONCLUDING REMARKS

The power of molecular techniques has led to revisions of bacterial systematics. For an overview of molecular approaches and techniques used to study entomopathogenic bacteria, please refer to Boemare & Tailliez (2009). However, one must keep in mind that a thorough approach for the identification and classification of unknown bacteria has to include a combination of both classical phenotypic tests as well as incorporation of molecular techniques. Good luck in finding the next generation of entomopathogenic Bacilli.

## ACKNOWLEDGMENTS

We would like to thank Drs. Meenal Vyas and Thomas Unruh for reviewing this book chapter. We would also like to thank Dr. Lerry Lacey for his reviews and editorial work.

## DISCLAIMER

Mention of a commercial or proprietary product does not constitute an endorsement of the product by the United States Department of Agriculture.

## REFERENCES

Ahmed, I., Yokota, A., Yamazoe, A., & Fujiwara, T. (2007). Proposal of *Lysinibacillus boronitolerans* gen. nov. sp. nov., and transfer of *Bacillus fusiformis* to *Lysinibacillus fusiformis* comb. nov. and *Bacillus sphaericus* to *Lysinibacillus sphaericus* comb. nov. *Int. J. Syst. Evol. Microbiol., 57*, 1117–1125.

Altschul, S. F., Gish, W., Miller, W., Myers, E. W., & Lipman, D. J. (1990). Basic local alignment search tool. *J. Mol. Biol., 215*, 403–410.

Anderson, C. M., & Haygood, M. G. (2007). α-Proteobacterial symbionts of marine Bryozoans in the genus. *Watersipora. Appl. Environ. Microbiol., 73*, 303–311.

Angus, T. A. (1954). A bacterial toxin paralysing silkworm larvae. *Nature, 173*, 545.

Ash, C., Priest, F. G., & Collins, M. D. (1993). Molecular identification of rRNA group 3 bacilli (Ash, Farrow, Wallbanks and Collins) using a PCR probe test. *Antonie van Leeuwenhoek, 64*, 253–260.

Ash, C., Priest, F. G., & Collins, M. D. (1994). *Paenibacillus* gen. nov. and *Paenibacillus polymyxa* comb. nov. In *Validation of the Publication of New Names and New Combinations Previously Effectively Published Outside the IJSB*, List no. 51. *Int. J. Syst. Bacteriol., 44*, 852.

Atashpaz, S., Khania, S., Barzegaria, A., Barara, J., Vahed, S. Z., Azarbaijani, R., & Omidi, Y. (2010). A robust universal method for extraction of genomic DNA from bacterial species. *Microbiology, 79*, 538–542.

Boemare, N., & Tailliez, P. (2009). Molecular approaches and techniques for the study of entomopathogenic bacteria. In S. P. Stock, I. Glazer, N. Boemare, & J. Vandenberg (Eds.), *Insect pathogens: Molecular approaches and techniques* (pp. 32–49). Wallingford, UK: CABI Publishing.

Crickmore, N., Zeigler, D. R., Feitelson, J., Schnepf, E., Van Rie, J., Lereclus, D., Baum, J., & Dean, D. H. (1998). Revision of the nomenclature for the *Bacillus thuringiensis* pesticidal crystal proteins. *Microbiol. Molec. Biol. Rev., 62*, 807–813.

Crickmore, N., Zeigler, D. R., Schnepf, E., Van Rie, J., Lereclus, D., Baum, J., Bravo, A., & Dean, D. H. (2011).

*Bacillus thuringiensis* Toxin Nomenclature *(2011)*. http://www.lifesci.sussex.ac.uk/Home/Neil_Crickmore/Bt/.

De Barjac, H. (1981). Identification of H-serotypes of *Bacillus thuringiensis*. In H. D. Burges (Ed.), *Microbial Control of Pests and Plant Diseases 1970–80* (pp. 35–43). London, UK: Academic Press.

De Barjac, H., & Bonnefoi, A. (1962). Essai de classification biochimique et sérologique de 24 souches de *Bacillus* du type *B. thuringiensis*. *Entomophaga, 7*, 5–31.

De Vos, P., Garrity, G., Jones, D., Krieg, N. R., Ludwig, W., Rainey, F. A., Schleifer, K.-H., & Whitman, W. B. (2009). *Bergey's Manual of Systematic Bacteriology* (2nd ed.). New York, NY: Springer.

Demkin, V. V., Edelstein, M. V., Zimin, A. L., Edelstein, I. A., & Suvorov, M. A. (2000). Detection of sequence variation in PCR-amplified fragments of *omp2* gene from 3 species of the family Chlamydiaceae using agarose gel electrophoresis containing bisbenzimide-PEG. *FEMS Microbiol. Lett., 184*, 215–218.

Devulder, G., Perriere, G., Baty, F., & Flandrois, J. P. (2003). BIBI, a Bioinformatics Bacterial Identification Tool. *J. Clin. Microbiol., 41*, 1785–1787.

Dingman, D. W. (2009). DNA fingerprinting of *Paenibacillus popilliae* and *Paenibacillus lentimorbus* using PCR-amplified 16S–23S rDNA intergenic transcribed spacer (ITS) regions. *J. Invertebr. Pathol., 100*, 16–21.

Dulmage, H. T., Correa, J. A., & Martinez, A. J. (1970). Coprecipitation with lactose as a mean of recovering the spore–crystal complex of *Bacillus thuringiensis*. *J. Invertebr. Pathol, 15*, 15–20.

Garrity, G. M., Bell, J. A., & Lilburn, T. G. (2004). Taxonomic outline of the Prokaryotes. In *Bergey's manual of systematic bacteriology* (2nd ed.). *Release 5.0*. New York, NY: Springer-Verlag.

Gols, R., Schutte, C., Stouthamer, R., & Dicke, M. (2007). PCR-based identification of the pathogenic bacterium, *Acaricomes phytoseiuli*, in the biological control agent *Phytoseiulus persimilis*. *Biol. Control, 42*, 316–325.

Gordon, R., Haynes, W., & Pang, C. (1973). *The genus* Bacillus. U.S. Dept. of Agriculture Handbook No. 427. pp. 291.

Heyndrickx, M., Vandemeulebroecke, K., Hoste, B., Janssen, P., Kersters, K., de Vos, P., Logan, N. A., Ali, N., & Berkeley, R. C. W. (1996). Reclassification of *Paenibacillus* (formely *Bacillus*) *pulvifaciens* (Nakamura 1984) Ash et al., 1994, a later subjective synonym of *Paenibacillus* (formely *Bacillus*) *larvae* (White, 1906) Ash et al., 1994, as a subspecies of *P. larvae* subsp. *larvae* and *P. larvae* subsp. *pulvifaciens*. *Int. J. Syst. Bacteriol., 46*, 270–279.

Höfte, H., & Whiteley, H. R. (1989). Insecticidal crystal proteins of *Bacillus thuringiensis*. *Microbiol. Rev., 53*, 242–255.

Holt, J. G., Krieg, N. R., Sneath, P. H. A., Staley, J. T., & Williams, S. T. (2000). *Bergey's Manual of Determinative Bacteriology* (9th ed.). Philadelphia, PA: Lippincott Williams & Wilkins.

Ji, N., Peng, B., Wang, G., Wang, S., & Peng, X. (2004). Universal primer PCR with DGGE for rapid detection of bacterial pathogens. *J. Microbiol. Meth., 57*, 409–413.

Kim, J. B., & Blackshaw, S. (2001). One-step enzymatic purification of PCR products for direct sequencing. *Current protocols in human genetics*. 11.6.1–11.6.4.

Kramer, M. F., & Coen, D. M. (2006). Enzymatic amplification of DNA by PCR: Standard procedures and optimization. *Current Protocols in Cytometry, 37*. A.3K.1–A.3K.15.

Krieg, A. (1981). The genus *Bacillus* insect pathogens. In M. P. Starr, H. Stolp, H. G. Trüper, A. Balows, & H. G. Schlegel (Eds.), *The Prokaryotes, a Handbook on Habitats, Isolation, and Identification of Bacteria, Vol. II* (pp. 1742–1755). Berlin, Heidelberg, New York: Springer-Verlag.

Lacey, L. A. (1984). Production and formulation of *Bacillus sphaericus*. *Mosq. News, 44*, 153–159.

Lacey, L. A., Unruh, T. R., Simkins, H., & Thomsen Archer, K. L. (2007). Gut bacteria of the Pacific coast wireworm, *Limonius canus*, inferred from 16s rDNA sequences and their implications for control. *Phytoparasitica, 35*, 479–489.

Lane, D. J. (1991). 16S/23S rRNA sequencing. In E. Stackebrandt, & M. Goodfellow (Eds.), *Nucleic Acid Techniques in Bacterial Systematics* (pp. 115–175). Chichester, New York, NY: John Wiley & Sons.

Laurent, P., Ripouteau, H., Cosmao Dumanoir, V., Frachon, E., & Lecadet, M.-M. (1996). A micromethod for serotyping *Bacillus thuringiensis*. *Lett. Appl. Microbiol., 22*, 259–261.

Lecadet, M. M., Frachon, E., Dumanoir, V. C., Ripouteau, H., Hamon, S., Laurent, P., & Thiery, I. (1999). Updating the H-antigen classification of *Bacillus thuringiensis*. *J. Appl. Microbiol., 86*, 660–672.

Leclerque, A., & Kleespies, R. G. (2008). Genetic and electron-microscopic characterization of *Rickettsiella tipulae*, intracellular bacterial pathogen of the crane fly, *Tipula paludosa*. *J. Invertebr. Pathol, 98*, 329–334.

Lindh, J. M., & Lehane, M. J. (2011). The tsetse fly *Glossina fuscipes fuscipes* (Diptera: Glossina) harbours a surprising diversity of bacteria other than symbionts. *Antonie Van Leeuwenhoek, 99*, 711–720.

Machado, A. P., Vivil, V. K., Tavares, J. R., Frederico, F. J. G., & Filho & Fischman, O. (2010). Antibiosis and dark-pigments secretion by the phytopathogenic and environmental fungal species after interaction *in vitro* with a *Bacillus subtilis* isolate. *Braz. Arch. Biol. Technol., 53*, 997–1004.

Nicholson, W. L. (2002). Roles of *Bacillus* endospores in the environment. *Cell. Mol. Life Sci., 59*, 410–416.

Pettersson, B., Rippere, K. E., Yousten, A. A., & Priest, F. G. (1999). Transfer of *Bacillus lentimorbus* and *Bacillus popilliae* to the genus *Paenibacillus* with emended descriptions of *Paenibacillus lentimorbus* comb. nov. and *Paenibacillus popillae* comb. nov. *Int. J. Syst. Bacteriol, 49*, 531–540.

Russell, B. L., Jelley, S. C., & Yousten, A. A. (1989). Carbohydrate metabolism in the mosquito pathogen *Bacillus sphaericus* 2362. *Appl. Environ. Microbiol., 55*, 294–297.

Schnepf, E., Crickmore, N., Van Rie, J., Lereclus, D., Baum, J., Feitelson, J., Zeigler, D. R., & Dean, D. H. (1998). *Bacillus thuringiensis* and its pesticidal crystal proteins. *Microbiol. Molecul. Biol. Rev., 62*, 775–806.

Shida, O., Takagi, H., Kadowaki, K., & Komagata, K. (1996). Proposal for two new genera, *Brevibacillus* gen. nov. and *Aneurinibacillus* gen. nov. *Int. J. Syst. Bacteriol, 46*, 939–946.

Soberón, M., Gill, S. S., & Bravo, A. (2009). Signaling versus punching hole: How do *Bacillus thuringiensis* toxins kill insect midgut cells? *Cell. Mol. Life Sci., 66,* (2009), 1337–1349.

Stahly, D., Andrews, R., & Yousten, A. (2006). The genus *Bacillus*: insect pathogens. In M. Dworkin, S. Falkow, E. Rosenberg, K.-H. Schleifer, & E. Stackebrandt (Eds.), *The Prokaryotes Vol. 4. Bacteria: Firmicutes, Cyanobacteria* (pp. 563–608). New York, NY: Springer.

Tailliez, P., Pages, P., Ginibre, N., & Boemare, N. (2006). New insight into diversity in the genus *Xenorhabdus*, including the description of ten new species. *Int. J. Syst. Evol. Microbiol., 56*, 2805–2818.

Tanada, Y., & Kaya, H. K. (1993). *Insect Pathology*. San Diego, CA: Academic Press.

Thiery, I., & Frachon, E. (1997). Identification, isolation, culture and preservation of entomopathogenic bacteria. In L. A. Lacey (Ed.), *Manual of Techniques in Insect Pathology* (pp. 55–90). San Diego, CA: Academic Press.

van Frankenhuyzen, K. (2009). Insecticidal activity of *Bacillus thuringiensis* crystal proteins. *J. Invertebr. Pathol, 101*, 1–16.

Wang, T. Y., Wang, L., Zhang, J. H., & Dong, W. H. (2011). A simplified universal genomic DNA extraction protocol suitable for PCR. *Genet. Molec. Res., 10*, 519–525.

Weisburg, W. G., Barns, S. M., Pelletier, D. A., & Lane, D. J. (1991). 16S ribosomal DNA amplification for phylogenetic study. *J. Bacteriol, 173*, 697–703.

Weiser, J. (1984). A mosquito virulent *Bacillus sphaericus* in adult *Simulium damnosum* from Northern Nigeria. *Zbl. Mikrobiol., 139*, 57–60.

Wilson, K. (2001). Preparation of genomic DNA from bacteria. *Current Protocols in Molecular Biology*. 2.4.1–2.4.5.

Wistreich, G. A., & Lechtman, M. D. (1988). *Microbiology* (5th ed.). New York, NY: Macmillan Publishing Company. pp. 913.

Yoon, J. H., Lee, K. C., Weiss, N., Kho, Y. H., Kang, K. H., & Park, Y. H. (2001). *Sporosarcina aquimarina* sp. nov., a bacterium isolated from seawater in Korea, and transfer of *Bacillus globisporus* (Larkin and Stokes 1967), *Bacillus psychrophilus* (Nakamura 1984) and *Bacillus pasteurii* (Chester 1898) to the genus *Sporosarcina* as *Sporosarcina globispora* comb. nov., *Sporosarcina psychrophila* comb. nov. and *Sporosarcina pasteurii* comb. nov., and emended description of the genus *Sporosarcina*. *Int. J. Syst. Evol. Microbiol., 51*, 1079–1086.

## APPENDIX

**1.** Nutrient broth

13 g commercially obtained nutrient broth in 1 l distilled water

Transfer 8 ml per screw-cap tube (17 × 150 mm) or 10 ml per Pyrex tube (22 mm diameter). Sterilize at 120°C for 15 min, 15 psi (103.4 kPa).
Alternatively:

| | |
|---|---|
| Peptone | 5.0 g |
| Beef extract | 3.0 g |
| Distilled water | 1000 ml |

Dissolve ingredients in distilled water. Adjust pH to 6.8. Transfer 8 ml per screw-cap tube (17 × 150 mm) or 10 ml per Pyrex tube (22 mm diameter). Sterilize for 15 min at 121°C, 15 psi (103.4 kPa).

**2.** $BNO_3$ (nitrate nutrient broth)

| | |
|---|---|
| Nutrient broth | 1 l |
| Potassium nitrate ($KNO_3$) | 10 g |

Adjust with sodium hydroxide to pH 7.6. Transfer 8 ml per screw-cap tube. Sterilize at 120°C for 15 min.

**3.** Nutrient agar

| | |
|---|---|
| Commercially obtained Nutrient agar | 28 g |
| Distilled water | 1 l |

Transfer ~200 ml into 250-ml flasks or 8 ml in screw-cap tube. Sterilize at 120°C for 15 min, 15 psi (103.4 kPa). Let tubes solidify at a slope position after autoclaving.
Alternatively:

| | |
|---|---|
| Peptone | 5.0 g |
| Beef extract | 3.0 g |
| Agar | 15.0 g |
| Distilled water | 1000 ml |

Dissolve ingredients in distilled water. Add magnetic stirbar; heat to boiling. Adjust pH to 6.8. Transfer 8 ml to sterile screw cap tubes or for plates autoclave in a 2-l flask. Sterilize for 15 min at 121°C, 15 psi (103.4 kPa).

For plates, pour 25 ml of autoclaved nutrient agar into Petri plates, flame top to remove air bubbles, allow to solidify at room temperature.

**4.** Nutrient agar pH 6

Same as Nutrient agar above except adjust to pH 6. Sterilize for 15 min at 121°C, 15 psi (103.4 kPa). Transfer 9 ml per screw-cap tube and cool on a slope after sterilization.

**5.** Gram stain

| | |
|---|---|
| Crystal violet | 1 g |
| Distilled water | 100 ml |

Gram iodine

| | |
|---|---|
| Potassium iodide | 2 g |
| Iodine | 1 g |
| Distilled water | 20 ml |

Adjust to 100 ml after complete dissolution.

Gram counterstain

| | |
|---|---|
| Basic Fuchsin | 0.1 g |
| Distilled water | 100 ml |
| Ethanol 95% | 50 ml |
| Acetone | 50 ml |

**6.** Craigies

| | |
|---|---|
| Nutrient broth | 13 g |
| Bacto agar | 2 g |
| Distilled water | 1 l |

Adjust to pH 7.2. Put a small cylinder glass into each screw-cap tube and dispatch 11 ml per tube.

**7.** Beef extract medium

| | |
|---|---|
| Beef extract | 4 g |
| Sodium chloride (NaCl) | 5 g |
| Pancreatic peptone or Bacto peptone | 10 g |
| Distilled water | 1 l |

Precipitate at 120°C for 30 min. Adjust to pH 7.3–7.4. Filter then sterilize at 110°C for 30 min.

**8.** Nutrient gelatin

| | |
|---|---|
| Beef extract Medium (see no. 7) (double concentrated) | 500 ml |
| Peptone | 10 g |
| NaCl | 5 g |
| Gelatin | 150 g |
| Distilled water | 500 ml |

Dissolve the peptone and NaCl in boiling beef extract medium. Add water and gelatin while stirring. Let it boil until dissolution then cool down to 40°C. Adjust to pH 7.4–7.6. Add either an egg white (previously mixed) or 75 ml horse-serum. Let precipitate for 30 min at 112/115°C. Filter on humid filter paper. Adjust to pH 7.0–7.2. Dispatch 10 ml per screw-tube and sterilize for 20 min at 120°C.

**9.** Nutrient gelatin without nitrate

| | |
|---|---|
| Meat extract medium (see no. 7) | 2 l |
| Gelatin | 75 g |
| Nutrient agar | 12 g |
| Trypsic peptone or Bacto peptone | 20 g |
| Potassium chloride (KCl) | 10 g |

Adjust to pH 7.6–7.8 and let cool until 50°C.
Add:

| | |
|---|---|
| Horse-serum | 100 ml |

Precipitate at 120°C for 15 min then filter when hot (heating funnel).

Add:

| | |
|---|---|
| Glucose | 20 g |

Dispatch 15 ml per screw-tube.
Sterilize at 110°C for 30 min.

**10.** Arginine dihydrolase medium

| | |
|---|---|
| L-Arginine hydrochloride | 5 g |
| Yeast extract | 3 g |
| Glucose | 1 g |
| Bromocresol purple solution (1.6%) | 1 ml |
| Distilled water | 1 l |

Adjust to pH 6.3–6.4. Distribute 5 ml per tube.
Autoclave at 120°C for 15 min. Make a control tube without arginine.

**11.** Basal salt medium

| | |
|---|---|
| Ammonium dibasic phosphate ($(NH_4)_2HPO_4$) | 1.0 g |
| Potassium chloride (KCl) | 0.2 g |
| Magnesium sulfate heptahydrate ($MgSO_4$, $7H_2O$) | 0.2 g |
| Yeast extract | 0.2 g |
| Distilled water | 1 l |

Dissolve ingredients, adjust to pH 7 then add

| | |
|---|---|
| Bromocresol purple | 0.05 g |

Distribute 10 ml into tubes and sterilize at 120°C for 15 min. For use: melt, cool to 50–60°C and add sterile carbohydrate solution to 1% final concentration.

**12.** Christensen

| | |
|---|---|
| Bacto peptone | 1 g |
| NaCl | 5 g |
| Monopotassium phosphate ($KH_2PO_4$) | 2 g |
| Agar | 20 g |
| Distilled water | 1 l |

Adjust to pH 6.8–7.0 then add

| | |
|---|---|
| Glucose | 1 g |
| Phenol red | 0.12 g |

Distribute in 10-ml amounts into screw-cap tubes (17 mm × 145 mm) and sterilize at 105°C for 30 min. Prepare a stock-solution of 20% urea in distilled water and sterilize on a 0.22-μm filter. Store at 4°C.

**13.** Stock solutions

*Stock solution 1*:

| | |
|---|---|
| Magnesium sulfate heptahydrate ($MgSO_4 \cdot 7H_2O$) | 12.3 g |
| Manganese monohydrate sulfate ($Mn\ SO_4 \cdot 1H_2O$) | 0.17 g |
| Zinc heptahydrate sulfate ($ZnSO_4 \cdot 7H_2O$) | 1.4 g |

Mix into 1 l-vial, dissolve gently by heating then adjust to 1 l with distilled water.
*Stock solution 2*:

| | |
|---|---|
| Ferric sulfate ($Fe_2(SO_4)_3$) | 2 g |
| Distilled water | 105 ml |
| Sulfuric acid ($H_2SO_4$) | 3 ml |

Heat for 4 min then filter.
Adjust to 1 l with distilled water.
*Stock solution 3*:

| | |
|---|---|
| Calcium chloride ($CaCl_2$, $H_2O$) | 14.7 g |
| Distilled water | 1 l |

These above stock-solutions can be kept for several weeks at room temperature

**14.** UG 'usual' medium

| | |
|---|---|
| Bacto peptone | 7.5 g |
| Potassium phosphate solution | 100 ml |
| Stock solution 1 (see no. 13) | 10 ml |
| Stock solution 2 (see no. 13) | 10 ml |
| Stock solution 3 (see no. 13) | 10 ml |
| Distilled water | 870 ml |

Potassium phosphate solution:

| | |
|---|---|
| Mono potassium phosphate ($KH_2PO_4$) | 68 g |
| Distilled water | 1 l |

Adjust to pH 7.4. Dispatch 110 ml per 1-l Erlenmeyer flask or 10 ml into tubes (22 mm diameter).
Sterilize at 120°C for 15 min.

**15.** UG 'usual' medium with nutrient agar

Usual medium + Bacto agar 15 g per 1 l

Bring to boil while stirring.
Dispatch in flasks and sterilize at 120°C for 15 min.
These flasks can be kept at room temperature, when used boil then cool until 55°C.

**16.** Starch nutrient agar

| | |
|---|---|
| Potassium phosphate solution | 100 ml |
| Stock solution 1 | 10 ml |
| Stock solution 2 | 10 ml |
| Stock solution 3 | 10 ml |
| Peptone | 7.5 g |
| Distilled water | 870 ml |

Potassium phosphate solution:

| | |
|---|---|
| Mono potassium phosphate ($KH_2PO_4$) | 68 g |
| Distilled water | 1 l |
| Adjust to pH 7.4 and add: | |
| Nutrient agar | 20 g |

Filter hot after autoclaving. Prepare 150 ml starch suspension (commercial starch) 7% in hot distilled water and add it to the filtered nutrient agar. Dispatch 100 ml into 125-ml flasks

**17.** M.B.S.

| | |
|---|---|
| Mono potassium phosphate ($KH_2PO_4$) | 6.8 g |
| Bacto-tryptose | 10.0 g |
| Yeast extract | 2.0 g |
| Magnesium sulfate heptahydrate ($MgSO_4 \cdot 7H_2O$) | 0.3 g |
| Calcium chloride dihydrate ($CaCl_2$, $2H_2O$) | 0.2 g |
| Stock solution | 10.0 ml |
| Distilled water | 1 l |

Stock solution:

| | |
|---|---|
| Manganese sulfate ($MnSO_4$, $H_2O$) | 2.0 g |
| Ferric sulfate ($Fe_2(SO_4)_3$) | 2.0 g |

| | |
|---|---|
| Zinc sulfate ($ZnSO_4$, $7H_2O$) | 2.0 g |
| Distilled water | 1 l |

Adjust to pH 7.2. Dispatch 110 ml per 1-l Erlenmeyer flask or 10 ml into tubes (22 mm diameter). Sterilize at 120°C for 15 min. For plating in Petri dish, add 1.5% nutrient agar to M.B.S. medium

**18.** MRVP medium

| | |
|---|---|
| Polypeptone or trypsic peptone | 5 g |
| Glucose | 5 g |
| NaCl | 5 g |
| Distilled water | 1 l |

Dissolve by gently heating. Adjust to pH 7 and dispatch per 5 ml in screw cap tube. Sterilize for 30 min at 105°C.

**19.** KOVACS reagent

| | |
|---|---|
| Para-dimethylaminobenzaldehyde | 5 g |
| Isoamyl alcohol | 75 ml |

Dissolve by gently heating in water-bath at 50°C then add slowly while stirring (use gloves, protective glasses and chemical hood).

| | |
|---|---|
| Hydrochloric acid 37% | 25 ml |

Store at 4°C for less than 3 months

**20.** Nitrate test reagents

*Reagent* $NO_3$ A

| | |
|---|---|
| Parasulfanilic acid | 0.8 g |
| Acetic acid (5 N solution in distilled water) | 100 ml |

*Reagent* $NO_3$ B

| | |
|---|---|
| Alpha-naphtylamine | 0.5 g |
| Acetic acid (5 N solution in distilled water) | 100 ml |

**Warning:** Alpha-naphtylamine is carcinogenic. Use for handling and disposing of the product.

**21.** VP reagents

*Reagent* VP A

| | |
|---|---|
| Potassium hydroxide | 40 g |
| Distilled water | 100 ml |

*Reagent* VP B

| | |
|---|---|
| Alpha-naphtol | 6 g |
| Absolute ethanol | 100 ml |

**Warning:** Alpha-naphtol is harmful.

**22.** Tryptone broth for indole test

| | |
|---|---|
| Tryptone | 10 g |
| NaCl | 5 g |

Dissolve in 1 l of deionized water dispense in 4-ml aliquots in capped tubes. Autoclave 15 min at 121°C.

**23.** TE buffer

| | |
|---|---|
| 1 M Tris-HCl (pH 8.0) | 1 ml |
| 0.5 M EDTA | 0.2 ml |

Bring to 100 ml with distilled water.

**24.** CTAB buffer

| | |
|---|---|
| CTAB (hexadecylcyltrimethyl-ammonium bromide) | 10 g |
| NaCl | 4.1 g |
| Distilled water | 100 ml |

Dissolve by gently heating in water bath at 65°C.

CHAPTER IV

# Bioassay of bacterial entomopathogens against insect larvae

MAUREEN O'CALLAGHAN*, TRAVIS R. GLARE[†] & LAWRENCE A. LACEY[‡]

*AgResearch, Private Bag 4749, Christchurch 8140, New Zealand
[†]Bio-Protection Research Centre, P.O. Box 84, Lincoln University, Lincoln 7647, New Zealand
[‡]IP Consulting International, Yakima, WA 98908, USA

## 1. INTRODUCTION

The discovery of the insecticidal activity of *Bacillus thuringiensis* (Bt) launched a new era in pest control, with Bt becoming the leading biopesticide used around the world to combat agricultural pests (predominantly Lepidoptera and Coleoptera) and insect vectors of human disease (predominantly aquatic dipteran species) (Beegle & Yamamoto, 1992; Glare & O'Callaghan, 2000; Lacey, 2007). 'Bt' is not a single entity; it is a collection of subspecies and hundreds of isolates that vary widely in their ability to produce a range of toxins and hence have diverse host ranges (Garczynski & Siegel, 2007). New strains and toxins are constantly being evaluated and tested for their efficacy against pests that have yet to be controlled by other measures. The identification of *Bt* subsp. *tenebrionis* (Btt) strains active against coleopteran pests, and *Bt* subsp. *israelensis* (Bti) and *Bacillus sphaericus* (serovarieties 5a5b and 25) active against pest Diptera such as mosquitoes and black flies, has led to further development of a wide range of products based on Bt. This chapter focuses mainly on methods for testing of Bt against these key agricultural pests and insect vectors of disease and presents some brief recent examples of bioassays of Bt against other insect orders. Several other bacterial genera have recently emerged as previously unrecognized insect pathogens and methods for testing some of these non-sporeforming bacteria are also briefly described.

A bioassay is the use of a living organism to assay, or measure the amount of a substance, such as a toxin, in an unknown sample. For this purpose a range of concentrations is required to precisely determine the quantity of bacterial toxin eliciting a mortality response in larval insects. A single discriminating concentration of bacterial suspensions may also be used to study the effect of abiotic and/or biotic factors on larvicidal activity of candidate isolates or formulations. In such cases the amount of toxin is known and some other biological effect is being tested.

Bioassays are performed for a range of purposes in insect pathology, including the assessment of activity of new isolates or formulations of bacterial pathogens, mode of action studies (for example assessment of the role of toxins as opposed to viable cells in insect toxicity) and quality control of commercial biopesticide products. Bioassays are used to measure the key parameters essential in selection of efficacious strains for microbial control, including factors such as host range, speed of kill, and activity against different life-stages. They are also an essential tool in studies examining development of pest resistance to insect pathogens.

A wide variety of different bioassay methods have been described and researchers must select the most appropriate technique based on knowledge of key factors including: the bacterial pathogen and its mode of action; and the test species (target pest and/or key non-target

MANUAL OF TECHNIQUES IN INVERTEBRATE PATHOLOGY
ISBN 9780123868992

species) and its feeding behavior and ease of laboratory culture). The most commonly used type of assay is the dose–mortality bioassay, in which mortality is measured with a range of bacterial concentrations after a specified time. A wide range of concentrations are necessary if the goal of the bioassay is to determine the $LC_{50}$ value (lethal concentration required to kill 50% of the target insects) for new bacterial strains. These assays are also often used to assess intra-specific variability among laboratory and field strains of test invertebrates, for example, when examining Bt resistance in invertebrate populations. A potency bioassay is a special case of the dose–mortality assay where mortality caused by the microbe of interest is compared against a standard. For example, the comparison of Bt strains against an international standard and expressed as international units per milligram. Other commonly used assays include time–response assays which are most suited to 2nd or 3rd instar larvae. Mortality is counted on successive days allowing calculation of the mean lethal time ($LT_{50}$). Feeding inhibition assays are generally conducted with 2nd and 3rd instar larvae and allow estimation of the concentration of the microbe of interest that causes a 50% reduction in larval weight in comparison with control larvae (effective concentration: $EC_{50}$). This type of assay can be used to compare, for example, activities of Bt preparations as the weight reduction in young larvae can be a more sensitive parameter than mortality in insect species susceptible to *Bt* subsp. *kurstaki* (Btk) (Navon, 2000). Amount of frass produced has also been used as a rapid method to determine specificity of Bt toxins against Lepidoptera (van Frankenhuyzen & Gringorten, 1991).

Methods for bioassay of strains and formulations of Bt are well established (Navon, 2000). Dulmage *et al.* (1971) first described a standardized methodology that became the official assay for formulations of Bt and includes the use of an 'international unit' (IU) of potency. Bioassay methods for Bt have continued to evolve and are outlined in more detail below. More specific methods are constantly under development to determine the susceptibility of insects to Bt toxins, especially for monitoring resistance development. Feeding disruption tests (FDT) use measures of frass production as a rapid toxicity indicator (Bailey *et al.* 1998, 2001). In FDT assays, the diagnostic dose of Bt toxin is fed in artificial diet meal pads containing a blue indicator dye, such as Trypan blue. Toxicity is measured by production of dyed fecal pellets at 24 and 48 h.

While there is a range of methods available, there are some basic principles and processes that are common to all bioassays. Prior to starting bioassays, consideration must be given to the criteria that will be used to assess activity; for example, mortality, morbidity, effect on development, and/or fecundity. The test insects used in bioassays must be uniformly healthy and vigorous and genetically homogeneous. These requirements may be easily met for laboratory reared species but may present significant challenges when insects must be collected from the field, for example for assays against non-target species. Careful attention must be paid to preparation and standardization of doses of microbial inoculum/toxin used in the assays.

Sometimes bioassays must be performed over several seasons, and where data are to be compared over different seasons, it is essential to perform bioassays under the same strict conditions, for example, using toxins from the same batch and using identical bioassay procedures. In studies looking at development of field resistance in the target pest, it is ideal to bioassay the control strain each time that field populations are assayed.

## 2. LABORATORY BIOASSAY OF *BACILLUS* PATHOGENS AGAINST LEPIDOPTERAN LARVAE

### A. Factors affecting activity

Differences between strains, difficulties in measuring the active components of Bt formulations, and environmental and host influences on toxicity all combine to complicate assessment of Bt toxicity to lepidopteran larvae. Selected biological factors that can significantly affect activity of microbial inocula/toxins and hence the results of bioassays are discussed below; methods for measurement of the impact of environmental parameters on efficacy are addressed in Section 2E.

#### *1. Biology and feeding behavior of test species*

Knowledge of the feeding behavior of the test species is essential in the design of bioassays. For example, *Heliothis virescens* feeds by grazing across the surface of the diet and is best assayed using the surface contamination bioassay, described in more detail in Section 2C3. In contrast, *Ostrinia nubialis* typically bores in to the diet, spending little time on the surface of the agar so would have reduced chance of exposure to Bt toxin applied to the diet surface. In this case, the diet incorporation method described in Section 2C4 is the most appropriate bioassay approach. Insects such as *O. nubialis* and *Trichoplusia ni* can be grouped together in bioassay containers as they will live in large numbers together without eating each other. In contrast, *H. virescens* is cannibalistic and must be held individually.

### 2. Larval age

In general, young larvae, particularly 1st instars, are more susceptible under the same conditions than older larvae but young larvae are fragile and require careful handling, so may not necessarily be the best choice for routine, high-volume bioassays. The age of larvae selected for use in bioassays depends to some extent on the criteria that are to be measured. For example, if the aim of the bioassay is to determine developmental response of larvae to various strains of Bt, then it may be appropriate to use neonate larvae or recently molted 3rd instar larvae, and measure ability/failure of neonates to reach 3rd instar, or 3rd instars to reach 4th instar. In most dose–mortality assays, strains or doses are screened against similarly aged larvae of an appropriate instar.

### 3. Anti-feedant effect of Bt

Some insects have been shown to move away from Bt-contaminated leaves as opposed to untreated or control-treated leaves and in some cases Bt formulations or crystals have been shown to repel feeding insects which may impair ability to dose larvae with the required dose (Glare & O'Callaghan, 2000). Feeding stimulants including molasses and sucrose have been used to increase larval intake of toxins and spores. Phagostimulants are often used in commercial formulations. These encourage pest insects to eat the maximum amount of pathogen before it deteriorates on foliage and with Bt, before the crystal toxin arrests feeding. Burges & Jones (1998) provide a comprehensive list of phagostimulants that have been used in association with Bt.

## B. Bioassay design and data analysis

Design of a robust bioassay involves the selection of sample sizes and concentrations which will achieve the required precision for the parameter estimates or confidence intervals for the parameters of interest, for example, the $LC_{50}$ (lethal concentration required to kill 50% of insects) or $LD_{50}$ (lethal dose required to kill 50% of insects). $LC_{50}$ is a standard measure to express virulence. A pilot study should give a general idea of the various lethal concentrations for mortalities of interest.

### 1. Concentrations/dosage

Bioassays to determine the efficacy of a bacterial pathogen will require testing of different concentrations/doses of the agent against the same stage/age of insect target. The standard approach is to use serial dilutions of the bacterium or active component, such as toxins. Dilutions are usually made in weak buffers, such as phosphate buffer, saline, or (less common with bacteria) weak wetting agent solutions such as Triton X-100 or Tween. It may be necessary to conduct preliminary assays to establish the correct range of dilutions needed, but this will depend to some extent on starting material and prior knowledge of the bacterium–insect system. Often five or more dilutions are used to establish an $LC_{50.}$ A good practice is to have several doses both above and below the $LC_{50}$ (McGuire *et al.*, 1997).

Dose–response may be represented as a parameter other than mortality. Sublethal effects, such as change in weight gain or fecundity are also used as measures of impact through bioassays. The concentration in these cases is often referred to as 'effective concentration' (EC) and can be expressed as an $EC_{50}$ or dose which causes 50% reduction in the attribute being measured.

Choice of controls is crucial for any successful bioassay. The control should include any carriers, dilution agents, or ingredients used in preparation of the treatments, lacking only the active agent.

### 2. Use of standards

One of the issues in measuring potency of bacterial pathogens or their toxins is comparing between assays conducted with different strains or in different laboratories, even if using the same insect. While some variation can be expected from bacteria (where the exact stage of development and growth is difficult to replicate between assays), it is generally easier to standardize between assays of toxins, such as the Cry toxins of Bt strains. The use of internationally available standards of the most common Cry toxins was first established for a Bt strain

E-61 with an assigned potency of 1000 International Toxic Units/mg (Glare & O'Callaghan, 2000). Subsequently, a more relevant standard based on the HD-1 strain of Btk was established and assigned a relative potency of 18,000 IU/mg. The supply of the original and subsequent standards is currently almost exhausted. Standard powders are also available for Bti (Dulmage *et al.*, 1985; Thiery & Hamon, 1998).

Another method to establish relative potency is to compare with the effect of a well-characterized chemical pesticide of known toxicity, such as a commercially available chemical. By establishing the $LC_{50}$ for both the chemical and toxic bacterial agent, comparison between experiments is possible.

### *3. Replication*

To allow reliable statistical analysis it is necessary to have sufficient replication, both within a bioassay and over time. The number of replications required is dependent on the level of variability encountered. To establish an $LC_{50}$, a minimum of three replicates is required, with control mortality below 5–10%. Sample size can be calculated prior to the start of the experiment, based on the level of accuracy required and expected variation. Marcus & Eaves (2000) suggested that a total of 750 insects provided reliable replication, spread over five bacterial concentrations and three replicates per dose. There are diminishing returns in terms of decreased variability with increasing sample size. In general, bioassays should be repeated at least three times as specific conditions such as health status of the insects and ambient conditions can influence the results.

### *4. Analysis of data*

In general, no corrections to the final mortality caused by each treatment/dilution are necessary if control mortality is less than 5–10%. Experiments where control mortality exceeds 20% should be repeated. If control mortality does not exceed 10%, the data are analyzed by probit analysis (Finney, 1971) after adjusting for control mortality using Abbott's formula (Abbott, 1925):

$$\text{Adjusted \% mortality} = \frac{\text{Observed \% mortality} - \text{average control \% mortality}}{100\% - \text{average control \% mortality}}$$

ANOVA and statistical tests for mean separation (such as Tukey's studentized multiple range test) can be used to analyze data after arcsine-square root transformations of percentage mortality data. For a detailed consideration of statistical analysis of bioassay data, see Chapter VII or Marcus & Eaves (2000).

## C. Diet-based laboratory bioassays

Bioassays using artificial diet are highly reproducible and efficient, particularly when only small volumes of the test material are available. Diet-based assays remove the need to collect or maintain a supply of plant material (which is inherently more variable than the prescribed artificial diet). Obviously this method can only be used where the test species can be reared on an artificial diet that supports normal growth of the insect and is not detrimental to the active agent being assayed. Hence an insect diet that contains antibiotic may not be suitable for assay of some microbial pathogens but might still be suitable if toxins were being assayed. Obviously, the composition of the diet should support optimal survival and development of the test larvae, thus minimizing levels of control mortality. Performance of different strains of larvae, for example, laboratory-adapted *versus* feral strains, may vary on artificial diet and some preliminary testing to optimize survival of the test species on diet may be required (Carpenter & Bloem, 2002).

Artificial diets contain nutrients that substitute for the natural food of the host insect and are typically quite complex. Initially diets were developed for bioassay of single insects, for example, Dulmage *et al.* (1971) proposed a diet for assay of *Trichoplusia ni*. The growing interest in application of Bt for control of a wide range of pests has necessitated the development of additional diets and a wide range of diets are now in use. Diet-based assays have been used most commonly in the assay of Bt strains against the lepidopteran pests *Spodoptera exigua* and *Helicoverpa (Heliothis) armigera* with various researchers using slight modifications of a basic diet recipe containing soybean/soy flour, yeast, vitamin C and other components. Navon *et al.* (1990) proposed a standardized bioassay diet (Appendix 1) that can be used for three lepidopteran species: *H. armigera, Spodoptera littoralis* and *Erias insulana*. Other diets that have been used successfully for bioassay of Bt against various insect larvae are also presented in Appendix 1. Similar diets have been used for other insects, such as *Lymantria* spp. (e.g., Keena *et al.*, 2010). A wide choice of diets that could be adapted for bioassays and general information on insect rearing on artificial diets is available in Carson (2003) and Schneider (2009). Some diets are now commercially available, such as for a range of caterpillars (e.g., beet armyworm, diamondback moth, cutworms, cornborers, gypsy moth) and Coleoptera (e.g., boll weevils, Colorado potato beetle) from BioServ (http://www.insectrearing.com/products/indiets.html).

Two approaches can be used in diet-based assays: (1) surface contamination, whereby the active agent is spread across the surface of solid artificial diet; or (2) diet incorporation, where the active agent is mixed with molten artificial diet before it is dispensed into the experimental arena, such that the Bt is dispersed evenly throughout the depth of the diet. As discussed earlier, the choice of method used depends mainly on the feeding behavior of the insect. *H. virescens* feeds by grazing on the surface of the diet and can be effectively assayed using the surface contamination method. In contrast *O. nubialis* will bore into diet, spending little time on the surface and is best assayed by the diet incorporation method.

### *1. Separation of bacterial developmental stages and toxins*

Many of the insect pathogenic bacteria kill through action of toxins. The active toxin(s) may be produced in a separate body within the sporangium (e.g., parasporal crystals of Bt), in varying amounts in spores and vegetative states or may be produced at only certain times during bacterial development. Separation of the toxin may be necessary for bioassays examining effects of toxins alone. Many methods have been used to separate components of bacteria, such as separation of vegetative cells from spores and parasporal bodies. The simplest approach has been to use young *versus* old cultures. For *Bacillus* and related genera, bacteria sporulate as they age. With some species, leaving a culture to grow for > 48 h allows over 95% of bacteria to form spores. More complex methods are required to separate toxin-containing crystals from spores. As the concentration of parasporal crystals in fermentation broth is usually about 1%, a large amount of water also has to be removed.

Braun (2000) presented several methods used to separate Bt spores from crystal toxins, such as methods based on polyethylene glycol (PEG), solubilization using KOH and isolation of crystals using ion exchange chromatography. Spores have been separated from parasporal crystals using other methods such as froth flotation and density gradient centrifugation using either CsCl, Renografin (Bristol–Myers Squibb Company, New Jersey), NaBr, Ludox (Sigma, St Louis), sucrose, or phase separation (Lin *et al.*, 2003). There are problems with each method, including inefficient recovery, and the need for expensive reagents and equipment. No method appears to work with all strains (Braun, 2000).

### *2. Preparation of inoculum for addition to diet*

Preparation of the inoculum is the same for any of the bioassay methods described below. All suspensions of inoculum for bioassay should be prepared from formulated product or fresh cultures just prior to each bioassay. The test material should be prepared in sterile distilled water or in a buffered saline solution, with the volume prepared depending on the size of the bioassay and amount of test material available. The initial suspension must be well homogenized and the addition of a wetting agent such as (0.1%) Tween 80 can assist in wetting and mixing of dry powders into suspension. Subsequent dilutions can then be made using routine dilution procedures. Six or seven dilutions of the test material should be made; each suspension should be homogenized with a vortex mixer just prior to making the next dilution in the series and again just before adding it to the diet.

### *3. Surface contamination assay*

For the surface contamination assay, prepare aliquots of each dilution as described above and place onto the surface of the diet, which has previously been prepared and dispensed into appropriate-sized containers. A good example of a surface contamination assay of Bt toxins against *Helicoverpa armigera* and *H. punctigera* is described by Bird & Akhurst (2007). A stock suspension of spore/crystal mix (Cry1Ac or CryAb) was diluted with distilled water to produce six two-fold dilutions and distilled water was used as the control. Aliquots (in this case 50 μl, but volume will depend on area/volume of artificial diet to be treated) of the series of doses were then applied to the surface of the artificial diet held in 24-well plates. The toxin suspension should be spread evenly across the surface of the diet and left to air dry before test larvae are added.

*4. Diet incorporation*

For this method (Figure 4.1), the diet must be prepared and autoclaved as per instructions for that particular diet, cooled to 55°C and maintained in a molten state in a water bath until used. The required dilutions of the test materials are then diluted 1:10 into aliquots of diet and mixed thoroughly in a blender. Timers should be used to precisely control the amount of mixing that occurs. The diet/dilution can then be poured into cups, or wells of a tissue culture plate, and allowed to solidify and dry before addition of larvae. The amount of diet in each cup should be sufficient to maintain the insect throughout the incubation phase but is not otherwise critical because the concentration of Bt is constant throughout the diet. An amount of approximately 5 ml in a 9-ml plastic cup is usually sufficient for most species.

*5. Incubation of larvae and evaluation of effects*

Healthy uniformly aged larvae must be carefully transferred to individual cups or wells of microtiter plates, such that damage to larvae is avoided. Careful handling is required in particular when using neonate larvae. A small camel-hair brush is an excellent tool for this process. Larvae should be transferred to the controls and most dilute sample first. Cups or trays can be covered with a Mylar film or similar material which maintains humidity but is perforated to allow aeration and prevent condensation. Conditions for incubation vary with test species and will be the same as the conditions prescribed for the rearing of each species. In general, 25–30°C and 40–60% RH are acceptable. The amount of time that larvae are allowed to feed on the diet must be considered. The longer the feeding period, the greater the variability in dosage consumed, reflecting differences in feeding rate and in intrinsic susceptibility with increasing age of test larvae.

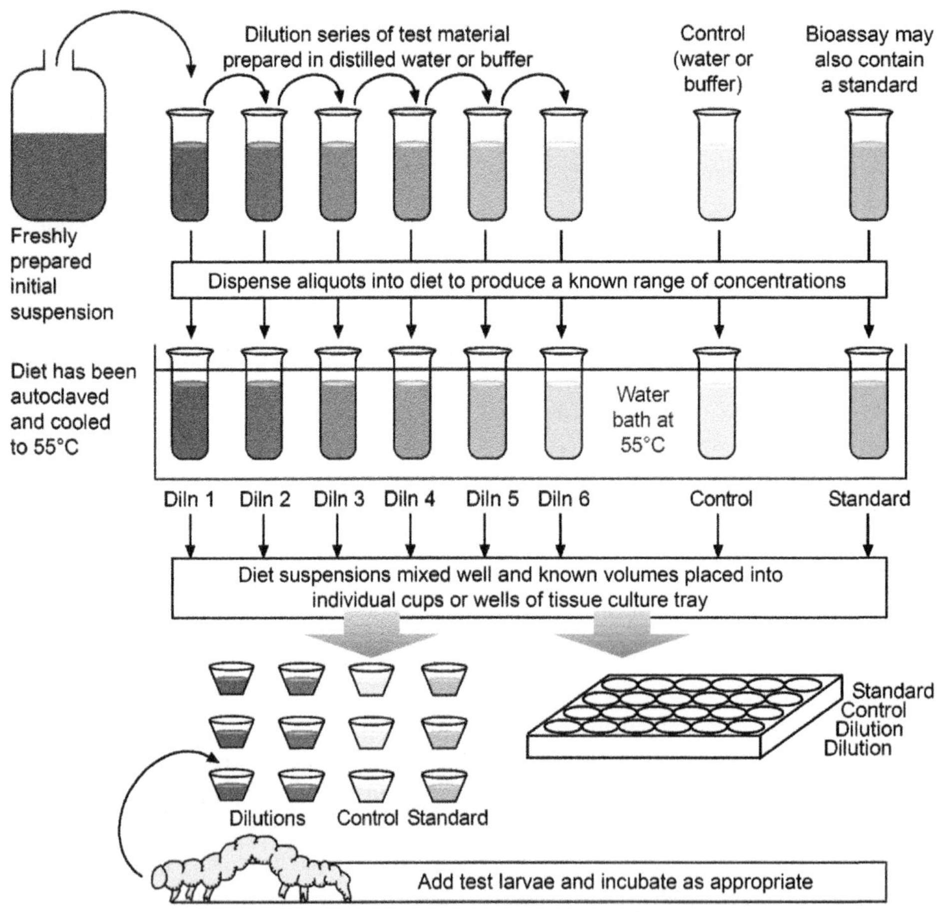


**Figure 4.1** Bioassay procedure for incorporation of suspensions of known concentrations of Bt/toxin into diet.

Evaluation of assays should take place at pre-determined time(s); for most bioassays this will be after 4–7 days' incubation. Mortality can be assessed visually but if there is any doubt, the larvae should be prodded to ascertain mortality. Lack of response or only very weak response (i.e. the larva is moribund) is scored as dead. Calculations of percentage mortality should be based on the number of larvae initially placed in containers and the number of living larvae after the incubation period, as survivors of some species will consume dead larvae.

In addition to simply measuring mortality, there may be a need to examine developmental responses of the test species to the microbe/toxin or to make age-specific mortality comparisons. For example in the bioassays of *Helicoverpa armigera* and *H. punctigera* described above by Bird & Akhurst (2007), one unfed neonate larva (< 24 h old) or one recently molted 3rd instar larva that had been raised on artificial diet was placed in each well. Mortality and larval growth inhibition (failure of neonates to reach 3rd instar) or failure of 3rd instars to reach 4th instar) were assessed after 7 days at conditions used for insect rearing.

Larval inhibition assays are often better measured by performing bioassays on artificial diet into which the Bt toxin has been incorporated. This ensures that a more consistent dose can be delivered to larvae compared with surface application of toxin where, once the surface of the diet has been penetrated by feeding activities, larvae are able to grow at normal rates. However, the diet incorporation method requires larger amounts of toxin and may not always be feasible, for example, in studies of populations with high levels of resistance.

## D. Application methods for bioassay of bacteria on foliar surfaces

For test insects that cannot be maintained on artificial diet, bioassays will need to be carried out on leaf surfaces. Leaf bioassays can be considered midway between bioassays on artificial diet and field assays, with leaf surfaces being more representative of the chemical and physical barriers present in an agricultural crop. Bioassay on leaf surfaces is also an essential step in screening programs to test activity of new Bt isolates or assess efficacy of formulations designed to improve environmental performance, such as those containing UV protectants, surfactants, rain-fasting materials and stickers. Some of these formulation components can affect the feeding behavior and hence ingestion of the active microbe.

The choice of plant depends on the test species and some screening may be necessary prior to bioassay to ensure that the foliage is palatable to the target and can be maintained at a quality level sufficient to support the survival and/or growth of the test species for the duration of the experiment. The plant species used can significantly impact on measurement of toxicity, with some plants shown to inhibit or enhance toxicity. For example the response of *Lymantria dispar* to Btk toxin has been shown to vary when larvae were fed on oak or aspen leaves (Glare & O'Callaghan, 2000). Cotton is ideal as it lasts well in sealed dishes and provides plenty of food and moisture and supports insects such as *O. nubialis* for up to a week (McGuire *et al.*, 1997). Maize, tomato, and cruciferous species are also commonly used. Depending on the plant chosen, it may be necessary to maintain foliage quality by placing leaf stems in water, agar or a suitable nutrient medium.

Laboratory bioassays on foliage surfaces can be carried out using whole leaves or leaf discs but in general similar methods for application of inocula are used for both types of assay. Assays using whole potted-plants and caged larvae may be appropriate in some circumstances (Navon, 2000).

### *1. Droplet method on foliage*

The droplet method involves the careful application of a known concentration (usually calculated to simulate expected field dose rate) onto the leaf surface. Leaves, or leaf discs, should be the same size and age. The droplet (typically of volume around 100 μl) may be spread with a sterile glass rod to fill a certain space marked onto the leaf surface, which ensures a consistent dose is provided to the test larvae such that $LC_{50}$ determinations can be made. The use of Tween 80 (approximately 0.1% v/v) in the inoculum suspension can aid even dispersal of the droplet within the required space. While providing an accurate dose that allows subtle differences in activity to be detected or comparisons of formulations to be made, the method is very laborious. Treated leaves are left to air dry before use in bioassays. Use of the droplet method in bioassay of *Plutella xylostella* is shown in Figure 4.2.

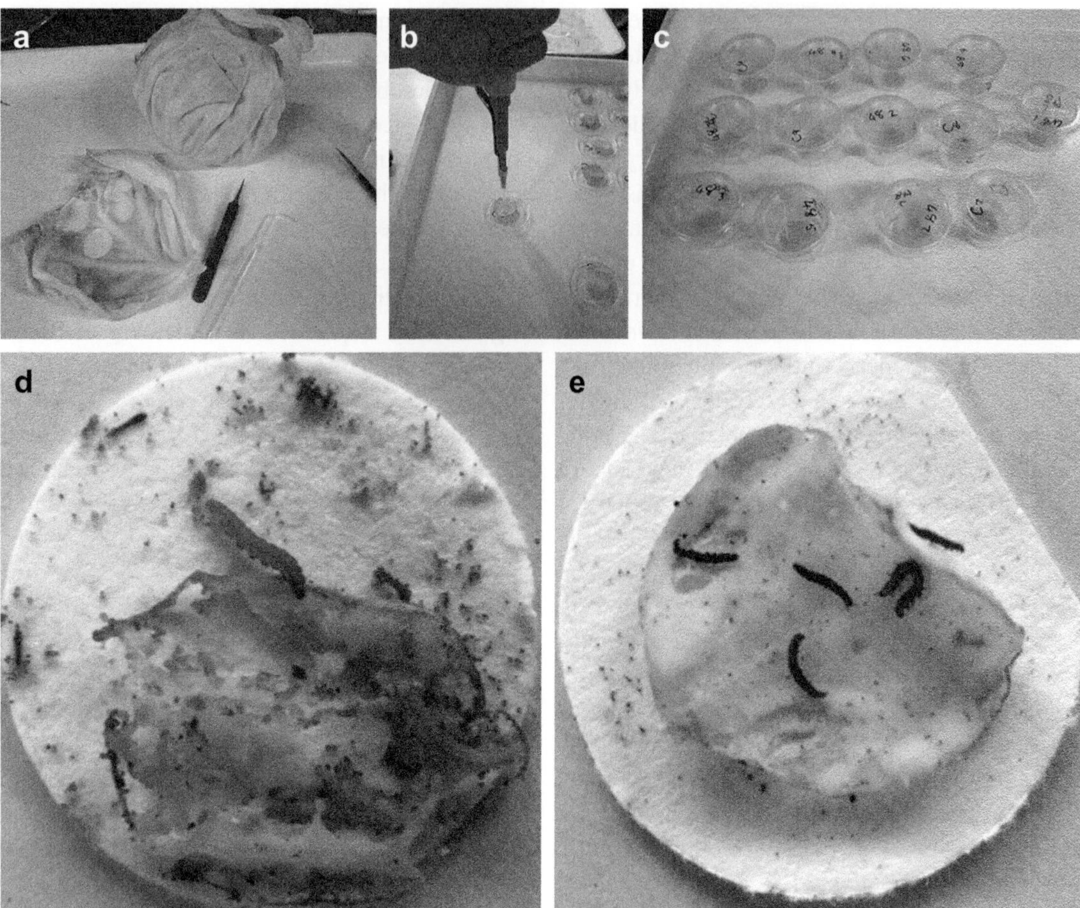

**Figure 4.2** Bioassay of *Yersinia entomophaga* against diamondback moth larvae. a. Discs are cut from cabbage leaves. b. Leaf discs are inoculated with aliquots of known concentration of *Y. entomophaga* toxin or live cells. c. Cups containing six larvae are incubated for 4 days at 22°C. d. Untreated larvae remain healthy and feeding 4 days after treatment. e. Dead larvae 4 days after treatment. (Courtesy of Mark Hurst and Sandra Jones, AgResearch, New Zealand.) Please see the color plate section at the back of the book.

*2. Leaf dip*

An alternative, slightly less labor-intensive method is leaf dipping in which detached leaves or even whole plants are immersed in a suspension of Bt/toxin and then allowed to dry. In comparison with the droplet method, much greater volumes of test material (Bt/toxin) must be used and coverage of the foliage tends to be much less uniform so there is less accurate control of the dose that the test species will be exposed to in subsequent bioassays. The type of foliage used can significantly affect the outcome. For example the waxy nature of cabbage leaves tends to repel liquids, leading to lower than expected levels of toxin being delivered in bioassays. Even leaves of various ages collected from the same plant can vary such that dose rates accumulating on leaves vary to an unsatisfactory level. McGuire *et al.* (1997) recommend that leaf/plant dipping only be used when no alternative method is available.

*3. Track sprayer*

The laborious nature of laboratory methods such as the droplet method can be overcome by using specialized equipment that delivers a consistent dose to foliage. One example of this type of equipment is the research track sprayer (Devries Manufacturing, Hollandale, MN, USA; www.devriesmfg.com; Figure 4.3). Primarily designed for research on delivery of herbicides and pesticides, the track sprayer is ideal for the treatment of foliage with uniform coverage of microbial inocula or toxins. The chamber contains a spray head with adjustable nozzles

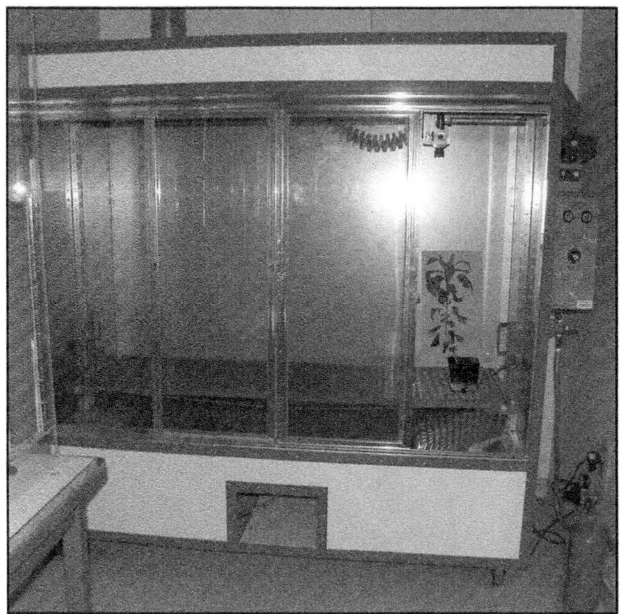

**Figure 4.3** A research track sprayer. The control panel on the right allows modification of nozzle speed and direction while pressure is generated from a regulated compressed air supply. To modify the chamber for rain, the air supply line is turned off and a water line connected to a second nozzle provides water for spraying plants or other substrates. A continually traversing nozzle provides rain simulation to test rain fastness of formulations.

and speed adjustment, spray pressure, and volumes can be varied, such that field application conditions can be effectively simulated. Using this equipment, large numbers of Bt isolates or formulations can be screened; in particular the technique can provide useful data on performance of formulations applied through an agricultural-type nozzle and patterns of spray splashing and dripping that may lead to inconsistent deposition of toxin on foliage, closely simulating the field situation. Care must be taken to ensure that all foliage is contacted by the spray suspension. The method is most useful where absolute accuracy of dose–response is not required and provides a useful halfway step between laboratory and field experimentation. The track sprayer can also apply water continuously for rain simulation and testing of the rainfastness of Bt formulations.

*4. Potter Tower*

The Potter Tower (Burkhard Agronomic Instruments, UK; www.burkardscientific.co.uk; Figure 4.4) is a useful alternative to the track sprayer. This air-operated spraying apparatus applies a precise and uniform deposit of spray over a circular area of 9 cm diameter. This apparatus is recognized as the reference standard for researching chemical spraying techniques and is frequently used in bioassay of insect pathogens (e.g., Morris, 1973; Vidal *et al.*, 1997; Bateman, 1999). In most cases, the apparatus is placed in a Plexiglas® cabinet. Operation of the Potter Tower is relatively simple. First a known volume (usually 1–5 ml) of test sample is placed into a vial located at the top of the tower. The test subject, often on a plant leaf, leaf discs, or only the target insect(s) in a suitable dish, is placed onto the platform at the base of the tower. Once the cabinet containing the tower is closed, the tower is activated by either a switch or lever that raises the platform holding the test subject into the base of the tower while simultaneously aspirating the test sample from the vial into the tower. The amount of sample sprayed per unit area onto the test subject is controlled by the volume placed in the vial. Calibration of the tower is required in order to establish the relationship between volume placed in the vial and the mass of material deposited onto a known area.

## E. Factors affecting larvicidal activity on leaf surfaces

A number of abiotic factors including solar radiation, temperature, and humidity (moisture) have been shown to affect the persistence and efficacy of Bt. Commercial formulations are constantly being developed to overcome these environmental constraints and these new products can be assessed in the first instance in laboratory bioassay systems.

*1. Solar simulation*

The inactivation of Bt and many potentially useful non-sporeforming bacterial entomopathogens by sunlight is well known and is a key factor limiting their persistence and efficacy in the field. To effectively compare formulations for their ability to protect microbial inocula or toxins, or to screen potential UV protectants, methods that allow uniform repeatable exposure to sunlight must be used. Laboratory screening using a solar simulator is an ideal method for preliminary screening of candidate microbes or formulations, so that only the more promising

**Figure 4.4** a. The Potter Tower. b. Placement of a leaf sample onto the treatment platform. c. Apple leaf discs in place ready for treatment. d. A 9-cm diameter Petri dish containing artificial diet prior to treatment.

candidates need to be subjected to more laborious and expensive field testing. It also enables year-round testing.

Solar simulators use a xenon light source and specific filters that provide irradiance that closely approximates the quality and quantity of natural sunlight and typically take the form of a solar chamber that consists of a power supply, light source and enclosure (Figure 4.5). The enclosure or chamber provides a work surface and shields the operator from radiation. Portable or handheld solar simulators and light sources are also available. The use of the Atlas Suntest CPS + solar simulator (SDL Atlas, USA) has been described previously (McGuire *et al.*, 2000; Lacey & Arthurs, 2005) and the method is amenable to test sensitivity of any bacterial strain or formulation. Sometimes the simulator is placed on top of a custom (but easily made) cabinet that provides a larger work surface area and provides greater separation between the test substrate and the heat generated by the light source.

Where bacterial formulations are being tested on plant tissue, the use of Tefcel plastic (American Durafilm,

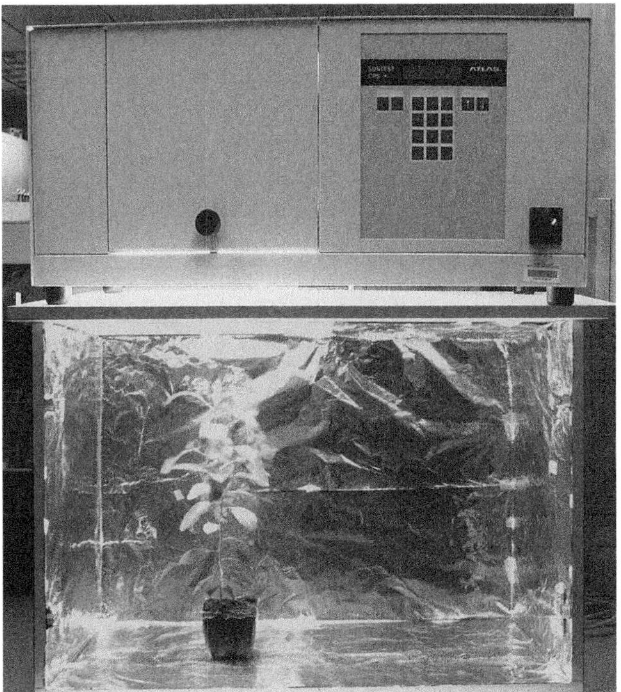

**Figure 4.5** Solar simulator in use for determination of solar sensitivity of formulations of Bt. The figure shows the solar simulator with a plant placed below the light source. In this modification, the top of the chamber has been removed to allow light to pass onto the plant. Tefcel plastic is placed across the opening, to prevent burning of the leaves.

Holliston, MA) is recommended, to reduce the heat to which plants are exposed. The system described above can be used to treat detached plant foliage or whole plants if the purpose-built cabinet is large enough, but whole plants must be carefully staked to ensure that leaves exposed are equidistant from the light source. Foliage should be treated using one of the methods described above and exposed to the light source for a designated period of time. McGuire *et al.* (2000) have described in detail the calibration of a solar simulator to test the stability of the Bt formulation Dipel on cotton plants; for 50% loss in activity of Bt, cotton plants would need to be exposed to the light for 8 h, at a distance of 20 cm from the light source. Required exposure times will vary with the pathogen (e.g., non-sporeforming bacteria will require much shorter exposure times), the host plant, formulation, and the particular experimental set-up used. Following exposure, plant leaves or discs are bioassayed using standard protocols, including the use of the necessary controls.

*2. Rainfastness*

Rainfall has been shown to cause a sharp decline in Bt activity on leaves (Glare & O'Callaghan, 2000) and commercial formulations often contain adjuvants to improve the adherence of spores and crystals to leaves. Methods which simulate rainfall in measurable doses allow relatively easy screening of strains and formulations.

Simulation of rainfall can be achieved using the track sprayer system described earlier by supplying water into the line that provides air pressure to the spray nozzle. The spray head can then be adjusted such that the head traverses back and forth across the plants or leaf discs. Various treatment conditions (e.g., light or heavy rainfall) can be achieved by varying the speed, nozzle, pressure, etc. Foliage treated with Bt or other pathogens are dried before being placed in the calibrated rain simulator and spaced within the chamber so that they receive even amounts of rainfall. McGuire *et al.* (1997) recommend testing rainfastness by exposing treated foliage to 5 cm of rain over a 50-min period, which simulates relatively heavy rain and gives a good indication of the ability of a formulation to resist rain. Leaves should be dried and leaf discs cut if necessary before use in bioassays.

*3. Temperature*

While temperature can affect survival of Bt, extreme temperatures are required to inactivate spores. Temperature does, however, affect the efficacy of Bt with higher temperatures typically increasing toxicity (Glare & O'Callaghan, 2000). This occurs largely through increasing the insect metabolic rate and the rate at which insects feed and consume Bt. Raising the temperature also accelerates bacterial development within the insect. Generally Bt is more toxic at higher temperatures and is not effective at field temperatures below around 15°C, possibly because of reduced feeding of the target insect. Increased larval mortality was seen in bioassays of Btk against *Choristoneura rosaceana* as incubation temperatures were increased up to 25°C (Li *et al.*, 1995). $LT_{50}$ values for *C. fumiferana* decreased from 12–17 days at 13°C to 2–4 days at 25°C (van Frankenhuyzen, 1990). The choice of temperature at which bioassays should be run will be influenced by the environment in which the pathogen will eventually be used but bioassays against key insect pests *H. armigera* and *H. punctigera* are typically carried out at 25°C and *O. nubilalis* at 27°C.

## 3. BIOASSAY OF *BACILLUS THURINGIENSIS* AGAINST DIPTERAN LARVAE

Due to their roles in the transmission of the causal agents of malaria, dengue, onchocerciasis, and a variety of other diseases of humans and domestic animals, mosquitoes (Culicidae) and black fly (Simuliidae) are considered some of the most medically important insect species. Effective larvicides can contribute to the integrated management of these pests by reducing or eliminating the emergence of the adults, the blood feeding stage responsible for transmitting disease-causing agents and pestiferous activity. Success of environmentally friendly means of control for black flies and mosquitoes was greatly increased with the discovery of varieties of *Bacillus thuringiensis* with elevated larvicidal activity against these insects (Lacey & Undeen, 1986; Lacey, 2007). Their activity against certain families of flies in the suborder Nematocera is principally due to the production of Cry4A, Cry4B, Cry11A and Cyt1A toxins. These are predominantly found in *B. thuringiensis* subsp. *israelensis* (Bti) and the PG-14 isolate of *B. thuringiensis* subsp. *morrisoni*. Isolates of certain varieties of *Bacillus sphaericus* (Serotypes 5a5b and 25) also have good demonstrated larvicidal activity for the microbial control of pest and vector mosquitoes (Lacey, 2007).

In order to assess and compare the efficacy of any bacteria against both mosquito and black fly larvae, the bacterial toxins responsible for larvicidal activity must be ingested by the larvae. Comparative efficacy testing must be conducted under repeatable conditions that permit normal or close to normal feeding rates and under conditions that do not produce abnormally high mortality (in excess of 10%) in control larvae. Although there are other aquatic Nematocera that are susceptible to bacteria and/or their toxins, we will emphasize methods for the testing of bacteria against larvae of mosquitoes and black flies. Readers interested in the bioassay of Bti against chironomid larvae and other aquatic Nematocera can refer to procedures used by Ali *et al.* (1981) and other authors cited by Lacey & Mulla (1990). A multitude of aquatic non-target organisms are not directly affected by ingesting bacteria or bacterial toxins that are active against mosquitoes and/or black flies, but they may be affected by formulation components or the consequence of removing target species from the ecosystem. Results of testing and a review of the literature on the effects of entomopathogenic bacteria on non-target aquatic insects are presented by Lacey & Mulla (1990) and Lacey & Merritt (2003).

### A. Bioassay of *Bacillus* pathogens against mosquito larvae

The bioassay of bacteria against larvae of most mosquito species is a relatively straightforward procedure. Larvae of most pest and vector mosquitoes are filter feeders and readily ingest particulates in the size range of bacteria and bacterial inclusions (Merritt & Craig, 1987). Some notable exceptions are predatory mosquitoes that consume aquatic insects including the larvae of other mosquito species.

#### *1. Factors affecting activity*

A number of factors, such as mosquito species and age, number of mosquito larvae per bioassay container, water quality and temperature, volume and depth of water, presence or absence of food and other particulates, shape of the bioassay container, and factors related to the bacterial preparation (size of particles, effect of adjuvants) can significantly influence toxin activity and the results of bioassays. Many of these factors also influence feeding rate and hence the amount of toxin ingested. In the next sections of this chapter we will present information on various bioassay parameters and how they might influence bioassay results.

##### *a. Effect of species*

The major difference in standardized bioassays for isolates of *B. thuringiensis* and *B. sphaericus* will be the target species. All filter feeding mosquito species thus far tested are susceptible to Bti. Susceptibility to *B. sphaericus* is strongly dependent upon the mosquito species used for bioassay. *Culex* and *Psorophora* species are highly susceptible to *B. sphaericus*, but many *Aedes* and *Ochlerotatus* species, especially *Aedes aegypti* are not susceptible or considerably less susceptible. Because *Ae. aegypti* is relatively easy to rear and its eggs can be stored dry between rearings, it is an ideal test animal for bioassays of isolates of *B. thuringiensis* subspp. *Culex quinquefasciatus* is the preferred test species for *B. sphaericus*. Although it it is normally very susceptible, high levels of resistance to this bacterium have been reported in certain populations of *Cx. quinquefasciatus* (Lacey, 2007).

*b. Feeding behavior*

In addition to other susceptibility factors related to species, the feeding behavior of a particular species will influence the amount of inoculum the larvae will come into contact with. Some species may feed predominantly at the surface of the water (*Anopheles* species), within the water column (some *Culex* spp.) or predominantly off the substratum (some *Aedes* spp.) or a combination of column and substrate feeding. Species that feed mostly at the surface usually consume less toxin due to settling of toxic moieties from their feeding zone and thus appear to be less susceptible to entomopathogenic bacteria.

The parasporal body of Bti ($\approx 0.75\ \mu m$), where the toxic moieties are located, are efficiently captured along with other fine particles (Merritt & Craig, 1987). Depending on its nature and quantity, particulate matter can reduce or enhance the feeding rate of mosquito larvae (Dadd, 1970) and hence the amount of toxic inclusions that are ingested by larvae (Sinègre *et al.*, 1981a). Even the addition of excess food that is normally desirable for species such as *Cx. quinquefasciatus* could inhibit feeding or interfere with the activity of toxins. During the exposure/incubation period, the absence of food will enable an accurate and repeatable bioassay of bacterial pathogens. Late 3rd and early 4th instars can withstand starvation with little or no effect on control mortality. In assays involving younger larvae that require more than 48 h exposure and incubation, a small amount of food (5 mg/l), such as finely ground lab chow, could be added to the assay cups in the second 24 h of exposure. The type of food (i.e. rich in protein or carbohydrate) could significantly affect the outcome of the experiment. For example, Skovmand *et al.* (1998) observed rapid development of 3rd instars to less susceptible 4th instars fed on a diet rich in protein, whereas the same instars fed on carbohydrate diet required a longer period of time to reach late 4th instars. As late 3rd and early 4th instar, the larvae were significantly more susceptible to Bti.

*c. Number of larvae/container, volume and quality of water, and features of the bioassay container*

Because mosquito larvae, especially those feeding within the water column and off the substratum can be very efficient in removing particulates, the number of larvae per container will influence the amount of toxic moieties that will be available to each individual (Sinègre *et al.*, 1981a). The amount of toxin per larva will also be a function of the volume of water used for the bioassay. Five milliliters of water per larva has enabled consistent and repeatable bioassays in studies conducted at the USDA-ARS Center for Medical, Agricultural, and Veterinary Entomology (CMAVE); Gainesville, FL, USA) on both *B. sphaericus* and *B. thuringiensis* subspp. against several mosquito species. The shape and depth of the container and water depth may also affect the availability of bacterial toxin (Skovmand *et al.*, 1997; 1998). Small disposable cups (depth *ca.* 4 cm) filled with 100 ml of chlorine-free water (deionized water, well water, or as a last resort, aged tap water), are used for bioassays at CMAVE. The presence of chlorine can significantly reduce the larvicidal activity of Bti and other *B. thuringiensis* subspp. (Sinègre *et al.*, 1981b) and be detrimental to larvae. The quantity of water used for bioassay at the Pasteur Institute (Paris, France) is 150 ml of water. Twenty larvae per container are used at CMAVE and 25 larvae per container are used at the Pasteur Institute.

*d. Larval age*

The age group that enables good survival in controls and consistent assay results are late 3rd and early 4th instars. Younger larvae (1st and 2nd instar) provide more sensitive targets, but tend to be less hardy and unable to survive much more than 24 h without food. Older fourth instars may pupate before the end of the exposure period. With the exception of first instars, mosquito larvae imbibe very little water. If solubilized toxins are to be assayed, it will be necessary to use 1st instars or to encapsulate the toxin(s).

*e. Temperature*

Within the range of temperatures tolerated for a given species, the activity of bacterial toxins in mosquito larvae is, for the most part, positively correlated with temperature (Sinègre *et al.*, 1981b; Wraight *et al.*, 1987; Lacey, 2007). Many of the species most utilized for laboratory bioassay (e.g., *Ae. aegypti, Cx. quinquefasciatus, An. albimanus,* and *An. stephensi*) are tropical in origin and thrive at 27°C. Lower temperatures may be required for other species from more temperate climates.

*2. Selection of dosage (concentrations)*

Almost invariably one observes a dose-dependent mortality response in mosquito larvae to Bti and *B. sphaericus* varieties. The exception being when bacterial preparations contain such low amounts of Nematocera-active toxin that the particulate load necessary to elicit a response may also inhibit feeding. A range of concentrations of more promising candidate bacteria should be selected (five to seven concentrations), that produce mortality from 10 to 90% with two concentrations above and two below the 50% mortality level.

Dosage is usually expressed in milligrams of spore powder or formulation per liter of water or in parts per million (ppm). The use of international toxicity units (ITU) to compensate for the day to day fluctuations in larval response to Bti has been proposed (de Barjac & Larget, 1979). ITUs are calculated by comparing primary powders to a standard and using the ratio of the $LC_{50}$ values of the standard and test sample times the assigned potency of the standard. For example:

$$\text{Potency (ITU) of sample} = \frac{LC_{50}\text{ for standard} \times \text{potency (ITU) of standard}}{LC_{50}\text{ for sample}}$$

The potency of IPS-78, the first standard prepared by de Barjac, for example, was arbitrarily assigned a potency of 1000 ITU/mg (de Barjac & Larget, 1979). If the $LC_{50}$ for the test sample is greater than that of the standard, the ITU rating will be inversely proportional to the rating of the standard. Primary powders of Bti were also proposed by Dulmage *et al.* (1985) as US standards.

Several reasons for avoiding the use of spore count as a measure of dosage are presented by Dulmage *et al.* (1990). Nevertheless, if suspensions of fresh cultures are used in assays, such as in preliminary screening of recently isolated strains, spore count (after heat shock at 80°C for 12 min) may be the only indication of dosage available. Ultimately, preparation of primary powders would enable comparative bioassays against a standard.

*3. Number of replicates and tests*

Each concentration and control should be replicated a minimum of four times per test date. Although a test conducted on a single date with replicated assays of each dosage will usually provide a more homogeneous set of data, it is advantageous to run replicate tests over time to account for the variations in susceptibility that are observed in different cohorts of mosquito larvae obtained from the same colony. Three replicate tests are the usual minimum.

*4. Bioassay protocol*

In order to have repeatable results, the various bioassay parameters should be standardized. Suggested parameters for standardized bioassay of varieties and formulations of Bti and *B. sphaericus* against mosquito larvae are presented in Table 4.1. Several similar protocols have been recommended for the repeatable bioassay of Bti (de Barjac & Larget, 1979; McLaughlin, *et al.*, 1984; Dulmage *et al.*, 1990) against mosquito larvae. The following is an eclectic combination of procedures for the preparation of inoculum and bioassay against mosquito larvae.

*a. Preparation of inoculum*

Suspensions of inoculum for bioassay should be freshly prepared from primary powders, formulated product, or fresh cultures just prior to each bioassay. Initial homogenates are prepared with standards and primary powders by suspending 50 mg of powder in 10 ml of deionized or distilled water and agitating for 10 min using a bead mill (20-ml penicillin flask with several 6-mm glass beads) or similar method. Do not use a blender to prepare suspensions. Reduction in larvicidal activity is possible due to shearing effects on toxic moieties. Subsequent dilutions can then be made from the homogenate using routine dilution procedures. Dulmage *et al.* (1990) describe making a 'stock' suspension from the above homogenate by adding 0.1 ml of the homogenate to 9.9 ml of water. The stock is then resuspended using a Vortex agitator. Following the procedures of de Barjac & Larget (1979) as used by Thiery & Hamon (1998), subsequent dilutions are made in the container in which the assay will take place by adding 15–120 µl of stock

**Table 4.1** Suggested parameters for bioassay of varieties of *Bacillus thuringiensis* and *Bacillus sphaericus* against mosquito larvae.

| | |
|---|---|
| Dosage of bacteria | mg/l (ppm)* |
| No. of concentrations | 5–7[†] |
| Replicate/control & concentration/test | 4 |
| Replicate tests (separate dates) | 3 |
| Age of larvae | Late 3rd early 4th |
| Duration of test | 24 h |
| Volume of water | 100 ml |
| Source of water | Nonchlorinated[‡] |
| Food | None[§] |

*Weight of primary powder or formulation; with fresh cultures of spores, spore count (after heat shock; 80°C for 12 min) may be the only indication of dosage available; several reasons for avoiding the use of spore count as a measure of dosage are presented by Dulmage *et al.* (1990).

[†]Concentrations should be chosen such that at least two produce mortality between 10 and 50% and at least two produce mortality between 50 and 90%.

[‡]Where deionized water is not available, well water or de-chlorinated (aged and/or aerated) tap water can be substituted.

[§]If very young larvae are used, addition of food may be necessary. Finely ground and suspended (mg/l) Purina Lab Chow, Tetramin, and similar have been used for larval diet. See Skovmand *et al.* (1998) for the effect of diet on susceptibility.

suspension directly into the assay cups. In order to minimize measurement errors, McLaughlin *et al.* (1984) recommended making a series of dilutions to be used for treatments using a minimum of 10 ml of bacterial suspensions for making subsequent dilutions and for applying the diluted suspensions to the bioassay cups.

*b. Handling of insects*

Materials and methods for rearing various species of mosquitoes used for bioassays are presented by Gerberg *et al.* (1994) and will not be covered here. When larvae are in the late 3rd to early 4th instar, they are removed from rearing trays by sieving and placed in chlorine-free water without food. If deionized water is not available, tap water that has been aerated for 24 h will suffice. Twenty late 3rd or 4th instars are transferred to each bioassay container using a Pasteur pipette with the narrow tip removed or an eye dropper with a sufficiently wide opening.

*c. Bioassay containers*

Small (*ca.* 4–5 cm deep) cups holding 100 ml of water will provide an adequate container in which to bioassay bacteria against most species of mosquitoes. Four replicate cups per concentration for each isolate and control are recommended. It is easier to first add larvae and then fill with the balance of the 100 ml of water minus the amount of water in which the application will be made, rather than adding the water first and then subtracting the amount that was added along with the larvae. If 10 ml of bacterial suspension will be added to each container rather than microliter amounts, the containers should only be filled to 90 ml prior to adding 10 ml treatment suspensions.

*d. Application of treatments and incubation of larvae*

Applications of bacterial suspensions should be made in at least 10 ml of water per container starting with the lowest concentration. Between applications of different concentrations agitate containers in which the inoculum is held to avoid setting. Food should not be added to the containers. After the appropriate treatments are added to each container, the larvae are then incubated for 24 h at 27°C in the case of *Ae. aegypti* and *Cx. quinquefasciatus*. Temperature is evenly maintained at CMAVE by placing the cups with larvae in large trays containing 2–3 cm of water. The trays are then set on thermostatically controlled heat tapes within a cabinet closed with a clear plastic curtain. The sensor for the thermostat is placed in one of the trays under water. The protocol for evaluation of *B. sphaericus* at CMAVE used 2nd instar *Cx. quinquefasciatus*. Because some strains of the bacterium require more time to kill, the exposure period was extended to 48 h with the addition of a small amount of larval diet (aquarium fish food) after the first 24 h.

*e. Assessment of mortality*

After 24 h of exposure to Bti, larvae are assessed for mortality. If larvae fail to respond to tapping the side of the assay cup, they are probed with a needle. Lack of response or only very weak response (i.e. the larva is moribund) is scored as dead. It is best to base calculations of % mortality on the number of larvae originally placed in the containers and number of living larvae after the exposure/incubation period due to the fact that dead larvae are often consumed by the survivors.

*f. Considerations for predatory larvae*

The bioassay of bacteria against predatory species of mosquitoes, notably *Toxorhynchites* species and certain species of *Psorophora* and others will necessitate first feeding the bacteria to a prey species, such as *Ae. aegypti*, and then adding the predator. Procedures used for bioassay of *B. sphaericus* and Bti against *Toxorhynchites* spp. are presented by Larget & Charles (1982) and Lacey (1983).

### B. Bioassay of Bti against black fly larvae

All black fly species require a lotic habitat for the larval stage. Their habitats range from small creeks to huge rivers with stream velocities as low as 2 cm/s to more than 2 m/s, respectively. Although they are normally thought of as filter feeders, a range of feeding strategies is observed for the family including filter feeding, grazing (scrapers), deposit feeders, and predators (Currie & Craig, 1987). The larvae of most vector or pestiferous species (i.e. those that warrant control) are filter feeders for a significant amount of the time. They feed on seston by filtering particulates in the size range of 0.09–350 μm from the flowing water with cephalic fans (Ross & Craig, 1980; Currie & Craig, 1987). The parasporal bodies of Bti ($\approx$ 0.75 μm), where the toxic moieties are located, are efficiently captured along with other particles. The combination of their filtering efficiency and susceptibility make black fly larvae ideal targets for Bti, even when the period of exposure is relatively brief (Lacey & Undeen, 1984, 1986).

*1. Factors affecting activity*

Several of the factors that affect the susceptibility of black fly larvae to Bti are the same as those that affect susceptibility of mosquito larvae (species, larval age, temperature, presence of food and other particulates) (Lacey *et al.*, 1978; Molloy *et al.*, 1981; Coupland, 1993). The major difference between bioassays of bacteria against black flies and mosquitoes is the requirement for running water for black fly larvae.

*a. Effect of bioassay system*

A number of bioassay systems have been proposed for the laboratory evaluation of bacteria against larval black flies. The current required for bioassay can be artificially created in a closed system by using a magnetic stirrer and stir bar in a beaker of water (Undeen & Nagel, 1978), or with air bubbles (Lacey & Mulla, 1977) (Figures 4.6 and 4.7) or other artificial methods (Molloy, 1982) or more naturally in systems which utilize flowing water (Gaugler *et al.*, 1980 [Figure 4.8]; Guillet *et al.*, 1985; Atwood *et al.*, 1992; Coupland, 1993). As with mosquito larvae, the most important factor is to enable normal larval feeding rates and avoid excess control mortality. Closed systems in which current is artificially created (Lacey & Mulla, 1997; Undeen & Nagel, 1978) are the least expensive, but the current is not as laminar as that obtained with flowing water. The systems utilized by Gaugler *et al.* (1980) and Coupland (1993) can be operated in both recirculation and flow through modes, thereby prolonging exposure if needed.

A portable system of mini-gutters for evaluation of larvicides including Bti against *Simulium damnosum* s.l. under semi-field conditions is described by Guillet *et al.* (1985). Conceivably this method could also be used in the laboratory, but a constant source of water would be required. Atwood *et al.* (1992) also described a mini-gutter

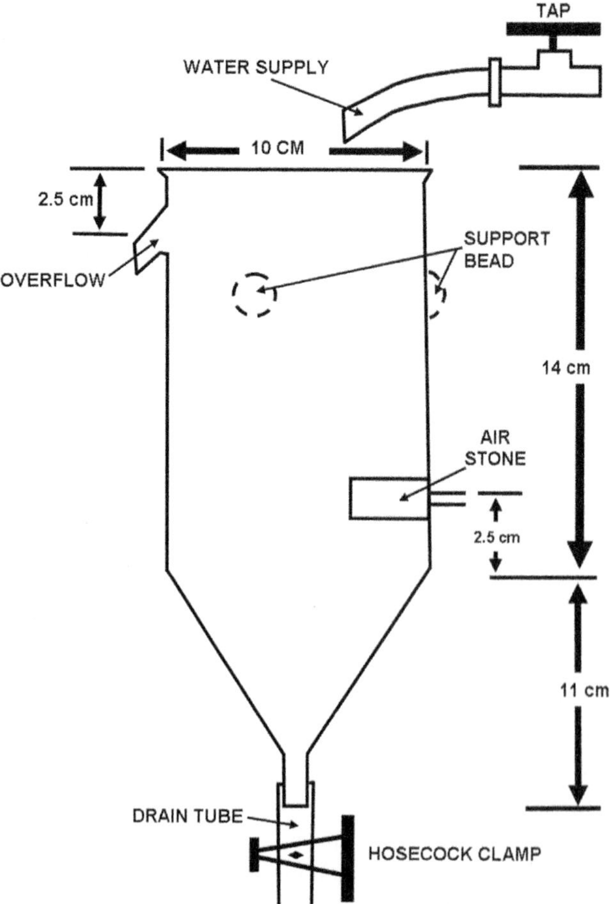


**Figure 4.6** Schematic drawing of a flushing bioassay system. (From Lacey & Mulla, 1977, with permission.)

**Figure 4.7** Assembled flushing bioassay system for evaluation of entomopathogens against black fly larvae. (From Lacey & Mulla, 1977, with permission.)

system for evaluation of Bti against *Cnephia pecuarum*. In all of the systems employed by Gaugler *et al.* (1980), Atwood *et al.* (1992), and Guillet *et al.* (1985), the authors used sources of water from natural habitats. Although this method would provide the microseston on which the larvae feed, different sources of stream water could introduce variable water conditions and comparison of data between different laboratories could be confounded by these variables. In addition to current, other environmental conditions may also have to be met. Species that are normally found in pristine, cold, fast-flowing streams may be difficult to use for laboratory bioassays unless their environmental conditions are adequately simulated.

*b. Effect of larval feeding*

Larval feeding rates can be highly influenced by particle concentration in the water. Feeding efficiency increases at lower particulate concentrations (Kurtak, 1978) and increasing particle concentrations up to an optimum will increase the quantity of food consumed, but an excessive amount of particulates will actually inhibit feeding (Gaugler & Molloy, 1980). If feeding is inhibited before or during exposure to Bti, fewer toxic inclusions will be consumed. If inhibition occurs following ingestion of toxin, mortality can be increased due to the fact that ingested particulates remain stationary in the midgut and hence prolong contact time with receptors on the target epithelium (Gaugler & Molloy, 1980).

*c. Exposure duration*

Length of exposure is another factor that will differ from mosquito bioassays. Under most operational conditions, the time interval that Bti formulations are added to larval breeding sites is fairly brief (for example when applied by aircraft or from 10–30 min when applied at ground level). To more realistically approximate natural conditions, exposure of the larvae to the bacterium in the laboratory should not be too protracted. For a standard bioassay, we suggest 30 min of exposure. Varieties of Bt that produce low amounts of toxins that are active against black flies will require a higher dosage and longer exposure time to produce significant mortality (Lacey *et al.*, 1978).

*2. Bioassay protocol*

The preparation of bacterial suspensions and number of replicates is as described above for mosquito bioassays. Table 4.2 presents some suggested standardized parameters for bioassay of Bti and other Bt varieties against black fly larvae.

*a. Collection of larvae*

Unless one is in one of the few laboratories with a black fly colony, larvae will have to be collected from a breeding site. Getting from the field to the laboratory with larvae of some species may require transport of the larvae in aerated containers of water under cool or

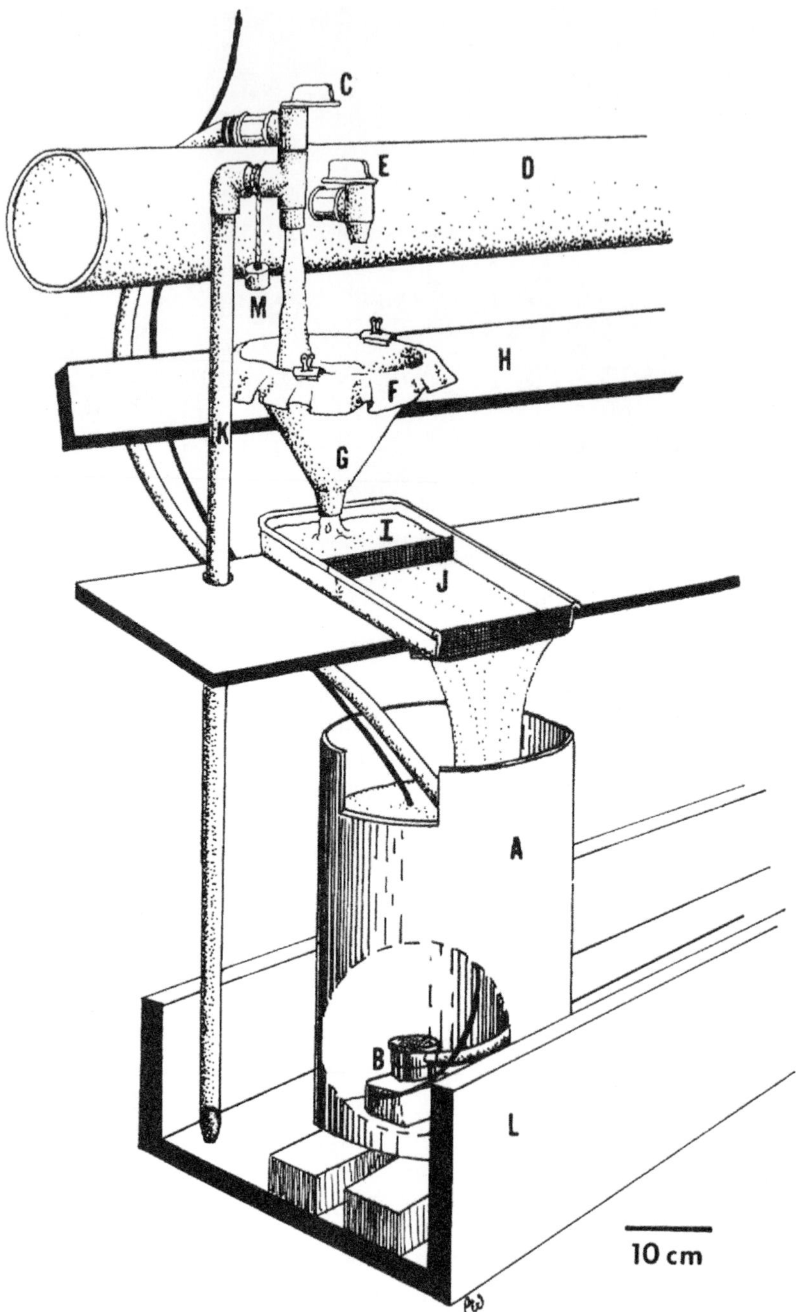


**Figure 4.8** Schematic drawing of a flow through bioassay system. A, reservoir tub; B, recirculation pump; C, recirculation valve; D, water supply pipe; E, supply valve; F, organdy cloth filter; G, delivery funnel; H, funnel support; I, tray reservoir section; J, tray attachment section; K, standing waste pipe; L, waste trough; M, cap. (From Gaugler *et al.*, 1980, with permission.)

cold conditions. Several species, including *Simulium damnosum* s.l. and *S. vittatum* (see Adler *et al.*, 2004 for a revision of this taxon), can be transported for short periods of time on the damp vegetation on which they were collected. Remove excess water from the plants and place loosely in plastic bags in a cool place for transport to the laboratory. An insulated box, such as an ice chest works well. A layer of ice separated from the bags of

**Table 4.2** Parameters for bioassay of *Bacillus thuringiensis* varieties against black fly larvae.

| | |
|---|---|
| Concentration of bacteria | mg/l (ppm)* |
| Duration of exposure | 30 min |
| No. of concentrations | 5–7† |
| Replicate/control & concentration/test | 4 |
| Replicate tests (separate dates) | 3 |
| Age of larvae | Penultimate and early last instar |
| Duration of test | 24 h |
| Volume of water | 1000 ml |
| Source of water | Chlorine free |
| Food | 5 mg/l after exposure‡ |

*Weight of primary powder or formulation; with fresh cultures of spores, spore count (after heat shock; 80°C for 12 min) may be the only indication of dosage available; several reasons for avoiding the use of spore count as a measure of dosage are presented by Dulmage *et al.* (1990).

†Concentrations should be chosen such that at least two produce mortality between 10 and 50% and at least two produce mortality between 50 and 90%.

‡Following the 30-min exposure period, finely ground and suspended (5 mg/l) Purina Lab Chow, Hog Chow, Tetramin, and similar have been used for larval diet.

larvae by a sheet of styrofoam is ideal for ensuring cool, but not excessively cold temperatures.

*b. Preparation of larvae for testing*

In the laboratory, the larvae should be removed from the bags and placed in trays in chlorine-free water. Penultimate instar larvae or young last instars [those which do not have melanized 'gill spots' (organs that will become the pupal respiratory filaments)] can be gently removed using a soft latex eye dropper by nudging the larvae with the dropper near the anal gills at the same time as drawing water into the dropper. Immediately transfer the larva to the container or system in which bioassay will be conducted. Using 20 larvae per replicate container provides sufficient test animals and will avoid the entanglement of larvae in their own silk that one sees when putting a large number of larvae in a container.

Allow larvae to acclimate for approximately 3 h before exposure to candidate bacteria. Prior to adding bacterial suspensions, dead, detached, and pupating larvae should be removed. Usually if healthy penultimate instars are used, the number of larvae that are not attached and actively feeding will be low or zero. If larvae are removed, add additional larvae to bring the count to the desired number for each bioassay container.

*c. Treatment and incubation*

After acclimation, and when larvae are feeding, add bacterial suspensions. In closed bioassay systems, including recirculating systems, the desired concentrations can be added all at once. If a flow-through system is used, the suspension will have to be dripped continuously in the flowing water for the desired exposure period. With isolates containing high amounts of Nematocera-active toxin(s), such as Bti, an exposure period of 30 min is recommended.

After exposure in closed non-flow-through systems, it will be necessary to change the water in which the larvae are exposed. This usually causes detachment of the larvae and care must be taken not to lose any of the larvae when the water is decanted. Detachment will also result in cessation of feeding, which can result in increased mortality. In the system used by Lacey & Mulla (1977) (Figure 4.6), the inoculum is flushed from the bioassay container by adding water to each container from individual taps while it drains from the bottom and overflows (Figure 4.7). Larvae are not exposed to air or cessation of current and hence, do not detach. Flushing of inoculum without larval detachment is ideal when using the flow-through systems such as that of Gaugler *et al.* (1980).

After changing the water, it will be important to add food to enable as close to normal feeding rates as possible. Five parts per million of finely ground lab chow or similar food added as a suspension should enable normal feeding rates. If feeding inhibition is noted (labral fans are held closed for longer intervals than normal) a lower concentration of food should be used.

Larvae are then incubated for at least 24 h before assessment of mortality. A lower temperature (20°C), than that used for mosquitoes is recommended to more realistically approximate stream conditions. After incubation, larvae are observed for signs of life. When in doubt regarding the condition of the larvae, use a probe. Data are treated as described above for mosquito larvae.

One exception may be necessary. Because black fly larvae do not always respond favorably to collection, handling, and the laboratory environment, it may be necessary to accept a higher level of control mortality. Twenty per cent control mortality may have to be the cut-off point for accepting test results. Because of lower temperatures and shorter exposure times, it may be necessary to take a second reading 24 h after the first.

Readers interested in field assessment of bacterial pathogens against simuliid larvae are referred to Molloy (1982) and Skovmand *et al.* (2007). A combination of field exposure and laboratory assessment of Bti formulations against black fly larvae is presented by Lacey & Undeen (1984). The assay of bacteria against other aquatic organisms especially those found in running water can be quite complex depending on their requirements. The system that is chosen should simulate as close as possible the system in which the organism is normally found.

## 4. BIOASSAY OF *BACILLUS THURINGIENSIS* AGAINST OTHER ORDERS

The general descriptions of bioassays above have concentrated on the use of *B. thuringiensis* against Lepidoptera and some Diptera. However, the toxins produced by various strains of Bt can be toxic to a wider range of invertebrates, many of which require different approaches for successful bioassay. Some relevant techniques are covered in Chapter XIII, which covers bioassay of bacteria against slugs and snails. Herein we consider Bt against Coleoptera, Diptera, Orthoptera, Homoptera and Acari, as examples of the other groups containing pest species.

### A. Coleoptera

The finding of Bt strains active against Coleoptera (Krieg *et al.* 1983) was of significance given the large number of pest species in the order. The subspecies was named *Bt tenebrionis* (Btt) and has been shown to produce a Cry3 protein. More coleopteran active Bt have been described since (e.g., Nazarian *et al.*, 2009).

Bioassay of Btt against Coleoptera has included use of leaf dip assays and other standard assays described for Lepidoptera to measure mortality. However, these methods have not produced reliable results for all beetles and their larvae and alternative approaches may be needed. It can be difficult to get uniform coverage of leaves. In addition bioassays that concentrate on mortality ignore the decrease in larval growth from ingestion of toxins. To overcome some of the limitations, de Leon & Ibarra (1995) used an assay that measured the growth rate of *Leptinotarsa texana* larvae (a relative of the Colorado potato beetle, *Leptinotarsa decemlineata*), to bioassay Cry 3 proteins. They found that growth rate was inversely related to ingested dose, so used linear regression of growth rate to estimate a mean effective dose ($ED_{50}$.) The benefit of using weight gain is that it is a continuous variable. In each assay, six inocula were used, differing in concentration by a factor of 0.5. Ten early 3rd instar larvae were used per dose, individually confined in wells of a 24-well plate, to prevent cannibalism.

Individually maintained larvae were first starved, weighed and then fed 2 µl of pure crystal dilution suspensions in 20% sucrose and 0.02% Tween 20 by placing the drops on wax paper and allowing larvae to feed. The insects consumed all the drops, so dose could be closely controlled. After treatment, larvae were moved to plants for monitoring and weighed after 2 days. A growth factor per dose was calculated, using mean final weight over the mean initial weight. No result was used unless the $ED_{50}$ was between the 2nd and 4th of the six doses, and control mortality under 10%. As with most bioassays the authors recommended a minimum of three replicates.

A leaf surface contamination assay was used by Costa *et al.* (2000) to assay to test Btt against Colorado potato beetle. To standardize, they selected 4th instar larvae weighing between 45 and 75 mg. Small leaf discs were given to each individual larva in a Petri dish daily. The amount of leaf eaten was assessed for each day. Btt-treated leaf discs were given to larvae for a 24-h period, or continuously. Leaf discs were contaminated by immersion in Btt solutions for 5 s, then strained and air-dried. By measuring the amount of leaf eaten, and multiplying by the Btt residue on the leaf, the specific dose consumed could be calculated. Residue was calculated by the weight of water loss. Changes in larval weight, mortality, fecundity could be assessed in this assay.

### B. Diptera (Sciaridae)

A bioassay for toxicity testing of Bti against fungus gnats (Diptera: Sciaridae), utilized the attraction of fungus gnat, *Bradysia impatiens*, to potato starch as a food (Taylor *et al.*, 2007). A medium made from dehydrated potato was mixed with suspension of Bti, divided into 30-ml plastic cups containing a wedge of potato on one side of the bottom edge, and the cup was covered. Fungus gnat larvae were transferred to the potato surface and maintained for 24 h after which mortality was assessed under a dissecting microscope.

### C. Orthoptera and Homoptera

Grasshoppers and locusts cause severe damage in many parts of the world, but few tests have used Bt toxins against these species. Peng *et al.* (2003), cited in Song *et al.* (2008) developed a toxicity test for grasshoppers based on immersion of corn seedlings in suspensions of toxins. For locusts, spore and trypsin-digested sporulating suspensions were used. Spore and protein concentration estimates were used to determine dose, respectively, then the samples were diluted in 2% Tween 80. Corn seedlings were immersed in suspensions for about 2 h, before the inoculated seedlings were fed to 30 locusts per treatment. Locusts were then maintained in cages at 28°C.

Sucking insects, such as Homoptera, present specific challenges when dosing with bacteria or toxins. Mora *et al.* (2007) used a bioassay method to dose the rice delphacid *Tagosodes orizicolus;* insects sucked solutions through parafilm to mimic natural feeding on plants. Insects were fed Bt solutions through perforated parafilm membranes. The bacteria and/or toxins were suspended in 10% sucrose and honey (1:48 vol/vol) solutions. Six-centimeter Petri dishes were filled to the top with sugar solution and then covered with parafilm. Insects were placed on the parafilm. Better results were obtained if the parafilm was perforated using a needle before the assay began. A similar bioassay of Bt toxins against the whitefly, *Bemisia tabaci* type B, was presented by Davidson *et al.* (1996).

### D. Acari

Erban *et al.* (2009) developed a method for bioassay of Btt for pests of stored grain, *Tyrophagus putrescentiae* (Acaridae), *Dermatophagoides farinae* (Pyroglyphidae), and *Lepidoglyphus destructor* (Glycyphagidae) using toxins of Btt. They used a diet-based feeding assay, where the commercially available Btt product Novodor FC (Valent BioSciences) was incorporated into a fish food rearing diet. The diet was composed of dog food, wheat germ, dried fish food, pangamin, and gelatin in ratios of 10:10:3:2:1 (w/w). Concentrations of Btt product ranged from 0.001 to 100 mg/g of diet, which was mixed, lyophilized, and ground to a powder. The powder was rehydrated for 24 h before use and 50 mg was fed to 50 mites at 75−85% RH. The number of mites was assessed after 25 days, so the bioassay used population suppression as the measure of efficacy.

### E. β-Exotoxin bioassay

β-Exotoxin is produced by some strains of Bt. It is a secreted thermostable low-molecular-weight adenine nucleotide analog that has some insect and mammalian toxicity. Because of the mammalian toxicity, it is widely regarded as unsuitable for inclusion in biopesticide formulations and the World Health Organization (1999) has recommended that commercial products are free of β-exotoxin. While there are methods to detect β-exotoxin through chromatography, bioassay against fly (*Musca domestica*) larvae is widely used. The test uses autoclaved culture supernatant of Bt fermentations (e.g., Ohba *et al.*, 1981). Diet-based bioassay methods are recommended. Dry diet ingredients are mixed with completely lysed cell cultures, reduced in volume to 10% through rotary evaporation and autoclaved to release exotoxin. The effect is measured as the percentage of larvae reaching pupation.

## 5. BIOASSAY OF OTHER ENTOMOPATHOGENS

### A. Spore-forming bacteria

The basic principles of bioassay used for Bt are often applicable to related insect pathogens, especially the spore-forming bacilli in genera such as *Lysinibacillus* and *Brevibacillus. Brevibacillus laterosporus*, for example, also produces vegetative cells, spores and sometimes parasporal bodies, so studies have used the methods described above to separate the stages for subsequent assay. Bioassay techniques used with *Brevibacillus* include: dilution of cell fractions, supernatant or spores in water containing mosquito larvae (Favret & Yousten, 1985); surface contamination of diet for larvae of the cabbage looper, *Trichoplusia ni* (Favret & Yousten, 1985; Ruiu *et al.* 2006); incorporation in sucrose solutions (e.g., 30% sucrose) for inoculation of larvae and adult houseflies (Ruiu *et al.*, 2006, 2007); diet incorporation for caterpillars (Oliveira *et al.*, 2004); mixing bacterial suspensions with wheat meal for the coleopteran *Tenebrio molitor* (Oliveira *et al.*, 2004); and for the mollusc *Biomphalaria glabrata*, treatment involved the addition of dried biomass (final concentration 50 mg/l) to plastic cups, with each cup containing 100 molluscs (Oliveira *et al.*, 2004).

### B. Non-spore-forming bacteria

Methods for assay of whole non-spore-forming bacteria against insects are simpler than those for spore-formers, in that only one stage needs to be considered. However, conditions of growth and age of cells can influence the virulence and production of toxins. Two basic methods have been used to determine the insecticidal potential of non spore-forming bacteria: the oral application of viable cells and the subcutaneous injection of protein extracts or living bacterial cells (Fuchs *et al.*, 2008).

#### *1. Hemolymph injection*

Injection of a known quantity of bacterial cells or toxin protein directly into the hemolymph is a standard technique to determine activity against insects (Jackson & Saville, 2000). Generally the technique is used with large caterpillars, such as *Manducta sexta* or *Galleria mellonella*. It can be useful to identify insecticidal activity but there are limitations as insect defense barriers such as the gut lining and peritrophic membrane are bypassed. Small aliquots, typically 5 µl of suspension containing cells or proteins of interest, are injected subcutaneously into larvae. A glass microsyringe capable of autoclaving is often used. Specialized apparatus are available to control precisely the amount of inoculum being injected (see Chapter V).

#### *2. Oral inoculation*

The methods for oral inoculation (also called the droplet method) are the same or similar to those described for Bt above, such as surface contamination or diet incorporation. Hurst *et al.* (2011) applied 20 µl of bacterial solution or proteins into an air dried 300-mm$^3$ section of general-purpose laboratory diet, and allowed the solution to absorb and rehydrate the matrix before providing it to larvae of several caterpillar species. Similarly, oral inoculation of *M. sexta* with bacteria was achieved by allowing 50-µl cultures to soak into an agar-based artificial diet before air drying the diet and providing it to larvae (Fuchs *et al.*, 2008).

Where diet is not available, leaf surfaces can be contaminated with bacterial or protein solutions and provided to larvae. Hurst *et al.* (2011) found that the addition of a spreader (0.02 % Dut, Elliot Chemicals Ltd) improved the coverage of bacterial solution on cabbage leaf discs for inoculation of diamondback moth larvae. These methods do not generally allow precise quantification of dose imbibed. A simpler version is to deprive larvae of diet or moisture and then place it in the centre of a ring of inoculation drops for which the exact quantity of inoculum is known. By calculating the amount of drops consumed by a larva (from number of drops remaining), dose can be estimated. Incorporation of food coloring or analine blue mixed with the suspension can improve visualization that an insect has consumed dose.

## 6. CONCLUSIONS

Data from laboratory bioassays can provide useful information on species susceptibility, (including development of field resistance), and for comparison of efficacy of bacterial isolates, particularly when new isolates are compared to an established standard. However, care must be taken when using laboratory-derived data to make predictions of relative activity under field conditions. For example, ITU ratings based on bioassays against *Ae. aegypti* do not always correspond to relative efficacy of Bti formulations against field populations of mosquito larvae (Dame *et al.*, 1981). Similarly, the ITU ratings for Bti formulations as determined with mosquito larvae are often not correlated with activity against black fly larvae in laboratory bioassays (Molloy *et al.*, 1984) nor under field conditions (Lacey & Undeen, 1984). Even laboratory bioassays of formulated Bti conducted with black fly larvae could provide misleading information related to relative efficacy under field conditions. Molloy *et al.* (1984) observed a positive correlation between particle size and efficacy, yet under field conditions, formulations with smaller particle sizes provide greater effective carry (i.e. kill black fly larvae for a greater distance downstream) due to settling of larger particles.

It is difficult to generalize about the advantages and disadvantages of bioassay methods because of the range of agents to test and insects to target. Toxin-producing bacteria provide different challenges to bacteria that kill through infection. Larvae and adult stages need different methods and the requirement to ingest cells or toxins requires knowledge of the insect feeding.

There are many sources of variation in bacterial bioassays against insects. Even where conditions are apparently similar, variation in outcome has been found. Robertson *et al.* (1995) repeated bioassays using Btt against *L. decemlineata* for 83 weeks and Btk against *Plutella xylostella* for 37 weeks and found significant variation between bioassays. Variation in the insect condition or intraspecific genetic differences (including resistance development) can confound results. Measuring the actual dose ingested as opposed to the offered dose is also a major difficulty. Many techniques outlined use inoculation methods which measure the amount of bacteria and/or toxins only prior to introduction of insects, not the level consumed. Also, precise quantification of the lethal component of the bacterial suspension is also difficult to measure and methods such as spore concentration to estimate dose are sometimes not related directly to toxin levels. An alternative method for assessment of toxin quantity has been suggested by Oestergaard *et al.* (2007), who quantified Bti toxin content using monoclonal antibodies as a means of quality control of commercial products. However, Shisa *et al.* (2002) showed discrepancies between cry gene-predicted and bioassay-determined insecticidal activities in Bt natural isolates.

Despite differences in the manner in which toxicity of bacteria are assessed, data from laboratory bioassays provides useful information on species susceptibility, including development of resistance, and allows comparison of bacterial isolates, particularly when primary powders of bacterial strains are compared to a standard. However, once effective isolates are identified in laboratory bioassays, the most realistic evaluation of formulated bacterial microbial control agents will always be under field conditions.

## ACKNOWLEDGMENTS

We thank Mark Hurst, Sandra Jones, Pauline Hunt, Richard Townsend, and Heather Headrick for their assistance with the figures.

## REFERENCES

Abbott, W. S. (1925). A method of computing the effectiveness of an insecticide. *J. Econ. Entomol., 18*, 265–267.

Adler, P. H., Currie, C., & Wood, D. M. (2004). *The Black Flies (Simuliidae) of North America*. Ithaca, NY: Cornell University Press. pp. 941.

Ali, A., Baggs, R. D., & Stewart, J. P. (1981). Susceptibility of some Florida chironomids and mosquitoes to various formulations of *Bacillus thuringiensis* serovar. *israelensis*. *J. Econ. Entomol., 74*, 672–677.

Atwood, D. W., Robinson, J. V., Meisch, M. V., Olson, J. K., & Johnson, D. R. (1992). Efficacy of *Bacillus thuringiensis* var. *israelensis* against larvae of the southern buffalo gnat, *Cnephia pecuarum* (Diptera: Simuliidae), and the influence of water temperature. *J. Amer. Mosq. Control Assoc., 8*, 126–130.

Bailey, W. D., Brownie, C., Bacheler, J. S., Gould, F., Kennedy, G. G., Sorenson, C. E., & Roe, R. M. (2001). Species diagnosis and *Bacillus thuringiensis* resistance monitoring of *Heliothis virescens* and *Helicoverpa zea* (Lepidoptera: Noctuidae) field strains from the southern United States using feeding disruption bioassays. *J. Econ. Entomol., 94*, 76–85.

Bailey, W. D., Zhao, G., Carter, L. M., Gould, F., Kennedy, G. G., & Roe, R. M. (1998). Feeding disruption bioassay for species and *Bacillus thuringiensis* resistance diagnosis for *Heliothis virescens* and *Helicoverpa zea* in cotton (Lepidoptera: Noctuidae). *Crop Prot., 17*, 591–598.

Barjac, H., & de Larget, I. (1979). Proposals for the adoption of a standardized bioassay method for the evaluation of insecticidal formulations derived from serotype H.14 of *Bacillus thuringiensis*. *Wld. Hlth. Org. mimeo. doc., WHO/VBC/79. 744*. pp. 16.

Bateman, R. (1999). Delivery systems and protocols for biopesticides. In F. R. Hall, & J. J. Menn (Eds.), *Biopesticides, Use and Delivery* (pp. 509–528). Totowa, NJ: Humana Press.

Beegle, C. C., & Yamamoto, T. (1992). History of *Bacillus thuringiensis* Berliner research and development. *Can. Entomol., 124*, 587–616.

Bird, L. J., & Akhurst, R. J. (2007). Variation in susceptibility of *Helicoverpa armigera* (Hubner) and *Helicoverpa punctigera* (Wallengren) (Lepidoptera: Noctuidae) in Australia to two *Bacillus thuringiensis* toxins. *J. Invertebr. Pathol., 94*, 84–94.

Braun, S. (2000). Production of *Bacillus thuringiensis* insecticides for experimental uses. In A. Navon, & K. R. S. Ascher (Eds.), *Bioassays of Entomopathogenic Microbes and Nematodes* (pp. 49–72). Wallingford, UK: CABI Publishing.

Burges, H. D., & Jones, K. A. (1998). Formulation of bacteria, viruses and protozoa to control insects. In H. D. Burges (Ed.), *Formulation of Microbial Biopesticides: Beneficial Microorganisms, Nematodes and Seed treatments* (pp. 33–127). Dordrecht, The Netherlands: Kluwer Academic Publishers.

Carpenter, J. E., & Bloem, S. (2002). Interaction between insect strain and artificial diet in diamondback moth development and reproduction. *Entomol. Exper. Appl., 102*, 283–294.

Carson, A. C. (2003). *Insect Diets: Science and Technology,* CRC Press. pp. 344.

Coupland, J. B. (1993). Factors affecting the efficacy of three commercial formulations of *Bacillus thuringiensis* var. *israelensis* against species of European black flies. *Biocontrol Sci. Technol., 3*, 199–210.

Costa, S. D., Barbercheck, M. E., & Kennedy, G. G. (2000). Sublethal acute and chronic exposure of Colorado potato beetle (Coleoptera: Chrysomelidae) to the delta–endotoxin of *Bacillus thuringiensis. J. Econ. Entomol., 93*, 680–689.

Currie, D. C., & Craig, D. A. (1987). Feeding strategies of larval black flies. In K. C. Kim, & R. W. Merritt (Eds.), *International Symposium on Ecology and Population Management of Black Flies* (pp. 155–170). University Park, PA: Pennsylvania State University Press.

Dadd, R. H. (1970). Comparison of rates of ingestion of particulate solids by *Culex pipiens* larvae: phagostimulant effect of water-soluble yeast extract. *Entomol. Exp. Appl., 13*, 407–419.

Dame, D., Savage, K., Meisch, M., & Oldacre, S. (1981). Assessment of industrial formulations of *Bacillus thuringiensis* var. *israelesis. Mosq. News, 41*, 540–546.

Davidson, E. W., Patron, R. B. R., Lacey, L. A., Frutos, R., Vey, A., & Hendrix, D. L. (1996). Activity of natural toxins against the silverleaf whitefly, *Bemisia argentifolii*, using a novel feeding bioassay system. *Entomol. Exp. Appl., 79*, 25–32.

Dulmage, H. D., Boening, O. P., Rehnborg, C. S., & Hansen, D. G. (1971). A proposed standardized bioassay for formulations of *Bacillus thuringiensis* Berliner. *Environ. Entomol., 19*, 182–189.

Dulmage, H. T., McLaughlin, R. E., Lacey, L. A., Couch, T. L., Alls, R. T., & Rose, R. I. (1985). HD-968-S-1983, a proposed U. S. standard for bioassays of preparations of *Bacillus thuringiensis* subsp. *israelensis*-H-14. *Bull. Entomol. Soc. Am., 31*, 31–34.

Dulmage, H., Yousten, A., Singer, S., & Lacey, L. (1990). Guidelines for the production of *Bacillus thuringiensis* (H-14) and *Bacillus sphaericus. Wld. Hlth. Org. TDR Prog.*, 88.

Erban, T., Nesvorna, M., Erbanova, M., & Hubert, J. (2009). *Bacillus thuringiensis* var. *tenebrionis* control of synanthropic mites (Acari: Acaridida) under laboratory conditions. *Exper. Appl. Acarol., 49*, 339–346.

Favret, M. E., & Yousten, A. A. (1985). Insecticidal activity of *Bacillus laterosporus. J. Invertebr. Pathol., 45*, 195–203.

Finney, D. J. (1971). *Probit Analysis*. Cambridge, UK: Cambridge University Press.

Fuchs, T. M., Bresolin, G., Marcinowski, L., Schachtner, J., & Scherer, S. (2008). Insecticidal genes of *Yersinia* spp.: taxonomical distribution, contribution to toxicity towards *Manduca sexta* and *Galleria mellonella*, and evolution. *BMC Microbiol., 8*, 214.

Garczynski, S. F., & Siegel, J. P. (2007). Bacteria. In L. A. Lacey, & H. K. Kaya (Eds.), *Field Manual of Techniques in Invertebrate Pathology: Application and Evaluation of Pathogens for Control of Insects and other Invertebrate Pests* (2nd ed.). (pp. 175–197) Dordrecht, The Netherlands: Springer Scientific Publishers.

Gaugler, R., & Molloy, D. (1980). Feeding inhibition in black fly larvae (Diptera: Simuliidae) and its effects on the pathogenicity of *Bacillus thuringiensis* var. *israelensis. Environ. Entomol., 9*, 704–708.

Gaugler, R., Molloy, D., Haskins, T., & Rider, G. (1980). A bioassay system for the evaluation of black fly (Diptera: Simuliidae) control agents under simulated stream conditions. *Can. Entomol., 112*, 1271–1276.

Gerberg, E. J., Barnard, D. R., & Ward, R. A. (1994). Manual for mosquito rearing and experimental techniques. *Amer. Mosq. Control Assoc. Bull., 5*, 1–98.

Glare, T. R., & O'Callaghan, M. (2000). *Bacillus thuringiensis: Biology, Ecology and Safety.* Chichester, UK: John Wiley and Sons. 350 pp.

Guillet, P., Hougard, J. M., Doannio, J., Escaffre, H., & Duval, J. (1985). Evaluation de la sensibilité des larves du complexe *Simulium damnosum* à la toxine de *Bacillus thuringiensis* H 14. 1. Méthodologie. *Cah. ORSTOM, sér. Ent. méd. Parasitol., 23*, 241–250.

Hurst, M. R. H., Jones, S. A., Binglin, T., Harper, L., & Glare, T. R. (2011). The main virulence determinants of *Yersinia entomophaga* MH96 is a broad host range insect active, toxin complex. *J. Bacteriol., 193*, 1966–1980.

Jackson, T. A., & Saville, D. J. (2000). Bioassays of replicating bacteria against soil-dwelling insect pests. In A. Navon, & K. R. S. Ascher (Eds.), *Bioassays of Entomopathogenic Microbes and Nematodes* (pp. 73–93). Wallingford, UK: CABI Publishing.

Keena, M. A., Vandel, A., & Pultar, O. (2010). Phenology of *Lymantria monacha* (Lepidoptera: Lymantriidae) laboratory reared on spruce foliage or a newly developed artificial diet. *Ann. Entomol. Soc. Am., 103*, 949–955.

Krieg, V. A., Huger, A. M., Langenbrunch, G. A., & Schnetter, W. (1983). *Bacillus thuringiensis* var. *tenebrionis*, a new pathotype effective against larvae of Coleoptera. *Z. Angew. Entomol., 96*, 500–508.

Kurtak, D. C. (1978). Efficiency of filter feeding of black fly larvae (Diptera: Simuliidae). *Can. J. Zool., 56*, 1608–1623.

Lacey, L. A. (1983). Larvicidal activity of *Bacillus* pathogens against *Toxorhynchites* mosquitoes (Diptera: Culicidae). *J. Med. Entomol., 20*, 620–624.

Lacey, L. A. (2007). *Bacillus thuringiensis* serovariety *israelensis* and *Bacillus sphaericus* for mosquito control. In T. G. Floore (Ed.), *Biorational Control of Mosquitoes. Bull. Amer. Mosq. Control Assoc., 7* (pp. 133–163).

Lacey, L. A., & Arthurs, S. P. (2005). New method for testing solar sensitivity of commercial formulations of the granulosis virus of codling moth (*Cydia pomonella*, Tortricidae: Lepidoptera). *J. Invertebr. Pathol., 90*, 85–90.

Lacey, L. A., & Merritt, R. W. (2003). The safety of bacterial microbial agents used for black fly and mosquito control in aquatic environments. In H. M. T. Hokkanen, & A. E. Hajek

(Eds.), *Environmental Impacts of Microbial Insecticides: Need and Methods for Risk Assessment* (pp. 151–168). Dordrecht, The Netherlands: Kluwer Academic Publishers.

Lacey, L. A., & Mulla, M. S. (1977). A new bioassay unit for evaluating larvicides against blackflies. *J. Econ. Entomol., 70*, 453–456.

Lacey, L. A., & Mulla, M. S. (1990). Safety of *Bacillus thuringiensis* (H-14) and *Bacillus sphaericus* to non-target organisms in the aquatic environment. In M. Laird, L. A. Lacey, & E. W. Davidson (Eds.), *Safety of Microbial Insecticides* (pp. 169–188). Boca Raton, FL: CRC Press.

Lacey, L. A., & Undeen, A. H. (1984). Effect of formulation, concentration and application time on the efficacy of *Bacillus thuringiensis* (H-14) against black fly (Diptera: Simuliidae) larvae under natural conditions. *J. Econ. Entomol., 77*, 412–418.

Lacey, L. A., & Undeen, A. H. (1986). Microbial control of black flies and mosquitoes. *Annu. Rev. Entomol., 31*, 265–296.

Lacey, L. A., Mulla, M. S., & Dulmage, H. D. (1978). Some factors affecting the pathogenicity of *Bacillus thuringiensis* Berliner against blackflies. *Environ. Entomol., 7*, 583–588.

Larget, I., & Charles, J.-F. (1982). Étude de l'activité larvicide de *Bacillus thuringiensis* variété *israelensis* sur les larves de Toxorhynchitinae. *Bull. Soc. Pathol. Exp., 75*, 121–130.

Leon, de, T., & Ibarra, J. E. (1995). Alternative bioassay technique to measure activity of Cry III proteins of *Bacillus thuringiensis*. *J. Econ. Entomol., 88*, 1596–1601.

Li, S. Y., Fitzpatrick, S. M., & Isman, M. B. (1995). Effect of temperature on toxicity of *Bacillus thuringiensis* to the oblique banded leafroller (Lepidoptera: Tortricidae). *Canad. Entomol., 127*, 271–273.

Lin, D.-Q., Yao, S.-J., Mei, L.-H., & Zhu, Z.-Q. (2003). Collection and purification of parasporal crystals from *Bacillus thuringiensis* by aqueous two-phase extraction. *Separ. Sci. Technol., 38*, 1665–1680.

Marcus, R., & Eaves, D. M. (2000). Statistical and computational analysis of bioassay data. In A. Navon, & K. R. S. Ascher (Eds.), *Bioassays of Entomopathogenic Microbes and Nematodes* (pp. 249–293). Wallingford, UK: CABI Publishing.

McGuire, M. R., Behle, R. W., Goebel, H. N., & Fry, T. C. (2000). Calibration of a sunlight simulator for determining solar stability of *Bacillus thuringiensis* and *Anagrapha falcifera* nuclear polyhedrosis. *Environ. Entomol., 29*, 1070–1074.

McGuire, M. R., Galan-Wong, L. J., & Tamez-Guerra, P. (1997). Bacteria: Bioassay of *Bacillus thuringiensis* against lepidopteran larvae. In L. Lacey (Ed.), *Manual of Techniques in Insect Pathology* (pp. 91–99). San Diego, CA: Academic Press.

McLaughlin, R. E., Dulmage, H. T., Alls, R., Couch, T. L., Dame, D. A., Hall, I. M., Rose, R. I., & Versoi, P. L. (1984). U.S. standard bioassay for the potency assessment of *Bacillus thuringiensis* serotype H-14 against mosquito larvae. *Bull. Entomol. Soc. Amer., 30*, 26–29.

Merritt, R. W., & Craig, D. A. (1987). Larval mosquito (Diptera: Culicidae) feeding mechanisms: mucosubstance production for capture of fine particles. *J. Med. Entomol., 24*, 275–278.

Molloy, D. (1982). Biological control of black flies (Diptera: Simuliidae) with *Bacillus thuringiensis* var. *israelensis* (Serotype 14): A review with recommendations for laboratory and field protocols. *Misc. Publ. Entomol. Soc. Am., 12*, 30.

Molloy, D., Gaugler, R., & Jamnback, H. (1981). Factors influencing efficacy of *Bacillus thuringiensis* var. *israelensis* as a biological control agent of black fly larvae. *J. Econ. Entomol, 74*, 61–64.

Molloy, D., Wraight, S. P., Kaplan, B., Geradi, J., & Peterson, P. (1984). Laboratory evaluation of commercial formulations of *Bacillus thuringiensis* var. *israeliensis* against mosquito and black fly larvae. *J. Agric. Entomol., 1*, 161–168.

Mora, R., Ibarra, J. E., & Espinoza, A. M. (2007). A reliable bioassay procedure to evaluate per os toxicity of *Bacillus thuringiensis* strains against the rice delphacid, *Tagosodes orizicolus* (Homoptera: Delphacidae). *Rev. Biol. Trop. (Int. J. Trop. Biol.), 55*, 373–383.

Morris, O. N. (1973). Dosage–mortality studies with commercial *Bacillus thuringiensis* sprayed in a modified Potter's tower against some forest insects. *J. Invertebr. Pathol., 22*, 108–114.

Navon, A. (2000). Bioassays of *Bacillus thuringienis* products used against agricultural pests. In A. Navon, & K. R. S. Ascher (Eds.), *Bioassays of Entomopathogenic Microbes and Nematodes* (pp. 1–24). Wallingford, UK: CABI Publishing.

Navon, A., Klein, M., & Braun, S. (1990). *Bacillus thuringiensis* potency bioassays against *Heliothis armigera, Earias insulana*, and *Spodoptera littoralis* larvae based on standardized diets. *J. Invertebr. Pathol., 55*, 387–393.

Nazarian, A., Jahangari, R., Jouzani, G. S., Seifinejad, A., Soheilivand, S., Bagheri, O., Keshavarzi, M., & Alamisaeid, K. (2009). Coleopteran-specific and putative novel *cry* genes in Iranian native *Bacillus thuringiensis* collection. *J. Invertebr. Pathol., 102*, 101–109.

Ohba, M., Tantichodok, A., & Aizawa, K. (1981). Production of heat-stable exotoxin by *Bacillus thuringiensis* and related bacteria. *J. Invertebr. Pathol., 38*, 26–32.

Oliveira, E. J., Rabinovitch, L., Monnerat, R. G., Passos, L. K., & Zahner, V. (2004). Molecular characterization of *Brevibacillus laterosporus* and its potential use in biological control. *Appl. Environ. Microbiol., 70*, 6657–6664.

Oestergaard, J., Voss, S., Lange, H., Lemke, H., Strauch, O., & Ehlers, R. U. (2007). Quality control of *Bacillus thuringiensis* ssp. *israelensis* products based on toxin quantification with monoclonal antibodies. *Biocontrol Sci. Technol., 17*, 295–302.

Robertson, J. L., Preisler, H. K., Ng, S. S., Hickle, L. A., & Gelernter, W. D. (1995). Natural variation: a complicating factor in bioassays with chemical and microbial pesticides. *J. Econ. Entomol., 88*, 1–10.

Ross, D. H., & Craig, D. A. (1980). Mechanisms of fine particle capture by larval black flies (Diptera: Simuliidae). *Can. J. Zool., 58*, 1186–1192.

Ruiu, L., Delrio, G., Ellar, D. J., Floris, I., Paglietti, B., Rubino, S., & Satta, A. (2006). Lethal and sublethal effects of

*Brevibacillus laterosporus* on the housefly (*Musca domestica*). *Entomol. Exper. Appl., 118*, 137–144.

Ruiu, L., Floris, I., Satta, A., & Ellar, D. J. (2007). Toxicity of a *Brevibacillus laterosporus* strain lacking parasporal crystals against *Musca domestica* and *Aedes aegypti*. *Biol. Control, 43*, 136–143.

Schneider, J. C. (Ed.). (2009). *Principles and Procedures for Rearing High Quality Insects*. Mississippi State University. pp. 352.

Shisa, N., Wasano, N., & Ohba, M. (2002). Discrepancy between *cry* gene-predicted and bioassay-determined insecticidal activities in *Bacillus thuringiensis* natural isolates. *J. Invertebr. Pathol., 81*, 59–61.

Sinègre, G., Gaven, B., & Jullien, J. L. (1981a). Contribution à la normalisation des épreuves de laboratoire concernant des formulations expérimentales et commerciales du sérotype H-14 de *Bacillus thuringiensis*. III. Influence séparée ou conjointe de la densité larvaire, du volume ou profondeur de l'eau et de la présence de terre sur l'efficacité et l'action larvicide résiduelle d'une poudre primaire. *Cah. ORSTOM, sér. Ent. méd. Parasitol., 19*, 157–163.

Sinègre, G., Gaven, B., & Vigo, G. (1981b). Contribution à la normalisation des épreuves de laboratoire concernant des formulations expérimentales et commerciales du sérotype H-14 de *Bacillus thuringiensis*. II. Influence de la température, du chlore résiduel, du pH et de la profondeur de l'eau sur l'activité biologique d'une poudre primaire. *Cah. ORSTOM, sér. Ent. méd. Parasitol., 19*, 149–155.

Skovmand, O., Hoegh, D., Pedersen, H. S., & Rasmussen, T. (1997). Parameters influencing potency of *Bacillus thuringiensis* var. *israelensis* products. *J. Econ. Entomol., 90*, 361–369.

Skovmand, O., Kerwin, J., & Lacey, L. A. (2007). Microbial control of mosquitoes and black flies. In L. A. Lacey, & H. K. Kaya (Eds.), *Field Manual of Techniques in Invertebrate Pathology: Application and Evaluation of Pathogens for Control of Insects and other Invertebrate Pests* (2nd ed.). Dordrecht, The Netherlands: Springer. pp. 735–750.

Skovmand, O., Thiery, I., Benzon, G. L., Sinegre, G., Monteny, N., & Becker, N. (1998). Potency of products based on *Bacillus thuringiensis* var. *israelensis*: interlaboratory variations. *J. Amer. Mosq. Control Assoc., 14*, 298–304.

Song, L., Gao, M., Dai, S., Wu, Y., Yi, D., & Li, R. (2008). Specific activity of a *Bacillus thuringiensis* strain against *Locusta migratoria manilensis*. *J. Invertebr. Pathol., 98*, 169–176.

Taylor, M. D., Willey, R. D., & Noblet, R. (2007). A 24-h potato-based toxicity test for evaluating *Bacillus thuringiensis* var. *israelensis* (H-14) against darkwinged fungus gnat *Bradysia impatiens* Johannsen (Diptera: Sciaridae) larvae. *Intern. J. Pest Manag., 53*, 77–81.

Thiery, I., & Hamon, S. (1998). Bacterial control of mosquito larvae: investigation of stability of *Bacillus thuringiensis* var. *israelensis* and *Bacillus sphaericus* standard powders. *J. Amer. Mosq. Control Assoc., 14*, 472–476.

Undeen, A. H., & Nagel, W. L. (1978). The effect of *Bacillus thuringiensis* ONR-60A (Goldberg) on *Simulium* larvae in the laboratory. *Mosq. News, 38*, 524–527.

van Frankenhuyzen, K. (1990). Effect of temperature and exposure time on toxicity of *Bacillus thuringiensis* Berliner spray deposits to spruce budworm, *Choristoneura fumiferana* Clemens (Lepidoptera: Tortricidae). *Canad. Entomol., 122*, 69–75.

van Frankenhuyzen, K., & Gringorten, J. L. (1991). Frass failure and pupation failure as quantal measurements of *Bacillus thuringiensis* toxicity to Lepidoptera. *J. Invertebr. Pathol., 58*, 465–467.

Vidal, C., Lacey, L. A., & Fargues, J. (1997). Pathogenicity of *Paecilomyces fumosoreus* (Deuteromycotina: Hyphomycetes) against *Bemisia argentifolii* (Homoptera: Aleyrodidae) with a description of a bioassay method. *J. Econ. Entomol., 90*, 765–772.

World Health Organisation. (1999). *Guideline specifications for bacterial larvicides for public health use. WHO Document WHO/CDS/CPC/WHOPES/99.2*. Geneva, Switzerland: World Health Organization.

Wraight, S. P., Molloy, D., & Singer, S. (1987). Studies on the Culicine mosquito host range of *Bacillus sphaericus* and *Bacillus thuringiensis* var. *israelensis* with notes on the effects of temperature and instar on bacterial efficacy. *J. Invertebr. Pathol., 49*, 291–302.

## APPENDIX: ARTIFICIAL DIET RECIPES

### A. General purpose diet for rearing and bioassay of lepidopteran insects (Dulmage *et al.*,1971)

**Group 1:** Distilled water (heated to boiling), 1200 ml; 4 M KOH, 18 ml; casein, 126 g; alfalfa meal (entomological grade), 54 g; sucrose, 126 g; Wesson salts mixture, 36 g; alphacel, 18 g; wheatgerm, 108 g; ascorbic acid, 14.5 g; aureomycin (250 mg/capsule), 0.5 g; 15% methyl-*p*-hydroxybenzoate solution, 35 ml; 10% choline chloride solution, 36 ml; 10% formaldehyde solution, 13 ml. For modifications of this recipe, please see McGuire *et al.*, 1997.

**Group 2:** Agar, granulated, dissolved in 2500 ml boiling distilled water, 90 g.

**Group 3:** Vitamin solution A: nicotinic acid amide, 12.0 g; calcium pantothenate, 12.0 g; thiamine hydrochloride, 3.0 g; pyridoxine hydrochloride, 3.0 g; biotin, 0.24 g; vitamin $B_{12}$, 0.024 g; distilled water, 1000 ml.

**Group 4:** Vitamin solution B: riboflavin, 6.0 g; folic acid, 3.0 g; Dissolve in solution of 2.24 g of KOH in 1000 ml distilled water.

Prepare Group 1 in a 1-gallon Waring blender equipped with variable speed transformer; add the components in the order listed with the blender operating at very low speed. Add group 2, cool to about 60–65°C, and add groups 3 and 4. Adjust transformer to maximum output and continue blending at slow speed for 2 min. Final pH of diet will be about 5.2.

### B. Diet for bioassay of third instar larvae of *Helicoverpa armigera, Earias insulana* and *Spodoptera littoralis* (Navon *et al.*,1990)

**Group 1:** Seed products/condensed food*, 16%; B vitamins solution†, 10.5%; cholesterol, 0.05%; choline chloride 0.09%; cellulose powder, 1.67%; sorbic acid, 0.03%; homogenize ingredients in half of the deionized water needed to make the volume of diet needed.

**Group 2:** Agar, 16.7%; methyl-*p*-hydroxybenzoate (nipagin), 0.19%; dissolved together in half of the water by heating in an autoclave.

The hot agar/nipagin solution (Group 2) and the nutrient homogenate (Group 1) are mixed together and formaldehyde (0.22%) and L-ascorbic acid (0.38%) are added to the diet maintained at 50°C.

**Note:** slightly different amounts of some ingredients are used when the diet is made for bioassay of neonate larvae (see Navon *et al.*, 1990, for details).

### C. General purpose diet suitable for rearing of many insect species (Singh, 1983)

**Group 1:** Agar, 2.5%; casein (80 mesh), 3.5 %; cellulose powder, 10%; Wesson's salt, 1%; wheatgerm (ground into a fine powder), 3%. Ingredients mixed together in the order listed in a Waring blender to give a fine mixture.

**Group 2:** Cholesterol, 0.05%; linoleic acid, 0.25% dissolved in dichloromethane at approximately 10 ml/g. Group 2 ingredients are then mixed well with the powdered ingredients of Group 1and the dichloromethane is then allowed to evaporate off under a fumehood for 24 h.

*Types and amounts of seed product/condensed food used in the diet varies depending on insect species. For *Spodoptera littoris* and *Helicoverpa armigera*, beans and whole milk powder are used at a ratio of 1:6; For *Earias insulana*, beans, whole milk powder and cotton seed protein were used at a ratio of 4.5:1:1.5.

†B vitamin solution: *i*-inositol, 4600 mg; pantothenic acid, 850 mg; niacin 300 mg; *p*-amino-benzoic acid, 154 mg; riboflavin, 24 mg; pyridoxine, 38 mg; thiamine, 28 mg; folic acid, 12 mg; biotin 4 mg; vitamin B12, 2 mg; all dissolved in 1 l distilled water.

**Group 3:** Distilled water, 65.0%; potassium hydroxide, 4 M 0.50%.

**Group 4:** Vanderzant vitamin mix*, 2%; sucrose (commercial grade), 3%; glucose, 0.5%; distilled water, 7.17%; streptomycin sulphate BP, 0.015%; benzylpenicillin (sodium) BP, 0.015%. Vitamins and sugars are dissolved in distilled water to make volume of 1000 ml.

**Group 5:** Methyl-parahydroxylbenzoate 15 g, sorbic acid 20 g, and 175 ml commercial-grade ethanol mixed together; 15 ml of this solution used to make 1 kg diet.

To prepare 1 kg of diet, 200 g of ingredients listed in Groups 1 and 2 are placed in a large beaker. Distilled water (650 ml) is then mixed with the dry ingredients and mixed thoroughly. The pH of the diet is adjusted to 6.5 through the addition of 5 ml 4M KOH at this stage. The diet is then covered and autoclaved for 20 min. The diet mixture is cooled to 70°C before the addition of 71.7 ml of the sugar and vitamin solution (Group 4), followed by 15 ml of the mould inhibitor (Group 5). Following thorough mixing, the diet can then be dispensed into appropriate containers.

### D. *Helicoverpa armigera* (Akhurst *et al.*, 2003)

Soy flour, 85 g; wheatgerm, 60 g; yeast, 50 g; ascorbic acid, 3 g; nipagin, 3 g; sorbic acid, 1 g; agar, 12 g; water 875 ml.

### E. *Plutella xylostella* (Carpenter & Bloem, 2002)

*Soy-B diet*

Casein, 17.50 g; soy flour, 15.0 g; cabbage flour, 15.0 g; Brewer's yeast, 8.10 g; sucrose, 17.50 g; vitamin premix, 5.0 g (purchased from Bio-Serv®); ascorbic acid, 2.0 g; I-inositol, 0.10 g; choline chloride, 0.50 g; Wesson salts, 5.0 g; aureomycin, 0.50 g; sorbic acid, 1.00 g: methyl-P, 1.0 g; cholesterol, 1.25 g; soy oil, 3.50 ml; KOH solution, 2.50 ml; alphacel, 2.50 g; agar, 11.25 g; water, 420 ml.

Note that Carpenter & Bloem (2002) reported that survival and growth of strains of *Plutella xylostella* varied on different diets. The diet listed above performed well for the three strains tested.

*Vanderzant vitamin mix available from Bio-Serv (www.insectrearing.com/index.html).

CHAPTER V

# Bacteria for use against soil-inhabiting insects

ALBRECHT M. KOPPENHÖFER*, TREVOR A. JACKSON[†] & MICHAEL G. KLEIN[‡]

*Department of Entomology, Rutgers University, New Brunswick, NJ 08901, USA
[†]AgResearch, Biocontrol, and Biosecurity, Lincoln Research Centre, Lincoln, Canterbury, Private Bag 4749, New Zealand
[‡]Department of Entomology, Ohio State University, OARDC, Wooster, OH 44691, USA

## 1. INTRODUCTION

Despite over 90% of insects spending at least part of their time in the soil, there are very few bacteria for use against soil-inhabiting pests. Over time, these insects have evolved resistance against the multitude of pathogenic microbes that can be isolated from soil and will kill leaf-feeding pests. It appears that only highly co-evolved or unique pathogenic bacteria are able to overcome the defenses of soil-dwelling pests. The bacteria that are available are almost exclusively used against white grubs, larvae of beetles in the family Scarabaeidae. Scarab adults and larvae are serious pests of crops, ornamentals, and turf throughout the world and are likely to increase in significance (Jackson, 1992; Jackson & Klein, 2006). In addition, scarabs like the Japanese beetle, *Popillia japonica*, are quarantined to prevent introduction and damage as has occurred in the United States and on Terceira Island, Azores (Lacey *et al.*, 1994). Jackson and Klein (2006) surveyed the general status of pathogens against scarabs, and reviews (Klein & Jackson, 1992; Klein, 1995; Koppenhöfer, 2007) have examined the status of bacteria against turf and soil pests in general.

This chapter concentrates on bacteria from the genera *Bacillus*, *Paenibacillus*, and *Serratia*, and describes methods of isolation, identification, propagation, bioassay, and preservation. *Bacillus thuringiensis* is a naturally occurring soil bacterium which produces protein crystals toxic to many insects. Among the vast number of *B. thuringiensis* strains available in collections, few have shown activity against soil pests. *Bacillus thuringiensis* subsp. *japonensis* Buibui strain which contains the Cry8Ca1 delta endotoxin has been found to be highly effective against larvae in the scarab subfamily Rutelinae, but it is less effective against pests in the Melolonthinae and Dynastinae subfamilies (Bixby *et al.*, 2007). After ingestion of spores and toxin crystals, the larva stops feeding while the *B. thuringiensis* spores germinate and multiply, and toxins degrade the midgut epithelium. Once the epithelial barrier is broken, bacteria enter the hemocoel producing a septicemia, and the insect dies within a few days. More recently, a strain of *Bacillus thuringiensis* (SDS-502) containing the Cry8Da toxin has been isolated with activity against larvae of the cupreous chafer, *Anomala cuprea* (Asano *et al.*, 2003) and both larval and adult Japanese beetle (Yamaguchi *et al.*, 2008), while a strain bearing a Cry8Ga toxin has been found with activity against melolonthine scarabs (Yu *et al.*, 2006; Shu *et al.*, 2009). Interestingly, these Cry8 toxin bearing strains appear to be specific to subfamilies within the Scarabaeidae and suggest that further isolation and testing would be useful to expand the range of scarab active isolates. *Bacillus thuringiensis* subsp. *israelensis* containing the Cry4A, Cry4B, Cry11A, and Cyt1A delta endotoxins has shown promise for the control of larvae of crane flies in the genus *Tipula* (Oestergaard *et al.*, 2006) and of larvae of various species of fungus gnats. Many of the techniques for dealing with scarab- or *Tipula*-active *B. thuringiensis* differ little from those for other *B. thuringiensis* strains and are covered in detail in Chapter III.

*Paenibacillus* (formerly *Bacillus*) *popilliae* and *P. lentimorbus* are obligate pathogens that cause 'milky

MANUAL OF TECHNIQUES IN INVERTEBRATE PATHOLOGY
ISBN 9780123868992

disease' in scarab larvae in the subfamilies Melolonthinae, Rutelinae, Aphodinae, and Dynastinae. The history and use of milky disease bacteria has been reviewed (Klein, 1992) and will not be covered in detail here. Milky disease bacteria strains are highly specific, showing little to no cross-infectivity to scarab species other than the one they were isolated from. Only the strain infecting *P. japonica* has been commercialized. Infection is initiated when the host insect ingests bacterial spores. The spores germinate in the gut and release a parasporal crystal protein that may facilitate entry of the vegetative rods into the hemolymph by damaging the midgut wall. The bacteria transverse the midgut epithelium and go through one multiplication cycle on the luminal side of the basal lamina in immature replacement cells and regenerative cells. Then they move into the hemolymph where they multiply through several cycles, and eventually sporulate and form parasporal bodies. The high concentration of refractile spores and parasporal bodies during the final infection stages gives the hemolymph the characteristic milky-white color (Figure 5.1 and book cover). Milky disease is inevitably fatal but the exact cause of host death is not fully understood; it seems to involve the depletion of fat body reserves. When the infected larva dies, typically after several weeks or even months, up to $5 \times 10^9$ spores may be released into the soil from the disintegrating host cadaver, leading to a buildup of the disease.

Soil insects are frequently found dead with a softened black or brown cadaver caused by a bacterial septicemia, often by 'potential pathogens'. These bacteria have little ability to consistently infect and kill their hosts but can grow once within the insect hemocoel. Exceptions are

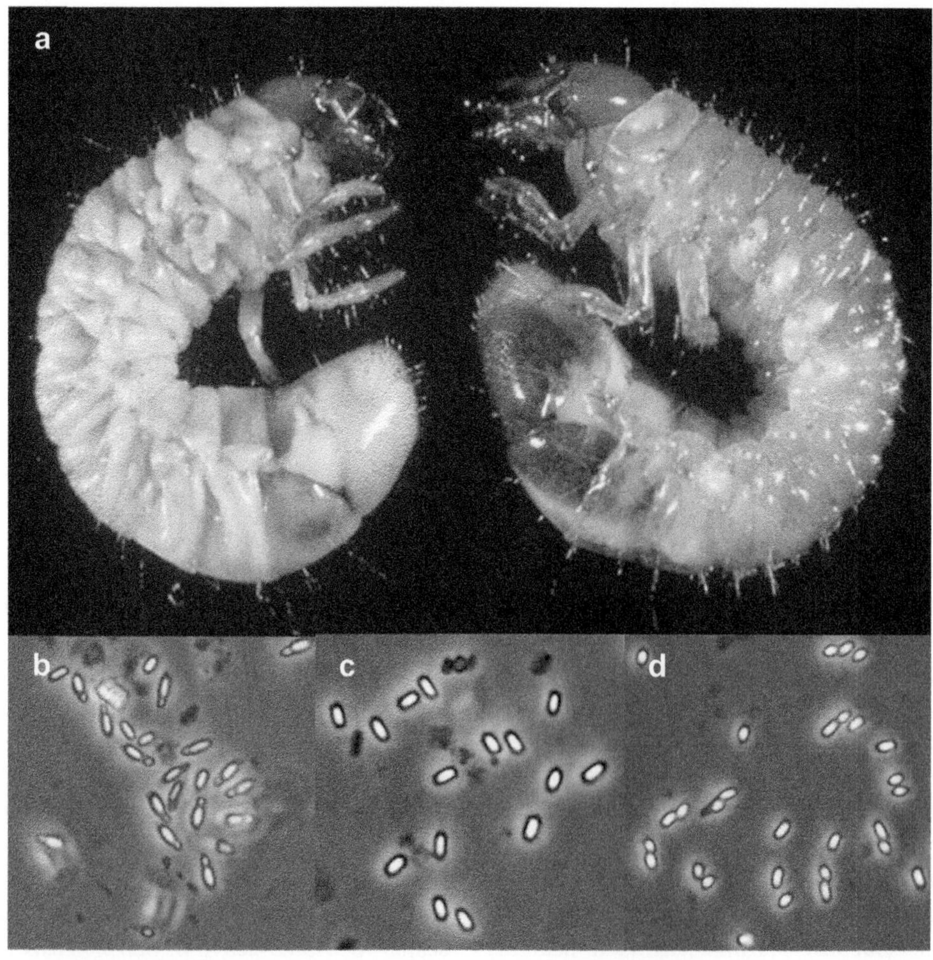


**Figure 5.1** Milky disease of the Japanese beetle (*Popillia japonica*). a. Healthy larva with clear hemolyph (right), diseased larva with milky haemolymph (left). b–d. Spore types from left to right, *Paenibacillus popilliae* (Type A1) from Japanese beetle; *P. lentimorbus* from Japanese beetle; *P. popilliae* from New Zealand grass grub.

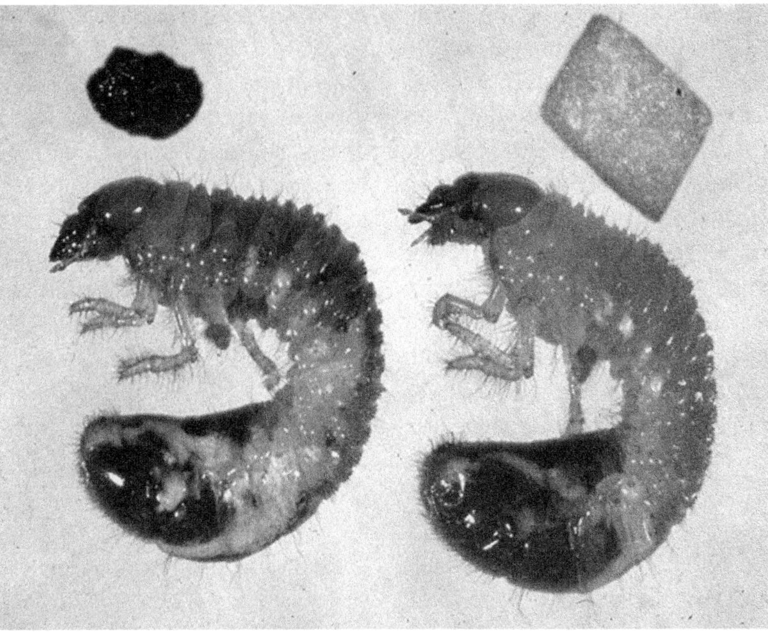

**Figure 5.2** Amber disease of the New Zealand grass grub, *Costelytra zealandica*. Healthy, feeding larva with darkened gut (left), non-feeding, amber diseased larva with cleared gut (right). Carrot pieces show consumption in the previous 24-h period.

*Serratia* species carrying a specific plasmid that causes amber disease in the New Zealand grass grub, *Costelytra zealandica*. The disease has been described by Jackson *et al.* (1993, 2001) and Gatehouse *et al.* (2009) and causes the infected insect to cease feeding and purge its gut, taking on an amber appearance due to the translucence of the tissues around the empty gut (Figure 5.2). The amber disease infected larva undergoes a long chronic disease period as fat bodies are reduced and tissues weaken until bacteria break through into the hemocoel and produce a septicemia. Non-feeding, amber colored larvae have also been found in other species, but, to date, no other amber disease producing bacteria have been isolated. *Yersinia entomophaga* (Hurst *et al.*, 2011) is another non-sporeforming bacterium pathogenic to soil insects, but in this case death is rapid from septicemia.

## 2. ISOLATION

When an insect is suspected of sickness or death due to a pathogen, standard bacteriological techniques (see Chapter III) can be used to isolate bacteria from the insect. Hemolymph, gut sections, or other infected tissue can be dilution streaked under aseptic conditions on to nutrient agar or similar media to produce single colonies which can be isolated and purified. However, many contaminant bacteria may also be isolated from dead or diseased insects. Before reaching the conclusion that an isolated bacterium is the disease causing organism, Koch's postulates must be proven—a microbe can not be described as a pathogen unless it can be isolated from a diseased host, purified, multiplied and, when applied to healthy hosts, produce the same disease symptoms. When a particular pathogen is suspected, selective agars can be employed (see below). Preliminary differentiation of bacteria can be made on the basis of cell morphology, results of Gram-staining (Chapter III), and utilizing biochemical tests as outlined in *Bergey's Manual of Systematic Bacteriology*, Vol. 2 (Garrity, 2005).

After determining cell shape and size and Gram stain, bacteria can be categorized and identified according to their growth requirements on standard or specialized media listed in Bergey's Manual. Rapid and convenient identification systems based on multi-test, substrate utilization tests have been developed for clinical diagnostics and can be used for a preliminary taxonomic determination of bacterial pathogens [e.g., API® strips (www.biomerieux-diagnostics.com) or Enterotube® II or Oxi/Ferm Tube® II (www.bd.com) systems]. The Biolog MicroStation™ System (Biolog, Hayward, CA, USA) can also be used. However, all systems have limitations and variable or critical reactions should be confirmed on specific media before a final identification is given. Certain fastidious bacteria, such as the milky disease bacteria (*P. popilliae* and *P. lentimorbus*), require more specialized media as discussed below.

Molecular identification provides another alternative for taxonomic determination. Sequence analysis of the 16S rRNA gene can provide a unique signature for bacterial species. Ribosomal genes are conserved as they are essential for cellular function, but small mutations in the gene will accumulate through bacterial evolution with variation in the gene sequence providing a basis for determining phylogeny (Clarridge III, 2004). Sequences from isolated organisms or putative pathogens can be compared with standards in Genebank or other databases and a taxonomic name assigned. However, for tracking of strains for epidemiological purposes or detection of strains with particular virulence factors, 16S rRNA gene analysis is not sufficient and other techniques, such as Pulsed-Field Gel Electrophoresis (PFGE) (Correa and Yousten, 2001) or Q-PCR (Monk *et al.*, 2010) should be used. 16S rRNA is particularly useful for non-cultured organisms but, in general, a combination of molecular methods, cultivation on discriminating media and bioassay should be used to determine the identity of a pathogenic bacterium.

## A. Isolation of milky disease bacteria

Milky disease bacteria, *P. popilliae* and *P. lentimorbus*, appear to be obligate pathogens. In nature, they are only found associated with diseased scarabs, or as spores in the surrounding soil, and require specialized conditions (bodies of their hosts) for growth and sporulation.

New isolates of milky disease bacteria can be obtained from the hemolymph of infected scarab larvae (Figure 5.1). Stahly *et al.* (1992a) reviewed procedures used in isolating these bacteria.

- Surface-disinfect grub with 0.5% sodium hypochlorite (NaClO, dilute chlorine bleach) solution in water.
- Puncture the cuticle with a needle (or snip off leg) and apply gentle pressure to the cadaver.
- Collect emerging droplets with spores in sterile water.

Spores from hemolymph can also be collected onto glass slides for later examination and culture (see Section 5A). After isolation, spores or vegetative rods of milky disease bacteria can be diluted and directly plated on J-Medium or MYPGP agar (see Appendix). St Julian *et al.* (1963) recommend 0.1% tryptone as a standard diluent, particularly if vegetative rods are involved. Generally, spores of milky disease bacteria germinate poorly in the laboratory with a maximum of 1–5% of the spores likely to produce colonies on agar (Stahly *et al.*, 1992a). Numerous attempts to increase germination with heat treatments or chemical additives have given variable results. The most successful procedures involved heating spores at 60°C for 15 min in calcium chloride ($CaCl_2$, 1 mM) at pH 7, and suspending them in cabbage looper hemolymph and tyrosine at an alkaline pH (Stahly *et al.*, 1992a). Krieger *et al.* (1996) improved germination after heating for 30 min at 75°C and applying pressure through a French press.

Ironically, poor spore germination may be beneficial to strain isolation. Milner (1977) has taken advantage of the poor germination in development of the following method for isolation of *P. popilliae* var. *rhopaea* from soil.

- Soil suspension (2 g wet weight of soil in 20 ml water) mixed with a germinating medium (0.5% yeast extract, 0.1% glucose, at pH 6.5) at a 1 : 50 ratio and subjected to a series of seven heat shocks (70°C for 20 min) at hourly intervals.
- Following that procedure, which kills vegetative cells and germinating spores of contaminant bacteria, an aliquot is spread on J-Medium and incubated anaerobically at 28°C for 7 days.

There is no information on how this procedure works with other varieties of *P. popilliae* or *P. lentimorbus*. Stahly *et al.* (1992b) note this is a very time-consuming procedure and is not an accurate method for quantifying spores in soil. An alternative procedure that they suggested utilizes natural resistance to vancomycin which is shown by many strains of *P. popilliae* (but to date no *P. lentimorbus* strains) to isolate the bacteria from spore powder or from inoculated soil.

- 1 g of spore powder (see Section 4A2d), or 1 g each of powder and soil, placed in 5 ml of water and heated at 60°C for 15 min to kill vegetative cells.
- Dilutions made in water and 0.1 ml plated on MYPGP agar containing vancomycin (150 μg/ml).
- Mold growth, a particular problem when soil samples are analyzed, prevented by the addition of cycloheximide (1 mg/ml).

- Plates incubated at 30°C for up to 3 weeks. Maximum *P. popilliae* colony counts occurred in about 9 days.

In these experiments, vancomycin did not interfere with germination or colony-forming ability of *P. popilliae* of the selected strains and is a valuable tool for isolation of those strains. However, subsequent studies found that not all strains of *P. popilliae* are vancomycin resistant.

## B. Isolation of *Serratia* spp.

The genus *Serratia* comprises species which are ubiquitous in nature and commonly found in soil and water which leads to frequent isolation of contaminatory *Serratia* isolates from diseased and dead insects. Isolation of putative pathogenic strains can be aided by selective media. Amber disease of the New Zealand grass grub is caused by specific strains of *S. entomophila* and *S. proteamaculans* (Grimont *et al.*, 1988) and septicemias can be produced by strains of *S. marcescens*.

A rapid typing system has been developed for isolating putative disease causing *Serratia* spp. based on the use of selective agars (O'Callaghan & Jackson, 1993).

- Soil or larval extracts can be dilution plated on caprylate-thallous agar (CTA) (Appendix) to allow *Serratia* spp. to grow, while repressing other microorganisms (Starr *et al.*, 1976).
- Alternatively, the tract from larvae with signs of amber disease can be extracted by holding each end of a larva with a forceps and pulling in opposite directions (Figure 5.3).
- The gut, which normally stays attached to the anterior end, can be crushed in water and spread on CTA, or directly streaked onto the agar.
- Cream-colored colonies, 2 mm in diameter, form after 4–6 days at 27°C and can be transferred to selective media to complete the identification process (see Section 3B). By contrast, a related species, *Serratia marcescens*, produces red colonies (Figure 5.4).

## C. Isolation of *Bacillus thuringiensis*

*Bacillus thuringiensis* is commonly isolated from soil and insects and can be readily isolated taking advantage of its heat-resistant spore and characteristic morphology from the combination of spore and toxin. Claus & Berkeley (1986) suggest the use of 50% (v/v) ethanol for at least 1 h to clean up *Bacillus* spores before isolating the bacterium. This procedure will kill vegetative cells but not increase the chance for mutations found with heating the culture above 70°C for 10 min. They also described media for selective isolation of *B. cereus* that can also be used to isolate *B. thuringiensis*.

- Commercially available medium, polymyxin pyruvate (to suppress Gram-negative organisms), egg yolk, mannitol, bromothymol blue agar (PEMBA) (Appendix) can be used.
- On this mixture, *B. cereus* colonies are slightly rhizoid, have a distinct turquoise blue color, and are usually surrounded by an egg yolk precipitate of similar color.
- Alternatively, a liquid enrichment broth (5 g peptone, 5 g meat extract, 10 g $KNO_3$, distilled water to 1000 ml) can be used.

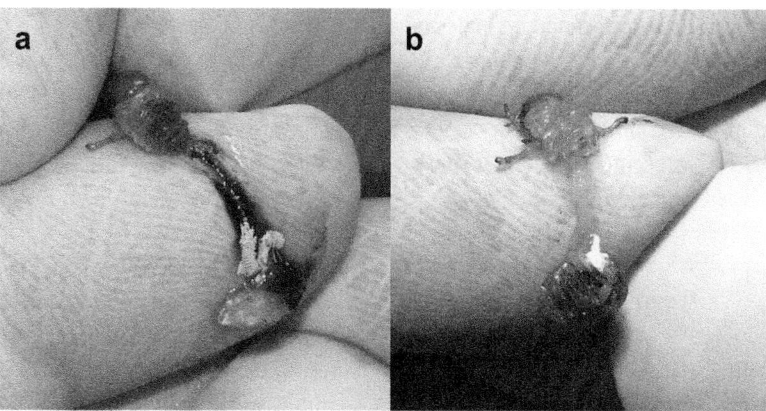


**Figure 5.3** In-field diagnostics for gut clearance due to amber disease. Larvae are pulled apart to reveal the condition of the midgut. a. Healthy larva with dark midgut, b. amber diseased larva with empty clear gut.

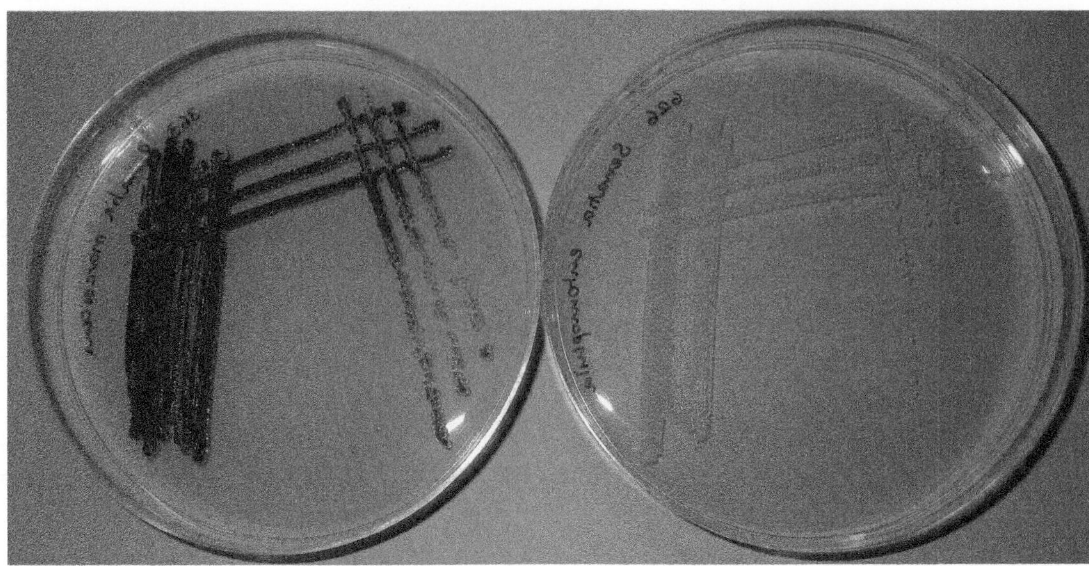

**Figure 5.4** Streak plate cultures of *Serratia marcescens* with red pigmentation (left) and *Serratia entomophila* (right) in 24-h culture on nutrient agar. Please see the color plate section at the back of the book.

- After the medium has been sterilized at 121°C for 15 min, pasteurized (10 min, 80°C) soil samples (5 g/100 ml) are added to the enrichment broth.
- After 24 h incubation at 30°C, the suspension is streaked on to nutrient agar plates and held at 37°C for colony formation.

Isolates of *B. thuringiensis* can be identified microscopically by the presence of parasporal bodies, or crystals, in the sporangium. Asano *et al.* (2003) used a simplified system for isolation by preparing soil suspensions in distilled water. Samples were mixed thoroughly and coarse matter allowed to settle. An aliquot of the clear liquid was removed, heated to 80°C for 10 min and dilution plated on to nutrient agar. Characteristic colonies were examined microscopically for the presence of spores and crystals and positive isolates selected. Other isolation procedures for *B. thuringiensis* are presented in Chapter III. A series of biochemical tests are outlined by Claus & Berkeley (1986) and in Chapter III, and the keys presented in Chapter III can help identify *Bacillus* species isolated from scarabs.

## 3. IDENTIFICATION

### A. Identification of milky disease bacteria

The general inability of the milky disease bacteria to grow and sporulate on standard microbiological media has resulted in confusion of the taxonomy of the group. Standard biochemical reactions utilized in the identification of most bacteria (Krieg & Holt, 1984; Sneath *et al.*, 1986) have not been very useful with milky disease bacteria. Milner (1981) found insignificant differences among four varieties (var. *popilliae*, *lentimorbus*, *melolonthae*, and *rhopaea*) of *P. popilliae* (then considered a single species) after 20 standard tests. The physiology of one milky disease bacterium, *P. popilliae* var. *popilliae* (presently *P. popilliae* Group 1) has been extensively examined. Those interested in exploring this subject are directed to the review of Stahly *et al.* (1992a) to obtain the extensive list of references available.

Based on molecular methods (phylogenetic tree construction based on 16S rRNA gene sequences), milky disease bacteria were transferred from the genus *Bacillus* to the genus *Paenibacillus* (Pettersson *et al.*, 1999) and are currently organized into two groups within the species *P. popilliae* and three groups within the species *P. lentimorbus* (Dingman, 2008, 2009). In the past, however, milky disease bacteria have been treated as varieties of *P. popilliae* (Milner, 1981; Ellis *et al.*, 1989; Klein & Jackson, 1992) or separated into the two species *P. popilliae* and *P. lentimorbus* (Gordon *et al.*, 1973; Claus & Berkeley, 1986; Stahly *et al.*, 1992a).

Initially, Dutky (1940) described two obligate pathogens, *B. popilliae* and *B. lentimorbus*, as the causative agents of Type A and B milky disease in Japanese beetle.

The distinguishing feature separating the two bacteria was considered the presence of a crystal, or parasporal body, in *P. popilliae* (Figure 5.1) but absence thereof in *P. lentimorbus*. The sporangium has a 'footprint' shape with the spore forming the sole and the parasporal body the heel. Milner (1981) suggested a single species but with four subgroups, A1, A2, B1, and B2, based on presence/absence of parasporal body and size and position of the spore. As distinguishing characters between the two species, differences in their lipid composition, spore surface topography, their antigen antibody reactions (Claus & Berkeley, 1986; Stahly *et al.*, 1992a), and their DNA GC (guanine, cytosine) ratios have in the past also been suggested. To help clear up the taxonomic confusion, Milner (1981), Claus & Berkeley (1986), and Stahly *et al.* (1992a) suggested obtaining information on host specificity, spore size and morphology, parasporal body structure, type of peptidoglycan, major type of quinone, GC ratios, nucleic acid hybridization, fatty acid composition, polar lipids, serology, menaquinones, and electrophoresis of enzymes.

However, more recent studies, using molecular methods to determine species and groupings and including different or new strains than in previous studies, found that many morphological or physiological characteristics previously used to separate species, varieties, or groups had limited utility. Thus, absence of a parasporal body in the genetic grouping defining *P. lentimorbus* strains seems to be the exception rather than the rule (Rippere *et al.*, 1998; Harrison *et al.*, 2000). While not found in any *P. lentimorbus* strains to date, vancomycin resistance may be common only among North American isolates of *P. popilliae* (Harrison *et al.*, 2000). Hence, more reliable identification of milky disease bacteria should also involve molecular methods. Which method or combination of methods is most appropriate will also be affected by the level at which separation between isolates is sought (e.g., general classification, intra-generic or intra-species separation) and how much other information about to-be-identified/separated isolates is already available.

DNA–DNA similarity analysis (DNA reassociation) combined with random amplified polymorphic DNA analysis were used by Rippere *et al.* (1998) and Harrison *et al.* (2000) to study the relationship of numerous strains/isolates of milky disease bacteria; these methods clearly segregated the isolates into *P. popilliae* or *P. lentimorbus*. However, more recent studies (Pettersson *et al.*, 1999; Dingman, 2008, 2009) have used comparative analysis of 16S rDNA for species identification and classification, a method that has become the standard for identification and classification of bacteria in general. While 16S rRNA has successfully defined milky disease causing species, strain differentiation requires other methods. Macdonald & Kalmakoff (1995) used PFGE to differentiate New Zealand isolates of *P. popilliae* from the USA biocontrol strain, and studies of North and South American strains showed that the technique would be used for monitoring of individual isolates (Correa & Yousten, 2001). RFLP and 16S rDNA sequence comparison has further been used (Dingman, 2009) to examine variation according to geographic region of isolation.

## B. Identification of *Serratia* spp.

Amber disease in *C. zealandica* is caused by specific isolates of *S. entomophila* and *S. proteamaculans* (Enterobacteriaceae) (Grimont *et al.*, 1988). However, bacteria of this genus are widespread in the environment and the main problem for identification is separating pathogenic isolates from the non-pathogenic bacterial pool in the soil. The study of *Serratia* spp. in the environment has been facilitated by development of a semi-selective CTA (Starr *et al.*, 1976). Thalium salts in the medium suppress many bacterial species but allow growth of small white colonies of *Serratia* spp. after 2–3 days at 25–30°C. To determine identity to species level, bacterial colonies conforming in appearance to *Serratia* spp. are transferred by spotting on to a series of three differentiating media (see Appendix) to determine whether they are species known to cause amber disease (O'Callaghan & Jackson, 1993). The key reactions are as follows: production of a halo on DNase–Toluidine Blue agar (Appendix) after 24 h incubation at 30°C indicates that the bacterium is from the genus *Serratia*; after 24 h incubation at 30°C on adonitol agar, *S. entomophila* produces yellow colonies, and *S. proteamaculans* produces blue/green colonies; as other *Serratia* species also produce yellow colonies, growth on itaconate agar after 96 h incubation at 30°C confirms the identity as *S. entomophila*.

While species can be determined by biochemical tests, identification of strain and pathotype requires further tests. *S. entomophila* can also be differentiated from other *Serratia* spp. by serotyping which has been useful for safety studies (Allardyce *et al.*, 1991). A number of techniques can be used to define strains of the bacterium.

*Serratia entomophila* is host to a wide range of bacteriophages which vary in their ability to enter and lyse *S. entomophila* strains. A phage-typing system, based on sensitivity spectrum to a range of phages, has been developed (O'Callaghan *et al.*, 1997) and is used in monitoring programs of applied strains. Strains can also be differentiated by molecular methods such as RFE (Rotating Field Electrophoresis) (Claus *et al.*, 1995) and PFGE which can be used to confirm the identity of recovered strains after application. Not all strains of *S. entomophila* and *S. proteamaculans* are pathogenic because the Sep Toxin Complex (Tc) and the Afp anti-feeding genes are carried on a specific plasmid present in the pathogenic isolates (Hurst *et al.*, 2000). Presence of the large pADAP plasmid is an indicator of pathogenicity (Glare *et al.*, 1993) and is used as a quality control indicator for *S. entomophila* products. A DNA probe targeting a section of the plasmid has been used with colony blot analysis for quantification of disease in epizootiology studies (Jackson *et al.*, 1997). A species-specific probe has also been developed as an alternative to laborious biochemical typing in the identification of *S. entomophila* (Hurst *et al.*, 2008). These methods may be superseded by Q-PCR which has shown the potential for culture independent rapid determination of species and pathotype from soil samples (Monk *et al.*, 2010).

Unfortunately there are few other media with the discriminating power of CTA for other bacterial entomopathogens. *Yersinia entomophaga* MH96 (Hurst *et al.*, 2011) has a profile similar to *Serratia marcescens* using the API 20E system, but subsequent culture tests on DNAse and CTA clearly rule out this identification. A unique 16S rRNA sequence defines this species. It is likely that molecular typing will increasingly take over from biochemical identifications as methods and databases are further developed.

### C. Identification of *Bacillus thuringiensis*

Identification of *B. thuringiensis* is covered in Chapter III and will not be detailed here. Scarab active strains have primarily been identified by screening of environmental isolates and bioassay (Ohba *et al.*, 1992; Asano *et al.*, 2003; Yu *et al.*, 2006). Scarab active strains have been found among *B. thuringiensis* subspecies *japonensis* and *galleriae* and contain spherical to ovoid parasporal bodies. They also contain *cry8* genes which appear to be the basis of their scarab activity (Yamaguchi *et al.*, 2008). Yu *et al.* (2006) have defined a PCR-RFLP method for recognition of *cry8*-type genes which will aid screening of new isolates but the ultimate test will be pathogenicity against the target host.

## 4. PROPAGATION AND PRODUCTION

### A. Propagation of milky disease bacteria

*Paenibacillus* spp. bacteria are extremely fastidious in their growth requirements and do not sporulate well in liquid media in fermenters or on agar-solidified nutrient media. The inability to produce highly infective *P. popilliae* spores on artificial media has severely restricted the use of milky disease spore powder as a biopesticide and limited milky disease products to those made from collected naturally infected larvae or by laboratory infection of healthy larvae for bacterial multiplication and extraction of spores.

#### *1.* In vitro *culturing*

Vegetative cultures of milky disease bacteria can be produced on complex artificial media. St Julian *et al.* (1963) developed what became the standard medium (J-Medium, see Appendix) for obtaining growth of milky disease bacteria cells. The final pH was 7.3–7.5, and as with all the media discussed here, the carbohydrate source was sterilized by filtration and added after the rest of the medium had been autoclaved. Other workers have used modifications of the basic J-Medium. Lingg & McMahon (1969) observed better growth when the proportions of tryptone and yeast extract were reversed to provide 1.5% tryptone, 0.5% yeast extract, 0.6% $K_2PO_4$, 0.2% glucose, and 1.5% agar with a final pH of 7.0–7.2. Costilow & Coulter (1971) suggested the more complex MYPGP medium (Appendix). Ellis *et al.* (1989) replaced the glucose in the J-Medium with trehalose (the major sugar in hemolymph) resulting in a medium with 1% tryptone, 0.5% yeast extract, 0.3% $K_2HPO_4$, 0.2% trehalose, and 1.5% agar. Milky disease bacteria have been found to grow on Difco® Brain Heart Infusion and Difco® Todd–Hewitt Broth media (D. Dingman, personal communication).

An *in vitro* method of milky disease spore production was patented by Ellis *et al.* (1989) and used to produce bacteria for a commercial product. Under test, bacteria reported to be *P. popilliae* were identified as *B. polymyxa* and *B. amylolyticus* (Stahly & Klein, 1992) which led to the withdrawal of all commercially *in vitro*-produced milky disease spore powder.

### 2. In vivo *production*

Dutky (1942) provided the following procedure for production of milky disease bacteria in larvae, based on his experience in the mass production program in the Eastern USA. Films of dried hemolymph with spores from diseased larvae on glass microscope slides (see Section 6A for details) may be kept as stock cultures for over 30 years.

#### *a. Preparation of inoculum*

- A standardized spore suspension is obtained by removing the spores from the slides with 0.5 ml of distilled water. This is facilitated by stroking the moistened film with a sterile pipette to bring the spores into suspension.
- The suspension is allowed to run into a sterile test tube by holding the tip of the pipette against one corner of the slide and tilting the slide toward the pipette tip.
- Fresh distilled water is flooded over the slide and the procedure is repeated.
- Spore counts on the suspension are made with a counting chamber, and the number of spores is adjusted to approximately $3 \times 10^8$ per ml.

To prepare sufficient spore suspension to inoculate 500–1000 larvae, the following will be helpful.

- Prepare a test tube (A) containing 1 ml of sterile distilled water and a second tube (B) containing 2 ml of sterile distilled water.
- Wash spores from two culture slides into the inoculum (B) with an additional 1-ml portion of sterile distilled water so that total volume of the suspension is 3 ml.
- Fill a capillary pipette to the 0.01-ml mark with a loopful of this suspension and discharge the capillary pipette into the sample tube (A).
- Shake the tube, withdraw a loopful of the suspension from the sample tube and fill a counting chamber.
- Place the counting chamber on the mechanical stage of a microscope and allow the chamber to stand for 2 min to permit settling of the spores in the chamber.
- Using a 65 × objective, count the number of spores in five large squares of a Levy, or similar, counting chamber (80 small squares).

The spore density can be calculated using formulae from Chapter III. Spore density should be adjusted to $3 \times 10^8$ spores per ml of suspension by proportional concentration or dilution to achieve the correct concentration for inoculation.

#### *b. Inoculating larvae*

- A hypodermic syringe with a 27–30-gauge needle is filled with the adjusted suspension and the entrapped air is expelled from the needle. Care must be exercised to ensure that no air bubble is present in the needle. Such a bubble will result in the failure of the small volumes to be injected into the grub.
- The loaded syringe is then put into a manually operated microinjector, and the micrometer screw is turned forward until a droplet is forced from the end of the needle. Alternatively, a motorized microinjector may be used (Figure 5.5). The expelled droplet is removed with a piece of absorbent cotton.
- To quantify the spore dosage, an inoculating dosage is discharged into a dry, sterile test tube and sterile distilled water (1 ml) is then pipetted into the tube.
- Counts are made on this sample with a hemocytometer to confirm the spore dosage.

Larvae are injected as follows:

- The grub is held firmly, but lightly, between thumb and forefinger, the dorsal posterior is positioned outward and the grub is forced on to the needle point so that the needle enters in the dorsal portion of the

**Figure 5.5** Motorized microinjector for intracoelomic and oral inoculation of measured quantities of microorganisms in suspension into insect larvae.

suture between the second and third posterior abdominal segments (Figure 5.5). Care must be taken that the needle enters horizontally so as not to puncture the intestine. In case of accidental puncture of the intestine of a grub during injection, the hypodermic needle should be sterilized immediately by swabbing with 0.5% sodium hypochlorite solution before proceeding with further inoculation. Otherwise, larvae may be infected with septicemia-producing bacteria contaminating the needle. Care must be also taken to ensure that the site of puncture is free from soil particles.
- The larva is then allowed to hang suspended on the needle during the injection.
- The injection is made by depressing the pressure bar of the ratchet mechanism of the microinjector which forces an inoculating dosage into the body of the grub.

It is preferable to use full-grown larvae, although small 3rd instar and even 2nd instar larvae may be used. By volume, the dosage approximates 3.3 μl. Since the inoculum is adjusted to $3 \times 10^8$ spores per ml, the resulting spore dosage approximates $10^6$ spores. Diseased grubs can also be produced by microinjection of vegetative cells of *P. popilliae* into grubs. The procedure is similar except that only 10–100 cells are injected in each larva.

*c. Incubating larvae*

- Boxes having a capacity of 500 larvae (approximately 60 × 60 × 30 cm) are used to hold the injected larvae during the period of incubation. These are equipped with metal cross-section separators which divide each of five layers into 100 compartments.
- After injection, the inoculated larvae are each dropped into a separate compartment. As soon as all compartments in a layer are occupied, soil (heated at 100°C for 24 h and brought up to 60% of the ball point) is added to fill the compartments.
- A flat, metal separator is placed on top of the filled layer, and a new cross section is put in the box. The compartments in the new layer are partially filled with soil, injected larvae are placed in these compartments, and the whole process is repeated until all five layers have been filled.
- The soil should contain 225 g of grass seed for each 40 kg of soil which, when sprouted, will serve as food for the larvae during incubation.
- The grass seed should contain a high percentage of ryegrass (perennial or annual, *Lollium* spp.) which germinates quicker than most other grass types.
- The boxes are incubated at 30°C for 10–12 days. High humidity should be maintained in the incubation chamber to prevent excessive drying out of the soil.

The use of soil at the proper moisture level is essential for satisfactory survival of larvae and spore production. Too-low soil moisture will limit grass seed germination and with that larval growth, which will reduce the spore yield. Excessive moisture will increase larval mortality by interfering with the normal gas exchange in the soil. The optimal moisture value for spore production and larval survival is dependent on the soil and has been found, for a large number of soil types, to be that which just prevents adherence of soil particles to the cuticle of the larvae. This is approximately 60% of the 'ball point' of the soil.

To determine the ball point of the soil, 100 g soil is weighed into containers of approximately 250 ml capacity, and water added in 2-ml increments. The soil is mixed thoroughly after each addition until it forms a plastic mass with just an excess of free moisture. When this point is reached, the soil ball formed will re-form when the ball is broken up and the mass agitated by rotating the container.

The amount of water added to bring the 100-g sample to this state is recorded as the 'ball point' value of the soil. Between 60 and 65% of this value is the moisture content which the soil should contain for use in the incubation boxes. These values are determined and expressed on the basis of ml of water per 100 g air-dried soil.

*d. Preparation of spore powder*

- After incubation, boxes are broken down, and the grubs with milky disease signs are screened out of the soil and dropped into a battery jar of ice water. The ice water inactivates the larvae, permitting thousands of them to be placed in the jar without danger of loss of spores due to larvae nipping one another.
- The larvae are then washed in a colander to remove adhering soil particles, returned to the glass jar, packed with ice and held in a refrigerator at 1–2°C until used.

- When sufficient numbers of diseased larvae have accumulated, the grubs are crushed by running them through a meat chopper or blender. After all the larvae have been run through the machine, it is washed out with a small quantity of water to remove the adhering grub tissue.
- The resulting suspension is placed in a graduated cylinder, brought up to an even volume, and spore counts are made on this suspension.
- The suspension is then added to the carrier ($CaCO_3$, calcium carbonate, precipitated, USP) so that the mixture will contain $10^9$ spores/g of dry material.
- The moist paste ($CaCO_3$ plus spore suspension) is mixed thoroughly by running it through a mixing device, such as a blade cutter or trowel mixer.
- After thorough mixing has been accomplished, the moist paste is passed through a high-speed impeller-type blower which shears the agglomerated particles.
- Drying of the dust is accomplished either by drawing heated air through the blower and exposing the finely divided particles to the warm air blast, or the paste can be spread out in a thin layer in front of a dehumidifier and pulverized when dry.

  Ellis *et al.* (1989) suggested that freeze-drying of the concentrate is preferable to the original air-drying procedure. When dry, this material is the stable concentrated spore–dust preparation.
- The concentrate is then mixed with nine parts of a dry carrier (talcum powder, marble flour, etc.) and stored until used. The final mixture contains $10^8$ spores/g.

The spore content of the dry spore–dust concentrate may be checked as follows:

- To 10 g of the powder in a volumetric flask (200 ml capacity) is added approximately 100 ml of distilled water, the flask is shaken to wet the particles, and 20 ml of concentrated hydrochloric acid is added to the suspension.
- The flask is gently rotated with the stopper out until gas evolution ceases.
- Distilled water is added until the volume is 200 ml, the flask is stoppered and shaken vigorously.
- A loopful of this suspension is then used to fill a cell counting chamber. The count on the suspension should be *ca.* $5 \times 10^7$ spores/ml. The actual count in millions is then divided by 200 yielding spore counts in $10^9$/g of the concentrated dust mixture.

Syringes should be flushed with distilled water and then with alcohol immediately after use to avoid a plugged needle or 'freezing' of the plunger in the syringe. Careful records should be made of numbers of larvae injected, spore and volume dosages, the number of larvae living at the end of the incubation period, the number diseased or not infected, and the yields. From such records, it is easy to control the process fully and correct any serious faults in procedure as they occur.

## B. Production of *Serratia* spp.

*Serratia* spp. can be grown on many common media, but to achieve high-density cultures media must be selected and a fermentation strategy determined. In fermenters, bacteria grow through the lag, and exponential phase and reach maximum cell density in the stationary phase. For *S. entomophila* a batch production system is used. Visnovsky *et al.* (2008) found that highest cell densities were produced on a sucrose, yeast extract, and urea medium with aeration sufficient to maintain 20% dissolved oxygen through the lag and exponential phases. The objective of *Serratia* cultivation is to produce robust cells for formulation and application. Because cells harvested before reaching the stationary phase showed poor storage stability, a standard fermentation is concluded 4–6 h into the stationary phase. Fermenter cultures can reach high densities ($4–6 \times 10^{10}$ cells/ml) and are as stable as unmodified fermenter broth when stored under refrigeration (4°C). The *S. entomophila* product Invade® for grass grub control was an unmodified fermenter broth containing $4 \times 10^{10}$ bacteria/ml which could be maintained for more than 3 months prior to application with minimal losses (Pearson & Jackson, 1995). To improve stability and facilitate distribution and application, a thermostable granule was developed using biopolymers and clays (Johnson *et al.*, 2001). The stabilization technology was developed into a biopolymer coated zeolite granule, Bioshield®, that is stable in ambient conditions and can be applied through standard farmers' machinery. *Serratia* products are quality-control tested for purity, cell density, pathogenic stability, and survival (Pearson & Jackson, 1995). Production of other non-sporeforming entomopathogenic bacteria will require modification of production and formulation systems. For example,

*Y. entomophaga* fermentation is carried out at 25°C to maintain toxin production (Hurst *et al.*, 2011).

### C. Production of *Bacillus thuringiensis*

The mass propagation and fermentation of scarab active strains of *B. thuringiensis* is the same as for other *B. thuringiensis* products and is presented in Chapter III. *B. thuringiensis* will grow and sporulate readily on nutrient agar (Appendix). Suzuki *et al.* (1992) outlined their procedure for producing spores and crystals for laboratory tests. Bacteria which had been purified on NYS agar for 24 h at 30°C were cultured in NYS broth medium overnight on a rotary shaker at 350 r.p.m. NYS medium (200 ml) was placed in a 1-l baffled flask and inoculated with 2 ml of the seed culture. Cells were cultured for 3 days, and the spore/crystal complex was collected by centrifugation for 15 min at 15,000 × *g*, and rinsed three times with distilled water. In later tests, Suzuki *et al.* (1994) used 10,000 × *g* for 10 min and added potassium sorbate (0.01%) to prevent bacterial growth. Asano *et al.* (2003) used centrifugation in Percoll gradients to collect pure crystals for testing.

## 5. BIOASSAY

### A. Bioassay for milky disease bacteria

There has been considerable interest in establishing the virulence and host range of the many varieties of *P. popilliae* and *P. lentimorbus* discussed above. Dutky (1941) outlined a series of tests of milky disease organisms (by injection and lab feeding) that has been used and slightly modified by many researchers, but is still generally valid today. He recommended injection of spores as the first test of susceptibility to *P. popilliae* infection as follows:

- Use an inoculating dose of $1{-}2 \times 10^6$ spores per larva, as outlined in Section 4A2.
- Use un-injected larvae and those injected with sterile water as controls.
- Hold larvae in individual tins or vials, in soil at 60% of the ball point with sprouted grass for food (Section 4A2), and examine daily for 15 days.

Poprawski & Yule (1990) replaced the soil–grass mixture with an artificial diet. Milner (1981) noted that the age of spores used in infectivity tests should be standardized since fresh spores may germinate poorly. St Julian *et al.* (1963) found that vegetative cells of var. *popilliae* could be tested by dilution in 0.01% tryptone. However, injections should be completed within 1 h of dilution. The cause of larval mortality and all macroscopic symptoms must be confirmed by microscopic examination. The low infectivity with injected free spores of *P. popilliae* produced *in vitro* (Ellis *et al.*, 1989) suggests that feeding tests are a more reliable indication of bacterial virulence than is injection.

For his feeding tests, Dutky (1941) used a standard spore–talc preparation described above.

- The spore–talc preparation was mixed with soil at $2 \times 10^9$ spores/kg air-dried soil, and the mixture was brought up to 60% of the ball point (Section 4A2).
- A rate of $2 \times 10^9$ spores/kg of soil or higher should result in about 85% milky larvae after 1 month (Dutky, 1963).
- Rates of 0.25 and $0.5 \times 10^9$ per kg have resulted in 50% milky Japanese beetle larvae (Dutky, 1963). However, it should be remembered that those values were obtained by pooling the result of many tests and that several factors including age of spores, nutritional status of the host and incubation temperature can influence the results (Milner, 1981).
- To obtain a better idea of the infectivity of the spore preparation, a range of four to five rates between $0.25 \times 10^9$ and $1{-}2 \times 10^{10}$ spores/kg should be used. Spore concentrate must be used for rates above $8 \times 10^9$ spores/kg soil.
- Soil for these tests should be at about 60% of the ball point of the soil.
- Larvae are incubated in the soil–spore powder medium with 10 g of grass seed per kilogram of soil added for food. For optimal results and comparison with old standards, the larvae should be incubated at 30°C. For a more realistic representation of natural soil conditions lower temperatures will generally be more appropriate (i.e. 20–25°C).
- Twenty or more larvae should be used for each concentration and rates should be replicated a minimum of three times.
- Examination of larvae twice a week for 1 month is needed to establish the infectivity of the milky disease variety or strain being tested.

- Inclusion of an untreated control, as well as a standardized spore preparation of known virulence, is essential for interpreting the results.

Other methods of feeding *P. popilliae* spores have been used. Poprawski & Yule (1990) used a pipetting gun fitted with a blunt glass tip to force-feed 0.1 ml of bacteria in 0.1% tryptone at rates of 2–3 × $10^7$ spores per ml to *Phyllophaga* larvae. The placement of the bacteria on pieces of carrot or grass roots (Milner, 1981) and sugarcane roots (David & Alexander, 1975) has also been employed. Milner (1981) reported a $LD_{50}$ of about $10^7$ spores per larva in his direct-feeding studies. Dingman (1996) noted that natural feeding of spores to grubs did not quantitatively or, consistently, introduce the pathogen into the host and developed a technique for force feeding *P. popilliae* spores to Japanese beetle and European chafer larvae which has been used to induce and study milky disease. Peroral injection was best done with a blunt 33-gauge needle and volumes of about 5 μl. Larvae were anesthetized with a stream of $CO_2$ gas, held ventral side up, and the needle introduced into the alimentary tract by moving the mandibles against the needle, and gently pushing the larva on to the needle before release of a measured dose of spores. The blunted needle could be inserted into the oral aperture and slid about 0.75 cm into the ventriculus without damage to the insect gut (Figure 5.6). The procedure was carried out using a stereomicroscope at 6 × magnification.

Results of these bioassay studies have shown that, in general, there is very little cross-infectivity of the strains of *P. popilliae* isolated from one grub species for grubs of another species. However, as noted above, the host range for varieties of *P. popilliae* is much greater when spores are injected rather than being fed. Milner (1981) notes that spore germination, host defense reactions, and host nutrition, all contribute to the specificity of milky disease bacteria.

No greenhouse tests on *P. popilliae* and *P. lentimorbus* have been reported. Thurston *et al.* (1994) tested the interaction between *P. popilliae* and entomopathogenic nematodes in a greenhouse study; however, the *Cyclocephala hirta* larvae used in the test had already been infected with *P. popilliae* under laboratory conditions

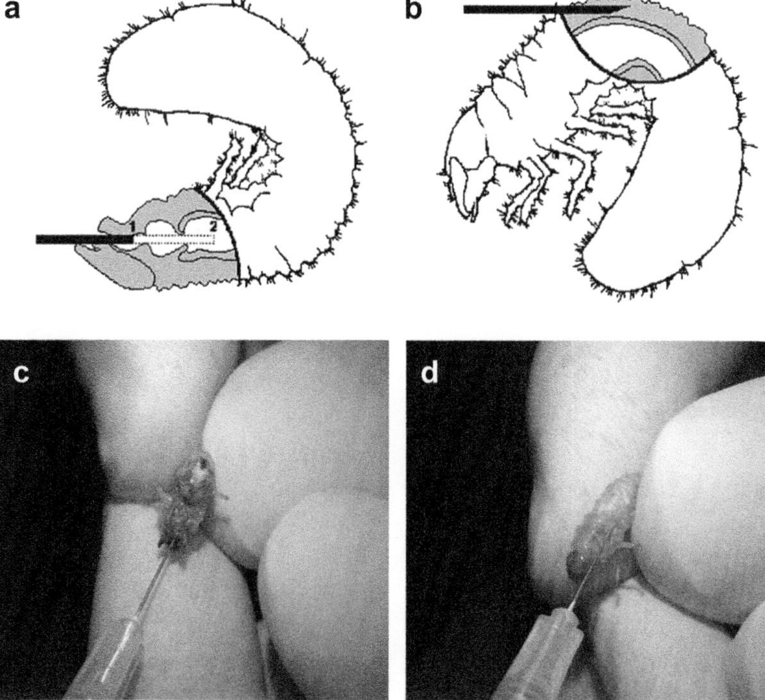


**Figure 5.6** Inoculation of scarab larvae. a and c. oral inoculation of a grass grub larva using a bluntened 30-gauge needle inserted into the oral cavity (1) or midgut (2). b and d. Intracoelomic injection of a grass grub larva with a 30-gauge needle inserted dorsally between the second and third abdominal segments. (a and b from Jackson & Saville 2000).

before being released into pots with soil and grass in the greenhouse. To test milky disease bacteria under greenhouse conditions a similar setup to that described for *B. thuringiensis* below (Section 5C) could be used. A spore–talc preparation would be added to the experimental units either onto the soil surface and then watered in or mixed into the soil (preferably in top 5 cm of soil or as appropriate for the system tested). Larvae would then be recovered after a period of time and examined for infection with exposure length depending on experimental conditions and objectives. This is the method by which *P. popilliae* is applied in the field for control of *P. japonica* (Klein *et al.*, 2007).

## B. Bioassay of *Serratia* spp.

Bioassays to detect amber disease bacteria are based on the ability of pathogenic bacteria to cause a cessation of feeding and produce the distinctive symptoms of the disease in infected larvae. As bioassays are time consuming and test insects are often difficult to obtain, it is important that the desired result, differentiation into pathogenic and non-pathogenic isolates, is obtained with the minimum of effort. In order to establish a screening program it is important to establish a clear-cut testing protocol. The dose of microbe should be calculated to provide a 'maximum challenge' but with minimal effects caused by non-pathogenic bacteria. The conditions of the assay and expected outcomes should be clearly defined. The design of the assay can then be established to meet these criteria. As an example, the protocol for detection of amber disease causing strains of bacteria is set out below.

### 1. *Testing protocol for detection of amber disease causing isolates of bacteria (from Jackson & Saville, 2000)*

- Scarab larvae should be collected from the field (ideally early 3rd instar, which can be identified by head capsule width, relatively small body size and a darkened midgut).
- Larvae are placed in individual compartments (2.5 $cm^3$) in trays with small ($\approx$ 20 mg) cubes of carrot for food and maintained in humid conditions in a dark incubator at 15°C for 24–48 h.
- Test bacteria are applied to carrot cubes either in suspension by micropipette (5 µl containing at least $1 \times 10^7$ cells) or the carrot cubes can be rolled across a growth of bacterial cells on an agar plate to provide a coating of bacteria on the carrot.
- Treated and control carrot cubes are placed in trays according to a predetermined randomized block design (see below).
- Healthy, feeding larvae are added to cells in the trays in the random order specified by the design.
- Assessment of feeding is made after 48 h and larvae transferred to fresh trays with untreated carrot.
- After a further 48–72 h, larvae are assessed for feeding and expression of amber disease symptoms.

To determine amber disease, larvae that do not feed on the fresh untreated carrot offered in the 2–5-day period and show the gut clearance characteristic of amber disease within 5–7 days are categorized as diseased while feeding larvae with a darkened gut are noted as healthy. Dead larvae (usually very few) are omitted from analysis as treatment is not expected to result in death within the time of the experiment. When large numbers of strains are being screened for pathogenicity, the key distinction is between the disease level observed for each strain and that observed in the control treatment, so multiple replicates of the control treatment should be included in the assay. It is also important that true randomization and replication are used in the experimental design (for a discussion of these points, see Jackson & Saville, 2000). The bioassay described above can be modified to compare strains in dosage response assays and evaluate different bacteria. To deliver exact doses of bacteria to the target host oral inoculation with a blunted needle can be carried out as described above (Figure 5.5).

## C. *Bacillus thuringiensis*

Sharpe (1976) found that a preparation of *B. thuringiensis* var. *galleriae* applied with a small spatula directly to the mouthparts of *P. japonica* larvae was able to invade the hemocoel and the parasporal crystal was found to be toxic. About 20% of diseased larvae contained sporulating *B. thuringiensis* as the only septicemic agent. Later, Sharpe & Detroy (1979) showed that *B. thuringiensis* preparations were only effective when given to actively feeding Japanese beetle larvae.

Ladd (1988) added spores and/or crystals of *B. thuringiensis* directly to agar medium to which 5%

sucrose had been added as a feeding stimulus. Small larvae were placed directly in holes made in the agar, or agar plugs were removed and fed to larvae (A. Slaney, personal communication).

Oestergaard *et al.* (2006) tested the virulence of *B. thuringiensis* var. *israelensis* (*Bti*) against 1st instars of the European crane fly, *Tipula paludosa*, another lawn pest.

- Cell wells of 24-well plates were half-filled with 1.2% agar on which first instars were individually kept and fed *Bti* treated leaves discs (chickweed, *Stellaria media*).
- *Bti* was applied to the discs at 8 μg/cm$^2$ using a Potter spray tower and transferred into the wells.
- Well plates were kept at 4, 8, 15, or 20°C with mortality being assessed daily for 4 days.
- There were 120 larvae for each temperature.

*Bti* caused over 90% mortality even at 4°C but the time to kill the larvae increased with decreasing temperature.

To test the virulence of *B. thuringiensis* var. *japonensis* (Buibui strain) (*Btj*) Suzuki *et al.* (1992) used laboratory reared scarab larvae. They collected eggs by allowing adults to oviposit in sand, and used the follow-up procedure.

- Eggs were transferred to compost medium and allowed to hatch.
- First to 3rd instars of various scarabs were fed in individual cups containing 1 g of sterilized compost with various amounts of *Btj* crystals for up to 21 days at 25°C in the dark.
- Mortality was assessed at weekly intervals.
- Ten larvae were used for each dosage and each experiment was conducted three times.

*Btj* caused 65−100% mortality at 25 μg toxin/g compost to a range of scarab species with 100% mortality of *Anomala cuprea* at 0.5 g toxin/g compost.

Later, Suzuki *et al.* (1994) suggested rearing scarab larvae at 25°C, 60% relative humidity, and 12 h/12 h dark/light.

Koppenhöfer & Kaya (1997) used field-collected 3rd instar *C. hirta* to determine the virulence of a liquid formulation of *Btj*.

- To reduce the possibility of the larvae being already infected by pathogens, only healthy looking larvae were first incubated for 4−6 days in 30-ml plastic cups filled with a sterilized moist [−6 kPa soil water potential; 23% (w/w) moisture] sandy loam soil with grass seed at 21−25°C. Only actively feeding and still-healthy looking larvae were used in the ensuing experiments.
- For experiments, the same cups, soil, grass, moisture, and temperature conditions were used except that the soil in treatments had been moistened with *Btj* suspension at the appropriate concentration (final concentrations of 3.13−100 μg δ-endotoxin per gram of soil) and was thoroughly mixed.
- After addition of a larva, the cups were kept on trays in plastic bags with moist paper towels to reduce moisture loss.
- Grub mortality was assessed after 7, 14, and 21 days.
- Treatments were replicated four times with 10−12 individual grubs per replicate.

The $LC_{50}$ for 3rd instar *C. hirta* was determined at 35 μg toxin per gram of soil after 21 days of exposure.

Greenhouse pot experiments were conducted by Koppenhöfer *et al.* (1999) to test the virulence of *Btj* against field-collected 3rd instar *C. hirta* and *C. pasadenae*.

- Plastic pots (14 cm diameter × 11 cm height) were filled to 9 cm height with a loamy sand and grass grown in the pots in the greenhouse (average temperature 23°C) for 3−5 weeks. The grass was cut weekly and watered two to three times per week.
- Nine healthy-looking larvae were placed on the soil surface 3 days before the start of the experiment; larvae that did not dig into the soil within 24 h were replaced.
- Experiments started with an application of 50 ml water or 50 ml *Btj* suspension containing the appropriate amount of a liquid *Btj* formulation to obtain the target rate of δ-endotoxin (0.25−10 kg/ha).
- The pots were destructively sampled 24 days after *Btj* application to determine larval survival.
- Treatments were replicated six to seven times.

To test the effect of *B. thuringiensis* SDS-502 on adult Japanese beetles, Yamaguchi *et al.* (2008) used the following procedure.

- Spore and crystal preparations of bacteria were suspended in 6% sucrose.
- Leaf discs of Virginia creeper were cut and coated with the bacterial suspension.
- Adult Japanese beetles were captured and introduced into containers with the leaf discs.

- Mortality of beetles and leaf area consumed were recorded after 3 days.

Japanese beetle and Alder leaf beetle (*Agelastica menetriesi*) were both found to be susceptible to the bacterium.

## 6. PRESERVATION

### A. Milky disease bacteria

Both *P. popilliae* and *P. lentimorbus* spores may be collected on glass slides and retained for future use in infectivity tests (Dutky, 1963). This method for preparing milky disease stock culture slides involves the following:

- Wash about 20 milky larvae in running water, followed by thorough rinsing in 40–50% alcohol; transfer to 45–50% alcohol.
- Transfer to 45–50% alcohol for 5 min.
- Transfer larvae to water at 45–50°C for 5 min.
- Remove larvae and place on clean cheesecloth or blotting paper until used.
- Bleed each larva by puncturing behind the head capsule with a needle, on to a clean slide (do not puncture or rupture gut).
- Place a second slide over the first leaving about 1.25 cm at the ends and separate the slides with a quick pulling movement in opposite directions.
- Allow slides to dry and store in a slide box for future use.

This procedure is designed to keep the blood from coagulating and to allow the maximum recovery of spores from each larva. Longer conditioning times may be required for larger larvae. In addition, the hemolymph from larger infected larvae can be placed on more than one set of slides, so that four to six slides may be made from a single larva. Spores isolated on microscope slides by Dr. S. Dutky in 1945 using the above procedure, and subsequently stored at room temperature, were as infective as spores from any other source when tested against Japanese beetle larvae in 1983 (Klein & Jackson, 1992).

For vegetative cells, lyophilization of bacteria in the early stationary phase in a medium with less than 0.2% fermentable carbohydrate has been recommended. Vegetative cells of *P. popilliae* and *P. lentimorbus* have also been maintained by transferring to fresh J or MYPGP agar every seven to 10 days (Stahly *et al.*, 1992a). Gordon *et al.* (1973) used monthly transfers of *P. popilliae* in tubes of J-Medium without glucose, but with 0.1% agar. Milky disease bacteria can also be preserved by adding glycerol (to 15%) to mid-log phase cells in MYPGP broth and storing the cell suspensions at −70°C (Dingman, personal communication). Lyophilization of vegetative cells, in either bovine serum or a solution of 5% sodium glutamate and 0.5% gum tragacanth, has also been successful (Stahly *et al.*, 1992a). Spores can be washed and added to dry, sterile soil, placed on microscope slides and dried, or kept as suspensions in water or alcohol (Milner, 1981; Stahly *et al.*, 1992a).

### B. *Serratia* spp.

Detailed suggestions for the maintenance of bacterial cultures from different genera are presented in Bergey's Manual (Garrity, 2005). In general, once purified, bacterial isolates should be stored for short-term preservation on agar slants or stabs under refrigeration. New isolates should be stored in an established reference collection. As with all microorganisms, subculture of bacterial strains should be kept to a minimum from the original reference to avoid loss of virulence and the selection of mutant strains in working cultures. For long-term storage, bacteria should either be kept frozen in glycerol, preferably at −80°C, under liquid nitrogen, or be lyophilized. Convenient systems for frozen storage have been developed as the Microbank (www.pro-lab.com) and Protect (www.tscswabs.co.uk) systems.

### C. *Bacillus thuringiensis*

As with non-spore-forming bacteria, the spore-forming species should not be maintained by subculturing at short intervals (Claus & Berkeley, 1986). Rather, sporulated cultures can be kept for at least 1 year at room temperature on agar slants protected from drying. Both sporulating and vegetative strains can be preserved by freeze-drying in a medium of skim milk powder (20% w/v) supplemented with 5% (w/v) meso-inositol, and subsequently stored under vacuum. As an alternative, cells can be preserved in liquid nitrogen using nutrient broth supplemented with 10% (v/v) glycerol or 5% (v/v) dimethyl sulfoxide as a cryoprotective medium. This latter method is recommended if vegetative cells are to be protected (Claus & Berkeley, 1986). Other methods for long-term storage of *Bacillus* spp. can be found in Chapter III.

## REFERENCES


Allardyce, R. A., Keenan, J. I., O'Callaghan, M., & Jackson, T. A. (1991). Serological identification of *Serratia entomophila*, a bacterial pathogen of the New Zealand grass grub (*Costelytra zealandica*). *J. Invertebr. Pathol., 57*, 250–254.

Asano, S., Yamashita, C., Iizuka, T., Takeuchi, K., Yamanaka, S., Cerf, D., & Yamamoto, T. (2003). A strain of *Bacillus thuringiensis* subsp. *galleriae* containing a novel *cry8* gene highly toxic to *Anomala cuprea* (Coleoptera: Scarabaeidae). *Biol. Control, 28*, 191–196.

Berkowitz, D. M., & Lee, W. S. (1973). A selective medium for the isolation and identification of *Serratia* marcescens. *Abstr. Ann. Meet. Am. Soc. Microbiol., 1973*, 105.

Bixby, A., Alm, S. R., Power, K., Grewal, P., & Swier, S. R. (2007). Susceptibility of four species of turfgrass-infesting scarabs (Coleoptera: Scarabaeidae) to *Bacillus thuringiensis* serovar *japonensis* strain Buibui. *J. Econ. Entomol., 100*, 1604–1610.

Claus, D., & Berkeley, R. C. W. (1986). Genus *Bacillus* Cohn 1872. In P. H. A. Sneath, N. S. Mair, M. E. Sharpe, & J. G. Holt (Eds.), *Bergey's Manual of Systematic Bacteriology* (Vol. 2) (pp. 965–1599). Baltimore: Williams and Wilkins.

Claus, H., Jackson, T. A., & Filip, Z. (1995). Characterization of *Serratia entomophila* strains by DNA fingerprints and plasmid profiles. *Microbiol. Res., 150*, 159–166.

Clarridge, J. E., III (2004). Impact of 16S rRNA gene sequence analysis for identification of bacteria on clinical microbiology and infectious diseases. *Clinical Microbiol. Rev., 17*, 840–862.

Costilow, R. N., & Coulter, W. H. (1971). Physiological studies of an oligosporogenous strain of *Bacillus popilliae*. *Appl. Microbiol., 22*, 1076–1084.

Correa, M. M., & Yousten, A. A. (2001). Pulsed-field gel electrophoresis for the identification of bacteria causing milky disease in scarab larvae. *J. Invertebr. Pathol., 78*, 278–279.

David, H., & Alexander, K. C. (1975). Natural occurrence of milky disease bacterium *Bacillus popilliae* Dutky on white grubs in India. *Current Sci., 44*, 819–820.

Dingman, D. W. (1996). Description and use of a peroral injection technique for studying milky disease. *J. Invertebr. Pathol., 67*, 102–104.

Dingman, D. W. (2008). Geographical distribution of milky disease bacteria in the eastern United States based on phylogeny. *J. Invertebr. Pathol., 97*, 171–181.

Dingman, D. W. (2009). DNA fingerprinting of *Paenibacillus popilliae* and *Paenibacillus lentimorbus* using PCR-amplified 16S-23S rDNA intergenic transcribed spacer (ITS) regions. *J. Invertebr. Pathol., 100*, 16–21.

Dutky, S. R. (1940). Two new spore-forming bacteria causing milky disease of the Japanese beetle. *J. Agric. Res., 61*, 57–68.

Dutky, S. R. (1941). Testing the possible value of milky diseases for control of soil-inhabiting larvae. *J. Econ. Entomol., 34*, 217–218.

Dutky, S. R. (1942). Method for the preparation of spore–dust mixtures of type A milky disease of Japanese beetle larvae for field inoculation. *USDA Bur. Entomol. Pl. Quar., ET-192*, 15.

Dutky, S. R. (1963). The milky diseases. In E. A. Steinhaus (Ed.), *Insect Pathology an Advanced Treatise* (Vol. 2) (pp. 75–115). New York: Academic Press.

Ellis, B-J., Obenchain, F. & Mehta, R. (1989) *In vitro* method for producing infective bacterial spores and spore-containing insecticidal compositions. U.S. Patent, 4,824,671.

Garrity, G. M. (2005). *Bergey's Manual of Systematic Bacteriology The Proteobacteria* (2nd ed.) (Vol. 2). New York: Springer. pp. 2816.

Gatehouse, H. S., Tan, B., Christeller, J. T., Hurst, M. R. H., Marshall, S. D. G., & Jackson, T. A. (2009). Phenotypic changes and the fate of digestive enzymes during induction of amber disease in larvae of the New Zealand grass grub (*Costelytra zealandica*). *J. Invertebr. Pathol., 101*, 215–221.

Glare, T. R., Corbett, G. E., & Sadler, A. J. (1993). Association of a large plasmid with amber disease of the New Zealand grass grub, *Costelytra zealandica*, caused by *Serratia entomophila* and *S. proteomaculans*. *J. Invertebr. Pathol., 62*, 165–170.

Gordon, R. E., Haynes, W. C., & Pang, C. H.-N. (1973). The genus *Bacillus*. *USDA Agric. Handbook, 427*, 283.

Grimont, P. A. D., Jackson, T. A., Ageron, E., & Noonan, M. J. (1988). *Serratia entomophila* sp. nov. associated with amber disease in the New Zealand grass grub, *Costelytra zealandica*. *Intern. J. Syst. Bacteriol., 38*, 1–6.

Harrison, H., Patel, R., & Yousten, A. A. (2000). *Paenibacillus* associated with milky disease in Central and South American scarabs. *J. Invertebr. Pathol., 76*, 169–175.

Holbrook, R., & Anderson, J. M. (1980). An improved selective and diagnostic medium for the isolation and enumeration of *Bacillus cereus* in foods. *Can. J. Microbiol., 26*, 753–759.

Hurst, M. R. H., Becher, S. A., Young, S. D., Nelson, T. L., & Glare, T. R. (2011). *Yersinia entomophaga* sp. nov., isolated from the New Zealand grass grub *Costelytra zealandica*. *Intern. J. System. Evol. Microbiol., 61*, 844–849.

Hurst, M. R. H., Glare, T. R., Jackson, T. A., & Ronson, C. W. (2000). Plasmid-located pathogenicity determinants of *Serratia entomophila* the causal agent of amber disease of grass grub show similarity to the insecticidal toxins of *Photorhabdus luminescens*. *J. Bacteriol., 182*, 5127–5138.

Hurst, M. R. H., Young, S. D., & O'Callaghan, M. (2008). Development of a species-specific probe for detection of *Serratia entomophila* in soil. *NZ Plant Prot., 61*, 222–228.

Jackson, T. A. (1992). Scarabs—pests of the past or future? In T. A. Jackson, & T. R. Glare (Eds.), *Use of Pathogens in Scarab Pest Management* (pp. 1–10) Andover: Intercept Ltd.

Jackson, T. A., & Klein, M. G. (2006). Scarabs as pests: a continuing problem. Scarabaeoidea in the 21st Century: A Festschrift Honoring Henry F. Howden. *Coleopt. Bull., 60*, 102–119.

Jackson, T. A., Boucias, D. G., & Thaler, J. O. (2001). Pathobiology of amber disease, caused by *Serratia* spp., in the New Zealand grass grub, *Costelytra zealandica*. *J. Invertebr. Pathol., 78*, 232–243.

Jackson, T. A., & Saville, D. J. (2000). Bioassays of replicating bacteria against soil-dwelling insect pests. In A. Navon, & K. R. S. Asher (Eds.), *Bioassays of Entomopathogenic Microbes and Nematodes* (pp. 73–93). Wallingford, UK: CABI Publishing.

Jackson, T. A., Huger, A. M., & Glare, T. R. (1993). Pathology of amber disease in the New Zealand grass grub *Costelytra zealandica* (Coleoptera: Scarabaeidae). *J. Invertebr. Pathol., 61*, 123–130.

Jackson, T. A., Townsend, R. J., Nelson, T. L., Richards, N. K., & Glare, T. R. (1997). Estimating amber disease in grass grub populations by visual assessment and DNA colony blot analysis. *Proc. NZ Plant Prot. Conf., 50*, 165–168.

Johnson, V. W., Pearson, J. F., & Jackson, T. A. (2001). Formulation of *Serratia entomophila* for biological control of grass grub. *NZ Plant Prot., 54*, 125–127.

Klein, M. G. (1992). Use of *Bacillus popilliae* in Japanese beetle control. In T. A. Jackson, & T. R. Glare (Eds.), *Use of Pathogens in Scarab Pest Management* (pp. 179–189). Andover: Intercept Ltd.

Klein, M. G. (1995). Microbial control of turfgrass insects. In R. L. Brandenburg, & M. G. Villani (Eds.), *Handbook of Turfgrass Insect Pests* (pp. 95–100). Lanham: Entomological Society of America.

Klein, M. G., & Jackson, T. A. (1992). Bacterial diseases of scarabs. In T. A. Jackson, & T. R. Glare (Eds.), *Use of Pathogens in Scarab Pest Management* (pp. 43–61). Andover: Intercept Ltd.

Klein, M. G., Grewal, P. S., Jackson, T. A., & Koppenhöfer, A. M. (2007). Lawn, turf and grassland pests. In L. A. Lacey, & H. K. Kaya (Eds.), *Field Manual of Techniques in Invertebrate Pathology: Application and Evaluation of Pathogens for Control of Insects and Other Invertebrates* (2nd ed.). (pp. 655–675). Dordrecht, The Netherlands: Springer.

Koppenhöfer, A. M. (2007). Microbial control of turfgrass insects. In M. Pessarakli. (Ed.), *Handbook of Turfgrass Management and Physiology* (pp. 299–314). Boca Raton, FL: CRC Press.

Koppenhöfer, A. M., & Kaya, H. K. (1997). Additive and synergistic interaction between entomopathogenic nematodes and *Bacillus thuringiensis* for scarab grub control. *Biol. Control, 8*, 131–137.

Koppenhöfer, A. M., Choo, H. Y., Kaya, H. K., Lee, D. W., & Gelernter, W. D. (1999). Increased field and greenhouse efficacy against scarab grubs with a combination of an entomopathogenic nematode and *Bacillus thuringiensis*. *Biol. Control, 14*, 37–44.

Krieg, N. R., & Holt, J. G. (1984). *Bergey's Manual of Systematic Bacteriology*, (Vol 1). Baltimore, MD: Williams and Wilkins. 964.

Krieger, L., Franken, E., & Schnetter, W. (1996). *Bacillus popilliae* var *melolontha* H1, a pathogen for the May beetles, *Melolontha* spp. In T. A. Jackson, & T. R. Glare (Eds.), *Proceedings of the 3rd International Workshop on Microbial Control of Soil Dwelling Pests* (pp. 79–87). Lincoln, New Zealand, AgResearch.

Lacey, L. A., Amaral, J. J., Coupland, J., & Klein, M. G. (1994). The influence of climatic factors on the flight activity of the Japanese beetle (Coleoptera: Scarabaeidae): implications for use of a microbial control agent. *Biol. Control, 4*, 298–303.

Ladd, T. L., Jr. (1988). Japanese beetle (Coleoptera: Scarabaeidae); influence of sugars on feeding response of larvae. *J. Econ. Entomol., 81*, 1390–1393.

Lingg, A. J., & McMahon, K. J. (1969). Survival of lyophilized *Bacillus popilliae* in soil. *Appl. Microbiol., 17*, 718–720.

Macdonald, R., & Kalmakoff, J. (1995). Comparison of pulsed-field gel electrophoresis DNA fingerprints of field isolates of the entomopathogen *Bacillus popilliae*. *Appl. Environ. Microbiol., 61*, 2446–2449.

Milner, R. J. (1977). A method for isolating milky disease, *Bacillus popilliae* var *rhopaea*, spores from the soil. *J. Invertebr. Pathol., 30*, 283–287.

Milner, R. J. (1981). Identification of the *Bacillus popilliae* group of insect pathogens. In H. D. Burges (Ed.), *Microbial Control of Pests and Plant Diseases 1970–1980* (pp. 45–59). London, UK: Academic Press.

Monk, J.Y.S., Vink, C.J., Winder, L.M., Hurst, M.R.H., & O'Callaghan, M. (2010). DNA Q-PCR and high-resolution DNA melting analysis for simple and efficient detection of biocontrol agents. In H.J. Ridgway, T.R. Glare & S.A. Wakelin (Eds.), *Paddock to PCR: demystifying molecular technologies for practical plant protection*. Proceedings of an International Symposium held 9 August 2010. (pp. 117–124). New Plymouth, New Zealand. Lincoln, New Zealand, New Zealand Plant Protection Society.

O'Callaghan, M., & Jackson, T. A. (1993). Isolation and enumeration of *Serratia entomophila*—a bacterial pathogen of the New Zealand grass grub, *Costelytra zealandica*. *J. Appl. Bacteriol., 75*, 307–314.

O'Callaghan, M., Jackson, T. A., & Glare, T. R. (1997). *Serratia entomophila* bacteriophages: host range determination and preliminary characterization. *Can. J. Microbiol., 43*, 1069–1073.

Oestergaard, J., Belau, C., Strauch, O., Ester, A., van Rozen, K., & Ehlers, R.-U. (2006). Biological control of *Tipula paludosa* (Diptera: Nematocera) using entomopathogenic nematodes (*Steinernema* spp.) and *Bacillus thuringiensis* subsp. *israelensis*. *Biol. Control, 39*, 525–531.

Ohba, M., Iwahana, H., Asano, S., Suzuki, N., Sato, R., & Hori, H. (1992). A unique isolate of *Bacillus thuringiensis* serovar *japonensis* with a high larvicidal activity specific for scarabaeid beetles. *Lett. Appl. Microbiol., 14*, 54–57.

Pearson, J. F., & Jackson, T. A. (1995). Quality control management of the grass grub microbial control product, Invade®. *Proc. Agron. Soc. NZ, 25*, 51–53.

Pettersson, B., Rippere, K. E., Yousten, A. A., & Priest, F. G. (1999). Transfer of *Bacillus lentimorbus* and *Bacillus popilliae* to the genus *Paenibacillus* with emended descriptions of *Paenibacillus lentimorbus* comb. nov. and *Paenibacillus popilliae* comb. nov. *Intern. J. System. Bacteriol., 49*, 531–540.

Poprawski, T. J., & Yule, W. N. (1990). Bacterial pathogens of *Phyllophaga* spp. (Col., Scarabaeidae) in southern Quebec, Canada. *J. Appl. Entomol., 109*, 414–422.

Rippere, K. E., Tran, M. T., Yousten, A. A., Hilu, K. H., & Klein, M. G. (1998). *Bacillus popilliae* and *Bacillus lentimorbus*, bacteria causing milky disease in Japanese beetles and related scarab larvae. *Intern. J. System. Bacteriol., 48*, 395–402.

Sharpe, E. S. (1976). Toxicity of the parasporal crystal of *Bacillus thuringiensis* to Japanese beetle larvae. *J. Invertebr. Pathol., 27*, 421–422.

Sharpe, E. S., & Detroy, R. W. (1979). Susceptibility of Japanese beetle larvae to *Bacillus thuringiensis*: associated effects of diapause, midgut pH, and milky disease. *J. Invertebr. Pathol., 34*, 90–91.

Shu, C., Yan, G., Wang, R., Zhang, J., Feng, S., Huang, D., & Song, F. (2009). Characterisation of a novel *cry8* gene specific to Melolonthidae pests: *Holotrichia oblita* and *Holotrichia parallela*. *Appl. Microbiol. Biotechnol., 84*, 701–707.

Sneath, P. H. A., Mair, N. S., Sharpe, M. E., & Holt, J. G. (1986). *Bergey's Manual of Systematic Bacteriology*, (Vol. 2). Baltimore, MD: Williams and Wilkins, pp. 1599.

Stahly, D. P., & Klein, M. G. (1992). Problems with *in vitro* production of spores of *Bacillus popilliae* for use in biological control of the Japanese beetle. *J. Invertebr. Pathol., 60*, 283–291.

Stahly, D. P., Andrews, R. E., & Yousten, A. A. (1992a). The genus *Bacillus*: insect pathogens. In A. Balows, H. G. Truper, M. Dworkin, W. Harder, & K. H. Schliefer (Eds.), *The Procaryotes* (Vol. 2) (pp. 1697–1745). New York, NY: Springer.

Stahly, D. P., Takefman, D. M., Livasy, C. A., & Dingman, D. W. (1992b). Selective medium for quantitation of *Bacillus popilliae* in soil and in commercial spore powders. *Appl. Environ. Microbiol., 58*, 740–743.

Starr, M. P., Grimont, P. D. A., Grimont, F., & Starr, P. B. (1976). Caprylate-thallous agar medium for selectively isolating *Serratia* and its utility in the clinical laboratory. *J. Clinical Microbiol., 4*, 270–276.

St Julian, G., Pridham, T. G., & Hall, H. H. (1963). Effect of diluents on viability of *Popillia japonica* Newman larvae, *Bacillus popilliae* Dutky, and *Bacillus lentimorbus* Dutky. *J. Insect Pathol., 5*, 440–450.

Suzuki, N., Hori, H., Ogiwara, K., Asano, S., Sato, R., Obha, M., & Iwahana, H. (1992). Insecticidal spectrum of a novel isolate of *Bacillus thuringiensis* serovar *japonensis*. *Biol. Control, 2*, 138–142.

Suzuki, N., Hori, H., Tachibana, M., & Asano, S. (1994). *Bacillus thuringiensis* strain Buibui for control of cupreous chafer, *Anomala cuprea* (Coleoptera: Scarabaeidae), in turfgrass and sweet potato. *Biol. Control, 4*, 361–365.

Thurston, G. S., Kaya, H. K., & Gaugler, R. (1994). Characterizing the enhanced susceptibility of milky disease-infected scarabaeid grubs to entomopathogenic nematodes. *Biol. Control, 4*, 67–73.

Visnovsky, G. A., Smalley, D. J., O'Callaghan, M., & Jackson, T. A. (2008). Influence of culture medium composition, dissolved oxygen concentration and harvesting time on the production of *Serratia entomophila*, a microbial control agent of the New Zealand grass grub. *Biocontrol Sci. Technol., 18*, 87–100.

Yamaguchi, T., Sahara, K., Bando, H., & Asano, S. (2008). Discovery of a novel *Bacillus thuringiensis* Cry8D protein and the unique toxicity of the Cry8D-class proteins against scarab beetles. *J. Invertebr. Pathol., 99*, 257–262.

Yu, H., Zhang, J., Huang, D., Gao, J., & Song, F. (2006). Characterization of *Bacillus thuringiensis* strain Bt185 toxic to the Asian cockchafer: *Holotrichia parallela*. *Current Microbiol., 54*, 13–17.

## APPENDIX 1

## Some media for isolation and culture of entomopathogenic bacteria for soil insects

### *Media for isolation of* Serratia *species*

1. Caprylate-thallous agar (CTA) for isolation of Serratia (Starr *et al.*, 1976)

CTA is made by preparation and mixing of two solutions. This recipe makes 1 l agar.

Solution A:

| | |
|---|---|
| Distilled water | 486.0 ml |
| Magnesium sulphate heptahydrate ($MgSO_4.7H_2O$) | 0.15 g |
| Potassium dihydrogen orthophosphate ($KH_2PO_4$) | 0.68 g |
| Dipotassium hydrogen orthophosphate anhydrous (Potassium phosphate) ($K_2HPO_4$) | 2.61 g |
| Calcium chloride solution (1%) ($CaCl_2$) | 1.0 ml |
| Trace Element Solution (see below) | 10.0 ml |
| *n*-Octanoic acid (caprylic acid) [$CH_3(CH_2)_6.COOH$] | 1.1 ml |
| **Thallium(I) sulphate** ($Tl_2SO_4$) | 0.25 g |
| Yeast extract (Difco) 5% w/v solution | 0.1 g |

**Warning:** Thallium(I) sulphate is extremely toxic if inhaled. Avoid contact with skin, eyes, and clothing. Add each ingredient to the distilled water, in the order given, dissolving each ingredient *completely* before adding the next. Adjust pH to 7.20 (use $K_2HPO_4$ to raise pH; $KH_2PO_4$ to lower).

Solution A can be made up to 1 week in advance, sterilized and stored at 4°C.

Solution B:

| | |
|---|---|
| Distilled water | 500.0 ml |
| Sodium chloride (NaCl) | 7.0 g |
| Ammonium sulphate [$(NH_4)_2SO_4$] | 1.0 g |
| Difco agar | 15.0 g |

Dissolve in distilled water. Adjust pH to 7.2 as above. Add agar and heat to boiling while stirring.

Sterilize Solution A and Solution B separately for 15 min at 120°C, 103 kPa (15 psi). Add Solution A to Solution B aseptically, stir vigorously. To prevent precipitation from occurring, pour agar while still hot. Agar plates keep well for several weeks at 4°C but the medium loses effectiveness if remelted.

CTA agar is useful for isolation of most *Serratia* species. Colonies are apparent after 3 days growth at 30°C and will grow to 2–5 mm with further incubation. Normally pigmented species often produce white colonies on CTA. Pure isolates should be confirmed as *Serratia* spp. after growth on DNase agar as a small number of non-*Serratia* spp. can grow on CTA.

Trace element solution for CTA:

| | |
|---|---|
| Trihydrogen phosphate ($H_3PO_4$) | 1.96 g |
| Ferrous sulphate heptahydrate ($FeSO_4.7H_2O$) | 0.055 g |
| Zinc sulphate heptahydrate ($ZnSO_4.7H_2O$) | 0.0287 g |
| Manganese(II) sulphate monohydrate ($MnSO_4.H_2O$) | 0.0223 g |
| Copper(II) sulphate pentahydrate ($CuSO_4.5H_2O$) | 0.0025 g |
| Cobalt(II) nitrate hexahydrate [$Co(NO_3)_2.6H_2O$] | 0.003 g |
| Boric acid powder ($H_3BO_3$) | 0.0062 g |

Dissolve in 1 l distilled water. Store at 4°C.

2. Deoxyribonuclease agar (DNAse) (from Berkowitz & Lee, 1973)

Recipe for 1 l agar:

| | |
|---|---|
| DNase agar (Difco) | 37.8 g |
| distilled water | 900.0 ml |
| L-arabinose | 10.0 g |
| Toluidine blue 0.1% w/v solution | 90.0 ml |

Dissolve agar in water; heat to boiling while stirring. Add the remaining components and sterilize before pouring.

*Serratia* colonies hydrolyze DNA clearing the dark color of the medium around the bacterial colonies. *S. entomophila* and *S. marcescens* are unable to ferment L-arabinose and produce a red halo around the colonies while other *Serratia* spp. will produce a yellow halo.

3. Adonitol agar (from O'Callaghan & Jackson, 1993)

Recipe for 1 l agar:

| | |
|---|---|
| Peptone | 8.33 g |
| Sodium chloride (NaCl) | 4.17 g |
| Bromothymol blue solution (see below) | 10.0 ml |
| Adonitol (dissolved in 20ml distilled water) | 5.0 g |
| Bacto agar | 12.5 g |

Bromothymol blue solution:

| | |
|---|---|
| Bromothymol blue | 0.2 g |
| Sodium hydroxide (NaOH), 0.1 M | 5.0 ml |
| Distilled water | 95.0 ml |

Dissolve peptone and NaCl completely in distilled water. Adjust pH to 7.4 before adding the bromothymol blue solution and the dissolved adonitol. Bring the volume of the solution to 1 l using distilled water. Heat to boiling while stirring to dissolve the agar. Sterilize for 30 min at 120°C, 103 kPa. Pour while still hot. Store at room temperature.

After 24 h incubation at 30°C on adonitol agar, *S. entomophila* produces yellow colonies, and *S. proteamaculans* produces blue/green colonies allowing differentiation between the two species.

4. Itaconate agar (from O'Callaghan & Jackson, 1993)

Recipe for 1 l agar:

| | |
|---|---|
| Sodium phosphate (dibasic) ($Na_2HPO_4$) | 6.0 g |
| Potassium phosphate (monobasic) ($KH_2PO_4$) | 3.0 g |
| Sodium chloride (NaCl) | 0.5 g |
| Ammonium chloride ($NH_4Cl$) | 1.0 g |
| Difco agar | 15.0 g |
| 0.01 M $CaCl_2$ solution (sterile) | 10.0 ml |
| 1 M $MgSO_4 \cdot 7H_2O$ solution (sterile) | 1.0 ml |
| 20% itaconate solution (filter sterilized) | 10.0 ml |

Dissolve the first four ingredients in 1 l distilled water in the order given. Adjust pH to 7.0. Add agar and heat to boiling while stirring. Sterilize for 30 min at 120°C, 103 kPa. Allow to cool, add the sterile solutions and mix thoroughly before pouring the agar plates. *S. entomophila* is the only *Serratia* spp. to grow on itaconate agar.

*Media for isolation and culture of* Paenibacillus *spp*

1. J-medium (from St Julian *et al.*, 1963)

Recipe for 1 l agar:

| | |
|---|---|
| Tryptone | 5.0 g |
| Yeast extract | 15.0 g |
| Dipotassium phosphate ($K_2HPO_4$) | 3.0 g |
| Glucose (filter sterilized) | 2.0 g |
| Distilled water | 1000.0 ml |

Dissolve ingredients in distilled water. Adjust pH to 7.3–7.5. If a solidified medium is desired, add 20 g agar and a magnetic stirbar; heat to boiling. Sterilize for 15 min at 121°C, 103 kPa. Add glucose after mixture cools.

2. MYPGP medium

Recipe for 1 l agar:

| | |
|---|---|
| Mueller–Hinton broth | 10.0 g |

| | |
|---|---|
| Yeast extract | 10.0 g |
| Dipotassium ($K_2HPO_4$) | 3.0 g |
| Sodium pyruvate ($C_3H_3O_3Na$) | 1.0 g |
| Glucose (filter sterilized) | 2.0 g |
| Distilled water | 1000 ml |

Dissolve ingredients in distilled water. Adjust pH to 7.1. If a solidified medium is desired, add 20 g of agar and heat to boiling while stirring. Sterilize for 15 min at 120°C, 103 kPa. Add glucose after mixture cools. Vancomycin may be added for isolation of tolerant strains of *P. popilliae* (Stahly *et al.*, 1992)

**3.** PEMBA (Holbrook & Anderson, 1980)

Recipe for 1 l agar:

| | |
|---|---|
| Peptone | 1.0 g |
| D-Mannitol | 10.0 g |
| Magnesium sulphate heptahydrate ($MgSO_4.7H_2O$) | 0.1 g |
| Sodium chloride (NaCl) | 2.0 g |
| Sodium diphosphate (dibasic) ($Na_2HPO_4$) | 2.5 g |
| Potassium dihydrogen orthophosphate ($KH_2PO_4$) | 0.25 g |
| Bromothymol blue (water soluble) | 0.12 g |
| Agar | 18.0 g |
| Distilled water | 1000 ml |

Dissolve ingredients in distilled water. Adjust pH to 7.4. Dispense 90-ml amounts into bottles; sterilize for 15 min at 120°C, 103 kPa.

Before using, add the following solutions to *each* bottle of molten and cooled (50°C) agar:

5.0 ml 20% w/v $C_3H_3O_3Na$, Sodium pyruvate, filter sterilized
100 units/ml Polymyxin, filter sterilized
5.0 ml Egg yolk emulsion (Oxoid SR 47)

If samples suspected of containing large numbers of fungi are to be examined, 1 ml 0.4% (w/v) actidione, filter sterilized, may also be added.

CHAPTER VI

# Identification of entomopathogenic fungi

RICHARD A. HUMBER

USDA-ARS Biological Integrated Pest Management Research Unit, Robert W. Holley Center for Agriculture and Health, 538 Tower Road, Ithaca, NY 14853-2901, USA

## 1. INTRODUCTION

This chapter provides the means for confident identifications of most genera of fungal pathogens affecting insects, spiders, or mites with little or even no formal mycological training. Despite its very broad applicability, this chapter is not a comprehensive guide for identifying all fungal genera affecting these invertebrates; these fungi are too diverse and there have been too many recent genomically based changes in their taxonomies to consider them all here.

While many important entomopathogenic fungi have diagnostic characters making them quickly identifiable, a great deal of rigorously documented, phylogenetically based taxonomic revision has fundamentally transformed the taxonomy of fungi (James *et al.*, 2006; Hibbett *et al.*, 2007) and, incidentally, had some profound effects for many major entomopathogenic fungal taxa.

The key provides multiple paths to a correct identification for fungi with multiple, distinctly different spore types (e.g., the conidial and resting spore states of entomophthoraleans). This chapter also discusses the preparation of slide mounts for light microscopy. Similar techniques are covered in other chapters, but good techniques for both microscopy and making good slides are essential skills for observing key taxonomic characters.

Before photomicrography, microscopic observations were recorded with drawings (freehand or aided by a camera lucida) that sometimes include extraordinary detail. Drawings are still an important and artistically pleasing way to illustrate complex images even though photographic documentation of microscopic images is more universally used. Digital photography has replaced its film-based predecessor and offers its own distinct challenges for most effective communication. The images used here are high-resolution images with minimal software adjustments. Images requiring a high depth of focus to convey essential information—e.g, images with detailed content presented at very low or high magnifications (where focal planes may be too thin to show key details in a single image)—are computer-generated montages of multiple focal planes combined into single images with greatly enhanced depth of focus.

## 2. QUICK GUIDE TO LIGHT MICROSCOPY

The identification of most entomopathogenic fungi necessarily depends on the observation of microscopic characters. Fortunately, however, many common entomopathogens can, with relatively little experience, be easily identified to the genus or, in some instances, the species by observation with either the unaided eye or low magnifications from hand lenses or stereo microscopes. Species identifications usually require confirmation of essential microscopic characters.

The ease with which key microscopic characters can be seen is directly affected by the quality of one's microscopy and slide preparative techniques. The following sections outline the few major skills needed to use a microscope well or to make good slide preparations.

MANUAL OF TECHNIQUES IN INVERTEBRATE PATHOLOGY
ISBN 9780123868992

2012 Published by Elsevier Ltd.

## A. Köhler illumination: the first critical step

The key to observing fine details in a microscope is not magnification; it is optical resolution, the ability to distinguish two adjacent objects. Many factors can affect image resolution, but the first and most important is to maintain Köhler illumination when using bright-field or differential interference optics. Phase-contrast images are much less sensitive to the physical settings of a microscope, but it is always a good idea to maintain Köhler illumination at all times.

The following steps to achieve Köhler illumination should be repeated for each objective used: focus sharply on any object in a slide and then:

1. Close down the field diaphragm (at the light source) and adjust the height of the condenser so that both the inner edge of this iris diaphragm and the object in the slide are sharply focused when seen through the eyepieces.
2. Open the field diaphragm until its image nearly fills the field of view and then center the field diaphragm image in the field of view with the condenser's centering screws.
3. Adjust the opening of the condenser diaphragm. The image of this diaphragm is seen by removing an eyepiece and looking down the inside of the microscope body; a focusing telescope can be useful but is not truly necessary for this step. The condenser diaphragm should be adjusted so that its opening fills some 80–90% of the diameter of the image in this back focal plane.

The condenser diaphragm should never be opened wider than the full diameter of the back focal plane; the resulting 'glare' of too much uncollimated light in the system severely degrades the image resolution. A frequent error in light microscopy is to close down the condenser diaphragm too far to increase the image contrast, but the resulting interference effects (seen as increasing graininess and darkening of object edges) also dramatically reduces image resolution.

## B. Coverslips

Microscopic image resolution is also affected by the type and thickness of coverslips used in slide preparations. The optics of microscope lenses are calculated to allow maximal resolution with #1½ coverslips (0.16–0.19 mm thick); maximal resolution is lower with either #1 and #2 coverslips (with thicknesses of 0.13–0.17 and 0.17–0.25 mm, respectively). Use glass coverslips for diagnostic work. Plastic coverslips are too thick, optically inferior, cause intolerable image degradation, and should be avoided for routine microscopic observations.

Full-sized 18- or 22-mm square or round coverslips may not be the most practical size for diagnostic purposes or whenever one must make large numbers of mounts in a short time. The total amount of glass and mounting medium to be used can be greatly reduced by scribing square coverslips into quarters with a diamond or carbide pencil and a slide edge as a straightedge, and then gently breaking those coverslips along the scratches if they do not break during the scribing. Ten or 12 such miniature coverslips can fit on a standard slide. Not only is less material consumed in this process, but the smaller area under each coverslip makes it easier to locate the fungus to be observed.

## C. Mounting media

Regardless of the mounting medium used, it is important to use no more than is needed to fill the volume under the coverslip. It is alright to use too little mounting medium, but using too *much* floats the coverslip, does not flatten the material to be examined, and prevents any later sealing with nail polish or other slide sealants. Mounting medium can be removed and a preparation further flattened without spreading mounting medium all over the slide (or microscope) and without lateral movement on the specimen by placing the slide into a pad of bibulous paper and applying whatever pressure is needed.

The choice of mounting medium and the means of preparing slide mounts can profoundly affect the apparent sizes of taxonomically important structures (Humber, 1976).

Recipes for some useful mounting media are given below in Section 2G. These include pure lactic acid to which acidic stains (e.g., aniline blue or aceto-orcein) may be added, lactophenol (which is better for semi-permanent mounts than lactic acid and is also compatible with acidic stains), and aceto-orcein (a highly useful general mounting medium that can hydrate even dried specimens and is indispensible for work with the Entomophthorales).

## D. Handling material to be observed

Novice slide-makers often include too much material in a slide in the hope that 'more is better'. In fact, the best, most useful slides usually include a tiny amount of material that has been carefully teased apart and spread. Using only small amounts of material in each mount may force repeated preparations to see specific structures, but the effort required is often distinctly rewarded by the results. In all practicality, most slides for initial diagnoses can be prepared quickly since the main characters are often readily seen regardless of the care in preparation. Mounts meant for photography and or archival preservation, however, benefit greatly from fastidious preparative attention.

The most useful tools to prepare slides of many fungi are not standard dissecting needles whose points are much too large to tease apart delicate fungal structures. The best tools may be '0' and 'minuten' insect pins mounted in soft wood sticks (e.g., the thick wooden match sticks available in the U.S. or wooden chopsticks). The blunt ends of stainless steel '0' (whose heads have been cut off) or 'minuten' insect pins should be pushed into the sticks. The points of both of these types of pins remain small and distinctly pointed even when viewed at high magnification (see Figure 6.1). The '0' pins are superb for coarse operations or teasing apart leathery or hard structures; 'minuten' pins are unexcelled for manipulating hyphae, conidiophores, or other delicate structures. These insect pins are also versatile tools for manipulating cultures. The points of '0' pins can be pounded out into very useful microspatulas. Standard or flattened points of '0' needles can be flame sterilized but the points of 'minuten' pins may melt and even burn if flamed; autoclaving in glass Petri dishes or in groups in folded foil packets is also a convenient way to sterilize these pins.

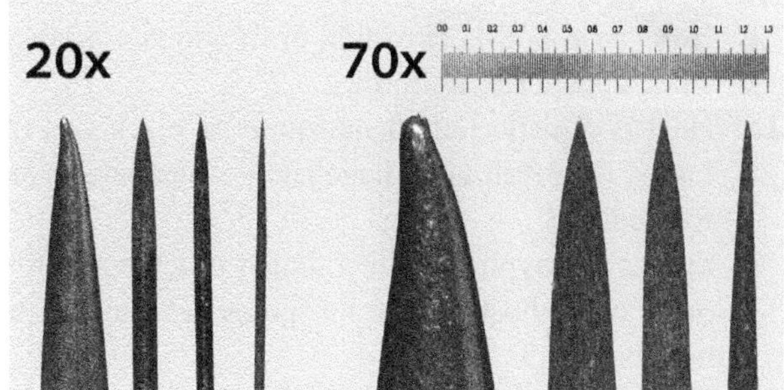


**Figure 6.1** Comparative appearances in a stereo microscope of dissecting needle points (L to R: standard dissecting needle, '0' insect pin, and 0.15 and 0.10 mm diam. 'minuten' insect pins). The higher magnification set includes a 1.3—mm stage micrometer scale.

To make good slides consistently is an art that benefits from plenty of practice and using common sense. Most taxonomically important structures can be detected well enough at magnifications of 50−75 × to know if a slide merits examination on the compound microscope. Virtually all microscopic examination of entomopathogenic fungi for diagnostic purposes can be done at a magnification of 400−450 ×; oil immersion is only rarely needed.

## E. Semi-permanent slide mounts

Most slide mounts are made strictly for immediate observation rather than for long-term storage for later reference. Many differing techniques can be used to make semi-permanent slides, but those most useful for invertebrate pathogens involve means to seal slides prepared with aqueous mounting media.

A very short-term seal may be obtained by painting a melted mixture of roughly equal amounts of paraffin and petroleum jelly around the coverslip. Extreme caution must be used if melting this mix over any open flame since paraffin vapors are highly flammable.

Coverslips are most often sealed by ringing them with fingernail polish, Canada balsam, or another slide-making resin. Apply a relatively narrow and thin first layer; once the sealant is dried, a thicker and more secure seal can be built up by later applications of the sealant, but always be sure that the edges of the subsequent layer(s) cover the inner and outer edges of earlier layer(s). Such preparations may remain sealed for several months but should not be relied on to last for years. No sealing method is likely to work unless only a *minimal* of mounting medium is included under the coverslip; slides on which *any* amount of mounting medium protrudes from under the coverslip will probably fail to seal.

More secure, longer-lasting aqueous mounts can be prepared with methods using two coverslips of unequal sizes (Kohlmeyer & Kohlmeyer, 1972). The basic method shown in Figure 6.2 is simple: the material is spread in a *minimal* drop of mounting medium on the small coverslip; the large coverslip is then lowered on to

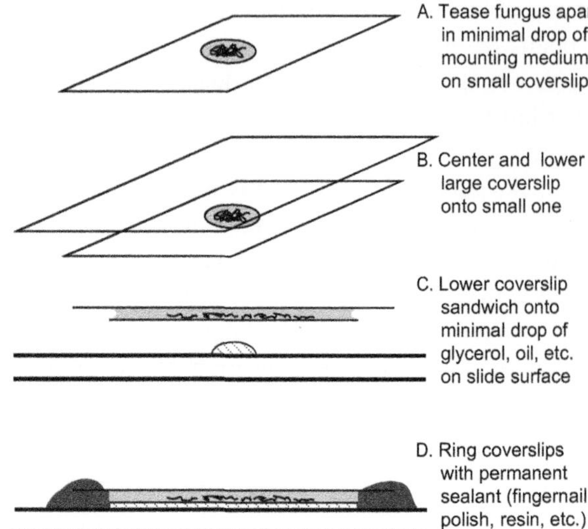


**Figure 6.2** Outline of the procedure to make coverslip 'sandwiches' and semipermanent slides.

the small one; the smaller coverslip of this sandwich is then attached to the standard microscope slide by a drop of glycerol, immersion oil, or resin; and the space under the edge of the large coverslip is filled with a permanent sealant. Kohlmeyer & Kohlmeyer (1972) modified this basic procedure with a preliminary sealing of the small coverslip on to the large one and allowing this first ring to dry before attaching the sandwich to the slide. Such a procedure is easier to describe than to execute flawlessly. Several points should be heeded to increase the likelihood of success:

- The relative size differences of the coverslips should be small. Pairing 18-mm and 22-mm square coverslips is suitable; mixing square and round coverslips should be avoided.
- It takes practice to get small enough drops of mounting fluids.
- It is easiest to use a small paint brush to apply the sealant.
- Adjusting the viscosity and solvent concentration in the sealant is the most difficult problem in this technique. Too much solvent tends to create bubbles in the sealing ring and may destroy the longevity of the mount. Inadequately thinned sealant may be too viscous to fill the space under the large coverslip.
- Excess (hardened) sealant can be cut away with a razor blade to improve the cosmetic appearance of the preparation.

## 3. KEY TO MAJOR GENERA OF FUNGAL ENTOMOPATHOGENS

This key should be used together with the taxonomic treatments and photos in Section 4; some generic entries include information about morphologically similar, related genera but that are identified separately in the key. Genera appear in the key for each distinctly different spore-bearing state formed (e.g., conidial *versus* sexual or thick-walled resistant spore states). More entomopathogenic fungi are illustrated and keyed, although in less detail, in Samson *et al.* (1988). The vegetative states of most fungi have little taxonomic value and are not treated in this key.

If no reproductive structures or spores are seen on first examination, incubate any culture for more time or transfer it to a nutritionally weaker medium; incubate fresh collections or desiccated specimens in a humid chamber at 100% RH for 24–48 h but check closely for fast-growing fungal and bacterial saprobes, and for mite and nematode contaminants, that can quickly overwhelm a fungal entomopathogen.

It is assumed that this key will be used primarily with infected specimens but most of the included fungi should also be identifiable from sporulating cultures so long as the user is aware of the host's identity and has a general idea about the appearance of the fungus on an infected host.

A brief glossary of terms used in the key and generic discussions is presented following the references cited. Textbooks of mycology and the *Dictionary of the Fungi* (Kirk *et al.*, 2008) provide more detailed definitions of terms and concepts.

**1.** Host arthropod is terrestrial or, if aquatic, no flagellate zoospores are formed...2
**1a.** Host is aquatic (larval mosquito or black fly) or, rarely, terrestrial; flagellate zoospores are formed...41
**2.** Spores and hyphae or other fungal structures visible on or cover host; few or no spores form inside the host body...3
**2a.** Fungal growth and sporulation wholly (or mostly) confined inside the host body...33
**3.** Elongated macroscopic structures (synnemata or club-like to columnar stromata) project from host...4

**3a.** Fungal growth may cover all or part of the host but large, projecting structures are absent…11

**4.** Conidia form on synnemata and/or on mycelium on the host body…5

**4a.** Flask-like to laterally flattened fruiting structures (perithecia) borne on or submersed in an erect, dense to fleshy, club-like to columnar stroma or on body of host; if mature, containing elongated asci with thickened apical caps…10

**5.** Conidia formed in short to long chains…6

**5a.** Conidia produced singly on many separate denticles on each conidiogenous cell or singly on conidiogenous cell, dry (or, less often, with a thin slime coat) or in small groups in a slime droplet…8

**6.** Conidiogenous cells flask-like, with swollen base and a distinct neck, borne singly or in loose clusters; chains of conidia often long and divergent (when borne on clusters of conidiogenous cells)… *Isaria* (or *Paecilomyces*)

**6a.** Conidiogenous cells short, with rounded to broadly conical apices (*not* having a distinctly narrowed and extended neck)…7

**7.** Conidiogenous cells clustered on ± swollen vesicle on short to long conidiophores projecting laterally from synnemata and/or the hyphal mat covering the host; conidia pale to yellow or violet in mass; affecting spiders…*Gibellula*

**7a.** Conidiogenous cells apical on broadly branched, densely intertwined conidiophores forming a compact hymenium; conidia in parallel chains forming columns or plate-like masses, usually green in mass…*Metarhizium*

**8.** Conidiogenous cell with swollen base and elongated, narrow to spine-like neck; conidia formed singly (usually with a distinct slime coating) or small groups in a slime droplet…*Hirsutella*

**8a.** Conidiogenous cell producing several to many conidia, each formed singly on separate denticles…9

**9.** Conidiogenous cell with an extended, denticulate apex (growing apex repeatedly forms a conidium and regrows [rebranches] just below the new conidium)…*Beauveria*

**9a.** Conidiogenous cell short, compact, cylindrical to broadly clavate; apex studded by many denticles…*Hymenostilbe*

**10.** Erect stroma bears perithecia superficial to partially or fully immersed (with only small circular opening raised above stromatic surface); perithecia scattered or aggregated into more or less differentiated, apical or lateral fertile part; asci (if present) with thickened apical cap perforated by narrow canal and filiform ascospores (that usually dissociate into one-celled part spores); conidia, if simultaneously present, being formed on host body, on lower portion of stroma, or on separate synnemata…[see *Cordyceps*]

**10a.** Perithecia on or partially immersed in a cottony to woolly hyphal layer covering host…*Torrubiella* (in the broad sense)

**11.** Fungus covering larval whitefly or scale host is a stroma (fleshy to hard mass of intertwined hyphae); sporulation occurs in cavities below the stromatic surface…12

**11a.** Host partially to completely covered by wispy, cottony, woolly, or felt-like growth or by a dark-colored, extensive patch having columns and chambers below its surface but *not* forming a dense stroma…13

**12.** Spores (conidia) fusoid, one-celled, in a slime mass oozing from chambers (with no differentiated wall, lined by conidiogenous cells) immersed in the stroma; affecting whiteflies and scales… *Aschersonia*

**12a.** Globose to flask-like perithecia (with a distinct wall) immersed in stroma, contain elongated asci with thickened apices or, at maturity, a (non-slimy) mass of globose, ovoid or rod-like spores (formed by dissociation of multiseptate ascospores); *Aschersonia* anamorph often present on same stroma…*Hypocrella*

**13.** Fungus forming pale to dark brown or black patch, ± extensive, on woody plant parts; surface dense to felt-like, with elongated or clavate thick-walled cells (teleospores) sometimes present; underneath forming columns and chamber-like spaces sheltering live scale insects; some scales penetrated by tightly coiled haustorial hyphae [Basidiomycota]… *Septobasidium*

**13a.** Fungal hyphae emerging from or covering host are colorless to light colored, wispy to cottony, woolly, felt-like, or waxy-looking mat…14

**14.** Flask-like to laterally compressed perithecia, superficial to partially immersed in fungal mat on the host; asci elongate, with thick apical cap and (if mature) filiform multiseptate ascospores tending to dissociate into one-celled part-spores; conidial state(s) may also occur on host body or synnemata; especially on spiders or hemipterans…*Torrubiella* (in the broad sense)

**14a.** Spores (conidia only) form on external fungal surfaces; no sexual structures (perithecia) are present…15

**15.** Conidia form on cells with elongated denticulate necks bearing multiple conidia or on awl- to flask-shaped or short blocky conidiogenous cells; conidia form singly or successively in dry chains or slime drops (anamorphic forms of ascomycetes)…16

**15a.** Conidia forcibly discharged and may rapidly form forcibly or passively dispersed secondary conidia (Entomophthorales)…25

**16.** Conidiogenous cell with an extended, denticulate apex (growing apex *repeatedly* forms a conidium, branches, and regrows below the previous conidium)…*Beauveria*

**16a.** Conidiogenous cells are awl- to flask-shaped, with or without an obvious neck; conidia single, in chains, or in multispored slime drops…17

**17.** Conidia single or in dry chains on apices of conidiogenous cells…18

**17a.** Conidia aggregate in slime drops at apices of conidiogenous cells…21

**18.** Conidia borne singly on conidiogenous cell with swollen base and one or more narrow, elongated necks; conidia globose or, if not, usually having an obvious slime coat; especially on mites…*Hirsutella*

**18a.** Conidia borne in dry chains, not covered by any obvious slime…19

**19.** Conidiophores much branched in a candelabrum-like manner but very densely intertwined, and forming nearly wax-like fertile areas; conidiogenous cells short, blocky, without apical necks; conidial chains long and, usually, laterally adherent in prismatic columns or continuous plates…*Metarhizium*

**19a.** Conidiophores individually distinct and unbranched or with a main axis and short side branches bearing single or clustered conidiogenous cells…20

**20.** Conidiogenous cells flask-like, with swollen base and a distinct neck, borne singly or in loose clusters; chains of conidia often long and divergent (when borne on clusters of conidiogenous cells)…*Isaria* (or *Paecilomyces)*

**20a.** Conidiogenous cells short and blocky with little obvious neck, borne in small clusters on short branches grouped in dense whorls on (otherwise unbranched) conidiophores; conidial chains short; especially on noctuid lepidopteran larvae…*Nomuraea*

**21.** Conidia aggregating in slime droplets or single on conidiogenous cells; conidia either (a) elongated, gently to strongly curved with ± pointed ends, one or more transverse septa and usually a short (basal) bulge or bend ('foot') and/or (b) one- to two-celled, variously shaped; hyphae may form apical or intercalary thick-walled, smooth or roughened, colorless to gold–brown, ± globose swellings (chlamydospores), single or few in linear groups…*Fusarium*

**21a.** Conidiogenous cells little thicker than hyphae, occurring singly or grouped into regular clusters and/or whorls; conidia one-celled, aggregated in slime drops; mycelium highly uniform in diam…22

**22.** Conidiogenous cells narrow and awl-like or tapering uniformly from slightly swollen base to narrow tip; occurring singly, paired, or in whorls on hyphae or in terminal clusters…23

**22a.** Conidiogenous cells with a swollen to flask-like base and a (usually short) neck often bent out of axis of the conidiogenous cell; conidiogenous cells borne singly, clustered, or in whorls aggregating in loose 'heads' on erect apically branching conidiophores poorly differentiated from vegetative hyphae…*Tolypocladium*

**23.** Multicellular, thick-walled, ± globose chlamydospores formed apically or on short lateral hyphae; usually on nematodes, but can affect insects…*Pochonia* [see *Lecanicillium*]

**23a.** No such chlamydospores present…24

**24.** Conidiogenous cells mostly in whorls but some occurring singly or in pairs…*Lecanicillium*

**24a.** Conidiogenous cells mostly single, few in pairs, with whorls absent or rare…*Simplicillium* [see *Lecanicillium*]

**25.** Primary conidia obviously uninucleate (in aceto-orcein) and sometimes seen to be bitunicate (with outer wall layer lifting partially off of spores in liquid mounts)...26

**25a.** Primary conidia obviously multinucleate or nuclei not readily seen (in aceto-orcein)...29

**26.** Conidia long clavate to obviously elongated (length/width ratio usually $\geq$2.5), papilla broadly conical, often with a slight flaring or ridge at junction with basal papilla...27

**26a.** Conidia ovoid to clavate; papilla rounded and frequently laterally displaced from axis of conidium...28

**27.** Conidia readily forming elongate secondary capilliconidia attached laterally to and passively dispersed from capillary conidiophores; rhizoids and cystidia not thicker than hyphae; rhizoids numerous, often fasciculate or in columns... *Zoophthora*

**27a.** Conidia never forming secondary capilliconidia; conidia often strongly curved and/or markedly elongated; rhizoids and/or cystidia 2–3× thicker than hyphae; especially on dipterans (or other insects) in wet habitats (on wetted rocks, in or near streams, etc.)...*Erynia*

**28.** Conidia never producing secondary capilliconidia; rhizoids 2–3× thicker than hyphae, terminating with prominent discoid holdfast; cystidia at base 2–3× thicker than hyphae, tapering towards apex...*Pandora*

**28a.** Conidia never producing secondary capilliconidia; rhizoids not thicker than hyphae, numerous, solitary to fasciculate, with weak terminal branching system or sucker-like holdfasts; cystidia as thick as hyphae, often only weakly tapered...*Furia*

**29.** In aceto-orcein, nuclei staining readily, with obviously granular contents...30

**29a.** In aceto-orcein, nuclei not readily visible or not staining...32

**30.** Conidia with apical point and broad flat papilla; after forcible discharge, embedded in halo-like zone of clear material...*Entomophthora*

**30a.** Conidia without apical projection and discharged by eversion of a rounded (not flat) papilla...31

**31.** Conidia pyriform with papilla merging smoothly into spore outline; formed by direct expansion of tip of conidiogenous cell (with no narrower connection between conidiogenous cell and conidium); rhizoids never formed...*Entomophaga*

**31a.** Conidia globose with papilla emerging abruptly from spore outline; formed on conidiogenous cells with a narrowed neck below the conidium; if present, rhizoids 2–3× thicker than hyphae, with discoid terminal holdfast...*Batkoa*

**32.** Conidia globose to pyriform, papilla rounded, nuclei numerous but inconspicuous (not stained in aceto-orcein); secondary conidia (a) single and forcibly discharged, (b) single and passively dispersed from apex of capillary conidiophore, or (c) multiple, small, forcibly discharged from single primary conidium; short to long, hair-like villi may cover old conidia...*Conidiobolus*

**32a.** Conidia globose to pyriform, papilla flattened, usually 4-nucleate; secondary conidia (a) forcibly discharged, resembling primary or (b) almond- to drop-shaped, laterally attached to a capillary conidiophore with a sharp subapical bend; especially on aphids or mites...*Neozygites*

**33.** Affecting larval bees (causing chalkbrood in Apidae or Megachilidae); fungus in cadavers is white to black, forming large spheres (spore cysts) containing smaller walled groups (asci) holding (asco)spores...*Ascosphaera*

**33a.** Affecting insects other than bees; ± thick-walled spores forming individually or not as above...34

**34.** Thick-walled spores formed *inside* a ± loosely fitted outer (sporangial) wall...35

**34a.** Thick-walled spores formed directly at apices of hyphae or hyphal bodies by budding or intercalary; not formed loosely inside another (obvious) fungal cell...36

**35.** Spores (oospores) smooth, colorless; formed inside irregularly shaped cell (oogonia); some cells in broad mycelium dividing internally, producing a narrow tube penetrating host cuticle and then forming evanescent terminal vesicle from which many biflagellate zoospores are released; affecting mosquitoes [Chromista: Peronosporomycetes] ...*Lagenidium*

**35a.** Spores (resistant sporangia) ± globose, golden-brown with hexagonally reticulate surface; formed inside close-fitting thin (but evanescent) outer wall; zoospores posteriorly uniflagellate, released through cracked wall on germination [Blastocladiomycota]...*Myiophagus*

**36.** In gregarious (periodical) cicada adults; terminal abdominal exoskeleton sements drop off to expose loose to compact fungal mass; spores thin-walled, 1–4 nucleate or, if thick-walled, with strongly sculptured surface...*Massospora*

**36a.** Not affecting cicadas, with obviously thick-walled spores occurring throughout body (not confined to terminal abdominal segments)...37

**37.** Spores (zygospores or azygospores) smooth or surface irregularly roughened, warted, or spinose; colorless to pale or deeply colored (various colors possible), brown, gray, or black; wall ± obviously two-layered...38

**37a.** Spores (thick-walled resistant sporangia) with surface regularly decorated with ridges, pits, punctations, striations, reticulations; yellow–brown to golden–brown; wall not obviously two-layered...40

**38.** Resting spore outer wall obviously melanized (gray, brown or black; inner wall is hyaline), with smooth or rough surface; binucleate (but nuclei may not stain in aceto-orcein unless wall is cracked); any conidia discharged from infected hosts produce almond- to drop-shaped secondary capilliconidia; affecting aphids, scales, or mites...*Neozygites*

**38a.** Resting spores colorless, colored, or dark, surfaces smooth or rough; any conidia discharged from infected hosts are not as above...38

**39.** When spores are gently crushed in aceto-orcein, nuclear staining is poor or absent; if nuclei are seen, contents are not strongly granular (Ancylistaceae)...*Conidiobolus*

**39a.** Nuclei in spores crushed gently in aceto-orcein have obviously granular (stained) contents...Entomophthoraceae (genus undetermined)

**40.** Sporangia ellipsoid (not globose), with a (±obvious) preformed dehiscence slit; wall thick, golden-brown, pitted to deeply sculptured; affecting larvae/pupae of mosquitoes (or midges) [Blastocladiomycota]...*Coelomomyces*

**40a.** Sporangia ± globose, with no visible dehiscence slit; wall relatively thin; surface with low (hexagonally) reticulate ridges; affecting terrestrial insects [Blastocladiomycota]...*Myiophagus*

**41.** Zoospores biflagellate [Peronosporomycetes]; affecting mosquito larvae...42

**41a.** Affecting terrestrial insects, forming ± globose sporangia with reticulately ridged surface inside host body; when dispersed into water, sporangia release posteriorly uniflagellate zoospores. Rarely collected....*Myiophagus*

**42.** Contents of zoosporangia in host body (mosquito larva) are transferred through narrow discharge tube to a gelatinous vesicle formed outside body, complete development of zoospores; vesicle swells and dissolves rapidly, releasing all zoospores simultaneously...*Lagenidium*

**42a.** Zoospores develop fully in single file in a narrow zoosporangium produced outside host (mosquito larva) and are released singly through the tip of zoosporangium...*Leptolegnia*

## 4. MAJOR GENERA OF FUNGAL ENTOMOPATHOGENS

This section is organized by the currently recognized phylogenetic sequence and its corresponding classification (James *et al.*, 2006; Hibbett *et al.*, 2007). The entomopathogenic watermolds include starting with the conidial fungi that are the most commonly encountered fungal entomopathogens and moving through the ascomycetes and basidiomycetes, Zygomycetes, Peronosporomycetes, and Chytridiomycetes that are progressively less common and may have narrower host ranges. Generic treatments include a brief diagnosis, and lists of major (but not all) diagnostic characters, characterizations of some common and important species, references to taxonomic literature useful for species identification, and, in some instances, further comments.

Labels on the figures correspond to the lettered diagnostic characters of the genera and species.

### A. Blastocladiomycete and Peronosporomycete watermolds

These phylogenetically diverse organisms produce uni- or biflagellate zoospores as the dispersive/infective units. Flagellate zoospores may be released from two possible sorts of sporangia, those with either thin or thick walls (usually propagative sporangia *versus* meiosporangia in which meiosis occurs, respectively). It would be unusual to detect hosts infected by these fungi during their vegetative states; these fungi are usually only detected when sporangia have been formed or are releasing zoospores. The Oomycetes are no longer accepted as being true fungi—although they are still studied by mycologists—but are now classified as 'fungal allies' in the kingdom variously referred to as Chromista or Straminipila for all organisms that produce a tinsel-type flagella (Alexopoulos *et al.*, 1996; Dick, 2001).

### *1.* Coelomomyces *Keilin (Figure 6.3)*

[Blastocladiomycota: Blastocladiales] Mycelium parasitic in hemocoel of aquatic dipteran larvae, coarse, wall-less in early development, budding off thick-walled resistant meiosporangia with thick deep golden or yellow-brown walls decorated by folds, ridges, warts, pits, etc.; sporangia releasing posteriorly uniflagellate zoospores on germination; zoospores infecting copepods or other aquatic crustaceans; haploid mycelium in crustacean hemocoele is wall-less, cleaving off posteriorly uniflagellate gametes that fuse in pairs; biflagellate zygotes encysting on and infecting Diptera, especially Culicidae (see Kerwin & Petersen, 1997).

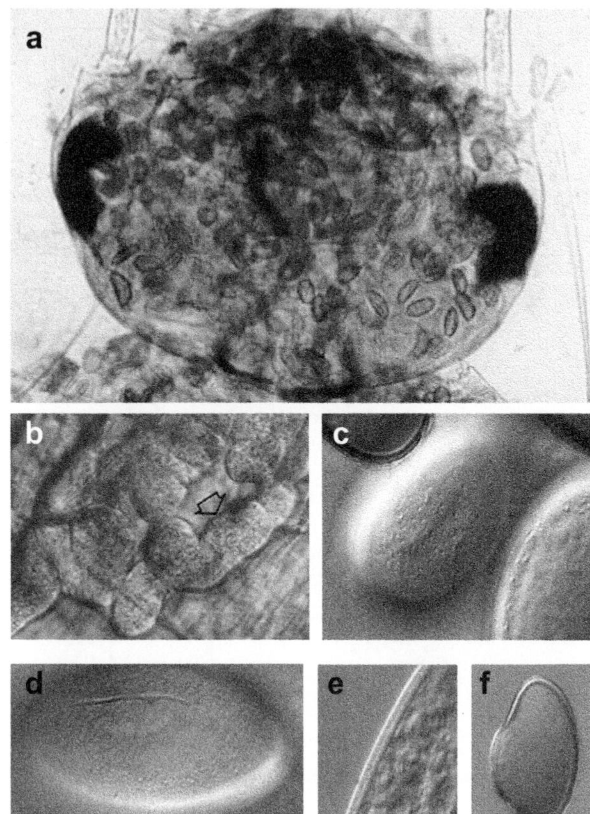


**Figure 6.3** *Coelomomyces*. a. Resistant sporangia in mosquito head (photo: DW Roberts). b. Wall-less hyphal bodies in mosquito hemocoel (from Padua *et al.*, 1986, *J. Invertebr. Pathol.*, *48*, 286). c–e. Resistant sporangia with punctate surface texture (c), preformed germination slit (d), and bilayered wall (e). f. Early germination of resistant sporangium showing gelatinous plug bulging through the germination slit.

#### *a. Key diagnostic characters*

(1) Mosquito (or other aquatic dipteran) larval body filled with coarse, wall-less hyphae that later convert almost completely to golden-brown resistant sporangia. (2) Resistant sporangia: ovoid, thick-walled, golden-brown, with obvious sculpturing or punctation on surface and a pre-formed germination slit. (3) Zoospores: posteriorly uniflagellate when released from germinating resistant sporangia or (as gametes) from haploid thalli in copepod or cladocerans. *Note: Gametes fuse in pairs; biflagellate zygotes swim to and encyst on dipteran host.*

#### *b. Major species*

*C. indicus* Iyengar—resistant sporangia 25–65 × 30–40 μm, with longitudinal ridges frequently anastomosing; affecting *Anopheles* spp. in Africa, India, Australasia, and Philippines. *C. dodgei* Couch & Dodge—resistant sporangia 37–60 × 27–42 μm with longitudinal slits (striae) 3–4 μm apart (giving banded appearance to sporangia); affecting *Anopheles* spp. in North America. *C. psorophorae* Couch—resistant sporangia 55–165 × 40–80 μm, surface with closely spaced punctation; affecting aedine and culicine mosquito larvae in the Northern Hemisphere.

#### *c. Main taxonomic literature*

Bland *et al.*, 1981; Couch & Bland, 1985.

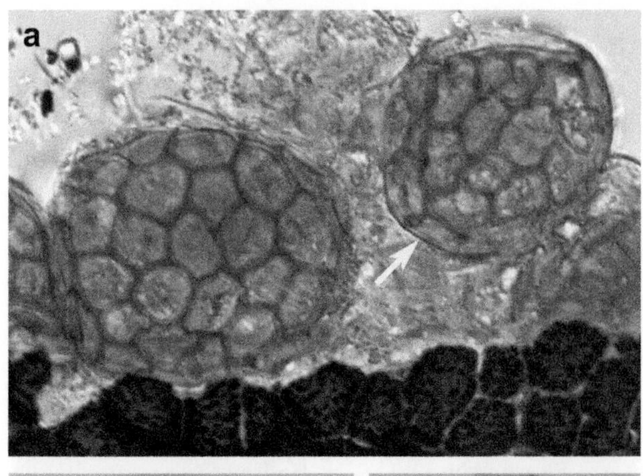


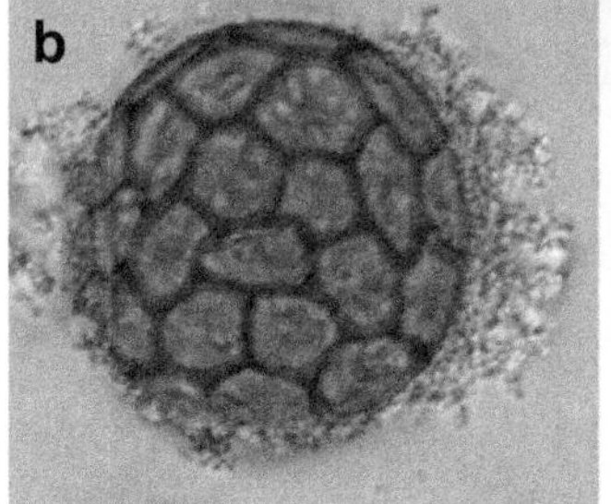


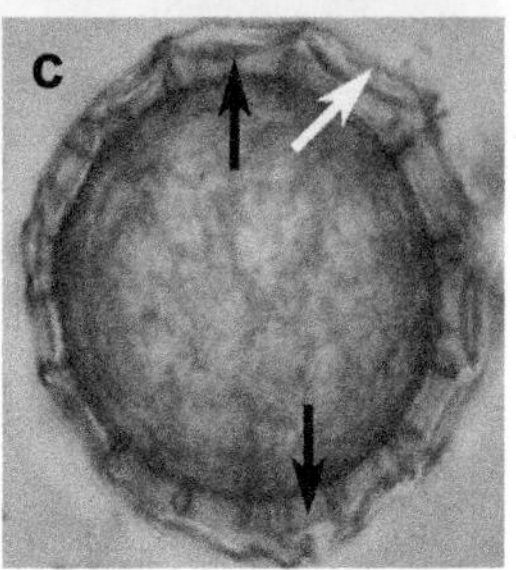


**Figure 6.4** *Myiophagus ucrainicus*. a–b. Reticulate resistant sporangia. c. Two walls are visible; the smooth wall (white arrow) is easily lost and of sporangial origin; the reticulate inner wall (black arrows) is the spore wall.

*2.* Myiophagus *Thaxter (Figure 6.4)*

[Blastocladiomycota: Blastocladiales] Monotypic: *M. ucrainicus* (Wize) Sparrow. Vegetative thallus endozoic, coenocytic, branched with frequent constrictions becoming empty of cytoplasm as sporangia form, and dissociating to produce free sporangia at maturity; sporangia globose (ovoid to ± fusiform), with slightly thickened wall, forming one to five exit papillae; zoospores posteriorly uniflagellate, uninucleate, oval to elongated (4–5 × 7–7.5 μm), with yellow to orange granules in anterior; resting sporangia mostly spherical, with gold-colored outer wall decorated with polygonal reticulation, cracking open upon germination to extrude globose sporangium and releasing zoospores; affecting scales, weevils, and lepidopterans (*not* known from mosquitoes or insects with aquatic stages).

*a. Key diagnostic characters*

(1) Resistant sporangia: globose, thick-walled, 20–30 μm diameter, golden-brown, with prominent hexagonal reticulation of surface. (2) Zoospores (*rarely* observed): posteriorly uniflagellate, released upon germination of resistant sporangia or from globose, thin-walled zoosporangia.

*b. Main taxonomic literature*

Sparrow, 1939; Karling, 1948.

*3.* Lagenidium *Schenk (see figures in Kerwin & Petersen, 1997)*

[Chromista/Straminipila: Peronosporomycetes: Lagenidiales] Mycelium in hemocoel of host (mosquito larva), coarse, thick, coenocytic but becoming septate, cells becoming oval to spherical and serving as zoosporangia or sex organs; partially differentiated zoosporangial contents extrude through evacuation tube (7–10 × 50–300 μm) to form gelatinous-walled vesicle outside host body; zoospores laterally biflagellate, reniform, cleaving in vesicle and released on mass when vesicle wall dissolves; oospores (zygotes) thick-walled, form between segments of same or adjacent mycelial strand.

*a. Key diagnostic characters*

(1) Mycelium becomes cellular; cells acting as zoosporangia or gametangia. (2) Zoosporogenesis: partially cleaved cytoplasmic blocks with thin connecting strands are transferred via a long evacuation tube through host cuticle to an external vesicle; zoospores complete differentiation in vesicle and disperse suddenly when gelatinous vesicle wall dissolves. (3) Zoospores: laterally biflagellate, kidney- or bean-shaped. (4) Resistant spores: thick-walled oospores, globose, formed by conjugations of adjacent cells (gametangia).

*b. Major species*

*L. giganteum* Couch—affecting mosquitoes (only entomopathogenic species).

*c. Main taxonomic literature*

Sparrow, 1960; Bland *et al.*, 1981.

*d. General comment*

*L. giganteum* can be a major mosquito pathogen; *Pythium* species are closely related oomycetes that may also affect mosquitoes but much less frequently. The biflagellate zoospores of both genera disperse in a similar way from externally formed vesicles but

*Pythium* hyphae are thin (≤3–4 μm diameter) and retain obvious cytoplasm after zoosporogenesis; *L. giganteum* hyphae are coarse (generally ≥ 5 μm diameter) whose cytoplasmic contents convert completely into zoospores or oospores.

#### *4.* Leptolegnia *de Bary*

[Chromista/Straminipila: Peronosporomycetes: Saprolegniales] Mycelium parasitic in hemocoel of mosquito larvae, slender, sparingly branched, producing lateral swellings; zoosporangia apical, mostly filamentous, no broader than hyphae (occasionally centrally wider), zoospores formed mostly in a single row. Oogonia usually on short lateral branches, densely ornamented with many finger-like projections, eventually containing one to three oospores with smooth to similarly decorated walls, filling most of oogonial volume.

##### *a. Key diagnostic characters*

(1) Mycelium relatively thin. (2) Zoosporangia: individual cells (zoosporangia) develop a thin exit tube penetrating host cuticle; partially cleaved cytoplasmic blocks are transferred through this tube into a growing vesicle at apex of the tube; zoospores complete differentiation in the vesicle and are dispersed upon the disappearance of the evanescent gelatinous vesicle wall. (3) Zoospores: laterally biflagellate (with anterior tinsel flagellum and posterior whiplash flagellum). (4) Resistant spores: thick-walled oospores, globose, formed by conjugations of adjacent cells (gametangia).

##### *b. Major species*

*L. chapmanii* Seymour—affecting mosquitoes (only entomopathogenic species).

##### *c. Main taxonomic literature*

Seymour, 1984.

##### *d. General comments*

*Leptolegnia chapmanii* is a mosquito pathogen that can kill its hosts with unusual speed. This fungus is also extraordinary since it infects its host either through the cuticle (by germinating zoospore cysts) or through the gut (by germination of ingested zoospore cysts) (Pelizza *et al.*, 2008).

### B. Entomophthoromycotina: Entomophthorales

The phylogenetic reclassification of fungi (James *et al.*, 2006; Hibbett *et al.*, 2007) validated what many mycologists have long known, that the phylum Zygomycota was a heterogeneous assemblage. The new phylogenic classification rejected this phylum, placed mycorrhizal zygomycetes into the new phylum Glomeromycota, and distributed the remaining zygomycetes into the new subphyla Entomophthoromycotina, Kickxellomycotina, Mucormycotina, and Zoopagomycotina without assigning these taxa to any phylum. Depending on future research, these subphyla may each may be recognized as a new phylum (or partially recombined in another new phylum) but none of their taxa belong in Glomeromycota.

The Entomophthorales comprise the most significant entomopathogenic zygomycetes (here used as a descriptive rather than a taxonomic term). The only other zygomycetes notably associated with arthropods are the endocommensal trichomycetes (again used descriptively rather than taxonomically) in the guts of insects and crustaceans (Lichtwardt *et al.*, 2001); these highly diverse fungi are not treated further here.

Entomophthoralean taxonomy is based primarily on the primary (and secondary) conidia that are their main infective units; this group's thick-walled resting spores—zygospores or azygospores (which differ *only* in whether gametangial conjugations do or do not occur, respectively, before their production) generally have little taxonomic value except at the rank of family (Humber, 1989). Whether entomophthoralean resting spores are sexual spores is a complex, unresolved question discussed by Humber (1981).

Entomopathogenic species generally develop vegetatively in the host hemocoel by forming small, rod-like to irregularly shaped, readily circulated hyphal bodies instead of extensive hyphae. In some genera, the vegetative cells are wall-less, hypha-like, or highly variable and often highly mobile protoplasts; many of these protoplastic cells closely resemble the host's hemocytes. 'Rhizoids' and 'cystidia' (defined in the glossary below) are terms applied with many meanings for structures found in many diverse cryptogams, and are treated as taxonomically signficant characters for the Entomophthorales. Cultures of these fungi may produce cystidia (which help to perforate the host cuticle before fungal sporulation) but not rhizoids (which anchor mycotized hosts to the substrate). Humber (1981, 1989) analyzed the value of all major characters treated as taxonomically significant for these fungi.

Despite using several different taxonomies, some useful general works to identify many entomopathogenic species include MacLeod & Müller-Kögler (1973), Keller (1987, 1991, 1997, 2002), and Bałazy (1993). After a long and spirited debate about the familial and generic taxonomy of Entomophthorales (summarized in Humber 1981, 1989), the most widely accepted and used current taxonomy for this order recognizes six families in this order (Humber 1989), at least two of which (Neozygitaceae and Entomophthoraceae) are obligatorily pathogenic for arthropods. Work in several laboratories worldwide is seeking to provide a robust and phylogenetically based reclassification of these fungi that seems likely to adjust—but not to overturn—the basic outlines of this current classification. The most problematic entomopathogenic genera in these new studies continue to be *Conidiobolus*, *Erynia*, *Furia*, and *Pandora*; it is still too early to know how these genera may be treated in a phylogenetic reclassification of the order.

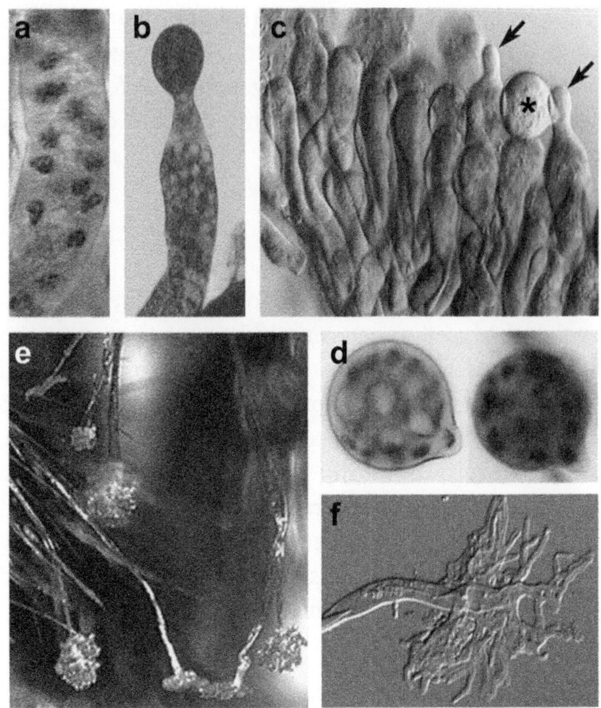


**Figure 6.5** *Batkoa*. a. Hyphal nuclei stained by aceto-orcein. b. Developing conidium with narrow neck between conidiogenous cell and conidium. c. Extended tips of conidiogenous cells (arrows) before conidia develop, and a discharged conidium (★). c. Multinucleate conidia stained in aceto-orcein. e–f. Rhizoids with discoid terminal holdfasts.

### *1.* Batkoa *Humber (Figure 6.5)*

[Entomophthoraceae] Hyphal bodies elongated, walled (not protoplastic); conidiophores simple with a narrow 'neck' between conidium and conidiogenous cell; primary conidia globose, multinucleate, discharged by papillar eversion; rhizoids (if present) obviously thicker than vegetative hyphae, with terminal discoid holdfast; resting spores bud laterally from parental cell; unfixed nuclei have granular contents stained by aceto-orcein.

#### *a. Key diagnostic characters*

(1) Nuclei: granular contents stain in aceto-orcein. (2) Conidiogenous cells: apex narrows and elongates before forming conidium. (3) Conidia: globose, multinucleate. (4) Rhizoids (if present): thick, with discoid terminal holdfast.

#### *b. Major species*

*B. apiculata* (Thaxter) Humber—conidia *ca.* 30–40 μm diameter; papilla often with pointed extension (apiculus); on hemipterans and flies. *B. major* (Thaxter) Humber—conidia *ca.* 40–50 μm diameter; papilla often with pointed extension; on diverse insects.

#### *c. Main taxonomic literature*

MacLeod & Müller-Kögler, 1973 (as *Entomophthora* spp.); Keller, 1987, 1991 (as *Entomophaga* spp.); Humber, 1989; Bałazy, 1993.

### *2.* Conidiobolus *Brefeld (Figure 6.6)*

[Ancylistaceae] Mycelium initially coenocytic, becoming irregularly septate, often forming walled, elongate hyphal bodies; conidiophores unbranched; primary conidia globose to pyriform with rounded apex and prominent papilla, multinucleate, forcibly discharged by papillar eversion; secondary conidia form (1) singly, resembling primaries, forcibly discharged, (2) multiply, forcibly discharged (microconidia; in subgenus *Delacroixia*), or (3) as cylindrical capilliconidia passively dispersed from capillary conidiophore (in subgenus *Capillidium*); resting spores (mostly zygospores) forming in from hyphal axis or apically (not budding off laterally); unfixed nuclei not

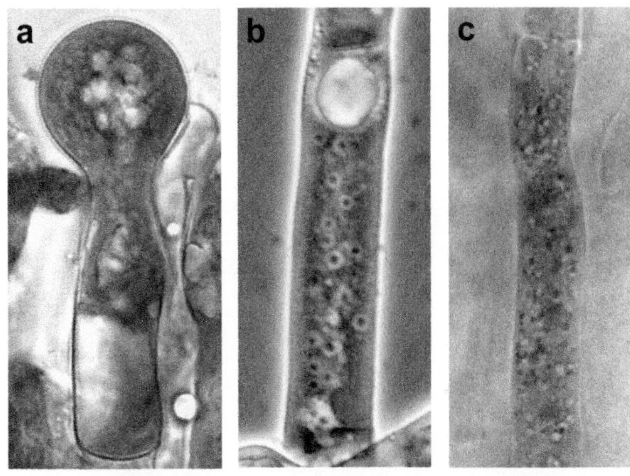


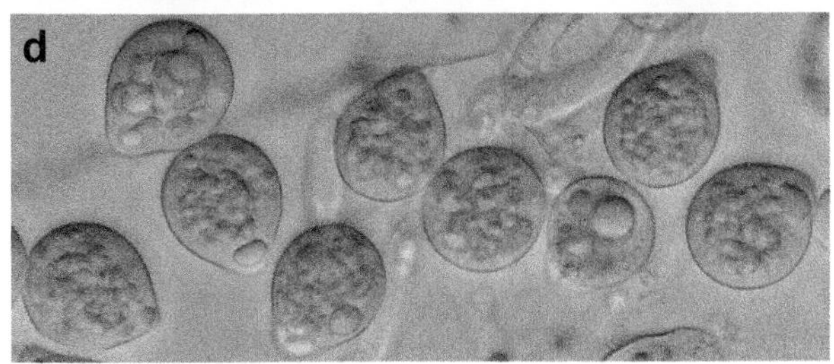


**Figure 6.6** *Conidiobolus*. a. Conidiogenous cell showing developing conidium; cytoplasm is being pushed into the conidium by a growing vacuole. b–c. Nuclei in hyphae with clear nucleoplasm and dense central nucleolus (a, phase contrast in live hypha; b, aceto-orcein). d. Conidia with rounded papillae.

(or poorly) stained in aceto-orcein, without coarsely granular contents.

*a. Key diagnostic characters*

(1) Nuclei: ± undifferentiated in aceto-orcein; contents not obviously granular. (2) Conidiophores: simple (rarely bifurcate at base). (3) Conidia: globose to pyriform, multinucleate. (4) Resting spores: formed in axis of parental cell or apically (not budded off). (5) Vegetative cells: walled (not protoplastic). (6) Subgenera are currently defined by types of secondary conidia formed (Ben-Ze'ev & Kenneth, 1982).

*b. Major species*

*C. coronatus* (Costantin) Batko—old conidia become covered by villose spines (1 to many many μm long); often forming secondary microconidia; weak pathogen of insects or vertebrates. *[Note: Positive identification requires presence of villose conidia]*. *C. obscurus* (Hall & Dunn) Remaudière & Keller Hall—conidia globose, hemispherical papilla emerges abruptly from spore outline, 30–40 μm diameter; nuclei may show faint (finely granular) peripheral staining in aceto-orcein; no capilliconidia or microconidia formed; especially on aphids. *C. thromboides* Drechsler—conidia mostly pyriform, papilla merges gradually into spore outline, 17–30 μm diameter; no capilliconidia or microconidia formed; especially on aphids.

*c. Main taxonomic literature*

King, 1976, 1977; Keller, 1987; Bałazy, 1993.

*d. General comment*

Gene-based studies do not support the current generic or subgeneric taxonomy but do not yet suggest how these fungi should eventually be reclassified.

*3.* Entomophaga *Batko (Figure 6.7)*

[Entomophthoraceae] Hyphal bodies fusoid to bead-like amoeboid protoplasts, later rod-like to spherical; conidiophores simple; primary conidia pyriform to ovoid,

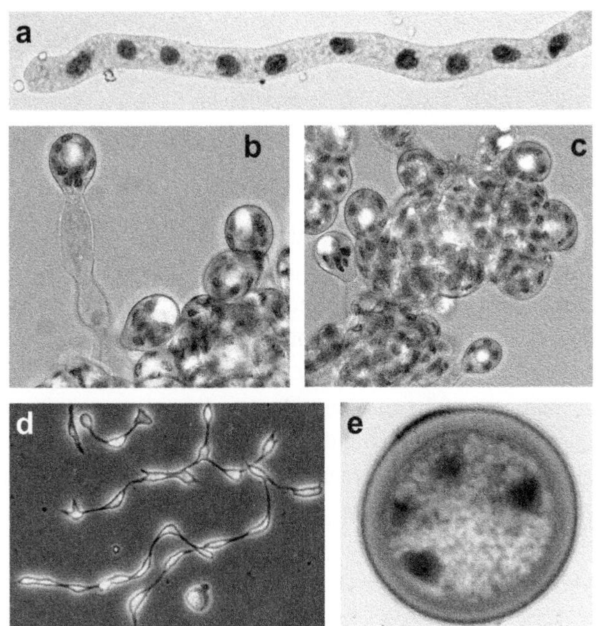


**Figure 6.7** *Entomophaga*. a. Hyphal nuclei (in aceto-orcein). b–c. Pear-shaped conidia and conidiogenous cell (b). d. Amoeboid vegetative protoplasts (in culture). e. Maturing resting spore; mature resting spores are usually binucleate, with a thicker and more obviously 2-layered wall.

multinucleate, discharged by papillar eversion; rhizoids and cystidia not formed; resting spores bud laterally from parental hypha; unfixed nuclei have granular contents staining in aceto-orcein.

*a. Key diagnostic characters*
(1) Nuclei: with coarsely granular contents in aceto-orcein. (2) Conidiogenous cells: no elongated, narrow neck below conidium. (3) Conidia: pyriform, multinucleate. (4) Vegetative growth: protoplastic, bead-like to fusoid or irregularly hyphoid; walled only late in development in host.

*b. Major species*
*E. aulicae* (Hoffman in Bail) Batko [species complex]—affecting diverse lepidopterans; *E. maimaiga* Humber, Soper & Shimazu (Soper *et al.*, 1988), affects gypsy moths (*Lymantria dispar*) in Japan, North America. *E. grylli* (Fresenius) Batko [species complex]—affecting diverse acridids (Orthoptera); *E. calopteni* (Bessey) Humber (Humber, 1989) specifically affects melanopline (spur-throated) grasshoppers and forms resting spores but not primary conidia in North America.

*c. Main taxonomic literature*
Keller, 1987; Soper *et al.*, 1988; Bałazy, 1993.

### 4. Entomophthora *Fresenius (Figure 6.8)*

[Entomophthoraceae] Vegetative cells short, rod-like (with or without cell walls); conidiophores simple; conidiogenous cells club-shaped; primary conidia with prominent apical point and broad, flat basal papilla, with two to 12 + nuclei, forcibly discharged by cannon-like mechanism with discharged conidia surrounded on substrate by a clear halo of undetermined material; rhizoids (if present) numerous, isolated or fasciculate, or inconspicuous (emerging only from fly mouthparts); resting spores bud laterally from parental hypha; unfixed nuclear contents coarsely granular in aceto-orcein.

*a. Key diagnostic characters*
(1) Primary conidia: with a broad, flat papilla and pointed apical projection; for species: conidial size, number, and size of nuclei. (2) Primary conidial discharge: cannon-like; discharged conidia surrounded by halo-like droplet. (3) Secondary conidia more broadly clavate, nonapiculate. (4) Rhizoids: present or absent; location, number, morphology if present.

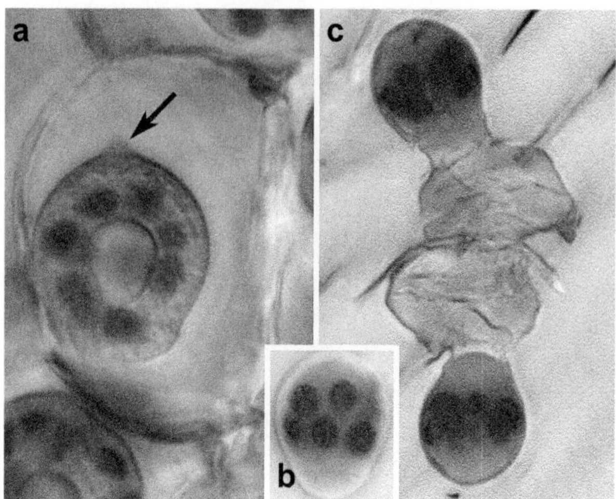


**Figure 6.8** *Entomophthora*. a. Discharged primary conidium with apiculus (arrow) and a broad, flat basal papilla and central vacuole, surrounded by a halo-like droplet of cytoplasm. b–c. Secondary conidia are broadly obovoid, not apiculate, and form rapidly on primary conidia (empty cells below the secondary conidia). Granular contents of nuclei are prominent in aceto-orcein.

*b. Major species*
*E. culicis* (Braun) Fresenius—affecting mosquitoes and black flies; conidia binucleate. *E. muscae* (Cohn) Fresenius [species complex; see Keller, 2002]—affecting muscoid flies; species are distinguished by conidial characters (size, number/size of nuclei), hosts affected, and biogeography (e.g., *E. schizophorae* is the primary species on flies in North America). *E. planchoniana* Cornu—affecting aphids (especially in Europe).

*c. Main taxonomic literature*
MacLeod *et al.*, 1976; Keller, 1987, 2002; Bałazy, 1993.

### 5. Erynia *Nowakowski [sensu Humber (1989)] (Figure 6.9)*

[Entomophthoraceae] Hyphal bodies rod-like to filamentous, walled; conidiophores digitately branched; primary conidia pyriform (clavate) to elongate, curved or straight with rounded to acute apices, uninucleate, basal papilla conical (often with slight flare at junction with spore) or rounded, bitunicate (outer wall layer may separate in liquid mounts), discharged by papillar

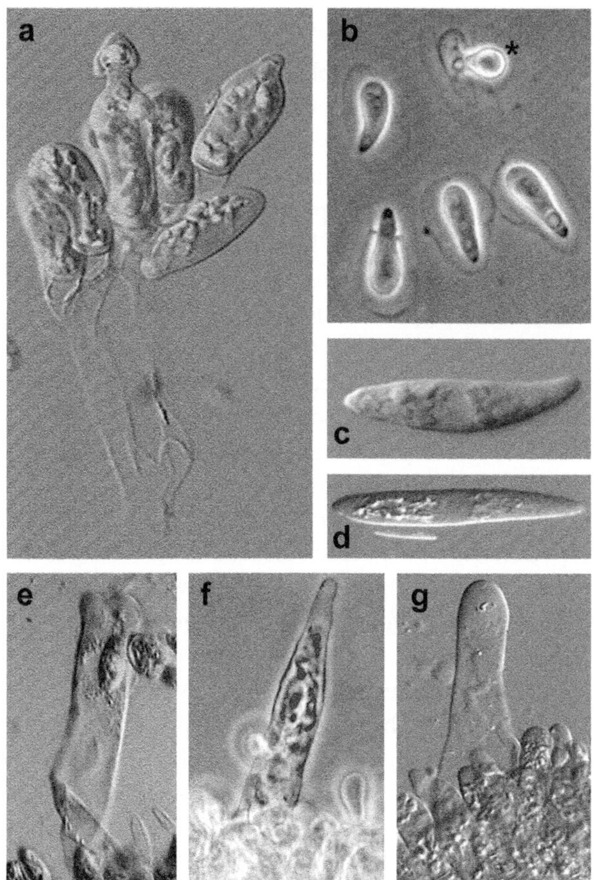


**Figure 6.9** *Erynia*. a. Branched conidiophore with short, blocky conidiogenous cells. b. Bitunicate primary conidia with outer wall layer completely separated with one forming a more globose secondary conidium (★). c–d. Primary conidia with conical papilla (to left). e-g. Cystidia emerging above level of conidial hymenium; note their large diameter and, in (e), apical branching.

eversion; secondary conidia like primaries or more globose, discharged by papillar eversion; rhizoids obviously thicker than hyphae, without disoid terminal holdfasts; cystidia columnar (some may branch at tip), obviously thicker than hyphae; resting spores bud laterally from parental cell; unfixed nuclei with coarsely granular contents in aceto-orcein.

*a. Key diagnostic characters*

(1) Conidiophores: digitately branched at apices; much entwined, difficult to separate. (2) Conidia: uninucleate, bitunicate, clavate to elongate, apices rounded or tapered to a blunt point; papillae broadly conical (± flared at junction with conidium) or rounded. (3) Rhizoids: thicker than hyphae, with no discoid terminal holdfast. (4) Cystidia: thicker than hyphae, not strongly tapered.

*b. Major species*

*E. aquatica* (Anderson & Anagnostakis) Humber—on Culicidae and Chironomidae (Diptera), especially in temporary, cold snow-melt pools; conidia long clavate 30–40 × 15–18 μm. *E. conica* (Nowakowski) Remaudière & Hennebert—on Culicidae and Chironomidae (Diptera); conidia 30–80 × 12–15 μm, often gently curved, tapering to subacute apex. *E. ovispora* (Nowakowski) Remaudière & Hennebert—on nematoceran flies; conidia broadly ovoid to ellipsoid, 23–30 × 12–14 μm. *E. rhizospora* (Thaxter) Remaudière & Hennebert—on Trichoptera; conidia lunate to straight, 30–40 × 8–10 μm, broadest in mid-length, tapering to blunt apex; resting spores wrapped by fine brown hyphae, often both outside as well as within host body.

*c. Main taxonomic literature*

Keller, 1991 (*Erynia* as defined by Remaudière & Hennebert, 1980); Bałazy, 1993 (as *Zoophthora* subgenus *Erynia*).

*d. General comment*

Current gene-based studies may not support *Erynia*, *Furia*, and *Pandora* as separate genera.

### 6. Furia *(Batko) Humber (Figure 6.10)*

[Entomophthoraceae] Hyphal bodies hypha-like, walled; conidiophores branched (± digitately) at apices; primary conidia clavate to obovoid, uninucleate, basal papilla rounded, bitunicate (outer wall layer may separate in liquid mounts), discharged by papillar eversion; rhizoids numerous, may be fasciculate, with diameter of vegetative hyphae, no discoid terminal holdfast; cystidia with diameter of hyphae; resting spores bud laterally from parental hypha; unfixed nuclei have granular contents staining in aceto-orcein.

*a. Key diagnostic characters*

(1) Conidiophores: branching close to conidiogenous cells. (2) Conidia: uninucleate, bitunicate, obovoid to clavate; apices and papillae rounded. (3) Rhizoids: as thick as hyphae, with sucker-like attachments or weak terminal branching systems but no discoid terminal

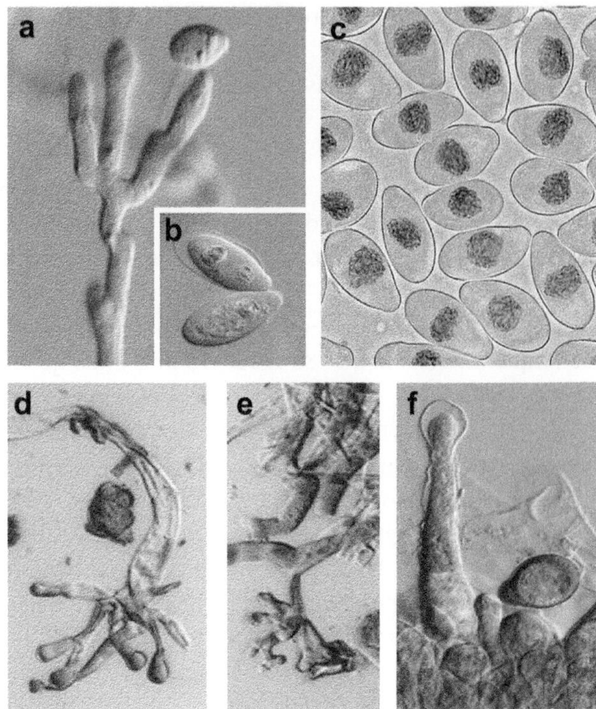

**Figure 6.10** *Furia*. a. Branched conidiophore (most are more highly branched). b. Bitunicate primary conidium with outer wall layer separating (upper). c. Primary conidia showing single nucleus with highly granular contents. d—e. Rhizoid holdfasts are branched but not discoid. f. Cystidium projecting from hymenium (note relative size of conidium at right).

holdfast; numerous, single, fasciculate, or in pseudo-rhizomorphs. (4) Cystidia: as thick as hyphae.

*b. Major species*

*F. americana* (Thaxter) Humber—conidia obovoid, 28—35 × 14—16 μm; on cyclorrhaphan (muscoid) flies. *F. virescens* (Thaxter) Humber—conidia broadly clavate to obovoid, 20—30 × 9—13 μm; especially on Noctuidae (Lepidoptera).

*c. Main taxonomic literature*

Li & Humber, 1984; Keller, 1991 (as *Erynia* spp.); Bałazy, 1993 (as *Zoophthora* subgenus *Furia* spp.).

*d. General comment*

Current gene-based studies may not support *Erynia*, *Furia*, and *Pandora* as separate genera.

### 7. Massospora *Peck (Figure 6.11)*

[Entomophthoraceae] Hyphal bodies rod-like to elongate, initially wall-less, filling and confined mainly to host's terminal abdominal segments; conidiophores line small cavities in abdomen; conidia one to six (mostly two) nucleate, passively dispersed when exposed after abdominal exoskeleton of live cicada sloughs off; resting spores thick-walled, surface deeply reticulate, budded off from parental hyphal bodies; conidia and resting spores not present in same host individual; restricted to emergences of gregarious cicadas (Hemiptera: Cicadidae), especially in North and South America; unfixed nuclear contents obviously granular in aceto-orcein.

*a. Key diagnostic characters*

(1) Affecting gregarious cicadas (high fungus/host species specificity). (2) Fungal growth restricted to terminal three to four abdominal segments. (3) Spore dispersal from live cicadas when internal spore mass is exposed by shedding of terminal abdominal exoskeleton. (4) Conidia: size, shape, number, and position of nuclei. (5) Resting spores: size, morphology of surface decorations.

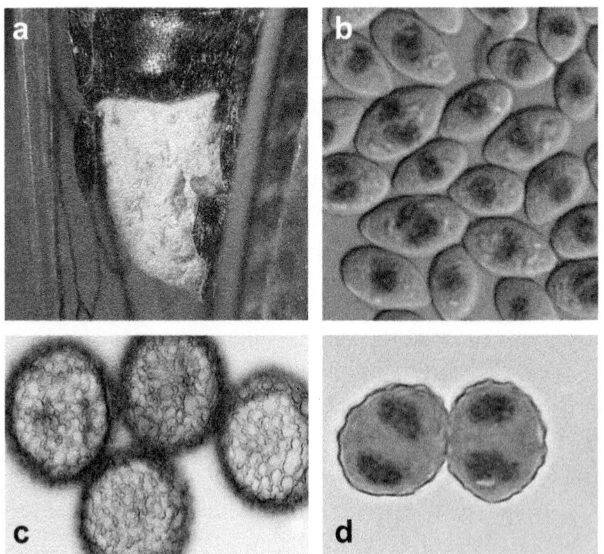

**Figure 6.11** *Massospora*. a—b. Abdomen of cicada with terminal segments fallen away to expose conidial mass of conidia of *M. levispora;*. b. Conidia with 1—3 nuclei and smooth wall. c—d. *M. cicadina* resting spores with deeply reticulate surface decoration c. and conidia d. that are binucleate and have a warted surface.

*b. Major species*

*M. cicadina* Peck—affecting 17-year cicada (*Magicicada septemdecim*).

*c. Main taxonomic literature*

Soper, 1974, 1981.

### *8.* Neozygites *Witlaczil (Figure 6.12)*

[Neozygitaceae] Hyphal bodies irregularly shaped, rod-shaped or spherical, usually three to five nucleate; conidiophores simple; primary conidia round, ovoid or broadly fusoid, with relatively flattened basal papilla, mostly four nucleate, forcibly discharged a short distance by papillar eversion; secondary conidia usually (more or less almond-shaped) capilliconidia passively dispersed from capillary conidiophores; resting spores bud laterally from short bridge between gametangia (hyphal bodies), black to smoke-gray, binucleate; nuclei (unfixed) staining poorly in aceto-orcein except during mitosis; especially on Hemiptera, thrips, and mites.

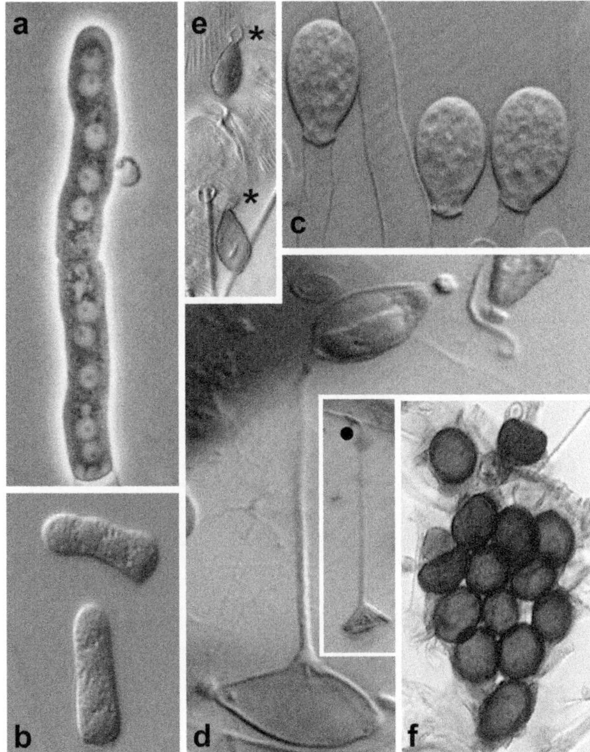


**Figure 6.12** *Neozygites* (from mites). a. Hypha with two 5-nucleate cells. b. Hyphal bodies. c. Primary conidia with flat papilla. d. Tertiary conidium with mucoid haptor on apex of capillary formed on a capilliconidium; inset shows apical bend of conidiophore (•). e. Capilliconidia attached to mite leg by mucoid haptor (★). f. Darkly melanized ovoid resting spores.

*a. Key diagnostic characters*

(1) Hyphal bodies: rod-like to spherical, routinely four nucleate. (2) Nuclei: staining poorly in aceto-orcein (unless mitotic); in mitosis, chromosomes stain well, vermiform, on central metaphase plate. (3) Conidia: four (± 1) nuclei, papilla flattened. (4) Secondary conidia: like primaries and forcibly discharged or drop- to almond-shaped capilliconidia with apical mucoid drop (haptor), dispersed passively from capillary conidiophores usually with apical knee-like bends. (5) Resting spores: zygospores with conspicuously dark outer wall, arising from short bridge between gametangia, binucleate, ovoid/smooth or round/usually roughened.

*b. Major species*

*N. fresenii* (Nowakowski) Remaudière & Keller—conidia subglobose, 17–20 μm diameter; zygospores ovoid, 25–50 × 15–30 μm; on aphids. *N. floridana* (Weiser & Muma) Remaudière & Keller—conidia 10–14 μm diameter; zygospores subglobose, dark brown, outer wall roughened, 14–25 μm in long dimension; on tetranychid mites. *N. parvispora* (MacLeod, Tyrrell & Carl) Remaudière & Keller—conidia small, 9–15 μm diameter; zygospores black, spherical to flattened, 15–20 μm diameter; especially on thrips.

*c. Main taxonomic literature*

Keller, 1991, 1997, 2007; Bałazy, 1993.

### *9.* Pandora *Humber (Figure 6.13)*

[Entomophthoraceae] Hyphal bodies filamentous, protoplastic or walled; conidiophores digitately branched; primary conidia clavate to obovoid, uninucleate, basal papilla rounded, bitunicate (outer wall layer may lift away in liquid mounts), discharged by papillar eversion; secondary conidia like primary or more globose; rhizoids clearly thicker than hyphae, with discoid terminal hold-fast; cystidia taper, thicker at base than hyphae; resting spores bud laterally from parental hypha; unfixed nuclei with obviously granular contents in aceto-orcein.

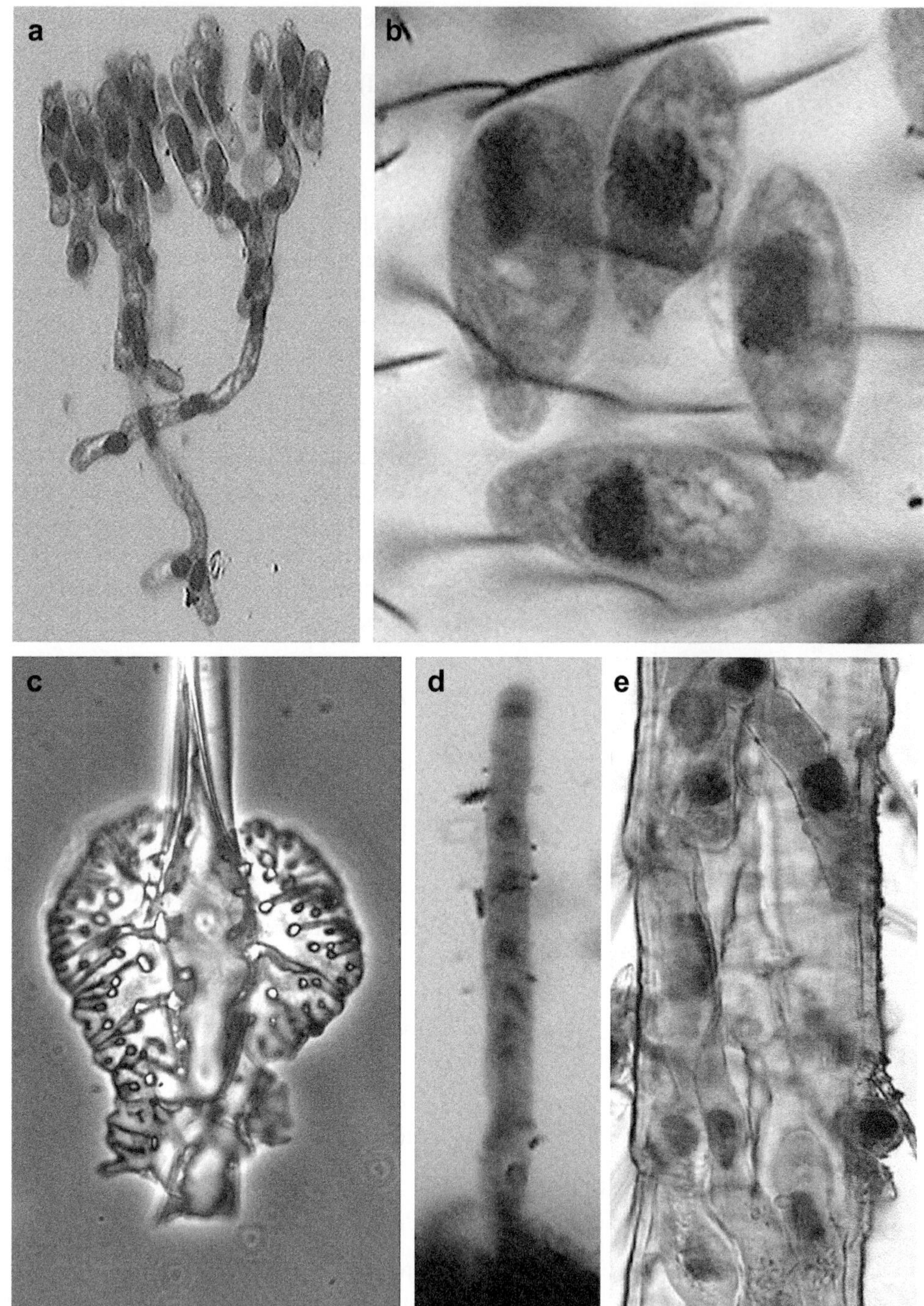


**Figure 6.13** *Pandora*. a. Digitately branched conidiophores. b. Uninucleate primary conidia (not showing bitunicate nature). c. Discoid terminal holdfasts of rhizoid. d. Cystidium projecting from hymenium. e. Hyphal bodies inside leg of aphid. All images (except c) stained in aceto-orcein to show nuclei.

*a. Key diagnostic characters*

(1) Conidiophores: branching digitately. (2) Conidia: uninucleate, bitunicate, obovoid to clavate; apices and papillae rounded. (3) Rhizoids: two to three times thicker than hyphae, relatively sparse, with prominent terminal discoid holdfast. (4) Cystidia: two- to three-times thicker than hyphae at base, tapering toward bluntly pointed apex. (5) Vegetative growth: hyphoid, protoplastic or walled in host.

*b. Major species*

*P. blunckii* (Bose & Mehta) Batko—on lepidopterans (esp. diamondback moth, *Plutella xylostella*); conidia pyriform, 15–20 × 7–11 μm. *P. delphacis* (Hori) Humber—especially on planthoppers (Hemiptera: Delphacidae); conidia broadly clavate, 30–35 × 12–18 μm; growing and sporulating well *in vitro* on diverse media. *P. neoaphidis* (Remaudière & Hennebert) Humber—on aphids; conidia broadly clavate, 15–40 × 9–16 μm, often with papilla laterally displaced; often difficult to isolate or reluctant to grow and sporulate well *in vitro*.

*c. Main taxonomic literature*

Humber, 1989; Keller 1991 (as *Erynia* spp.); Bałazy, 1993 (as *Zoophthora* subgenus *Neopandora* spp.).

*d. General comment*

Current gene-based studies may not support *Erynia*, *Furia*, and *Pandora* as separate genera. *P. neoaphidis* has been somewhat mysterious since it was first characterized as *Empusa aphidis* by Thaxter (1888) for the lack of any observations of naturally formed resting spores; Scorsetti *et al.* (2012) illustrated and used molecular techniques to confirm the identity of the resting spores of this species in field-collected aphids.

### *10.* Zoophthora *Batko [sensu Humber (1989)] (Figure 6.14)*

[Entomophthoraceae] Hyphal bodies rod-like to hyphoid, walled; conidiophores digitately branched; primary conidia clavate to obovoid, uninucleate, basal papilla rounded, bitunicate (outer wall layer may separate in liquid mounts), discharged by papillar eversion; secondary conidia (1) resembling primaries or more globose and discharged by papillar eversion or (2) elongate capilliconidia passively dispersed from capillary conidiophore; rhizoids as thick as vegetative hyphae, numerous, individual or fasciculate, discoid terminal holdfast absent; cystidia as thick as hyphae; resting spores bud laterally from parental hypha; unfixed nuclei have granular contents staining in aceto-orcein.

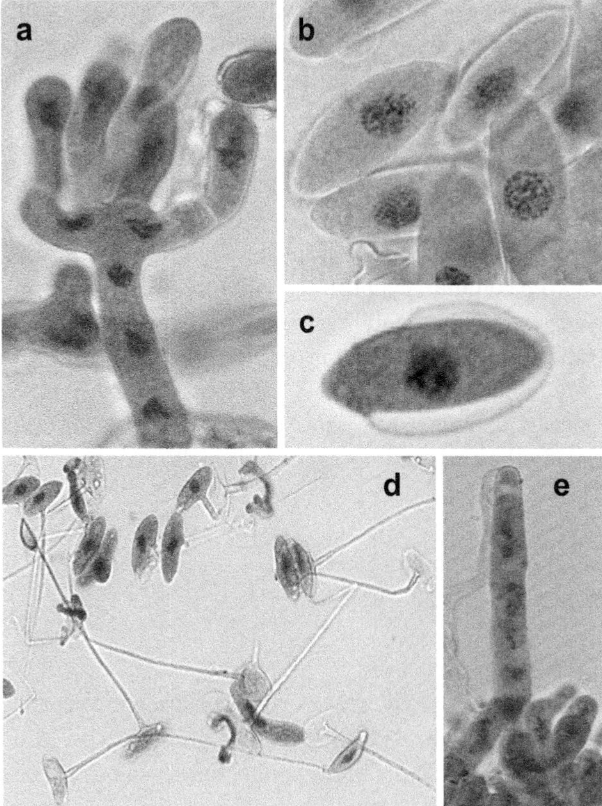


**Figure 6.14** *Zoophthora*. a. Branched conidiophore. b. Primary conidia showing conidal papilla and prominently granular nuclear contents. c. Conidium showing bitunicate nature (with outer and inner wall layers becoming separated in liquid mounts). d. Secondary capilliconidia on capillary conidiophores. e. Cystidium is about as thick as conidiogenous cells. All fungi were stained with aceto-orcein to show granular nuclear contents.

*a. Key diagnostic characters*

(1) Conidiophores: branching digitately at apices. (2) Conidia: papilla broadly conical (± flared at junction with spore), rarely rounded, elongate (L/D usually ≥ 2.5), straight or curved, apex rounded or tapering to point. (3) Secondary conidia: elongated capilliconidia passively dispersed from capillary conidiophore or resembling primary, forcibly discharged from short, thick conidiophore. (4) Rhizoids: ± thick as hyphae, holdfasts (if present) weakly differentiated, sparse terminal

branching systems, and/or small and sucker-like; usually numerous, single or fasciculate or forming thick pseudorhizomorphs. (5) Cystidia: ± thick as hyphae, untapered or ± tapered apically.

*b. Major species*

*Z. phalloides* Batko—conidia elongated, curved, with bluntly rounded apices; on aphids. *Z. phytonomi* (Arthur) Batko—conidia cylindrical, straight; resting spores colorless (or, possibly, darkened with roughened outer wall); especially affecting *Hypera* spp. (Coleoptera: Curculionidae) on alfalfa. *Z. radicans* (Brefeld) Batko [species complex; see Bałazy, 1993]—conidia more or less bullet-shaped, tapering to bluntly pointed apex, 15–30 μm long, with L/D ratio ~2.5–3.5.

*c. Main taxonomic literature*

Keller, 1991, 2007; Bałazy, 1993.

*d. General comment*

Current gene-based studies do consistently separate *Zoophthora* from a poorly resolved group of species in *Erynia*, *Furia*, and *Pandora*; the future state of these latter genera will depend on further studies. The monotypic *Orthomyces aleyrodis* (Steinkraus *et al.*, 1998), which has been reported only once, resembles *Zoophthora*: it forms short ovoid primary conidia that form capillary secondary conidiophores that are wider than in *Zoophthora*; the secondary conidia on these capillaries are globose and have a small conical papilla, and may be forcibly dislodged rather than passively dispersed.

## C. Basidiomycota: Pucciniomycetes: Septobasidiales

The world of entomopathogenic basidiomycetes is surprisingly narrow in comparison to the rich diversity of Entomophthorales and Ascomycota affecting insects and other invertebrates. There is a considerable diversity of basidiomycetes (not otherwise covered here; but see Cooke & Godfrey, 1964; Barron, 1977) that utilize nematodes as possible alternate sources of nutrients. The only notable basidiomycete pathogens or parasites obligately associated with insects are the rust-like fungi of Septobasidiales, and especially the dozens of species of *Septobasidium* (Couch, 1938). These fungi are, effectively, zoophilic rusts whose nourishment derives wholly from partial parasitism of scale insect populations underlying crust-like fungal thalli. The global knowledge of these fungi depends heavily on a classic monograph by Couch (1938); later major studies of this genus include those by Azema (1975) and the validation (Gómez & Henk, 2004) of Couch's (1938) many new but invalidly published species of *Septobasidium*. The biology and systematics of *Septobasidium* and its relatives exceed the scope of this chapter except for including *Septobasidium* in the key.

The mention of basidiomycetes before the ascomycete fungi may surprise some readers, but the recent phylogenetic re-evaluation of fungi (James *et al.*, 2006) clearly indicates that the Basidiomycota preceded the Ascomycota.

## D. Ascomycete Anamorphs and Teleomorphs

A very large, diverse range of ascomycete fungi are obligatorily associated with insect and other arthropod hosts, although the pathogenic/parasitic taxa are not uniformly distributed through the Ascomycota: the most diverse entomogenous ascomycetes (and perhaps the most amazing, too) may be the minute ectoparasites of the Laboulbeniomycetes (Tavares, 1985); this amazing group comprises more than 2000 species in more than 140 genera and is not treated further here. Among the so-called bitunicate ascomycetes (formerly Loculoascomycetes, now the Dothideomycetes) *Myriangium* and *Podonectria* are among the less common genera attacking scale insects; neither is treated here. The greatest number and diversity of entomopathogenic ascomycetes are the sexual and conidial forms in the order Hypocreales (Sordariomycetes; formerly class Pyrenomycetes); the vast majority of these fungi are assigned to the three families—Clavicipitaceae, Cordycipitaceae, and Ophiocordycipitaceae—recently split on phylogenetic grounds (Sung *et al.*, 2007) from the Clavicipitaceae in its older and vastly inclusive sense. It is among these clavicipitoid taxa that almost *all* major conidial entomopathogens are found, as well as the hundreds of species originally described in *Cordyceps*.

Whether conidial fungi (Fungi Imperfecti, or the Coelomycetes and Hyphomycetes) should be retained as useful taxonomic categories was one of the most hotly debated mycological issues of decades past. That debate is over and was settled quietly and without any further opposition by the rise of gene-sequence-based taxonomic approaches. Most conidial fungi that are entomopathogenic have now been placed confidently into the existing classifications of the Ascomycota; the most significant unresolved issues involve genera that are now known to be phylogenetically heterogeneous, with their traditional generic concepts including fungi allied to two or more different orders or families (e.g., the phylogenetically based reclassifications of the entomopathogenic species of *Paecilomyces* and *Verticillium* discussed further below but where, for many possible reasons, the correct gene-based affinities of some species remain to be determined).

### *1.* Aschersonia *Montagne and* Hypocrella *Saccardo (Figure 6.15)*

[Sordariomycetes: Hypocreales: Clavicipitaceae] *ANAMORPH: Aschersonia*, with stroma hemispherical or cushion-shaped (sometimes indistinct), superficial, usually light to brightly colored (yellow, orange, red, etc.), covering host insect, with one or more conidia-forming zones (locules) sunken into stroma and opening by wide pore or irregular crack; conidia hyaline, one-celled, spindle-shaped, extruded onto stromatic surface from locules in slime masses. *TELEOMORPH: Hypocrella*, with perithecia (walled structures containing asci and ascospores) globose to pyriform, immersed in stroma with opening protruding from stroma; asci cylindrical, with prominent hemispherical apical thickening penetrated by a narrow canal; ascospores filiform, with numerous transverse septa, dissociating at maturity to produce numerous cylindrical part-spores but sometimes remaining intact. Hosts: coccids and aleyrodids.

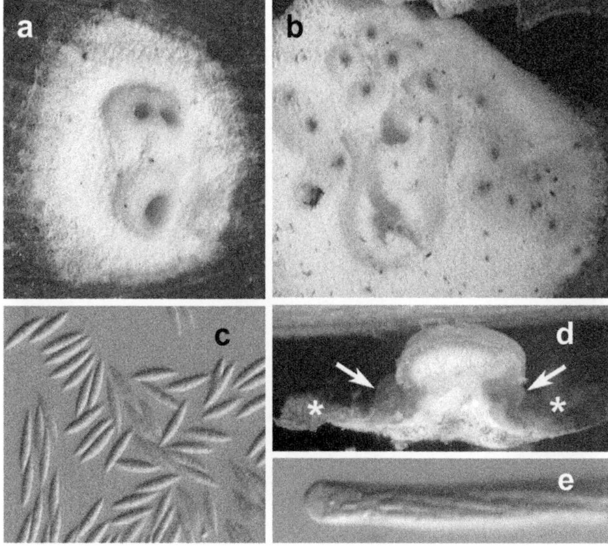


**Figure 6.15** *Aschersonia* (a–d) and *Hypocrella* (e). a. Stroma with sunken conidial zones. b. Conidial zones sunken into stroma. c. Conidia. d. Stroma (detached) with conidial masses (arrows), and thin hypothallus (★) extending into substrate. e. Thick ascus cap (left) with long, multiseptate ascospores in ascus.

#### *a. Key diagnostic characters*

(1) Stroma: presence, size, cross-sectional profile, color. (2) (*Aschersonia*) Conidiogenous locules: sunken in stroma, arrangement on stroma, release of conidia in slime. (3) (*Aschersonia*) Conidia elongated, fusoid, aseptate. (4) (*Hypocrella*) Perithecia: embedded in stroma. (5) (*Hypocrella*) Asci: long, with apical thickening penetrated by a narrow channel.

#### *b. Major species*

*Aschersonia aleyrodis* Webber—stromata *ca.* 2 mm diameter × 2 mm high, orange to pink or cream-colored, surrounded by thin halo of hyphae spreading on leaf surface. Conidia bright orange in mass, 9–12 × 2 μm.

#### *c. Main taxonomic literature*

Petch, 1914, 1921; Mains, 1959a,b; Liu *et al.*, 2005, 2006; Chaverri et al., 2008.

#### *d. General comments*

*Aschersonia* species are the conidial states of the less frequently found *Hypocrella* states; both genera are widespread in the tropics and subtropics. *Hypocrella* perithecia are immersed in the surface of stromata on which *Aschersonia* state may also occur. The most used taxonomy for both genera is that of Petch (1914, 1921) but more recent studies show that there can also be a *Hirsutella*-like synanamorph (Liu *et al.*, 2005), and molecular revisions now appearing may tend to lead to a degree of phylogenetically based fragmentation of both of these genera (Chaverri *et al.* 2005, 2008; Liu *et al.*, 2006).

### *2.* Ascosphaera *Spiltoir & Olive (Figure 6.16)*

[Eurotiomycetes: Onygenales] Affecting larval bees; mycelium septate, with sex organs (globose ascogonia and papillate trichogynes) formed on separate mycelia; fertilized ascogonia swell to form a large 'nutriocyst' (spore cyst) containing many individual asci; each ascus

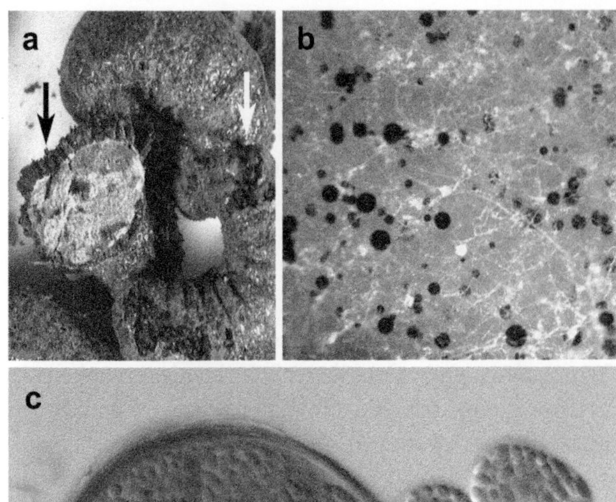


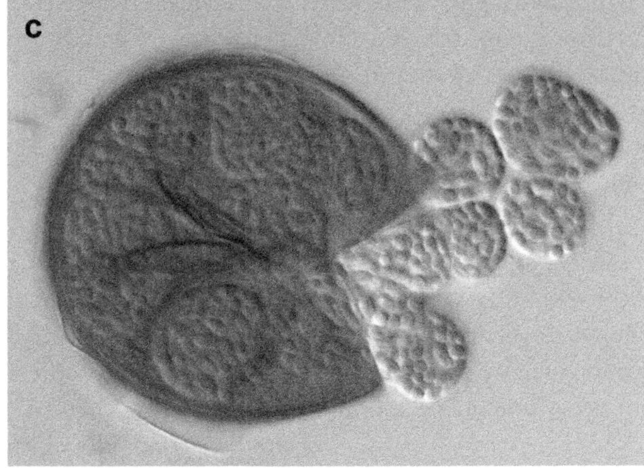


**Figure 6.16** *Ascosphaera*. a. Melanized nutriocysts (arrows) of *A. aggregata* under cuticle of infected bee larva. b. Melanized nutriocysts in culture of *A. proliperda*. c. Ruptured nutriocyst wall releasing globose asci and their many ascospores.

internally producing numerous ascospores. Appearance of mature infection is a blackened (or white) larva filled with a dense mass of balls (nutriocysts) containing balls (asci) containing ovoid to elongate (asco)spores.

*a. Key diagnostic characters*

(1) Pathogenic to bee larvae. (2) Mature infection is notable for the presence of balls (ascospores) within balls (asci) within balls (nutriocysts).

*b. Major species*

*A. aggregata* Skou—affecting leafcutting bees (Megachilidae) in North America; nutriocysts 140–550 × 100–400 μm; ascospores long ovoid to cylindrical, 4–7 μm long. *A. apis* (Maassen ex Claussen) Olive & Spiltoir—affecting honeybees (*Apis mellifera*) worldwide; nutriocysts 50–120 μm diameter; ascospores short ovoid or allantoid (bean-shaped), 2–3.5 μm long.

*c. Main taxonomic literature*

Skou, 1972, 1988; Rose *et al.*, 1984; Bissett, 1988; Anderson *et al.* 1998.

*3.* Beauveria *Vuillemin (Figure 6.17)*

[Sordariomycetes: Hypocreales: Cordycipitaceae] Forming a dense white covering on host exoskeleton, occasionally synnematous (forming erect fascicles of hyphae); conidiogenous cells usually densely clustered (or whorled or solitary), colorless, with globose or flask-like base and denticulate (toothed) apical extension (rachis) bearing one conidium per denticle; conidia aseptate. *TELEOMORPH: Cordyceps*. Hosts: extremely numerous and diverse.

*a. Key diagnostic characters*

(1) Conidiogenous cells: extending apically into sympodial rachis. (2) Rachis: denticulate, with one

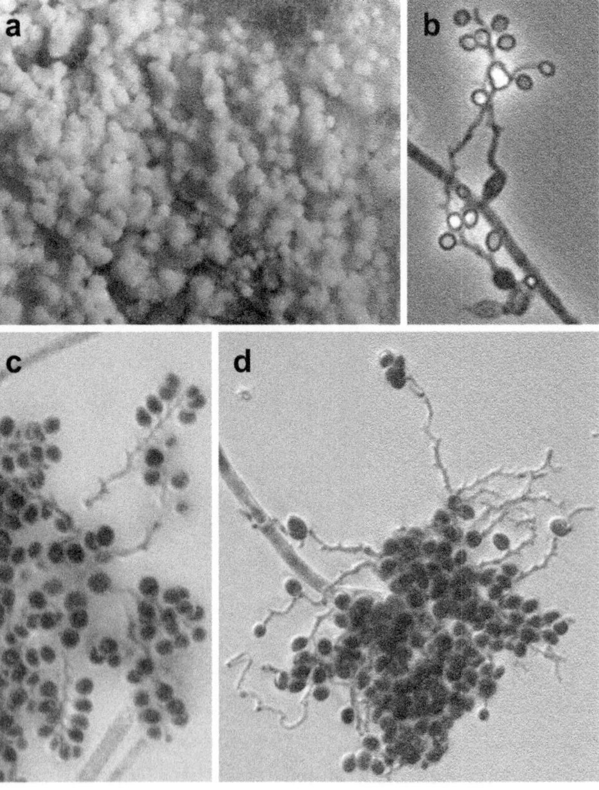


**Figure 6.17** *Beauveria*. a. Typical dense balls of conidia. b–d. Conidiogenous cells with swollen bases and denticulate raches extending apically with a conidium formed successively on each denticle (sterigma).

conidium per denticle. [Note: rachis *must* be denticulate to be *Beauveria.*] (3) Conidia: size, shape, and surface characteristics.

*b. Major species*

*B. bassiana* (Balsamo) Vuillemin—conidia nearly globose, ≤ 3.5 μm diameter *B. brongniartii* (Saccardo) Petch—conidia long ovoid to cylindrical, 2.5–4.5 (6) μm long; mostly on Scarabaeidae (Coleoptera). *For genotypically defined species*, see Rehner *et al.* (2011).

*c. Main taxonomic literature*

Hoog, 1972; Samson & Evans, 1982; Rehner *et al.*, 2011.

*d. General comments*

The taxonomy of *Beauveria* is being transformed by gene-based studies: teleomorphic connections to *Cordyceps* species have been confirmed (Shimazu *et al.*, 1988; Li *et al.*, 2001). Even though teleomorphs are exceedingly rarely found, mating type genes are univerally present and apparently functional so that outcrossing events may be occurring (infrequently) in this genus (Meyling *et al.*, 2009). *B. bassiana* and *B. brongniartii* are confirmed to be partially resolved species complexes, and a number of new species (mostly identifiable only by genotypic data) have been described after a four-gene analysis of a massive and globally inclusive study of this genus (Rehner *et al.*, 2011).

*4.* 'Cordyceps' *in the broad sense (Figure 6.18)*

[Sordariomycetes: Hypocreales: Clavicipitaceae—*Metacordyceps* G.H. Sung, J.M. Sung, Hywel-Jones & Spatafora; Cordycipitaceae—*Cordyceps* Fries; Ophiocordycipitaceae—*Ophiocordyceps* Petch, *Elaphocordyceps* G.H. Sung & Spatafora] Forming one or more erect stromata on a host, with perithecia confined to an apical (or subapical) fertile portion or with scattered on stromatic surface; perithecia flask-shaped, superficial to fully immersed in stroma; asci elongated, with thickened apical cap penetrated by a fine pore, with 8 filiform, multiseptate ascospores which fragment to form one-celled part-spores in most taxa. *ANAMORPHS (incomplete list):* Clavicipitaceae—*Metarhizium, Nomuraea*; Cordycipitaceae—*Beauveria, Isaria, Lecanicillium*; Ophiocordycipitaceae—*Hirsutella, Hymenostilbe, Tolypocladium.* Hosts: numerous, diverse insects.

**Figure 6.18** *Cordyceps* (sensu lato). a–b. Erect stromata bearing superficial perithecia (esp. in inset, b). c–d. Asci cap of ascus with distinct central canal through which ascospores are discharged; ascospores inside these asci are multiseptate and may dissociate into 1-celled part-spores.

*a. Key diagnostic characters*

(1) Erect stroma(ta) bearing asci in perithecia. (2) Asci: filiform, with a thickened apical cap having a central channel. (3) Ascospores: at first eight filiform spores per ascus with many transverse septa; in most species, dissociating at maturity to yield one-celled part spores.

*b. Major species*

*Cordy. militaris* Link: Fries—on lepidopterans; stromata single (rarely multiple) per host, < 10 cm high, thickly clavate and unbranched, orange, with swollen fertile part at apex. *Cordy. tuberculata* (Lebert) Mains—on lepidopterans; stromata off-white, several per host, with partially immersed, sulfur to bright yellow perithecia scattered toward apices. *Oph. lloydii* (Fawcett) G.H. Sung, J.M. Sung, Hywel-Jones & Spatafora—on ants; stromata white to cream-colored, ≤ 1 cm high, with discoid apical fertile part; perithecia partially immersed. *Oph. unilateralis* (Tulasne & C. Tulasne) Petch—on ants; fertile portion a swollen pad borne below apex of stroma; ascospores remain filiform (not dissociating to part-spores); on ants. *Oph. sinensis* (Berkeley) G.H. Sung, J.M. Sung, Hywel-Jones & Spatafora—on larval lepidopterans; stromata dark brown to black, < 8 cm high, with elongated and little swollen apical fertile portion;

used in Chinese herbal medicine as a general tonic and against many diverse ailments.

*c. Main taxonomic literature*

Kobayasi, 1941, 1982; Mains, 1958; Kobayasi & Shimizu, 1983; Sung, 1996; Sung *et al.*, 2007, Kepler *et al.*, 2012; also see *http://cordyceps.us*

*d. General comments*

The unworkably complex traditional taxonomy of *Cordyceps* (Kobayasi, 1941, 1982) led to species identifications based more often on magnificent water-colored illustrations (Kobayasi & Shimizu, 1983) or photographs (Sung, 1996) than on taxonomic descriptions. The phylogenetic reclassification of fungi has transfigured *Cordyceps*: two new families were segregated from the Clavicipitaceae, and hundreds of *Cordyceps* species reassigned among four genera and in each of these families (Sung *et al.*, 2007). *Cordyceps* (*sensu stricto*) includes species with pale to brightly colored fleshy (soft) stromata, perithecia on or immersed in (and perpendicular to the surface of) the stroma or subiculum. *Metacordyceps* species form fibrous to tough stromata with a stipe expanding into a cylindrical or clavate apical fertile part, pale to yellow–green or greenish or purplish, and hosts that are usually buried in soil. Few *Elaphomyces* species attack insects (cicada nymphs or other subterranean insects), and these form fibrous, pale to brownish stromata with a fertile (apical) knob or (lateral and subapical) pads or cushions containing perithecia. *Ophiocordyceps* species usually form darkly pigmented stromata arising from hosts buried in soil or decaying wood, stromata may be wiry to clavate/fibrous, and most perithecia occur on the stromatic surface or in patches or lateral pads below the tip of the stroma. Among the genera into which *Cordyceps* (*sensu lato*) was split, *Ophiocordyceps* now includes the most species followed by *Cordyceps* in its newly restricted sense.

5. Fusarium *Link (Figure 6.19)*

[Sordariomycetes: Hypocreales: Nectriaceae] Fruiting body (if present) a stromatic pad (sporodochium) with conidial hymenium on its surface, pale tan to yellow to orange or red. Conidiophores solitary or aggregated, simple or branched, bearing apical conidiogenous cells. Conidiogenous cells (phialides) short, cylindrical to much elongated, awl-like; forming one or two conidial types: *macroconidia* curved to canoe-shaped with sometimes prominent foot-like appendage on basal cell, with one or more transverse septa, usually released in slime heads or spore masses, and/or *microconidia* aseptate, small, ovoid to cylindrical, produced in slime or dry and in chains from elongate, awl-shaped conidiogenous cells. *TELEOMORPH: Nectria*. Hosts: scales and many other insects.

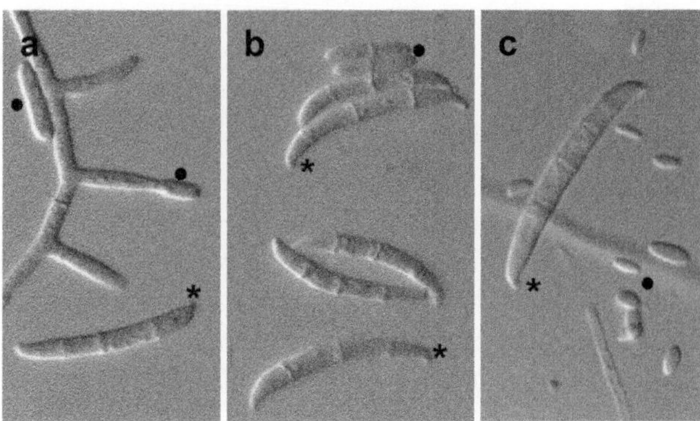


**Figure 6.19** *Fusarium*. a. Conidiogenous cells and microconidia (•); arrows indicate the slightly bent 'foot' (★) at base of macroconidium. b. Transversely septate macroconidia and 2-celled microconidia. c. Macroconidium and variably shaped unicellular microconidia.

*a. Key diagnostic characters*

(1) Macroconidia (the key diagnostic character) are highly variable in morphology, but are elongated, more or less curved, frequently described as boat- or canoe-shaped, and include one to several transverse septa, with a slight basal extension (foot). (2) Macroconidia: morphology, number of septa, color. (3) Colors of mycelium and/or exudations into culture media. (4) Microconidia (if formed): morphology of spores and conidiogenous cells. (5) Chlamydospores (if present): color, size, apical and/or intercalary formation, surface texture.

*b. Main taxonomic literature*

Booth, 1971; Nelson *et al.*, 1983; O'Donnell *et al.*, 2011.

*c. General comments*

*Fusarium* species delimitations have long been disputed, and their identifications now depend strongly on DNA sequence data (see O'Donnell *et al.*, 2011) but molecular species recognition approaches are still not definitive.

A few Fusaria may be specific pathogens for scale insects (Booth, 1971) but O'Donnell *et al.* (2011) found nearly 160 entomogenous *Fusarium* isolates to be distributed among many species complexes and that some of them differed little from phytopathogenic Fusaria.

### *6.* Gibellula *Cavara (Figure 6.20)*

[Sordariomycetes: Hypocreales: Cordycipitaceae] Synnemata (if present) white, yellowish, grayish to violet when fresh; becoming brownish with age, with conidial heads in a compact hymenium or more or less isolated; conidiophores septate, rough-walled, with an apical swelling (vesicle); conidiogenous cells (phialides) clustered on short swollen cells on vesicle, without obvious necks, with apical wall becoming progressively thickened; conidia aseptate, smooth, single or in short, dry chains. *TELEOMORPH: Torrubiella.* Hosts: spiders.

#### *a. Key diagnostic characters*

(1) Synnemata: presence, color, conidiophore arrangement on synnema (or on host body). (2) Wall texture on hyphae or conidiophores. (3) Conidiophores: size, surface texture, presence/absence of narrow isthmus below terminal vesicle.

#### *b. Major species*

*G. pulchra* (Saccardo) Cavara—conidiophores long, projecting from surface of white, lilac, yellow or orange synnemata. *G. leiopus* (Vuillemin) Mains—conidiophores short, crowded, forming dense fertile areas on lilac to purple synnemata.

#### *c. Main taxonomic literature*

Samson & Evans, 1992; Tzean *et al.*, 1997.

#### *d. General comments*

The apical vesicles of *Gibellula* conidiophores resemble those of *Aspergillus* but *Aspergillus* has (usually) globose conidia borne on phialides with a short neck. *Pseudogibellula formicarum* (Mains) Samson & Evans closely resembles *Gibellula pulchra* but is an insect rather than spider pathogen and has conidiogenous cells whose apices multiply denticulate (Samson & Evans, 1973). A second conidial state of *Gibellula/Torrubiella* species, *Granulomanus* spp. (Hoog, 1978), forms long, thin cylindrical conidia on polyphialidic conidiogenous cells.

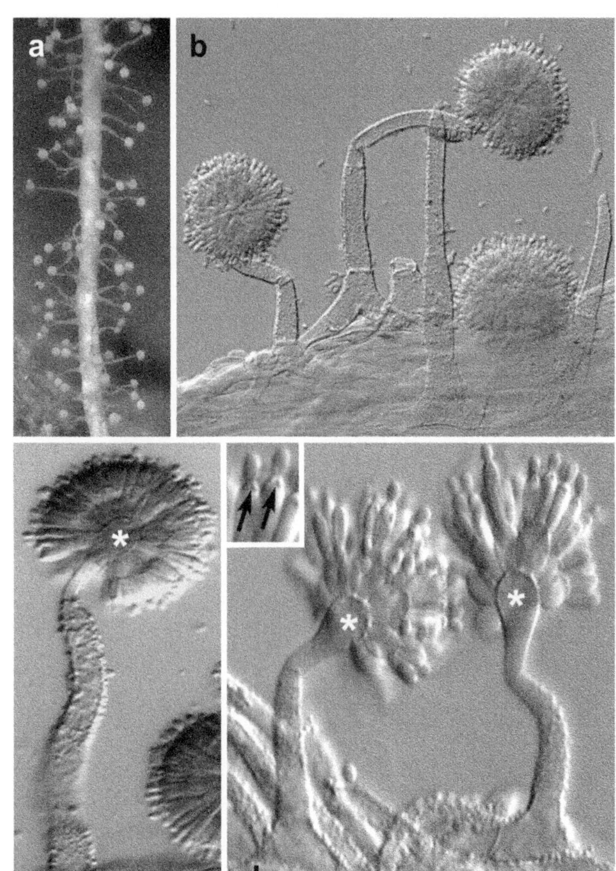


**Figure 6.20** *Gibellula.* a. Synnema bearing lateral conidiophores. b–d. Rough-surfaced conidiophores with inverted T- or Y-like bases (b) narrow to an apical vesicle (★) bearing densely clustered conidiogenous cells. d. Metulae (short cells) on vesicles bear clusters of conidiogenous cells (phialides) whose apices thicken (arrows in inset) as successive conidia form.

### *7.* Hirsutella *Patouillard (Figure 6.21)*

[Sordariomycetes: Hypocreales: most in Ophiocordycipitaceae; some taxa in Clavicipitaceae, Cordycipitaceae] Synnemata (if present) erect, often prominent, thin, compact, hard or leathery; conidiogenous cells (phialides) scattered to crowded, projecting laterally from synnema or from hyphae on host body, swollen basally and narrowing into one or more slender necks; conidia aseptate (rarely two-celled), hyaline, round, rhombic, elongate or like segments of citrus fruit, covered by persistent mucus, borne singly or two or more in droplets of mucus. *TELEOMORPHS:* mainly *Cordyceps* or

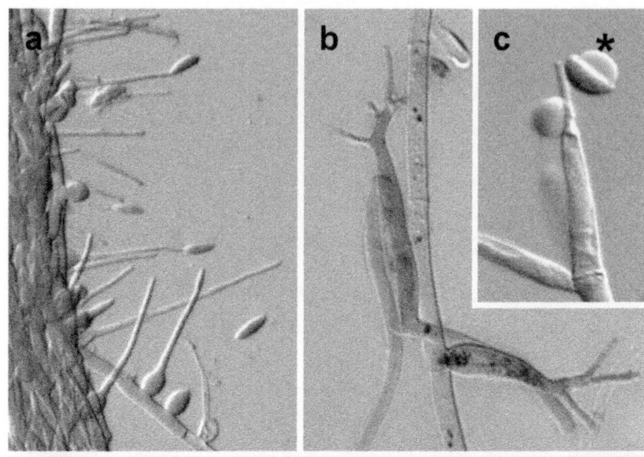


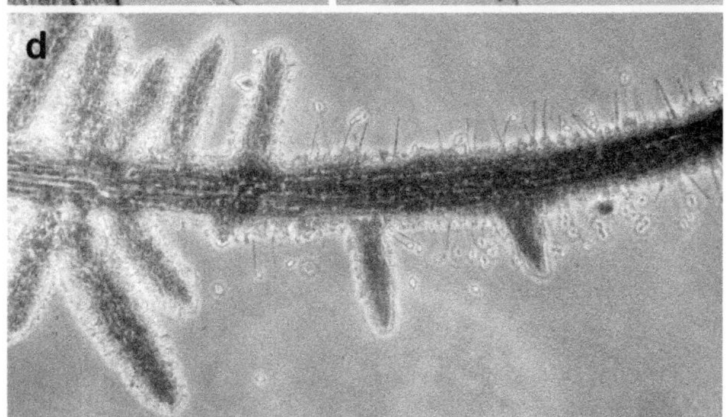


**Figure 6.21** *Hirsutella.* a–c. Conidiogenous cells with swollen bases and long, narrow, necks and conidia usually in an obvious slime drop (★ in c); long conidiogenous cells of *H. lecaniicola* b. are polyphialides (with several necks per conidiogenous cell). d. Synnema of *H. citriformis* have many lateral branches and prominent phialides.

*Torrubiella.* Hosts: many diverse insects (one species affecting nematodes).

*a. Key diagnostic characters*

(1) Conidiogenous cell (generic character): with swollen to flask-like base and one or more elongated, narrow necks. (2) Conidia: borne singly or in small groups, covered by persistent slime (slime absent from some species). (3) Synnemata: present in most species; not formed by several (mononematous) species. (4) Conidiogenous cell (specific characters): shape, size, with a single neck or polyphialidic.

*b. Major species*

*H. citriformis* Speare—on leaf- and planthoppers (Hemiptera: Cercopidae, Delphacidae); synnemata long, numerous, gray or brown, with many short lateral branches. *H. rhossiliensis* Minter & Brady—on nematodes and mites; not forming synnemata; conidiogenous cells with a single short, narrow neck; conidia resemble orange segments (straight on one side, curved on the other) or ellipsoid. *H. saussurei* Patouillard (type species)—synnemata long, numerous on various hymenopterans (wasps); *H. thompsonii* Fisher—on mites; conidiogenous cells with a distinctly swollen base and one short, narrow neck or polyphialidic; conidia globose with a smooth or wrinkled surface and no obvious slime layer.

*c. Main taxonomic literature*

Mains, 1951; Minter & Brady, 1980; Samson *et al.*, 1980; Minter *et al.*, 1983; Evans & Samson, 1982; Rombach & Roberts, 1989; Hodge, 1998.

*d. General comments*

Despite its many species and the importance this genus, *Hirsutella* has never been monographed, and its literature is dispersed; the closest substitute for a published monograph is a doctoral dissertation by Hodge (1998). Identifying *Hirsutella* species can be difficult due to this lack of a monograph and because the morphologies of *Hirsutella* species seem to intergrade with species from several other genera such as *Lecanicillium* and *Tolypocladium* (Humber & Rombach, 1987). Recent molecular studies (Sung *et al.*, 2007) have placed many *Hirsutella* species in Ophiocordycipitaceae, but other named and unnamed taxa of *Hirsutella* belong in Cordycipitaceae or Clavicipitaceae.

### 8. Hymenostilbe *Petch emend. Samson & Evans (Figure 6.22)*

[Sordariomycetes: Hypocreales: Ophiocordycipitaceae] Synnemata cylindrical or slightly tapered apically, covered by compact layer of conidiogenous cells; conidiogenous cells polyblastic, bearing solitary conidia on short denticles; conidia aseptate, hyaline, smooth or roughened. *TELEOMORPH: Cordyceps.* Hosts: diverse insects.

*a. Key diagnostic characters*

(1) Synnemata present. (2) Conidiogenous cells form compact hymenium. (3) Conidiogenous cells polyblastic. (4) Teleomorphic connections with *Cordyceps* spp. (which may occur together with *Hymenostilbe* synnemata).

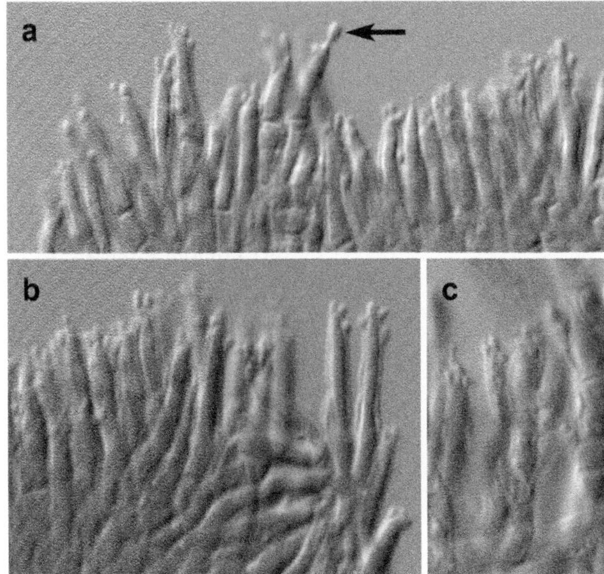


**Figure 6.22** *Hymenostilbe.* a–c. Conidiogenous cells with irregularly denticulate apices in dense hymenium on synnema; no conidia (except at arrow) remain on denticles.

*b. Major species*

*H. dipterigena* Petch—on flies; associated with *Cordyceps dipterigena;* synnemata brown, long.

*c. Main taxonomic literature*

Samson & Evans, 1975; Hywel-Jones 1995a, b, 1996.

*9.* Isaria *Persoon and Paecilomyces Bainier (Figure 6.23)*

[Sordariomycetes: Hypocreales: Cordycipitaceae for *Isaria* spp.] Conidiophores usually well developed, synnematous in many species, bearing whorls of divergent branches and conidiogenous cells (phialides), colorless to pigmented (but not black, brown, or olive); conidiogenous cells with a distinct neck and base flask- to narrowly awl-shaped or nearly globose, borne singly or in groups in whorls on conidiophores, on short side branches or in apical whorls; conidia aseptate, hyaline to colored, in dry divergent chains. *TELEOMORPHS: Cordyceps* or *Torrubiella.* Hosts: numerous, diverse insects.

*a. Key diagnostic characters*

(1) Conidiophores often synnematous. (2) Conidiogenous cells (phialides) single or whorled, with swollen (flask-like, clavate to globose) base and a distinct neck; orientation of phialides gives cluster of spore chains a feathery to cottony appearance. (3) Conidia: in long, divergent chains, one-celled, ovoid to elongate (rarely globose).

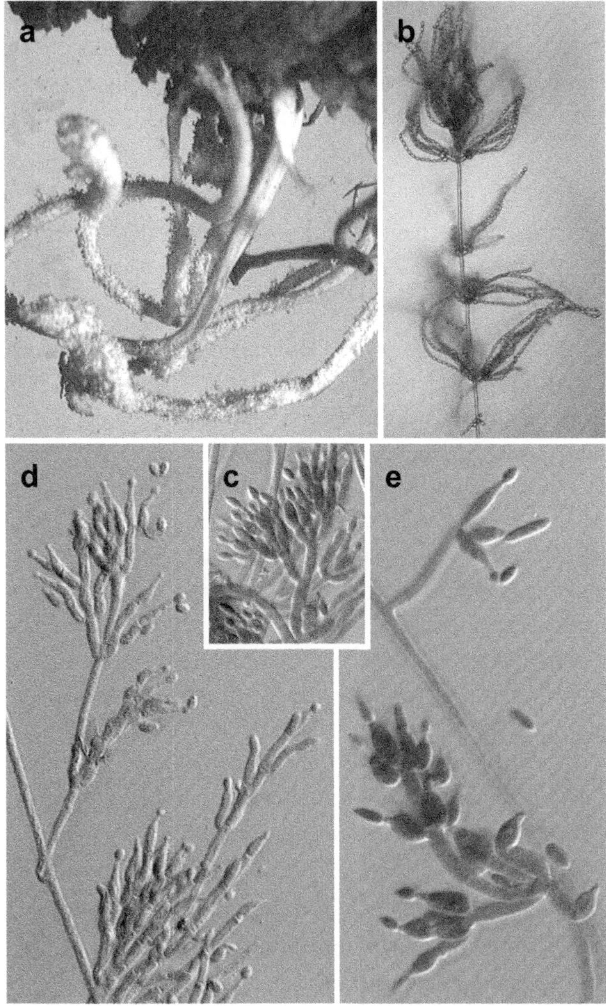


**Figure 6.23** *Isaria.* a. Synnemata. b. Clusters of divergent conidial chains. c–e. Conidiogenous cells (phialides) with swollen bases and prominent necks bearing conidia.

*b. Major species*

*I. farinosa* (Holm) Fries—synnemata often present; conidia short fusoid to lemon-shaped, $\leq 3\ \mu m$ long, smooth walled, white to cream-colored in mass. *I. fumosorosea* Wize—synnemata usually present; conidiophores and phialides with smooth uncolored walls; conidia long ovoid, $\leq 4\ \mu m$ long, rosy–tan to smoky pink (or gray) in mass. *Paec. lilacinus* (Thom)

Samson—conidial mass gray to tan with indistinct lavender shading; conidiophores slightly colored (in comparison to conidia), with roughened walls; conidia ellipsoid to fusoid, 2–3 μm long.

*c. Main taxonomic literature*

Samson, 1974; Luangsa-ard *et al.*, 2004, 2005.

*d. General comments*

Samson (1974) recognized two sections in *Paecilomyces*, with the type and all other species in section *Paecilomyces* were anamorphs of cleistothecial ascomycetes (Eurotiomycetes: Eurotiales). All entomopathogens were treated in *Paec.* section *Isarioidea* and are now known to be anamorphs of perithecial ascomycetes in the clavicipitoid families of Hypocreales (Luangsa-ard *et al.*, 2004) for which the further use of *Paecilomyces* is not phylogenetically supported. Luangsa-ard *et al.* (2005) reclassified most of these species in Cordycipitaceae in the genus *Isaria*; the few residual species from sect. *Isarioidea* placed in Clavicipitaceae (*P. carneus*, *P. marquandii*) and Ophiocordycipitaceae (*P. lilacinus*) are still referred to as *Paecilomyces* species but must eventually be assigned to other genera than *Paecilomyces* and *Isaria.*

*10.* Lecanicillium *W. Gams & Zare,* Simplicillium *W. Gams & Zare, and* Pochonia *Batista & O.M. Fonseca (Figure 6.24)*

[Sordariomycetes: Hypocreales: Cordycipitaceae—*Lecanicillium*, *Simplicillium*; Clavicipitaceae—*Pochonia*] Conidiophores little differentiated from vegetative hyphae; conidiogenous cells (phialides) in whorls (verticils) of two to six, paired, or solitary on hyphae or apically on short side branches; conidia hyaline, aseptate, borne in slime droplets or dry chains. Chlamydospores (in *Pochonia*) multicellular, ± globose, thick-walled, mostly on short lateral branches. TELEOMORPHS*: Cordyceps, Torrubiella* for entomopathogens. Hosts: scales, aphids, other insects (or nematodes).

*a. Key diagnostic characters*

(1) Conidiogenous cells (phialides): arrangement in whorls or pairs (or singly); elongated and usually tapering uniformly from the base. (2) Conidia: aggregating in slime drops at phialide tips (*rarely* single or in dry chains).

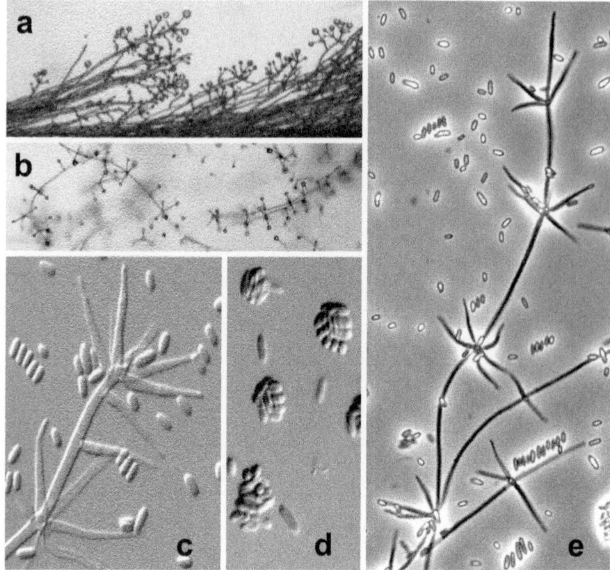


**Figure 6.24** *Lecanicillium*. a–b. Conidiophores showing whorled and single phialides bearing apical slime balls. c,e. Tapered conidiogenous cells form singly or in whorls. d. Balls of conidia in mucus.

*b. Major species*

*Lec. lecanii* (Zimmermann) Zare & W. Gams—conidia 2.5–3.5 × 1–1.5 μm, ± consistent in size, phialides 11–20 μm long. *Lec. longisporum* (Petch) Zare & W. Gams—conidia 5–10.5 × 1.5–2.5 μm, phialides 20–40 μm long. *Lec. muscarium* (Petch) Zare & W. Gams—conidia 2.5–5.5 × 1–1.5–(1.8) μm, highly variable in size, phialides 20–35 μm long. *V. fusisporum* W. Gams—conidia fusoid (spindle-shaped, 4–6 μm long).

*c. Main taxonomic literature*

Gams, 1971; Zare & Gams 2001; Zare *et al.*, 2001.

*d. General comments*

As with the switch to *Isaria* from *Paecilomyces* (see above), Gams (1971) classified *Verticillium* in two sections, *Verticillium* (including the type species) and *Prostrata* (which included insect-associated species). Species in section *Prostrata* proved to be clavicipitoid hypocrealeans whereas the species of section *Verticillium* were also phylogenetically mixed, but not clavicipitaceous (Sung *et al.*, 2001). The reclassification of sect. *Prostrata* assorted its species among a number of genera (Gams & Zare, 2001), only some of which included entomopathogens; these genera are, in turn, distributed among all three clavicipitoid families (Sung *et al.*, 2007) Clavicipitaceae (*Pochonia*, *Rotiferophthora* [not treated here]), Cordycipitaceae (*Lecanicillium*, *Simplicillium*), and Ophiocordycipitaceae (*Haptocillium* [not treated here]). *Verticillium lecanii* was moved to *Lecanicillium* as a three-species complex (*lecanii*, *muscarium*, and *longisporum*) along with most of the entomopathogenic species with strongly whorled phialides. *Simplicillium* differs from *Lecanicillium* in having its conidiogenous cells mostly single, but some in pairs and very few in whorls.

### *11.* Metarhizium *Sorokin (Figure 6.25)*

[Sordariomycetes: Hypocreales: Clavicipitaceae] Mycelium often covering affected hosts; conidiophores in dense patches; individual conidiophores broadly branched (candelabrum-like), densely intertwined; conidiogenous cells with rounded to conical apices, forming a dense hymenium; conidia aseptate, cylindrical or ovoid, in long chains usually adhering laterally and forming prismatic or cylindrical columns or solid plates of spores, pale to bright green to yellow–green, olivaceous, sepia or white in mass. *TELEOMORPH: Metacordyceps*. Hosts: extremely numerous and diverse.

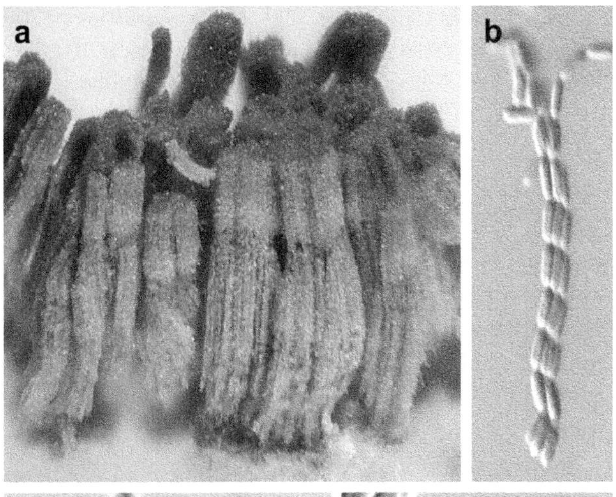


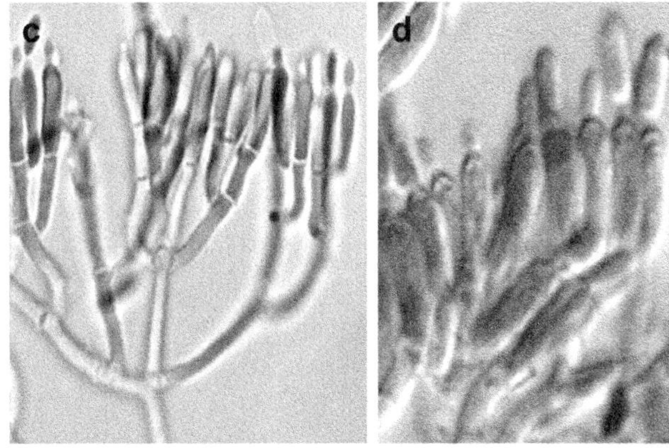


**Figure 6.25** *Metarhizium*. a. Masses of long, parallel conidial chains forming prismatic columns. b. Conidial chains showing side-by-side placement of spores in adjacent chains. c. Broadly branched conidiophores intermesh tightly to form dense, flat fertile surfaces (hymenia). d. Blunt apices of phialides thicken progressively as conidia form.

#### *a. Key diagnostic characters*

(1) Conidiogenesis occurring in dense hymenia. (2) Conidiophores branching repeatedly at broad angles and resembling candelabra (although observation of individual conidiophores is difficult). (3) Conidiogenous cells clavate or cylindrical, with a rounded to conical apex, no obvious neck; apical wall progressively thickening as conidia are produced. (4) Conidia produced in long chains; chains often adhere laterally to form prismatic columns or solid plates.

#### *b. Major species*

*M. anisopliae* (Metschnikoff) Sorokin—conidiogenous cell cylindrical; conidia $\leq 9\ \mu m$ long, cylindrical and often with a slight central narrowing, forming very long, laterally adherent chains, usually some shade of green. *M. majus* (Johnston) J.F. Bischoff, Rehner & Humber—conidia $\geq 11\ \mu m$; host usually a scarabaeid. *M. acridum* (Driver & Milner) J.F. Bischoff, Rehner & Humber—conidiogenous cells clavate to broadly ellipsoid; conidia light gray–green in mass, ovoid (*not* cylindrical), 7–11 μm long; relatively slow to develop, affecting Orthoptera.

#### *c. Main taxonomic literature*

Rombach *et al.*, 1986, 1987; Liang *et al.*, 1991; Driver *et al.*, 2000; Bischoff *et al.*, 2009; Kepler *et al.*, 2012.

#### *d. General comments*

*Metarhizium* is a sufficiently important genus of entomopathogens that its taxonomy has received two molecularly based revisions, both of which basically validated (and expanded) the morphological taxonomy of Rombach *et al.* (1987). Driver *et al.* (2000) used two genes to divide *M. anisopliae* and *M. flavoviride* into a series of new varieties and moved a commercialized pathogen of grasshoppers and locusts from *M. flavoviride* to a variety of *M. anisopliae*. Bischoff *et al.* (2009) included new species and taxa not treated by the Driver classification in a four-gene analysis that raised the *M. anisopliae* varieties from the earlier revision to species rank and retained several other species; the similar multilocus study of the *M. flavoviride* complex has still to be published but is expected to raise the varieties in that taxon to species rank. Identification of *Metarhizium* species will now need to be based on sequence data, especially for the translation-elongation factor (EF-1α) gene (Bischoff *et al.*, 2009), since the morphologies of currently accepted taxa are only poorly differentiated.

### *12.* Nomuraea *Maublanc (Figure 6.26)*

[Sordariomycetes: Hypocreales] Mycelium septate, white, with flocculent overgrowth, sparse in culture to dense on insects (often completely covering the host), usually becoming green or purple–gray to purple; conidiophores single or (rarely) synnematous (if synnematous, with a sterile base), erect, bearing whorls of short and blocky branches (metulae) with clusters of short phialides on metulae; conidiogenous cells short, with blunt apices and little if any distinct neck; conidia aseptate, smooth, round to ovoid or elongate and slightly curved, in short, divergent chains, pale to dark green, purple–gray to purple, or (rarely) white in mass. *TELEOMORPH: Metacordyceps, Ophiocordyceps.* Hosts: Noctuoid lepidopterans or spiders.

#### *a. Key diagnostic characters*

(1) Conidiophores: with conigenous cells in dense, individually distinct whorls. (2) Conidiogenous cells: short, blocky, with no distinct neck (but seemingly papillate at apex). (3) Conidia: one-celled, in short, divergent chains.

#### *b. Major species*

*N. rileyi* (Farlow) Samson (Clavicipitaceae)—on Noctuidae (especially larvae); conidial mass light (gray–green), covering host; conidia ovoid, in short chains. *N. atypicola* Yasuda (Ophiocordycipitaceae)—on spiders; conidia lavender–gray to purple.

#### *c. Main taxonomic literature*

Samson, 1974; Hywel-Jones & Sivichai, 1995; Sosa-Gómez *et al.*, 2009; Kepler *et al.*, 2012.

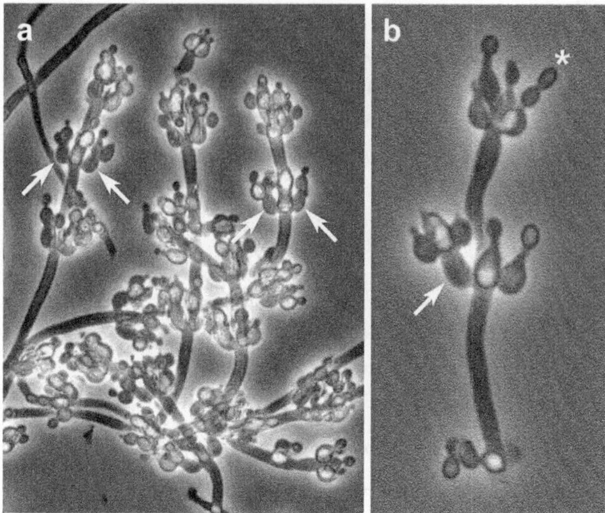


**Figure 6.26** *Nomuraea rileyi*. a. Note beaded appearance of whorls of conidiogenous cells borne directly on conidiophores or on short blocky cells (metulae; arrows). b. Ovoid conidia are formed in short chains (★); note the short, narrow necks of conidigenous cells.

#### *d. General comments*

It is clear that *Nomuraea* as traditionally circumscribed is phylogenetically heterogeneous. The most common species, *N. rileyi*, is a clavipitaceous pathogen affecting a very broad range of lepidopteran hosts worldwide, and may exert significant natural control over these insects. The ophiocordycipitaceous spider pathogen, *N. atypicola*, is much less common. A study of mitochondrial ribosomal gene sequences (Sosa-Gómez *et al.*, 2009) is notable since its analysis of a large number of *N. rileyi* isolates and other hypocrealean entomopathogens show show patterns of relationships that are directly parallel with those shown by the nuclear genes (e.g., Sung *et al.*, 2001, 2007).

### *13.* Tolypocladium *W. Gams (Figure 6.27)*

[Sordariomycetes: Hypocreales: Ophiocordycipitaceae] Conidiophores irregularly branched; conidiogenous cells (phialides) single or clustered, often forming terminal 'heads' with aggregated clusters of conidiogenous cells; conidiogenous cells with globose to flask-like base, narrowing abruptly to distinct neck that often bends away from axis of conidiogenous cell; conidia globose to cylindrical, aseptate, colorless, in slime heads; especially from nematoceran dipteran hosts. *TELEOMORPH: Elaphocordyceps* (for *T. inflatum* [= *T. niveum*]; Hodge

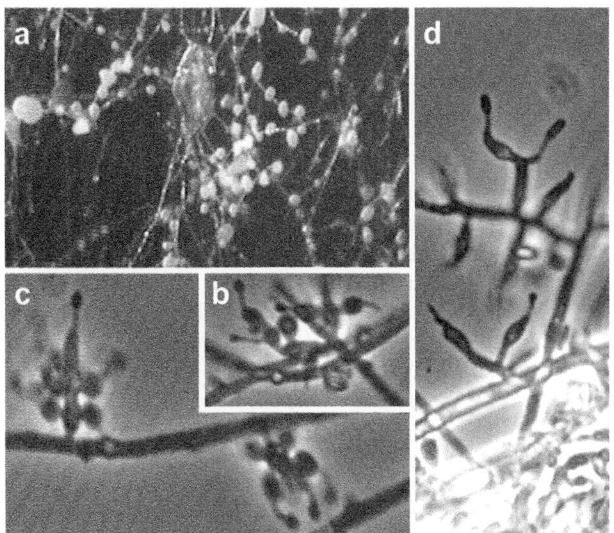


**Figure 6.27** *Tolypocladium*. a. Mucoid conidial balls formed on clusters of conidiogenous cells. b–d. Phialides with swollen bases and thin necks (usually bent away from cell axis) usually occur in small clusters.

*et al.*, 1996; Sung *et al.*, 2007). Hosts: mostly small dipterans.

*a. Key diagnostic characters*
(1) Conidiogenous cells (phialides): with flask-like to subglobose base, short narrow necks often bent out of the axis of the base; occurring singly or in whorls on vegetative cells. (2) Conidia: aseptate, released in slime drops.

*b. Major species*
*T. cylindrosporum* W. Gams—conidia cylindrical, straight or slightly curved. *T. extinguens* Samson & Soares—conidia subglobose, short ovoid or slightly curved (bean-shaped).

*c. Main taxonomic literature*
Bissett, 1983; Samson & Soares, 1984; Hodge *et al.*, 1996.

*d. General comments*
This genus has conidiogenous cells that can resemble those of species of *Lecanicillium*, *Hirsutella,* and other genera bearing conidial slime balls. The various conidia shapes of *Tolypocladium* differ from those usually seen in *Hirsutella* but may recall those of *Lecanicillium*. *Tolypocladium* phialides differ from those of verticillioid fungi by their distinct necks that usually bend to the side. Except for releasing conidia into slime droplets, *Tolypocladium* phialides also resemble the conidiogenous cells of *Beauveria* with a single conidium but before the extended denticulate rachis forms. *Tolypocladium* still remains a poorly circumscribed genus despite its use as the commercial source of cyclosporin antibiotics.

*14.* Torrubiella *Boudier (Figure 6.28)*

[Sordariomycetes: Hypocreales: Cordycipitaceae; some species reassigned to other genera in Clavicipitaceae] Stroma absent or poorly developed as a light- to brightly colored mycelial mat (subiculum) on the host; perithecia elongate, white to yellow to orange or red, superficial to immersed; asci elongated, with thickened apical cap penetrated by a fine pore, with eight filiform, multiseptate ascospores filiform, multiseptate, fragmenting at maturity to form one-celled part-spores; especially on spiders or scale insects. *ANAMORPHS: Gibellula*, *Granulomanus*, *Hirsutella* and other genera. Hosts: spiders, Hemiptera.

*a. Key diagnostic characters*
(1) Subiculum (mycelium covering host) compact to woolly, *not* organized into distinct stroma (either erect as in *Cordyceps* or covering host). (2) Perithecia: immersed to superficial. (3) Host: generally spiders (or Hemiptera).

*b. Major species*
None of the more than 50 species and varieties in this genus is especially common. See the general comments above for *Cordyceps*. Most species are found on spiders and often occur together with an anamorphic (conidial) state; most nonaraneous species of *Torrubiella* affect scale insects but are easily distinguished from *Hypocrella* by the absence of a compact, dense stroma.

*c. Main taxonomic literature*
Kobayasi 1982; Kobayasi & Shimizu 1982, 1983; Humber & Rombach, 1987, Johnson *et al.*, 2009, Kepler *et al.*, 2010, Sato *et al.*, 2010.

*d. General comment*
*Torrubiella* was traditionally distinguished from *Cordyceps* based on whether perithecia formed directly on the host's body or on an erect stroma, respectively. This distinction is challenged, however, by *T. ratticaudata* (Humber & Rombach, 1987, with an erect stroma but perithecia only on the host spider's body) and by the gene-based transfer of *T.*

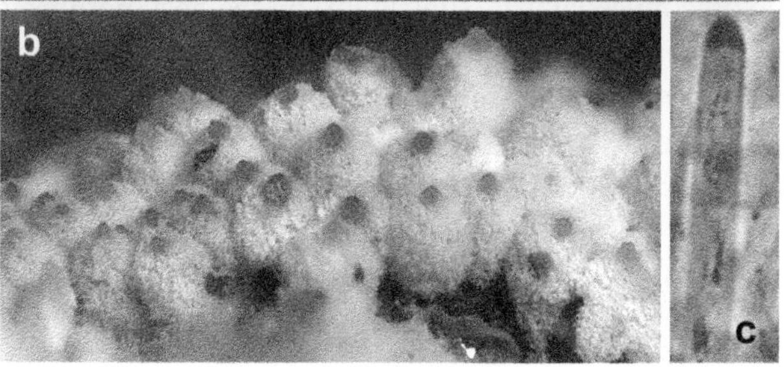


**Figure 6.28** *Torrubiella*. a–b. Body of affected spider covered by dense clusters of perithecia embedded in loose hyphal webs; with emergent (uncovered) darker ostioles or perithecial openings, especially in (b). c. Thickened apical cap of developing ascus, still too immature to show central pore through which ascospores would be discharged.

*pulvinata* (with perithecia on the host body but forming no stroma at all) to *Ophicocordyceps* (Kepler *et al.*, 2010). All species originally classified as *Torrubiella* but that are now placed by their genes in Clavicipitaceae *sensu stricto* will probably have to be reclassified in another genus; many *Torrubiella* species are confirmed to belong in Cordycipitaceae (although there is still no molecular confirmation for the correct familial placement of the type species, *T. aranicida*; see Sung *et al.*, 2007).

## REFERENCES


Alexopoulos, C. J., Mims, C. W., & Blackwell, M. (1996). *Introductory Mycology* (4th ed.). New York, NY: John Wiley & Sons. pp. 868.

Anderson, D. L., Gibbs, A. J., & Gibson, N. L. (1998). Identification and phylogeny of spore-cyst fungi (*Ascosphaera* spp.) using ribosomal DNA sequences. *Mycol. Res., 102*, 541–547.

Azema, R. (1975). Le genre *Septobasidium* Patouillard. *Documents Mycologiques, 6*, 1–24.

Bałazy, S. (1993). *Entomophthorales. Flora of Poland (Flora Polska), Fungi (Mycota), 24*. Botany, Kraków: Polish Acad. Sci., W. Szafer Inst. 1–356.

Barron, G. L. (1977). *The Nematode-Destroying Fungi (Topics in Mycobiology No. 1)*. Guelph: Canadian Biological Publ. pp. 140.

Ben-Ze'ev, I., & Kenneth, R. G. (1982). Features-criteria of taxonomic value in the Entomophthorales: I. A revision of the Batkoan classification. *Mycotaxon, 14*, 393–455.

Bischoff, J. F., Rehner, S. A., & Humber, R. A. (2009). A multilocus phylogeny of the *Metarhizium anisopliae* lineage. *Mycologia, 101*, 512–530.

Bissett, J. (1983). Notes on *Tolypocladium* and related genera. *Can. J. Bot., 61*, 1311–1329.

Bissett, J. (1988). Contribution toward a monograph of the genus *Ascosphaera*. *Can. J. Bot., 66*, 2541–2560.

Bland, C. E., Couch, J. N., & Newell, S. Y. (1981). Identification of *Coelomomyces*, Saprolegniales, and Lagenidiales. In H. D. Burges (Ed.), *Microbial Control of Pests and Plant Diseases 1970–1980* (pp. 129–162). London, UK: Academic Press.

Booth, C. (1971). The Genus *Fusarium*. Kew, UK: Commonw. Mycol. Inst., pp. 237.

Carmichael, J. W., Kendrick, W. B., Connors, I. L., & Sigler, L. (1980). *Genera of Hyphomycetes*. Edmonton, Canada: Univ. of Alberta Press. pp. 386.

Chaverri, P., Bischoff, J. F., Evans, H. C., & Hodge, K. T. (2005). *Regiocrella*, a new entomopathogenic genus with a pycnidial anamorph and its phylogenetic placement in the Clavicipitaceae. *Mycologia, 97*, 1225–1237.

Chaverri, P., Liu, M., & Hodge, K. T. (2008). A monograph of the entomopathogenic genera *Hypocrella, Moelleriella*, and *Samuelsia* gen. nov. (Ascomycota, Hypocreales, Clavicipitaceae), and their *Aschersonia*-like anamorphs in the Neotropics. *Stud. Mycol., 60*, 1–66.

Cooke, R. C., & Godfrey, B. E. S. (1964). A key to the nematode-detroying fungi. *Trans. Brit. Myc. Soc., 47*, 61–74.

Couch, J. N. (1938). *The Genus* Septobasidium. Chapel Hill: University of North Carolina Press, pp. 480.

Couch, J. N., & Bland, C. E. (Eds.). (1985). *The Genus* Coelomomyces. Orlando, FL: Academic Press, pp. 399.

Dick, M. W. (2001). *Straminipilous Fungi: Systematics of the Peronosporomycetes Including Accounts of the Marine Straminipilous Protists, the Plasmodiophorids, and Similar Organisms*. Dordrect, The Nether lands: Kluwer Academic Publ. pp. 670.

Driver, F., Milner, R. J., & Trueman, J. W. H. (2000). A taxonomic revision of *Metarhizium* based on a phylogenetic analysis of rDNA sequence data. *Mycol. Res., 104*, 134–150.

Evans, H. C., & Samson, R. A. (1982). Entomogenous fungi from the Galapagos Islands. *Can. J. Bot., 60*, 2325–2333.

Gams, W. (1971). Cephalosporium-*artige Schimmelpilze (Hyphomycetes)*. Stuttgart, Austria: Gustav Fischer Verlag. pp. 262.

Gams, W., & Zare, R. (2001). A revision of *Verticillium* sect. *Prostrata*. III. Generic classification. *Nova Hedwigia, 72*, 329–337.

Gómez, L. D., & Henk, D. A. (2004). Validations of the species of *Septobasidium* (Basidiomycetes) described by John N. *Couch. Lankesteriana, 4*, 75–96.

Hibbett, D. S., Binder, M., Bischoff, J. F., Blackwell, M., Cannon, P. F., Eriksson, E. I., Huhndorf, S., James, T., Kirk, P. M., Lücking, R., Lumbsch, H. T., Lutzoni, F., Matheny, P. B., McLaughlin, D. J., Powell, M. J., Redhead, S., Schoch, C. L., Spatafora, J. W., Stalpers, J. A., Vilgalys, R., Aime, M. C., Aptroot, A., Bauer, R., Begerow, D., Benny, G. L., Castlebury, L. A., Crous, P. W., Dai, Y.-C., Gams, W., Geiser, D. M., Griffith, G. W., Gueidan, C., Hawksworth, D. L., Hestmark, G., Hosaka, K., Humber, R. A., Hyde, K. D., Ironside, J. E., Koljalg, U., Kurtzman, C. P., Larsson, K.-H., Lichtwardt, R., Longcore, J., Miadlikowska, J., Miller, A., Moncalvo, J.-M., Mozley-Standridge, S., Oberwinkler, F., Parmasto, R., Reeb, V., Rogers, J. D., Roux, C., Ryvarden, L., Sampaio, J. P., Schüßler, A., Sugiyama, J., Thoron, R. G., Tibell, L., Untereiner, W. A., Walker, C., Wang, Z., Weir, A., Weiss, M., White, M. M., Winka, K., Yao, Y.-J., & Zhang, N. (2007). A higher-level phylogenetic classification of the *Fungi. Mycol. Res., 111*, 509–547.

Hodge, K. T. (1998). Revisionary studies in *Hirsutella* (anamorphic Hypocreales: Clavicipitaceae). Ithaca, NY: Ph.D. dissertation, Cornell. University.

Hodge, K. T., Krasnoff, S. B., & Humber, R. A. (1996). *Tolypocladium inflatum* is the anamorph of *Cordyceps subsessilis. Mycologia, 88*, 715–719.

Hoog, G. S. de (1972). The genera *Beauveria, Isaria, Tritirachium*, and *Acrodontium* gen. nov. *Stud. Mycol., 1*, 1–41.

Hoog, G. S. de (1978). Notes on some fungicolous Hyphomycetes and their relatives. *Persoonia (Leiden), 10*, 33–81.

Humber, R. A. (1976). The systematics of the genus *Strongwellsea* (Zygomycetes: Entomophthorales). *Mycologia, 68*, 1042–1060.

Humber, R. A. (1981). An alternative view of certain taxonomic criteria used in the Entomophthorales (Zygomycetes). *Mycotaxon, 13*, 191–240.

Humber, R. A. (1989). Synopsis of a revised classification for the Entomophthorales (Zygomycotina). *Mycotaxon, 34*, 441–460.

Humber, R. A., & Rombach, M. C. (1987). *Torrubiella ratticaudata* sp. nov. (Pyrenomycetes, Clavicipitales) and other fungi from spiders on the Solomon Islands. *Mycologia, 79*, 375–382.

Hywel-Jones, N. L. (1995a). Notes on *Cordyceps nutans* and its anamorph, a pathogen of hemipteran bugs in Thailand. *Mycol. Res., 99*, 724–726.

Hywel-Jones, N. L. (1995b). *Hymenostilbe ventricosa* sp. nov., a pathogen of cockroaches in Thailand. *Mycol. Res., 99*, 1201–1204.

Hywel-Jones, N. L. (1996). *Cordyceps myrmecophila*-like fungi infecting ants in the leaf litter of tropical forest in Thailand. *Mycol. Res., 100*, 613–619.

Hywel–Jones, N. L., & Sivichai, S. (1995). *Cordyceps cylindrica* and its association with *Nomuraea atypicola* in Thailand. *Mycol. Res., 99*, 809–812.

James, T. Y., Kauff, F., Schoch, C. L., Matheny, P. B., Hofstetter, V., Cox, C. J., Gelio, G., Gueidan, C., Fraker, E., Miadlikowska, J., Lumbsch, H. T., Rauhut, A., Reeb, V., Arnold, A. E., Amtoft, A., Stajich, J. E., Hosaka, K., Sung, G.-H., Johnson, D., O'Rourke, B., Crockett, M., Binder, M., Curtis, J. M., Slot, J. C., Wang, Z., Wilson, A. W., Schüßler, A., Longcore, J. E., O'Donnell, K., Mozley-Standridge, S., Porter, D., Letcher, P. M., Powell, M. J., Taylor, J. W., White, M. M., Griffith, G. W., Davies, D. R., Humber, R. A., Morton, J. B., Sugiyama, J., Rossman, A. Y., Rogers, J. D., Pfister, D. H., Hewitt, D., Hansen, K., Hambleton, S., Shoemaker, R. A., Kohlmeyer, J., Volkmann-Kohlmeyer, B., Spotts, R. A., Serdani, M., Crous, P. W., Hughes, K. W., Matsuura, K., Langer, E., Langer, G., Untereiner, W. A., Lücking, R., Büdel, B., Geiser, D. M., Aptroot, A., Diederich, P., Schimitt, I., Schultz, M., Yahr, R., Hibbett, D. S., Lutzoni, F., McLaughlin, D. J., Spatafora, J. W., & Vilgalys, R. (2006). Reconstructing the early evolution of Fungi using a six-gene phylogeny. *Nature, 443*, 818–822.

Johnson, D., Sung, G.-H., Hywel-Jones, N. L., Luangsa-ard, J. J., Bischoff, J. F., Kepler, R. M., & Spatafora, J. W. (2009). Systematics and evolution of the genus *Torrubiella* (Hypocreales, Ascomycota). *Mycol. Res., 113*, 279–289.

Karling, J. S. (1948). Chytridiosis of scale insects. *Amer. J. Bot., 35*, 246–254.

Keller, S. (1987). Arthropod-pathogenic Entomophthorales of Switzerland. I. *Conidiobolus, Entomophaga*, and *Entomophthora. Sydowia, 40*, 122–167.

Keller, S. (1991). Arthropod-pathogenic Entomophthorales of Switzerland. II. *Erynia, Eryniopsis, Neozygites, Zoophthora*, and *Tarichium. Sydowia, 43*, 39–122.

Keller, S. (1997). The genus *Neozygites* (Zygomycetes, Entomophthorales) with special reference to species found in tropical regions. *Sydowia, 49*, 118–146.

Keller, S. (2002). The genus *Entomophthora* (Zygomycetes, Entomophthorales) with a description of five new species. *Sydowia, 54*, 157–197.

Keller, S. (2007). Arthropod-pathogenic Entomophthorales from Switzerland. III. First additions. *Sydowia, 59*, 75–113.

Kepler, R. M., Kaitsu, Y., Tanaka, E., Shimano, S., & Spatafora, K. J. W. (2010). *Ophiocordyceps pulvinata* sp. nov., a pathogen of ants with a reduced stroma. *Mycoscience, 52*, 39–47.

Kepler, R. M., Sung, G.-H., Ban, S., Nakagiri, A., Chen, M.-J., Huang, B., Li, Z., & Spatafora, J. W. (2012). New teleomorph combinations in the entomopathogenic genus *Metacordyceps. Mycologia, 104*. 182–197.

Kerwin, J. L., & Petersen, E. E. (1997). Fungi: Oomycetes and Chytridiomycetes. In L. Lacey (Ed.), *Manual of Techniques in Insect Pathology* (pp. 251–268). London: Academic Press.

King, D. S. (1976). Systematics of *Conidiobolus* (Entomophthorales) using numerical taxonomy. II. Taxonomic considerations. *Can. J. Bot., 54*, 1285–1296.

King, D. S. (1977). Systematics of *Conidiobolus* (Entomophthorales) using numerical taxonomy. III. Descriptions of recognized species. *Can. J. Bot., 55*, 718–729.

Kirk, P. M., Cannon, P. F., Minter, D. W., & Stalpers, J. A. (2008). *Dictionary of the Fungi* (10th ed.). Wallingford, UK: CABI Publisher. pp. 784.

Kobayasi, Y. (1941). The genus *Cordyceps* and its allies. *Sci. Rep. Tokyo Bunrika Daig., Sect. B, 5*, 53–260.

Kobayasi, Y. (1982). Keys to the taxa of the genera *Cordyceps* and *Torrubiella*. *Trans. Mycol. Soc. Japan, 23*, 329–364.

Kobayasi, Y., & Shimizu, D. (1982). Monograph of the genus *Torrubiella*. *Bull. Nat. Sci. Museum, Ser. B (Botany), 8*, 43–78.

Kobayasi, Y., & Shimizu, D. (1983). *Iconography of Vegetable Wasps and Plant Worms*. Osaka: Hoikusha Publ. Co., Ltd. pp. 280.

Kohlmeyer, J., & Kohlmeyer, E. (1972). Permanent microscopic mounts. *Mycologia, 64*, 666–669.

Li, Z., & Humber, R. A. (1984). *Erynia pieris* (Zygomycetes: Entomophthoraceae): description, host range and notes on *Erynia virescens*. *Can. J. Bot., 62*, 653–663.

Li, Z., Li, C., Huang, B., & Fan, M. (2001). Discovery and demonstration of the teleomorph of *Beauveria bassiana* (Bals.) Vuill., an important entomogenous fungus. *Chin. Sci. Bull., 46*, 751–753.

Liang, Z.-Q., Liu, A.-Y., & Liu, J.-L. (1991). A new species of the genus *Cordyceps* and its *Metarhizium* anamorph. *Acta Mycol. Sinica, 10*, 257–262.

Lichtwardt, R. W., Cafaro, M. J., & White, M. M. (2001). *The Trichomycetes: Fungal Associates of Arthropods*. Revised online edition. http://www.nhm.ku.edu/~fungi/monograph/text/mono.htm.

Liu, M., Chaverri, P., & Hodge, K. T. (2006). A taxonomic revision of the insect biocontrol fungus *Aschersonia aleyrodis*, its allies with white stromata and their *Hypocrella* sexual states. *Mycol. Res., 110*, 537–554.

Liu, M., Rombach, M. C., Humber, R. A., & Hodge, K. T. (2005). What's in a name? *Aschersonia insperata*: a new pleoanamorphic fungus with characteristics of *Aschersonia* and *Hirsutella*. *Mycologia, 97*, 246–253.

Luangsa-ard, J. J., Hywel-Jones, N. L., & Samson, R. A. (2004). The polyphyletic nature of *Paecilomyces sensu lato* based on 18S–generated rDNA phylogeny. *Mycologia, 96*, 773–780.

Luangsa-ard, J. J., Hywel-Jones, N. L., Manoch, L., & Samson, R. A. (2005). On the relationships of *Paecilomyces* sect. *Isarioidea* species. *Mycol. Res., 109*, 581–589.

MacLeod, D. M., & Müller-Kögler, E. (1973). Entomogenous fungi: *Entomophthora* species with pear-shaped to almost spherical conidia (Entomophthorales: Entomophthoraceae). *Mycologia, 65*, 823–893.

MacLeod, D. M., Müller-Kögler, E., & Wilding, N. (1976). *Entomophthora* species with *E. muscae*-like conidia. *Mycologia, 68*, 1–29.

Mains, E. B. (1951). Entomogenous species of *Hirsutella, Tilachlidium*, and *Synnematium*. *Mycologia, 43*, 691–718.

Mains, E. B. (1958). North American entomogenous species of *Cordyceps*. *Mycologia, 50*, 169–222.

Mains, E. B. (1959a). North American species of *Aschersonia* parasitic on Aleyrodidae. *J. Insect Pathol., 1*, 43–47.

Mains, E. B. (1959b). Species of *Hypocrella*. *Mycopathol. Mycol. Appl., 11*, 311–326.

Meyling, N. V., Lubeck, M., Buckley, E. P., Eilenberg, J., & Rehner, S. A. (2009). Community composition, host range, and genetic structure of the fungal entomopathogen *Beauveria* in adjoining agricultural and seminatural habitats. *Molec. Ecol., 18*, 1282–1295.

Minter, D. W., & Brady, B. L. (1980). Mononematous species of *Hirsutella*. *Trans. Br. Mycol. Soc., 74*, 271–282.

Minter, D. W., Brady, B. L., & Hall, R. A. (1983). Five hyphomycetes isolated from eriophyid mites. *Trans. Br. Mycol. Soc., 81*, 455–471.

Nelson, P. E., Toussoun, T. A., & Marasas, W. F. O. (1983). Fusarium *Species: an Illustrated Manual for Identification*. University Park: Penn. State Univ. Press. pp. 193.

O'Donnell, K., Humber, R. A., Geiser, D. M., Kang, S., Park, B., Robert, V. A. R. G., Crous, P. W., Johnston, P. R., Aoki, T., Rooney, A. P., & Rehner, S. A. (2011). Phylogenetic diversity of insecticolous fusaria inferred from multilocus DNA sequence data and their molecular identification via the Internet at FUSARIUM-ID and *Fusarium MLST*. *Mycologia, 103*. in press.

Pelizza, S. A., López Lastra, C. C., Becnel, J. A., Humber, R. A., & García, J. J. (2008). Further research on the production, longevity, and infectivity of the zoospores of *Leptolegnia chapmanii* Seymour (Oomycota: Peronosporomycetes). *J. Invertebr. Pathol., 98*, 314–319.

Petch, T. (1914). The genera *Hypocrella* and *Aschersonia*. *Ann. R. Bot. Gdns. Peradeniya, 5*, 521–537.

Petch, T. (1921). Studies in entomogenous fungi. II. The genera of *Hypocrella* and *Aschersonia*. *Ann. R. Bot. Gdns. Peradeniya, 7*, 167–278.

Rehner, S. A., Minnis, A. M., Sung, G.-H., Luangsa-ard, J. J., Devotto, L., & Humber, R. A. (2011). Phylogeny and systematics of the anamorphic, entomopathogenic genus *Beauveria*. *Mycologia, 103*, 1055–1073.

Remaudière, G., & Hennebert, G. L. (1980). Révision systématique de *Entomophthora aphidis* Hoffm. in Fres. Description de deux nouveaux pathogènes d'aphides. *Mycotaxon, 11*, 269–321.

Rombach, M. C., & Roberts, D. W. (1989). *Hirsutella* species (Deuteromycotina; Hyphomycetes) on Philippine insects. *Philip. Entomol, 7*, 491–518.

Rombach, M. C., Humber, R. A., & Roberts, D. W. (1986). *Metarhizium flavoviride* var. *minus* var. nov., a pathogen of plant- and leafhoppers of rice in the Philippines and Solomon Islands. *Mycotaxon, 27*, 87–92.

Rombach, M. C., Humber, R. A., & Evans, H. C. (1987). *Metarhizium album* Petch, a fungal pathogen of leaf- and planthoppers of rice. *Trans. Br. Mycol. Soc., 88*, 451–459.

Rose, J. B., Christensen, M., & Wilson, W. T. (1984). *Ascosphaera* species inciting chalkbrood in North America and a taxonomic key. *Mycotaxon, 19*, 41–55.

Samson, R. A. (1974). *Paecilomyces* and some allied Hyphomycetes. *Stud. Mycol, 6*, 1–119.

Samson, R. A., & Evans, H. C. (1973). Notes on entomogenous fungi from Ghana. I. The genera *Gibellula* and *Pseudogibellula*. *Acta Bot. Neerl, 22*, 522–528.

Samson, R. A., & Evans, H. C. (1975). Notes on entomogenous fungi from Ghana. III. The genus *Hymenostilbe*. *Proc. Koninkl. Nederl. Akad. Wetensch., Ser. C, 78*, 73–80.

Samson, R. A., & Evans, H. C. (1982). Two new *Beauveria* spp. from South America. *J. Invertebr. Pathol., 39*, 93–97.

Samson, R. A., & Evans, H. C. (1992). New species of *Gibellula* on spiders (Araneida) from South America. *Mycologia, 84*, 300–314.

Samson, R. A., & Soares, G. G., Jr. (1984). Entomopathogenic species of the hyphomycete genus *Tolypocladium*. *J. Invertebr. Pathol., 43*, 133–139.

Samson, R. A., Evans, H. C., & Latgé, J.-P. (1988). *Atlas of Entomopathogenic Fungi*. Berlin: Springer-Verlag. pp. 187.

Samson, R. A., McCoy, C. W., & O'Donnell, K. L. (1980). Taxonomy of the acarine parasite *Hirsutella thompsonii*. *Mycologia, 72*, 359–377.

Sato, H., Ban, S., Masuya, H., & Hosoya, T. (2010). Reassessment of type specimens of *Cordyceps* and its allied described by Dr. Yosio Kobayasi preserved in TNS. Part 1. The genus *Torrubiella*. *Mycoscience, 51*, 154–151.

Scorsetti, A. C., Jensen, A. B., López Lastra, C., & Humber, R. A. (2012). First report of *Pandora neoaphidis* resting spore formation *in vivo* in aphid hosts under field conditions. *Fung. Biol., 116*. 196–203.

Seymour, R. L. (1984). *Leptolegnia chapmanii*, an oomycete pathogen of mosquito larvae. *Mycologia, 76*, 670–674.

Shimazu, M., Mitsuhashi, W., & Hashimoto, H. (1988). *Cordyceps brongniartii* sp. nov., the teleomorph of *Beauveria brongniartii*. *Trans. Mycol. Soc. Japan, 29*, 323–330.

Skou, J. P. (1972). Ascosphaerales. *Friesia, 10*, 1–24.

Skou, J. P. (1988). Japanese species of *Ascosphaera*. *Mycotaxon, 31*, 173–190.

Soper, R. S. (1974). The genus *Massospora*, entomopathogenic for cicadas, Part, I, Taxonomy of the genus. *Mycotaxon, 1*, 13–40.

Soper, R. S. (1981). New cicada pathogens: *Massospora cicadettae* from Australia and *Massospora pahariae* from Afghanistan. *Mycotaxon, 13*, 50–58.

Soper, R. S., Shimazu, M., Humber, R. A., Ramos, M. E., & Hajek, A. E. (1988). Isolation and characterization of *Entomophaga maimaiga* sp. nov., a fungal pathogen of gypsy moth, *Lymantria dispar*, from Japan. *J. Invertebr. Pathol., 51*, 229–241.

Sosa-Gómez, D. R., Humber, R. A., Hodge, K. T., Binneck, E., & da Silva-Brandão, K. L. (2009). Variability of the mitochondrial SSU rDNA of *Nomuraea* species and other entomopathogenic fungi from Hypocreales. *Mycopathologia, 167*, 145–154.

Sparrow, F. K., Jr. (1939). The entomogenous chytrid *Myrophagus* Thaxter. *Mycologia, 31*, 439–444.

Sparrow, F. K., Jr. (1960). *Aquatic Phycomycetes* (2nd ed.). Ann Arbor: University of Michigan Press. pp. 1187.

Steinkraus, D. C., Oliver, J. B., Humber, R. A., & Gaylor, M. J. (1998). Mycosis of bandedwinged whitefly (*Trialeurodes abutilones*) (Hemiptera: Aleyrodidae) caused by *Orthomyces aleyrodis* gen. and sp. nov. (Entomophthorales: Entomophthoraceae). *J. Invertebr. Pathol., 72*, 1–8.

Sung, G.-H., Hywel-Jones, N. L., Sung, J.-M., Luangsa-ard, J. J., Shrestha, B., & Spatafora, J. W. (2007). Phylogenetic classification of *Cordyceps* and the clavicipitaceous fungi. *Stud. Mycol., 57*, 5–59.

Sung, G.-H., Spatafora, J. W., Zare, R., Hodge, K. T., & Gams, W. (2001). A revision of *Verticillium* sect. *Prostrata*. II. Phylogenetic analyses of SSU and LSU nuclear rDNA sequences from anamorphs and teleomorphs of the Clavicipitaceae. *Nova Hedwigia, 72*, 311–328.

Sung, J.-M. (1996). *The Insects-Born Fungus of Korea in Color*. Seoul: Kyohak Publ. pp. 315.

Tavares, I. (1985). *Laboulbeniales (Fungi, Ascomycetes)*. Lehre: Mycologia Memoir No. 9. J. Cramer. pp. 627.

Thaxter, R. (1888). The Entomophthoreae of the United States. *Mem. Boston Soc. Nat. Hist., 4*, 133–201.

Tzean, S. S., Hsieh, L. S., & Wu, W. J. (1997). The genus *Gibellula* on spiders from Taiwan. *Mycologia, 89*, 309–318.

Zare, R., & Gams, W. (2001). A revision of *Verticillium* section *Prostrata*. IV. The genera *Lecanicillium* and *Simplicillium*. *Nova Hedwigia, 73*, 1–50.

Zare, R., Gams, W., & Evans, H. C. (2001). A revision of *Verticillium* section *Prostrata*. V. The genus *Pochonia*, with notes on *Rotiferophthora*. *Nova Hedwigia, 73*, 51–86.

## NOTES

1. *Paecilomyces lilacinus* was recently reassigned as the type and only species of the new genus by Luangsa-ard, J. J., Houbraken, J., van Doorn, T., Hong, S. B., Borman, A. M., Hywel-Jones, N. L., & Samson, R. A. (2011). *Purpureocillium*, a new genus for the medically important *Paecilomyces lilacinus*. *FEMS Microbiol. Lett., 321*, 1411–149.
2. Article 59 of the International Code of Botanical Nomenclature, which allowed separate and valid names for both sexual and conidial states of pleomorphic fungi (such as the entomopathogenic Hypocreales), was eliminated during the 2011 International Botanical Congress in Melbourne, Australia. Without regard for how variable the different forms of fungi might be, in the future any one fungal organism will be allowed only a single valid name. The choice of which generic name will be applied among sexual and conidial morphs for any group of fungi will be determined by a complex and

lengthy committee-based process. This major rule change will dramatically affect many fungal entomopathogens, but it will take years before the final, official decisions between competing names may be made. Until then, mycologists will need to continue using the rejected dual nomenclatural system for naming these fungi.

## BRIEF GLOSSARY OF MYCOLOGICAL TERMS

Irregular plurals of terms follow a slash (/). Terms listed here apply to the bracketed fungi at the end of definitions.

**Anamorph.** The conidial (asexual) state of an ascomycete fungus, usually formed in the complete absence of any sexual state. Anamorphs may have nomenclaturally valid names different from the sexual state (*teleomorph*) of the same fungus; two or more distinctly different (and separately nameable) conidial states formed by a single teleomorphic species are referred to as *synanamorphs*.

**Ascus/asci.** Cell in which a single nucleus undergoes meiosis, after which one or more (usually eight) *ascospores* are cleaved out of the cytoplasm. [Ascomycota.]

**Capilliconidium/capilliconidia.** A passively dispersed conidium produced apically on a long, slender (capillary) *conidiophore* arising from another conidium. [Entomophthorales, *e.g.*, *Neozygites*, *Zoophthora.*]

**Conidiogenous cell.** The cell on which a conidium forms, usually with only a single place (locus) on which a conidium forms; some conidiogenous cells have two or more conidiogenous loci. [anamorphic Ascomycota, Entomophthorales.]

**Conidiophore.** A simple or branched hypha or hyphal system bearing conidiogenous cells and their conidia. [anamorphic Ascomycota, Entomophthorales.]

**Conidium/conidia.** Fungal mitospore formed externally on a *conidiogenous cell*; conidia are *not* formed wholly inside any other cell (*ascus*, *sporangium*, etc.,) nor as external meiospores (basidiospores on a basidium, the cell in basidiomycetes in which both karyogamy and meiosis occurs prior to basidiosporogenesis). [anamorphic Ascomycota, Entomophthorales.]

**Cystidium/cystidia.** In the Entomophthorales, more or less differentiated hyphae that precede and facilitate the emergence of the developing *conidiophores* through the host cuticle; cystidia usually project above the *hymenium*, but soon lose their turgor and collapse. Cystidia are rarely seen on any but very fresh specimens. [Entomophthorales; *e.g.*, *Pandora* spp.]

**Denticle.** One of several to many small, conical to spike- or thorn-like or truncate projections on a conidiogenous cell, each of which bears a single conidium. [anamorphic Ascomycota; *e.g.*, *Beauveria* or *Hymenostilbe* spp.]

**Hymenium/hymenia.** A compact palisade layer of sporulating cells (conidiogenous cells, asci, etc.). [Ascomycota; Entomophthorales.]

**Papilla/papillae.** The basal portion of an entomophthoralean conidium by which conidia attach to conidiogenous cells and which is usually involved in forcible discharge of conidia. [Entomophthorales.]

**Perithecium/perithecia.** A globose, ovoid or pear-shaped walled structure in which *asci* and *ascospores* form; perithecia may be superficial or partially to fully immersed in the fruiting body. Each perithecium has an apical opening (ostiole) through which the *ascospores* are discharged. [Ascomycota: Sordariomycetes.]

**Polyphialide.** A conidiogenous cell having more than one neck, each of which produces one or more conidia; relatively common in *Hirsutella* species that do not form synnemata. [anamorphic Ascomycota.]

**Rachis/raches.** A geniculate (or sometimes zig-zag) apical extension of a conidiogenous cell produced by sympodial branching of the elongating extension beneath each successive conidium formed. [*Beauveria* spp.]

**Rhizoid.** In the Entomophthorales, more or less differentiated hyphae that contact and anchor a host to the substrate; they may or may not have differentiated terminal holdfasts. [Entomophthorales.]

**Sporangium/sporangia.** A cell or 'spore sac' inside of which (mitotic or meiotic) spores form; this is a very general term that can be correctly applied to diverse structures in nearly every class of fungi.

**Stroma/stromata.** A loose to fleshy or dense mass of vegetative hyphae on or in which spores (conidia or ascospores) are produced. Conidial stromata are usually very dense and compact, not extending very far above the host or substrate) (*e.g.*, *Aschersonia* spp.); ascomycetous stromata bearing *perithecia* may be either low and compact (*e.g.*, *Hypocrella* spp.) or upright and club- to column-like (*e.g.*, *Cordyceps* spp.). [anamorphic and teleomorphic Ascomycota.]

**Subiculum.** A woolly to crust-like mycelial growth (a) on the substrate under a fungal fruiting body or (b) on or in which perithecia are borne on the surface of a host.

**Synanamorph.** See anamorph.

**Synnema/synnemata.** An erect, branched or simple (unbranched) aggregation of hyphae; loosely fasciculate to compact, leathery or brittle in consistency, bearing conidiogenous cells and conidia. [anamorphic Ascomycota; *e.g.*, *Hirsutella.*]

**Teleomorph.** The sexual (ascus-forming) state of an ascomycete fungus that can produce one or more conidial (anamorphic) states.

**Zoospore.** A uni- or biflagellate motile spore produced in a *zoosporangium*. [Blastocladiomycetes; Peronosporomycetes.]

**Zoosporangium/zoosporangia.** The *sporangium* in which flagellate *zoospores* develop; *zoospores* and *zoosporangia* are formed only by watermolds. [Blastocladiomycetes; Peronosporomycetes.]

## APPENDIX: RECIPES OF STAINS AND REAGENTS

### 1. Aceto-orcein [nuclear stain/mounting medium]

| | |
|---|---|
| Orcein (natural or synthetic) | 1.0 g |
| Acetic acid, glacial | 45.0 ml |

Dissolve the orcein in hot glacial acetic acid, dilute 1 : 1 with distilled water and reflux or boil for at least 5 min. If boiled, replace lost volume with 50% glacial acetic acid. Filter at least twice to remove undissolved particulates. This stain continues to throw a precipitate over time and requires periodic clarification by filtration. There are many other ways of preparing aceto-orcein, most of which recommend refluxing; this simplified procedure yields a very satisfactory stain.

**Note**: both aceto-orcein and 50% acetic acid (with or without addition of an acidic stain) are excellent general mounting media for diagnostic slide preparations because of their low viscosity and ability to wet even hydrophobic material.

**2.** Lactophenol [mounting medium]

| | |
|---|---|
| Phenol (crystals) | 20 g |
| Lactic acid | 20 g |
| Glycerol | 40 g |
| Distilled water | 20 ml |

The addition of phenol (which acts as a preservative for long-term retention of slides) to this mixture is strictly optional; its omission has no effect on the utility of the mixture. Any available acidic stain can be added to lactophenol.

**3.** Lactic acid [mounting medium]

Anhydrous lactic acid with or without the addition of stains such as acid fuchsin, aniline blue or other acidic dyes can be used as an outstanding temporary to semi-permanent mounting medium.

CHAPTER VII

# Laboratory techniques used for entomopathogenic fungi: Hypocreales

G. DOUGLAS INGLIS*, JUERG ENKERLI† & MARK S. GOETTEL*

*Agriculture and Agri-Food Canada Research Centre, Lethbridge, Alberta, Canada T1J 4B1
†Agroscope Reckenholz-Tanikon, Reckenholzstrasse 191, Switzerland

## 1. INTRODUCTION

Recent taxonomic studies on fungi using molecular techniques have provided important information on their phylogenies and new classifications (Chapter VI). Many genera of entomogenous fungi previously handled under the formed anamorphic class Hyphomycetes, have been reclassified in the Order Hypocreales in the Phylum Ascomycota, and coupled to their teleomorphs (e.g., *Cordyceps* spp.). The anamorphic states of these fungi are filamentous and reproduce asexually by conidia generally formed aerially on conidiophores arising from the substrate. The most common route of host invasion is through the external integument, although infection through the digestive tract is possible. Conidia attach to the cuticle, germinate, and penetrate the cuticle. Once in the hemocoel, hyphae colonize tissues throughout the host, forming yeast-like hyphal bodies often referred to as blastospores. Host death is often due to a combination of the action of a fungal toxin, physical obstruction of blood circulation, nutrient depletion, and invasion of organs. After host death, hyphae usually emerge from the cadaver and, under appropriate conditions of temperature and humidity, produce conidia on the exterior of the host. Conidia are then dispersed by wind or water. Under dry or cool conditions, insects often remain intact as fungus-filled 'mummies' and the fungus only emerges and sporulates once the cadaver is brought under appropriate conditions for fungal growth and conidiation. Consequently, entomopathogenic Hypocreales are among the most commonly encountered insect pathogens as cadavers often remain intact for long periods and the external mycelium is conspicuous (see Chapter I).

Most entomopathogenic Hypocreales are facultative pathogens and are relatively easily grown in pure culture on defined or semi-defined media. However, many non-pathogenic microorganisms quickly colonize insect cadavers, especially if they are not already colonized by a pathogen. Consequently, it is often a challenge to determine whether a hypocrealean fungus isolated from an insect cadaver was responsible for the insect's death.

Entomopathogenic Hypocreales have one of the widest spectra of host ranges among entomopathogens. A variety of factors may determine or influence the susceptibility of a host to infection by these fungi. These include the genetics of the fungal strain, the host's physiological state, nutrition, defence mechanisms, presence of other microorganisms as well as a number of other diverse factors such as environmental parameters. Prediction of the ecological host range based on laboratory bioassay results remains a challenge, as in most cases, insects are much more susceptible to infection under laboratory or caged situations than they are in their field environment. To be useful in predicting field virulence, laboratory bioassays must attempt to mimic those conditions most likely to be expected in nature.

In this chapter we provide techniques for the study of entomopathogenic hypocrealean fungi. Due to the

MANUAL OF TECHNIQUES IN INVERTEBRATE PATHOLOGY
ISBN 9780123868992

diversity of entomopathogenic fungi and the very wide range of hosts that they infect, we provide information both in the form of generalizations and via specific examples. In many cases, readers will need to adapt these techniques to specific requirements. Recent and on-going taxonomic revisions make it difficult to provide the latest epithets for all general For instance, the former *Verticillium lecanii* now represents several species in the genus *Lecanicillium* while the former *Metarhizium anisopliae* now represents nine species within *Metarhizium* including *M. anisopliae.* Where possible, we provide examples using the most recent taxonomy; however, where ambiguous, we retain former taxonomic names.

The reader is referred to Tanada & Kaya (1993), Boucias & Pendland (1998), Butt *et al.* (2001), Upadhyay (2003), Vestergaard *et al.* (2003), Ekesi and & Maniania (2007), Wraight *et al.* (2007), Goettel *et al.* (2010), Roy *et al.* (2010), and Vega & Kaya (2012) for more recent reviews on entomopathogenic hypocrealean biology, epizootiology, pathogenesis and development as microbial control agents.

## 2. DETECTION AND CHARACTERIZATION

Traditionally, classical microbiological methods have been used to isolate and identify hypocrealean fungi. While classical methods are still heavily utilized, the application of molecular methods, alone or in combination with classical methods, have provided valuable insights into the biology of entomopathogenic Hypocreales.

### A. Classical methods

Entomopathogenic Hypocreales can most often be isolated from insect cadavers or from soil. However, methods that favor isolation of these fungi may be required when isolating them from soil or when contaminants are a problem. Special care must be taken when isolating these fungi from within plant tissues as they often occur on plant surfaces (e.g., on the phylloplane).

#### *1. Isolation from cadavers*

Entomopathogenic Hypocreales may be harvested directly from insect cadavers on which the fungus has already sporulated. If external sporulation has not occurred, cadavers may be placed in suitable environments and the fungus allowed to produce hyphae or conidia externally. Alternately, cadavers may be homogenized and the homogenate plated on an appropriate agar medium. In cases where isolation is difficult, it may be necessary to experimentally infect more insects, before finally obtaining a pure culture (see Section 5 below).

In many instances, it is desirable to surface-sanitize cadavers to remove potential contaminants on their integument. However, this is possible only if the fungus has not yet emerged and sporulated on the host surface. The most common 'disinfectants' used are sodium hypochlorite (1–5%) and/or ethanol (70%) (Chapter I). The efficacy of the sanitation procedure can be confirmed by swirling the treated cadaver in a nutrient broth (with or without antibacterial agents) and checking for bacterial growth after 1–3 days. This method may yield unsatisfactory results if the integrity of the cadaver integument is poor.

If the identity of the fungus is known, it is preferable to use a selective medium developed for the particular fungus in question (Section 2 below). However, if a selective medium does not exist or if the identity of the fungus is unknown, virtually any medium used for propagation of entomopathogenic Hypocreales can be used (see Section 3A). We routinely use sabouraud dextrose agar with yeast extract (SDAY) (Appendix: 2) supplemented with streptomycin sulphate (0.08%) and penicillin (0.03%).

It may be necessary to first induce external growth of entomopathogenic Hypocreales by placing the cadaver in an environment with high relative humidity (e.g., on a water agar medium amended with antibacterial agents, on moistened filter paper, or adjacent to moistened cotton batten in a sealed sterile container such as a Petri dish sealed with parafilm or plastic film). For insects that have died of mycosis, vegetative growth and/or sporulation on the surface of the integument usually occurs within a few days at 20–25°C under conditions of high humidity. Once adequate growth is observed on cadavers, the fungus can be transferred to an appropriate agar medium with a needle. If prolific conidiogenesis is observed on the cadaver, conidia can be streaked on to an agar medium directly or conidia can be suspended in buffer or water before being streaked. Addition of an antimicrobial agent such as chloromycetin (50 μg/ml) to the buffer or water may be useful. It may also be possible to obtain relatively pure colonies of entomopathogenic Hypocreales by attaching the cadaver to the lid of the Petri dish with an adhesive such as double-sided sticky tape. Conidia released from the cadaver which land on the medium may produce mycelial colonies that are relatively free of contaminants.

Another isolation method requires the homogenization of cadavers followed by dilution plating of the homogenate on to an appropriate semi-selective medium. The surface-sanitized cadaver can be homogenized using a variety of methods. Potter–Elvehjem tubes may be used for larger insects and micropestles in 1.5-ml microcentrifuge tubes for smaller insects. Mechanical tissue grinders and blenders may also be used to homogenize insects, particularly those with hard exoskeletons, but it is more difficult to maintain a sterile environment and preclude cross-contamination of samples using this equipment.

### *2. Selective media*

Selective media are frequently used for the isolation of entomopathogenic Hypocreales. Most bacteria are inhibited by low pH and, in general, fungal growth will be favored on media where the pH is less than 5. In most instances, however, inhibition of bacteria is achieved by amending media with antibacterial agents. Frequently used wide-spectrum antibacterial agents include chloramphenicol, tetracycline, and streptomycin. These agents may be used alone or in combination with agents with a more specific mode of action (e.g., penicillin, novobiocin, and chloramphenicol). While crystal violet is commonly used as a coloring agent to provide contrast for visualizing light-colored colonies, it also inhibits the growth of gram-positive bacteria. Antimicrobial agents must be used with caution as they may be inhibitory to some species or strains of fungi. Furthermore, care must be taken to appropriately dispose of cultures (e.g., by autoclaving) to ensure that bacteria possessing resistance to antimicrobial agents are not released into the environment.

Inhibition of contaminant fungi is more problematic than bacteria, and fungal contaminants are invariably a problem when attempting to isolate entomopathogenic Hypocreales from soil. Fungi in the genera *Trichoderma*, *Mucor* and *Rhizopus* are fast growing and can rapidly (*ca.* 2–3 days) obscure colonies of entomopathogenic Hypocreales making isolation difficult to impossible. Species of *Penicillium* and *Aspergillus* can also be problematic because they are prolific sporulators and are common in soil. Undesirable fungi can be inhibited by amending media with fungicides but other materials such as Rose Bengal, oxgall, O-phenyl-phenol, and/or sodium desoxycholate have also been used. In addition to facilitating isolation of entomopathogenic Hypocreales, selective media have advanced our knowledge of host targeting, population dynamics, and their saprotrophic behavior.

#### *a.* Beauveria *and* Metarhizium *spp.*

Media based on the differential activity of the fungicide, dodine (N-dodecylguanidine monoacetate) have been successfully used to isolate *Beauveria bassiana* and *Metarhizium* spp. from soil and insect cadavers. An oatmeal agar medium amended with 650 μg/ml dodine suppressed growth of *Penicillium* spp., *Trichoderma viride* and Mucorales species but supported growth of *B. bassiana* and *Metarhizium* spp. (Beilharz *et al.*, 1982). Subsequently, Chase *et al.* (1986) confirmed that an oatmeal medium amended with 600 μg/ml of dodine resulted in good isolation of *B. bassiana*; crystal violet was added to the medium to enhance the contrast with fungal colonies. However, this concentration of dodine was inhibitory to *M. anisopliae*. When the dodine concentration was reduced to 500 μg/ml, and 400 μg/ml of the fungicide benomyl (Benlate) was added to the medium, both *B. bassiana* and *M. anisopliae* were effectively isolated from soil (Chase *et al.*, 1986). Liu *et al.* (1993) observed that even 300 μg/ml of dodine inhibited isolation of some isolates of *M. anisopliae*. They found that a reduced concentration of dodine (10 μg/ml) in combination with 500 μg/ml of cyclohexamide increased the efficacy of recovery of *Metarhizium* from soil. Since dodine is not readily available and expensive, Rangel *et al.* (2010) developed a medium using more economical amounts of this compound. Fernandes *et al.* (2010) developed a dodine-free medium which favored isolation of *Metarhizium acridum*, however it is not effective for all strains of the fungus (S. Jaronski, personal communication). Alternately, a medium based on low sugar content and high pH supplemented with $CuCl_2$, and crystal violet was developed for selective isolation of *B. bassiana* by Shimazu & Sato (1996). Strasser *et al.* (1996) developed a dodine-based medium for selective isolation and maintenance of *Beauveria brongniartii*. Recipes for these selective media are provided in Appendix: 1.

#### *b.* Culicinomyces clavisporus

Panter & Frances (2003) reported that low concentrations of thiabendazole and dichloran inhibited growth of mucoralean fungi but permitted growth of *C. clavisporus* (Appendix: 1g).

*c.* Paecilomyces lilacinus

*Paecilomyces lilacinus* is more tolerant of high salt concentrations than many other fungi. The amendment of an agar medium with sodium chloride, pentachloronitrobenzene, benomyl, and Tergitol was found to facilitate recovery of *P. lilacinus* from soil by inhibiting growth of contaminant fungi such as *Rhizopus* and *Trichoderma* (Mitchell *et al.*, 1987; Appendix: 1h).

*d.* Lecanicillium *spp.*

A selective medium for *Lecanicillium* spp. which excludes bacteria and actinomycetes and holds in check the growth of contaminating fungi was developed by Kope *et al.* (2006; Appendix: 1i).

*3. Isolation from soil*

Entomopathogenic fungi are usually heterogeneously distributed in soil, putatively in or near insect cadavers. Thus, pooling soil samples usually increases the frequency of isolation. Soil can be collected with any number of tools but soil corers are most frequently used since a specified volume can be removed. The depth is usually limited to the top 10–15 cm of the organic and/or A horizon soil zone. The collection tool should be surface-sanitized between samples to avoid cross-contamination. Upon collection, the soil samples are usually placed in a cool environment (≈ 5°C). Although entomopathogenic fungi may remain viable in soil for relatively long periods, it is recommended that samples be processed as quickly as possible, usually within 5 days of collection.

The methods described below favor the isolation of propagules. To isolate fungi present in soil as actively growing or dormant hyphae, a number of methods have been devised (e.g., immersion methods, direct hyphal isolation, and washing methods). For a summary of these, see Parkinson (1994).

*a. Dilution spread plating*

Entompathogenic hypocrealean fungi are considered to be poor competitors in soil relative to other soilborne fungi. Thus, the direct isolation of entomopathogenic fungi from soil usually relies on the use of a selective medium. The most popular isolation method is the soil dilution plate method [see Section 4A2e(1)]. A commonly used procedure is to place 10 g of soil into 90 ml of sterile water or buffer (usually ≈ pH 6–7). The sample is then homogenized to release propagules from the soil matrix. The preferred technique is to use a commercial blender but stirring the slurry with a magnetic stir bar or on a mechanical shaker for 20–60 min may be sufficient.

Following homogenation, the aliquots of 100–200 µl are spread on to an appropriate medium using a glass or plastic spreader. In cases where the density of entomopathogenic fungi in soil is high, it may be necessary to dilute the homogenate further. However, densities of entomopathogenic fungi in soil are usually low, and larger volumes (≈ 1 ml) of the original sample may be spread on to the agar medium using an inclined rotary motion of the Petri dish; an agar concentration of 2–3% may facilitate absorption of the solution into the medium. When propagules are associated with soil particles, it may be necessary to increase the viscosity of the dilution solution. Increasing the viscosity will prolong the sedimentation of particles which will reduce the variation in the number of particles transferred. A number of materials may be used, but the most common and easy to prepare is a low concentration of agar (≤ 0.2%).

Once the homogenate is spread over the agar medium, cultures are incubated at an appropriate temperature (20–25°C for most taxa) for 3–7 days. Individual colonies can then be transferred to a suitable nutrient medium. For prolific sporulators, if possible, make the transfers before sporulation occurs.

*b. Direct plating*

In some instances, it may be desirable to plate soil directly which is faster than soil dilution plating. The simplest method is to sprinkle soil particles on to the surface of an agar medium and allow fungal hyphae to radiate out from the particles. Unless a selective medium is used, this usually provides unsatisfactory isolation of entomopathogenic fungi due to overgrowth of contaminant fungi.

A better strategy is to embed the particles in an agar medium (Warcup, 1950). For this technique, ≈ 5–15 g of soil is placed in a sterile Petri dish. The soil may be sieved prior to its placement in the dish or the large soil aggregates can be disrupted once in the dish. Approximately 15 ml of an appropriate molten (≤ 50°C) agar medium is added to the dish and the soil particles dispersed using a swirling motion. After incubation of the cultures (usually between 20 and 25°C), isolates can be transferred to suitable medium. Disadvantages are that colonies embedded in the medium may be difficult to retrieve, and it is relatively arduous to dilute the original sample with sterile soil or sand if densities of colony-forming units (CFU) in the soil are high.

*c. Insect baiting*

Insect baits may be used to indirectly isolate fungi from soil. In general, entomopathogenic Hypocreales are considered to be weak saprotrophs but since they possess the ability to infect living insects, they can gain access to a living insect relatively free of competitors. Although larvae of the greater wax moth (*Galleria mellonella*) are most commonly used, larvae of other insects such as the large flour beetle (*Tribolium destructor*) and the pine bark beetle (*Acanthocinus aedilis*) may be also be utilized (Zimmermann, 1986). Klingen *et al.*, (2002) found that using larvae of the anthomyid, *Delia floralis*, isolated *T. cylindrosporum* more frequently than did *G. mellonella.*

Soil samples are placed in containers, the soil is moistened, and larvae are added to the soil and incubated for ≈ 14 days in conditions that favor development of the target fungus. Although the quantity of soil may limit the larval density, in general, five to 15 larvae per container are used. Larvae are placed on the soil surface, and the soil is agitated or the containers are gently inverted or shaken periodically to ensure that the larvae remain exposed to the soil. Cadavers are collected at intervals and processed for internal fungi (see Section 1A1 above). Non-soil dwelling insects (e.g., *G. mellonella*) are very susceptible to infection by entomopathogens in soil which increases the sensitivity of this method. Zimmermann (1998) recommends first air-drying the soil prior to re-moistening to prevent infections by entomopathogenic nematodes.

*4. Isolation from plant tissues*

Entompathogenic hypocrealean fungi can exist on the surfaces of plants (i.e. epiphytes) or within plant tissues (i.e. endophytes). For the isolation of endophytic entomopathogenic Hypocreales, extreme precautions must be taken to ensure that fungi which may exist as epiphytes or contaminants on the plant surfaces are not isolated. For this reason, when attempting to isolate endophytes, intact plant tissues must be surface sanitized to kill potential contaminants on the plant surfaces. Plant tissues can include leaves, roots, stems or fruit.The most common sanitizers used are sodium hypochlorite (1−5%) and/or ethanol (70%). The efficacy of sanitization can be confirmed by swirling the treated plant tissues in a nutrient broth and checking for microbial growth after several days. As sanitizers can be quickly absorbed into injured or cut plant tissues, thereby killing the endophyte, only intact tissues should be used. For instance, Ownley *et al.* (2008) recommend using intact whole seedlings. It may be possible to use whole excised leaves as long as the petiole is not immersed in the sanitizer and is asceptically removed below the sanitized portion following the sanitization procedure. The sanitized plant tissue can then be asceptically cut into pieces and plated on to the desired agar media. Researchers must be cognisant that no sanitizer will be 100% effective, and that a number of factors will adversely affect the efficacy of sanitizers. For example, polar sanitizers are less effective if the plant tissue to be santizied possesses high amounts of cuticular waxes or is pubescent. In such cases, the use of surfactant or two or more sanitizers in combination may prove useful. It must also be realized that isolation of a putative endophyte from surface-santized plant tissue should be confirmed using direct evidence (e.g., microscopy) before it can be definitively concluded that the fungus was indeed a true endophyte.

*5. Purification*

Following isolation, it is usually necessary to ensure that the isolated fungus is free from contaminant microorganisms (excluding mycoviruses) and, if desired, that it represents a single genotype. Most entomopathogenic Hypocreales are relatively prolific sporulators and conidia can be satisfactorily streaked on an agar medium containing an antibacterial agent(s) (see Section 2A2 above). Individual colonies free of bacteria can then be collected.

CFUs do not necessarily originate from a single propagule and therefore additional steps may be desirable to ensure that a CFU represents a single genotype. The two most common methods used to achieve genotype purity are the isolation of individual conidia or the isolation of hyphal tips. Entomopathogenic Hypocreales are heterokaryons, and heterokaryotic mycelium may give rise to uninucleate or homokaryotic conidia, or to homokaryotic hyphal tips (Webster, 1986). Therefore, one must be cognisant that both of the above methods do not necessarily ensure that a unique karyotype will be obtained.

Although a micromanipulator can be used, a single conidium or germling (i.e. a germinated condium consisting of the conidium and a germ-tube) can be isolated without the aid of a manipulator. Conidia of entomopathogenic Hypocreales are generally too small to isolate with a dissecting microscope so a compound light microscope should be used. Either a non-inverted or inverted compound microscope can be used. A number of methods have been described for the isolation of fungal propagules using compound microscopes, but the light location method requires no specialized equipment (Tuite, 1969). For the light location method, a suspension of conidia is spread on to a relatively transparent agar medium (12–15 ml per dish); care must be taken to ensure that the density of conidia is low enough to allow isolation of individual propagules. For the isolation of non-germinated conidia, the cultures should be examined soon after the carrier has absorbed into the medium. For germlings, the cultures should be maintained at 20 to 25°C for *ca.* 6–12 h before being examined. The Petri dish is placed on the stage and the culture is examined under bright field illumination. The 10 × objective lens is preferable to larger magnification lenses since it provides the largest field of view with an adequate level of resolution. All other objective lenses should be removed to allow for more working room. Once a segregated conidium or germling has been located within a discrete area, the stage is adjusted so that the target propagule is in the center of the field. The objective lens is then swung to one side, the microscope diaphragm is closed until only the target area is illuminated, and a disk of agar (e.g., 4–5 mm diameter) is excised and aseptically transferred to an appropriate agar medium in a Petri dish or slant tube.

Hyphal tip isolations are, in general, easier to conduct than single conidium isolations, but it is not always possible to isolate a single hypha. Mycelium or a suspension of conidia is placed centrally on an agar medium. To promote diffuse hyphal growth which will facilitate isolation, the medium used should be relatively low in available nutrients. The margin of the colony should be examined with a dissecting microscope or compound light microscope, a section of agar containing a hyphal tip is then removed using a sterile spatula, knife, or needle, and aseptically transferred on to an appropriate agar medium.

## *6. Identification using morphological characters*

The vast majority of Hypocreales can be presumptively identified using asexual morphological characters. Successful identification of entomopathogenic Hypocreales relies on adequate observation of both conidia and conidiogenous cells. A number of methods are used to prepare Hypocreales for microscopy. Whole mount slide preparations (Chapter VI) can be quickly prepared, and, if done correctly, are very useful for identifying Hypocreales; whole mounts are the method of choice for phialidic taxa that are prolific sporulators, such as *Aspergillus* and *Penicillium* species. However, considerable disruption to conidiogenous cells and dehiscence of conidia may occur during this type of preparation. Therefore, we routinely use slide culture preparations for more critical observations of conidiogenesis (Figure 7.1). Small blocks of an appropriate agar medium (≈ 1 cm square) are placed on a sterile glass coverslip (22 × 30 mm) situated on 2% water agar medium in a Petri dish (Harris, 1986). The agar block is then inoculated with the fungal tissue, and a second sterile coverslip is placed on top of the block. The lid of the Petri

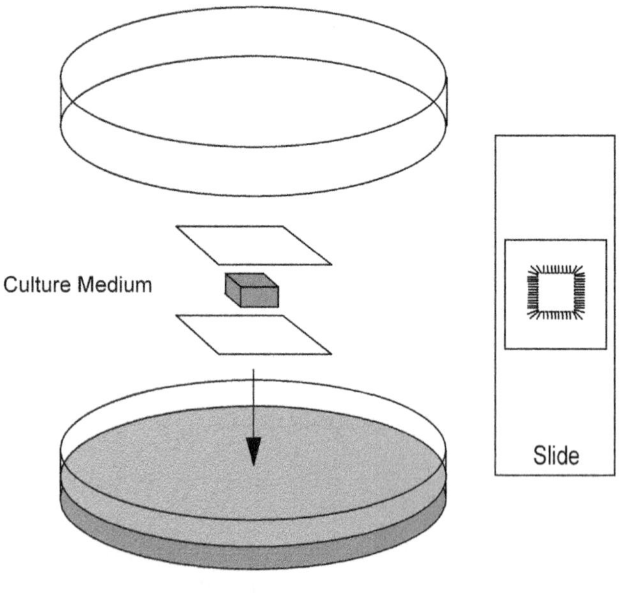


**Figure 7.1** Slide culture. A small block of an appropriate agar medium is inoculated, sandwiched between two sterile cover slips, and incubated on 2% water agar in a Petri dish. Once the fungus has started to sporulate, the agar is removed, and the coverslips are mounted on a microscope slide.

dish is replaced and the culture is incubated at an appropriate temperature. Vegetative growth may be visible on the surface of the coverslip(s) within a few days, but the culture must be incubated long enough to allow conidiogenesis. Sporulation can usually be observed using a dissecting microscope. The top coverslip and, if adequate, the bottom coverslip, are removed from the agar block and prepared for mounting. It is recommended that replicate cultures be established to allow for situations where the coverslip is removed before adequate sporulation has occurred.

A number of methods are available for preparing semi-permanent and permanent microscope slides (Chapter VI). Polyvinyl alcohol mounting medium (PVA; Appendix: 3a) is easy to prepare and use, has a long shelf-life, and provides high-quality and stable slide preparations. To mount coverslips from slide cultures, the material is first wetted by flooding the coverslip with a PVA wetting agent (Appendix: 3b). Excess wetting agent is removed by blotting and the coverslip is then mounted in a drop of PVA. The PVA is allowed to harden in an oven or on a slide warmer at 40°C for *ca.* 24–36 h. Slide preparations should be examined under phase-contrast, but if bright-field illumination is preferred, lacto-fuchsin can be added to the PVA solution (Appendix: 3c). See Chapter VI for a dichotomous key for the identification of the genera of entomopathogenic fungi.

## B. Molecular methods

Since the first edition of the book was published in 1997, the application of molecular-based methods has become commonplace in microbiology including the study of entomopathogenic hypocrealean fungi. Molecular methods are powerful tools that facilitate detection, identification/characterization, and/or quantification of fungi. They allow direct identification, detection, and analysis of molecular genetic traits (e.g., genotyping), which are not, or only indirectly possible using morphological, physiological, and biochemical methods. Molecular methods have become an essential tool for many biological disciplines including, for instance, population and evolutionary biology, ecology, taxonomy, and systematics. However, molecular methods require an equipment infrastructure, can be expensive, and the development and careful evaluation of specific tools for each target is mandatory. Furthermore, it is inevitable that a certain degree of trouble-shooting will be necessary for each situation. Molecular-based analyses rely on polymerase chain reaction (PCR) as the core technology, and the various analytical steps involved should be examined as a "full cycle approach" as illustrated in Figure 7.2 (Amann *et al.*, 1995). In this

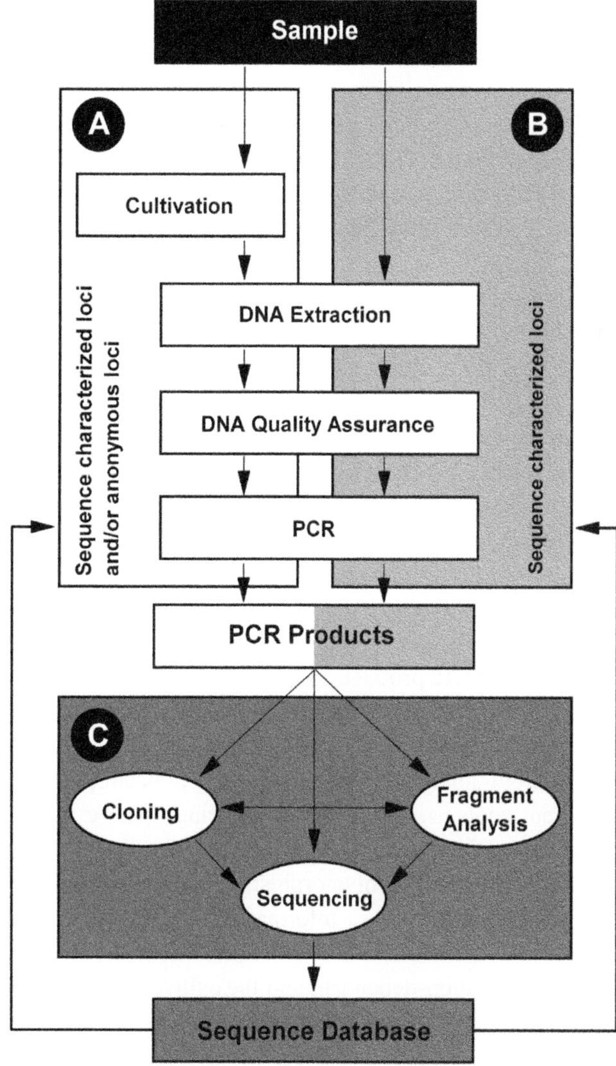


**Figure 7.2** Flow chart illustrating the principles applied to study entomopathogenic hypocrealean fungi using PCR-based molecular techniques. Cultivation-dependent analyses (A) are indicated in white boxes, and cultivation-independent analyses (B) are highlighted in light gray boxes. Analytical procedures (C) used to assess PCR products generated with cultivation-dependent and cultivation-independent methods are indicated by dark gray boxes. Details are described in the text. (Modified from Enkerli & Widmer (2010).)

chapter, we divide molecular methods into two broad categories: (1) cultivation-dependent; and (2) cultivation-independent methods (Enkerli & Widmer, 2010). For the first category, molecular analyses are performed on pure cultures of fungal isolates obtained through traditional cultivation, whereas cultivation-independent molecular analyses are performed directly (i.e. without cultivation) on DNA extracted from environmental samples (e.g., insect cadavers, soil samples). In cultivation-dependent analyses, two basic types of loci can be targeted by PCR (Figure 7.2): (1) sequence-characterized loci consisting of defined genes or DNA regions (e.g., ribosomal RNA genes and flanking spacer regions); and (2) anonymous loci which are undefined regions or genes that are distributed within genomes. Anonymous loci usually are amplified simultaneously (e.g., randomly amplified polymorphic DNA and amplified fragment length polymorphism) using PCR primers that hybridize to multiple loci, while sequence characterized loci are amplified individually using locus specific primer pairs. In contrast, only sequence characterized loci can be targeted in cultivation-independent analyses, where samples represent mixtures of organisms (metagenomic DNA extract). Primers used to amplify anonymous loci do not allow distinction of individual genomes present in such samples and it is therefore not possible to determine the origin of a particular PCR product. PCR products generated with cultivation-dependent as well as cultivation-independent methods are subsequently analyzed by applying a variety of different analytical procedures (e.g., electrophoretic fragment analysis, cloning or sequencing). These procedures allow the detection of specific sequence characteristics like nucleotide polymorphisms or length variations, which represent the basic information for the discrimination, identification and detection of individual fungal taxa.

The cultivation-dependent and the cultivation-independent approaches each possess strengths as well as weaknesses, and it is important that the researcher is cognisant of the relative merits of each approach as well as their limitations. It is important to stress that these two approaches can be complementary, and depending on the research question to be answered, one or the other may be favored.

### 1. Cultivation-dependent analyses

Cultivation-dependent analyses combine molecular methods with traditional microbiological methods (e.g., isolation techniques using plating or baiting techniques). A major advantage of cultivation-dependent analysis is that the researcher possesses the living culture, which is often required for subsequent experimentation (e.g., virulence assessments). Furthermore, certain molecular methods also require a viable culture (e.g., identification/characterization of a particular strain). In this chapter, we limit our discussion of cultivation-dependent molecular methods to diagnostic PCR, sequence-based identification, and genotyping. The cultivation-dependent approach is also applied when performing indepth genetic characterization of hypocrealean fungi such as phylogenetic analyses, whole genome sequencing, and functional analyses of specific genes and proteins, but these topics are beyond the scope of the chapter and the reader is referred to a number of excellent references on molecular methods for additional information (e.g., Current Protocols in Molecular Biology).

#### a. Preliminary steps

As the acquisition of a fungal strain is required, the initial step is to isolate a fungus from an insect or other substrate using traditional microbiological methods (see Section 2A above). The next step is to isolate a single genotype or strain. Many researchers assume that a single colony on the surface of a medium originated from a single propagule and thus represents a unique genotype. This can not be assumed. Thus, it is important to take the appropriate steps to ensure the isolation of a unique genotype. The hyphal growth form of hypocrealean fungi complicates this. The most commonly used method is to isolate a single propagule (e.g., a conidium or hyphal body), although the isolation of an individual hyphal tip may be used as well (see Section 2A5 above). One must also be cognisant that hyphae of many hypocrealean fungi exist as heterokaryons. Once a unique strain has been isolated, sufficient biomass must be propagated and the strain stored for subsequent evaluation. As the possibility of genetic changes to the strain in storage should be guarded against, storage of the biomass at low temperature (e.g., −80°C or below) or in the absence of water under vacuum (i.e. lyophilized) is preferred. For long-term storage, lyophilization is recommended.

#### b. Genomic DNA extraction

Effective extraction of genomic DNA is a pre-requisite for all subsequent molecular analyses and a variety of extraction methods can be used. For all the steps involved, it is critical that the equipment and material used (including plastic- and glassware) is free of potentially contaminating DNA or DNA degrading enzymes (DNases). This should be assured by the manufacturer (i.e. certified nucleic acid, and DNase/RNase-free plastics and tips) or in the case of materials that are re-used, they must be pretreated to ensure that they do not contain residual DNA or DNase activity. Methods for removing residual DNA and inactivating DNases include autoclaving, soaking in 10% bleach (i.e. 0.525% sodium hypochlorite) for 30 min, or the use of a number of commercially available surface cleaners designed for this purpose such as DNA Away® (Molecular Bioproducts) or Ultra Clean® (Mo-Bio). Care must be taken to ensure that materials are well rinsed using purified water. Plastics can be checked for cleanliness using PCR.

*(1). Homogenization.* The first step typically involved in DNA extraction is homogenization of the fungal cells (also referred to as cell disruption and/or lysis) using mechanical, chemical and/or enzymatic treatments to release cytoplasmic components into an appropriate buffer. As a rule of thumb, extraction of DNA from rapidly growing cells is favored over extraction from older cells as older hyphae may no longer be viable and thus may contain no or reduced amounts of DNA. DNA is relatively stable within cells, and short- to medium-term storage of biomass without any adverse impact on DNA quality can be accomplished at −20°C, although storage at −80°C is preferred. The quality and amounts of DNA required are dependent on the method(s) to be applied. For example, the quality of DNA (e.g., sheared DNA) is less of an issue for diagnostic PCR than it is for genotyping.

Given the extensive cell wall containing a chitin scaffold of hypocrealean fungi, effective extraction of DNA typically requires an initial mechanical homogenization step. The efficacy of this procedure may be enhanced by prior lyophilization of the fungal biomass. A variety of mechanical homogenization methods can be applied ranging from the utilization of bead beaters to tissue tearors to the use of mortar and pestles. Regarding the latter, a commonly used strategy is to grind the tissue in liquid nitrogen with or without a grinding agent such as DNA-free sand. The biomass is typically ground until it is a powder. Liquid nitrogen serves to make the cells brittle as well as to ensure that heat build-up is minimized; heat build-up will cause shearing of DNA. Heat build-up during extraction is a particular issue when using bead beaters (i.e. glass or steel beads placed in a tube with the fungal biomass are rapidly oscillated generating mechanical shear forces that disrupt the cells). For this reason, time of oscillation and maintainance of tubes at low temperatures is imperative when using the bead beating method. Furthermore, it may be possible to disrupt cells (e.g., hyphal bodies) using ultrasound or extreme pressure (e.g., using a French press), but these methods typically are less effective and also require specialized equipment. Mechanical homogenization is often followed or combined with a chemical lysis step. Chemical lysis includes incubation of the biomass in a lysis buffer containing reagents like sodium dodecyl sulfate (SDS), cetyltrimethylammonium bromide (CTAB), or phenol (Figure 7.3). Furthermore, in certain protocols, enzymatic treatments (e.g., protease K, chitinase) are included in the homogenization step to aid cell lysis and the release of the DNA.

To facilitate extraction of high-quality nucleic acids from entomopathogenic hypocrealean fungi, protoplasts (i.e. individual cells consisting of the protoplasm and cell membrane with the cell wall removed by enzyme treatment) can be generated from young mycelia, and nucleic acids extracted from the protoplasts (St Leger & Wang, 2009).

*(2). Purification.* After completion of the homogenization step, the DNA has to be purified from the crude extract (Figure 3). In this step, cell wall fragments, contaminating lipids, and proteins are removed by applying techniques like centrifugation, phenol/chloroform extraction, column purification and/or precipitation (e.g., with potassium acetate, ethanol, isopropanol, and/or polyethylene glycol). RNA can be removed by adding RNases to the crude or purified DNA extract. A number of commercially available kits can be used to extract fungal DNA (e.g., Qiagen DNeasy® Plant Mini Kit, Qiagen Inc; NucleoSpin® Plant II, Machery−Nagel Inc). Many researchers choose to use commercial kits for logistical reasons (e.g., ease and consistency) and because of the possibility of adapting such kits to automated large-scale extraction. The primary disadvantage of using commercial kits is the increased cost. However, savings in labor often offset the higher cost of using kits.

Once DNA has been extracted, it is recommended that it be evaluated for quality and quantity. This can be accomplished by electrophoresis in an agarose gel using an appropriate size marker (e.g., lambda-*Hind*III DNA

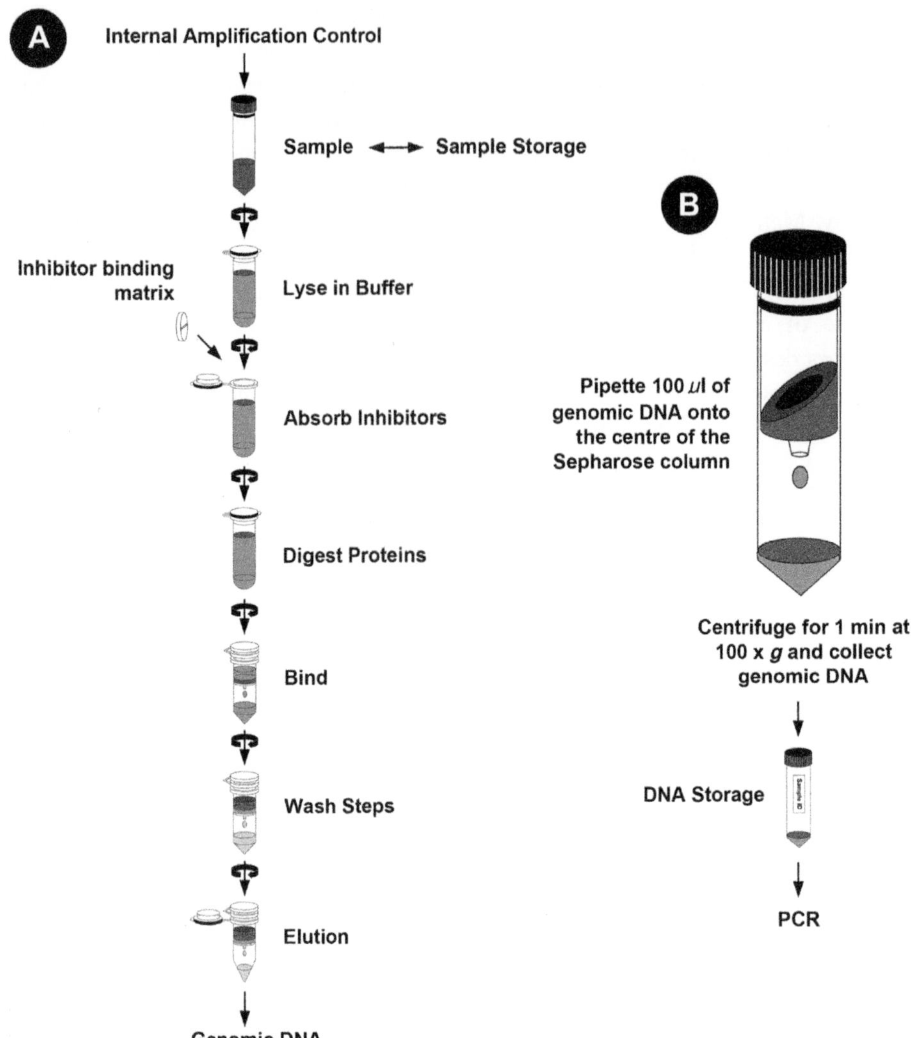


**Figure 7.3** An example of a DNA extraction procedure based on the QIAamp DNA Stool Mini Kit (Qiagen QIAamp® DNA Stool Handbook 2010). Step A includes addition of an Internal Amplification Control, and extraction of DNA using the commercial kit. After addition of the IAC, the sample can be stored at low temperature. The sample is then lysed using a lysis buffer step. Although not shown in this figure, a prior mechanical homogenization step is often required to facilitate DNA release (see text for additional information). In this example, PCR inhibitors (e.g., polyphenolic compounds) are scavenged using an inhibitor binding matrix and contaminating proteins are removed by digestion with proteinase K. Subsequently, DNA is bound to a silica membrane, impurities are removed by washing, and DNA is eluted from the spin column. DNA can then be stored at −20°C or colder. In instances where excessive polyphenolic compounds are present (see text), a post-clean-up step may be utilized (Step B). In this example, DNA is purified applying sepharose gel filtration in 1-ml Centriprep spin column tubes. DNA (100 µl) is slowly pipetted on to the center of the column, the column is centrifuged, and the eluate containing the purified DNA collected and stored.

ladder) and buffer system, and subsequent staining and visualizing of the DNA. Traditionally, the fluorescing intercalating dye, ethidium bromide, has been used to visualize DNA. However, ethidium bromide is a mutagen, and increasingly non-toxic dyes such as GelRed® (Biotium, Inc.) or SYBR® Green (Molecular Probes, Invitrogen Life Science) are favored. DNA of high quality is visible in the gel as a single genomic band of high molecular weight, whereas sheared DNA appears as a smear. Most often, quantities of DNA are estimated by visualizing electrophoresed genomic DNA in an agarose gel relative to a size marker of known concentration such as the lambda-*Hind*III DNA ladder. However, other specialized equipment can also be used. For example, DNA concentrations can be estimated using a spectrophotometer or fluorometer (e.g., Nanodrop), or through the use of capillary electrophoresis (e.g., QIAxcel®). Once extracted, DNA should be stored at −20°C or lower, and care should be taken to avoid repeated freeze–thaw cycles. As a rule of thumb, higher concentrations of DNA will store for a longer period at low temperatures than will DNA at low concentrations.

*c. Cultivation-dependent diagnostic PCR*

PCR is a powerful and versatile method that can be used to identify and characterize entomopathogenic hypocrealean fungi. Here we define diagnostic PCR as the application of PCR to amplify a sequence characterized locus (or region) in a taxon-specific manner (i.e. amplification products [amplicon] are obtained only from target and not from non-target taxa). Typically, diagnostic PCR is utilized at the species level of resolution (Entz *et al.*, 2005), but detection at the subspecies (Castrillo *et al.*, 2003; Destefano *et al.*, 2004) or subgenus (Schneider *et al.*, 2011) levels may also be possible. Sequence-characterized loci targeted with diagnostic PCR either are loci that are unique for the selected taxon (i.e. not present in non-target taxa), or what is more often the case, they are loci that are conserved at different levels among related taxa or even all fungi. For a conserved locus to be effective for diagnostic PCR, it must contain variable regions next to conserved regions that allow discrimination between target and non-target taxa and provide taxon specific sequence signatures. For diagnostic PCR, primers are designed that hybridize to such specific sequence signatures, thus allowing taxon-specific PCR amplification of the conserved locus. The ribosomal RNA genes and flanking regions (i.e. internal transcribed spacer and intergenic regions) are examples of such loci (Figure 7.4).

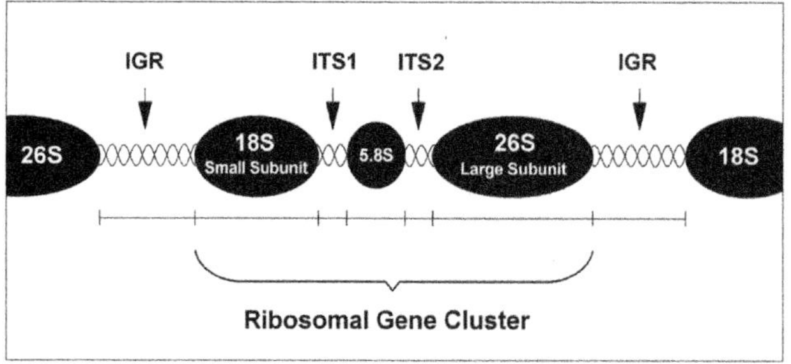


**Figure 7.4** A schematic diagram of the ribosomal RNA gene cluster found in entomopathogenic hypocrealean fungi. The rRNA gene cluster consists of the 18S rRNA gene (small subunit), the 5.8S rRNA gene, the 26S rRNA gene (large subunit), and the internal transcribed spacer regions (ITS1 and ITS2) flanking the genes. Also shown is the intergenic region (IGR) spanning the regions between individual rRNA gene clusters units.

For the development of diagnostic PCR approaches, the discovery or identification of the target locus is of major importance. Unique loci can be identified using techniques like Randomly Amplified Polymorphic DNA (RAPD) or Amplified Fragment Length Polymorphism (AFLP) (see Section 2B1e below). As whole genome sequences of entomopathogenic hypocrealean fungi become available, it will be possible in the future to also identify such loci using comparative genomic analysis. However, at this point whole genome sequences of entomopathogenic fungi are available only for one isolate of *Metarhizium robertsii* and one isolate of *Metarhizium acridum* (Gao *et al.*, 2011). The evaluation of conserved loci for diagonsitc PCR involves indepth sequence analysis and comparison of target and non-target taxa (see Section 2B1d below). Conserved loci are only valuable for diagnostic PCR when target taxon-specific sequence signatures can be identified. Therefore, the effort necessary to identify conserved loci that are suitable for diagnostic PCR can be substantial.

For diagnostic PCR to be effective, primer design and validation is a critical step. Although a detailed description of primer design is beyond the scope of this chapter, the reader must be cognisant of a number of salient issues. When designing primers for diagnostic PCR, the general rules for primer design should be followed. Ideally, primers should have a size of 20–24 nucleotides, an annealing temperature (Ta) of 55–60°C and yield a fragment of at least 100 base pairs in size to allow for sufficient resolution in agarose gel electrophoresis. However, primer characteristics may deviate from these values depending on the situation. The Achilles heel of diagnostic PCR is the absolute requirement for primer comprehensiveness (i.e. all genotypes within the target taxon) and specificity (i.e. only the target taxon is amplified).

To address primer comprehensiveness, evaluation typically involves conducting PCR on characterized strains of the target taxon with emphasis placed on evaluating as many strains as is logistically possible. To evaluate primer specificity, DNA obtained from a variety of non-target fungal taxa are included, and such evaluations should include phylogenetically similar or related taxa. Furthermore, primer specificity can be evaluated by performing identity searches in public sequence databases (see Section 2B1d below). When conducting diagnostic PCR, the inclusion of negative (i.e. PCR reaction mix minus DNA template) and positive (i.e. PCR reaction mix with DNA from a known positive added) controls should always be included. The sensitivity (i.e. the detection limit) of a PCR is also a consideration, but this is less of an issue for cultivation-dependent approaches (i.e. relative to cultivation-independent PCR) because DNA template is usually available in sufficient amounts. The issues discussed above are important even when approaching diagnostic PCR based on published reports. Differences in equipment and reagents (e.g., *Taq* polymerase) between laboratories can affect PCR and therefore it is almost always necessary to conduct in-house validation of published primers and PCR conditions. In this regard, Ta and reagent concentrations (particularly the concentration of MgCl) are most important. Thus, if specificity is deemed insufficient, altering the concentration of the PCR reagents, particularly the concentration of DNA and MgCl, and/or altering the cycling conditions (e.g., increasing Ta) may result in increase specificity. As a variety of factors may have to be altered in concert, optimization of PCR must be approached in a methodical manner and carefully evaluated in stages. For additional information on PCR optimization, the reader should consult any number of references dedicated to PCR.

Following amplification, PCR products must be analyzed by electrophoretic sizing or sequencing, with the former being most commonly used for routine diagnostic PCR. Most laboratories utilize agarose gel electrophoresis (e.g., 1% TAE-agarose) and stain DNA with ethidium bromide or alternate dyes such as GelRed® or SYBR® Green, and visualize the stained DNA under ultraviolet (UV) light. The size of the amplicons is ascertained relative to an appropriate size maker (e.g., 100 base pair ladder). Automated high-resolution capillary electrophoresis can be used as an alternative to agarose gel electrophoresis (e.g., QIAxcel). This technique can reduce time and labor when handling multiple samples, deliver automatic sizing of amplicons, and save data in an electronic format. When developing or validating diagnostic PCR, it is always a good idea to verify the sequence of arbitrarily selected amplicons. If the PCR results in a single band, the PCR product can be purified using a commercial kit (e.g., a PCR clean-up kit to remove excess primers, nucleotides, DNA polymerase, and salts) and sequenced directly. If the PCR reaction results in multiple bands (for a possible indication of unspecific amplification), the bands may have to be separated using electrophoresis, the band of interest carefully excised, and purified using a commercial kit (e.g., a gel purification kit). To excise a band, the gel is laid on top of a sheet of plastic and placed on to the transilluminator for visualization. The band is cut out using a DNA-free scalpel and placed into a microfuge tube. Personnel should wear protective clothing and a face shield for this procedure. Purified amplicons can be sequenced directly or indirectly (i.e. the amplicon is cloned prior to sequencing). In the latter case, the excised and purified DNA fragment is ligated into an appropriate plasmid vector, transformed into an *Escherichia coli* host, which is then grown on selective media. Following plasmid extraction, the cloned fragment can be sequenced and compared to the expected target sequence.

*d. Sequence-based identification*

Another commonly used method to identify or characterize entomopathogenic hypocrealean fungi is to sequence conserved loci (sometimes also referred to as phylogenetic genes because of their use in phylogenetic analyses) and to compare obtained sequences with reference sequences. Targeted loci should contain conserved as well as variable sequence regions. The conserved regions can be used as 'anchors' for sequence alignments (to guarantee that homologous sequences are compared), whereas the variable regions provide sequence differences (polymorphisms, sequence signatures) that allow the discrimination of taxa. In contrast to diagnostic PCR, primers are designed that bind to a conserved sequence region of the targeted gene which allows the use of the same primers to amplify the same locus from related taxa as well. Following amplification, the resultant amplicon can be sequenced using the dye termination method (i.e. Sanger) and a capillary genetic analyzer. Depending on the conditions and the chemistry used (e.g., the type of polymer in the analyzer capillaries), sequence reads in excess of 800−1000 nucleotides are attainable. Thus, loci or regions to be targeted in sequence-based identification ideally are less than 800 nucleotides in length for practical reasons (i.e. only one sequencing reaction is required, and it is not necessary to generate contigs). If sequence data greater than 800 nucleotides is required, multiple staggered sequencing reactions are required, and the researcher must subsequently generate a contig (i.e. form a contiguous set of overlapping DNA segments from the same gene) using appropriate software such as Sequencher (Gene Codes Corporation) or Geneious (Biomatters Ltd). The rRNA genes and flanking regions (Figure 7.4) are the most commonly targeted loci for sequence-based identification of entomopathogenic hypocrealean fungi. However, other loci like the largest subunit of nuclear RNA polymerase II (RPB1), the second largest subunit of nuclear RNA polymerase II (RPB2), the elongation factor 1-$\alpha$ (EF-1$\alpha$) or β-tubulin are used as well.

Once the sequence of the target locus has been obtained, it can be compared with reference sequences to allow identification or characterization of the fungus. The simplest form of analysis is to apply a Basic Local Alignment Search Tool comparison of nucleotide sequences (BLASTN) within sequence databases such as GenBank of the National Center of Biotechnology Information (NCBI). This analysis retrieves the most similar sequences present in the database. The names associated with these sequences ideally allow identification of the organism under investigation or they may at least provide the names of close relatives. However, successful identification is dependent on the accuracy of the information present in the database, and the researcher should always scrutinize information within NCBI and other genetic databases (e.g., fungal identities associated with sequence data) with a critical eye.

More informative analyses involve the application of phylogenetic algorithms to obtain a measure of relatedness (e.g., using PHYLIP, PAUP, Bionumerics, etc.) between selected reference sequences and the sequence of the organism under investigation. Reference sequences may be obtained from the NCBI sequence database or determined in-house. For many, it will be desirable to create their own sequence database. The first step is to align sequences using an appropriate multi-sequence-alignment program (e.g., Clustal X), and then refine the alignment manually (e.g., BioEdit, GeneDoc). Phylogenetic analyses can then be performed using a number of algorithms such as maximum parsimony, maximum likelihood, Bayesian phylogenetic inference, and distance matrix based methods. The latter method (e.g., Neighbor-joining) is often preferred because it is computationally less complicated. Results of such analyses are represented in phylogenetic trees. Support for internal branches within the tree (i.e. as a measure of clade robustness) can be obtained by bootstrap analysis. Finally, it is important that sequence data generated and verified in such analyses is made available to the scientific community, and submitted to the NCBI sequence database. The constantly increasing information in such databases represents a crucial step in the 'full cycle approach' (Figure 7.2) as it steadily improves the base for design of specific primers or accurate identification and characterization of fungi. An example of a majority rule consensus phylogram using a Bayesian analysis of the 5′-end of the EF-1$\alpha$ gene to resolve the *M. anisopliae* complex is presented in Figure 7.5 (Bischoff *et al.*, 2009).

*e. Genotyping isolates*

The application of molecular techniques to study genetic variability within taxa is a very powerful and useful method available for studying mitosporic hypocrealean fungi. Here we restrict our discussion to an overview of the most commonly used methods. The methods can be divided into: (1) analysis of anonymous loci (i.e. RAPD, universally primed PCR, repetitive PCR, AFLP); and (2) analysis of sequence characterized loci (i.e. simple sequence repeat, single nucleotide polymorphism) (Figure 7.2). All available methods possess strengths and weaknesses, and here we emphasize the relative strengths and weaknesses of each method to allow the reader to decide which method may be most appropriate for their research. For more information on the methods as well as information on additional methods, we refer the reader to references such as Caetano-Anollés & Gresshoff (1997), Avis (2004), Semagn *et al.* (2006), and Castrillo & Humber (2009).

*(1). Randomly amplified polymorphic DNA.* RAPD, also referred to as 'Arbitrary Amplified PCR', has been a popular method in the past to examine genetic variation in entomopathogenic fungi (Williams *et al.*, 1990; Welsh & McClelland, 1990). RAPD analysis relies on the simultaneous amplification of anonymous loci using short primers (eight to 12 nucleotides in length). RAPD primers have also been referred to as random primers because they amplify sequences of DNA that are random (i.e. amplicons are not derived from a known gene

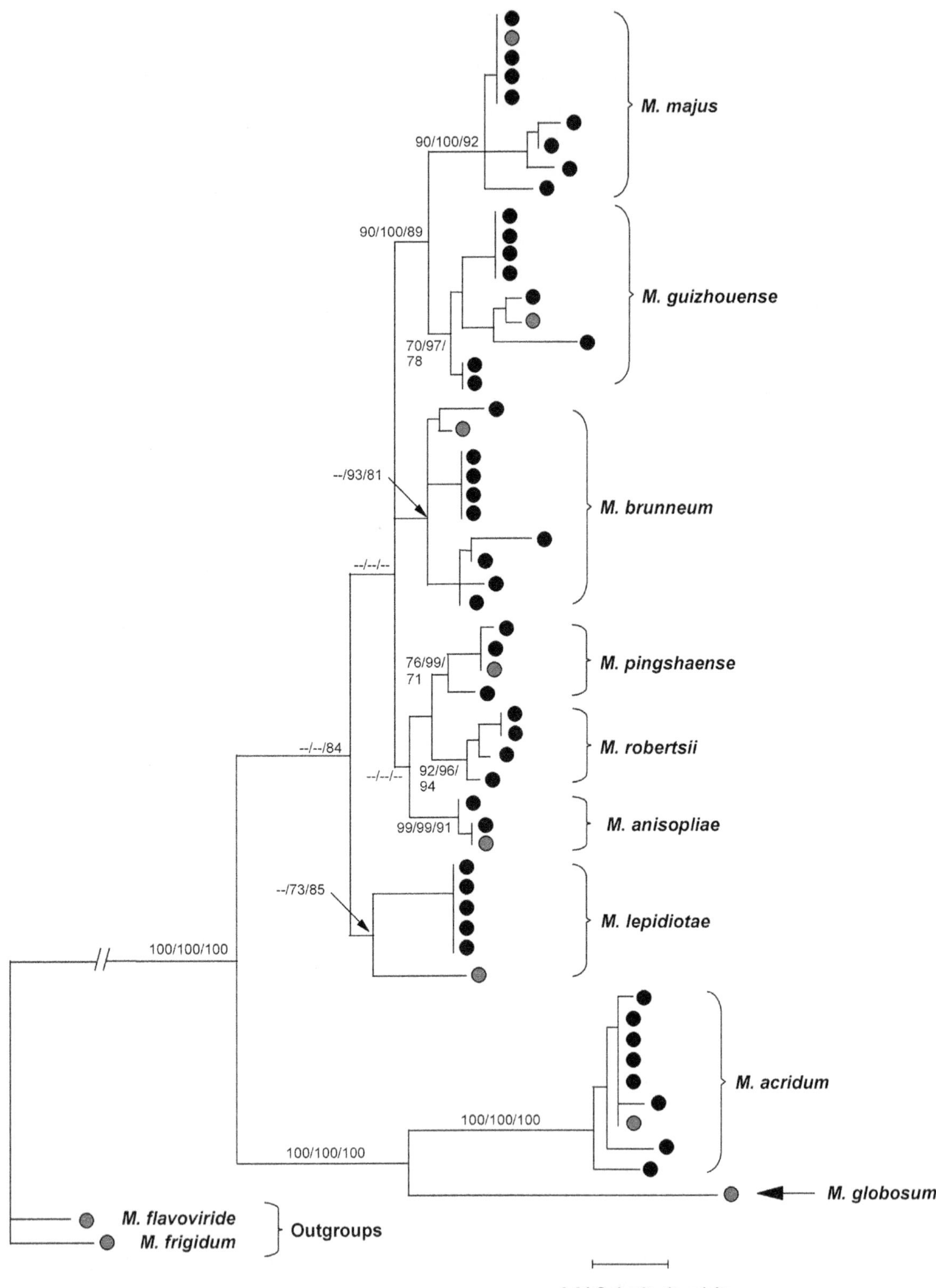


**Figure 7.5** A majority rule consensus phylogram from the Bayesian analysis of the 5′-end of the EF-1α resolving the phylogenetic relationships within the *Metarhizium anisopliae* species complex. Bootstrap values greater than 70% are shown at nodes, and values for the backbone and terminal nodes are presented together to prevent overcrowding. Ex-type isolates are denoted as gray circles. [Phylogram recreated from Bischoff *et al.* (2009).]

sequence). RAPD-PCR is performed with a single primer (one sequence), and a PCR product is generated if the primer is able to bind to the template DNA at close distance (< 3 kb) and in opposite orientation (3′-ends of the primer face each other). Due to the short length of the primer, there are numerous annealing sites throughout the whole genome and a variety of products are generated in one reaction. It is almost always necessary to screen a variety of primers to determine ones that deliver informative amplicons. Amplicons are typically resolved and visualized using gel electrophoresis. Resulting banding patterns (profiles) can be analyzed for the presence or absence of specific bands that allows comparisons of different fungal isolates or taxa.

RAPD was one of the first methods available to assess genetic variability among fungal isolates. It has been applied to study genetic variation within entomopathogenic hypocrealean species including *B. bassiana* (Maurer *et al.*, 1997; Fernandes *et al.*, 2006), *B. brongniartii* (Cravanzola *et al.*, 1997; Piatti *et al.*, 1998), *Hirsutella thompsonii* (Mozes-Koch *et al.*, 1995; Aghajanzadeh *et al.*, 2007), *L. lecanii* (Mor *et al.*, 1996), *M. anisopliae* (Fegan *et al.*, 1993; Leal *et al.*, 1994; Velásquez *et al*, 2007), *N. rileyi* (Boucias *et al.*, 2000; Vargas *et al.*, 2003), and *Isaria farinosa* (Chew *et al.*, 1998). In other studies, RAPD analysis has been used to ascertain associations among genotypes with specific hosts (Bridge *et al.*, 1997; Maurer *et al.*, 1997) or correlations with geographical origin (Leal *et al.*, 1994; Boucias *et al.*, 2000). The RAPD technique is a fast and simple method and it has been used extensively. However, it has the salient disadvantage of low reproducibility. The quality and concentration of template DNA, the concentration and source of PCR components, the cycling conditions as well as the type of equipment used all can affect reproducibility. As a result, we do not advocate the use of RAPD other than as a preliminary method. If a decision is made to use RAPDs, it is imperative that careful attention be paid to quality control (e.g., contamination prevention) and that independent runs are employed to ensure reproducibility of the method.

Another application of the RAPD technique is its use as a tool for identification and development of sequence characterized amplified region (SCAR) markers. Such markers can be applied for diagnostic PCR and they ideally represent unique loci (see Section 2B1c above and Section 2B2c below). RAPD bands that have been identified as specific for a particular taxon or isolate are excised from an agarose or polyacrylamid gel (for higher resolution), purified and sequenced (see Section 2B1c and Section 2B1d above). Primers are then designed from the sequence data, and carefully validated for comprehensiveness and specificity as described above (see Section 2B1c above). Two examples in which SCAR analysis was applied to study entomopathogenic hypocrealean fungi include those of Castrillo *et al.* (2003) and Takatsuka (2007).

*(2). Universally primed PCR.* The universally primed PCR (UP-PCR) technique is closely related to RAPD analysis and relies on the same principles (Bulat & Mironenko, 1990). Like RAPD-PCR, UP-PCR amplifications are based on the use of single primers. However, UP-PCR primers are longer (15–20 base pairs) and a higher annealing temperature is used relative to RAPDs. As a result, UP-PCR amplifications are more specific and reproducible than RAPD analyses. This technique provides a simple and fast way to assess genetic variability among fungal isolates. Like RAPD-PCR, the accuracy of UP-PCR is dependant on accurate band scoring, and a good practice is to conduct multiple runs and only include bands that are consistently produced. UP-PCR has been used to discriminate strains of *B. bassiana* (Meyling & Eilenberg, 2006) and *L. lecanii* (Mitina *et al.*, 2007).

*(3). Repetitive element PCR.* The underlying principle of reitive element PCR (Rep-PCR) is based on the use of repetitive DNA elements that are distributed throughout the genome as annealing sites for specific primers. DNA fragments located between two repeated DNA elements (i.e. inter fragments) are amplified if the repeated DNA elements are located at close distance (< 3 kb) and are in the opposite direction. Like in RAPD analysis, generated PCR products are resolved using gel electrophoresis and resulting profiles are analyzed for presence or absence of specific bands. Rep-PCR approaches have been developed for a variety of different types of repetitive elements. These include BOX-PCR (Versalovic *et al.*, 1994), Enterobacterial Repetitive Intergenic Consensus-PCR (ERIC-PCR) (Versalovic *et al.*, 1994), Inter Simple Sequence Repeats (ISSR-PCR) (Zietkiewicz *et al.*, 1994), and Inter Retrotransposon Amplified Polymorphism PCR (IRAP-PCR) (George *et al.*, 1998). Rep-PCR approaches are fast and simple to perform. An

attractive feature of the Rep-PCR is that the primers work in a variety of organisms and no previous knowledge of the genomic structure is necessary. Furthermore, Rep-PCR genomic fingerprinting also circumvents the requirement to identify suitable arbitrary primers by trial and error that is inherent to RAPD genotyping. Similarly to UP-PCR, Rep-PCR approaches are more reproducible and reliable than RAPD analyses as the primers are longer than RAPD primers and therefore enable amplification at PCR conditions that are more stringent. However, the potential for contamination and false results still exists with Rep-PCR, and the use of appropriate controls is imperative. The resolution of Rep-PCR may be lower than for other genotyping methods, and researchers should be cognisant that genotypes deemed to be identifical using Rep-PCR may not be identical (i.e. clonal). Rep-PCR has been used to examine genetic diversity and population structure of *B. bassiana* (Aquino de Muro *et al.*, 2005; Wang *et al.*, 2005; Estrada *et al.*, 2007), and *N. rileyi* (Han *et al,* 2002).

*(4). Amplified fragment length polymorphism.* AFLP is a robust whole-genome method that can be readily applied to genotyping hypocrealean fungi. Like RAPD, this method targets anonymous loci and does not require any knowledge of the DNA target. However, this method is more specific and reproducible than RAPD. The AFLP method is specifically designed to detect polymorphic DNA and consists of four basic steps: (1) digestion of genomic DNA with one or more restriction enzymes; (2) ligation of adaptor sequence to the cohesive ends of the restriction sites; (3) amplification of the fragments using primers that anneal specifically to the adapter and restriction site; and (4) electrophoretic separation and visualization of the resultant PCR products (Figure 7.6). It is often desirable to use two different restriction enzymes to maximize fragment diversity. In the case of fungi, two amplification steps should be conducted to assure that only a subset of the restriction fragments are amplified (Vos & Kuiper, 1997). Amplification step one is commonly referred to as 'pre-amplification' (Figure 7.6D). Amplification step two involves the use of 'selective AFLP primers' to reduce fragment complexity (Figure 7.6F); these primers correspond to the adapter and restriction site sequences and have additional nucleotides at the 3′-ends extending into the restriction fragments. The 3′-extensions are called the 'selective nucleotides'. Alternatively, primers of the pre-amplification may have a selective nuleotide attached to provide a first reduction of fragment complexity. AFLP profiles can be visualized using agarose or polyacrylamide electrophoresis, or using an automated capillary genetic analyzer. In the latter case, the 5′-end of one of the selective primers has to be tagged with a fluorescent label to allow visualization of AFLP amplicons.

AFLP analyses have been performed to investigate the genetic diversity of various entomopathogenic fungi including *B. bassiana* (Aquino de Muro *et al.*, 2003, 2005), *H. thompsonii* (Tigano *et al.*, 2006), *N. rileyi* (Boucias *et al.*, 2000; Devi *et al.*, 2007), and *M. anisopliae* (Inglis *et al.*, 2008). This technique has a number of advantages over other genotyping methods. For example, large numbers of polymorphic bands can be detected in one experiment (Meudt & Clarke, 2007), the method can be automated using capillary genetic analyzers (Inglis *et al.*, 2008), and it typically provides very reproducible results due to the use of long and specific primers (Taylor *et al.*, 1999). Disadvantages of AFLP include its relative high labor requirements and the need for a relatively high degree of technical expertise and equipment infrastructure. Also, there can be some variability in AFLP results between different laboratories as a result of variation in technique, chemicals, and equipment.

Similarly to the RAPD technique, the AFLP technique can be used to identify and develop SCAR markers (see Section 2B1e above). Potentially unique bands (Figure 7.6H; arrows) are excised and sequenced, and primers can be designed and carefully validated. Despite its robustness, a major drawback of AFLP to identify SCAR for use in diagnostic PCR is that unique AFLP amplicons are typically based on polymorphisms at restriction sites. Thus, very limited sequence variation may be represented within the amplicon (e.g., only one polymorphic nucleotide, *versus* unique regions) which can hinder efficient primer design.

*(5). Simple sequence repeats.* PCR based analysis of simple sequence repeat (SSR) loci, also referred to as 'microsatellites', is currently the most widely applied method for analyzing genetic variation, population structure, and relatedness (Avis, 2004). SSR loci are sequence characterized (see Section 2B1d above), consist of short tandemly repeated sequences of one to six nucleotides, and they are scattered throughout the genome of most organisms (Goldstein & Schlötterer, 1999) including

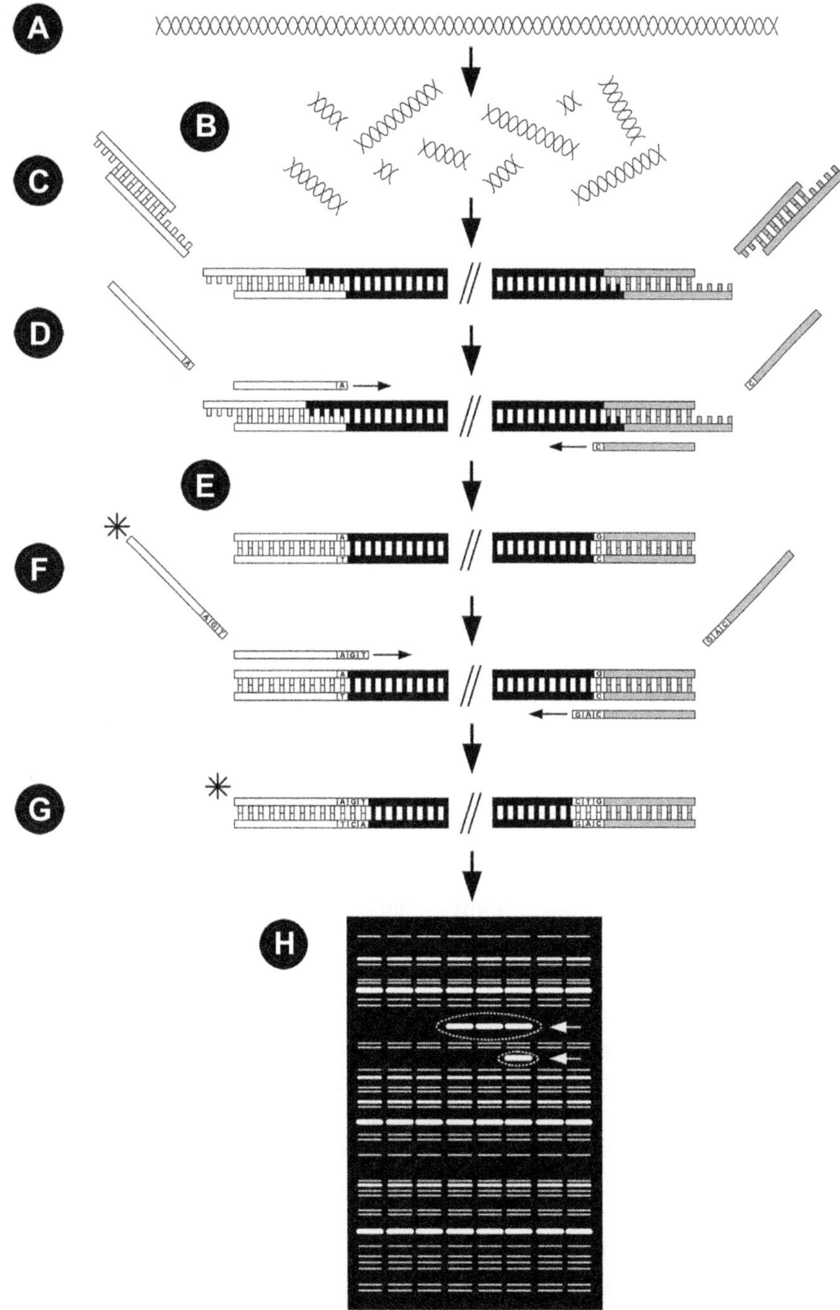


**Figure 7.6** The amplified fragment length polymorphism process (Vos & Kuiper, 1997): A. extracted genomic DNA; B. DNA restricted with one or more restriction enzymes; C. attachment of the adapter sequences to the restriction fragments; D. the pre-amplification step with primers that anneal to the adapter and restriction site sequences; E. resultant amplicon with the primer adapters incorporated; F. secondary amplification with selective primers (i.e. primers that correspond to the adapter and the restriction site sequences and that have additional nucleotides at the 3′-ends extending into the restriction fragment), one of which may be tagged with a fluorescent marker at the 5′-end; G. the resultant amplicon from secondary amplification (i.e. containing the fluorescent marker); and H. the electropherogram showing unique amplicons (white arrows). Details are described in the text.

hypocrealean fungi (Rehner & Buckley, 2003; Dalleau-Clouet *et al.*, 2005; Enkerli *et al.*, 2005). Alleles of a particular locus represent length polymorphisms which originate from different numbers of repeats. SSR loci are PCR amplified individually using primer pairs that bind specifically to the unique sequences flanking the SSR. Allele sizes (i.e. size of PCR product) are subsequently determined using electrophoresis techniques, such as capillary electrophoresis. SSR markers tend to be highly polymorphic (up to 20 alleles) and they are potentially independently segregating, which are important criteria when discriminating closely related organisms and performing population genetic analyses.

SSR analysis is very reliable and reproducible, and compared to other genotyping methods, it is relatively simple and cost efficient. SSR-based genotyping typically relies on analysis of multiple loci and application of multiplex PCR approaches (up to five loci amplified in one reaction). Furthermore, the use of a capillary genetic analyzer (i.e. sequencer) allows for a high degree of automation of the procedure. However, the isolation and characterization of individual SSR markers is time consuming, cost intensive and demands considerable expertise in molecular techniques (e.g., Rehner & Buckley, 2003; Enkerli *et al.*, 2005). Furthermore, SSR primers developed for a particular species often can not be applied to different species. This specificity does allow analysis of SSR loci in complex DNA samples such as DNA obtained from infected insects or soil. Compared to AFLP, which is based on a genome-wide assessment, SSR analyses rely on a limited number of loci and therefore its power to distinguish closely related genotypes may be lower.

SSR markers are available for a variety of entomopathogenic hypocrealean species including *B. bassiana* (Rehner & Buckley, 2003), *B. brongniartii* (Enkerli *et al.*, 2001), *M. anisopliae* (Enkerli *et al.*, 2005; Oulevey *et al.*, 2009), and *I. fumosorosea* (Dalleau-Clouet *et al.*, 2005). The markers have been applied to identify and characterize strains (Leland *et al.*, 2005; McGuire *et al.*, 2005), to monitor strains released for biological control (Enkerli *et al.*, 2004; Wang *et al.*, 2004), and to assess population structures and genetic diversity (Enkerli *et al.*, 2001; Gauthier *et al.*, 2007; Velásquez *et al.*, 2007; Meyling *et al.*, 2009; Freed *et al.*, 2011).

*(6). Single nucleotide polymorphism.* Single Nucleotide Polymorphism (SNP) markers are widely applied in biomedical fields and are becoming increasingly popular for studies in evolution, conservation biology, and ecology (Morin *et al.*, 2004). A SNP represents a single base pair position in a specific locus (i.e. sequence characterized locus) for which different sequence signatures (alleles) are found in individuals of a population. SNP analysis typically requires initial PCR amplification of the locus containing the SNP(s). Subsequently, the PCR product is purified (see Section 2B1c above) and submitted to a SNP detection assay. Various approaches have been developed (Kim & Misra, 2007); however, the most widely used strategy is based on single-base extension of primers that bind immediately adjacent to the SNP. Extensions are performed with fluorescently labeled dideoxynucleotides and resulting products are visualized and the size determined using capillary electrophoresis. Like SSR analysis, SNP based genotyping involves analysis of multiple SNPs. Typically, SNP loci display two different alleles (biallelic) and, in rare cases, three or even four alleles can be found (Kim & Misra, 2007). Once developed, SNP analysis represents a fast and efficient method for genotyping strains.

SNP discovery and development relies on comprehensive sequence comparisons using alignments of sequences of the target locus of representative strains (see Section 2B1d above). The necessary sequences may be obtained by sequencing the target locus in selected strains, which requires substantial sequencing efforts, or they may be obtained from public sequence databases. SNPs have been detected in genomes of many organisms (Kim & Misra, 2007) including fungi (Bain *et al.*, 2007; Kristensen *et al.*, 2007; Lambreghts *et al.*, 2009; Munoz *et al.*, 2009; Broders *et al.*, 2011). However, as far as entomopathogenic fungi are concerned SNPs have only been developed for the entomophthoralean species, *Pandora neoaphidis* (Fournier *et al.,* 2009).

## *2. Cultivation-independent analyses*

Cultivation-independent approaches can be used to detect species and/or strains in environmental samples (i.e. cultivation-independent diagnostic PCR) and they can be applied for analysis and comparison of fungal communities (i.e. community profiling). Here we restrict our discussion to cultivation-independent diagnostic PCR. Analyses of fungal community structures typically target fungi at the kingdom, phylum, or class levels, and they are not specifically applied to investigate insect pathogenic fungi. For further reading on cultivation-independent methods, we refer the reader to Kirk *et al.* (2004) and Enkerli & Widmer (2010).

Cultivation-independent approaches for detection of species and/or strains of hypocrealean fungi in environmental samples like insect tissues or cadavers, plant material, feces, water, or soil are often more efficient than cultivation-based approaches as they allow circumvention of time-consuming cultivation steps. Furthermore, this approach allows the study of species/strains that are difficult to isolate and/or cultivate, and/or are morphologically difficult to identify. As well, samples can be archived until it is convenient to process them, thereby providing a considerable logistical advantage over cultivation-based methods. As indicated previously, a significant limitation of cultivation-independent methods is that isolates do not exist for subsequent experimentation, and the molecular tools to be used must be carefully validated to ensure the accuracy of results. Various issues need to be considered when developing and applying cultivation-independent diagnostic PCR. In the following paragraphs we address issues involved in the preparation of DNA samples as well as those related to the PCR methodology.

*a. Effective extraction of genomic DNA*

A critical aspect of cultivation-independent methods is reliable and efficient extraction of high-quality DNA from environmental samples such as insect tissue or cadavers, plant material, feces, water, or soil. A variety of protocols and commercial kits are available to extract DNA from such samples. Regardless of the protocol to be used, all methods require the same basic steps which are homogenization of the sample followed by DNA purification (see Section 2B1b above). Environmental samples are homogenized in the presence of an appropriate buffer by applying mechanical, chemical or enzymatic procedures individually or in combination. Mechanical procedures typically involve grinding (e.g., using a mortar and pestle or tissue tearor), bead beating, freezing/thawing, and/or sonication. As indicated previously, care must be taken to prevent excessive exposure to heat during mechanical procedures, and it is often necessary to keep samples cool during the process. Depending on the type of sample, protocols may have to be adapted and optimized to achieve optimal yield and quality. This is particularly the case for extraction of DNA from soil samples. Soil has a highly complex chemical and physical composition and protocols optimized for one soil type may not be efficient for DNA extraction from another soil type (Frostegård *et al.*, 1999; Kabir *et al.*, 2003).

Efficient DNA extraction requires an exhaustive cell lysis which depends on the structural state of the fungus. For instance, the extraction of DNA from some types of fungal spores may require harsher lysis procedures than extractions from mycelium (Castrillo *et al.*, 2007). Performing successive extractions from the same sample may also be necessary as this can result in more efficient DNA extraction (Bürgmann *et al.*, 2001; Feinstein *et al.*, 2009). Given the variety of DNA extraction methods available, we advocate that pilot studies be undertaken to ascertain which method is most effective for the sample and fungal state to be targeted. If the end goal is to obtain a quantitative measure of biomass (see Section 4B1), it is essential that the DNA extraction method is consistently applied across samples. In this regard, the use of a quantitative Internal Amplification Control (IAC) is recommended (see Section 2B2c below).

*b. PCR inhibition*

Extraction of DNA with high-quality and purity from environmental samples is challenging because materials that are highly inhibitory to PCR are often co-extracted with nucleic acids. This is particularly a concern when extracting DNA from fecal or soil samples where polyphenolic compounds such as humic, fluvic, and tannic acids, and other inhibitory materials commonly occur. A number of strategies can be used to prevent or minimize issues caused by inhibitory substances, and the most effective strategy or combination of strategies should be determined in pilot studies. We categorize the different strategies into: (1) DNA dilution; (2) post clean-up procedures; (3) use of robust *Taq* DNA polymerases; and (4) minimization of inhibitor effects during PCR using chemical additives. The simplest strategy employed is to dilute DNA extracts sufficiently to lower the impact of inhibitory substances on PCR amplification (Wilson, 1997; Arbeli & Fuentes 2007); however, this strategy can only be used in instances where concentration of the target is not limiting. Various different methods can be used to post clean-up DNA extracts for which inhibition of PCR has been encountered. This includes precipitation of DNA with isopropanol (LaMontagne *et al.*, 2002) or polyethylene glycol 8000 (Widmer *et al.*, 1996) followed by the removal of inhibitory compounds remaining in the liquid phase. Other methods include spin column purification based on micro-filtration (Widmer *et al.*, 1999), gel filtration with Sepharose (Inglis *et al.*, 2010) or Sephadex resins (Jackson *et al.*, 1997), DNA binding with silica (Schneider *et al.*, 2011), or selective binding of polyphenolic inhibitors for instance with polyvinylpolypyrrolidone (Widmer *et al.*, 1996).

A variety of DNA spin column systems are commercially available, and a number of commercial DNA extraction kits utilize spin columns within the kit. In some instances, it may be desirable to use supplemental columns in conjunction with commercial kits (Figure 7.3; Inglis *et al.*, 2010). Some commercial kits utilize proprietary materials designed to bind inhibitors. An example is the Qiagen QIAamp DNA Stool Mini Kit® that utilizes Inhibidex® tablets. Although often overlooked, the type of *Taq* polymerase used can significantly abrogate the impacts of PCR inhibitors (Katcher & Schwartz, 1994). For example, we have had some success using HotStart KAPA® *Taq* DNA polymerase (Kappa Biosystems) to reduce the effects of PCR inhibition. The final strategy that is commonly used to address PCR inhibition is the addition of chemicals that minimize the effects of inhibitory substances on PCR. Examples include proteins like bovine serum albumin (BSA; Romanowski *et al.*, 1993; Comey *et al.*, 1994), Phage T4 gene 32 protein (Kreader, 1996), and skim milk (Arbeli & Fuentes, 2007) that scavenge inhibitory substances and thereby reduce the effects of inhibitory substances on DNA polymerases. It is important to realize that it is almost always necessary to employ a combination of strategies in concert.

When extracting DNA from complex substrates such as cadavers, feces, or soil, it is unlikely that all PCR inhibitors will be completely removed. Therefore, it is very important to guard against false negative results when applying diagnostic PCR, and to ascertain the degree of PCR inhibition for quantitative analyses. Inhibition can be quantified by spiking known amounts of recombinant template DNA into serial dilutions of sample DNA and subsequent quantification by qPCR and statistical analyses (Schneider *et al.*, 2009). Another common approach is to use an internal amplification control (IAC, also referred to as internal positive control [IPC]), which is amplified simultaneously (in the same PCR reaction) with the target. IACs can be relatively simple (e.g., exogenous DNA possessing unique primer

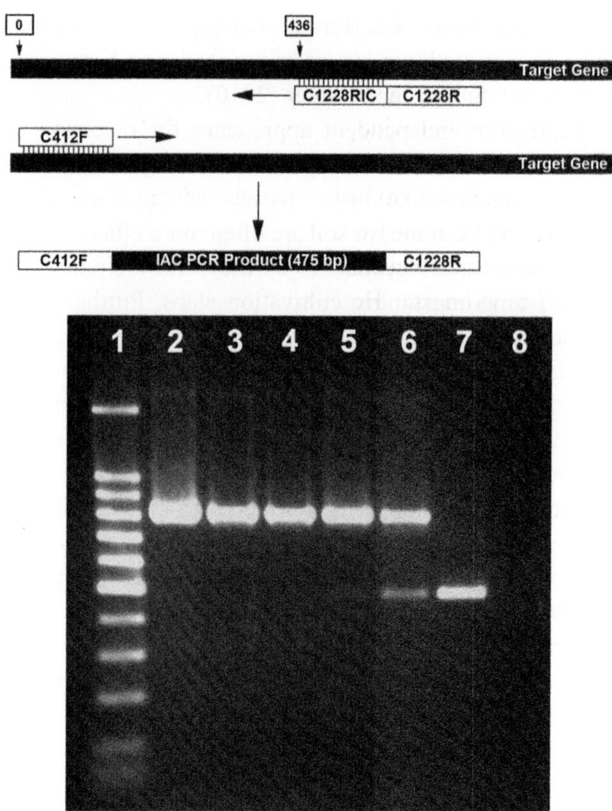


**Figure 7.7** Construction and use of an internal amplification control (IAC). In this example, an IAC was constructed using mutagenic primer strategy in which the IAC was designed to amplify under the same PCR conditions as those of the target gene (Inglis & Kalischuk, 2003). The deletion in the target gene was achieved by PCR amplification of the 16S rRNA gene with mutagenic primer C1228RIC to yield a 475-base-pair product instead of the 816-base-pair product. The 475-base-pair product was cloned into the pGEM-T EASY vector and transformed into *Escherichia coli* JM109 cells, the plasmid DNA was extracted with a QIAprep spin miniprep kit, was linearized by digestion with *Nco*I enzyme, the enzyme removed, and the concentration of linearized IAC was adjusted to 700 copies/μl. Agarose gel demonstrating the impact of cell density on detection of the target (i.e. upper band) relative to the IAC (i.e. lower band) using the same primer set. Lane 1, 100-base-pair molecular weight marker (the dark band is 500 base pairs); lane 2, $10^1$ dilution template; lane 3, $10^2$ dilution; lane 4, $10^3$ dilution; lane 5, $10^4$ dilution; lane 6, $10^5$ dilution; lane 7, no template and IAC added; lane 8, no template or IAC. Note that the IAC was added to all samples with exception of the sample run in lane 8. Also note that competition occurred for primer binding between the template and IAC illustrating the importance of optimizing concentration of the IAC when using this strategy.

sites that can be amplified with a separate primer set) or quite sophisticated (e.g., exogenous DNA possessing the same primer annealing sites as the taxon-specific primers, but designed to provide an amplicon of smaller size) (Denis *et al.*, 1999; Inglis & Kalischuk, 2003). For the latter approach, an IAC has to be constructed that has a deletion between the two primers used for amplification (Figure 7.7). Such constructs can be created by the use of mutagenic primers. However, when applying this approach, care must be taken to use the proper concentration of IAC DNA in order to guard against primer competition with sample DNA. The advantage of this method is that a single primer set can be used and the sensitivity for detection of the target and the IAC is the same, which may not be the case when using separate primer pairs.

*c. Cultivation-independent diagnostic PCR*

Cultivation-independent diagnostic PCR in its basic steps is identical to cultivation-dependent diagnostic PCR (see Section 2B1c above). The main difference between the two approaches is that, in cultivation-independent diagnostic PCR, the amount of target template DNA is small and part of a complex DNA extract (metagenomic DNA extract) while in cultivation-dependent diagnostic PCR, the template DNA (obtained from pure culture) is not limited. In addition, the total amount of DNA extracted from a sample can be very small, for instance when extracting DNA from an insect cadaver. The development of cultivation-dependent diagnostic PCR basically involves all the issues discussed above for the development of cultivation-dependent diagnostic PCR (i.e. locus selection, primer design, and validation). In fact, a first step for most is to develop a cultivation-dependent diagnostic PCR, which subsequently is adapted for cultivation-independent diagnostics. The presence of complex DNA in a PCR can affect primer comprehensiveness and specificity particularly when working with soil DNA extracts. Therefore, reconsidering these aspects is crucial when adapting cultivation-dependent diagnostics to cultivation-independent analysis. This can be done by assessing and comparing amplification products obtained from DNA samples (metagenomic DNA) containing the target and/or target-free DNA samples (metagenomic DNA) spiked with DNA of the target taxon. Non-spiked target-free DNA samples as well as samples spiked with non-target DNA should be included as controls. Resulting PCR products are subsequently assessed by gel electrophoresis (see Sections 3B1 and 3B2 above). The PCR product should represent a single band as multiple bands may indicate lack of specificity. Furthermore, obtained PCR products should be cloned and sequenced to confirm product identity (see Section 2B1c above). The sequence identity of multiple clones should be verified to allow detection of possibly different sequences of similar or even identical size.

Cultivation-independent diagnostic PCR approaches for detection of target taxa in environmental samples such as soil or insects have been developed for various entomopathogenic hypocrealean fungi. They include detection of *M. acridum* (Entz *et al.*, 2005) and *B. brongniartii* (Schwarzenbach *et al.*, 2007) at the species level, *B. bassiana* (Castrillo *et al.*, 2003, 2008; Bell *et al.*, 2009), *M. acridum* (Bell *et al.*, 2009), and *M. anisopliae* var. *anisopliae* (Destefano *et al.*, 2004) at the sub-species level, and *Metarhizium* clade 1, which includes the six species *M. majus*, *M. guizhouense*, *M. pingshaense*, *M. anisopliae*, *M. robertsii* and *M. brunneum* at the sub-genus level (Schneider *et al.*, 2011).

It is also important that one is cognisant that diagnostic PCR is subject to a detection threshold also referred to as PCR sensitivity, that will vary depending on the target locus (copy number), and the PCR conditions (e.g., the Ta, PCR reagents, and cycling conditions used) (see Section 2B1c above). Most primers developed for cultivation-independent detection of entomopathogenic hypocrealean fungi target conserved loci that are present in multiple copies in the genome such as rRNA genes or ITS regions which may be present at 100 copies or more per genome (Herrera *et al.*, 2009; Schneider *et al.*, 2011). Such target loci allow for a lower detection limit as compared to target loci that are present in a single copy in the genome (e.g., a SCAR marker). To achieve the lowest-possible detection limit, diagnostic PCR has to be performed under optimal amplification conditions. For instance, a lower Ta or additional PCR cycles will allow for more sensitive detection. However, at the same time, this may affect comprehensiveness and specificity (see Section 2B1c). The detection limit of a diagnostic PCR can be determined, for instance, by performing PCR on target-free DNA samples (metagenomic DNA) spiked with serial dilutions of DNA of the target taxon. Obtained PCR products are separated by gel electrophoresis and analyzed for presence or absence of the target band (see Section 2B1c above), which allows determination of the lowest target DNA concentration yielding a PCR product (detection limit). Cultivation-independent diagnostic PCR may detect target copy numbers as low as 10 copies per PCR reaction (Schwarzenbach *et al.*, 2007). However, limits for reliable detection are typically around 100 copies per reaction (see also Section 4B1). As outlined previously, reliable and sensitive detection of a target taxon is also dependent on effective DNA extraction (e.g., effective lysis of target structures) and avoidance of PCR inhibition. Therefore, the considerations made should always be kept in mind when performing cultivation-independent diagnostic PCR.

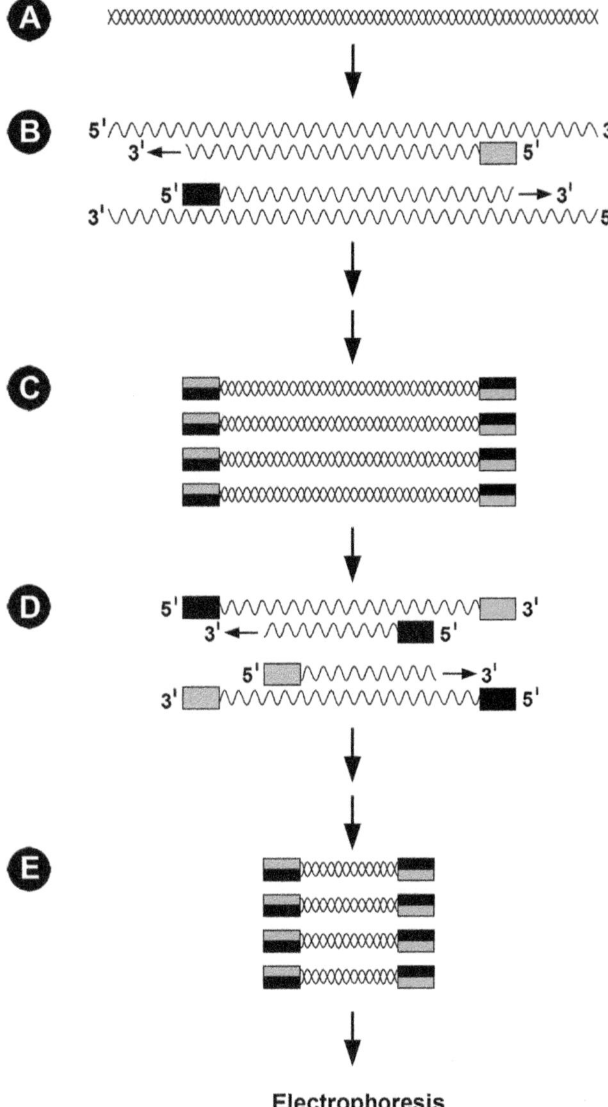


**Figure 7.8** A schematic illustrating the use of nested PCR for direct detection of an entomopathogenic hypocrealean fungus where: A. is genomic DNA; B. annealing of the first set of primers (i.e. outer primers) to the target and DNA extension; C. amplified DNA from the first PCR reaction which may contain non-target products due to non-specific amplification; D. annealing of the second set of primers to the PCR products from the first PCR reaction (i.e. inside primers) and DNA extension; and E. uncontaminated final product. Amplicons are then visualized by electrophoresis.

Nested and/or semi-nested PCR are two strategies that are commonly used to achieve a low detection limit. Both strategies involve two rounds of PCR using two primer sets (Figure 7.8). The first set of primers is designed to allow amplification from a wider range of related taxa while the second set is specific to the target taxon. In the case of nested PCR, both primers used in the second round of PCR are new and bind to the first round product internally to the first primer set (internal primers), whereas for semi-nested PCR, one new internal primer is combined with the corresponding first round primer. By using two primer sets in succession, the detection range can be lowered and at the same time the specificity can be increased. Thus, nested PCR provides comprehensiveness as well as specificity and sensitivity. The major limitations of nested PCR are that the number of amplifications, and thus labor and reagents, are doubled. As mentioned before, it is important to always include negative controls to detect contamination.

### *d. Future methods*

*(1). Metagenomics.* A type of cultivation-independent approach which is increasing in importance in various fields of ecology is metagenomics. Metagenomics is the study of metagenomes (i.e. all of the genetic information present in a particular environment or sample). Although metagenomics has not been applied to the study of entomopathogenic hypocrealean fungi to date, it has applications to examine the impacts of entomopathogenic fungi on soil microbial communities and/or to detect specific organisms or genes in complex environmental samples. A detailed description of the methods used in metagenomics is beyond the scope of this chapter, and the reader is referred to the metagenomics literature for additional information.

*(2). Diagnostic arrays.* Another approach which is increasing in importance in ecological research is the use of diagnostic microarrays (Sessitsch *et al.*, 2006; Xu,

2006). Microarrays (i.e. gene arrays) are glass slides to which several thousand DNA probes have been attached in defined arrays. The DNA probes are designed to specifically bind to nucleic acids of different target organisms or taxonomic groups including specific strains within a species, all strains within a species, all species within a genus, or even multiple genera or higher phylogenetic groupings. These probe arrays are exposed to fluorescent-labeled DNA samples of unknown composition. Unbound DNA is then washed from the array surface, and only strongly bound DNA (i.e. complementary) remains hybridized. Labeled DNA bound to the probes fluoresce and generate a signal which is detected using specifically designed array readers. This allows determination of the composition of the DNA sample (i.e. the presence or absence of individual targets represented on the microarray). Microarray analysis constitutes a powerful high throughput approach for simultaneous detection and identification of thousands of target organisms in complex environmental samples and has great potential for use in research and diagnostics of entomopathogenic fungi. However, the development of microarray probes is demanding and there are currently no microarrays available that allow specific detection of entomopathogenic fungi.

## 3. INOCULUM PROPAGATION AND PREPARATION

### A. Inoculum propagation

Detailed studies on fungal pathogenesis, virulence, physiology and evaluation as microbial control agents require methods for the propagation of mycelia and/or infectious propagules. Most entomopathogenic fungi are easily propagated on defined or semi-defined media containing suitable nitrogen and carbon sources. Although conidia of most species are commonly stored at 4–5°C, several simple precautions can be taken to extend the shelf-life of conidia stored under refrigeration. Mycelia and hyphal bodies can also be stored refrigerated or dehydrated after treatment with a cryoprotectant.

Submerged culture is used for production of mycelia and hyphal bodies. Some species of entomopathogenic Hypocreales may also produce conidia under submerged culture, but these conidia are usually short-lived. Conidia of entomopathogenic hypocrealean fungi are most often produced aerially using surface culture techniques. In most cases, the choice of culture method will be dictated by the type of propagule desired.

Continuous serial subculturing on artificial media should be avoided as this can result in genetic changes that can affect the phenotype. In some cases phenotypic changes resulting from the selection for specific genetic traits as a result of heterokaryosis (e.g., attenuation of virulence) can be restored after a single or several passages through an insect host. However, a superior strategy is to utilize proper storage methods in order to preserve the original genotype. Methods for preservation of cultures are presented in Chapter X.

#### *1. Surface culture*

Surface culture is used for the routine maintenance of isolates and for production of conidia. One of the most commonly used media by insect pathologists is SDAY (Appendix: 2a); however, many other media such as cornmeal, czapeck-dox, malt extract, nutrient, potato dextrose agar (PDA), and Sabouraud's maltose agar are also often used. Some researchers advocate more complex media such as mixed cereal agar (Appendix: 2b) to maintain strain 'vigor'. However, this has not been substantiated and these media have not been widely accepted in insect pathology, although most entomopathogenic Hypocreales including *B. bassiana*, *C. clavisporus*, *Metarhizium* spp., *T. cylindrosporum* and *Lecanicillium* spp. will grow and sporulate well on complex media (Goettel, 1984).

Surface culture is usually accomplished on agar media within glass bottles or disposable plastic Petri dishes. The agar surface is inoculated under sterile conditions with a suspension of either conidia or hyphal bodies; initial growth is faster when hyphal bodies are used. The dishes are normally incubated between 25 and 27°C under varying light conditions. To reduce water loss from media, Petri dishes can be sealed with Parafilm. Within 7–10 days, the cultures will normally have sporulated and the conidia are harvested either by direct scraping from the surface using a sterile rubber policeman or by washing off with an appropriate buffer or distilled water. For larger scale production of conidia, economical nutritive substrates such as rice, bran or cereal grains are typically used (Chapter VIII).

The use of inert substrates as a base for the aerial growth and sporulation of hypocrealean fungi enables the complete separation of the pathogen from the nutritive medium. Goettel (1984) used cellophane (i.e. cellulose film) as a semi-permeable membrane which allows harvest of non-cellulolytic fungi virtually free of nutritive substrate contamination. Approximately one part bran (w) and 10 parts distilled water (v) are combined in autoclavable pans (e.g., tin cookware roasting pans). Sheets of presoaked cellophane are then layered over the bran mixture. Each pan is then placed into an autoclavable plastic bag and autoclaved for 1 h at 138 kPa (20 psi, ≈ 125°C). Extreme care is taken in cooling the autoclave as slowly as possible to prevent the bran mixture from boiling, thereby causing the cellophane to lift off the bran surface. The pans are allowed to cool and the surface of the cellophane is inoculated with the fungus using a hypodermic needle; inoculum is injected on to the surface of the cellophane in one corner of the pan and spread over the surface by moving the pan in a swirling motion. The amount of inoculum required must be adjusted according to the size of pans used. After incubation for 1–2 weeks, the pans are removed from the autoclave bags and the cellophane with adhering fungus is gently lifted off the bran surface. The fungal mat, which consists almost exclusively of conidia, is scraped off the cellophane surface using a spatula. Up to $2.3 \times 10^8$ conidia/cm$^2$ cellophane surface were obtained with several entomopathogenic Hypocreales after a 14-day incubation period at 20°C (Goettel, 1984).

Inert substrates can be used as a base for the aerial growth and sporulation of entomopathogenic Hypocreales. For instance, Bailey & Rath (1994) used a nutrient-impregnated membrane for production of *M. anisopliae* conidia. Strips of absorbent membrane (i.e. Superwipe, an absorbent fibrous material), are dipped into a conidial-nutrient solution (e.g., ultra-high temperature treated milk (1.4%), sucrose (0.2%) and $2.1 \times 10^5$ conidia/ml). The membranes are then incubated in bottles or steel boxes. Yields are in the order of $10^7$ conidia/cm$^2$.

Aerial conidia can also be produced on the surface of still liquid media. Kybal & Vlcek (1976) used a polyethylene hose (≈ 300 cm diameter), sealed at both ends and partially filled with a nutrient broth inoculated with the fungus. Sterile air is then pumped through the hose. While mycelial growth occurs within the liquid medium, sporulation occurs at the liquid–gas interface. After incubation, the liquid is drained and the conidia are harvested. Using this system, after 12 days' incubation, Samšiňáková *et al.* (1981) obtained $10^9$ conidia of *B. bassiana* per cm$^2$ liquid surface [peptone (0.8%) and sorbitol (1%)].

### 2. *Submerged culture*

Submerged culture is used for the production of mycelia, hyphal bodies, or sometimes, conidia. The type of material to be obtained (i.e. mycelium, hyphal body or conidium) can usually be controlled by selecting specific fungal strains and varying culture conditions (Chapter VIII). At the laboratory level, submerged culture is usually accomplished in flasks with magnetic stirrer bars, shake flasks on rotary or wrist shakers, or in larger plastic or glass containers as air-lift fermenters.

Under the proper conditions, certain species and/or strains produce conidia in submerged culture. For instance, out of 14 isolates of *Hirsutella thompsonii*, only one was found to be capable of producing submerged conidia, and only if corn steep liquor (1%) and Tween 80 (0.2%) were added to the culture medium (van Winkelhoff & McCoy, 1984). Some isolates of *M. acridum* (as *Metarhizium flavoviride*) can be induced to switch from vegetative growth to conidiation by varying the ratio of yeast to sucrose; more than $1.5 \times 10^9$ conidia/ml were produced in a medium containing brewers' yeast (2–3%) and sucrose (2–3%) (Jenkins & Prior, 1993). Similarly, submerged conidia of a *B. bassiana* isolate were produced in a sucrose (2%)–yeast extract (0.5%)–basal salts medium ($1.7 \times 10^8$ conidia/ml), whereas mostly hyphal bodies were produced in a sucrose (2.5%)–yeast extract (2.5%) medium ($7.4 \times 10^8$ hyphal bodies/ml) (Rombach, 1989).

For production of larger quantities of biomass, air-lift fermentation in 10- or 20-l autoclavable plastic carboys can be used (McCoy *et al.*, 1975; Pereira & Roberts, 1990). In this system, autoclaved nutrients and an antifoaming agent (V-30 at 25 ppm, Dow Corning or vegetable oil at 0.1%) are added to the flasks in up to 1 l of water. Antimicrobial agents, fungal inoculum, and filter-sterilized water (using an inline autoclavable filter with a 0.5-μm pore size) are then added until the desired volume is reached. Compressed filtered air (using the same inline filters described above) is injected into the flask through a glass pipette extending to the bottom and is allowed to escape through an opening at the top of the flask. A modification of this system is to extend a plastic tube to the bottom of the flask. To obtain adequate aeration, reduce foaming, and obtain superior mixing, the bottom of the hose is plugged and small perforations are made in the bottom 10 cm of the hose (R. Pereira, personal communication). The air is first warmed and humidified by passing through water in an Erlenmeyer flask kept on a hot plate set at low heat or wrapped with heating tape. The mycelium is harvested by filtering after 3–4 days' incubation. Using this method, yields of *H. thompsonii* were 30 g wet weight/l of culture medium (0.5% each of dextrose and yeast extract and 0.05% peptone) supplemented with basal salts (McCoy *et al.*, 1975). Air-lift fermentation is often used for the production of biomass which is used for the inoculation of solid substrates such as rice. For instance, a simple medium of sucrose (1.5%) and nutritional yeast (1.5%) is used to produce inoculation cultures of *B. bassiana* and *M. anisopliae* using this method (R. Pereira, personal communication).

### 3. *Propagule storage*

Once the fungus is cultured, it must be stored unless it is used immediately. Most of the same methods used for preservation of fungal cultures (Chapter X) can also be used for short or long-term storage of fungal propagules. Conidia, hyphal bodies, and mycelia can be stored at 4°C for up to several weeks or even months. However, this depends much on the species and/or strain in question. Conidia usually store best if they are kept under dry cool conditions (Chapter VIII). If cooling is not possible, then the moisture content can be critical. For instance, Hedgecock *et al.* (1995) obtained gradual decline of viability of *M. acridum* (as *M. flavoviride)* conidia possessing a moisture content of 5% after 4 months of storage at 38°C, whereas a rapid loss of viability occurred in conidia possessing a moisture content of 15%. In contrast, Daoust & Roberts (1983) reported that conidia of *M. anisopliae* survived best when RH was high (97%) at moderate (19–27°C) or low (0–4°C) temperatures. For storage under dry conditions, conidia should be harvested, air dried for several days preferably in a class 2A or higher biosafety cabinet, and then maintained over anhydrous silica gel crystals. Laminar flow hoods **should not** be used to ensure worker safety and prevent contamination of the immediate enivronment. Hyphal bodies can be stored wet or dry at 4–5°C. For instance, Gardner & Pillai (1987) stored hyphal bodies of *T. cylindrosporum* in distilled water for 5 months at 4°C without appreciable loss of viability. Blachère *et al.* (1973) developed a method for drying and storage of hyphal bodies of *B. brongniartii*; hyphal bodies were centrifuged, mixed with formulating ingredients (Appendix: 4), and dried at 4°C. Essentially no loss of viability occurred after 8 months of storage at 4°C in vacuum-sealed plastic bags.

Mycelium can usually be stored wet for short periods at 4°C, but precise information on this method of storage is currently lacking. Mycelium can also be stored dry. Mycelium is harvested during the active growth phase of submerged cultivation, vacuum filtered, washed, sprayed with a cryoprotectant sugar solution, air dried and stored at 4°C. As indicated previously, older hyphae may not be viable and care must be exercised to ensure that mycelium is harvested at a time in which viability is at a maximum. Upon rehydration, mycelia may produce conidia. Mycelia of *M. anisopliae* and *B. bassiana* treated with 10% solutions of maltose or sucrose produced more conidia upon rehydration than did mycelia treated with a solution of dextrose alone (Pereira & Roberts, 1990). In contrast, treatment of mycelia with maltose and dextrose in combination provided superior conidiation of *B. bassiana* from dried mycelia stored at room temperature. Roberts *et al.* (1987) used 10% sucrose for drying mycelium of *C. clavisporus*.

### 4. *Inoculum preparation*

Propagules are usually formulated in a water carrier but any number of carriers may be used (e.g., solid substrates). Although propagules with hydrophilic cell walls are readily suspended in water (e.g., *T. cylindrosporum* and *Lecanicillium* spp.), this is not the case for conidia possessing hydrophobic cell walls (e.g., *B. bassiana* and *M. anisopliae*). To uniformly suspend hydrophobic propagules in water, it is necessary to sonicate and/or use mechanical suspension methods. Mechanical suspension of propagules using micropestles (Figure 7.9) can provide excellent suspension of *B. bassiana* and *Metarhizium* spp. conidia in water without causing cell damage. Although a surfactant may facilitate suspension of propagules, it is generally not necessary and may interfere in the adherence of the

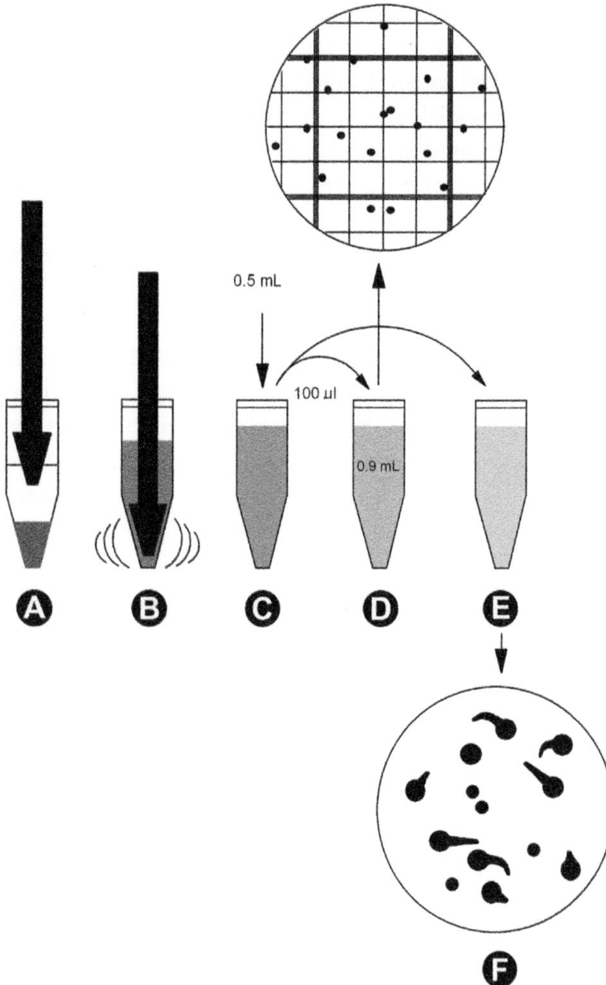


**Figure 7.9** Suspension, enumeration and density adjustment of propagules of entomopathogenic Hypocreales: A. propagules at the bottom of a 1.5-ml microcentrifuge tube are gently agitated in ≈ 0.5 ml of carrier with a sterile micropestle; B. clumps of propagules are separated by vigorous agitation of the suspension aided by the use of a pestle motor; C. the volume of the carrier is increased to ≈ 1 ml by the addition of 0.5 ml of the carrier; D. the suspension is diluted 10 ×, 10 µl is loaded into a hemocytometer, and the number of propagules per ml calculated; E. the original suspension is diluted accordingly to obtain the desired target concentration; and F. viability of conidia is ascertained.

propagule to the host insect. To suspend hydrophobic conidia, harvested conidia are placed in a 1.5-ml microcentrifuge tube, ≈ 0.5 ml of sterile water is added to the tube, the micropestle is inserted into the tube, and the conidial mass is gently agitated with the micropestle by hand (prevents liberation of conidia into air). The micropestle is then attached to the motor (e.g., Kontes, Concept Inc, Clearwater FL) and the suspension is vigorously agitated while moving the pestle in an up and down, and side to side motion (*ca.* 30 s).

Although fresh conidia harvested from an agar medium can usually be rehydrated by placing directly into water as described above, care must be taken when rehydrating conidia. Rapid imbibition of water by dehydrated hypocrealean conidia can result in damage to cell membranes. Such damage can be avoided by using warm water or by prior rehydration under high humidity. For instance, the viability of *M. acridum* exhibited over 70% germination with prior rehydration in water-saturated atmosphere compared to less than 25% when immersed directly into water (Moore *et al.*, 1997). Faria *et al.* (2009) demonstrated the importance of water temperature to imbibitional damage to conidia of *M. acridum* and *M. anisopliae*; viability of conidia of both species was less than 5% if immersed in water at 0.5°C compared to over 88% after immersion in water at 33°C.

Evidence now suggests that oil can be a more efficacious carrier than water for propagules of some entomopathogenic hypocrealean fungi (Bateman *et al.*, 1993; Inglis *et al.*, 1996a), and conidia with hydrophobic cell walls are readily suspended in oil using micropestles. In addition, Faria *et al.* (2009) reported that paraffinic oil provided considerable protection from imbibitional damage.

## 4. QUANTIFICATION

It is often desireable to quantify hypocrealean fungi (e.g., to assess targeting and fate). Given the inherent difficulties in accurately measuring biomass of hyphal fungi, most efforts at quantifying entomopathogenic Hypocreales have relied on classical microbiological methods to enumerate infectious propagules applied in both laboratory and field experiments. It is well recognized that the inherent limitations of classical microbiological methods grossly underestimate hyphal biomass, and thus are not effective in this regard. Although they have not been extensively used to date, molecular-based quantification methods may be used to obtain a more accurate measure of hypocrealean fungal biomass (Section 4B below).

## A. Classical methods

Quantification of inoculum for bioassays or field application is usually accomplished through direct counts. Enumeration of propagules from insects or infested substrates (e.g., leaves or soil) is based primarily on both direct counts and indirect recovery methods (e.g., dilution plating). Direct counts should normally be adjusted to take into account the proportion of propagules that are viable.

### *1. Direct enumeration*

*a. Inoculum*

*(1). Hemocytometer quantification.* A hemocytometer is commonly used to quantify numbers of propagules per unit volume or weight. The propagule suspension is vortexed and approximately 10 μl of the suspension is then loaded on to each side of the hemocytometer. Propagules must first be allowed to settle; for a water carrier, propagules settle rapidly (*ca.* 5 min) but in oil, adequate settling may require more than an hour. It is also imperative that the excessive volumes of oil are not loaded on to the hemocytometer as oil infiltrating between the lip of the counting chamber and the coverslip will lift the coverslip and increase the volume of the counting chamber thereby providing an overestimate of propagule density. When using water, care must be taken to ensure that the water carrier does not evaporate before measurements are taken. Consequently, measurements using a water carrier should be taken as soon as possible after the propagules settle. The 'cell' type (be it 'A', 'B' or 'C') on the hemocytometer that contains a countable number of propagules ($\approx$ 20−100) is selected, and the number of the propagules within each of five 'cells' on each of the two sides of the hemocytometer (total of 10 'cells') are counted. The five 'cells' of each side of the hemocytometer should comprise the four corner 'cells' and the middle 'cell' the fifth 'A' 'cell' consists of all 25 'C' 'cells' (Figure 7.10). If greater than 10% of the propagules appear clustered, the entire procedure should be repeated, making sure that the propagules are dispersed by vigorous pipetting or vortexing of the original suspension.

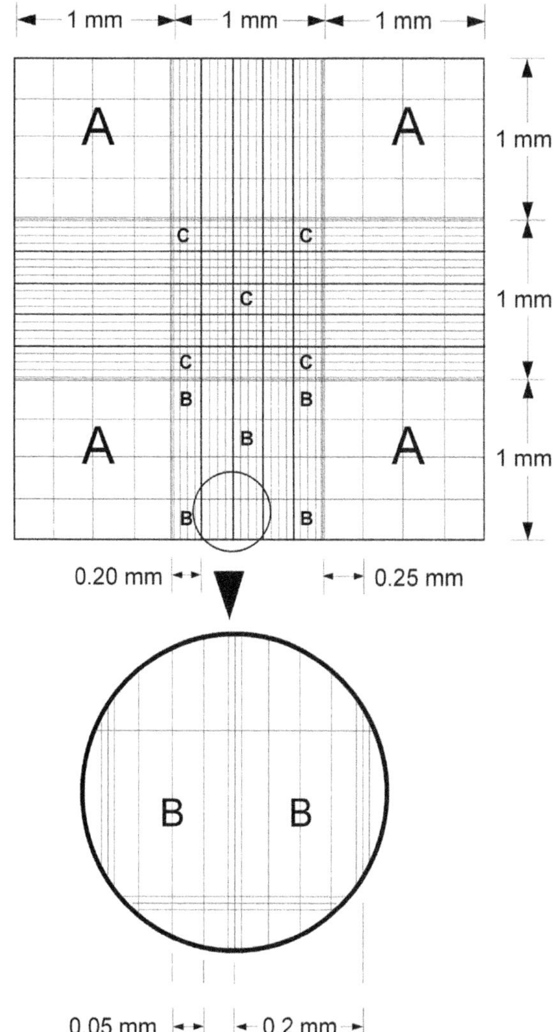


**Figure 7.10** An Improved Neubauer hemocytometer used to enumerate propagule densities. The number of propagules in each of five 'cells' ('A', 'B' or 'C') are counted; the fifth 'A' 'cell' consists of all 25 'C' 'cells'.

The reference squares within each hemocytometer 'cell' are used to facilitate counting; the boundary of each 'cell' is determined by the center line in the group of three. To avoid counting propagules twice, count those that are touching the center line only when they are on the top and left center line; those that touch the bottom and right side center lines of the 'cell' are not counted. The average number of propagules per 'cell' is calculated and a good measure of count uniformity is when the standard deviation (SD) is within 10−15% of the mean. To obtain an estimation of the number of propagules per milliliter in the original suspension, the average number of propagules per 'cell' is then multiplied by the volume conversion factor. For Improved Neubauer hemocytometers, the total volume of each 'C' 'cell' is $0.02\ \text{cm} \times 0.02\ \text{cm} \times 0.01\ \text{cm} = 4.0 \times 10^{-6}\ \text{cm}^3$ (Figure 7.10). The average number of propagules per 'C' 'cell' is divided by this number or multiplied by its inverse ($2.5 \times 10^5$) to obtain the number of propagules per ml; for 'A' and 'B' cells, means are multiplied by $1.0 \times 10^4$ and $2.0 \times 10^5$, respectively. For example, a mean count of 50 propagules per 'C' 'cell' with an SD of $\leq 10$ ($n = 10$) is considered to be a good count, and 50 is multiplied by the conversion factor, $2.5 \times 10^5$ to obtain a value of $1.25 \times 10^7$ propagules /ml.

The original suspension is then diluted to obtain the desired target concentration using the formula $C_1V_1 = C_2V_2$ where $C_1$ is the initial concentration, $V_1$ is the required volume for dilution of the original aliquote, $C_2$ is the desired concentration, and $V_2$ is the desired final volume. For instance, if the initial concentration ($C_1$) obtained through a hemocytometer count is $1.25 \times 10^7$ propagules/ml and the desired concentration ($C_2$) is $1 \times 10^6$ propagules/ml in 1 ml ($V_2$), then the equation is:

$$V_1 = [1 \times 10^6\ \text{propagules/ml} \times (1.0\ \text{ml})]/1.25 \times 10^7 = 0.08$$

Therefore, 0.08 ml of the original suspension is added to 0.92 ml of an appropriate carrier to obtain a total volume of 1 ml containing a suspension of $1 \times 10^6$ propagules/ml (i.e. the desired concentration). The total volume of inoculum required is then determined, and the appropriate dilution is made. For instance, if 10 ml of inoculum at the target concentration of $1 \times 10^6$ propagules/ml is desired, then $V_1$ is multiplied by 10, and 0.8 ml of the original suspension is added to 9.2 ml of the appropriate carrier to obtain 10 ml of suspension containing $1 \times 10^6$ propagules/ml.

In instances where the concentration of propagules in the original suspension is too high to get an accurate count on the hemocytometer (i.e. > 100/'cell'), it is necessary to first dilute the suspension prior to enumeration. For example, if an insect is to be treated with $10^5$ propagules in 0.5 μl, the concentration of propagules per unit volume that is required is $2 \times 10^8$ propagules/ml. At this density, each 'C' 'cell' of the hemocytometer would contain ≈ 800 propagules (i.e. $2 \times 10^8$ propagules per ml/$2.5 \times 10^5$ per ml) which are too numerous to count. Therefore, the original solution is diluted 10 times (e.g., 100 μl of the propagule suspension into 900 μl of water) and the concentration is reduced to ≈ 80 propagules per 'C' 'cell'. The mean number of propagules per ml in the diluted sample is then multiplied by the dilution factor, in this case by 10, to obtain the concentration of propagules in the original sample. When dilution of the original sample is necessary, it is desirable to obtain counts from two independent dilutions. For dose–mortality experiments, suspensions for each dose can be prepared from independent samples or the original sample can be diluted (see Section 5A for experimental design information). However, when dilution of the original sample is used, it is important that the accuracy of the dilutions be confirmed.

*(2). Other quantification methods.* Although not commonly employed, a number of other methods can be used to estimate concentrations of propagules. Turbidimetric methods have been used to estimate propagule concentrations of entomopathogenic Hypocreales. This method relies on the transmittance of light through the propagule suspension and the fact that the percentage light transmitted will diminish in proportion to the turbidity. Turbidity readings must be standardized in terms of numbers of propagules per unit volume. The relationship between turbidity and propagules per unit volume is usually determined using a hemocytometer. Although the turbidimetric method is relatively simple and rapid, it is subject to errors due to variation in size, shape, and clumping of propagules. Two additional methods used are dilution plating and the most probable number (MPN) method [see Sections 4A2e(1) and 4A2e(2) below].

*(3). Viability—germination assessments.* In contrast to hemocytometer counts and the turbidimetric methods, plating techniques [see Sections 4A2e(1) and 4A2e(2) below] provide a measure of viable propagules per unit volume. However, they usually take 3–7 days before data is obtained and generally provide a conservative estimate of propagule concentrations. Since hemocytometer and turbidimetric methods do not distinguish between viable and non-viable propagules, it is necessary to determine spore viability so that doses can be prepared on the basis of viable propagules. Germination assessments typically provide results within 24 h. In contrast, vital dye assessments of propagule viability can provide results within an hour [see Section 4A1a(4) below].

For germination assessments, propagules are typically distributed over the surface of a translucent agar medium such as SDAY or PDA, and the prevalence of propagules that germinate after a specified period is measured. Care should be exercised when selecting an appropriate culture medium, as some species or isolates of fungi may germinate more readily on some media. In addition, it may be important to slowly hydrate some species such as *Metarhizium* in warm water before spread plating for a germination test. Propagules from the same batch to be used in the bioassay should be processed immediately prior to the preparation of inoculum (*ca.* 1 or 2 days). A commonly used procedure is to prepare a suspension of propagules in water and to spread the suspension on to the surface of the medium at a density sufficiently high to permit rapid observation, yet sparse enough to limit obscuration of propagules from overgrowth by hyphae.

A satisfactory density of spores is usually obtained by spreading $\approx 10^6$ propagules in 100 μl on to media in an 8.5-cm diameter Petri dish. Propagules are incubated in the dark at an appropriate temperature for a specified period (usually 18–24 h) and the area to be observed is fixed (e.g., lactophenol, Chapter VI); it is usually sufficient to place a couple of drops of the fixative on to the surface of the medium and then place a glass coverslip(s) over the area. Once the fixative has absorbed into the medium, the area can be examined microscopically. Phase-contrast microscopy is usually preferred but if bright field examination is used, the propagules should be stained (e.g., lacto-fuchsin, Appendix: 3c or lactophenol cotton blue, Chapter VI) after fixation and prior to placement of the coverslip. An inverted compound microscope may also be used.

Viability of conidia in an oil carrier may also be determined as outlined above. However, problems occur when attempting to enumerate propagules of species with small round conidia (e.g., *B. bassiana*) as the oil can form tiny emulsion droplets on the agar surface which are difficult to distinguish from ungerminated conidia. To circumvent this, propagules in an oil emulsion can be stained with a drop of lacto-fuchsin (Appendix: 3c) which is miscible in oil, prior to placement of the coverslip on the medium. Alternately, the oil suspension can be diluted in kerosene and applied directly on to the culture medium. A drop of kerosene is added to the culture medium and covered with a cover slip. As it takes several minutes for water droplets to move from the agar into the kerosene, there is sufficient time to conduct viability counts prior to water drop formation (S. Jaronski, personal communication).

Propagules are usually considered viable if germ-tube lengths are $\geq$ two times the diameter of the propagule in question. For some taxa, conspicuous swelling (e.g., *B. bassiana*) or formation of a barbell shape (e.g., *T. cylindrosporum*) of germinating conidia may be used to indicate viability. Numbers of germinated and non-germinated propagules in arbitrarily selected fields of view or in parallel transects, usually defined with an ocular micrometer are counted. Although a minimum of 200 propagules may be sufficient, counting 500 or more can increase accuracy. It is desirable to determine the viability of propagules on replicate cultures and at various positions on the medium within the same Petri dish.

A major limitation to this technique is that the rate of propagule germination is usually normally distributed and hyphae from early germinating propagules can obscure non-viable and later-germinating propagules, thereby affecting the accuracy of the counts. To circumvent this problem, benomyl has been used to inhibit the hyphal development of *B. bassiana* and *M. anisopliae* (Milner *et al.*, 1991; Inglis *et al.*, 1996a). The mode of action of benomyl is inhibition of spindle formation during mitosis as a result of binding to tubulin. Ascomycetous fungi are highly sensitive to benomyl (Edgington *et al.*, 1971) at relatively low concentrations (*ca.* 0.005% Benlate 50 WP per ml); benomyl permits germ-tube formation but prevents hyphal development thereby preventing overgrowth (Figure 7.9F). However, caution should be exercised when using benomyl. For instance, Faria *et al.* (2009) found that the inclusion of benomyl at 0.005 and 0.1 g/l in a yeast extract agar-based solid medium (YEA) containing 0.5 g yeast extract, 100 mg gentamicin, 0.1 g Tween 80, and 16 g agar per liter substantially delayed germination of *M. acridum* conidia as compared to conidia of *M. anisopliae* and *B. bassiana*.

To calculate viable propagules per unit volume, total counts estimated with the hemocytometer are multiplied by the germination percentage. For example, if the viability of propagules was 75% and the total count from the hemocytometer was $2 \times 10^8$ propagules per ml, the number of viable propagules per ml is $2 \times 10^8 \times 0.75 = 1.5 \times 10^8$ viable propagules/ml.

*(4). Viability—vital stain assessments.* Vital stains can be used to rapidly determine conidial viabilities. Several

fluorochromes have been used successfully to determine viability of conidia of entomopathogenic Hypocreales. The optical brightener Tinopal BOPT can be used to differentiate between viable and non-viable conidia of *M. anisopliae* and *B. bassiana* (Jimenez & Gillespie, 1990). Conidia are stained with a 0.05% Tinopal (bi-striazinyl amino stilbene) solution in 0.05% Triton X-100 for 30 min and are viewed using a fluorescence microscope under UV light. Viable conidia fluoresce only weakly whereas non-viable conidia fluoresce brightly. However, this technique was unsuitable for conidia of *M. majus*.

Fluorescein diacetate (FDA) can be used alone, or in conjunction with propidium iodide (PI), to determine viabilities of *Isaria fumosorosea* and *B. bassiana* conidia (Schading *et al.*, 1995). Aqueous suspensions of conidia (4 µl) are mixed with equal amounts of freshly prepared working solutions of FDA and PI (optional) (Appendix: 3) in a dark room illuminated with a 40-W photographic safelight. The mixture is stirred with a pipette tip and covered with a coverslip. The slides are then viewed under epifluorescence using a 450–490-nm (blue light) exciter filter and a barrier filter in conjunction with a DM500 dichromatic mirror. Viable conidia fluoresce bright green, and, if PI is also used, nuclei of dead conidia fluoresce red. Only PI will fluoresce when green light (515–560 nm) is used. If PI is not used, after counting the green conidia using epifluoresence, just enough substage light (phase contrast) is added to identify and count all conidia present within the same field. Viable conidia maintain their fluorescence for only short periods (10–30 s) once normal substage light is applied. Concentrations of the fluorochromes used are critical; by experimenting with concentrations, these staining techniques can probably be adapted to assess conidial viability of most species of entomopathogenic Hypocreales. Further information on using fluorescent techniques is presented in Butt (1997) and Chapter XV.

The use of live–dead staining strategies has become relatively commonplace in microbiology. A number of commercial companies produce live–dead kits for fungi (e.g., Invitrogen Life Science). For example, the LIVE/DEAD® Yeast Viability Kit from Invitrogen Corporation combines a proprietary two-color fluorescent probe for yeast viability (FUN® 1) with a fluorescent fungal surface labeling reagent Calcofluor® White M2R; plasma membrane integrity and metabolic function of fungi are required to convert the yellow–green-fluorescent intracellular staining of FUN® 1 into red–orange intravacuolar structures, whereas Calcofluor White M2R labels cell-wall chitin with blue-fluorescence regardless of metabolic state. Assessment of viability can be done manually using microscopy. Alternatively, microtiter or flow cytometry can be used to enumerate numbers of viable *versus* dead cells in a population. Although flow cytometry is readily used to enumerate viable and non-viable bacteria and yeasts, to our knowledge it has not been extensively used to enumerate filamentous fungi to date.

*b. On insect surfaces*

Direct quantification of propagules on the surface of insect integuments has been successfully accomplished in relatively few instances. On very small insects (e.g., thrips and aphids), conidia can be directly visualized by mounting the insect in a drop of lactophenol blue or lactophenol acid fuchsin. Scanning electron microscopy (SEM) can be used to obtain information on the spatial distribution of propagules, but preparation of specimens for electron microscopy can be very labor intensive and electron microscopes are not readily available. However, the growing availablity of environmental SEMs has greatly simplified sample preparation, and these microscopes may prove useful in estimating propagule densities of entomopathogenic hypocrealean fungi.

The use of tissue specific and non-specific dyes offers enormous potential for direct enumeration of hypocrealean propagules. Conidia of *Lecanicillium* spp. stained with the fluorescent dye Uvitex were enumerated on the integument of aphids using epifluorescence microscopy in conjunction with image analysis (Girard & Jackson, 1993) and FDA was used to observe conidia of *I. fumosorosea* germinating directly on sweet potato whitefly nymphs or on foliage itself (R.L. Schading, unpublished observations cited in Schading *et al.*, 1995). Stains can be rendered tissue-specific by conjugation to tissue-specific antibodies. The most common method is to produce polyclonal antibodies in rabbits or chickens to protein or carbohydrate antigens on the surface of cell walls of propagules and/or hyphae. Briefly, the antibodies are bound to cell wall antigens, unbound antibodies are removed, the tissue is exposed to anti-rabbit or anti-chicken IgG conjugated with the fluorochrome, fluorescein isothiocyanate (FITC), the tissue is re-washed and examined microscopically. However, the cuticle of many insects autofluoresce thereby obscuring the propagules stained with fluorescent dyes. *In situ* quantification of hypocrealean fungi with fluorescent dyes has been most successful on small, soft-bodied insects (e.g., aphids). More details on use of fluorochromes are presented by Butt (1997).

*2. Indirect enumeration*

While direct enumeration techniques may increase the sensitivity of enumeration and provide information on the spatial distribution of propagules, these methods are considerably more labor intensive than indirect methods. Furthermore, direct enumeration can not be used in situations where physical or microbial contaminants obscure the propagules being enumerated. For indirect enumeration, the propagules must first be recovered. Recovery methods differ according to the substrates on which the propagules occur.

*a. Recovery—insects*

Normally, living insects should be killed before they are processed; mechanical injury, exposure to $CO_2$, or freezing have all been used to kill insects, but the method selected should have a negligible effect on the fungus. Propagules must first be either washed from the surface of the insect or obtained through homogenization of the host. Sonication has been shown to efficaciously remove bacteria from leaf surfaces but the use of sonication for the recovery of entomopathogenic Hypocreales from insect integuments has not been well studied.

*(1). Washing.* Although propagules of entomopathogenic fungi may be strongly attached to the cuticle of insects, they can be dislodged by vigorous washing and/or homogenization. To recover propagules by washing, insects are shaken in the wash solution using a vortexer or mechanical shaker. The wash solution may be water or buffer amended with a surfactant (e.g., 0.1% Silwet L77 which is much more efficient than Tween 80, S. Jaronski, personal communication) but caution must be used in selecting a surfactant since they can be toxic to fungi. Both reciprocal and rotary shakers are used but the former is preferred since it provides a more vigorous agitation of insects. If a rotary shaker or vortexer is employed, it may be necessary to add coarse sterile sand or small glass beads to the wash solution to aid in the mechanical detachment of propagules from the integument. The speed at which the shakers are operated should be as high as possible (> 200 rpm) without disrupting the insect integument or killing fungal cells. The duration of the washing step using shakers also varies, but times in excess of 1 or 2 h are typical.

The volume of the wash solution and the type of vessel used also influence the efficacy of the wash procedure. Wash volumes in excess of 5 ml in relatively wide-diameter containers result in the greatest turbulence of the wash solution. Since relatively large wash volumes are required, it may be necessary to include several insects to increase the density of propagules released into the wash solution. The differential recovery of propagules from the external integument and the relatively low labor requirements are the primary advantages to the wash method. With insects that readily release their foregut contents, it may be necessary to coat the mouthparts (e.g., with molten paraffin) prior to washing (Inglis *et al.*, 1996a). The primary disadvantage to the wash method is the decreased sensitivity of the technique as a result of the large wash volumes required and recovered colonies usually represent a conservative estimate of inoculum densities due to clumping of propagules and/or inadequate detachment from the integument.

*(2). Homogenization.* The homogenization method may be more efficacious than the wash method for recovery of propagules that are strongly attached to the cuticle or that are located in relatively inaccessible locations (e.g., cuticular folds). Since insects may be homogenized in relatively small volumes, the sensitivity of the homogenization method is generally greater than the wash method; the increased sensitivity facilitates measurements of inoculum densities between individual insects in the population. Furthermore, the homogenization method is conducive to the recovery of propagules from specific tissues (e.g., excised alimentary canals).

Homogenization of insects is usually accomplished using pestles or mechanical grinders (see Section 2A1), but bead beaters may also be used; when using bead beaters care must be taken to ensue that fungal cells are not disrupted and to avoid excessive heat build-up. When homogenizing insects, the effect of antimicrobial agents released in insect hemolymph must be considered. In this regard, a chilled buffer should always be used and the homogenate processed as rapidly as possible after disruption of the insect.

Disadvantages to the homogenization method include the inability to distinguish between propagules on the integument surface with those in the hemolymph and/or alimentary canal following the homogenization of whole insects; the possible inactivation of propagules due to encapsulation and/or melanization resulting from liberation of the hemolymph; conservative estimates of populations due to clumping and strong attachment of propagules to the integument; and the inability to distinguish between propagule types (e.g., conidia, hyphal bodies, or hyphal fragments).

*b. Recovery—foliage and spore traps*

Quantification of propagule density per unit area is often required in laboratory and field experiments, particularly following spray application. Propagules can be recovered from a number of substrates that possess a defined or measurable area. Two examples of substrates used to assess propagule densities include glass coverslips and leaves. Round coverslips ($\approx$ 10–15 mm diameter) are suitable since they are relatively robust and readily fit into containers used for washing. One to several coverslips can be attached to an object such as a plastic Petri dish lid with double-sided tape or other adhesive. The lids with attached coverslips can be placed at the base of a spray tower in the laboratory or on the soil surface in the field. If desirable, the height of the coverslips can be varied (e.g., relative to the plant canopy) by attaching the Petri dish lids to the tops of stakes with glue. The dishes should be removed as soon after application as possible and stored at low temperature ($\approx 5°C$) until they can be processed.

To recover propagules, coverslips are either pooled or washed individually. The wash solution is diluted, spread on to an agar medium and the number of CFU counted at the appropriate dilution. Populations are calculated as: ($n \times$ dilution factor)/area; where $n$ is the number of propagules at the desired dilution; area is usually reported in $cm^2$. In field settings, it is recommended that coverslips be placed at defined sites within each plot (subplots) to obtain a measure of spatial variation.

Measures of propagule density per unit area can also be obtained from foliage. Leaves are arbitrarily collected from a defined position in the canopy and stored at low temperatures ($\approx 4°C$) prior to being processed. If a leaf area meter is available, leaves can be cut into pieces of approximately the same size and shape, and total area measured after the wash step. If a leaf area meter is unavailable, a cutting device of defined area, (e.g., a cork borer) can be used to remove a disk of leaf tissue; this is considerably more time consuming and increases potential variation as a result of the smaller areas obtained from each leaf. The leaf pieces or disks are usually pooled prior to recovery. Propagules can be detached from the leaf segments using conventional washing, sonication, or homogenization. Once in suspension, propagules are diluted, CFU recovered on an appropriate medium, and propagule densities per unit area calculated as described below (see Section 4A2e below).

To obtain a measure of the spatial distribution of propagules on leaves, it may be possible to use a leaf imprint technique. Leaves are uniformly pressed against the surface of an appropriate agar medium, the outline of the leaf is marked, and the position of CFU recorded after incubation of the cultures.

It is often useful to relate propagule densities to droplet deposition data. To obtain a measure of droplet deposition, water- or oil-sensitive papers or strips can be used. They are usually placed in proximity to the coverslips or leaves from which propagule populations are quantified. Water-sensitive cards can not be used under conditions of high humidity, precipitation or where they may come into contact with water on vegetation or supports (e.g., dew). For qualitative assessments of size and density, cards can be visually compared to standards. For quantitative estimation of droplet density, the number of droplets per unit area (usually 1 $cm^2$) can be counted with the aid of a magnifying lens or stereomicroscope. In addition to droplet densities, image analysis systems can provide information on droplet area, size and percent coverage (Inglis *et al.*, 1997).

*c. Recovery—soil habitats*

The soil dilution spread-plate method is the most common technique used for quantifying propagules in soil. Soil should be collected to a defined depth with a soil corer. If possible (particularly in field experiments), it is recommended that samples be removed from a number of specific locations (e.g., subplots) within a sample area. Soil samples in plastic bags should be placed in a cool environment until they can be processed. The soil samples are usually mixed in bags and subsamples of soil are removed, weighed and homogenized in water or buffer (see Section 2A3).

Although any soil weight can be used, it is most common to place 10 g of soil into 90 ml of the homogenation liquid. The use of blenders to homogenize soil is relatively labor intensive and it is more difficult to maintain sterility. In a method designed to expedite the homogenation procedure, particularly when the objective is to recover conidia from soil, 1-g samples are vigorously vortexed in 9 ml of water or buffer for a specific period of time (usually between 30 and 60 s). Following homogenization, the slurry is diluted, aliquots spread on to a selective agar medium (see Section 2A2), and CFU are counted at the appropriate dilution.

*d. Recovery—aquatic habitats*

Water samples and/or insects can be collected with a variety of sampling devices (Atlas & Bartha, 1981). The water samples can then be concentrated (e.g., by centrifugation or filtration) or diluted as required, and CFU enumerated using the dilution spread plating method. Fresh water contains a number of potential fungal contaminants (e.g., Saprolegniales) and a selective medium should be used if possible. The relationship between inoculum density (in water or soil) and pathogenicity can also be quantified.

*e. Enumeration*

*(1). Dilution spread plating.* The most popular method for quantifying inoculum of entomopathogenic fungi on/in insect hosts or soil is based on dilution spread plating (Figure 7.11). Following dislodgement of propagules from insects, tissues or soil particles, the suspension is diluted as required (usually in a four- to 10-fold dilution series). The dilution method usually provides a conservative estimate of populations in soil primarily due to the attachment of propagules to soil particles and/or to conidial aggregation. When densities of entomopathogenic hypocrealean propagules are low, the dilution method may lack sensitivity primarily due the relatively large initial dilution (dilution factor = 50−100).

For spread plates, aliquots of 100−200 μl are generally spread on to an appropriate medium; for most applications the use of a selective medium is recommended, particularly one that prevents the growth of bacteria and fast-growing fungi (see Section 2A2). The liquid should be allowed to absorb into the medium before cultures are transferred to an appropriate temperature (20−25°C for most taxa). To obtain a measure of variation due to plating, it is recommended that two or more dishes be prepared at each dilution. Dishes are incubated for 3−7 days depending on the medium or organism, and since there is evidence that light exposure can inhibit colony development in some taxa (Chase *et al.*, 1986), cultures should be maintained in the dark as a precaution.

After an appropriate amount of time, colonies are counted at the dilution yielding 30−300 CFU per dish. To increase the accuracy of the counts (especially for smaller colonies), a 2- or 3 × stereoscope microscope may be used. Densities are usually presented as CFU/g of insect or per insect (standardized by weight) or as CFU/g of soil dry weight. An electronic moisture analyser can be used to determine the moisture content of soil. Alternatively, a weighed subsample (10 g) is dried at ≈ 110°C for a minimum of 12−24 h and the percentage of water in the sample is calculated. The densities of conidia per unit fresh weight are then adjusted accordingly.

In an attempt to reduce error that may occur during dilution and/or plating, and to expedite the procedure, mechanized plating systems have been developed. Spiral platers, initially shown to be effective for bacteria (Gilchrist *et al.*, 1973) have been used for enumeration of entomopathogenic Hypocreales (Fargues *et al.*, 1996). Image analysis systems can expedite the enumeration of colonies and provide a permanent record of cultures. However, as a result of the hyphal growth characteristics, and the relatively rapid growth rate exhibited by many entomopathogenic Hypocreales, it is often necessary to make 'judgement calls' when colonies grow together or are found in close proximity to each other. Also, obscurement of hypocrealean fungal colonies by other fast growing fungi (e.g., Mucorales) can be an issue when using spiral platers.

*(2). Most probable number.* An alternative to the spread plate method is the MPN technique (Woomer, 1994). The MPN method was developed for the enumeration of bacteria in water and has been adapted for use with filamentous fungi (Figure 7.11). As with the spread plate method, it is first necessary to separate the fungal propagules from the substrate. Once in suspension, the sample is diluted to the point of propagule extinction. Traditionally, dilutions were usually four- or 10-fold, with three to 10 replicate samples per dilution; however, a variety of dilution series and number of replicate samples can be used. Although a liquid medium may be used, a solid substrate is generally used for filamentous fungi. As with the spread plate method, it is desirable to use a selective medium (see Section 2A2). A physical barrier is usually required to prevent spreading of the suspension across the agar surface; this is accomplished

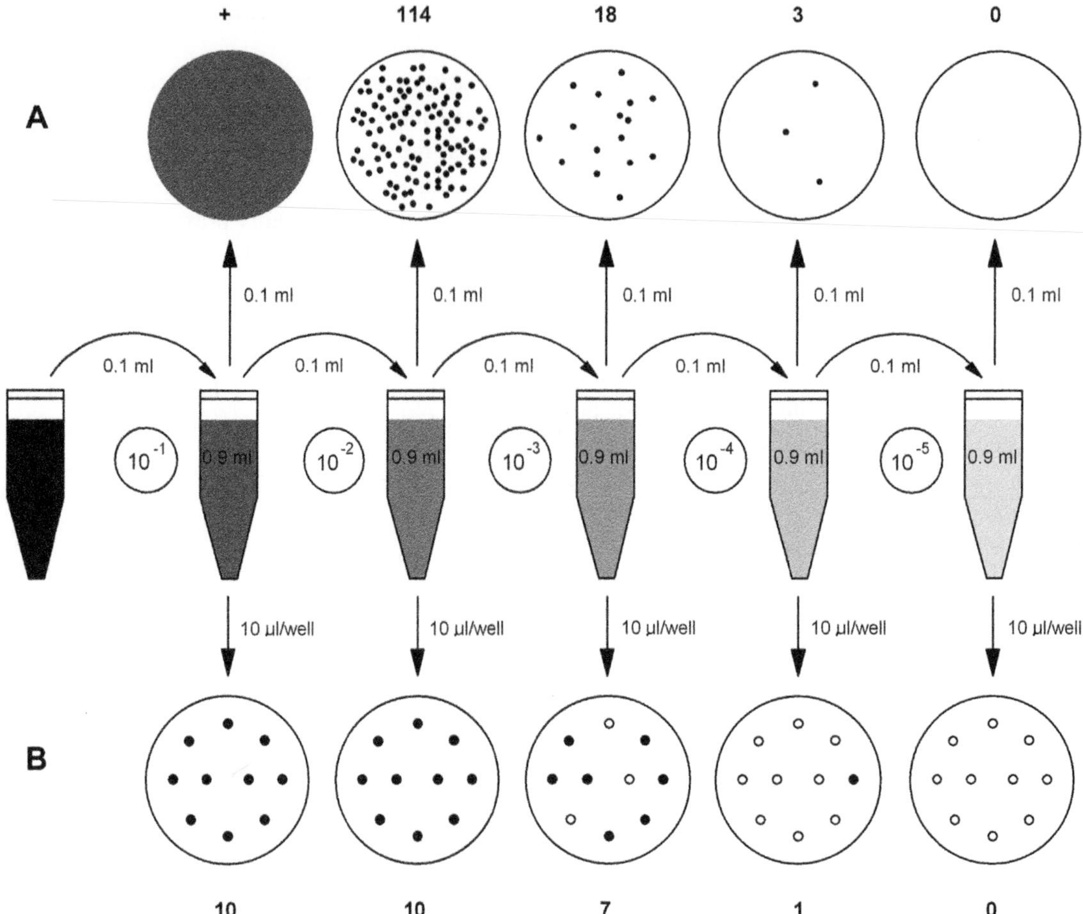


**Figure 7.11** Enumeration of propagules using the dilution plating and most-probable number (MPN) methods. The original suspension is diluted in a 10-fold dilution series by placing 100 µl in 0.9 ml of water or buffer; each dilution is vortexed ($\approx$ 10 s) before the transfer is made. For the dilution plating method A, 100 µl of the suspension is uniformly spread on the surface of an agar medium, and the culture is incubated. After an appropriate amount of time, colonies are counted at the dilution yielding 30–300 CFU per dish. The culture from the $10^{-1}$ dilution contains too many colonies to obtain an accurate count; colonies were confluent and crowded each other. The $10^{-2}$ dilution culture contained 114, well-distributed and easily counted colonies. The culture from the $10^{-3}$ dilution contained 18 colonies, seven more than would be predicted from the $10^{-2}$ culture. The percentage error in each colony in such a high dilution makes such a count unreliable but duplicate or triplicate cultures may be used to minimize this error. The number of viable propagules per ml = number of colonies × tube dilution factor × plating dilution factor. Using the $10^{-2}$ culture, propagules per ml = 114 × 100 × 10 = $1.14 \times 10^5$ propagules per ml. For the MPN method B, 10 replicate samples (10 µl) are placed on the agar medium from each dilution tube; wells are made with a cork borer to prevent spreading of the carrier. Three consecutive dilutions are selected at or immediately before propagule extinction and the number of positive wells are recorded. In this case the dilutions used may be $10^{-2}$–$10^{-4}$ (10–7–1) or $10^{-3}$–$10^{-5}$ (7–1–0). The MPN (per propagules in the first dilution used) are acquired from a MPN table (e.g., Meynell & Meynell, 1970). If $10^{-2}$ is used as the minimum dilution, the MPN is 11.6. If $10^{-3}$ is used, the MPN is 1.16. The number of viable propagules per ml = MPN × inverse of the minimum dilution × plating dilution factor. Since 10 µl was placed in each well, the plating dilution factor is 1 ml/0.01 ml or 100. Using the $10^{-2}$ as the minimum dilution, propagules per ml = $11.6 \times 10^2 \times 100 = 1.16 \times 10^5$ propagules per ml. In this example, the dilution plating and MPN methods provide a similar estimation of propagule density.

by placing a glass or polypropylene ring in the medium, or by making a well in the agar surface with a cork borer. Volumes placed within the wells will vary according to the size of the well and the absorptive properties of the medium. The cultures are incubated and after an appropriate amount of time, the wells are scored as positive or negative for growth of the target fungus. The number of positives and negatives at the dilutions before extinctions are used. From MPN tables or calculators based on a Poisson distribution, numbers of viable propagules can be obtained and density of propagules per unit volume or weight is then calculated (Meynell & Meynell, 1970); a number of MPN calculators are now freely available that conveniently allow researchers to vary the dilutions and the numbers of replicates employed. Although it is considered to be less accurate than the dilution spread plate method, the MPN method may provide comparable results (Harris & Sommers, 1968). The salient advantages of the MPN technique are its greater overall flexibility and scope of application. For example, microorganisms with specific attributes can be readily identified (e.g., fungi with chitinase activity if a chitin medium is used).

*(3). Host assays.* Another approach for enumeration of fungal propagules is through the use of bait insects. This technique is relatively labor intensive and is used principally for fungi for which there is no selective medium available or for aquatic samples. Susceptible insects are exposed to the propagule infested substrate (e.g., leaf surface, soil homogenate or water sample) for a defined period in a controlled environment, and the incidence of mortality is recorded periodically. Concentrations of propagules per unit area or volume of infested substrate/water can then be estimated according to dose—mortality data obtained under similar conditions with known numbers of propagules.

Another approach which can be used to quantify infective units in both water and soil is the MPN method [see Section 4A2e(2) above]; this technique is frequently used for assessing infective units of filamentous fungi that are obligate parasites (e.g., Ciafardini & Marotta, 1989). Although the MPN method is subject to many of the same limitations as the previously reported host assay method, it does not rely on a dose—mortality curve to estimate densities. Normally the insect pest of interest is used, but the sensitivity of the test is increased by using a highly susceptible insect host. For example, *G. mellonella* larvae are more sensitive to infection by entomopathogenic Hypocreales in soil than are soil-inhabiting insects (see Section 2A3c). The methods used are analogous to those described previously for enumeration of fungal propagules on an agar medium. The substrate (e.g., water or soil) is serially diluted and insects are exposed to the diluted substrates for a defined period of time. The dilution where no mortality is observed is selected, and the number of infective units (IU) is determined from MPN tables/calculators, and IU per volume of water or per unit weight of soil are calculated.

## B. Molecular methods

### *1. Quantitative PCR*

#### *a. Overview*

Quantitative real-time PCR (qPCR) is a very powerful technology that allows researchers to detect and quantify specific nucleic acid sequences in a sample. It is used for a variety of applications including detection and quantification of microorganisms in environmental samples, or quantification of gene expression. A major advantage of qPCR is the large dynamic range of quantification which may be up to six orders of magnitude. The technology is and has been applied extensively in many fields of biological research during the past decade. However, to date only a few studies have used qPCR to investigate the biology of entomopathogenic hypocrealean fungi. For instance, it has been used to detect and quantify specific strains or species in insect hosts, (Bell *et al.*, 2009; Anderson *et al.*, 2011), on plant surfaces (Castrillo *et al.*, 2008), and in soil; (Castrillo *et al.*, 2008; Schwarzenbach *et al.*, 2009), or to ascertain expression of genes involved in germination, conidiogenesis and pathogenesis (Fang & Bidochka, 2006; Peng *et al.*, 2009). As a result of the hyphal growth form of fungi, conventional methods such as the dilution spread-plate method provide a gross underestimation of fungal biomass. In this regard, the application of qPCR can provide a much more accurate measure of biomass and will provide key information on aspects of the biology of hypocrealean fungi. In this section, we provide a general overview on qPCR technologies and discuss the most relevant applications. For further reading and discussion of the topic, the reader should consult any number of references on qPCR (e.g., Bustin, 2004; Smith & Osborn, 2009; St Leger & Wang, 2009; Stock *et al.*, 2009).

*b. General strategies for quantitative PCR*

Real-time qPCR relies on the amplification of a target locus similar to conventional PCR. However, in real-time PCR quantification, the synthesis of DNA is monitored over time using fluorescence-based chemistries. qPCR requires the use of a specialized thermocycler that allows detection of fluorescence during amplification. The course of the reaction in qPCR and conventional PCR is identical and can be divided into two phases, the exponential phase in which the product accumulates exponentially, and the plateau phase in which the accumulation slows down because reaction components are consumed and become limited (Figure 7.12). During the first phase, fluorescence initially remains at background levels and the increase in signal is not detectable, even though the product accumulates exponentially. Eventually, a product concentration is reached that produces a fluorescence signal above the background. To allow comparison of individual reactions in a run, a threshold signal intensity is set. This threshold signal is determined by equipment-specific algorithms and represents the lowest reliable signal intensity exceeding the background in all samples. The number of cycles necessary to reach this threshold signal is defined as the threshold cycle (Ct) and it is determined for each sample. Small numbers of template at the start of the reaction will result in high Ct values and large numbers of template at the start of the reaction will result in small Ct values. It is this relationship that forms the basic concept for quantification with real-time qPCR.

There are two primary strategies employed to quantify the starting amount of template in a reaction: (1) absolute quantification; and (2) comparative quantification. Absolute quantification utilizes a standard curve from which the concentration of template in the sample is interpolated. The standard curve is generated in parallel to the sample amplification and consists of a serial dilution of a known concentration of a cloned target DNA (plasmid) or genomic DNA for instance. Comparing the Ct value obtained for each sample to the standard curve (i.e. Ct values plotted against log of the starting amount of standard template), allows calculation of the amount of starting template DNA in the sample. Results are

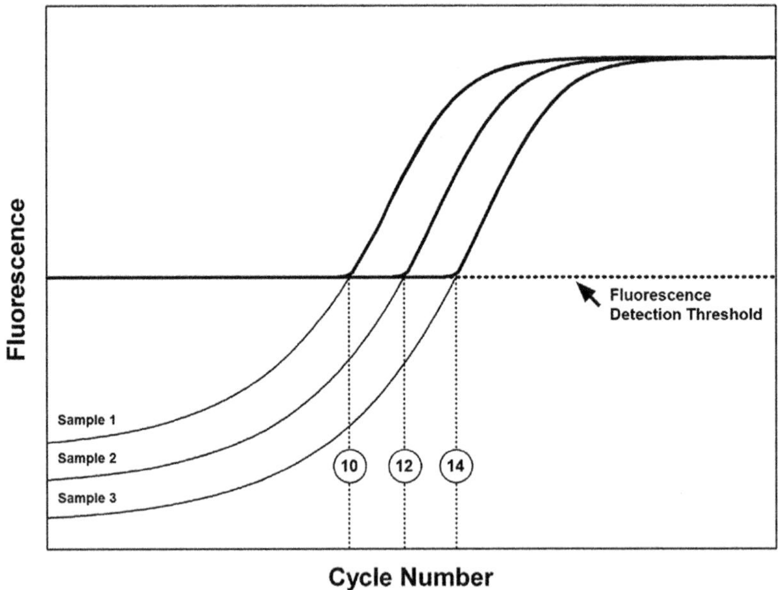


**Figure 7.12** Principles of real-time qPCR fluorescence quantification of target DNA in three samples possessing different target DNA concentrations (i.e. copy numbers). During the PCR process, DNA concentration of the target increases exponentially. The cycle number when the DNA concentration reaches the fluorescence detection threshold (i.e. the point at which the concentration can be reliably inferred from fluorescence intensity) is termed the threshold cycle (Ct). The Ct is inversely proportional to the log of the initial copy number of the target, and thus the lower the Ct value the higher the initial copy number in the sample; in this example, Ct values for samples 1, 2, and 3, are 10, 12, and 14, respectively. To convert Ct value to an actual value (e.g., copy number per unit volume), a standard curve of known concentrations is run simultaneously with the samples, and the copy number per unit volume interpolated from the standard curve.

typically expressed as copy numbers or micrograms per amount of sample (e.g., cell numbers, tissue or sample weight). Comparative quantification is applied, for instance, when comparing the expression of a gene in different strains or in different growth stages of a fungus (pathogenesis, conidiogenesis, germination). In this case, qPCR is used to determine the amount of mRNA of the target gene relative to the amount of mRNA of a calibrator gene, which in most cases is a housekeeping gene that is expressed in constant levels (e.g., actin, glyceraldehyde 3-phosphate dehydrogenase, translocation elongagation factor 1-$\alpha$). qPCR applied to quantify mRNA is also referred to as reverse transcription qPCR (RT qPCR, not to be confused with RT-PCR as the abbreviation for real-time PCR) because the initial step involved is the reverse transcription of the mRNA into cDNA. The cDNA is subsequently amplified in qPCR. Amplification of the gene of interest and the calibrator gene can be performed in separate reactions or in one reaction (multiplex PCR) if different dye systems are available to distinguish the different PCR products.

Two basic types of chemistries can be applied in real-time PCR quantification. The first is referred to as nonspecific chemistry. It relies on DNA intercalating dyes that emit fluorescent light when bound to double-stranded DNA. The amount of fluorescent signal obtained is proportional to the amount of DNA produced and it increases in parallel with product accumulation. SYBR® Green is the most common intercalating dye used. The second chemistry that is commonly used is referred to as probe-based chemistry. For this, an additional oligonucleotide (probe) is required, which hybridizes to the target between the two PCR primers. Various different probe-types have been devised and the most frequently used one is the TaqMan probe (Bustin, 2004). TaqMan probes are labeled with a fluorescent reporter at the 5′-end and a quencher at the 3′-end. No fluorescence is emitted as long as the probe is intact (i.e. the quencher is in close proximity to the reporter and quenches the fluorescence emitted by the reporter). The TaqMan probe hybridizes to the target during the annealing and extension step of the qPCR. The $5' \rightarrow 3'$ exonuclease activity of the *Taq* polymerase releases the reporter and the fluorescence is no longer quenched. The resulting increase in fluorescent signal is proportional to the amount of product accumulating in the reaction.

The primary advantages of a nonspecific chemistry like SYBR® Green over a probe-based approach like TaqMan are the relatively low cost, the simplicity of the assay as only two primers (no additional probe) are needed, less optimization is required, and the availability of the option to perform high resolution melting analyses. The probe-based chemistry is more complex to develop as it requires the development and validation of a specific probe. However, the use of a probe often results in increased specificity compared to SYBR® Green. Furthermore, it allows quantification of multiple targets in one reaction (multiplex PCR); probes with specificity to different target genes are labeled with different fluorophores emitting fluorescent light at different wavelengths which allows quantification of different target genes simultaneously.

The same issues that are important when developing and applying cultivation-independent diagnostic PCR are also critical when developing and applying qPCR to detect and quantify a target strain or taxon in complex environmental samples. In the following we focus on the issues that are particularly relevant to qPCR. As described above, effective and efficient DNA extraction and avoidance of PCR inhibition are crucial when working with environmental samples and all the procedures require careful optimization to allow accurate qPCR-based measurements. It is crucial to perform the necessary control steps using for instance spiking approaches and internal amplification controls as described above.

Another critical issue involves the primers used for amplification of the target locus. Primers should be designed and validated according to the considerations outlined above. Ideally the size of a qPCR amplicon should be 50–150 bases for optimal PCR efficiency. However, in many cases, primers have to be designed based on availability of sequence signatures, which may limit flexibility in selection of the product size. The primer Ta should be between 55 and 60°C, and with a TaqMan probe, the Ta of the probe should be ≈ 10°C above the Ta of the primers. Also, the Ta of both primers should be within 1°C. It is also important to carefully evaluate the comprehensiveness and specificity of the primers applying the procedures described above for the evaluation of primers used in cultivation-independent diagnostic PCR. In addition to these procedures, specificity of the amplification and product identity can be tested using high-resolution melting (HRM) analysis, a technique available when performing qPCR with nonspecific chemistry. In HRM analyses, the denaturation (melting) of a PCR product using slowly increasing temperature is monitored. DNA intercalating dyes used in nonspecific chemistry emit fluorescence when bound to double-stranded DNA, therefore the extent of melting can be monitored by the decrease in fluorescence. The resulting melting curve is characteristic for a PCR product of a specific sequence, and provides information on product identity and specificity of the amplification. HRM analyses are fast and simple to perform and can be routinely applied as a control. However, it is important to emphasize that HRM analysis alone is insufficient to prove specificity. Sequencing of PCR products obtained with amplifications from selected samples should always be part of the specificity assessments (see Section 2B1b).

When running qPCR assays, it is important to evaluate the performance, which includes assessment of the reproducibility, efficiency, sensitivity, and dynamic range of the PCR. The reproducibility can be assessed by comparing the Ct values of replicate reactions. Efficiency, sensitivity and dynamic range are assessed based on standard curves, which consist of serial dilutions of a known template (nanograms of genomic DNA or copies of plasmid DNA containing the target locus) either alone or spiked into target-free complex DNA samples when working with environmental samples. The standard curve (i.e. the plot of Ct values *versus* log of starting amount of template) should be linear with a correlation coefficient of $R^2 > 0.980$. From this curve, the PCR amplification efficiency can be calculated, which typically is done using equipment specific software. In a PCR with 100% amplification efficiency, the amount of product is doubled in every cycle of the exponential phase. With an optimized reaction, amplification efficiencies between 90 and 105% should be reached. Furthermore, it is important that the standard curve covers the entire range of target concentrations in the test samples to show that results obtained from the test samples are within the linear dynamic range of the assay and the assay provides the necessary sensitivity. In this context it is important to consider that PCR sensitivity of a specific assay also depends on the copy number of the target locus (i.e. does the locus occur in single or multiple copy per genome). This is the case when targeting genes or regions of the rRNA gene cluster which belong to the most preferred loci used for qPCR. The fact that the number of rRNA gene cluster copies can vary even among different isolates may add additional complications (Herrera *et al.*, 2009). Despite the limitations of targeting rRNA genes, quantitative data provided by qPCR is typically within one order of magnitude in accuracy and can provide valuable information that is unattainable by other methods.

*c. Viability quantitative PCR*

A considerable disadvantage of conventional PCR as well as qPCR is that these techniques do not allow distinction between viable or non-viable cells. However, such information might be crucial when monitoring entomopathogenic fungi in biological control. One approach to address this problem has been to apply RT qPCR to target mRNA, which in contrast to DNA, has a rapid turnover in viable cells and is quickly degraded in dead cells (Bleve *et al.*, 2003). An alternative to this approach is the use of the intercalating dyes, ethidium monoazide (EMA) (Rudi *et al.*, 2005) or propidium monoazide (PMA) (Vesper *et al.*, 2008) to distinguish between cells with an intact membrane (viable) and cells with a disrupted membrane (non-viable). Intercalating dyes such as EMA and PMA possess the ability to penetrate the cell membrane of dead cells and covalently bind to DNA following irradiation, thereby preventing the annealing of primers to the DNA. As a result, PCR products are obtained only from viable cells. However, there are two important aspects one must be aware of as they might affect reliability of the method. First, dyes to some extent may also penetrate membranes of viable cells, thereby providing an underestimate of viability. Secondly, particles potentially present in the sample may interfere with the irradiation step and as a result, PCR amplification may also occur from non-viable cells. Recent evidence suggests that PMA is superior to EMA as it penetrates cells possessing intact cell membranes to a lesser degree than EMA. Viability qPCR has proven itself to be an exceptionally powerful and useful high-throughput method to quantify living microorganisms in a variety of habitats. This method was originally developed to quantify viable bacteria, and to our knowledge, it has not yet been applied to entomopathogenic hypocrealean fungi. However, viability qPCR has been used to quantify yeasts (Andorrà *et al.*, 2010), and it is expected that the application of this method to entomopathogenic fungi will provide valuable information on their biology.

## *2. Fluorescence* in situ *hybridization (FISH)*

Fluorescence *in situ* hybridization is a non-sequencing-based method that relies on the hybridization between a target sequence of DNA or RNA with a fluorescently labeled probe (DNA or RNA) of complementary sequence to the target. This method allows detection/estimation of biomass in complex communities and substrates using epifluorescence and/or confocal laser scanning microscopy, or by flow cytometry. When targeting ribosomal RNA or DNA, in theory, each cell containing rDNA or rRNA will be stained by the probe molecule during the hybridization procedure. Ribosomal RNA is typically targeted as opposed to rDNA (rRNA gene) to increase sensitivity (i.e. due to higher copy numbers). A typical FISH procedure involves four steps: (1) fixation and permeabilization; (2) hybridization; (3) washing to remove unbound probe; and (4) detection/quantification of labeled cells. Samples must be fixed (e.g., on glass slides, on membrane filters, or in paraffin sections) to prevent degradation of nucleic acids, particularly RNA, and permeabilization of the cells is necessary to allow penetration of labeled probes. A number of strategies have been used to label probes. The most common strategy is to use direct fluorescent labeling because it is the fastest and cheapest, and it does not require additional sample processing. FISH allows researchers to visualize specific fungi within substrates, providing valuable information on biomass and spatial distributions. For microscopy, both cryohistology and paraffin histology can be used, but the former is preferred. Quantification of cells can be accomplished using image analysis software in conjunction with fluorescence microscopy or by flow cytometry (FCM-FISH).

As with any method, FISH possesses inherent weaknesses/limitations. The RNA or DNA sequence of the target must be known to allow the development of complementary probes. In this regard, it is essential that the specificity and sensitivity of the probes be carefully validated before they are used to avoid inaccurate results. Validation should involve *in silico*, *in vitro*, and *in vivo* evaluations. It is important to realize that probes are often limited in their capacity to distinguish single nucleotide changes complicating the ability to discriminate between closely related taxa. Additional issues include: (1) the target may be inaccessible to hybridization (e.g., due to rRNA–rRNA and rRNA–protein interactions); (2) not all sites within the ribosome are equally accessible (e.g., blocked by rRNA secondary structure); (3) the GC content of the probe influences stringency of hybridization; (4) large probes can cause high background fluorescence but small probes may lack hybridization specificity; (5) small probes more readily penetrate cells but might carry fewer labels thereby affecting the intensity of fluorescence; (6) not all cells are equally permeabilizable using standard fixation methods; (7) cells present in small numbers may have signal intensities that are below detection limits or lost in high levels of background fluorescence; (8) background fluorescence and/or autofluorescence can obscure detection; and (9) the method does not necessarily distinguish living from dead cells. A number of advances have helped mitigate some of these issues. Examples include the development of brighter fluorochromes, improved probe design software, use of agents that overcome the variable and sometimes insufficient penetration of probes (e.g., peptide nucleic probes), and substantially better instrumentation. While FISH can provide unique and valuable information on its own, it is a particularly valuable *in situ* tool when employed in conjunction with other methods (e.g., qPCR).

### *3. Enzyme-linked immunosorbent assay (ELISA)*

Although DNA- and RNA-based methods (e.g., qPCR and FISH) have reduced the popularity of antibody-based detection/quantification methods in microbiology, enzyme ELISA may still have applications to quantify fungal biomass. ELISA is used to quantify the presence of a particular antigen or an antibody in a sample. However, in the study of entomopathogenic hypocrealean fungi, the detection/quantification of antigens is of primary importance. Here we define an antigen as any molecule that binds specifically to an antibody (e.g., the immunoglobulin, IgG). Although antigens are primarily proteins, other types of molecules such as carbohydrates and secondary metabolites can also bind specifically to antibodies. There are four basic types of ELISA: (1) direct ELISA; (2) indirect ELISA; (3) sandwich ELISA; and (4) competitive ELISA; with variations within each type. In this chapter we restrict our description to direct and indirect ELISA, and the reader should consult the literature for information on the application of other ELISA strategies. For both direct and indirect ELISA, an antigen is affixed to a surface, typically individual well surfaces in 96- or 384-well microtiter plates. This is often referred to as antigen ‘immobilization’. The primary distinction between direct and indirect ELISA is that only a primary antibody is used in direct ELISA, whereas both a primary and secondary antibody are used in indirect ELISA. A primary antibody is an antibody that binds directly to the antigen. For direct ELISA, the primary antibody is conjugated with an appropriate reporter enzyme such as horseradish peroxidase or alkaline phosphatase (i.e. a labeled antibody). Other reporter enzymes can be used as well, but have not gained widespread acceptance. For indirect ELISA, the secondary antibody is conjugated with a reporter enzyme, and the secondary antibody is specific for the primary antibody (e.g., goat anti-rabbit). For both direct and indirect ELISA, it is critical that any unbound enzyme conjugate is removed and this is achieved by washing the microtitre plate with detergents. Furthermore, any plastic surface in the microtitre plate well that remains uncoated by the antigen must be 'blocked', and this is typically achieved by flooding the well with bovine serum albumin or casein after immobilization of the antigen. The quantitative feature of ELISA is dependant on the ability to detect how many reporter antibodies have bound directly to antigens in the case of direct ELISA or to primary antibodies bound to antigens in the case of indirect ELISA. This is typically accomplished spectrophotomically. To achieve this, an appropriate substrate is added to the microtitre plate containing the antigen and enzyme-labeled antibody. The enzyme conjugated to the antibody bound directly or indirectly to the antigen then catalyzes the breakdown of the substrate resulting in a quantitative color reaction that is measured using a spectrophotometer. There are a number of salient advantages of using indirect ELISA instead of direct ELISA. These include: a large diversity of labeled secondary antibodies are available commercially; the method is very versatile because the same secondary antibody can be used regardless of the primary antibody to be detected in a single species (e.g., a rabbit); maximal immunoreactivity of the primary antibody is achieved because it is not conjugated; sensitivity is increased because each primary antibody contains several epitopes that are available to bind the secondary antibody; and different visualization markers can be used with the same primary antibody. However, direct ELISA is easier to complete because there are fewer steps and cross-reactivity of the secondary antibody is not an issue. For this reason, direct ELISA has become the primary method used for immunohistochemical staining of tissues but is not extensively used to quantify microbial biomass.

The application of ELISA to detect/quantify entomopathogenic hypocrealean fungi primarily relies on the use of polyclonal antibiotics (i.e. as opposed to monoclonal antibodies). In the case of entomopathogenic fungi, very few antibodies are available commercially. Therefore, it will be necessary for the researcher to generate polyclonal antibodies; this can be done in-house or by using a commercial company that specializes in production of antibodies. To generate polyclonal antibodies, it is necessary to inject an appropriate animal with an antigen or antigens. While any antigen can be used, in the case of entomopathogenic hypocrealean fungi, crude antigen (e.g., whole cell homogenates) or specific antigen (e.g., cell wall-specific) preparations are used. Each of these strategies possesses relative advantages and disadvantages depending on the experimental objective. For example, crude antigen preparations may result in poor specificity and excessive cross-reactivity. Furthermore, crude antigen preparations may not provide meaningful quantitative results (e.g., if antigen concentrations are not consistent across cells). While specific antigens may provide superior specificity and quantitativeness, it is necessary to characterize targeted proteins in advance. For example, it is necessary to characterize the protein itself, ascertain expression and the temporal/spatial distribution of the protein, as well as confirm specificity experimentally (i.e. antibodies do not cross-react with antigens in/on non-target organisms). A number of 'clean up' procedures can be applied to purify antisera. For example, antisera can be purified using a saturated ammonium sulfate method, dialysis, and/or immunoaffinity chromatography. Undefined antigen targets, a requirement to utilize sentient animals to produce antibodies, the difficulty in distinguishing viable from non-viable cells are issues encountered when using ELISA as a quantitative method. As a result, many researchers have chosen to utilize DNA- and RNA-based methodologies such as qPCR as an alternative to ELISA (see Section 4B1). Although not used extensively, ELISA has been used to detect *M. anisopliae* at concentrations as low as $10^3$ spores/ml in soil (Guy & Rath, 1990).

## 5. INFECTION AND BIOASSAY

Methods for the rapid, systematic infection of insects for use in other studies (e.g., sporulation of fungi on cadavers), for studies on the mode of infection, or for the *in vivo* maintenance of a pathogen, seldom require precise dosing. The simplest method for such pathogen transmission is to expose healthy insects to cadavers from which the fungus has erupted and produced infectious propagules. This may be the only method to maintain fastidious, obligate pathogens, but entomopathogenic Hypocreales currently under consideration as microbial control agents are seldom obligate and fastidious. In contrast, in conducting bioassays, dose–time–mortality responses are necessary to compare differences between pathogen strains or species, or to evaluate the effects of factors such as environment, methods of targeting, formulation, or propagule type. Therefore, bioassays must be repeatable and strict control of the dose and dosing method (i.e. inoculation) must be exercised. A number of other factors such as host species, age and physiological condition, size of bioassay chamber, and incubation conditions (e.g., temperature, humidity, photoperiod) may also affect bioassay results and these factors should be measured/controlled in bioassays.

Although there is commonality amongst bioassays in general experimental design, no standardized bioassay systems exist for entomopathogenic Hypocreales. This is mostly due to the wide host range of many entomopathogenic hypocrealean fungi which necessitates the utilization of a variety of bioassay designs. Therefore, specific bioassays must be developed for most host–pathogen combinations. For a description of bioassay methods, see Butt & Goettel (2000).

## A. Experimental design and analyses

While bioassays can provide valuable information on the pathogen–insect–environment interaction, the value of the results obtained is dependent on the design, execution, analysis, and interpretation of results. The objective of this section is to summarize the salient aspects of bioassay design and analysis. For more detailed descriptions of the statistical analyses presented, the reader should consult the biometrics literature.

### *1. Definitions*

An *experimental unit* is the unit to which a treatment is applied (e.g., a group of insects); the term *plot* is synonymous with experimental unit. A *variable* is a measurable characteristic of an experimental unit and variables are either *dependent* (Y) or *independent* (X); independent variables are controlled by the experimenter. Dependent variables (i.e. measured response due to the effects of independent variables) are classed as *discrete* (i.e. real number values between specified limits) or *continuous* (i.e. any value between certain limits); discrete values are typically either counts or quantal (i.e. living or dead). A *treatment* is an agent or condition (independent variable) whose effect is to be measured and compared with other treatments. When a treatment is applied to more than one experimental unit, it is *replicated.* Replication is necessary to provide an estimate of experimental error (measure of variation among experimental units). The complexity of bioassay designs range from those that are relatively simple (e.g., one treatment is compared to another, usually a control treatment in a binary experiment) to those that are more complex (e.g., factorial designs). A *factor* is an independent variable and, in factorial experiments, each factor has at least two *levels* (several states within each factor). An *interaction* between two factors occurs if the level of one factor alters the impact of the other factor on the dependent variable.

### *2. Design and analysis*

The design of a bioassay is of paramount importance to the successful outcome, and one should consult a biometrician for advice before commencing the experiment. The initial step of any bioassay is to establish the objective(s) of the experiment and to formulate hypotheses to test (e.g., $H_0$ hypotheses). It is important that the concomitant attributes of the bioassay (e.g., environment variables) be pertinent and should be considered relative to the objectives and design of the experiment. Once hypotheses have been formulated, treatments and an appropriate design are selected to answer the questions posed. The statistical design employed should be the simplest design that provides an acceptable level of precision without compromising the need for replication, power, and appropriate tests. Proper randomization and replication are fundamental to appropriate experimental design, and extreme care must be exercised to ensure that replicates are not pseudoreplicates; pseudoreplicates are when the experimental unit is not an independent observation of a treatment. Inexperienced researchers often make the mistake of treating observations as replicates, thereby artificially increasing the degrees of freedom and the chance of committing a type I statistical error (i.e. falsely rejecting the $H_0$ hypothesis). Therefore, it is incumbent upon the researcher to ensure that replicates represent independent units.

Analysis of variance (ANOVA) is frequently used to evaluate the impacts of entomopathogenic Hypocreales on insects in bioassays. The two most common designs to which ANOVA is applied in bioassays are the completely randomized design (CRD) and the randomized complete block design (RCBD). There are two sources of variation in the CRD; among experimental units within a treatment (experimental error), and among treatment means. While the CRD may maximize degrees of freedom for estimating experimental error, it may be inefficient since the experimental error encompasses the entire variation among experimental units except that due to treatments. For this reason, the CRD is usually limited to bioassays that are conducted in uniform environments where experimental units are essentially homogeneous (e.g., controlled environment chambers). In most bioassays, the RCBD is preferred over the CRD since the RCBD incorporates a measure of variation among blocks (replicates). Variation accounted for by block differences can be removed from the total variation; the experimental error is typically the block by treatment interaction, which serves as a baseline against which variation among treatments can be assessed. Selection of the appropriate model is facilitated by listing all possible sources of variation and their associated degrees of freedom (e.g., ANOVA table), and identifying the tests to be conducted.

Most bioassays are influenced by a number of unmeasurable variables (e.g., dose estimation, the physiological status of the insects or of the fungus) and therefore, it is imperative that the bioassays be repeated at different times. One strategy is to repeat the entire experiment and compare results between trials. It is necessary to confirm homogeneity of variance (e.g., using Bartlett's test) before the trials are combined; variances associated with different treatment means should be homogeneous. Even when the variances between treatments are found to be homogeneous, the final decision on whether data from different trials should be combined rests with the researcher (i.e. whether the data between trials are consistent). In contrast to repeating the entire experiment, another strategy used in bioassays with entomopathogenic Hypocreales is to conduct each block of the experiment separately as a RCBD (single- or multi-factor experiment) with the blocks conducted at different times. Each of the blocks is carried out autonomously from each other and variation not attributable to the treatments can be removed by the model (e.g., block effect). However, if the experimental unit is not consistent, and this change influences the treatment effect, then this design may not be appropriate.

One of the assumptions of parametric statistics is that data are drawn from a normal distribution. To confirm this, data should be tested for normality, kurtosis and skewness (e.g., D'Agostino *et al.*, 1990; Univariate Procedure of SAS, SAS Institute Inc.). In conjunction with normality testing, residual *versus* predicted (ANOVA model) values should be plotted to determine that variances are homogeneous. In cases where the data are not normally distributed or variances are heterogenous, an appropriate transformation may be used. Log, square root, and arcsine transformations are frequently used to normalize distributions and/or homogenize data (Little & Hills, 1978). Data obtained from bioassays are often based on counts (e.g., number of dead insects) that are expressed as percentages or proportions of the total sample. These types of data are typically binomially distributed and variances tend to be larger in the middle ($\approx$ 50%) relative to the two ends of the range. Application of an arcsine transformation to such data may normalize the variances. However, the arcsine transformation will only be effective if the range of percentages is greater than 40% (Little & Hills, 1978) and most of the means do not lie between 30 and 70% (Snedecor & Cochran, 1989).

When evaluating quantal mortality data (i.e. percentage dead) it is useful to know if an insect died of mycosis or other causes. To remove effects not due to the entomopathogen, mortality observed in the control treatment (i.e. treated with the carrier alone) is considered using Abbott's formula (Abbott, 1925). Since some death would be expected in control treatment insects, the proportion of insects killed by the entomopathogen alone is:

$$P = [(C - T)/C] \times 100$$

where P is the estimated percentage of insects killed by the entomopathogen alone, C is the percentage of the control insects that are living, and T is the percentage of the treated insects that are living after the experimental period. Abbott's formula can also be applied to a variety of other types of data (e.g., propagule viability or germination tests). When applied to count data, Abbott's formula is also called the odds ratio or the cross products ratio (Johnson *et al.*, 1986; Schaalje *et al.*, 1986):

$$P = [1 - ((T_A/T_B)(C_A/C_B))] \times 100$$

where P is the estimated percentage of insects killed by the pathogen, T and C are count data from the treated and control treatments before$_B$ and after$_A$ treatment with the pathogen. A useful online tool for calculation of corrections for control mortalities using several methods is available at http://www.ehabsoft.com/ldpline/onlinecontrol.htm.

One may want to determine whether insects died of mycosis or other factors; colonization of cadavers by the fungus can be checked to confirm activity (see Section 5C below). While this may provide adequate results when the proportion of cadavers colonized is high, it generally provides a conservative estimate of mycosis; although non-colonized insects may have died from mycosis, the fungus could have been competitively excluded from the substrate by saprotrophic microorganisms. Extreme care should be exercised in interpreting these types of data.

Once the data have been collected and the assumptions for analysis of variance have been satisfied, the data may be analyzed as predetermined in the experimental design stage of the bioassay. Analysis of single factor experiments as either a CRD or RCBD is relatively straightforward. However, subsequent to a significant F-test for the factor in question (i.e. treatments), it may be necessary to determine which of the treatment means ($n \geq 3$) are significantly different. This is usually accomplished using a mean separation test at a selected $\alpha$-level (discrete variables) or with pre-planned comparisons using orthogonal contrasts; the least square means function of SAS is frequently used for unbalanced designs or for interactions between factors (e.g., in a factorial experiment). A mean separation test that controls both type I error (rejecting the $H_0$ when it is true) and II error (accepting the $H_0$ when it is false) should be used [see Jones (1984) for a comparison of means tests]. Even if mean differences are statistically significant, these differences should be deemed biologically "significant" by the researcher in order to be valid. Accurate interpretation of the experimental results is of paramount importance in the formulation of new hypotheses and in the implementation of subsequent experiments.

Factorial designs are used in many bioassays with entomopathogenic Hypocreales. This type of analysis allows the experimenter to determine the degree to which factors influence each other (i.e. whether an interaction exists between factors). Although the results obtained from factorial experiments may be more difficult to interpret than single-factor experiments, the information obtained on the interaction between factors is often important. For example, when comparing the efficacy of two entomopathogens across a number of doses, the interaction between taxa and dose may be more important to the researcher than the response of the individual factors alone. In the simplest type of factorial experiment, the same error term (residual error term) is used to test all factors.

In some experiments, a factor may not be independent. For example, if mortality is examined at time intervals in the same group of insects (i.e. time cannot be treated as an independent variable). In such instances, it is necessary to use a repeated measure statistic method. Advances in mixed model analysis such as those utilized in the mixed procedure of SAS (Littell) *et al.*, 2006; SAS Institute Inc.) are facilitating analysis of repeated measure data, and the use of mixed models have become the preferred method for analyzing temporal changes. The SAS mixed procedure handles variance heterogeneity among treatments and correlations among the data, and it assumes both a normal distribution and a linear model.

Regression analysis is frequently used to analyse the efficacy of entomopathogenic Hypocreales, particularly in instances where the relationship between dose and mortality is of interest to the researcher. Besides providing a measure of efficacy (e.g., lethal dose), this type of analysis may also provide important information on the mechanism of pathogenesis. The discrete data required for this type of analysis is quantal. Although probit, or logit transformations may be used to linearize the response, which is typically sigmoidal in its untransformed plot, a number of other models (e.g., log−log) can also be used (Robertson & Preisler, 1992). There is no evidence to indicate the superiority of the probit *versus* logit models, and both methods provide similar median lethal dose (LD) results (Robertson & Preisler, 1992). How well the data fit the assumptions of the model is called the goodness-of-fit and this is usually tested using a $\chi^2$ test; values predicted by the model are compared to actual values to derive this statistic. Additional information that is typically obtained from this type of analysis includes: $LD_{50}$ and $LD_{95}$ with 95% confidence intervals (95%); slope and standard error of the slope; and y-intercept of the regression. In bioassays with more than one treatment, the dose−response lines can be tested for parallelism and for a common y-intercept using log-likelihood ratio tests (Finney, 1971).

The two most important factors determining the power of dose−mortality analyses are dose selection and sample size. Selection of doses depends on the lethal dose of interest. For example, doses that provide a response between 25 and 75% are most useful for determinations of $LD_{50}$. The time at which data are collected is dependent on the researcher, and data collected at different times (e.g., day 5 and 6) can be analyzed separately. Sample size also influences the precision of the analyses. Robertson *et al.* (1984) concluded that 240 insects were required for a reliable response in a typical bioassay, although a sample size of 120 insects was adequate in most instances. As indicated earlier, it is important that dose−mortality experiments be repeated. Results from replicate bioassays can be compared using a 'common-line' model (Finney, 1971). A number of software packages (i.e. POLO, GLIM) that analyse dose−mortality data are available commercially or from non-profit organizations (Payne, 1978; Robertson *et al.*, 1984).

In time−dose−response experiments (i.e. disease progress), dose is kept constant and time is varied. This contrasts with dose−mortality experiments where the reverse is true (i.e. dose is varied but time at which mortality is assessed is constant). Time-course analysis provides a measure of lethal time, usually reported as the time at which 50% of the test insects have died. The use of probit- or logit-regression models to analyse time−mortality data is only valid if different groups of insects are used at each time. As indicated above, if the same group of insects is used, the data will be correlated and therefore analysis with standard probit techniques is invalid. In situations where it is not possible to obtain independent samples for each observation time (i.e. the number of insects is limited), methods that permit analysis of correlated response data must be used. Correlated data such as survivorship curves may be fitted to a Weibull function (Pinder *et al.*, 1978), and median lethal times with upper and lower 95% confidence limits estimated (e.g., using the SAS Lifereg Procedure, SAS Intstitute, Inc.).

Throne *et al.* (1995) described a method for analyzing correlated time−mortality data using complementary log−log, logit, or probit transformation of proportion insects killed (download program at http://www.ars.usda.gov/pandp/people/people.htm?personid=5643). In many instances, it is desirable to analyse both time− and dose−mortality data. Data from time−dose−response experiments are usually analysed by modeling time trends separately for each dose or by estimating dose trends separately for each time (Robertson & Preisler, 1992). However, Preisler & Robertson (1989) describe a method that estimates mortality over time in insects exposed to a series of increasing doses of insecticides (time−dose−mortality data); regression based on the complementary log−log model was used to analyse time trends for all dose levels simultaneously. Nowierski *et al.* (1996) applied the complementary log−log model to time−dose−mortality relationships for several entomopathogenic Hypocreales in grasshopper bioassays.

Analysis of covariance (ANCOVA) combines features of ANOVA and regression. Although ANCOVA is an extremely powerful technique, it has not been extensively applied to bioassays with entomopathogenic Hypocreales. Analysis of covariance can be used to remove variability associated with the dependent variable by including a concomitant variable in the model. The most common use of ANCOVA is to increase the precision in randomized bioassay experiments (Snedecor & Cochran, 1989). In such applications, the covariate (X) is a measurement (e.g., insect weight or dose) taken on each experimental unit before treatments are applied that predicts to some degree the final response of Y on the unit. By adjusting for the covariate, the experimental error is reduced and thus a more precise comparison among treatments is achieved. It is assumed that the slopes of the regression of Y and the concomitant variable or covariate do not differ significantly among the treatments. Analysis of covariance may also be used to adjust for sources of bias in bioassay experiments (Steel & Torrie, 1960; Snedecor & Cochran, 1989). For example, in studying the relationship between food consumption and dose, a measure of food consumption from insects treated with varying doses of the entomopathogen as well as the initial size of each insect are recorded and differences between the mean size of the insects exposed to the different doses are noted. If food consumption is linearly related to size, differences found in consumption among different dose treatments may be due, in part, to insect size. Size is consequently included in the ANCOVA model to remove bias.

Analysis of covariance can also be used to test for differences in regression relationships (intercepts and slopes) among treatments. Application of ANCOVA to dose−mortality data can be used as an alternative to traditional probit- or logit-analysis with $\chi^2$ tests. Using the general linear model (GLM) procedure of SAS, heterogeneity of slopes can be tested. The analyses can also test for differences in intercepts assuming a constant regression relationship among treatments (SAS Institute Inc.).

A less well-known statistical method that may be useful for analyzing bioassay data is the application of the competing risks theory. This theory deals with situations in which there is interest in the failure (or exit) times of individuals, where the subjects are susceptible to two or more causes of failure, and where the failure occurs over time (Johnson, 1992). Explanation of the models involved and several formulations of the theory with application to insect experimentation are provided by Schaalje *et al.* (1992).

## B. Inoculation

Presentation of a precise dose to the host is imperative for accurate, repeatable bioassay results. Whereas *per os* inoculation is required with most non-fungal entomopathogens, in contrast, most entomopathogenic Hypocreales require some form of inoculation of the integument. This, at times, can be difficult to accomplish with precision depending on the size and type of insect and the requirement to inoculate large numbers of insects. Therefore, rapid methods have been developed that simplify inoculation, yet still provide repeatable results even though the precise dose is not always known. Very indirect methods such as allowing the insect to walk on the surface of a sporulating culture have been used; however, such methods are crude and should be avoided, other than possibly for transmission requirements such as *in vivo* maintenance of cultures.

### *1. Injection*

Inoculation by injection is most often used when large numbers of infected insects are required, when studying internal immunological responses, or when attempting to maintain a pathogen in hosts. To inoculate insects internally, an aqueous suspension of propagules can be injected using a 1-ml tuberculin syringe fitted with a fine needle (e.g., 30 gauge) using a motorized microinjector to drive the syringe plunger. The volume of inoculum to be injected depends on the size of host insect; however, the smallest volume possible should be used to minimize any adverse impacts on the insect. The use of a motorized microinjector allows the inoculum to be delivered in a highly controlled manner (e.g., volume). Oil is often used as a carrier, and the microinjector should be calibrated by expelling the oil on to a pre-weighed filter paper and then weighing the paper and the weight of the oil is then divided by its specific gravity to determine the volume delivered. Hand-held or otherwise immobilized insects (e.g., adhered to sticky tape) are inoculated by piercing the intersegmental membrane and injecting the propagules directly into the hemocoel. Insects can first be immobilized with $CO_2$ or by chilling if required. Large numbers of insects can be injected, especially if the microinjector is equipped with a foot pedal. If a precise dose is required, care must be taken that the infective propagules do not settle within the syringe during inoculation.

Because the cuticle is an important barrier to infection, inoculation by injection has seldomly been used to measure virulence. However, Ignoffo *et al.* (1982) used inoculation through injection to demonstrate that resistance may not be solely at the integumental level; larvae of *Anticarsia gemmatalis*, a normally resistant species, injected with either hyphal bodies or conidia of *Nomuraea rileyi*, were much more resistant than larvae of *Trichoplusia ni*, a susceptible insect. In addition, analysis of the dose–time–mortality data suggested that this method of inoculation may be useful in bioassays of other insect–pathogen combinations.

### 2. Per os

Since entomopathogenic Hypocreales generally enter the host's body via direct penetration of the integument, *per os* (i.e. by mouth) inoculation is seldom used, unless the objective is to specifically infect the insect via the alimentary tract. For *per os* inoculation, microinjection devices have often been used; *per os* inoculation is achieved in a virtually identical manner to that described above for the injection method except that the end of the needle is 'blunted' with fine emery cloth, and inoculum is introduced directly into the foregut (e.g., pharynx, esophagus, crop, or proventriculus). However, extreme care must be exercised to ensue that the delivery device does not puncture or damage the foregut wall.

An alternative method to the injection method for *per os* inoculation is to present the inoculum *via* a bait or food source. The easiest method is to incorporate infective propagules directly onto the surface or within the bait substrate (e.g., propagules can be mixed in a sugar solution for presentation to flies or applied on to leaves for presentation to leaf eating insects). Further details on inoculating methods using baits are presented below (Section 5B3). Whatever the method of inoculation, it is virtually impossible to prevent surface contamination of the insect (see Chapter I). It must be noted that even with proper disinfestation, external infection can occur from infective propagules excreted in the frass (Allee *et al.*, 1990). Consequently, any conclusions on internal *versus* external infection based on *per os* inoculation must be validated (e.g., histologically; see Chapter XV).

### 3. *Topical*

For infection via the external integument, inoculum of entomopathogenic hypocrealean fungi is most commonly administered by some form of topical application. The type of topical inoculation method adopted usually depends on the size of insect, the number of insects to be inoculated, the formulation used, and the precision required. Fluorescent dyes (0.1% w/w; Day-Glo Color Corp.) in oil or aqueous formulations can be used to determine which insect body parts come into contact with formulated conidia (e.g., baits) (Inglis *et al.*, 1996c) after laboratory inoculation or field application.

#### *a. Direct application*

*(1). Immersion.* Although the dose cannot be measured precisely, immersion of insects into suspensions of propagules has been used successfully in bioassay of entomopathogenic Hypocreales. A series of aqueous suspensions are prepared with increasing concentrations of infective propagules. Insects are dipped singly into a suspension for a specified time. When using hydrophobic conidia, however, it would be expected that conidia would immediately adhere to the cuticle and much of the water would remain in the container (i.e. the concentration of conidia in the dipping suspension would change with every insect dipped). Consequently, in order to ensure that the dose received by each insect is as constant as possible, a separate suspension should be used for each insect dipped (i.e. insects should not be consecutively dipped into the same suspension). This is especially important if large insects are dipped into small volumes of aqueous inoculum. Insects can also be placed in small screened cages or bags that are dipped into propagule suspensions. Using dipping methods, exposure is usually expressed as the number of conidia/ml of suspension. This inoculation method is not recommended if precise information on the virulence of a particular strain of entomopathogenic hypocrealean fungus is the primary objective of the bioassay.

An alternative method used especially with small insects (e.g., aphids), is to flood propagule suspensions over the insect (Hall, 1976); the insects are placed on a filter paper in a Buchner funnel, and a spore suspension is gently poured in, immersing the insects. After a specified time (several seconds), the suspension is quickly drained off by suction. Alternatively, the insects are placed on detached leaf pieces or disks and flooded with the inoculum as described above.

*(2). Spraying.* One of the most commonly employed inoculation methods of topical inoculation is to spray the propagules directly on to the host integument. Several experimental spray devices are available commercially. The ones most commonly used for application of entomopathogenic Hypocreales are stationary sprayers such as the Potter spray tower. Track sprayers, where the spray nozzles are moved over the host at a controlled speed can also be used (see Chapter IV). Less expensive, yet very efficient systems can be easily developed using a plastic cylinder and an artist's air-brush. The equipment must be calibrated and care must be taken so that the delivery of the inoculum at the host level is uniform. During calibration and host treatment, droplet size, density and distribution pattern and propagule deposition should be monitored as previously described (Section 4A2). During the spray operation, insects are often immobilized on a sticky surface such as double-sided sticky tape, by chilling or by short exposure to $CO_2$. Insects that are not immobilized should be removed from the spray arena as soon as possible to prevent them from picking up additional inoculum from contaminated surfaces. Dosages are usually expressed as number of propagules/cm$^2$ surface area.

*(3). Droplet.* With larger insects, a precise droplet of the inoculum can be placed directly on to a specific region of an insect's integument. Microinjectors (see Section 5B1 above) or micropipettes can be used to apply volumes as small as 0.5 µl. However, this is usually not possible when applying aqueous suspensions, especially when using micropipettes, as the water droplet is difficult to deposit on the hydrophobic insect cuticle. Deposition of

the droplet in an area where it is absorbed by capillary motion may be helpful (e.g., at the pronotal shield of locusts). Care must be taken when using oil formulations as the oil itself can be toxic. Doses are usually expressed as the number of propagules/insect.

*b. Indirect application*

Rather than presenting the inoculum directly on to the insect surface, indirect methods can be used to present inoculum via a secondary substrate. The most common method is to inoculate a substrate and then transfer the insects on to it; insects pick up the inoculum by contact with the substrate as they feed or move across the substrate. A less common method is to present the inoculum within a food source; insects surface contaminate themselves in the process of eating (e.g., Inglis *et al.*, 1996c). The methods for deposition of the substrate are essentially the same as described above; inoculum is deposited on the surface of a substrate by dipping, spraying or direct deposition.

Application of inoculum on to a leaf surface is commonly used with plant feeding insects such as caterpillars. Leaves or leaf disks are treated and presented to insects in bioassay containers such as Petri dishes. This method has been used successfully to bioassay many entomopathogenic fungi against a variety of hosts including caterpillars and the Colorado potato beetle (Ignoffo *et al.*, 1983, and references therein).

Loss of inoculum and/or sticking of the leaf disk onto the surface of the bioassay container are often a problem, especially with oil formulations. To overcome this, Inglis *et al.* (1996c) presented leaf disks impaled on insect pins to grasshopper nymphs; 5-mm diameter lettuce disks were inoculated with 0.5 μl of an oil/conidial suspension of *B. bassiana*. Each inoculated disk was pierced with a pin and suspended ≈ 2 cm into the bioassay vial from a foam plug. Insects were allowed to feed for a certain period and only insects that had completely consumed the bait were used in the assay. Alternatively, the area of the leaflet or disk consumed can be recorded and the relative inoculum calculated accordingly; however, this is very time consuming.

For assay of *B. bassiana* against adult flies, Watson *et al.* (1995) treated 35-cm$^2$ sheets of plywood with either dry or wet formulations. Flies were exposed by placing $CO_2$-anesthetized individuals on the treated surface of the plywood and covering them with an inverted Petri dish bottom for 3 h.

*4. Aquatic*

Inoculation of aquatic insects is usually accomplished by introducing the insects directly into aqueous suspensions containing infective propagules. Exposures are expressed as number of propagules/ml of rearing medium. Stationary systems are usually adequate when insects such as mosquito or chironomid larvae are assayed (see Chapter IV). However, systems with running water are necessary when assaying black fly larvae (see Chapter IV). Simple stationary methods may be satisfactory with fast-acting pathogenic hypocrealean fungi such as *Culicinomyces clavisporus*. However, continuous exposure of larvae to various concentrations of conidia is not an ideal bioassay system with slower acting Hypocreales such as *Tolypocladium cylindrosporum*, (Goettel, 1987); the effective concentration can vary according to length of exposure, as mosquitoes are continually re-ingesting conidia that are still viable in excreta resulting in significant variability among replicates. An inoculation system limiting exposure time of larvae and infective propagules may reduce inter-replicate variability (Nadeau & Boisvert, 1994); larvae are placed in cups containing a suspension of infective propagules for a set period, and afterwards the desired exposure period larvae are harvested, rinsed, and placed into new containers with water without inoculum.

*5. Soil*

Conducting bioassays in soil requires cognisance of the physical characteristics of the soil to be used as these will influence the bioassay. Prior to inoculation of soil with hypocrealean fungi, the classification, texture, cation exchange capacity, organic matter content, pH, and moisture characteristics of the soil should be determined if not known. Although soil can be stored moist at low temperatures for a period of time with little effect on the soil microflora, it is desirable to use soil as soon as possible after collection to minimize possible storage effects. If it is necessary to store soil for long periods, the soil should be weighed (to determine per cent moisture content), dried to an appropriate level using a drying oven (e.g., $\geq$ 50% of the maximum water-holding capacity), and then stored dry until required. Just before use, the moisture content of the soil should be restored to the target level based on water potential. Researchers must be cognisant that drying soil will impact the biological characteristics of the soil (e.g., bacterial community structure will be affected). An alternative to dry storage of soil is to maintain soil at low temperatures (e.g., $\leq -20°C$). As with dry storage, freezing soil will affect microbial communities. Repeated freeze–thawing of the samples should be avoided.

Quantification of infective propagule density and distribution in soil is important. Propagules may be applied to soil as a dry preparation, in aqueous suspension, or formulated in/on a solid carrier (e.g., wheat bran or alginate pellets), and applied directly to the soil surface or uniformly mixed throughout the soil. The incorporation of dry propagules into soil can result in clumping of inoculum, either when the propagules are added to moistened soil or when they are added to dry soil which is subsequently moistened. Clumping can be exasperated when using soils high in clay content. A satisfactory distribution of propagules may be achieved by spraying conidia (e.g., with an air-brush) on to moistened soil while it is continuously mixed. Once incorporated, the soil moisture level (i.e. by weight) can then be increased to the desired level without affecting the distribution of propagules.

To test the distribution of infective propagules of hypocrealean fungi in soil, cores should be removed from various locations (e.g., using a soil corer or large cork borer), the samples weighed, propagules recovered on an agar medium, CFU counted, and the number of CFU per unit weight of soil calculated (see Section 4A2c). The number of CFU in soil at the different sample locations are compared with each other, and to theoretical populations, to obtain a measure of propagule uniformity.

Once propagules have been incorporated into the soil and the proper moisture level achieved, soil may be dispensed into containers that usually range in volume from 200 to 1000 ml, but volumes may be increased according to need. Insects are situated in the soil at a specific depth during soil placement or they are placed on the soil surface and permitted to move into the soil. The biological characteristics of the insect used will determine the best strategy in this regard. Although most bioassays have focused on insects that inhabit soil for a portion of their life-cycle (e.g., scarab beetle larvae), the influence of entomopathogenic Hypocreales on insects that are exposed to soil for a relatively short period of time (e.g., ovipositing grasshoppers) can also be tested using these protocols (e.g., Inglis *et al.,* 1995a).

## C. Incubation and mortality assessments

Following inoculation, the insects must be incubated, preferably under controlled environmental conditions. This is usually carried out in environmental chambers or cabinets controlling factors such as temperature, photoperiod and humidity. For information on methods to elucidate the effects of environmental parameters on pathogen efficacy, see Section 6. The choice of bioassay chamber, feeding regime and incubation method are all important in successful bioassays and will vary according to the needs of the host. Insects can be incubated singly or bulked in cages or assay chambers. In general, incubation conditions should be those that favor survival of non-inoculated insects. A rule of thumb is to ensure overall mortality of control treatment insects remains below 10% over the experimental period.

Mortality assessments are generally made daily and, due to the slow nature of most entomopathogenic hypocrealean fungi, may need to be carried out for up to 2 weeks or more after inoculation. Cadavers must be removed before the fungus sporulates to prevent horizontal transmission. When evaluating mortality data, it is useful to know whether an insect died of mycosis or other causes. To determine if insects died of mycosis, colonization of the cadaver by the hypocrealean fungus should be ascertained; to achieve this, cadavers are incubated in a high moisture environment (e.g., on moistened filter paper or water agar) and, if the cadavers are subsequently colonized by the fungus, these insects are considered to have died from mycosis. However, researchers must be cognisant that the inability of a fungus to colonize a cadaver does not preclude the possibility that the fungus caused or contributed death of the insect, and the importance of comparing mortality rates to an appropriate control treatment is essential (see Section 5A above).

### *1. Insects in epigeal habitats*

Larger insects are usually incubated singly in plastic containers, such as 500-ml food containers. They can also be bulked and incubated in small cages. Insects should be fed as required and conditions for proper growth and development provided. Larvae are usually transferred to individual compartments in plastic trays and reared on artificial diets. Care should be taken that the diets do not contain antimicrobial agents that could interfere with disease progression. For smaller insects inoculated directly on their host plant (i.e. detached leaf or leaf disk), incubation is often carried out in Petri dishes containing either water agar, moistened filter paper, or a soaked piece of cotton batten. Under such conditions of high humidity, detached leaves or leaf disks usually remain viable and are able to provide nutrients for their host for many days. Care should be taken not to place too many insects on a leaf surface, as this accelerates leaf deterioration. Insects should be transferred to fresh leaf surfaces as required. Alternately, bioassays can be developed that utilize rooted plants. For instance, Lacey *et al.* (1999) used a rooted cabbage bioassay system to evaluate *I. fumosorosea* against whitefly eggs and nymphs.

Escape by tiny insects such as thrips can pose unique challenges in bioassay cage design. For instance, Ugine *et al.* (2005) used detached bean leaves, within 30-ml polystytrene cups sealed with Parafilm to bioassay *B. bassiana* against western flower thrips while Pourian *et al.* (2008) used 3-cm diameter 35-mm film canisters and cucumber leaf disks for bioassay of *M. anisopliae* against onion thrips.

#### *2. Insects in aquatic habitats*

Insects in aquatic habitats, such as mosquito larvae, are commonly incubated in 200 ml of mosquito culture medium within 500-ml plastic food containers or beakers. Using this system, high inter-replicate variability in mortality often occurs, and Goettel (1987) attributed this to differences in microflora and fauna that establish in the water in the different replicate containers. He suggested that this could be overcome by inoculating each container with a defined suspension before introducing the larvae; however, this has yet to be tested.

#### *3. Insects in soil habitats*

Most bioassays with soil-inhabiting insects are conducted in soil placed in containers in a controlled-environment chamber. Temperature is the easiest variable to control. Although soil temperatures may be similar to air temperatures in controlled-environment chambers, where possible, it is recommended that the temperature in soil be recorded. There are numerous sensors or transducers that can be used to measure temperature (Livingston, 1993a). Water availability in soil is a more problematic parameter to control than temperature (see Section 6A2c). Since in most soil assays the insects are cryptic, daily mortality assessment is often not possible. Consequently, mortality assessments are made at the termination of the bioassay period.

## 6. ASSESSMENT OF ENVIRONMENTAL PARAMETERS

The ultimate challenge facing researchers studying entomopathogenic Hypocreales as microbial control agents is to predict field efficacy from laboratory-acquired data. *In vitro* assays can be used to determine pertinent variables affecting fungal growth, development and pathogenicity. Persistence of propagules can also be estimated through simulated field studies. Ultimately, bioassays using target hosts and incorporating the pertinent environmental variables will provide information most suited for prediction of efficacy under field conditions.

### A. Fungal tolerances

There is great genotypic and phenotypic variability among strains of entomopathogenic Hypocreales within the same species which affects, among other things, persistence of infective propagules in the field. To better predict efficacy under field conditions, variability amongst strains to persist in controlled environments should first be determined in laboratory assays. Such assays have been developed to study the three salient environment parameters: sunlight, temperature and humidity. Furthermore, it is essential that persistence evaluated in controlled settings be validated under field conditions.

#### *1. Sunlight*

Natural sunlight is one of the more important factors affecting survival of propagules under field conditions, and the UV-B (295–320 nm) components of sunlight are the most detrimental to microbial cells. However, irradiation at different wavelengths may be beneficial by promoting photo-reactivation, a phenomenon whereby the detrimental effects of UV irradiation are counteracted by the organism. Consequently, assays assessing tolerance to sunlight should preferably use polychromatic light at a temperature favorable to photo-reactivation (Fargues *et al.*, 1996).

To assess the impact of light radiation on entomopathogenic Hypocreales, propagules of the fungus are uniformly deposited on a substrate (a variety of substrates can be used including glass slides, Petri dishes, filter paper, or foliage) and then exposed to a source of simulated sunlight. Natural sunlight is variable and unpredictable and therefore should be avoided, especially if replicates are to be made on different days. Several artificial sunlight devices are available commercially. Long-pass filters are used to block short wavelengths under 295 nm to simulate natural sunlight (Rougier *et al.*, 1994). The substrates containing the propagules are irradiated for the desired period. The irradiance received can be varied by changing the distance of the substrate from the light source. Intensity of UV-B radiation should be measured with a radiometer. Adequate ventilation is paramount as irradiation can produce significant levels of ozone which, in itself, can be toxic to infective propagules. A non-irradiated (i.e. shaded) control treatment should always be included in experiments. Care must be taken to avoid UV-B diffuse radiation, as it can also be detrimental to propagule viability (Smits *et al.*, 1996).

Although it is often desirable to simulate natural sunlight as closely as possible and include polychromatic light, much information can still be gained using much simpler and cheaper light sources. For instance, Inglis *et al.* (1995b) tested the effects of UV protectants using a UV-B fluorescent bulb (Ultra-Lum, Carson, CA) which emits radiation from 260 to 400 nm with a peak at 300–310 nm. However, wavelengths under 295 nm which normally do not reach the earth's surface should be filtered using long pass filters if possible.

Following exposure, infective propagules are harvested and viability is assessed using any of the methods described above (see Section 4). If germination counts are used, it is preferable to use the benomyl method [see Section 4A1a(3)], as UV irradiation may delay germination in a proportion of the propagules, thereby increasing the problems caused by obstruction of counts due to hyphal growth of the early germinated propagules. Propagule survival is estimated by comparing the viability of the irradiated propagules with the viability of the shaded, control propagules (per cent survival = number of viable propagules following irradiation/ number of viable propagules in control × 100).

### 2. *Temperature*

Temperature is an important factor that determines the rate of germination, growth, sporulation and survival of entomopathogenic Hypocreales. Furthermore, temperature can impart indirect effects through impacts on the host physiology (e.g., efficacy of the immune response to infection). Elucidating effects of temperature typically rely on the use of controlled environment chambers to keep temperatures constant (± 1°C).

#### a. *Spore germination*

In determining the effects of temperature on spore germination, it is important to realize that even very short periods of changes in temperature can significantly affect responses. Therefore, destructive sampling should be used whenever possible. Adequate numbers of replicated cultures must be set up at several temperatures, usually between 5 and 40°C if upper and lower limits are sought. Periodically, several replicates are removed and germination rates determined. It is simplest if cultures are fixed with a drop of fixative (e.g., lactophenol cotton blue), covered with a coverslip and the spore germination rates determined later. The per cent germination is calculated for each culture (i.e. temperature/time combination). Mean percentages for each temperature/time combination are transformed to their logit values to obtain a straight-line relationship between germination and time. Maximum-likelihood methods are used to estimate lag phase, germination rate, and time to germination of a certain proportion of spores (Hywel-Jones & Gillespie, 1990).

#### b. *Vegetative growth*

Effects of temperature on growth of entomopathogenic hypocrealean fungi are most easily assessed on semi-synthetic media in Petri dishes by measuring the rate of colony growths. Petri dishes containing an adequate medium (see Section 3A1) are inoculated centrally, either by placing a suspension of propagules (e.g., 0.1 μl) or a small plug (e.g., 6 mm diameter) taken from an actively growing, non-sporulating culture (e.g., colony margin), and incubated for a period of time at several experimental temperatures. For most entomopathogenic Hypocreales, a range of temperatures between 4 and 40°C should be chosen. Cultures are incubated in total darkness for ≈ 2 weeks under conditions of high humidity. Three to five replicate dishes should be prepared for each temperature/ isolate combination. Surface radial growth is recorded daily using two perpendicular measurements which can be drawn at the bottom of each Petri dish at each interval if the culture medium is translucent. This alleviates the need for destructive sampling (i.e. if non-translucent medium is used). Since radial growth from day 3 to day 12 fits a linear model ($y = vt + b$) for most entomopathogenic hypocrealean fungi, where 'v' is the growth velocity, growth rates (velocity in mm/day) are used as the main parameter to evaluate the influence of temperature on fungal growth (Fargues *et al.*, 1992). In order to compare maximum growth rates between isolates, analyses (e.g., ANOVA) can be done on relative growth rates (%) calculated from the maximum growth rate for each isolate (see Section 5A2).

*c. Moisture*

Effects of moisture on germination, growth and sporulation of entomopathogenic Hypocreales are usually carried out using media adjusted to different water activities. For aerial studies, the relative humidity (RH) is usually maintained with glycerol or saturated salt solutions (Appendix: 4) at a constant temperature. Manipulation of water potentials of liquid and solid media is accomplished by adjusting the solute concentration in the medium and equilibrating the medium or test material in a closed chamber. Solutes used to generate specific humidities include various salts, glucose, sucrose, glycerol, or polyethylene glycol (PEG). Since the solute in the medium may have other effects, it is prudent to test several. The water potential, as affected by the solutes, can be easily calculated as follows:

$$\Psi_{ts} = -4.46 \times 10^{-3}\ T\Delta T$$

where $\Psi_{ts}$ is pressure in megapascals (MPa), T is absolute temperature in °K and $\Delta T$ is the freezing point depression (Griffin, 1994). The reader is referred to Rockland (1960) and Dhingra & Sinclair (1985) for more information on the use of solutes to control humidity and water potential.

Since aerial humidities occur at equilibrium only at the solution/air interface, it is important to include a method for air circulation for optimum humidity control. Such a system has been developed for study of effects of humidity on entomopathogenic Hypocreales (Fargues *et al.*, 1997). Air is constantly circulated with a membrane air pump over a saturated salt solution in one chamber (18 × 27 × 18 cm), containing 1 kg of salt in 0.5 l distilled water, into a second chamber (27 × 36 × 18 cm) in which the test materials are placed. Air exchange is approximately one complete air change in the test chamber every 4–5 min. Humidity is monitored within the test chamber with probes attached to data loggers.

To study effects of water on germination, growth, and sporulation of fungi, a series of agar or aqueous media are prepared using different aqueous solute solutions (e.g., glycerol or PEG) to obtain media with a range of water activities ($a_w$) (Magan & Lacey, 1984). The media are inoculated with propagules and placed in the humidity-controlled chambers, each treatment in a chamber corresponding to $a_w$ of the medium. For growth in liquid media, flasks need not be included in a controlled humidity environment; however, water lost due to evaporation must be replaced daily. Germination, growth and sporulation are then evaluated (see Sections 4A1a, 6A2a, and 6A2b). Inch & Trinci (1987) demonstrated that there was good correlation on effects of water activity on growth of *I. farinosus* between measurements from shake flasks and those from solid medium.

It is often desirable to determine the effect of RH on sporulation on the surface of the insect cadaver. Insects are experimentally infected with the pathogen (see Section 5). Immediately after death, cadavers are transferred to the controlled humidity chamber and incubated for the desired time, usually 10–15 days for most entomopathogenic Hypocreales. The cadavers are then either washed or homogenized and the propagules are enumerated (see Section 4).

The availability of water to microorganisms in soil is affected by a number of factors, the most significant of which is soil texture. For example, the availability of water will not be equal in two soils with different textures but the same percentage water content (v/w). By controlling the water potential of soil, it becomes possible to study the effects of water on the efficacy of entomopathogenic Hypocreales, independent of soil texture, and *vice versa*. The total water potential of soil is the sum of the component water potentials so that:

$$\Psi_t = \Psi_g + \Psi_m + \Psi_o + \ldots$$

where $\Psi_t$ is the total water potential, $\Psi_g$ is a gravitational potential constant, $\Psi_m$ is the matric potential, $\Psi_o$ is the osmotic potential, and other less significant potentials (i.e. pressure potential) are indicated by dots. Matric potential is the potential arising from the attraction of the soil matrix for water (adsorption and capillary). The osmotic potential of soil arises from dissolved solutes and lowered activity of water attributable to interaction with charged surfaces (Livingston, 1993b). Water potential possesses units of pressure, usually MPa or bars where 1 MPa is equal to 10 bars. Saturated soil has a water potential of ≈ 0 bars, and as the soil becomes drier, the water potential becomes increasingly more negative. Soil water desorption curves can be determined using a number of methods including pressure plates, resistance blocks, tensiometers, thermocouple psychrometers, neutron scattering, γ-ray attenuation, ultrasonic energy, and/or filter paper methods (Livingston, 1993b; Topp, 1993; Topp *et al.*, 1993). There are advantages and disadvantages to each of these methods, and the reader should consult the numerous references to obtain additional information on the quantification of soil water potentials.

Studdert & Kaya (1990) investigated the effect of water availability in two soils (organic and sandy loam texture) on the efficacy of *B. bassiana* against beet armyworm (*Spodoptera exigua*) pupae. Soil containing conidia was dispensed into 200-ml plastic containers containing pupae, and the containers were covered with polyethylene sheets leaving an air space of ≈ 1.5 cm. Containers with relatively moist soils (−3, −2 and −15 bars) were placed in plastic containers covered with damp towels and kept in the dark at a controlled temperature; the towels were wetted periodically. Less than 2% of the initial soil moisture was lost during the experiment (10 days). For drier soils (−37 and −200 bars), containers were maintained in desiccators over saturated salt solutions. Saturated salt solutions were used to control the water content of the atmosphere (Dhingra & Sinclair, 1985; Appendix: 4).

While it is possible to maintain a relatively constant water potential in soils kept in closed containers, this is not the case when soils are exposed to the atmosphere. Inglis *et al.* (1995a) studied the susceptibility of ovipositing grasshoppers to *B. bassiana* conidia in soil. Grasshoppers choose the depth at which they oviposit according to soil texture and moisture. They will readily oviposit into soil at or near field capacity. However, moisture is rapidly lost from soil in cups, particularly under conditions of low ambient humidity typical of arid agroecosystems. The daily addition of water to the soil surface is unsatisfactory due to the abnormal placement of egg pods near the top of the soil profile (≈ 1–2 cm) by females. To reduce saturation at the surface but permit the addition of water to soil, containers were fitted with a central watering tube (Figure 7.13). The porous gravel base acted as a water reservoir from which water could rapidly spread across the soil bottom, and then move upward in the profile by capillary action. Capillary and absorption forces associated with soil matrix determine the field capacity of soil; field capacity is the point at which the macropores are filled with air but water remains in micropores or capillary pores (≈ −0.1 to −0.3 bars). Although the maintenance of soil at field capacity (or relative to field capacity) is an imprecise measure of matric potential, in some instances it is used as an estimation of water availability.

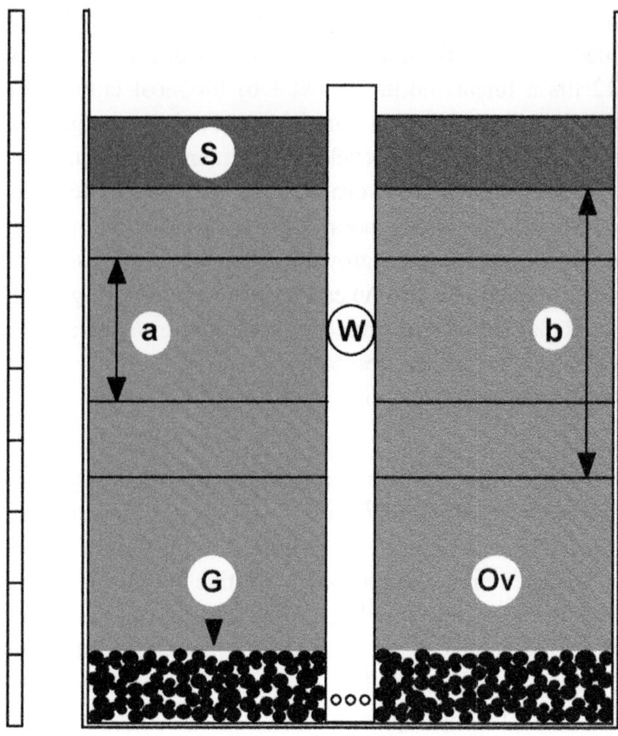


**Figure 7.13** Soil container used to test the efficacy of *Beauveria bassiana* conidia in soil (Ov) against ovipositing grasshoppers. The watering tube (W) allowed the addition of water to the bottom of the soil profile. The porous gravel (G) facilitated lateral movement of water and acted as a reservoir from which water moved upward into the profile by capillary action. Soil moisture was maintained near field capacity at the depth where eggs were deposited. A layer of sterile soil (S) was placed on the surface to prevent liberation of conidia due to oviposition activity. The bar adjacent to the container is 10 cm in length in 1-cm increments. Most egg pods were deposited between the lines indicated by the arrow marked 'a', but ranged between the lines indicated by the arrow marked 'b'.

The effects of rain on persistence of propagules on leaf or insect surfaces can be evaluated using rainfall simulators (e.g., Tossell *et al.*, 1987). The substrate in question (i.e. leaf or insect surface) is inoculated with a known quantity of propagules (see Section 5B3), and then the propagules are enumerated following exposure to simulated rain. The effects of the rain on propagule removal from leaves were assessed using analysis of covariance with propagule populations on leaves before exposure to rain used as the covariate. Per cent reduction in propagules due to rain exposure was determined as: number of propagules (e.g., CFU) prior to exposure – number of propagules after exposure)/number of propagules prior to exposure (Inglis *et al.*, 1995c).

## B. Inoculum persistence

It is often necessary to quantify rates of change in propagule populations over time, particularly when studying the effects of environmental parameters on the efficacy of entomopathogenic Hypocreales. To estimate persistence of entomopathogenic hypocrealean propagules in epigeal habitats, a number of substrates can be used (see Section 4A2). Most often, propagule survival on leaves is measured, since a measure of area can be obtained.

Fransen (1995) used a leaf imprint technique to directly assess survival of *Aschersonia aleyrodis* conidia on leaf surfaces; conidia were inoculated onto leaf surfaces and incubated. Leaf imprints were made on water agar periodically up to 20 days after conidial application. Immediately after each impression was made, the numbers of germinated and ungerminated conidia were recorded microscopically. The agar cultures were then incubated under conditions favorable for conidial germination (25°C) for 24 h and the numbers of germinated and ungerminated conidia were once again recorded and reduction in germination was assessed.

However, enumeration of most propagules associated with a number of substrates including leaves, other plant organs (e.g., flowers or fruit), soil, and insects must be assessed through indirect methods (see Section 4A2). Using such methods, it is often convenient to present populations per unit weight or insect (e.g., stadium), especially when it is difficult to accurately measure surface areas (e.g., insects). When significant variations in weight or size occur (e.g., individual insects or leaves), it is usually necessary to standardize weights. For example, CFU per mg can be calculated and then multiplied by the mean weight of the insects processed to obtain a measure of CFU per average insect. While weight measurements are useful for comparing populations within a specific substrate, it is less accurate than area for comparing propagule densities between substrates. While population densities or infective propagules per unit area, weight, or volume are most commonly used for assessing the persistence of propagules, the incidence or prevalence of insect mortality or of propagule survival over time (e.g., effect of environmental parameters on propagule germinability *in vitro*) have been used as measures of persistence.

For statistical comparisons of propagule persistence between treatments, it is usually necessary to normalize the data before analysis; a logarithmic transformation is usually required to normalize variance for data obtained from dilution plate counts [see Section 4A2e(1)]. Although it is desirable to sample from different populations at each collection date, in most instances it is logistically difficult to do so, and samples are often obtained from the same population (e.g., same plot or plant) at each time. Repeated measure statistics using the mixed procedure of SAS (SAS Institute Inc.) can be used to analyze correlated variables (see Section 5A2).

## C. Host–pathogen interactions

Bioassays using static conditions may be useful in comparing activity of different isolates, but they usually provide minimal information on the performance of the pathogen under field conditions. Consequently, bioassays must be developed that incorporate as many pertinent environmental parameters as possible. Inoculation techniques and the environmental conditions chosen should mimic as much as possible the natural situation and, more specifically, conditions at the level of the host microhabitat. By varying single or multiple factors, and by using information obtained on fungal tolerances, it should be possible to develop predictive models that would be indispensable in the development of entomopathogenic Hypocreales as microbial control agents. For instance, to study the effects of temperature on a thermoregulating host such as the grasshopper, Inglis *et al.* (1996b) used bioassay cages fitted with incandescent bulbs to allow for behavioral thermoregulation by the host (Figure 7.14). Cages are placed in a large controlled temperature chamber, and the periods of 'simulated sunshine' are varied by limiting the time the bulb is turned on, which

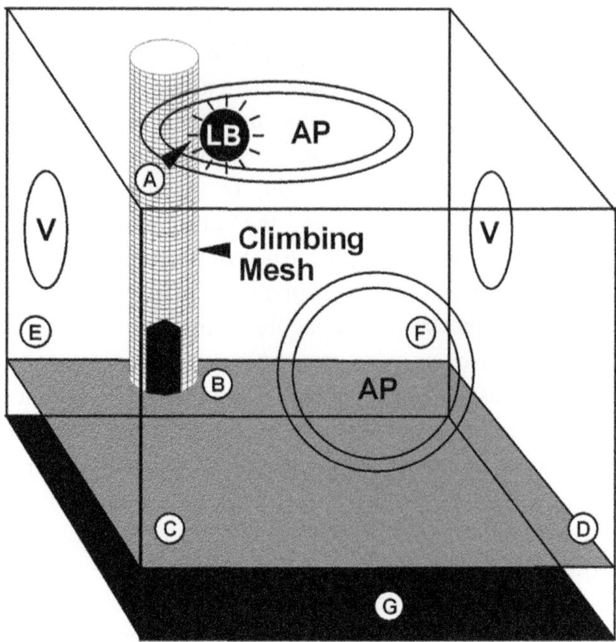


**Figure 7.14** A bioassay cage used to test the effect of grasshopper body temperature on the efficacy of *Beauveria bassiana*. To elevate their body temperature by basking, nymphs had to climb a plastic mesh (8 cm diameter × 28 cm high) towards a heat source (25-W incandescent light bulb; LB) in an aluminum cage (40 × 40 × 30 cm). The cage was equipped with a perforated metal floor, and a clear Plexiglas top and front. Air flow was enhanced by mesh vents (V) and access ports (AP), and temperatures were monitored at points A to F.

provides a heat gradient. Inoculated hosts are introduced into the system and monitored twice daily. Mortality of grasshoppers exposed to heat from the bulbs for various time periods per day are then compared to mortality in individuals not exposed to the heat gradient (i.e. in cages in which the bulbs are not turned on).

Studies on the effects of ambient humidity on infection of hosts can be accomplished using controlled humidity chambers (see Section 6A*2c*). Preferably, diurnal humidities and temperatures should be used to better mimic natural conditions.

Soil is very complex and the efficacy of entomopathogens is affected by a number of biotic and abiotic variables. Soil parameters that influence the efficacy of entomopathogenic Hypocreales include temperature, texture, water activity, pH, and exposure to agrochemicals. (McCoy *et al.*, 1992). However, other soil characteristics (e.g., compaction and physical structure) likely affect entomopathogenic Hypocreales as well, but have not been extensively tested. It is important to consider the abiotic and biotic characteristics of soil in the experimental design phase of the bioassay. The presence of a microflora can also influence the efficacy of entomopathogenic Hypocreales, especially in soil (e.g., Groden & Lockwood, 1991; Pereira *et al.*, 1993). The two most common methods used to study the effects of soil microorganisms on entomopathogenic hypocrealean activity are to compare efficacy in sterile or pasteurized soil *versus* non-sterilized soil, or stimulate microbial activity by amending soil with nutrients and subsequently comparing disease development in insects in the amended and unamended soils. Sterilization of soil is accomplished using heat (with or without steam), chemical sterilants or irradiation (Wolf & Skipper, 1994). All sterilization methods result in chemical alterations, primarily to the organic component. Heat is most commonly used to sterilize soil but high temperatures cause hydrolyzation of carbohydrates, coagulation of proteins, and influence a variety of other chemical properties of soil (Wolf & Skipper, 1994). Heat is more effective when used in combination with high-pressure steam (121°C > 103 kPa = 15 PSI). Although steam enhances conduction and penetration, the sterilization efficacy is further facilitated if soil is dry and exposed in relatively thin layers. Pasteurization of soil is accomplished by exposing soil to relatively low temperatures (≈ 60–75°C) for short periods of time. Although less severe (i.e. reduced chemical modifications to organic molecules), pasteurization generally does not affect thermophilic microorganisms, bacterial endospores, or cells in a desiccated state. *Bacillus* species and other spore-forming taxa (e.g., *Paenibacillus*, *Clostridium*) are cosmopolitan in soil, and in soils where endospores are prevalent, tyndalization may be used. Tyndalization involves the repeated (two to three times), short-term exposure of soil to high temperatures (e.g., 100°C) at intervals (usually 24 h). The rationale for this method is that high temperature exposure stimulates endospore germination and vegetative growth then occurs. Vegetative cells are vulnerable to high temperature, and are easily killed by exposure to high temperatures. The primary limitations to the use of sterilized or pasteurized soils include altered soil chemistry, a modified microbial community due to chemical changes, and the rapid colonization by pioneer colonizers following sterilization, which can result in the presence of large populations of a relatively small number of taxa.

The other strategy that is used to elucidate the effect of soil microorganisms on entomopathogenic Hypocreales is to add nutrients (carbon and nitrogen) to soil. At periods following the amendment of soil, vegetative growth (e.g., conidial germination and/or hyphal growth) and/or mortality assessments are conducted as an indication of fungistasis by soil-borne microorganisms. Although the soil amendment method may be useful for establishing the potential effect of microbial antagonists on entomopathogenic Hypocreales, the selection for certain types of microorganisms that results from the treatment may not be indicative of soils in nature. For example, the incorporation of simple carbohydrates (e.g., glucose) may favor fast-growing fungi with relatively simple nutritional requirements (e.g., Mucorales) to the detriment of taxa with more complex carbohydrate requirements (e.g., cellulolytic organisms). For both the sterilization and soil amendment methods, it is important that both total microbial biomass and microbial diversity are monitored (see Section 4).

In an effort to duplicate natural conditions, field-cage experiments can be used. However, results from field-cage assays must not be used as conclusive evidence of entomopathogenic hypocrealean efficacy in field settings. Cages provide a microclimate that is very different from that of the natural environment. For instance, cage screening provides shading and protection from wind. However, field-cages provide an excellent method to study host–pathogen interactions, especially if many of the factors have already been determined under laboratory conditions. For instance, Inglis *et al.* (1997) investigated the influence of environmental conditions on the efficacy of *B. bassiana* against grasshoppers in field environments using caged insects. Laboratory-acquired bioassay results showed that grasshoppers behaviorally respond to infection by thermoregulating and are able to recover from disease. In a field experiment, it was demonstrated that a higher incidence and more rapid development of disease occurred in grasshoppers placed in shaded cages than in cages exposed to direct sunlight or protected from UV-B radiation. However, methods for field-cage trials must be adapted for the different conditions and parameters being tested.

Once microclimatic constraints are better quantified and understood, it may be possible to overcome some of these inhibitory situations through improved formulation, strain selection, genotypic manipulation, and inoculum targeting. Identification of microlimatic constraints would also allow development of predictive models which would identify windows of opportunity, thereby optimizing efficacious use of these microbial control agents.

## ACKNOWLEDGMENTS

For the initial edition of this chapter, we thank Grant Duke (Lethbridge Research Centre, Lethbridge, Alberta), John Vandenberg (USDA, Ithaca, NY), Ann Hajek (Cornell University, Ithaca, NY) for their useful suggestions, Toby Entz (Lethbridge Research Centre, Lethbridge, Alberta), and Dan Johnson (University of Lethbridge, Lethbridge, Alberta) for their suggestions on experimental design and analyses, and Jacques Fargues and Nathalie Smits (INRA-Montpellier, France) for their suggestions on assessment of environmental parameters. We also thank Jenny Gusse (Lethbridge Research Centre, Lethbridge, Alberta), for providing comments on the use of molecular methodologies, Toby Entz for his comments on statistical analyses, and Reza Talaei Hassanloui (University of Tehran, Karaj), Ann Hajek and Stefan Jaronski (USDA, Sydney, MT) for providing useful suggestions in the updated edition of this chapter.

## REFERENCES

Abbott, W. S. (1925). A method for computing the effectiveness of an insecticide. *J. Econ. Entomol., 18*, 265–267.

Aghajanzadeh, S., Prasad, D. T., & Mallik, B. (2007). Genetic diversity in *Hirsutella thompsonii* isolates based on random amplified polymorphic DNA analysis. *BioControl., 52*, 375–383.

Allee, L. A., Goettel, M. S., Gol'berg, A., Whitney, H. S., & Roberts, D. W. (1990). Infection by *Beauveria bassiana* of *Leptinotarsa decemlineata* larvae as a consequence of fecal contamination of the integument following *per os* inoculation. *Mycopathologia., 111*, 17–24.

Amann, R. I., Ludwig, W., & Schleifer, K. H. (1995). Phylogenetic identification and *in situ* detection of individual microbial cells without cultivation. *Microbiol. Rev., 59*, 143–169.

Anderson, R. D., Bell, A. S., Blanford, S., Paaijmans, K. P., & Thomas, M. B. (2011). Comparative growth kinetics and virulence of four different isolates of entomopathogenic fungi in the house fly (*Musca domestica* L.). *J. Invertebr. Pathol., 107*, 179–184.

Andorrà, I., Esteve-Zarzoso, B., Guillamón, J. M., & Mas, A. (2010). Determination of viable wine yeast using DNA binding dyes and quantitative PCR. *Intl. J. Food Microbiol., 144*, 257–262.

Aquino de Muro, M., Elliott, S., Moore, D., Parker, B. L., Skinner, M., Reid, W., & El Bouhssini, M. (2005). Molecular characterisation of *Beauveria bassiana* isolates obtained from overwintering sites of Sunn Pests (*Eurygaster* and *Aelia* species). *Mycol. Res., 109*, 294–306.

Aquino de Muro, M., Mehta, S., & Moore, D. (2003). The use of amplified fragment length polymorphism for molecular analysis of *Beauveria bassiana* isolates from Kenya and other countries, and their correlation with host and geographical origin. *FEMS Microbiol. Lett., 229*, 249–257.

Arbeli, Z., & Fuentes, C. L. (2007). Improved purification and PCR amplification of DNA from environmental samples. *FEMS Microbiol. Lett., 272*, 269–275.

Atlas, R. M., & Bartha, R. (1981). Microbial Ecology*: Fundamentals and Applications*. Reading, UK: Addison-Wesley Publishing Co., pp. 560.

Avis, J. C. (2004). *Molecular markers, natural history, and evolution*. Sunderland, MA: Sinauer Associates, pp. 541.

Bailey, L. A., & Rath, A. C. (1994). Production of *Metarhizium anisopliae* spores using nutrient-impregnated membranes and its economic analysis. *Biocontrol Sci. Technol., 4*, 297–307.

Bain, J. M., Tavanti, A., Davidson, A. D., Jacobsen, M. D., Shaw, D., Gow, N. A. R., & Odds, F. C. (2007). Multilocus sequence typing of the pathogenic fungus *Aspergillus fumigatus*. *J. Clin. Microbiol., 45*, 1469–1477.

Bateman, R. P., Carey, M., Moore, D., & Prior, C. (1993). The enhanced infectivity of *Metarhizium flavoviride* in oil formulations to desert locusts at low humidities. *Ann. Appl. Biol., 122*, 145–152.

Beilharz, V. C., Parbery, D. G., & Swart, H. J. (1982). Dodine: A selective agent for certain soil fungi. *Trans. Brit. Mycol. Soc., 79*, 507–511.

Bell, A. S., Blanford, S., Jenkins, N., Thomas, M. B., & Read, A. F. (2009). Real-time quantitative PCR for analysis of candidate fungal biopesticides against malaria: Technique validation and first applications. *J. Invertebr. Pathol., 100*, 160–168.

Bischoff, J. F., Rehner, S. A., & Humber, R. A. (2009). A multilocus phylogeny of the *Metarhizium anisopliae* lineage. *Mycologia, 101*, 512–530.

Blachère, H., Calvez, J., Ferron, P., Corrieu, G., & Peringer, P. (1973). Etude de la formulation et de la conservation d'une préparation entomopathogène a base de blastospores de *Beauveria tenella* (Delacr. Siemaszko). *Ann. Zool. Ecol. Anim, 5*, 69–79.

Bleve, G., Rizzotti, L., Dellaglio, F., & Torriani, S. (2003). Development of reverse transcription (RT)-PCR and real-time RT-PCR assays for rapid detection and quantification of viable yeasts and molds contaminating yogurts and pasteurized food products. *Appl. Environ. Microbiol., 69*, 4116–4122.

Boucias, D., & Pendland, J. C. (1998). *Principles of Insect Pathology*. Boston: Kluwer Academic Publishers, pp. 537.

Boucias, D. G., Tigano, M. S., Sosa-Gomez, D. R., Glare, T. R., & Inglis, P. W. (2000). Genotypic properties of the entomopathogenic fungus. *Nomuraea rileyi. Biol. Control, 19*, 124–138.

Bridge, P. D., Prior, C., Sagbohan, J., Lomer, C. J., Carey, M., & Buddie, A. G. (1997). Molecular characterization of isolates of *Metarhizium* from locusts and grasshoppers. *Biodivers. Conserv, 6*, 177–189.

Broders, K. D., Woeste, K. E., SanMiguel, P. J., Westerman, R. P., & Boland, G. J. (2011). Discovery of single-nucleotide polymorphisms (SNPs) in the uncharacterized genome of the ascomycete *Ophiognomonia clavigignenti-juglandacearum* from 454 sequence data. *Molec. Ecol. Res., 11*, 693–702.

Bulat, S. A., & Mironenko, N. (1990). Species identity of the phytopathogenic fungi *Pyrenophora teres* Dreschsler and *P. graminea* Ito & Kuribayashi. Mikol. *Fitopatol, 24*, 435–441.

Bürgmann, H., Pesaro, M., Widmer, F., & Zeyer, J. (2001). A strategy for optimizing quality and quantity of DNA extracted from soil. *J. Microbiol. Methods, 45*, 7–20.

Bustin, S. A. (2004). *A – Z of Quantitative PCR*. LaJolla, CA: International University Line, pp. 882.

Butt, T. M. (1997). Complementary techniques: Fluorescence microscopy. In L. A. Lacey (Ed.), *Manual of Techniques in Insect Pathology* (pp. 355–365). London, UK: Academic Press.

Butt, T. M., & Goettel, M. S. (2000). Bioassays of Entomogenous Fungi. In A. Navon, & K. R. S. Ascher (Eds.), *Bioassays of Entomopathogenic Microbes and Nematodes* (pp. 141–195). Wallingford, U.K: CABI Publishing.

Butt, T., Jackson, C., & Magan, N. (Eds.). (2001). *Fungal Biocontrol Agents—Progress, Problems and Potential*. Wallingford, UK: CABI Publishing, pp. 390.

Caetano-Annollés, G., & Gresshoff, P. M. (1997). *DNA Markers: Protocols, Applications, and Overviews*. New York, NY: Wiley-VCH, p. 364.

Carmichael, J. W. (1955). Lacto-fuchsin: A new medium for mounting fungi. *Mycologia, 47*, 611.

Castrillo, L. A., & Humber, R. A. (2009). Molecular methods for identification and diagnosis of fungi. In S. P. Stock, J. Vandenberg, I. Glazer, & N. Boemare (Eds.), *Insect Pathogens. Molecular Approaches and Techniques* (pp. 50–70). Wallingford, UK: CABI Publishing.

Castrillo, L., Griggs, M. H., & Vandenberg, J. D. (2008). Quantitative detection of *Beauveria bassiana* GHA (Ascomycota: Hypocreales), a potential microbial control agent of the emerald ash borer, by use of real-time PCR. *Biol. Control, 45*, 163–169.

Castrillo, L. A., Thomsen, L., Juneja, P., & Hajek, A. E. (2007). Detection and quantification of *Entomophaga maimaiga* resting spores in forest soil using real-time PCR. *Mycol. Res., 111*, 324–331.

Castrillo, L. A., Vandenberg, J. D., & Wraight, S. P. (2003). Strain-specific detection of introduced *Beauveria bassiana* in agricultural fields by use of sequence-characterized amplified region markers. *J. Invertebr. Pathol., 82*, 75–83.

Chase, A. R., Osborne, L. S., & Ferguson, V. M. (1986). Selective isolation of the entomopathogenic fungi *Beauveria bassiana* and *Metarhizium anisopliae* from an artificial potting medium. *Florida Entomol., 69*, 285–292.

Chew, J. S. K., Strongman, D. B., & MacKay, R. M. (1998). Comparisons of twenty isolates of the entomopathogen *Paecilomyces farinosus* by analysis of RAPD markers. *Mycol. Res., 102*, 1254–1258.

Ciafardini, G., & Marotta, B. (1989). Use of the most-probable-number technique to detect *Polymyxa beta* (Plasmodiophoromycetes) in soil. *Appl. Environ. Microbiol., 55*, 1273–1278.

Comey, C. T., Koons, B. W., Presley, K. W., Smerick, J. B., Sobieralski, C. A., Stanley, D. M., et al. (1994). DNA extraction strategies for amplified fragment length polymorphism analysis. *J. Forensic Sci., 39*, 1254–1269.

Cravanzola, F., Piatti, P., Bridge, P. D., & Ozino, O. I. (1997). Detection of genetic polymorphism by RAPD-PCR in strains of the entomopathogenic fungus *Beauveria brongniartii* isolated from the European cockchafer (*Melolontha* spp.). *Lett. Appl. Microbiol., 25*, 289–294.

D'Agostino, R. B., Belanger, A., & D'Agostino, R. B. (1990). A suggestion for using powerful and informative tests for normality. *Am. Statist., 44*, 316–321.

Dalleau-Clouet, C., Gauthier, N., Risterucci, A. M., Bon, M. C., & Fargues, J. (2005). Isolation and characterization of microsatellite loci from the entomopathogenic hyphomycete, *Paecilomyces fumosoroseus. Mol. Ecol. Notes., 5*, 496–498.

Daoust, R. A., & Roberts, D. W. (1983). Studies on the prolonged storage of *Metarhizium anisopliae* conidia: Effect of temperature and relative humidity on conidial viability and virulence against moquitoes. *J. Invertebr. Pathol, 41*, 143–150.

Denis, M., Soumet, C., Rivoal, K., Ermel, G., Blivet, D., Salvat, G., & Colin, P. (1999). Development of a m-PCR assay for simultaneous identification of *Campylobacter jejuni* and *C. coli. Lett. Appl. Microbiol., 29*, 406–410.

Destefano, R. H. R., Destefano, S. A. L., & Messias, C. L. (2004). Detection of *Metarhizium anisopliae* var. *anisopliae* within infected sugarcane borer *Diatraea saccharalis* (Lepidoptera, Pyralidae) using specific primers. *Genet. Mol. Biol., 27*, 245–252.

Devi, U. K., Reineke, A., Rao, U. C. M., Reddy, N. R. N., & Khan, A. P. A. (2007). AFLP and single-strand conformation polymorphism studies of recombination in the entomopathogenic fungus. *Nomuraea rileyi. Mycol. Res., 111*, 716–725.

Dhingra, O. D., & Sinclair, J. B. (1985). *Basic Plant Pathology Methods*. Boca Raton, FL: CRC Press, pp. 355.

Edgington, L. V., Khew, K. L., & Barron, G. L. (1971). Fungitoxic spectrum of Benzimidazole compounds. *Phytopathol., 61*, 42–44.

Ekesi, S., & Maniania, N. K. (Eds.). (2007). *Use of Entomopathogenic Fungi in Biological Pest Management*. Kerala, India: Research Signpost, pp. 333.

Enkerli, J., Kölliker, R., Keller, S., & Widmer, F. (2005). Isolation and characterization of microsatellite markers from the entomopathogenic fungus *Metarhizium anisopliae. Mol. Ecol. Notes., 5*, 384–386.

Enkerli, J., & Widmer, F. (2010). Molecular ecology of fungal entomopathogens: molecular genetic tools and their applications in population and fate studies. *BioControl., 55*, 17–37.

Enkerli, J., Widmer, F., Gessler, C., & Keller, S. (2001). Strain-specific microsatellite markers in the entomopathogenic fungus *Beauveria brongniartii. Mycol. Res., 105*, 1079–1087.

Enkerli, J., Widmer, F., & Keller, S. (2004). Long-term field persistence of *Beauveria brongniartii* strains applied as biocontrol agents against European cockchafer larvae in Switzerland. *Biol. Control, 29*, 115–123.

Entz, S. C., Johnson, D. L., & Kawchuk, L. M. (2005). Development of a PCR-based diagnostic assay for the specific detection of the entomopathogenic fungus *Metarhizium anisopliae* var. *acridum. Mycol. Res., 109*, 1302–1312.

Estrada, M. E., Camacho, M. V., & Benito, C. (2007). The molecular diversity of different isolates of *Beauveria bassiana* (Bals.) Vuill. as assessed using intermicrosatellites (ISSRs). *Cell Mol. Biol. Lett., 12*, 240–252.

Fang, W., & Bidochka, M. J. (2006). Expression of genes involved in germination, conidiogenesis and pathogenesis in *Metarhizium anisopliae* using quantitative real-time RT-PCR. *Mycol. Res., 10*, 1165–1171.

Fargues, J., Goettel, M. S., Smits, N., Ouedraogo, A., Vidal, C., Lacey, L. A., Lomer, C. J., & Rougier, M. (1996). Variability in susceptibility to simulated sunlight of conidia among isolates of entomopathogenic Hyphomycetes. *Mycopathologia, 135*, 171–181.

Fargues, J., Maniania, N. K., Delmas, J. C., & Smits, N. (1992). Influence de la température sur la croissance *in vitro* d'hyphomycètes entomopathogènes. *Agronomie, 12*, 557–564.

Fargues, J., Ouedraogo, A., Goettel, M. S., & Lomer, C. J. (1997). Effects of temperature, humidity and inoculation method on susceptibility of *Schistocerca gregaria* to *Metarhizium flavoviride. Biocontrol Sci. Technol., 7*, 345–356.

Faria, M., Hajek, A. E., & Wraight, S. P. (2009). Imbibitional damage in conidia of the entomopathogenic fungi *Beauveria bassiana, Metarhizium acridum* and *Metarhizium anisopliae. Biol. Control., 51*, 346–354.

Fegan, M., Manners, J. M., Maclean, D. J., Irwin, J. A. G., Samuels, K. D. Z., Holdom, D. G., & Li, D. P. (1993). Random amplified polymorphic DNA markers reveal a high degree of genetic diversity in the entomopathogenic fungus *Metarhizium anisopliae* var. *anisopliae. J. Gen. Microbiol., 139*, 2075–2081.

Feinstein, L. M., Sul, W. J., & Blackwood, C. B. (2009). Assessment of bias associated with incomplete extraction of microbial DNA from soil. *Appl. Environ. Microbiol., 75*, 5428–5433.

Fernandes, E. K. K., Costa, G. L., Moraes, A. M. L., Zahner, V., & Bittencourt, V. (2006). Study on morphology, pathogenicity, and genetic variability of *Beauveria bassiana* isolates obtained from *Boophilus microplus* tick. *Parasitol. Res., 98*, 324–332.

Fernandes, E. K. K., Keyser, C. A., Rangel, D. E. N., Foster, N., & Roberts, D. W. (2010). CTC medium: a novel dodine-free selective medium for isolating entomopathogenic fungi, especially *Metarhizium acridum*, from soil. *Biol. Control, 54*, 197–205.

Finney, D. J. (1971). *Probit Analysis*. Cambridge, UK: Cambridge University Press.

Fournier, A., Widmer, F., & Enkerli, J. (2009). Development of a single nucleotide polymorphism (SNP) assay for genotyping of *Pandora neoaphidis*. *Fungal Biol., 114*, 498–506.

Fransen, J. J. (1995). Survival of spores of the entomopathogenic fungus *Aschersonia aleyrodis* (Deuteromycotina: Coelomycetes) on leaf surfaces. *J. Invertebr. Pathol., 65*, 73–75.

Freed, S., Jin, L., & Ren, S. X. (2011). Determination of genetic variability among the isolates of *Metarhizium anisopliae* var. *anisopliae* from different geographical origins. *World J. Microbiol. Biotechnol., 27*, 359–370.

Frostegård, A., Courtois, S., Ramisse, V., Clerc, S., Bernillon, D., Le Gall, F., Jeannin, P., Nesme, X., & Simonet, P. (1999). Quantification of bias related to the extraction of DNA directly from soils. *Appl. Environ. Microbiol., 65*, 5409–5420.

Gao, Q., Jin, K., Ying, S. H., Zhang, Y., Xiao, G., Shang, Y., Duan, Z., Hu, X., Xie, X. Q., Zhou, G., Peng, G., Luo, Z., Huang, W., Wang, B., Fang, W., Wang, S., Zhong, Y., Ma, L. J., St. Leger, R. J., Zhao, G. P., Pei, Y., Feng, M. G., Xia2, Y., & Wang, C. (2011). Genome sequencing and comparative transcriptomics of the model entomopathogenic fungi *Metarhizium anisopliae* and *M. acridum PLoS Genetics, 7*, e1001264.

Gardner, J. M., & Pillai, J. S. (1987). *Tolypocladium cylindrosporum* (Deuteromycotina: Moniliales), a fungal pathogen of the mosquito *Aedes australis* II. Methods of spore propagation and storage. *Mycopathologia, 97*, 77–82.

Gauthier, N., Dalleau-Clouet, C., Fargues, J., & Bon, M. (2007). Microsatellite variability in the entomopathogenic fungus *Paecilomyces fumosoroseus*: genetic diversity and population structure. *Mycologia, 99*, 693–704.

George, M. L. C., Nelson, R. J., Zeigler, R. S., & Leung, H. (1998). Rapid population analysis of *Magnaporthe grisea* by using rep-PCR and endogenous repetitive DNA sequences. *Phytopathol., 88*, 223–229.

Gilchrist, J. E., Campbell, J. E., Donnely, C. B., Peeler, J. T., & Delanay, J. M. (1973). Spiral plate method for bacterial determination. *Appl. Microbiol., 43*, 149–157.

Girard, K., & Jackson, C. W. (1993). Using fluorescent microscopy and image analysis to assess distribution of *Verticillium lecanii* spores on *Rhopalosiphum padi*. *Proc. Soc. Invertebr. Pathol., Ashville, N.C.*55, (abstract).

Goettel, M. S. (1984). A simple method for mass culturing entomopathogenic hyphomycete fungi. *J. Microbiol. Methods, 3*, 15–20.

Goettel, M. S. (1987). Studies on bioassay of the entomopathogenic hyphomycete fungus *Tolypocladium cylindrosporum* in mosquitoes. *J. Am. Mosq. Control Assoc., 3*, 561–567.

Goettel, M. S., Eilenberg, J., & Glare, T. R. (2010). Entomopathogenic fungi and their role in regulation of insect populations. In L. I. Gilbert, & S. S. Gill (Eds.), *Insect Control: Biological and Synthetic Agents* (pp. 387–431). San Diego, CA: Academic Press.

Goldstein, D. B., & Schlötterer, C. (1999). *Microsatellites: Evolution and Applications*. New York, NY: Oxford University Press, pp. 368.

Griffin, D. H. (1994). *Fungal Physiology* (2nd ed.). New York: Wiley–Liss, pp. 458.

Groden, E., & Lockwood, J. L. (1991). Effects of soil fungistasis on *Beauveria bassiana* and its relationship to disease incidence in the Colorado potato beetle, *Leptinotarsa decemlineata*, in Michigan and Rhode Island soils. *J. Invertebr. Pathol., 57*, 7–16.

Guy, P. L., & Rath, A. H. C. (1990). Enzyme-linked immunosorbent assay (ELISA) to detect spore surface antigens of *Metarhizium anisopliae*. *J. Invertebr. Pathol., 55*, 435–436.

Hall, R. A. (1976). A bioassay of the pathogenicity of *Verticillium lecanii* conidiospores on the aphid, *Macrosiphoniella sanborni*. *J. Invertebr. Pathol., 27*, 41–48.

Han, Q., Inglis, G. D., & Hausner, G. (2002). Phylogenetic relationships among strains of the entomopathogenic fungus, *Nomuraea rileyi* as revealed by partial β-tubulin sequences and inter-simple sequence repeat (ISSR) analysis. *Lett. Appl. Microbiol., 34*, 376–383.

Harris, J. L. (1986). Modified method for fungal slide culture. *J. Clinc. Microbiol., 24*, 460–461.

Harris, R. F., & Sommers, L. E. (1968). Plate-dilution frequency technique for assay of microbial ecology. *Appl. Microbiol., 16*, 330–334.

Hedgecock, S., Moore, D., Higgins, P. M., & Prior, C. (1995). Influence of moisture content on temperature tolerance and storage of *Metarhizium flavoviride* conidia in an oil formulation. *Biocontrol Sci. Technol, 5*, 371–377.

Herrera, M. L., Vallor, A. C., Gelfond, J. A., Patterson, T. F., & Wickes, B. L. (2009). Strain-dependent variation in 18S ribosomal DNA Copy numbers in *Aspergillus fumigatus*. *J. Clin. Microbiol., 47*, 1325–1332.

Hywel-Jones, N. L., & Gillespie, A. T. (1990). Effect of temperature on spore germination in *Metarhizium anisopliae* and *Beauveria bassiana*. *Mycol. Res., 94*, 389–392.

Ignoffo, C. M., Garcia, C., & Kroha, M. J. (1982). Susceptibility of larvae of *Trichoplusia ni* and *Anticarsia gemmatalis* to intrahemocoelic injections of conidia and blastospores of *Nomuraea rileyi*. *J. Invertebr. Pathol., 39*, 198–202.

Ignoffo, C. M., Garcia, C., Kroha, M. J., Samšiňáková, A., & Kálalová, S. (1983). A leaf surface treatment bioassay for determining the activity of conidia of *Beauveria bassiana* against *Leptinotarsa decemlineata*. *J. Invertebr. Pathol., 41*, 385–386.

Inch, J. M. M., & Trinci, P. J. (1987). Effects of water activity on growth and sporulation of *Paecilomyces farinosus* in liquid and solid media. *J. Gen. Microbiol., 133*, 247–252.

Inglis, G. D., & Kalischuk, L. D. (2003). Direct detection of *Campylobacter* species in bovine feces using polymerase chain reaction. *Appl. Environ. Microbiol., 69*, 3435–3447.

Inglis, G. D., & Kalischuk, L. D. (2004). Direct quantification of *Campylobacter jejuni* and *Campylobacter lanienae* in feces of cattle by real-time quantitative PCR. *Appl. Environ. Microbiol., 70*, 2296–2306.

Inglis, G. D., Duke, G. M., Goettel, M. S., & Kabaluk, J. T. (2008). Genetic diversity of *Metarhizium anisopliae* var. *anisopliae* in southwestern British Columbia. *J. Invertebr. Pathol., 98*, 101–113.

Inglis, G. D., McAllister, T. A., Larney, F. J., & Topp, E. (2010). Prolonged survival of *Campylobacter* species in bovine manure compost. *Appl. Environ. Microbiol., 76*, 1110–1119.

Inglis, G. D., Feniuk, R. P., Goettel, M. S., & Johnson, D. L. (1995a). Mortality of grasshoppers exposed to soil-borne *Beauveria bassiana* during oviposition and nymphal emergence. *J. Invertebr. Pathol., 65*, 139–146.

Inglis, G. D., Goettel, M. S., & Johnson, D. L. (1995b). Influence of ultraviolet light protectants on persistence of the entomopathogenic fungus, *Beauveria bassiana*. *Biol. Control, 5*, 581–590.

Inglis, G. D., Johnson, D. L., & Goettel, M. S. (1995c). Effects of simulated rain on the persistence of *Beauveria bassiana* conidia on leaves of alfalfa and wheat. *Biocontrol Sci. Technol., 5*, 365–369.

Inglis, G. D., Johnson, D. L., & Goettel, M. S. (1996a). Effect of bait substrate and formulation on infection of grasshopper nymphs by *Beauveria bassiana*. *Biocontrol Sci. Technol., 6*, 35–50.

Inglis, G. D., Johnson, D. L., & Goettel, M. S. (1996b). Effects of temperature and thermoregulation on mycosis by *Beauveria bassiana* in grasshoppers. *Biol. Control, 7*, 131–139.

Inglis, G. D., Johnson, D. L., & Goettel, M. S. (1996c). An oil-bait bioassay method used to test the efficacy of *Beauveria bassiana* against grasshoppers. *J. Invertebr. Pathol., 67*, 312–315.

Inglis, G. D., Johnson, D. L., & Goettel, M. S. (1997). Effects of temperature and sunlight on mycosis (*Beauveria bassiana*) of grasshoppers under field conditions. *Environ. Entomol., 26*, 400–409.

Jackson, C. R., Harper, J. P., Willoughby, D., Roden, E. E., & Churchill, P. F. (1997). A simple, efficient method for the separation of humic substances and DNA from environmental samples. *Appl. Environ. Microbiol., 63*, 4993–4995.

Jenkins, N. E., & Prior, C. (1993). Growth and formation of true conidia by *Metarhizium flavoviride* in a simple liquid medium. *Mycol. Res., 97*, 1489–1494.

Jimenez, J., & Gillespie, A. T. (1990). Use of the optical brightener Tinopal BOPT for the rapid determination of conidial viabilities in entomogenous deuteromycetes. *Mycol. Res., 94*, 279–283.

Johnson, D. L. (1992). Introduction: biology, ecology, field experimentation and environmental impact. In C. J. Lomer, & C. Prior (Eds.), *Biological Control of Locusts and Grasshoppers* (pp. 267–278). Wallingford, UK: CABI Publishing.

Johnson, D. L., Hill, B. D., Hinks, C. F., & Schallje, G. B. (1986). Aerial application of the pyrethroid deltamethrin for grasshopper (Orthoptera: Acrididae) control. *J. Econ. Entomol., 79*, 181–188.

Jones, D. (1984). Use, misuse, and role of multiple-comparison procedures in ecological and agricultural entomology. *Environ. Entomol., 13*, 635–649.

Kabir, S., Rajendran, N., Amemiya, T., & Itoh, T. (2003). Quantitative measurement of fungal DNA extracted by three different methods using real-time polymerase chain reaction. *J. Biosci. Bioeng., 96*, 337–343.

Katcher, H. L., & Schwartz, I. (1994). A distinctive property of Tth DNA polymerase: enzymatic amplification in the presence of phenol. *Biotechniques, 16*, 84–92.

Kim, S., & Misra, A. (2007). SNP genotyping: technologies and biomedical applications. *Annu. Rev. Biomed. Eng., 9*, 289–320.

Kirk, J. L., Beaudette, L. A., Hart, M., Moutoglis, P., Khironomos, J. N., Lee, H., & Trevors, J. T. (2004). Methods of studying soil microbial diversity. *J. Microbiol. Methods, 58*, 169–188.

Klingen, I., Eilenberg, J., & Meadow, R. (2002). Effects of farming system, field margins and bait insect on the occurrence of insect pathogenic fungi in soils. *Agric. Ecosystems & Environment, 91*, 191–198.

Kope, H. H., Alfaro, R. I., & Lavallée, R. (2006). Virulence of the entomopathogenic fungus *Lecanicillium* (Deuteromycota: Hyphomycetes) to *Pissodes strobi* (Coleoptera: Curculionidae). *Can. Entomol., 138*, 253–262.

Kreader, C. A. (1996). Relief of amplification inhibition in PCR with bovine serum albumin of T4 gene 32 protein. *Appl. Environ. Microbiol, 62*, 1102–1106.

Kristensen, R., Berdal, K. G., & Holst-Jensen, A. (2007). Simultaneous detection and identification of trichothecene- and moniliformin- producing *Fusarium* species based on multiplex SNP analysis. *J. Appl. Microbiol., 102*, 1071–1081.

Kybal, J., & Vlček, V. (1976). A simple device for stationary cultivation of microorganisms. *Biotechnol. & Bioengineer, 18*, 1713–1718.

Lacey, L. A., Kirk, A. A., Millar, G., Mercandier, G., & Vidal, C. (1999). Ovicidal and larvicidal activity of conidia and blastospores of *Paecilomyces fumosoroseus* (Deuteromycotina: Hyphomycetes) against *Bemesia argentifolii* (Homoptera: Aleyrodidae) with a description of a bioassay system allowing prolonged survival of control insects. *Biocontrol Sci. Technol., 9*, 9–18.

Lambreghts, R., Shi, M., Belden, W. J., de Caprio, D., Park, D., Henn, M. R., Galagan, J. E., Bastürkmen, M., Birren, B. W., Sachs, M. S., Dunlap, J. C., & Loros, J. J. (2009). A high-density single nucleotide polymorphism map for *Neurospora crassa*. *Genetics, 181*, 767–781.

LaMontagne, M. G., Michel, F. C., Holden, P. A., & Reddy, C. A. (2002). Evaluation of extraction and purification methods for obtaining PCR-amplifiable DNA from compost for microbial community analysis. *J. Microbiol. Methods, 49*, 255–264.

Leal, S. C. M., Bertioli, D. J., Butt, T. M., & Peberdy, J. F. (1994). Characterization of isolates of the entomopathogenic fungus *Metarhizium anisopliae* by RAPD-PCR. *Mycol. Res., 98*, 1077–1081.

Leland, J. E., McGuire, M. R., Grace, J. A., Jaronski, S. T., Ulloa, M., Park, Y. H., & Plattner, R. D. (2005). Strain selection of a fungal entomopathogen, *Beauveria bassiana*, for control of plant bugs (*Lygus* spp.) (Heteroptera: Miridae). *Biol. Control, 35*, 104–114.

Littell, R. C., Milliken, G. A., Stroup, W. W., Wolfinger, R. D., & Schabenberger, O. (2006). *SAS for Mixed Models* (2nd ed.). Cary, NC: SAS Institute Inc.

Little, T. M., & Hills, F. J. (1978). *Agricultural Experimentation*. New York, NY: John Wiley and Sons, pp. 368.

Liu, Z. Y., Milner, R. J., McRae, C. F., & Lutton, G. (1993). The use of dodine in selective media for isolation of *Metarhizium* spp. from soil. *J. Invertebr. Pathol., 62*, 248–251.

Livingston, N. J. (1993a). Soil temperature. In M. R. Carter (Ed.), *Soil Sampling and Methods of Analysis* (pp. 673–682). Boca Raton: Lewis Publishing.

Livingston, N. J. (1993b). Soil water potential. In M. R. Carter (Ed.), *Soil Sampling and Methods of Analysis* (pp. 559–567). Boca Raton: Lewis Publishing.

Magan, N., & Lacey, J. (1984). Effect of temperature and pH on water relations of field and storage fungi. *Trans. Br. Mycol. Soc., 82*, 71–81.

Maurer, P., Couteaudier, Y., Girard, P. A., Bridge, P. D., & Riba, G. (1997). Genetic diversity of *Beauveria bassiana* and relatedness to host insect range. *Mycol. Res., 101*, 159–164.

McCoy, C. W., Hill, A. J., & Kanavel, R. F. (1975). Large-scale production of the fungal pathogen *Hirsutella thompsonii* in submerged culture and its formulation for application in the field. *Entomophaga, 20*, 229–240.

McCoy, C. W., Storey, G. K., & Tigano-Milani, M. S. (1992). Environmental factors affecting entomopathogenic fungi in soil. *Pesqui. Agropecu. Bras, 27*, 107–111.

McGuire, M. R., Ulloa, M., Park, Y. H., & Hudson, N. (2005). Biological and molecular characteristics of *Beauveria bassiana* isolates from California *Lygus hesperus* (Hemiptera: Miridae) populations. *Biol. Control, 33*, 307–314.

Meudt, H. M., & Clarke, A. C. (2007). Almost forgotten or latest practice? AFLP applications, analyses and advances. *Trends Plant Sci., 12*, 106–117.

Meyling, N. V., & Eilenberg, J. (2006). Isolation and characterisation of *Beauveria bassiana* isolates from phylloplanes of hedgerow vegetation. *Mycol. Res., 110*, 188–195.

Meyling, N. V., Lubeck, M., Buckley, E. P., Eilenberg, J., & Rehner, S. A. (2009). Community composition, host range and genetic structure of the fungal entomopathogen *Beauveria* in adjoining agricultural and seminatural habitats. *Mol. Ecol., 18*, 1282–1293.

Meynell, G. G., & Meynell, E. (1970). *Theory and Practice in Experimental Bacteriology* (2nd ed.). Cambridge, UK: Cambridge University Press, pp. 347.

Milner, R. J., Huppatz, R. J., & Swaris, S. C. (1991). A new method for assessment of germination of *Metarhizium* conidia. *J. Invertebr. Pathol, 57*, 121–123.

Mitchell, D. J., Kannwischer-Mitchell, M. E., & Dickson, D. W. (1987). A semi-selective medium for the isolation of *Paecilomyces lilacinus*. *J. Nematol., 19*, 255–256.

Mitina, G. V., Mikhailova, L. A., & Yli-Mattila, T. (2007). RAPD-PCR, UP-PCR and rDNA sequence analyses of the entomopathogenic fungus *Verticillium lecanii* and its pathogenicity towards insects and phytopathogenic fungi. *Arch. Phytopathol. Pflanzenschutz, 41*, 113–128.

Moore, D., Langewald, J., & Obognon, F. (1997). Effects of rehydration on the conidial viability of *Metarhizium flavoviride* mycopesticide formulations. *Biocontrol Sci. Technol., 7*, 87–94.

Mor, H., Gindin, G., BenZeev, I. S., Raccah, B., Geschtovt, N. U., & Ajtkhozhina, N. (1996). Diversity among isolates of *Verticillium lecanii* as expressed by DNA polymorphism and virulence towards *Bemisia tabaci*. *Phytoparasitica., 24*, 111–118.

Morin, P. A., Luikart, G., & Wayne, R. K. (2004). SNPs in ecology, evolution and conservation. *Trends Ecol. Evol., 19*, 208–216.

Mozes-Koch, R., Edelbaum, O., Livneh, O., Sztejnberg, A., Uziel, A., Gerson, U., & Sela, I. (1995). Identification of *Hirsutella* species, isolates within a species and intraspecific heterokaryons by random amplified polymorphic DNA (RAPD). *J. Plant Dis. Protect., 102*, 284–290.

Munoz, C., Talquenca, S. G., & Volpe, M. L. (2009). Tetra primer ARMS-PCR for identification of SNP in beta-tubulin of *Botrytis cinerea*, responsible of resistance to benzimidazole. *J. Microbiol. Methods, 78*, 245–246.

Nadeau, M. P., & Boisvert, J. L. (1994). Larvicidal activity of the entomopathogenic fungus *Tolypocladium cylindrosporum* (Deuteromycotina: Hyphomycetes) on the mosquito *Aedes triseriatus* and the black fly *Simulium vittatum* (Diptera: Simulidae). *J. Am. Mosq. Control Assoc., 10*, 487–491.

Nowierski, R. M., Zeng, Z., Jaronski, S., Delgado, F., & Swearingen, W. (1996). Analysis and modeling of time–dose–mortality of *Melanoplus sanguinipes, Locusta migratoria migratorioides*, and *Schistocerca gregaria* (Orthoptera: Acrididae) from *Beauveria, Metarhizium*, and *Paecilomyces* isolates from Madagascar. *J. Invertebr. Pathol., 67*, 236–252.

Oulevey, C., Widmer, F., Kölliker, R., & Enkerli, J. (2009). An optimized microsatellite marker set for detection of *Metarhizium anisopliae* genotype diversity on field and regional scales. *Mycol. Res., 113*, 1016–1024.

Ownley, B. H., Griffin, M. R., Klingeman, W. E., Gwinn, K. D., Moulton, J. K., & Pereira, R. M. (2008). *Beauveria bassiana*: Endophytic colonization and plant disease control. *J. Invertebr. Pathol, 98*, 267–270.

Padhye, A. A., Sekhon, A. S., & Carmichael, J. W. (1973). Ascocarp production by *Nannizzia* and *Arthroderma* on keratinous and non-keratinous media. *Sabouraudia, 11*, 109–114.

Panter, C., & Frances, S. P. (2003). A more selective medium for *Culicinomyces clavisporus*. *J. Invertebr. Pathol., 82*, 198–200.

Parkinson, D. (1994). Filamentous fungi. In S. H. Mickelson (Ed.), *Methods of Soil Analysis. Part 2. Microbiological and Biochemical Properties* (pp. 329–350). Madison, WI: Soil Science Society of America, Inc.

Payne, C. D. (Ed.). (1978). *The GLIM System Release 3.77 Manual*. Downers Grove: Numerical Algorithms Group.

Peng, G., Xie, L., Hu, J., & Xia, Y. (2009). Identification of genes that are preferentially expressed in conidiogenous cell development of *Metarhizium anisopliae* by suppression subtractive hybridization. *Curr. Genet., 55*, 263–271.

Pereira, R. M., & Roberts, D. W. (1990). Dry mycelium preparations of entomopathogenic fungi, *Metarhizium anisopliae* and *Beauveria bassiana*. *J. Invertebr. Pathol., 56*, 39–46.

Pereira, R. M., Stimac, J. L., & Alves, S. B. (1993). Soil antagonism affecting the dose–response of workers of the red imported fire ant, *Solenopsis invicta*, to *Beauveria bassiana* conidia. *J. Invertebr. Pathol., 61*, 156–161.

Piatti, P., Cravanzola, F., Bridge, P. D., & Ozino, O. I. (1998). Molecular characterization of *Beauveria brongniartii* isolates obtained from *Melolontha melolontha* in Valle d'Aosta (Italy) by RAPD-PCR. *Lett. Appl. Microbiol., 26*, 317–324.

Pinder, J. E., Wiener, J. G., & Smith, M. H. (1978). The Weibull distribution: a new method of summarizing survivorship data. *Ecology, 59*, 175–179.

Pourian, H. R., Ezzati-Tabrizi, R., & Talaei-Hassanloui, R. (2008). An improved cage system for the bioassay of *Metarhizium anisopliae* on *Thrips tabaci* (Thysanoptera: Thripidae). *Biocontrol Sci. Technol., 18*, 745–752.

Preisler, H. K., & Robertson, J. L. (1989). Analysis of time–dose–mortality data. *J. Econ. Entomol., 82*, 1534–1542.

Rangel, D. E. N., Dettenmaier, S. J., Fernandes, E. K. K., & Roberts, D. W. (2010). Susceptibility of *Metarhizium* spp. and other entomopathogenic fungal species to dodine-based selective media. *Biocontrol Sci. Technol., 20*, 375–389.

Rehner, S. A., & Buckley, E. P. (2003). Isolation and characterization of microsatellite loci from the entomopathogenic fungus *Beauveria bassiana* (Ascomycota: Hypocreales). *Mol. Ecol. Notes., 3*, 409–411.

Roberts, D. W., Dunn, H. M., Ramsay, G., Sweeney, A. W., & Dunn, N. W. (1987). A procedure for preservation of the mosquito pathogen *Culicinomyces clavisporus*. *Appl. Microbiol. Biotechnol, 26*, 186–188.

Robertson, J. L., & Preisler, H. K. (1992). *Pesticide Bioassays with Arthropods*. Boca Raton, FL: CRC Press, pp. 127.

Robertson, J. L., Russell, R. M. & Savin, N. E. (1980). *POLO, A User's Guide to Probit Or LOgit Analysis*, United States Department of Agriculture Forest Service, Pacific Southwest Forest and Range Experiment Station, General Technical Report PSW-38, pp. 15.

Robertson, J. L., Smith, K. C., Savin, N. E., & Lavigne, R. J. (1984). Effects of dose selection and sample size on the precision of lethal dose estimates in dose–mortality regression. *J. Econ. Entomol., 77*, 833–837.

Rockland, L. (1960). Saturated salt solutions for static control of relative humidity between 5° and 4°C. *Anal. Chem., 32*, 1375–1375.

Romanowski, G., Lorenz, M. G., & Wackernagel, W. (1993). Use of polymerase chain-reaction and electroporation of *Escherichia coli* to monitor the persistence of extracellular plasmid DNA introduced into natural soils. *Appl. Environ. Microbiol., 59*, 3438–3446.

Rombach, M. C. (1989). Production of *Beauveria bassiana* (Deuteromycotina. Hyphomycetes) sympoduloconidia in submerged culture. *Entomophaga, 34*, 45–52.

Rougier, M., Fargues, J., Goujet, R., Itier, B., & Benateau, S. (1994). Mise au point d'un dispositif d'étude des effets du rayonnement sur la persistance des microorganismes pathogènes. *Agronomie, 14*, 673–681.

Roy, H. E., Vega, F. E., Chandler, D., Goettel, M. S., Pell, J. K., & Wajnberg, E. (Eds.). (2010). *The Ecology of Fungal Entomopathogens*. Dordrecht, The Netherlands: Springer, pp. 198.

Rudi, K., Moen, B., Dromtorp, S. M., & Holck, A. L. (2005). Use of ethidium monoazide and PCR in combination for quantification of viable and dead cells in complex samples. *Appl. Environ. Microbiol., 71*, 1018–1024.

Samšiňáková, A., Kálalová, S., Vlček, V., & Kybal, J. (1981). Mass production of *Beauveria bassiana* for regulation of *Leptinotarsa decemlineata* populations. *J. Invertebr. Pathol., 38*, 169–174.

Schaalje, G. B., Charnetski, W. A., & Johnson, D. L. (1986). A comparison of estimators of the degree of insect control. *Communications in Statistics: Simulations and Computations, 15*, 1065–1086.

Schaalje, G. B., Johnson, D. L., & Van der Vaart, H. R. (1992). Application of competing risks theory to the analysis of effects of *Nosema locustae* and *N. cuneatum* on development and mortality of migratory locusts. *Environ. Entomol., 21*, 939–948.

Schading, R. L., Carruthers, R. I., & Mullin-Schading, B. A. (1995). Rapid determination of conidia viability for entomopathogenic Hyphomycetes using fluorescence microscopy techniques. *Biocontrol Sci. Technol., 5*, 201–208.

Schneider, S., Enkerli, J., & Widmer, F. (2009). A generally applicable assay for the quantification of inhibitory effects on PCR. *J. Microbiol. Methods, 78*, 351–353.

Schneider, S., Rehner, S. A., Widmer, F., & Enkerli, J. (2011). A PCR-based tool for cultivation-independent detection and quantification of *Metarhizium* clade 1. *J Invertebr. Pathol., 108*, 106–114.

Schwarzenbach, K., Enkerli, J., & Widmer, F. (2009). Effects of biological and chemical insect control agents on fungal community structures in soil microcosms. *Appl. Soil Ecol., 42*, 54–62.

Schwarzenbach, K., Widmer, F., & Enkerli, J. (2007). Cultivation-independent analysis of fungal genotypes in soil using simple sequence repeat markers. *Appl. Environ. Microbiol., 73*, 6519–6525.

Semagn, K., Bjornstad, A., & Ndjiondjop, M. N. (2006). An overview of molecular marker methods for plants. *African J. Biotechnol., 5*, 2540–2568.

Sessitsch, A., Hackl, E., Wenzl, P., Kilian, A., Kostic, T., Stralis-Pavese, N., Sandjong, B. T., & Bodrossy, L. (2006). Diagnostic microbial microarrays in soil ecology. *New Phytol., 171*, 719–736.

Shimazu, M., & Sato, H. (1996). Media for selective isolation of an entomogenous fungus, *Beauveria bassiana* (Deuteromycotina: Hyphomycetes). *Appl. Entomol. Zool., 31*, 291–298.

Smith, C. J., & Osborn, A. M. (2009). Advantages and limitations of quantitative PCR (Q-PCR)-based approaches in microbial ecology. *FEMS Microbiol. Ecol., 67*, 6–20.

Smits, N., Rougier, M., Fargues, J., Goujet, R., & Bonhomme, R. (1996). Inactivation of *Paecilomyces fumosorosea* by diffuse and total solar radiation. *FEMS Microbiol. Ecol., 21*, 167–163.

Snedecor, G. W., & Cochran, W. G. (1989). *Statistical Methods*. Ames, IA: The Iowa State University Press.

Steel, R. G. D., & Torrie, J. H. (1960). *Principles and Procedures of Statistics*. New York: McGraw–Hill Book Company, Inc.

St Leger, J., & Wang, C. (2009). Entomopathogenic fungi and the genomics era. In S. P. Stock, J. Vandenberg, I. Glazer, & N. Boemare (Eds.), *Insect pathogens. Molecular approaches and techniques* (pp. 365–400). Wallingford, UK: CABI Publishing.

Strasser, H., Forer, A., & Schinner, F. (1996). Development of media for the selective isolation and maintenance of *Beauveria brongniartii*. In T. A. Jackson, & T. R. Glare (Eds.), *Microbial Control of Soil Dwelling Pests* (pp. 125–130). Lincoln: AgResearch.

Stock, S. P., Vandenberg, J., Glazer, I., & Boemare, N. (Eds.). (2009). *Insect pathogens. Molecular approaches and techniques*. Wallingford, UK: CABI Publishing.

Studdert, J. P., & Kaya, H. K. (1990). Water potential, temperature, and clay-coating of *Beauveria bassiana* conidia: effect on *Spodoptera exigua* pupal mortality in two soil types. *J. Invertebr. Pathol., 56*, 327–336.

Takatsuka, J. (2007). Specific PCR assays for the detection of DNA from *Beauveria bassiana* F-263, a highly virulent strain affecting the Japanese pine sawyer, *Monochamus alternatus* (Coleoptera: Cerambycidae), by a sequence-characterized amplified region (SCAR) marker. *Appl. Entomol. Zool., 42*, 619–628.

Tanada, Y., & Kaya, H. K. (1993). *Insect Pathology*. London: Academic Press, pp. 666.

Taylor, J. W., Geiser, D. M., Burt, A., & Koufopanou, V. (1999). The evolutionary biology and population genetics underlying fungal strain typing. *Clin. Microbiol. Rev., 12*, 126–146.

Throne, J. E., Weaver, D. K., Chew, V., & Baker, J. E. (1995). Probit analysis of correlated data: multiple observations over time at one concentration. *J. Econ. Entomol., 88*, 1510–1512.

Tigano, M. S., Adams, B., Maimala, S., & Boucias, D. (2006). Genetic diversity of *Hirsutella thompsonii* isolates from Thailand based on AFLP analysis and partial β-tubulin gene sequences. *Genet. Mol. Biol., 29*, 715–721.

Topp, G. C. (1993). Soil water content. In M. R. Carter (Ed.), *Soil Sampling and Methods of Analysis* (pp. 541–557). Boca Raton: Lewis Publishing.

Topp, G. C., Galganov, Y. T., Ball, B. C., & Carter, M. R. (1993). Soil water desorption curves. In M. R. Carter (Ed.), *Soil Sampling and Methods of Analysis* (pp. 569–579). Boca Raton: Lewis Publishing.

Tossell, R. W., Dickson, W. T., Rudra, R. P., & Wall, G. J. (1987). A portable rainfall simulator. *Canad. Agric. Engineer., 29*, 155–162.

Tuite, J. (1969). *Plant Pathological Methods: Fungi and Bacteria*. Minneapolis: Burgess Publishing Company.

Ugine, T. A., Wraight, S. P., Brownbridge, M., & Sanderson, J. P. (2005). Development of a novel bioassay for estimation of median lethal concentrations ($LC_{50}$) and doses ($LD_{50}$) of the entomopathogenic fungus *Beauveria bassiana*, against western flower thrips. *Frankliniella occidentalis. J. Invertebr. Pathol., 89*, 210–218.

Upadhyay, R. K. (Ed.). (2003). *Advances in Microbial Control of Insect Pests*. New York, NY: Kluwer Academic/Plenum, pp. 330.

van Winkelhoff, A. J., & McCoy, C. W. (1984). Conidiation of *Hirsutella thompsonii* var. *synnematosa* in submerged culture. *J. Invertebr. Pathol., 43*, 59–68.

Vargas, L. R. B., Rossato, M., Ribeiro, R. T. D., & de Barros, N. M. (2003). Characterization of *Nomuraea rileyi* strains using polymorphic DNA, virulence and enzyme activity. *Braz. Arch. Biol. Technol., 46*, 13–18.

Veen, K. H., & Ferron, P. (1966). A selective medium for the isolation of *Beauveria tenella* and of *Metarrhizium anisopliae*. *J. Insect Pathol., 8*, 268–269.

Vega, F. E., & Kaya, H. K. (2012). *Insect Pathology* (2nd ed.). San Diego, CA: Academic Press.

Velásquez, V. B., Carcamo, M. P., Merino, C. R., Iglesias, A. F., & Duran, J. F. (2007). Intraspecific differentiation of Chilean isolates of the entomopathogenic fungi *Metarhizium anisopliae* var. *anisopliae* as revealed by RAPD, SSR and ITS markers. *Genet. Mol. Biol., 30*, 89–99.

Versalovic, J., Schneider, M., DeBruijn, F. J., & Lupski, J. R. (1994). Genomic fingerprinting of bacteria using repetitive sequence-base polymerase chain reaction. *Methods Mol. Cell Biol., 5*, 25–40.

Vesper, S., McKinstry, C., Hartmann, C., Neace, M., Yoder, S., & Vesper, A. (2008). Quantifying fungal viability in air and water samples using quantitative PCR after treatment with propidium monoazide (PMA). *J. Microbiol. Methods, 72*, 180–184.

Vestergaard, S., Cherry, A., Keller, S., & Goettel, M. (2003). Hyphomycete fungi as microbial control agents. In H. M. T. Hokkanen, & A. E. Hajek (Eds.), *Environmental Impacts of Microbial Insecticides* (pp. 35–62). Dordrecht, The Netherlands: Kluwer Academic Publishers.

Vos, P., & Kuiper, M. (1997). AFLP analysis. In G. Caetano-Annollés, & P. M. Gresshoff (Eds.), *DNA Markers: Protocols, Applications, and Overviews* (pp. 115–132). New York, NY: Wiley–VCH.

Wang, C., Fan, M., Li, Z., & Butt, T. M. (2004). Molecular monitoring and evaluation of the application of the insect-pathogenic fungus *Beauveria bassiana* in southeast China. *J. Appl. Microbiol., 96*, 861–870.

Wang, S. B., Miao, X. X., Zhao, W. G., Huang, B., Fan, M. Z., Li, Z. Z., & Huang, Y. P. (2005). Genetic diversity and population structure among strains of the entomopathogenic fungus, *Beauveria bassiana*, as revealed by inter-simple sequence repeats (ISSR). *Mycol. Res., 109*, 1364−1372.

Warcup, J. H. (1950). The soil-plate method for isolation of fungi from soil. *Nature, 166*, 117−118.

Watson, D. W., Geden, C. J., Long, S. J., & Rutz, D. A. (1995). Efficacy of *Beauveria bassiana* for controlling the house fly and stable fly (Diptera: Muscidae). *Biol. Control, 5*, 405−411.

Webster, J. (1986). *Introduction to Fungi*. Cambridge, UK: Cambridge University Press.

Welsh, J., & McClelland, M. (1990). Fingerprinting genomes using PCR with arbitrary primers. *Nucleic Acids Res., 18*, 7213−7218.

Widmer, F., Seidler, R. J., & Watrud, L. S. (1996). Sensitive detection of transgenic plant marker gene persistence in soil microcosms. *Mol. Ecol., 5*, 603−613.

Widmer, F., Shaffer, B. T., Porteous, L. A., & Seidler, R. J. (1999). Analysis of *nif*H gene pool complexity in soil and litter at a Douglas fir forest site in the Oregon cascade mountain range. *Appl. Environ. Microbiol., 65*, 374−380.

Williams, J. G. K., Kublik, A. R., Rafalski, J. A., & Tingcy, S. V. (1990). DNA polymorphisms amplified by arbitrary primers are useful as genetic markers. *Nucleic Acids Res., 18*, 6531−6535.

Wilson, I. G. (1997). Inhibition and facilitation of nucleic acid amplification. *Appl. Environ. Microbiol., 63*, 3741−3751.

Wolf, C., & Skipper, H. D. (1994). Soil sterilization. In S. H. Mickelson (Ed.), *Methods of Soil Analysis. Part 2. Microbiological and Biochemical Properties* (pp. 41−51). Madison, WI: Soil Science Society of America, Inc.

Woomer, P. L. (1994). Most probable number counts. In S. H. Mickelson (Ed.), *Methods of Soil Analysis. Part 2. Microbiological and Biochemical Properties* (pp. 329−350). Madison, WI: Soil Science Society of America, Inc.

Wraight, S. P., Inglis, G. D., & Goettel, M. S. (2007). Fungi. In L. A. Lacey, & H. K. Kaya (Eds.), *Field Manual of Techniques in Invertebrate Pathology* (2nd ed.). (pp. 223−248) Dordrecht, The Netherlands: Springer.

Xu, J. (2006). Microbial ecology in the age of genomics and metagenomics: concepts, tools, and recent advances. *Mol. Ecol., 15*, 1713−1731.

Zietkiewicz, E., Rafalskim, A., & Labuda, D. (1994). Genome fingerprinting by simple sequence repeat (SSR)-anchored polymerase chain reaction amplification. *Genomics, 20*, 176−183.

Zimmermann, G. (1986). The 'Galleria bait method' for detection of entomopathogenic fungi in soil. *J. Appl. Entomol., 102*, 213−215.

Zimmermann, G. (1998). Suggestions for a standardised method for reisolation of entomopathogenic fungi from soil using the bait method (G. Zimmermann. J. Appl. Ent. 102, 213−215, 1986). IOBC/WPRS Bulletin. *Insect Pathogens and Insect Parasitic Nematodes, 21*, 289.

## APPENDIX

### 1. Selective media

*(a).* Beauveria *medium (Chase* et al., *1986)*

2% oatmeal infusion
2% agar
550 μg/ml dodine (N-dodecylguanidine monoacetate)
5 μg/ml chlortetracycline
10 μg/ml crystal violet

*(b).* Beauveria *medium (Shimazu & Sato, 1996)*

0.3% bactopeptone
1.5% agar
0.2 mg/ml $CuCl_2$
2 μg/ml crystal violet

**Note:** pH adjusted to 10.

*(c).* Beauveria *medium (Rangel* et al., *2010)*

1. Stock: 1.5% Dodine
   2.3 g Cyprex/Syllitt 65 w in 97.7 ml $H_2O$, or 95−200% ETOH
2. PDA + 1g/l yeast extract
3. Add 10 ml stock to 990 ml medium

*(d).* B. brongniartii *medium (Strasser* et al, *1996)*

10% peptone
20% glucose
12% agar
600 μg/ml streptomycin
50 μg/ml tetracycline
100 μg/ml dodine (N-dodecylguanidine monoacetate)
50 μg/ml cycloheximide (actidione)

**Note:** pH is adjusted to 6.3 using 1M HCl.

*(e).* Metarhizium *medium (Veen & Ferron, 1966; Liu* et al., *1993)*

1% glucose
1% peptone
1.5% oxgall
3.5% agar

10 μg/ml dodine (N-dodecylguanidine monoacetate)
250 μg/ml cycloheximide (actidione)
500 μg/ml chloramphenicol

**Note:** cyclohexamide is extremely toxic and should be handled with great care.

*(f). Metarhizium acridum medium (Fernandes* et al., *2010)*

PDA supplemented with:

0.5 g/l cholamphenicol
0.001 g/l thiabendazole
0.25 g/l cycloheximide

*(g). Culicinomyces medium (Panter & Frances, 2003)*

2.8% nutrient agar
20 μg/ml neomycin
10 μg/ml streptomycin
2 μg/ml thiabendazole
2 μg/ml dichloran

*(h). Paecilomyces lilacinus medium (Mitchell* et al., *1987)*

3.9% potato dextrose agar
1–3% NaCl
0.1% Tergitol
500 μg/ml pentachloronitrobenzene
500 μg/ml benomyl
100 μg/ml streptomycin sulphate
50 μg/ml chlortetracycline hydrochloride

**Note:** Benomyl is generally not available any more; however, Benlate containing the active ingredient benomyl (e.g., 50%) can be used.

*(i). Lecanicillium medium (Kope* et al., *2006)*

2 g L-sorbose
2 g L-aparagine
1 g $K_2HPO_4$
1 g KCl
0.5 g $MgSO_4 \cdot 7H2O$
0.01 g FeNaEDTA
20 g agar
1 l water
0.3 g streptomycin $SO_4$
0.05 g chlorotetracycline HCl
0.8 g pentachloronitrobenzene
1 g $NaB_4O_7 \cdot 10H_2O$

**Note:** pH adjusted to 4.0 with 10% $H_3PO_4$.

## 2. General culture media

*(a). Sabouraud dextrose agar + yeast extract (SDAY)*

10 g neopeptone
40 g dextrose
2 g yeast extract
15 g agar
1 l distilled water

*(b). Mixed cereal agar (Padhye* et al., *1973)*

25 g Pablum baby mixed cereal
5 g agar
250 ml water

**Note:** Mix ingredients and boil in a sealed container as the baby cereal contains spore-forming bacteria. Let cool and autoclave in small amounts as this medium will boil over easily.

*(c). Liquid culture medium (Pereira & Roberts, 1990)*

1% dextrose
1% yeast extract
0.05% antimicrobial agents (200,000 units Penicillin, 250 mg streptomycin/ml)
0.1% sunflower oil

*(d). Liquid culture medium (Samšiňáková* et al., *1981)*

2.5% glucose
2.5% soluble starch
2% corn-steep
0.5% NaCl
0.5% $CaCO_3$

**Note:** pH adjusted to 5.

## 3. Stains and mounting media

*(a). Polyvinyl alcohol (PVA) mounting medium*

Dissolve 8.3 g PVA in 50 ml deionized water
Add 50 ml lactic acid
Add 5 ml glycerine and filter if necessary
Add 0.1 g acid fuchsin if desired
Keep at room temperature for 24 h before using

*(b). Polyvinyl alcohol wetting agent*

50 ml 95% ethanol
25 ml acetone
25 ml 85% lactic acid

*(c). Lacto-fuchsin mounting medium and stain (Carmichael, 1955)*

0.1 g acid fuchsin
100 ml lactic acid

*(d). Fluorescein diacetate (FDA) (Schading* et al.*, 1995)*

Mix 35 μl of a stock solution of FDA (4 mg FDA/ml of acetone) in 4 ml deionized water.

**Note:** keep on ice protected from light and use within 1 h of preparation.

*(e). Propidium iodide (PI) (Schading* et al.*, 1995)*

Mix 60 μl of a stock solution of PI (3 mg/ml of deionized water) in 5 ml deionized water.

**Note:** store as for FDA.

## 4. Miscellaneous

*(a). Blastospore storage formulation (Blachère* et al.*, 1973)*

1 kg blastospores (22% wet moisture)
1 kg silica powder
250 ml of 200 g sucrose and 5 g sodium glutamate in water
250 ml of liquid paraffin containing 10% polyoxyethylene glycerol oleate

*(b). Germination medium (Milner* et al.*, 1991; Inglis* et al.*, 1996b)*

0.1% yeast extract
0.1% chloramphenicol
0.01% Tween 80
0.001–0.005% Benlate

*(c). List of saturated salts used for regulation of relative humidites*

Humidities vary from 1 to 2% from those previously published (J. Virolleaud, unpublished).

| *Saturated salt solutions* | *Relative humidity (%) at different temperatures (°C)* | | | | | | | | *Solubility at 20°C (Change)*[a] |
|---|---|---|---|---|---|---|---|---|---|
| | *5* | *10* | *15* | *20* | *25* | *30* | *35* | *40* | |
| Lithium chloride ($LiCl \cdot XH_2O$) | 14 | 14 | 13 | 12 | 12 | 12 | 12 | 11 | 81% (+) |
| Magnesium chloride ($MgCl_2 \cdot 6H_2O$) | 35 | 34 | 34 | 33 | 33 | 33 | 32 | 32 | 40% (=) |
| Potassium carbonate ($K_2CO_3 \cdot 2H_2O$) | — | 47 | 44 | 44 | 43 | 43 | 43 | 42 | 52% (+) |
| Magnesium nitrate ($Mg(NO_3)_2 \cdot 6H_2O$) | 58 | 57 | 56 | 55 | 53 | 52 | 50 | 49 | 43% (+) |
| Sodium chloride (NaCl) | 76 | 76 | 76 | 76 | 75 | 75 | 75 | 75 | 36% (=) |
| Potassium chloride (KCl) | 88 | 88 | 87 | 86 | 85 | 85 | 84 | 82 | 37% (+) |
| Potassium sulphate ($K_2SO_4$) | 98 | 98 | 97 | 97 | 97 | 96 | 96 | 96 | 11% (+) |

[a]Change in solubility at temperatures above 20°C.

CHAPTER VIII

# Mass production of entomopathogenic Hypocreales

STEFAN T. JARONSKI* & MARK A. JACKSON[†]

*United States Department of Agriculture, Agriculture Research Service, Pest Management Research Unit, Northern Plains, Agricultural Research Laboratory (NPARL), 1500 N. Central Avenue, Sidney, MT 59270, USA

[†]United States Department of Agriculture, Agriculture Research Service, Crop Bioprotection Research Unit, National Center for Agricultural Utilization Research (NCAUR), 1815 N. University Street, Peoria, IL 61604, USA

## 1. INTRODUCTION

The entomopathogenic Ascomycetes, *Beauveria bassiana*, *Metarhizium anisopliae* sensu lato, *Nomuraea rileyi*, *Lecanicillium* spp., *Isaria fumosorosea*, and *I. farinosus*, have undergone ever-increasing attention during the past 50 years as microbial pest control agents. There are currently more than 170 commercial products available around the world (Faria & Wraight, 2007). As a result, there is a fairly substantial body of literature and patents, and a number of proprietary techniques practised by commercial entities relating to the production of these fungi.

This chapter presents three basic approaches for mass production of these Hypocreales fungi: solid substrate fermentation to produce aerial conidia; liquid fermentation to produce blastospores, particularly of *I. fumosorosea*; and a very recent, more specialized, liquid fermentation for *Metarhizium* spp. microsclerotia for use in granules.

## 2. CULTURE MAINTENANCE

The availability and development of stable, stock cultures is imperative for long-term studies on the production and use of entomopathogenic fungi as biological control agents. Solid-substrate and liquid culture methods share common beginnings which include the development of stable stock cultures of the fungal entomopathogen. Suitable long-term storage methods for stock cultures of entomopathogens include air-drying, freeze-drying, freezing at ultra-low temperatures, submersion in oil or sterile water, or maintenance in host material (Kirsop & Snell, 1984). The objectives for storing cultures under these conditions to create a state of suspended animation *via* greatly reduced metabolic activity are two-fold: (1) preservation of the genetic make-up of the culture; and (2) reliable availability of stock cultures of the organism for experimental use. Serial vegetative transfer of stock cultures as a maintenance technique is greatly discouraged since mutations may occur that change the genetic make-up of the stock culture.

Two techniques are routinely used to start stock cultures of fungal entomopathogens: isolation of single-spores or isolation of the hyphal tip of a growing culture. Details for doing so are given in Chapter VII of this manual. The isolates are transferred onto a nutritive agar plate containing Potato Dextrose agar (PDA), Sabouraud Dextrose agar (SDA), with or without 1% yeast extract, or any appropriate nutritive agar for fungi, and grown out on the plate as a monoculture. The isolate of the entomopathogen must be sufficiently stable to undergo the repeated growth cycles required for inoculum development and for culture growth. Genetic instability can often be detected by presence sectoring with different appearance in radially growing cultures.


ISBN 9780123868992

2012 Published by Elsevier Ltd.

Store fungal stock cultures made from single spore or hyphal tip isolates at −80°C in a sterile solution of 10% glycerol. The glycerol acts as a cryoprotectant that aids in maintaining cellular integrity during freezing and thawing. To obtain these stock cultures, the isolate of the entomopathogen is grown from single spore or hyphal tip isolates on nutrient agar plates (PDA or SDA plates) for 2−3 weeks at room temperature (~25°C). See Chapter VII of this manual for more information.

Once cultures are grown-out on the plate, mince the cultures into very small pieces (~4−5 mm$^2$) using a sterile scalpel. Add approximately 10 agar pieces to each cryovial containing 1 ml of 10% glycerol. Prepare multiple cryovials (50−200) from each agar culture plate of the entomopathogen and store at −80°C for future use.

Alternatively, allow the cultures to sporulate and then collect the conidia into sterile 30% glycerol and freeze aliquots in a number of sterile 1.5- or 2-ml cryovials.

On an industrial scale the initial culture is used to inoculate several hundred slants of the appropriate nutritive medium, sufficient to initiate a year's worth of production runs. When fully sporulated, these slants are then preserved by addition of sterile 10% glycerol and frozen at −25°C. For each production cycle the appropriate number of slants is thawed, the glycerol removed, and the conidia transferred to liquid fermentation medium.

When conducting experiments with a hypocrealean fungus, initially inoculate 5−10 nutritive agar plates with the contents from one cryovial stock culture and incubate the inoculated agar plates at an appropriate growth temperature, typically at room temperature, unsealed, in the laboratory. After 2−3 weeks of incubation, most entomopathogenic fungal cultures will have sporulated or grown-out across the agar plates.

Choose a specific culture age (2 weeks, 3 weeks, etc.) for using these cultures as inoculum for experiments. Using a consistent inoculum age for experiments is critical in obtaining consistent results in the subsequent culturing experiments. By inoculating cultures weekly, you always have same-age cultures available for starting experiments.

## 3. SOLID SUBSTRATE PRODUCTION OF AERIAL CONIDIA

With some exceptions, the principal and most practical infectious stage of the entomopathogenic Hypocreales is the aerial conidium. In nature this aerial conidium is borne on erect structures in ready contact with the atmosphere, e.g., on the surface of insect cadavers. Mass production of aerial conidia using solid substrate fermentation is the closest approximation to the natural process and can be the most efficient process. Historical approaches in mass production have been discussed by Bartlett & Jaronski (1984) and Jenkins & Goettel (1997). A number of authors have described liquid fermentation to produce microcycle conidia (Thomas *et al.*, 1987; Jenkins & Prior, 1993; Kassa *et al.*, 2004), but solid substrate fermentation remains the most efficient and amenable to lower technology situations and is the primary production method in industry today.

This section is oriented for a small laboratory needing to produce decagram or hectogram quantities of hypocrealean aerial conidia for use in field trials or other experiments. The procedures presented here are not to be considered comprehensive nor the last word, but rather as a starting point for an investigator to adapt to individual needs and situations. The information here is based on methods routinely practised by the staff at the Northern Plains Agricultural Research Laboratory (NPARL) in Sidney, MT, for the past 11 years and are based on the senior author's previous industrial experience with commercial production of Hypocreales. Emphasis is on simpler methods amenable for use in individual laboratories that do not have sophisticated equipment. The basic method presented here is the product of a training course implemented by the NPARL laboratory for USDA scientists and subsequently repeated with scientists in a number of countries. It is basically a biphasic fermentation, with production of fungus in liquid fermentation used to inoculate solid substrate on which the fungus conidiates. A schematic diagram of the process is shown in Figure 8.1.

There is considerable genetic variability within each species of fungus, one expression of which is the extent of conidiation of an isolate, especially *in vitro*. Some isolates have prolific conidiation; others clearly tend toward mycelial growth and inherently poor conidiation (Figure 8.2). Some isolates produce excessive heat during fermentation, which requires the installment of a cooling system to maintain the bed temperatures. Others produce a cement-like mycelium that binds the substrate together minimizing surface area and spore production. For example, Leland *et al.* (2005), in their multi-criterion screen of *B. bassiana* for use against *Lygus* spp.,

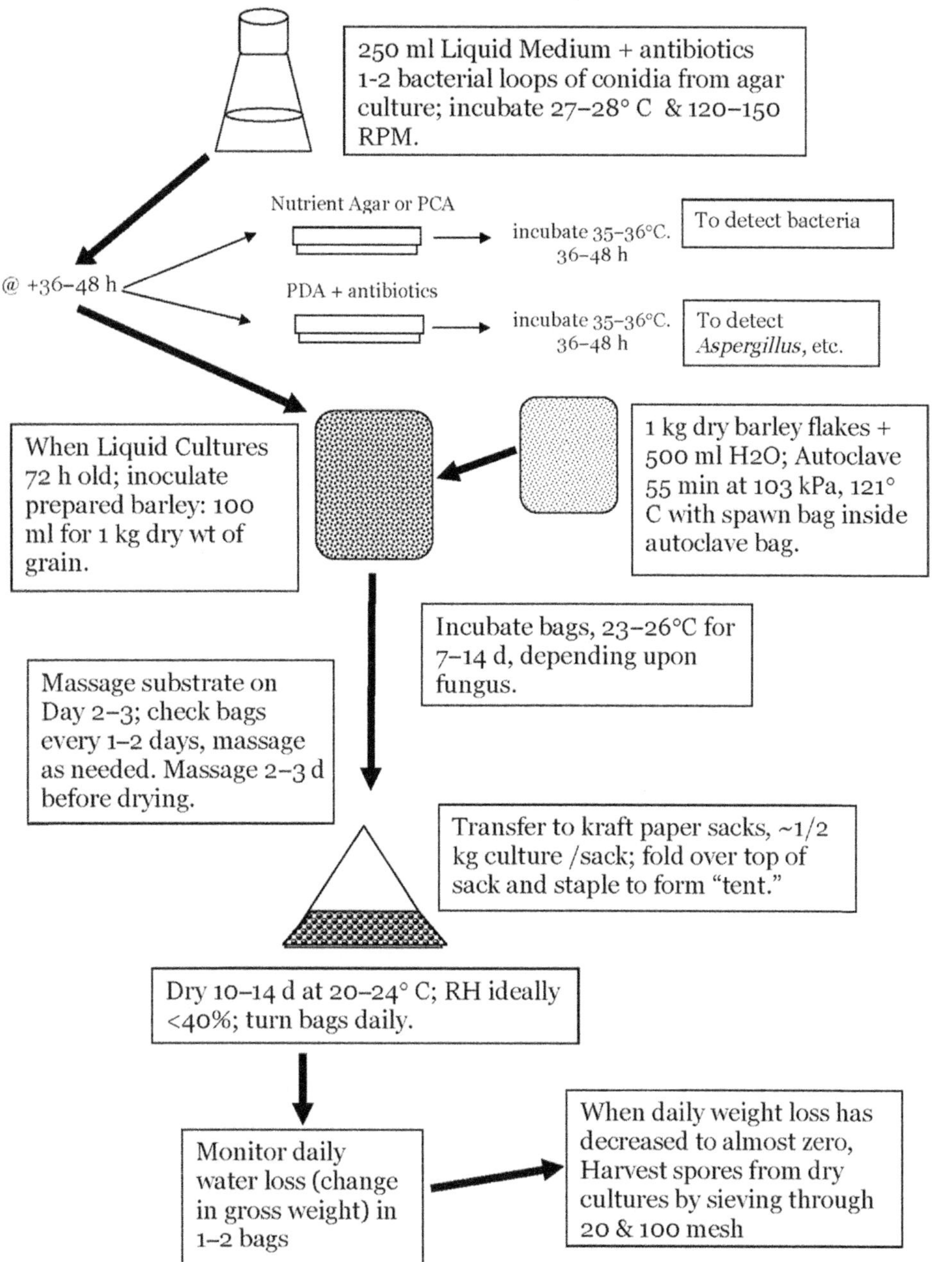


**Figure 8.1** Schematic diagram of solid substrate production of aerial conidia, including timing of phases and steps for contamination detection.

discovered that all of their efficacious and environmentally well-adapted isolates had very poor conidiation in solid substrate fermentation, even despite minor manipulation of fermentation variables (Jaronski, unpublished data). Consequently, a number of isolates could not be produced in sufficient quantity for small plot field trials and the remainder required an uneconomic level of production. Ultimately, conidial production (considered in terms of ha of use per unit of production (kg or l), is a critical determinant of the practical success of a fungus. Extensive modification of substrate and fermentation conditions could theoretically improve conidial yield, but can be a significant distraction from one's primary goals. Thus, we strongly advise that the ability of a candidate

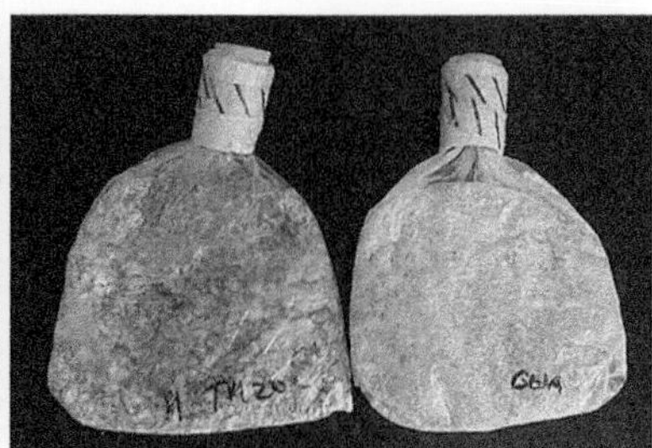

**Figure 8.2** (Left) Whole solid substrate cultures of a *Metarhizium robertsii*, showing desirable profuse spore production. Note the fine powder, which is concentrated conidia, between the substrate particles; (Right) differences between two isolates of *Beauveria bassiana*, TM20 and GHA, in spore production on solid substrate illustrate genetic limitations of this trait. The bag of GHA culture is white with profuse sporulation while that of TM20 is darker because much more of the substrate is visible because of poorer mycelial growth.

fungus to conidiate *in vitro* be evaluated during preliminary strain selection, along with virulence and adaptation of that fungus isolate for the environmental conditions in the target arena. Cultivation on the common agar media, such as PDA and SDA (with or without yeast extract supplementation), can be misleading. Even agar media based on the solid substrate of choice, e.g., rice or barley flour may not be truly indicative of performance during solid substrate fermentation. We have encountered numerous *B. bassiana* and *Metarhizium* spp. isolates that conidiated well on such media only to have extensive mycelial growth and poor conidiation on rice or barley substrate (Jaronski, unpublished data). Perhaps the wisest recourse is to first evaluate conidial production on the solid substrate to be later used, but in replicated small quantities, with elimination of isolates that perform poorly.

## A. Liquid fermentation phase

The first step in solid substrate fermentation is to produce a profuse blastospore culture with which to inoculate the sterilized solid substrate. Conidia harvested from agar media can be used directly as an inoculum for solid substrate. Introduce a small volume of sterile Tween 80® or similar nonionic wetting agent into a Petri plate or bottle of sporulated fungus culture, scrape the spores from the agar surface into suspension and then transfer the appropriately diluted conidial suspension to the solid substrate. Concentration of the conidial inoculum should be at least $1 \times 10^7$ conidia/ml but does not have to be rigorously controlled unless quantitative studies are being conducted. There are disadvantages to using conidia as inoculants, however. There is a danger that conidial harvest from agar media may become contaminated; even traces of bacterial contamination may subsequently adversely affect the solid substrate fermentation phase. Conidia take ~24 h to germinate and grow into hyphae, which then colonize the solid substrate. This lag phase allows any contaminating bacteria or saprophytic fungi, which generally multiply more quickly than the hypocrealean fungus, to get a head start and eventually overwhelm the fungus. Nevertheless, there are situations when aerial conidia must be used as inoculum. To reduce bacterial contamination an antibiotic, such as streptomycin (10 μg/ml) or chloramphenicol (25 μg/ml), can be used to make the conidial suspension fluid; yeast or fungal contamination cannot be controlled.

We recommend a liquid culture of blastospores for inoculation of solid substrate. An appropriate recipe is in the Appendix. Prepare 100 ml of liquid culture for every kg of dry grain used as solid substrate. Media that overwhelmingly favor production of blastospores over mycelium are desirable for the liquid fermentation phase. Mycelium does not distribute as readily over solid substrate in the volumes prescribed here and can result in zones of little fungal growth in the substrate. A blastospore suspension readily and easily mixes with the solid substrate to produce even growth throughout the substrate. For *B. bassiana, Metarhizium* s. l., *N. rileyi,* and *Isaria* spp., the basic medium contains glucose or sucrose as the carbohydrate, yeast extract (or killed brewers' yeast) as a nitrogen source, plus corn steep liquor. A basic recipe is given in the Appendix to this chapter. Quantities and ratio of glucose/sucrose to yeast extract can vary somewhat without effect. The corn steep liquor, a product of corn wet-milling process, is essential for good blastospore production by these fungi. Because corn steep liquor, having 50% solids, is very viscous, the recipe in the Appendix has a weight equivalent. An alternative to the corn steep is a dry form, SoluLys AST® (Rocquette), which can be substituted appropriately (8.5 g SoluLys equals 17 g corn steep liquor). We have found that the pH does not have to be adjusted (Jaronski, unpublished data). Use of a broad-spectrum antibiotic, such as gentamycin (10–15 μg/ml), streptomycin (10 μg/ml) or chloramphenicol (25 μg/ml) is recommended. The latter two antibiotics, however, must be added aseptically after autoclaving because they are not heat stable; gentamycin can be autoclaved in the medium.

Transfer one to two bacterial loops of conidia from an agar culture to 100 ml liquid medium in a 250-ml flask. One milliliter of an earlier blastospore culture can also be used. This scale allows the use of one flask per kilogram of solid substrate. Alternatively, larger or smaller volumes of liquid culture can be prepared in proportionally larger flasks, but these cultures will have to be subsequently dispensed aseptically to the solid substrate in appropriate volumes. Use of individual flasks for each amount of substrate also minimizes contamination of an entire production batch and makes inoculation of substrate more efficient.

Incubate the containers of liquid culture at 120–150 r.p.m. on a rotary shaker, at 24–28°C for 72–96 h. Agitation is required for good fungal growth and blastospore production. If available, fermentation flasks having baffles are preferable to regular flasks because the former greatly increase aeration of the liquid medium, but baffled flasks are not essential. Once the cultures are quite turbid with growth (48–72 h with blastospore inoculants; 72–96 h if conidia were used as inoculant) the liquid cultures are suitable for inoculation of solid substrate. The liquid cultures can be used immediately or stored up to a week at 4–6°C before use. While the inoculum can be quantified with a hemocytometer count of blastospores in the liquid culture, this step is not necessary. Blastospore concentrations of $1 \times 10^7$–$5 \times 10^8$ blastospores/ml are suitable for inoculation. As a semi-quantitative measurement, blastospore concentration in the liquid cultures should ideally be 30 + blastospores per microscope field when a bacterial loop of culture is examined under a coverslip with 400 × magnification. For routine production, a fairly turbid, 3- to 4-day-old liquid culture is sufficient without quantification.

Quality control procedures during liquid fermentation are very important in preventing production losses to contaminants. Once contaminants establish themselves in the solid substrate, the fungus cultures cannot be rescued. Inoculate Nutrient agar or Plate Count agar with a sample of each liquid culture after 24–36 h of incubation and incubate for 36–48 h at 35–36°C to detect bacteria. The incubation temperature prevents the hypocrealean fungi from growing, yet allows most bacteria to multiply and become manifest as tiny colonies. Contaminant fungi (*Aspergillus*, etc.) can be detected on PDA plus antibiotic within 2–3 days when incubated at 35–36°C. These temperatures do not greatly inhibit most species of *Aspergillus* spp., *Rhizopus* spp., *Penicillium* spp., etc. If necessary, liquid cultures can be refrigerated for up to a week while these contaminations checks are being conducted. In addition, you can sample the liquid culture aseptically and examine with 400 × phase-contrast magnification for common bacteria, which can be observed as motile rods or chains of cocci, as well as atypical fungal growth (based on your familiarity with the appearance of your fungi). The liquid and solid fermentation steps should be scheduled to facilitate these contamination detection steps (Figure 8.1).

### B. Solid substrate phase

#### *1. Selection of solid substrates*

The basic requirements for healthy growth and conidia production of any filamentous fungus are adequate nutrition (suitable carbon, phosphorous and nitrogen sources), temperature, pH, moisture, and gas exchange of $CO_2$ and $O_2$. While trace elements and vitamins can be added, these are not essential for adequate conidiation (S. Jaronski, personal observation). Substrates can be generally any agricultural product, or porous inorganic carrier. A wide variety of substrates have been described in the literature (Bartlett & Jaronski, 1984). Once hydrated and sterilized, the substrate can readily absorb further nutrients from liquid medium, providing the necessary requirements for healthy fungal biomass production. Filamentous fungi decompose starch by secreting highly concentrated hydrolytic enzymes from their hyphal tips to penetrate the substrate and access nutrients, promoting metabolic activity and rapid development. However, the physical heterogeneity of substrate beds presents a drawback to solid substrate fermentation. This is a challenge that can be decreased by periodic mixing of granules (manually massaging the plastic fermentation bags) to facilitate mycelial branching during early stages of vegetative growth in order to promote metabolism, which in turn optimizes spore production. In addition, the insulative nature of a solid substrate matrix imposes limitations to the volume and geometry of a solid substrate culture and bed thickness needs to be controlled. Generally, a solid substrate mass should be no more than 5–7 cm thick.

A wide variety of grains and inorganic substrates have been examined for use (Bartlett & Jaronski, 1984; Jenkins & Goettel, 1997; Sahayaraj & Namachivayam, 2008). A review of the literature will identify a wide range of potential substrates. A detailed discussion of advantages and disadvantages of these alternatives is beyond the scope of this chapter. In general, good carbohydrate and organic nitrogen sources are needed for good sporulation.

There are two commonly used grains: rice (*Oryza sativa*) and *flaked* barley (*Hordeum vulgare*). Flaked oats (*Avena sativa*), flaked rye (*Secale cereale*), and wheat (*Triticum* spp.) groats or bulghur are also suitable. Wheat bran is also suitable, but less so than the previously mentioned grains because it generally provides lower conidial yields and is not as appropriate for producing dry (*versus* fresh) conidia, even when the bran is mixed with organic or mineral inert to increase inter-particle spacing and substrate friability. We have experimented with many grains, beans, seeds, pulp, etc. Unprocessed barley grain itself is not suitable as a substrate because of its hard outer hull. The hull must be removed and the remaining grain must be milled or flaked. Flaked barley is readily available and cheap in North America and Europe as an animal feed and for making beer. Similarly, rolled or flaked oats are oat groats that have been rolled into flat flakes under heavy rollers and then steamed and lightly toasted, making them very suitable as a solid substrate. These are also widely available and cheap in North America and Europe. Rice is widely used as a substrate for production of *B. bassiana* and *Metarhizium* spp. Parboiled rice is much more preferable than regular uncooked rice. Parboiling is a special cooking process prior to milling, in which paddy rice is soaked and then steam cooked. The rice is then dried and subjected to the standard milling process to remove the hull and bran. This process changes the texture of the rice, making it firmer and less sticky, with more durable kernels. Parboiled rice is not readily available in the North American consumer market but can be obtained through food service suppliers. Parboiled rice is superior to regular white rice, because of differences in handling characteristics. Nevertheless, flaked barley (Figure 8.3) or oats are our substrates of choice for hypocrealean fungi because: (1) the fungi will penetrate and utilize the grain with ease; (2) hydrated barley and oats absorb all liquid inoculum with great efficiency during autoclaving without any previous preparation, whereas rice must be cooked first; (3) after hydration and steam sterilization, these grains crumble well into individual granules that provide a large surface area during fermentation and maintain friability better than rice; (4) barley and oats maintain moisture content well during the fermentation; (5) the two grains do not decompose into minute particulates that can mix with the end product; (6) barley and oats yield greater numbers of conidia than rice; and (7) both grains are readily obtained in North America and are cheaper than rice, especially parboiled rice. Ultimately the decision about which substrate to use will be dictated by grain availability and cost.

An alternative to grain is a small, inert substrate that can absorb liquid media. One such consists of clay beads, e.g., Seramis™ granules, used in European horticulture and hydroponics. Another is Celetom®, or Diatomite®, a mined, diatomaceous earth granule produced in the U.S. to absorb liquid spills. Perlite is not suitable because it does not absorb liquid medium evenly and can disintegrate during spore harvesting to produce fines, which can be a problem in sprayable formulations. Absorptive

**Figure 8.3** Magnified view of flaked barley, which is an ideal solid substrate for hypocrealean fungi.

granules from recycled paper, e.g. Biodac®, are not suitable, nor are clay granules used in cat litter. Zeolite granules are another potential inert mineral substrate, but zeolite from different sources can have subtly different properties that can affect fungal growth and sporulation. The Seramis and Celetom granules offer potential for flexible control of nutrients and for recycling (after washing and re-sterilization), but in general conidial yields are lower than with grains. A suitable liquid medium must also be developed for each fungus.

*2. Selection of solid substrate container*

A wide variety of rigid- and flexible-walled containers can be used—glass jars, glass or metal pans, plastic boxes, mushroom spawn bags, polypropylene autoclave bags, and even the ubiquitous high-density polyethylene or polypropylene shopping bag. A key characteristic is that the container should provide good gas exchange while retarding water loss. The size of the container is partially dictated by conidial needs. Typically, a hypocrealean fungus can produce 20–150 g conidia/kg, having a titer of $1 \times 10^{10}$–$2 \times 10^{11}$ conidia/g (depending on the fungus) of solid substrate.

Flexible container walls allow massage of the fungal cultures at inoculation and during fermentation. For small production batches, each involving 100–200 g of substrate, plastic autoclave bags or small, high-density polyethylene plastic shopping bags can be used (Figure 8.4). Autoclave bags, 25 × 30 cm (for 50–200 g substrate) or 30 × 60 cm (for 200–400 g substrate), can be readily obtained from major scientific or hospital supply houses. These bags can be closed by simply turning over the flap and securing with staples or clips, but this method can lead to frequent contamination of the cultures in handling and reduced gas exchange. Alternatively, bags may be constructed with necks that allow closure with a foam or cotton plug, or capped with a cloth membrane. Small fermentation bags can be constructed of polyvinylchloride (PVC) water pipe joints through which the open end of the bag is passed, draped back over the joint and taped in place with autoclave tape, creating a flexible-walled flask (Figure 8.4). Zipper-lock bags generally do not allow sufficient gas exchange for good fungal growth and sporulation, but Ziploc® Brand Fresh Produce Bags (S.C. Johnson Inc.) , with a nominal capacity of 3.8 l, do. These bags however, can not be autoclaved, but

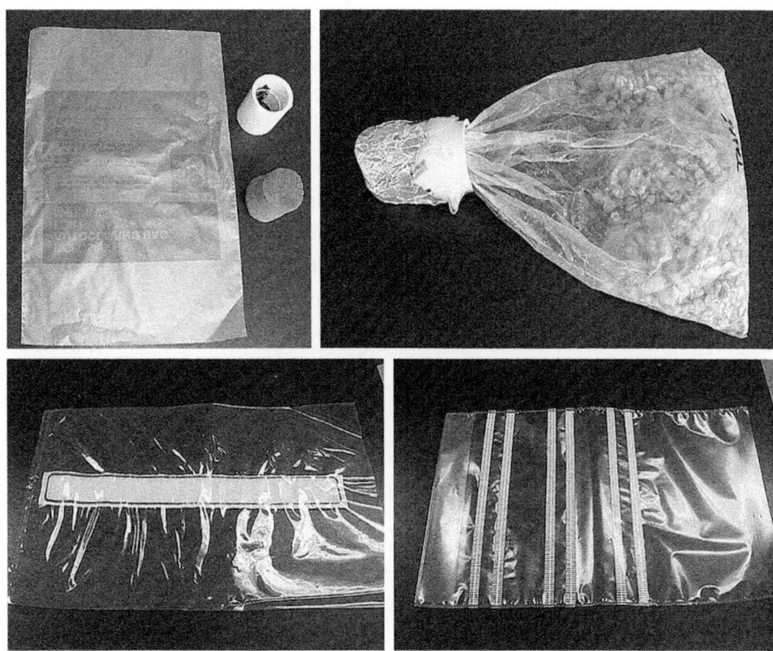

**Figure 8.4** (Top) Small, homemade solid substrate fermentation bag. (Left) the components being a small autoclave bag, plastic pipe joint and foam stopper; (right) the assembled bag, containing 100 g of autoclaved substrate. (Bottom) Two types of commercial mushroom spawn bags useful for production of hypocrealean fungi ((left) Unicorn bag, (right) SacO2 bag).

unopened bags approach sterility. The substrate must be autoclaved separately and aseptically transferred to these bags.

For greater production needs (> 1 kg of conidia) commercial mushroom spawn bags, which can each contain 1–1.5 kg substrate are preferable, although bags must be obtained in quantities that may be cost inhibitive. Spawn bags are readily autoclavable, rugged, have flexible walls, and allow low-pressure air exchange and maintenance of water activity levels conducive to good fungal growth and conidiation (Aw > 0.98). Two types of commercial bags are available: a traditional spawn bag with one or more gas-permeable vent patches (Unicorn Bags, Garland TX, USA), and SacO2 Microsac with multiple ventilation strips (SacO2, Combiness, Belgium; see Figure 8.4). If a mineral carrier is to be used in lieu of a grain, care must be taken to prevent the substrate from perforating plastic bag walls. Alternatively, one can make bags of Tyvek® (E.I. Dupont Corp.), a non-woven polyethylene, building sheathing membrane (Jaronski, unpublished data). Tyvek offers good gas permeability but reduced water vapor permeability and can be autoclaved. It comes in large sheets 1–3 m wide and 30–60 m long, which must be cut, folded and stapled to create bags. With Tyvek bags, initial fermentation should be carried out using a plastic outer bag to prevent moisture loss, but the outer bag is removed once the culture has begun to sporulate. Bag sizes can be tailored to substrate amounts; bags containing 5 kg substrate are possible (S. Jaronski, unpublished data).

### *3. Preparation of solid substrate*

#### *a. Flaked barley or oats*

Add 50 ml of reverse osmosis or distilled, not tap, water for every 100 g of dry grain within the fermentation bag, and mix well by shaking. Let the moistened grain rest for 15–30 min to properly absorb moisture before autoclaving; an occasional mixing by manual agitation of the bags may be beneficial for even water dispersal through the grain. Commercial spawn bags may be heat sealed at this point. With small autoclavable bags, place the dry grain into the bag, add the requisite amount of water and mix well. If a simple open bag is used, the open end should be folded over and loosely clipped. An outer autoclave bag is advisable for keeping the substrate uncontaminated after autoclaving. If a stoppered plastic bag is used, add a foam or cotton plug to close it. Lay the bags flat on racks or pans and autoclave at 103 kPa (121°C) for 30 min for substrate volumes of ~250 ml each to 55 min for substrate volumes of ~1 l or slightly greater.

With spawn bags, flatten out the moistened substrate within the bag to form a 3- to 5-cm-thick bed. Simply turn 5–10 cm of the open end of the bag over and down and place the bag, open end first, into a larger autoclave bag. This prevents any messy spills should the bag open during autoclaving. Alternatively, the open end may be heat sealed (see Section 3B4b) and the bag autoclaved without an outer bag. However, the outer bag allows storage of the sterilized substrate without contamination for days or weeks. Close the autoclave bag loosely, e.g., with a single piece of autoclave tape to allow steam to escape during the autoclave process. Place the bags horizontally on racks or in autoclave pans without stacking or packing. The substrate bed within bags must not be more than 3–5 cm thick, or sterilization will be insufficient. If autoclaving sealed bags, be sure that the bags are not inflated with air; they may explode during autoclaving if they are.

Autoclave spawn bags with 1 kg substrate at 121°C and 103 kPa for 55 min. Use a gravity-exhaust autoclave or cycle; vacuum-exhaust cycles can cause the bags to rupture. It is also important to arrange bags so that the heat can evenly penetrate the substrate from all sides. Solid substrate, especially moistened grain, has insulative properties that can cause problems in homogenous sterilization. Before autoclaving substrate for the first time, it is advisable to become very familiar with the autoclave to be used, as well as the principles and problems of autoclaves in general (they are not fool proof and have problems). The internet provides some good sources of information (for example, http://oomyceteworld.net/protocols/autoclave%20operation.pdf). Initial use of biological indicators inserted into test bags of substrate, then removed and incubated per manufacturer's instructions will reveal problems in complete sterilization of the medium using available sterilization equipment.

#### *b. Parboiled rice*

Add 30 ml of water per 100 g rice in the bags. A hydration medium of $KH_2PO_4$ (0.97 g/l); $H_2SO_4$ (0.41 g/l), and yeast extract (0.31 g/l) can be used with rice for *B. bassiana*, but it does not seem beneficial for *Metarhizium* spp. Because of the initially large volume of water, bags must be kept upright during autoclaving, during which time the excess water is absorbed by the rice. Some experimentation may be necessary to fine tune the necessary volume of water, depending on the type of rice used. There is variation among long- and short-grained rice and with broken *versus* whole rice; experimentation will be needed for whatever type is to be used. Autoclave the bags at 121°C for the periods described above for barley/oats.

*c. Polished or non polished rice (non parboiled)*

Most of the published methods for preparing rice are inadequate for polished rice. Some protocols call for precooking the rice before autoclaving, which adds extra, potentially cumbersome steps. A protocol suitable for smaller-scale production was developed by C. Hauxwell, Queensland University of Technology (personal communication), and has been adapted for our laboratory. Place the requisite amount of dry rice in a small spawn or other fermentation bag and autoclave as described elsewhere in this chapter without adding water. Following autoclaving, add sterilized, distilled, or reverse osmosis water through the bag opening at the rate of 75 ml per 100 g of rice. Loosely reseal the bag with autoclave tape or by re-stoppering. Then microwave the bag at full power until all the liquid is absorbed. As a general guideline, with 600 W power output, optimal microwaving should be 5 min for 100 g of rice. The exact time and power should be empirically determined; the appropriate endpoint is when all the water is absorbed by the rice, which is relatively soft after microwaving. It is possible to generate sufficient heat with high-power (> 1000 W) microwave ovens to melt the bags, or, if the bag is sealed, it may explode. The size of the fermentation bag will be dictated by the size of the microwave oven. It will not be possible to use full-size spawn bags with this approach in standard microwave ovens. After autoclaving, allow the substrate to cool completely before inoculating. The substrate in the bags should be subsequently broken up by hand massage. If you leave the bags of substrate overnight the substrate mass will crumble much more easily. Crumble the substrate well, so that it is as close to individual flakes as possible for maximum surface area and thus sporulation.

An alternative to steam autoclaving is tyndalization. While steam sterilization is much more preferable, tyndalization may be useful when an autoclave is inaccessible. Tyndalization requires that the liquid or solid substrate medium is thoroughly heated to 100°C (boiling water) and maintained at that temperature for 15 min, then cooled and incubated at room temperature for a day, then reheated. Three cycles are necessary to eliminate microbial contaminants.

The autoclaved bags of solid substrate can be stored in sealed outer bags for several weeks at room temperature; refrigeration causes excessive moisture condensation inside the bags and should be avoided. Storage of autoclaved substrate for 3–4 days is beneficial because it allows any contaminating fungi (*Aspergillus, Rhizopus*, etc.) to grow out and become visible, and thus preclude contaminated fermentations.

### *4. Inoculation of substrate*

In general, the volume of liquid inoculant should be 75–100 ml/kg of dry substrate. It has been our experience that lower volumes result in uneven growth through the substrate and opportunity for any contaminants to outcompete the fungus. Greater volumes do not increase spore yield and may even decrease it by creating excessively moist conditions, which interfere with gas transport through the substrate.

*a. Small bags*

Inoculation is simple and very similar to inoculation of a glass flask. In a sterile environment, open the bag by removing the closure and add the requisite amount of liquid inoculum (7.5–10 ml liquid culture/100 g dry substrate) with a sterile pipette. Replace the neck closure. Twist the bag below the neck or otherwise pinch the bag neck closed and massage the bag and/or shake it to disperse the inoculum through the substrate. The bags can then be placed for incubation by either laying on their sides or suspended from the neck rings. The key point is that gaseous transfer between atmosphere and culture should not be impeded by a constricted bag neck.

*b. Spawn bags*

Ideally, all inoculation procedures should be conducted in a laminar flow hood or other sterile environment to minimize contamination. In a worst-case situation, a 'clean room' can be created with HEPA-filter room air cleaners, use of sterile garments, gloves, shoe and head coverings, and stringent aseptic technique.

A thermal-impulse plastic bag sealer (Figure 8.5) will be necessary. A sealer with a 5-mm-wide seal which is long enough to seal across a spawn bag (40 cm) is

**Figure 8.5** Thermal impulse bag sealer, necessary for use with commercial mushroom spawn bags.

needed. These devices can be readily obtained from a number of companies that advertise on the internet, such as Uline Corp. Pleasant Prairie WI, USA. Place the sealer in the laminar flow hood and turn it on to a setting sufficient to seal the bag plastic but not completely melt the plastic. This may take some experimentation.

Remove flasks from the shaker to a location accessible to the sterile work area. Remove spawn bags one at a time from outer autoclave bags within the sterile work area. The spawn bag plastic may stick to the autoclave bag; if it does, gently separate the two manually and slide the spawn bag out. Stand the spawn bag up (the advantage of a gusseted spawn bag is that it readily stands vertically) and open it (Figure 8.6A). Remove the flask closure, flame the flask's rim and pour an appropriate volume of inoculum into the bag, taking care not to wet the upper surfaces of the bag. The inoculation rate is 75–100 ml of liquid culture per kg of dry substrate. At this point the open end of the bag should be folded over several times or twisted to temporarily close it and the substrate and inoculum within mixed together by shaking the bag vigorously by hand (Figure 8.6B).

Lay the bag flat on the work surface and seal the open end of the bag with the heat sealer (Figure 8.6C). The bag

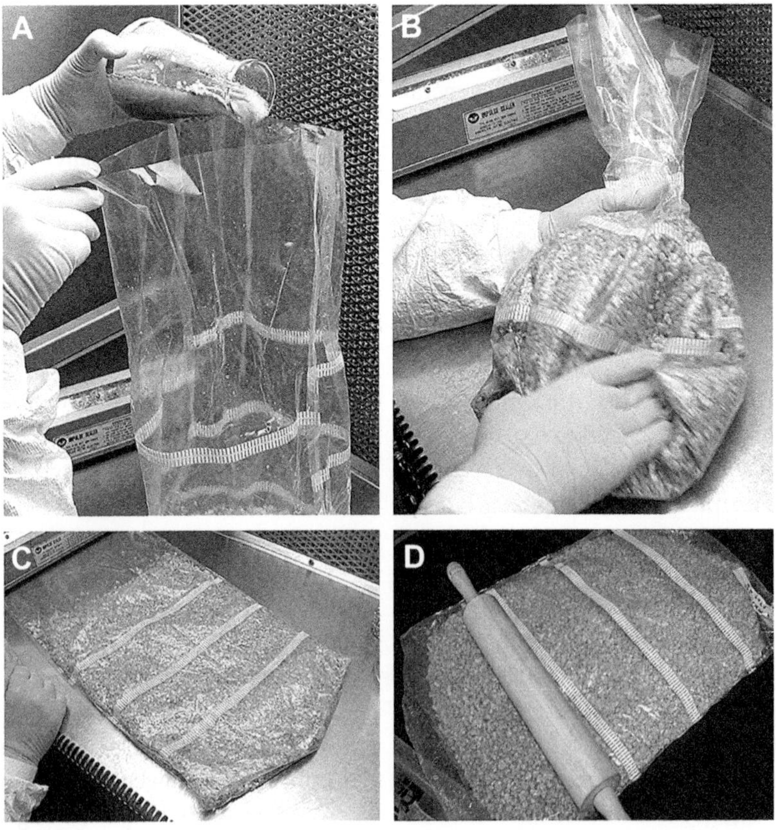


**Figure 8.6** A. inoculating substrate within a spawn bag; B. mixing the inoculum and substrate; C. heat sealing the bag opening; D. breaking up the whole culture with a pastry rolling pin.

should have only slight to moderate head space within it before sealing, but not be greatly inflated, because this latter condition will make later massaging of substrate difficult to impossible. However, some head space is needed for good gas exchange. It is advisable to make three parallel seals, about a centimeter apart and within a few centimeters of the bag end. This procedure securely seals the bag against any possible leaks. Flatten the inoculated substrate within each bag into 5- to 7-cm-thick beds. Thicker beds reduce gas transport and heat dissipation, leading to reduced conidial yields.

We have also used an alternative inoculation method whereby the spawn bags are sealed before autoclaving and the liquid inoculum is either added through an opened corner of the sealed bag end, or injected with a hypodermic syringe through the wall of the bag. With the former method, make a diagonal cut at one corner of the heat sealed end of the bag, creating an opening 8–10 cm wide. Add the inoculum through this opening and reseal. For syringe inoculation, use a 60-ml syringe with large bore needle (10–15 gauge). The liquid inoculum cannot have mycelial clumps that would clog the needle. First, sterilize a small surface of the bag wall with 70% ethanol, then introduce the inoculum by injection and immediately cover the hole with a piece of tape.

### 5. *Fermentation*

Incubate the bags of inoculated solid substrate at 24–27°C. An incubator is not necessary if you can use a room with relatively stable temperatures. Temperatures of 20–24°C will prolong the fermentation period. Temperatures below 20°C not only greatly slow mycelial growth but also reduce eventual conidial yield for many fungal isolates. Because the fermentation is exothermic, ambient temperatures greater than 28°C may reduce conidial yield in the center of the substrate mass, especially if the thickness of the substrate is greater than 7–10 cm. This fermentation phase may be conducted in the dark for *B. bassiana* and *Metarhizium* s.l.; light has no major effect on growth or conidiation. However, many isolates of *I. fumosorosea* and *I. farinosus* have a requirement of far blue or near UV light for good conidiation (Sakamoto *et al.*, 1985). Such light spectrum can be supplied with 'daylight spectrum' (a correlated color temperature of 5000–6500 K) fluorescent lighting suspended 10–20 cm over the bags of substrate. There are isolates of *I. fumosorosea* that will conidiate well in the dark (Jaronski, unpublished data); this characteristic may make them more preferable than others.

As soon as the substrate has visible mycelial growth and has begun to be bound up by the fungus, it must be broken up regardless of fermentation container. This stage usually occurs within 48 h of inoculation. Contrary to some published information, mechanical breaking up of the whole culture is not deleterious to good sporulation and is actually necessary. Failure to break the substrate up at this point leads to lower conidial yields by as much as 80% in our experience. Break up the substrate within each bag by manual massage and manipulation, taking care not to rupture the bag seams. Judicious use of a kitchen pastry rolling pin is also valuable in further disrupting the substrate mass (Figure 8.6D). The goal is to reduce the solid culture to individual particles as much as possible, maximizing surface area for subsequent growth and conidiation.

Allow the fermentation to continue for a total of 7–14 days for the solid substrate phase. Some isolates of *Metarhizium* spp. and *B. bassiana* require longer periods of fermentation and as much as 1 month is needed for good sporulation (Jaronski, unpublished data). With most isolates however, the shorter period is sufficient; by 7–10 days, sporulation of most strains has reached a plateau. A time-course study can be conducted with a desired isolate to optimize cycle length, if desired. Cultures of *Metarhizium* spp. should be watched for white vegetative overgrowth that can occur in the latter part of the solid substrate cycle. If or when this white growth is noticed, the fermentation should be discontinued immediately and the culture dried as described below. Allowing the white mycelial growth to continue will greatly diminish spore yield.

Contamination: while fungal contamination of *B. bassiana* cultures is readily visible (foci of green or black fungus amidst white *B. bassiana* mycelium), contamination of *Metarhizium* spp., particularly by green *Aspergillus* spp. or *Penicillium* spp., can be much more difficult to detect. In general the shade of green of a contaminant *Aspergillus* spp. is different from *Metarhizium* spp. and can be discerned with experience. Contamination by bacteria or yeasts is usually manifested by wet spots or poor mycelial growth in the substrate. If contamination is in isolated foci within the substrate, the physical manipulation of the substrate described earlier will spread it through the entire bag. It is best in such situations to autoclave and dispose of the contaminated bag(s). If the quality control techniques described earlier did not reveal any contamination, then lack of complete sterilization of the substrate, non-sterile inoculation, or other improper handling of materials is to be suspected.

### 6. *Culture drying phase*

If the spores are to be used within 1–2 weeks after production and only in an aqueous carrier for application, then drying is not necessary. See Section 3B7a for harvest of fresh spores. For good shelf-life and especially for formulation in oils or wettable powders, fungus spores have to be dried. The desired moisture of conidia is $< 7\%$ (w/w) by gravimetric determination or a water activity of $< 0.30$. Water activity ($a_W$), commonly used in the food industry, is a measure of vapor pressure of water in a substance relative to pure water and is therefore a measure of the biological availability of water, especially for microorganisms. The $a_W$ can be directly related to relative humidity and is expressed on a scale of 1.000 (= 100% RH) to 0.000 (= 0% RH) [see Wikipedia (2011) for a discussion of this topic].

In general, solid substrate cultures of *B. bassiana* may be dried quickly, within 2–3 days (although longer drying periods are satisfactory), whereas it has been our experience that *Metarhizium* spp. cultures need slower drying ($> 5$ days) to maintain high conidial viability. Some larger-scale production systems merely transfer the sporulated cultures to open pans, tubs, or bins and allow drying to proceed without containment. However, this method can lead to contamination of nearby areas, as well as exposure of workers to aerial spores and is not to be recommended. Drying within some sort of contained system is preferable.

The simplest method for slow, semi-controlled drying is to transfer the sporulated cultures to '57- to 66-lb' (26- to 30-kg) Kraft paper bags such as used in North American grocery and hardware stores. These bags, made of a heavy brown paper, have the appropriate moisture permeability for good drying and also serve to contain the conidia during the drying cycle. The large bags typically measure 305 × 178 × 432 mm (Figure 8.7); smaller, 'hardware' bags (15 × 9 × 30 cm) suffice for small quantities of substrate. These are suitable for spawn bag cultures. Bags or pouches may also be prepared from heavy shelf paper. *Beauveria bassiana* conidia can be rapidly dried using sacks of thinner paper, exposed to moving air in a relatively dry atmosphere ($< 40\%$ RH), or in plastic drying boxes described below in this section.

Prepare each paper bag by cutting one-third of its length horizontally off the top (Figure 8.7). This piece, which should be no wider than the bag, is placed as a liner on the bottom of the bag to increase tensile strength of the paper exposed to moist substrate. If determining yields, tare bags prior to adding substrate and again after substrate is added. In any case, determine gross weight of two filled bags in any production batch and record the weights. These bags will be used to monitor the drying cycle.

Ideally the work area for transfer of whole cultures to drying sacks should be well ventilated or at least contained to prevent contamination of adjacent areas. After donning appropriate protective wear (see Section 3B10) and surface sterilizing the work area, cut open each fermentation bag one at a time. Transfer 500–1000 g (fresh weight of whole culture) into each large paper bag, or up to 300 g into a smaller paper bag. The transfer should be slow and careful to minimize escape of spores into the air. With regular mushroom spawn bags it is often convenient to cut open the top, then cut down one seam to the level of the substrate. The culture can then be transferred with a large scoop with minimal escape of spores. When most of the culture has been transferred, the bag can be opened completely and the remaining culture

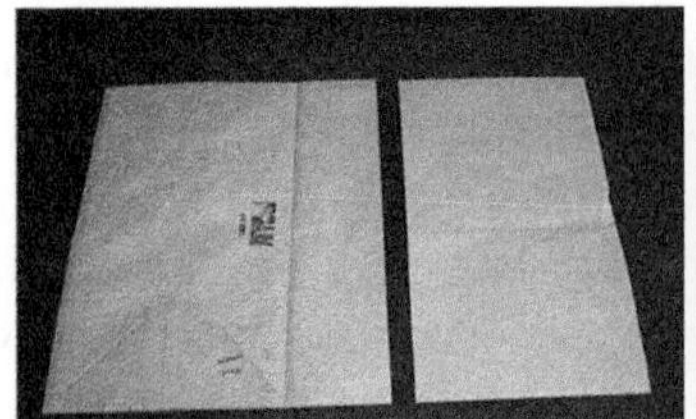

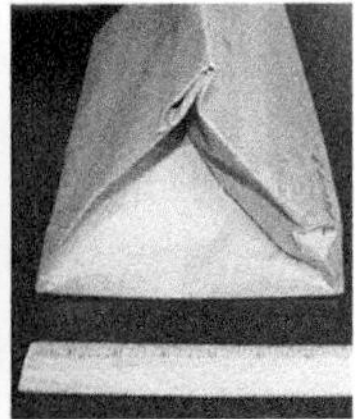

**Figure 8.7** Large 'kraft paper' grocery or hardware bag used for drying solid substrate cultures. (Left) the distal 17 cm of this bag is cut off for placement into the bottom of the bag as a liner; (Center) diagonal view of the filled and closed drying bag; (Right) side view of drying bag, showing how the top is folded and stapled shut.

transferred with a small scoop or large spoon. Move emptied sacks to an autoclave bag and dispose after autoclaving.

Crumble the substrate well but gently by hand to avoid excessive release of conidia into the air when transferring. Fold over the end of the bag twice and seal the sack with staples or binder clips (Figure 8.7). This procedure creates a triangular tent in which the culture will dry. Place bags on a wire drying rack (a solid surface will prevent even drying) (Figure 8.8).

Drying should occur at 20–26°C and low (30–40%) humidity. In situations where ambient humidity is high (> 70%), use of a dehumidifier in the drying room is advisable. Turn bags daily to move the substrate around so that the culture dries evenly. Failure to turn the substrate, especially during the initial 3–4 days, can result in uneven drying and even vegetative regrowth, lowering the harvestable yield of spores. Designate one or two drying bags for monitoring of moisture loss. Weigh each drying bag periodically during the drying cycle (whole gram accuracy is sufficient) and monitor the progress of the drying by plotting gross bag weight *versus* day of drying. Continue drying until there is little or no weight change between successive days (drying has then reached an asymptote). At 20–25°C and 30–45% RH drying is usually complete within 8–10 days. If ambient humidities are high (> 40–50%) use of a dehumidifier in the drying room is strongly recommended.

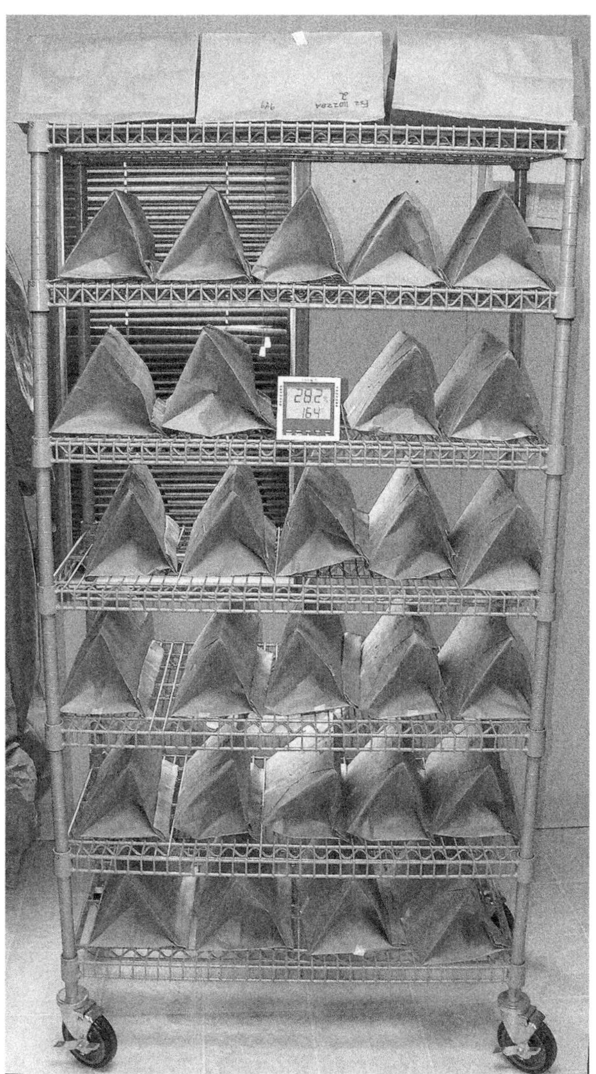

**Figure 8.8** Bags of fungal culture arranged for drying on a mobile wire rack.

When the drying phase is completed, determine water activity or gravimetric moisture content of the whole culture. Gravimetric moisture content requires destructive sampling of 1–2 g of conidia. In the simplest approach, place 1–2 g of whole culture in a tared container, then determine gross and net weights. Dry at 50°C for 24–48 h, then reweigh. The weight loss divided by the original net weight represents moisture content. Alternatively, moisture may be determined with an infrared moisture analyzer, e.g., Ohaus MB23, Mettlar Toledo MJ33, or IRMA-Lab 858 (MoistTech, Irma CA, USA). Ideally, moisture should be measured with a water activity meter, e.g., Aqualab® (Decagon Products, Inc., Pullman WA, USA), AW Labswift™ (Neutec Group Inc., Farmingdale NY, USA), or LabMASTER-aw (Sartorius, Bohemia NY, USA), or Hygrolab 3 (Rotronic AG, Bassersdorf, Switzerland).

For relatively rapid drying, such as for *B. bassiana*, air chambers may be fashioned from suitably sized plastic storage boxes (Figure 8.10). Each box is fitted with air-line connections, which are connected via tubing to an air manifold with flow control valves. Air is supplied to the manifold by an electric compressor and distributed to multiple boxes according to the flow valves. The lid of each box has two 5-cm-diameter holes covered with 150–200 thread count fabric attached with tape or hot glue. These act as exhaust vents. Within each box there is a raised platform fashioned from metal hardware cloth or screening. The screen is crimped to elevate the platform 2–4 cm off the bottom of the box. When ready to be dried, 500 g of whole culture are transferred to the mesh platform in a box, the lid replaced, and the air supply connected.

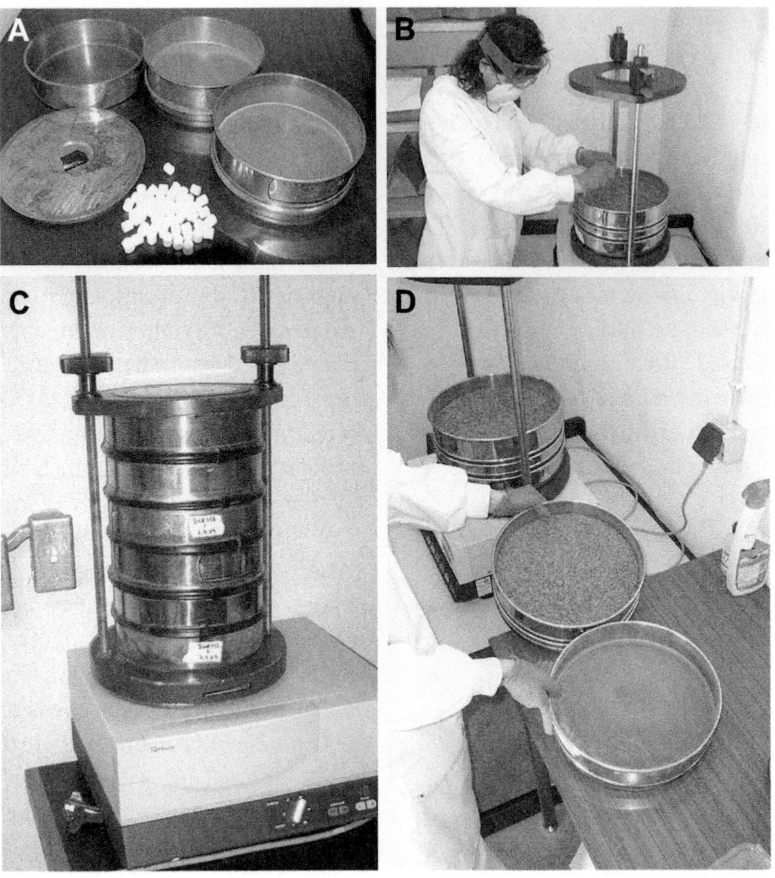

**Figure 8.9** Sieves for harvesting dry conidia from whole solid substrate culture. A. The components: an ASTM 20-mesh sieve, 100-mesh sieve, lid, collecting pan, and ceramic beads. B. Worker loading dry whole culture into sieves. C. Two sets of assembled sieves on a vibratory shaker. D. Harvested conidia from 1 kg of *Metarhizium robertsii* solid substrate culture.

This procedure should be conducted in a surface-sterilized work area with spore capture capability to prevent inadvertent spore escape into the surrounding environment (see Section 3B10). If yields are to be determined, tare the boxes prior to adding substrate and weigh again after substrate is added. Be sure to dry only one isolate at a time and clean the work area thoroughly between isolates. Crumble the substrate well when transferring and work slowly to minimize spore release into the air. It is essential to wear appropriate protective devices during this procedure (see Section 3B10). Use a Gilmont® flow meter or equivalent to adjust air flow (l/min) to equal 20 air exchanges per hour for *B. bassiana*. For this purpose you will need to calculate the volume of the box used. Weigh each tared drying box daily or every other day during the drying cycle. To monitor progress of drying, plot gross box weight *versus* day of drying. Continue drying until there is little or no weight change between successive days (drying has reached an asymptote). At this point determine water activity or gravimetric moisture content.

### 7. Harvest

#### a. Fresh conidia

In cases when fresh conidia can be used in aqueous suspension sprayed on to a target crop shortly after production, the spores do not have to be dried. The bags of sporulated substrate can be stored at 3–5°C until use. When you are ready to apply the fungus, suspend the whole culture in water with a wetting agent (e.g., 0.1% Tween 80, Tween 20, or Silwet L77®) in a large container that can be readily agitated to wash the conidia off the substrate. The spore suspension is then filtered through an ASTM 100 mesh (100 μm) screen which will remove particles large enough to clog most sprayer nozzles and screens. In North America, 100-, 200-, and 400-μm mesh strainers that fit common 11.4-, 18.9-, 26.5-, and 208-l (3-, 5-, 7-, and 55-gal) plastic buckets can be obtained (e.g., US Plastics Corp., Lima OH, USA). With such screens the spore suspension can be filtered and directly loaded into the buckets at one time; the buckets have water-tight lids with spouts which make facilitate storage and later transfer to a sprayer. In case such screens are unavailable, one to two layers of cheescloth or three to four layers of women's stocking hosiery can be used. The conidial concentration can then be determined by hemocytometer count and viability by germination on a suitable agar, both as described by Inglis *et al.* (Chapter VII). Such suspended conidia should be kept refrigerated as long as possible before use and should be sprayed within 1–2 days to avoid loss in viability.

*b. Dry conidia*

In most cases dry conidia are desired for storage or for incorporation into formulations. The ideal technical-grade conidial powder should have the highest concentration of conidia possible. Typically, a relatively pure, dry *B. bassiana* conidial powder will have $1.2–1.6 \times 10^{11}$ conidia/g, while a similar *Metarhizium* spp. powder will have $5–6 \times 10^{10}$ conidia/g. Harvesting dry conidia from substrate should also provide the greatest per cent recovery possible. It does little good to have a high conidial count per gram of substrate but only be able to recover < 50% of the conidia. Several methods with good recoveries are possible.

The simplest method is mechanical classification using nested sieves on a vibratory shaker (Figure 8.9). Typically, two sieves, an ASTM 20-mesh (0.850-mm opening) sieve is nested on an ASTM 100-mesh (0.100-mm opening) or ASTM 120-mesh (0.125-mm opening) sieve. A 20-cm diameter sieve can efficiently accommodate up to 500 g whole dry culture; a 30-mm-diameter sieve, 1 kg of culture. A lid on the ASTM 20 sieve is critical to prevent conidia from being released into the immediate environment. These sieves should be mounted on a vibratory sieve shaker such as one of the models made by Retsche GmbH, Haan, Germany, which have a throwing motion of up to 3 mm with angular momentum. If necessary, manual shaking of the sieves is possible although laborious.

Transfer of dry culture from spawn bag to sieve should be made in an exhaust hood or in a hood system that will

**Figure 8.10** Plastic storage boxes adapted for rapid drying of fungus cultures. For details see text.

capture escaping conidia and appropriate personal protection equipment should also be worn (see Section 3B10). Add several coins or metal washers weighing a total of a few grams to the 100-mesh sieve and place on the collection pan. These coins assist in the passage of conidia through the 100-mesh screen. In their absence, considerable retention of conidia with fine substrate particles can occur, causing reduced recovery. Place a 20-mesh sieve on top of the 100-mesh sieve. Transfer the dry whole culture to the 20-mesh sieve, being careful to generate as little spore escape as possible. Add 10–12 glass marbles, ceramic beads, or clean metal nuts to the culture, and then add a sieve lid. Ideally the sieve joints should be taped with electrical tape to prevent conidial dust from escaping. Place on the vibratory shaker and operate for 20–25 min, following the manufacturer's instructions for optimal separation and sieving. At the end of this period, remove the collection pan from the sieves and transfer the dry conidia to a suitable storage container. In our experience, recovery efficiencies range from 60–90% depending upon the fungus strain; strains that are very prolific spore producers have higher recoveries than those that are not. In the absence of a shaker the sieves can be swirled manually on a table top, with occasional tapping of the sieves on the table surface. This process is very laborious, but it does harvest the spores.

Another method, especially for production of larger amounts of a single fungus isolate, involving kilograms of whole culture, is to combine mechanical agitation of the dry whole culture with vacuum collection of the conidia. One such apparatus, developed by Dan Johnson,

University of Lethbridge, Lethbridge AB, Canada, consists of a cyclone-collector vacuum cleaner connected to a portable hand- or motor-operated concrete mixer (personal communication). The mixer has a Plexiglas covering attached over its opening, held to the mixer with four or five circular, 2- to 3-mm-thick magnets, creating a gap of like dimensions (Figure 8.11). The gap is necessary to allow air to flow into the mixer's chamber when the vacuum cleaner is operated. The vacuum cleaner is one of many models that has a bagless, cyclone collector. During harvest the whole dry culture is gently transferred from spawn bags to the mixer, the lid attached to the mixer opening using the magnetic clamps, and the vacuum hose attached to a port in the lid's center. As the mixer is rotated, the physical agitation dislodges the conidia from the substrate and they are rapidly sucked up by the vacuum cleaner. A disadvantage of this approach is that the cleanup/decontamination of the vacuum cleaner is difficult. One solution is a separate vacuum cleaner for each fungus strain.

**Figure 8.11** An alternative device for harvesting conidia from solid substrate consisting of an electrically powered concrete mixer with a Plexiglas covering coupled to a bagless, cyclone-type vacuum cleaner (not shown). For details see text. (Photograph courtesy of Dan Johnson, University of Lethbridge, Alberta Canada.)

A third, more sophisticated, method to harvest conidia is with a Mycoharvester™ (VBS Agriculture Ltd., Beaconsfield, UK, http://www.mycoharvester.info). This device is a lab-scale, stainless, cyclone dust collector, coupled to a fluid bed agitator. The Mycoharvester can be easily cleaned, decontaminated, and autoclaved. However, its cost may be prohibitive for the researcher wishing only to obtain enough conidia for a small-plot field trial.

Conidia should have a moisture of 5% or less. If the moisture of the conidia is greater than 7% w:w (water activity, $a_w > 0.4$) then the conidia should be dried further over silica or calcium sulfate desiccant for several days to a week to a satisfactory moisture level. In so doing, the conidial powder should be in a layer no thicker than 1−2 cm and should be turned several times during the drying period. The loss in moisture can be monitored periodically by changes in the net weight of the conidia.

## *8. Storage*

Conidia of *Beauveria bassiana*, *Metarhizium* spp. and *Isaria* are hygroscopic. Therefore, long-term storage (> 6 months) of conidia, especially those of many *Metarhizium* isolates, should be in air- and water-impermeable containers, such as a heat-sealed or zipper-lock bags, Mylar® foil pouch or bag, such as used for coffee (e.g., Sorbent Systems, IMPAK Corp., Los Angeles CA, USA), along with a desiccant packet (e.g., Humisorb™, Sorbent Systems). For best long storage, the dry conidial powders should be stored, as described above ($<-15°C$), in a manual defrost, not frost-free, freezer. The latter type will undergo periodic thaw and refreeze cycles that are deleterious to fungal conidia.

## *9. Quantification of production*

Calculation of total spore yield involves not only the harvested conidia but also those remaining on the substrate. Therefore, the 'spent substrate' from which the conidia have been removed should be weighed and a 10-g sample be retained. Suspend this sample in 0.05−0.1% Silwet L77, or 0.1% Tween 80 or Tween 20, and dilute appropriately for hemocytometer counts as per the methods outlined in Chapter VII. Similarly, a 0.1-g sample of conidial powder should be suspended in distilled water plus wetting agent, diluted and conidia counted. In sampling a conidial powder, heterogeneity of conidial concentration in the powder can occur, so it is important to make a composite sample of many small amounts randomly across the powder for an accurate conidial count.

It is also important to remember that such counts follow a Poisson distribution, so that sufficient conidia need to be counted for a desired level or precision. That number is

$$n = t^2/D^2$$

where n is the required number of conidia, t is the tabular t value for 95% confidence, and D is the desired precision. For example, to achieve a precision of ± 10% in 95% of replicate counts, a minimum of 384 conidia must be counted, regardless of the number of hemocytometer cells required.

The number of conidia per gram of harvested powder is multiplied by the number of grams of powder. The number of conidia washed from the substrate is multiplied by the total weight of spent substrate. The two values are then combined and divided by the initial dry weight of the substrate to determine conidia per kg or gram of substrate.

More often, it is more convenient to calculate the practical yield as harvested conidia per weight of starting substrate. But in doing so for comparative purposes, it is important to remember that slight variations in the harvesting process, especially when different people conduct the harvest, can affect the apparent yield.

It is also important to calculate the conidial viability before use. Methods for this measurement are detailed in Chapter VII. Viability of harvested conidia is rarely 100%; dry *Metarhizium* sp. conidia often have apparent viabilities as low as 60% using the standard germination test criteria.

### *10. Safety*

Whenever working with cultures when there is potential exposure to conidia, appropriate personal protection is highly advisable. Both the conidia and the beta-1,4 glucans of grains can be sensitizers and cause allergic reaction with repeated exposure. Wear a lab coat or Tyvek® clean suit, as well as latex gloves, eye protection, and a NIOSH N/P/R 95/100 respirator or equivalent. Alternatively, use a powered air purifier respirator. Sources of such equipment include 3M, St Paul MN, USA (Easy Air 7800 series, Air-Mate); Martindale Protection Ltd, London, UK (Mark V); Protector Safety Limited, Manchester, UK (RPFF 81); Racal Safety Limited, Wembley, UK (R.P.M. 81, Powerflow, Breathe Easy 7); or Willson Safety Products, Reading PA, USA (6783 respirator series).

In addition, when more than occasional spore production is anticipated, an exhaust system capable of efficient spore capture is advisable. Such can be found at amateur wood-working supply companies. One such system is illustrated in Figure 8.12. It consists of a 2- to 3-h.p. cyclone dust collector designed for the carpentry and woodworking industry, e.g., Tempest™ 2HP or Tempest 3HP (Penn State Industries, Philadelphia PA, USA), attached to a metal hood over the work area. Any escaping conidia are drawn into the exhaust system and collected in a drum for later disposal. The dust collector has a further HEPA filter to prevent escape of very small particles into the environment.

**Figure 8.12** Cyclone dust collector as used in the carpentry industry but used here to contain escaped conidial dust during harvesting. The unit is fed from a hood in an adjacent harvesting room behind the collector. Most conidia captured from the air in the harvesting room are contained in the plastic barrel beneath the cyclone collector. On the right of the unit is a HEPA filter on the exhaust to prevent conidia from being released into the environment.

## 4. LIQUID FERMENTATION PRODUCTION

### A. Overview of liquid culturing of entomopathogenic fungi

While the solid-substrate fermentation method for the production of aerial conidia on the surface of moistened grains mimics the natural conidia production process by entomopathogenic Hypocreales, the use of liquid culture fermentation for growing these entomopathogenic fungi presents a unique opportunity for producing other stable, infective fungal propagules such as 'yeast-like' blastospores for foliar applications, or the microsclerotia of *Metarhizium* spp., which have shown promise for use in spray and granular applications, respectively (Jaronski & Jackson, 2008).

Dimorphic entomopathogenic fungi that can be induced to grow in a 'yeast-like' fashion in liquid culture (Figure 8.13A) can be excellent candidates for liquid culture production. Clearly, economic considerations dictate that the production method developed must rapidly produce very high yields of stable,

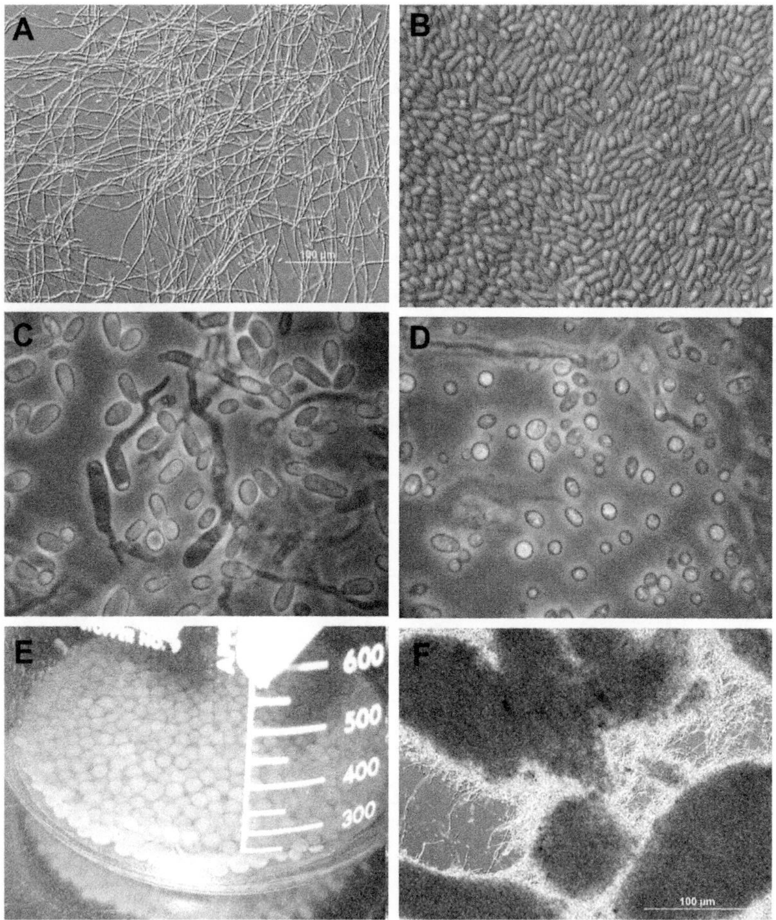


**Figure 8.13** Patterns of growth by fungal cultures cultivated using aerated, submerged culture. Dimorphic fungi are capable of hyphal-mycelial growth A. or "yeast-like" blastospore vegetative growth in the special medium described in this chapter B. Mixed blastospore-hyphal cultures of *Beauveria bassiana* C. and *Metarhizium anisopliae* D. also occur in liquid inocula for solid substrate fermentation. Culture conditions and inoculum size can be used to induce hyphal growth in a pellet form E. The hyphae of *Metarhizium anisopliae* and related *Metarhizium* species can differentiate to form melanized, hyphal aggregates termed microsclerotia which can be harvested by filtration, air dried and made into granules F.

infective propagules. The fungi *I. fumosorosea, I. farinosus, B. bassiana, Lecanicillium lecanii, L. muscarium, L. longisporum, N. rileyi* and *Metarhizium* spp. can be induced to grow in a yeast-like fashion in submerged culture; blastospores of *I. fumosorosea* and *Lecanicillium* spp. are both produced on a commercial scale. Studies with *I. fumosorosea* have demonstrated that 'yeast-like' blastospores of this fungus can be rapidly produced in high concentrations if appropriate concentrations of nitrogen are provided (Jackson *et al.*, 1997). The rapid germination rate of dried *I. fumosorosea* blastospores following rehydration make these propagules ideal candidates for use as a contact biopesticide as an alternative to aerial conidia. Blastospore production is especially valuable for *I. fumosorosea* isolates that do not conidiate readily without far blue/near UV light as discussed earlier.

Some of the Hypocreales, namely *B. bassiana* and *M. acridum*, also form what have been termed microcycle conidia (Thomas *et al.*, 1987; Jenkins & Prior, 1993; Bosch & Yantorno, 1999; Zhang & Xia, 2008). Formation of these microcycle conidia has not really been reduced to practical methodology, however. For microcycle conidial production in liquid culture, a vegetative fungal growth cycle is required followed by hyphal differentiation to form conidia. Because the existing processes do not produce pure microcycle conidia without mycelium and blastospores, separation of conidia from the latter is really needed for spray application.

For Hypocreales that are to be used as granular applications in soil, a very stable fungal propagule is required. Conidia or 'yeast-like' propagules are often not sufficiently stable or amenable to these applications. Conversely, fungal propagules such as microsclerotia or chlamydospores that often function as overwintering structures for many fungi are well suited to use as granular biopesticides. Recent studies showed that liquid fermentation can be used to produce microsclerotia of *Metarhizium* spp. (Jackson & Jaronski, 2009) (Figure 8.13F). Microsclerotial granules of *M. anisopliae* have shown excellent potential for control of soil-dwelling insects (Jaronski & Jackson, 2008). The further development of cost-effective liquid culture production techniques for fungal microsclerotia has potential to heighten commercial interest in using granular fungal agents as seed coatings, soil amendments, or even in spray applications.

## 1. Media development and screening for production of fungal propagules

A critical evaluation of the potential of your entomopathogenic fungus to produce a suitable infective propagule in liquid media is an important first step in determining the commercial potential of the fungus and, therefore, needs additional testing or development. Provided below are techniques that can be employed to determine the amenability of your fungal entomopathogen to production using liquid culture methods and further development as a bioinsecticide.

Entomopathogenic fungi are not only capable of infecting insects but can grow as saprophytes on a wide variety of simple and complex nutrients. In liquid culture, many of these fungi display dimorphic growth as hyphal filaments or as 'yeast like' blastospores. As growth proceeds and becomes unbalanced due to nutrient limitations, vegetative hyphal filaments of *Metarhizium* spp. may differentiate to form sclerotial bodies (Figure 8.13F). Manipulation of the nutritional and environmental conditions during the liquid culture growth of entomopathogenic fungi can be used to induce a desired growth pattern or to evaluate the potential of the selected organism to produce an appropriate propagule for use as a biocontrol agent.

While many nutritional factors play a role in the growth pattern or yields of fungal entomopathogens, carbon, nitrogen, and oxygen are of central importance. These three elements are the most abundant components of the cell and provide fungal cultures with metabolic energy and building blocks for growth. The source and/or concentration of the nitrogen and carbon components of the medium can have a dramatic impact on fungal propagule form and yield (Jackson & Schisler, 1994; Jackson & Jaronski, 2009). It has also been shown that these nutritional components of the growth medium can affect the desiccation tolerance, storage stability, and/or biocontrol efficacy of the fungal propagule produced (Inch *et al.*, 1986; Bidochka *et al.*, 1987; Jackson & Bothast, 1990; Jackson & Schisler, 1992; Hallsworth & Magan, 1995; Jackson *et al.*, 1997; Stephan & Zimmermann, 1997; Vidal *et al.*, 1998; Vega et al., 2003; Sheare & Jackson, 2006).

One approach to screening fungal entomopathogens for amenability to liquid culture production focuses on using various combinations of readily available carbon and nitrogen sources in a complete basal medium supplied with vitamins and trace metals (Appendix). Media are designed to provide either a nutritionally rich or nutritionally poor media with various carbon-to-nitrogen (C : N) ratios. The general scheme is shown in the Appendix. Acid-hydrolyzed casein is often used as a readily available nitrogen source. Over 80% of the milk protein, casein, is hydrolyzed to free amino acids, which are easily utilized by most fungi as a source of nitrogen and carbon. Glucose is a readily used carbon and energy source by most fungi. While glucose and acid hydrolyzed casein are provided as examples, numerous other carbon and nitrogen sources can be used for these studies. To supply sufficient oxygen to the cultures, baffled flasks are used and the agitation rate is set near the upper limits of the shaker incubator (typically 300–400 r.p.m.). Since shake incubators vary in their 'throw' or agitation radius, the agitation rate should be set at a speed that provides vigorous agitation without culture broth reaching the flask stopper.

The strategy behind using rich and weak media with differing C : N ratios evolved from our experience in optimizing production media for various fungal biocontrol agents. Conidial formation by fungi in aerated, liquid culture is often induced by the depletion of an essential nutrient. By using a nutritionally weak medium, differentiation by the fungal culture to form conidia can often be rapidly observed. Conversely, fungal cultures that produce blastospores, yeast-like vegetative propagules, produce highest numbers of these propagules in nutritionally rich media. Since blastospores are basically the yeast form of the fungus, a rich medium that increases biomass production generally increases blastospore yields. Nutritionally rich media with high carbohydrate concentrations have been shown to enhance the production of microsclerotia by *M. anisopliae* s.l. in liquid culture (Jackson & Jaronski, 2009). Thus, the diversity of nutritional environments in our media screening protocol allows evaluation of the potential of our fungal biocontrol agent to produce conidia, blastospores, or microsclerotia in liquid culture.

Another critical aspect of these screening media is the presence of sufficient minerals, vitamins, and trace metals for balanced growth. By supplementing the nitrogen and carbon sources in the liquid medium with sufficient quantities of minerals and growth factors and by using culture conditions that provide highly aerated growth conditions (baffled flasks, agitation rates of 300–400 r.p.m.), balance growth is assured regardless of whether these nutrients are provided in the complex carbon or nitrogen source. This is a critical consideration particularly when various carbon and nitrogen sources are being evaluated.

While myriad systems exist for organizing experimental data, it is imperative that the development of experimental media be performed in a very structured manner in order to avoid volume anomalies when all components are added. The media formulation template shown below is used to aid in the preparation of stock cultures of the various nutritional components and in the development of protocols for preparing medium optimization experiments with varied nutritional components. As an example, a sample media formulation sheet for *I. fumosorosea* blastospore production is shown in Figure 8.14. For further guidance on media development and on the composition of various complex substrates, refer to the media formulation guide provided by Trader's Protein (Zabriske *et al.*, 1980).

**MEDIA FORMULATION**

EXPERIMENT # E110724 DATE: July 24, 2011
MEDIUM: Basal + Trace Metals REPLICATES: 3
pH: 5.5, uncontrolled ACID: 2N HCl BASE: 2N NaOH
FLASK: 250 ml baffled, Erlenmeyer ORGANISM: Metarhizium anisopliae F52
RPM: 300 TEMP: 28 C INCUBATOR: New Brunswick 4230

| Ingredient | 1 | 2 | 3 | 4 | 5 | 6 | 7 | 8 | 9 | 10 | 11 | 12 |
|---|---|---|---|---|---|---|---|---|---|---|---|---|
| Basal Medium (2X) (ml) | 50 | 50 | 50 | 50 | 50 | 50 | | | | | | |
| Glucose (20%) (ml) | 5 | 8.3 | 9 | 22.5 | 37.5 | 40.5 | | | | | | |
| Acid Hydrolyzed Casein (g) | 1 | 0.34 | 0.2 | 4.5 | 1.5 | 0.9 | | | | | | |
| DH2O (ml) | 35 | 31.7 | 31 | 17.5 | 2.5 | - | | | | | | |
| Inoculum (5E+06) (ml) | 10 | 10 | 10 | 10 | 10 | 10 | | | | | | |
| | | | | | | | | | | | | |
| Carbon Concentration | 8 | 8 | 8 | 36 | 36 | 36 | | | | | | |
| Carbon to Nitrogen Ratio | 10:1 | 30:1 | 50:1 | 10:1 | 30:1 | 50:1 | | | | | | |
| | | | | | | | | | | | | |
| Medium # | 1 | 2 | 3 | 4 | 5 | 6 | | | | | | |
| | | | | | | | | | | | | |
| Total Volume | 100 ml | | | | | | | | | | | |

| BASAL MEDIUM (2X) | 1L | 20% GLUCOSE (w/v) |
|---|---|---|
| VITAMIN MIX (50X) | 40 ml | 500 ml |
| $KH_2PO_4$ | 4 g | 100 g |
| $MgSO_4$ | 0.6 g | -autoclave separately |
| $CaCl_2$ | 0.8 g | INOCULUM D-$H_2O$ 200 ml |
| $FeSO_4$ | 0.1 g | |
| Zn, Mn, Co (100X stock solutions) | 20 ml each | |
| D-$H_2O$ | 900 ml | |

**Figure 8.14** Sample formulation sheet used by NCAUR, Peoria IL, in preparing liquid fermentation media.

During initial studies, grow the fungal cultures for 7–8 days with samples being taken after 2, 4, 6, and 8 days of growth. Evaluate culture samples over time for the pattern of growth (filamentous or yeast-like), biomass accumulation, spore concentrations, and microsclerotia formation.

## B. Blastospore production

### 1. General methods

#### a. Inoculum

Inoculum concentration is a critical culturing parameter when conducting liquid-culture fermentation studies. The size of the inoculum affects culture parameters such as growth rate, nutrient utilization, and culture morphology. Stock cultures are used to inoculate shake flask starter cultures and then serially transferred to larger and larger vessels, depending on the scale at which liquid culture or solid-substrate production will be conducted. In the case of liquid culture production of fungal blastospores or microsclerotia, this scale-up process is repeated until a sufficiently large volume of product is produced. If the fungal culture being evaluated sporulates on nutritive agar plates, conidia can be used as inoculum. It is critical that the conidial concentration be consistent for all liquid culture cultivations. Use a conidial inoculum that delivers conidia concentrations of $5 \times 10^6$ conidia/ml of culture volume. This concentration of conidia ensures a reasonably dense fungal culture that has been shown to support dense growth and to inhibit pellet formation for hyphal cultures (Jackson, unpublished data). It is a good idea to experimentally determine the appropriate conidial inoculum for your specific entomopathogen culture prior to conducting experimental studies.

Stock cultures of fungus are preserved as sporulated agar chucks in 10% glycerol at −80°C as previously described. Weekly inoculation of PDA plates by stock cultures provides sporulated cultures for use as inoculum following 2–3 weeks' incubation at room temperature. Flood these plate cultures with 10 ml of sterile deionized $H_2O$ and scrape with a sterile inoculating loop to release and suspend conidia. Withdraw the conidial suspension from the plate with a sterile 10-ml pipette and dilute as appropriate with sterile deionized $H_2O$ in a sterile beaker to produce inoculum with a conidia concentration of $5 \times 10^7$ conidia/ml.

For fungal cultures that do not sporulate well or at all on nutritive agar plates, use small (~5 $mm^2$) diced agar pieces from the grown-out plate culture as the pre-culture inoculum for liquid cultures. Typically, use a quarter of a 9-cm-diameter agar plate culture to inoculate each liquid culture flask (100–200 ml volume). Inoculate larger liquid pre-cultures with additional agar pieces. Grow these agar-inoculated liquid cultures for 2–4 days to produce a homogenous hyphal culture that is then used as inoculum for liquid culture studies. Again, standardization and consistency is critical for inoculum development. Use of same-age, liquid pre-cultures derived from agar chunks as inoculum for liquid culture studies is required to ensure consistent culture growth and propagule formation.

#### b. Fermentation

It is critical that these cultures are highly aerated (300–400 r.p.m.) to ensure maximum 'yeast-like' growth (blastospores) with minimal hyphal growth.

Another key for developing reproducible cultures of fungal entomopathogens, in terms of growth and propagule formation, is inhibiting mycelial ring formation on the shake flask wall. The biomass in newly inoculated, vigorously agitated liquid cultures tend to adhere to the walls of the culture flask immediately above the liquid line. When this occurs, the culture biomass is no longer growing in 'liquid culture' but is growing more like a 'solid-substrate' culture on the side of the flask. This must be avoided or attended to on a regular basis to ensure consistent results and true submerged liquid culture. The tenacity of flask wall adherence by fungal biomass is dependent on the fungus, inoculum, and culture medium. It is imperative that this biomass ring be removed periodically during culture growth. This is particularly important early in the fermentation. During the first 3 days of growth, remove biomass rings from shake flask walls two to four times a day. Stopping the shaker incubator and shaking the flasks so that the force of the culture broth dislodges the biomass ring back into the liquid medium is often sufficient maintenance. If the fungal biomass is too tightly adhered to the flask, it must be pushed back into the medium using a sterile pipette or inoculation loop. Performing this operation in a biocontainment hood is recommended to reduce the chance of contaminating the culture. In addition, biomass ring removal should be performed on only two to three flasks at a time. It is critical to keep the rotary shaker incubator in operation for proper aeration of the cultures. Turning the shaker incubator off for even short periods of time stops aeration of the culture flasks leading to anaerobic growth conditions that may adversely affect the growth characteristics of the culture.

*c. Quantification*

*(1). Sampling.* When sampling liquid cultures of entomopathogenic fungi, care must be taken in using aseptic technique and, as mentioned above, removing the cultures from the shaker incubator for a minimum period of time to reduce problems associated with oxygenation of the cultures. The use of a biocontainment hood for sampling will reduce the risk of microbial contamination. Removing one to two flasks at a time from the shaker incubator and taking samples for analysis later, once all the flasks have been sampled, is recommended to help maintain oxygenated cultures. Our typical analyses of cultures during growth include measuring biomass accumulation (dry weights), propagule concentrations (spores or microsclerotia) and microscopically evaluating culture morphology and checking for the presence of unwanted microbial contaminants.

*(2). Dry weight determination.* Collect two, 1-ml samples of whole culture broth from culture flasks at various times, taking care to obtain a representative sample of culture biomass. The fungal biomass is separated from the spent medium by vacuum filtration onto pre-weighed filter disks (Whatman GF/A, Maidstone, UK). Dry weights are then determined by drying the biomass and filter disk at 60°C to a constant weight and reweighing of the filter disk. The difference in the disk weight represents the fungal biomass in the culture. Use duplicate whole-culture broth samples for each treatment variable.

*(3). Blastospore counts.* Spore concentrations are determined microscopically using a hemocytometer. Withdraw 1–2 ml of culture aseptically and transfer to a vial or test tube. Dilute appropriately for counting in a hemocytometer, using the methods described in Chapter VII. Constantly vortex suspensions to ensure sample homogeneity during sampling.

*d. Harvesting*

After the liquid fermentation process is complete, a method for stabilizing the fungal propagules for use at a later time is imperative. Collecting the fungal biomass and drying it to a moisture content that reduces metabolic activity and the potential for contamination by unwanted microorganisms is often the most practical approach to stabilization. Various methods for drying include freeze-drying, spray-drying, fluidized-bed drying and air-drying. While all these drying methods have benefits, air-drying is a preferable laboratory method for evaluating the potential for the fungal propagule to be stabilized as a dry preparation that can be easily accomplished in a laboratory setting. Air-drying also mimics commercial drying protocols that can be used for scale-up of the stabilization process. A description of the air-drying process is provided below.

*(1). Blastospore separation and drying.* Liquid culture produced blastospores are harvested from the fungal culture broth by first mixing whole cultures with a filter aid such as diatomaceous earth (DE). DE is typically added to whole cultures at a rate of 5% (w/v). This concentration can be varied with experimentation to determine appropriate levels of DE addition. The purpose of DE addition is two-fold: to aid in filtration and to separate the fungal propagules during drying so that they remain separate after drying and do not fuse together. Since liquid culture fungal fermentations often have biomass concentrations of ~25 g/l, the 5% DE level produces a fungal biomass : DE ratio of 1 : 2. In our experience, this ratio produces a friable filter cake that dries well and yields a final dry preparation with individual fungal propagules.

For air-drying, blastospores are harvested from the fermentation broth by adding 1 g of DE for every $2 \times 10^{10}$ blastospores. The DE-fungus biomass is then filtered from the spent culture supernatant on an appropriately sized Buchner funnel with a Whatman #1 filter. The resulting filter cake (~70% moisture) can be crumbled by hand or in a blender and then layered (~1 cm thick) on aluminum foil or in a shallow pan. The filter cake is air-dried overnight (14–20 h) in an air-drying chamber or in a laminar air flow. For long term storage survival, air-drying blastospores with moistened air (> 65% RH) is recommended (Jackson & Payne, 2007). This step can be accomplished using the controlled humidity air-drying chamber described in the next section, 4B1d(2). Once the spore formulations are less than 4% moisture, the viability and blastospore yield of dry formulation are evaluated using spore germination and yield measurements, as previously described. The dry blastospore formulations should be vacuum packaged and stored at 4°C for maximum shelf-life. If vacuum packaging is not available, zip-lock storage bags can be used with reasonable results.

*(2). Controlled humidity, air-drying chamber.* In order to dry blastospore preparations with air that contains consistent relative humidity (RH) levels, one needs to use a controlled-humidity drying chamber (Figure 8.15). This laboratory-scale chamber dries under controlled humidity conditions with air moving over the sample. In this drying chamber, the air source is compressed air which is very dry (RH 7−14%). In our system, 345 kPa air was provided by pressure regulation at an air flow of 10 l/min, as determined with an airflow meter. The airflow is split and regulated to provide various volumes of dry air or wet air. Moist air is obtained by bubbling the dry compressed air through a 11.4-l (3-gal) filter/pressure vessel containing 4 l of water. To achieve consistent RH values in the drying air, the wet air and dry air is metered to the drying chamber to achieve a RH value of more than 65% for the drying air. Uniform air flow over the blastospore/DE preparations is achieved for each shelf in the drying chamber by positioning a 1.27-mm perforated pipe at each shelf (Figure 8.15). The vents on the far side of the cabinet are fitted with/without filters depending on the nature of the sample being dried.

In general, stop drying when the moisture level of the DE : fungal biomass formulation is less than 5%. The moisture content of the dried blastospore preparations, expressed as ((wet weight minus dry weight)/dry weight) × 100, can be determined with a moisture analyzer as described in Section 3B6. Less automated gravimetric methods can be for moisture determination such as those used for dry weight determinations. After drying, store the fungal biomass : DE formulations in zipper-lock bags or vacuum packaging such as is available for home kitchens (at 4°C for optimal shelf-life).

*(3). Viability testing of dried fungal spore preparations.* Viability analyses should be conducted immediately after drying as a measure of 'desiccation tolerance' and at various intervals during storage to determine the 'storage stability' of the dried preparations. We typically evaluate the viability of dried fungal formulations monthly for 1 year or until the viability of the sample is below 10%. For dried fungal spore formulations, the viability of the formulation is evaluated by measuring the ability of the spores to germinate following rehydration using a previously described 6-h blastospore germination assay (Jackson *et al.*, 1997). Briefly, suspend ~ 50 mg of the air-dried spore formulation in 50 ml of sterile potato dextrose broth in a 250-ml, baffled Erlenmeyer flask. Incubate the suspension in a shaker incubator at 300 r.p.m. and an incubation temperature of 28°C for 6 h. A sample of the spore suspension is taken for microscopic analysis. Per cent germination is determined microscopically by

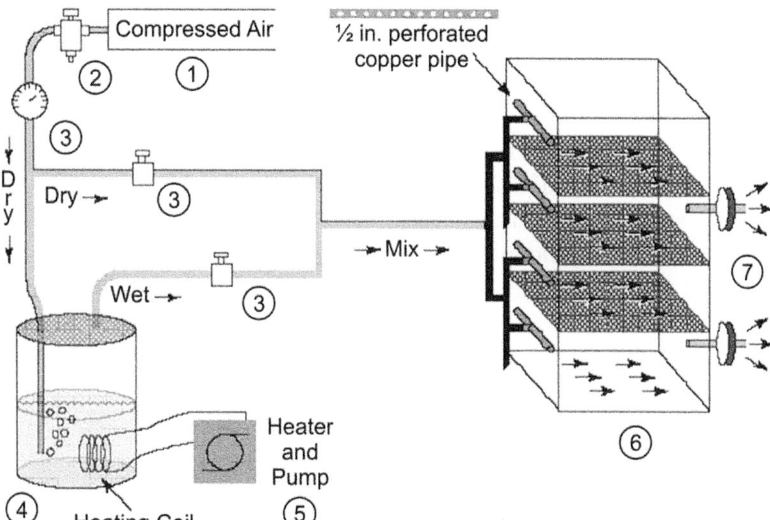


**Figure 8.15** Schematic of laboratory-scale air-drying chamber with constant airflow and relative humidity control. Low relative-humidity compressed air is used to supply dry air that can be humidified by passage through water. A mixture of the two air streams are metered to provide drying air with appropriate relative humidity.

evaluating 100 discrete blastospores for germ tube formation at 400 × magnification using either phase-contrast or bright-field microscopy. Do not assess germination of clumps of blastospores where discrete blastospores can not be seen. Spores with germ tube formation equal to half the length of the spore are considered germinated.

In addition to determining viability of the dry spore formulation, it is also important to be able to estimate the number of free spores that result from the resuspension of the dried formulation in water. Obviously this number will be dependent on the spore concentration in the fermentation broth, the amount of filter aid used to harvest and dry the spores, and the clumping of the spores following drying. To determine typical spore release after rehydration, add 1 g of the dried spore formulation to 100 ml in a 250-ml beaker. Place a stir bar in the beaker and mix on a magnetic stirrer for 30 min. Turn off the stirrer and allow the DE to settle to the bottom of the beaker for 10 s. Take a sample of the spore suspension and determine spore concentration microscopically using a hemocytometer. Spore concentration per ml × volume (100 ml) will provide you with spores released per gram of dried spore formulation.

#### 2. *Methods specific to* Isaria fumosorosea *and* I. farinosus

The entomopathogenic fungus *I. fumosorosea* (Ifr) rapidly produces high concentrations of stable, infective blastospores when grown in liquid culture under appropriate nutritional and environmental conditions (Jackson *et al.*, 1997, 2003). *Isaria farinosus* should also perform similarly. After rewetting, air-dried blastospores of Ifr rapidly infect and kill numerous 'soft-bodied' insect pests including whiteflies, aphids, psyllids, and subterranean termites. The following protocol will yield high concentrations of blastospores that survive drying with good shelf-life when stored at 4°C.

Shake flask cultures (100 ml) of Ifr are grown in 250-ml, baffled Erlenmeyer flasks under highly aerated conditions using the liquid culture production medium and growth conditions defined in the Appendix. The use of a conidial inoculum of Ifr that provides $5 \times 10^6$ conidia/ml in the culture will result in ~$0.8-1.2 \times 10^9$ blastospores/ml within a fermentation time of 96 h. As an alternative approach, 3-day-old blastospore cultures of Ifr can be used to inoculate blastospore production flasks or bioreactors. The use of blastospore pre-culture inocula that provide $5 \times 10^6$ blastospores/ml will reduce fermentation time from 96 h to less than 48 h while maintaining high blastospore yields. If the blastospores are to be used immediately (< 6 days), no drying step is necessary and the whole culture should be stored at 4°C until used to reduce unwanted microbial contamination. If the blastospores will be used for testing at a later date, air-drying and storage under vacuum at 4°C is recommended.

#### 3. *Methods specific for* Beauveria bassiana, Metarhizium *spp., and* Nomuraea rileyi

These three fungi readily produce blastospores in liquid culture with many of the same techniques described earlier. With the medium described for production of liquid inoculum for solid substrate fermentation (Section 3B7 and Appendix), consisting of glucose, yeast extract and corn steep liquor, most strains of *B. bassiana*, *Metarhizium* spp. and *N. rileyi* will produce prolific amounts of blastospores with minimal mycelial generation. The critical ingredient seems to be corn steep liquor; no other components beyond it, a carbohydrate, and an organic nitrogen source seem necessary.

### C. *Metarhizium anisopliae* microsclerotia

The development of a liquid culture production process for microsclerotia (small sclerotia) of the entomopathogenic fungus *M. anisopliae* s. l. provides a novel method for producing dry, granular formulations of the fungus for use in controlling soil-dwelling insects (Jackson & Jaronski, 2009). When these microsclerotial granules are incorporated into moistened soil, they germinate sporogenically to produce infective conidia (Jaronski & Jackson, 2008). Unfortunately *B. bassiana* and *I. fumosorosea* and *I. farinosus*. do not seem to form microsclerotia (Jaronski, unpublished data); microsclerotia production by *N. rileyi* has not been studied. Microsclerotia production studies with *Metarhizium* spp. cultures, encompassing all the major new species, except *M. acridum*, which remains untested, showed that highest yields of microsclerotia occurred when a liquid medium rich in carbon (36 g/l carbon) was used with a C : N ratio of 30 : 1 to 50 : 1 (Jackson & Jaronski, unpublished data). Recent studies have shown that the medium with the lowest nitrogen content (C : N ratio 50 : 1) produced the highest concentrations of stable microsclerotia in the shortest fermentation time (Jackson & Jaronski, 2009). The specific techniques for the production and stabilization of microsclerotia of *Metarhizium* s. l. will be described with reference to the methods previously described in this chapter. The following production and stabilization protocol will yield high concentrations of air-dried microsclerotia with excellent shelf-life.

Stock cultures of *Metarhizium* s. l., derived from single-spore isolates of the fungus, are grown for 3–4 weeks on PDA at room temperature. Conidia are harvested from sporulated cultures into 10 ml of a sterile, aqueous solution of 0.04% Tween 80 and scraped with a sterile inoculating loop to release and suspend conidia. The conidial suspension is withdrawn from the plate with a sterile 10-ml pipette and diluted with sterile deionized $H_2O$ to obtain a conidia concentration of $5 \times 10^7$ conidia/ml for use in liquid culture microsclerotia production experiments.

*1. Media for microsclerotia production*

Preparation of liquid medium for *Metarhizium* s. l. microsclerotia production is detailed in the Appendix. Note that the medium consists of a series of previously prepared components, assembled under aseptic conditions in the fermentation vessel, and inoculated.

*2. Main liquid culture*

Shake flask cultures (100 ml) of *Metarhizium* are grown in 250-ml, baffled Erlenmeyer flasks under highly aerated conditions using the liquid culture production medium detailed in the Appendix. Larger volume cultures can be obtained in baffled 3.4-l Fernbach flasks, containing 1–1.5 l liquid medium. It is critical that these cultures are highly aerated (300–400 r.p.m.) to ensure maximum growth rate and microsclerotia yields. Culture flasks with molded baffles and a liquid volume of no more than 40% of flask volume are critical for microsclerotia production. To combat medium foaming with the required agitation, an antifoam agent, e.g., Antifoam 204 (Sigma-Aldrich) can be used, typically at the rate of 0.1 ml/l. The use of a conidial inoculum of *Metarhizium* spp. that provides $5 \times 10^6$ conidia/ml in the final culture volume and the growth conditions described above will result in $5–29 \times 10^4$ microsclerotia/ml after a fermentation time of 6–8 days at 23–28°C. The replacement of conidial inoculum with hyphal pre-culture inoculum from 2-day-old cultures of *Metarhizium* will reduce fermentation times to 4–6 days. As microsclerotial production proceeds, the liquid culture will darken, even becoming almost black after 4–6 days. This coloration seems to be due to production of melanin during microsclerotial production.

*3. Monitoring microsclerotia production*

During initial studies, grow the liquid cultures for 7–8 days with samples being taken every few days. For microsclerotia concentration determinations, 100 μl of culture broth is placed on a glass slide, overlaid with a coverslip and the number of microsclerotia counted. The lower magnification range of a compound microscope (40 ×, 100 × magnification) should be used for this purpose, although some stereomicroscopes will have sufficiently high magnification to be also useful. The microsclerotia are counted when compact, sometimes melanized, hyphal aggregates larger than 50 μm are observed. Only well-formed microsclerotia are counted. The culture broth can be diluted with water, as appropriate, for ease of counting the microsclerotia. The endpoint is the number of microsclerotia per unit volume of fermentation broth. Dry weight determinations of biomass, as described earlier, should also be determined in parallel with microsclerotia counts.

*4. Harvest*

Harvest microsclerotia from the fungal culture broth by first mixing whole cultures with a filter aid such as diatomaceous earth or diatomite (DE), such as Celite™ Hyflo® (World Minerals Inc., Santa Barbara CA, USA), Diatomite™ (Diatomite, Burlington ON, Canada), or Celetom™ (EP Minerals, Reno NV, USA). DE may also be obtained from beer brewery supply and swimming pool supply companies. Diatomaceous earth is typically added to whole cultures at a rate of 5% (w/v). This concentration can be varied with experimentation to determine appropriate levels of DE addition. The purpose of DE addition is two-fold: to aid in filtration and to separate the fungal propagules during drying so that they remain separate after drying and do not fuse together. Since liquid culture fungal fermentations often have biomass concentrations of ~25 g/l, the 5% DE level typically produces a fungal biomass : DE ratio of 1 : 2. This ratio produces a friable filter cake that dries well and yields a final dry preparation with individual fungal propagules.

The microsclerotia-containing biomass can be used fresh, as wet inoculum, or more preferably as a dry granular formulation. Whole culture microsclerotia-containing fermentation broth can be stored for up to 3–4 months at 4°C without loss of viability. For longer shelf-life (more than 1 year), microsclerotia should be granulated with an inert material such as DE, air-dried to less than 4% moisture, and stored at 4°C. or vacuum packed for storage at room temperature. For air-drying, microsclerotia are harvested from the fermentation broth by adding DE at the rate of 50 g DE/l fermentation broth. The DE–*Metarhizium* biomass is then separated from the spent culture supernatant by filtration on an appropriately sized Buchner funnel with a Whatman #1 filter; paper towels can also be used in lieu of the filter paper. Regardless of the filtration media, it is expedient to place a mesh screen between the filter and the funnel to speed filtration. The resulting filter cake is granulated to the desired size for the application requirements.

Filter cakes have ~75% moisture and can be granulated or extruded to particles of 1 mm or larger at this time. Moist filter cake can be extruded, for example using a pasta maker, then dried and broken into small pieces. A method with smaller batches is to partially dry the filter cake to the point at which the mass is friable and slightly moist, then break it up in a small culinary food chopper, followed by complete drying. Alternatively, for smaller microsclerotial granules, additional granulation or extrusion can be conducted using smaller screen sizes as the moisture content decreases in the filter cake. Microsclerotial granules are finally air-dried overnight to less than 4% moisture as previously described in this chapter.

#### *5. Determining viability of dried fungal microsclerotia preparations*

For storage stability studies with air-dried microsclerotia preparations, precisely 25 mg of the air-dried microsclerotial preparation is sprinkled on to water agar plates that are then incubated, unsealed, at 28°C. Microsclerotia viability is determined by measuring hyphal germination from microsclerotial granules after 24 h incubation. The first 100 granules of the microsclerotial preparation found on each water agar plate are examined with a stereo microscope and counted as germinated whenever hyphal growth is observed. Results are presented as per cent germination. These same plates are incubated for an additional 7 days at 28°C to allow for conidia formation on the granules. Conidial production is measured by counting the total number of conidia produced on each plate. Briefly, each plate is flooded with 10 ml of sterile de-ionized water and the surface is scraped with a plastic 10-µl loop to dislodge conidia. The conidial suspension is pipetted from the rinsed plate and the suspension volume measured. The conidia concentration of each suspension is determined microscopically with a hemocytometer. Conidia production per gram of air-dried microsclerotial preparation is calculated and used for comparison between air-dried microsclerotial preparations from various cultures and from the same cultures after various storage times. Dry blastospore formulations are vacuum-packaged and stored at 4°C for maximum shelf-life. If vacuum packaging is not available, zipper-lock storage bags can be used with reasonable results.

### D. Microcycle conidia of *B. bassiana* and *M. acridum*

As mentioned earlier both of these fungi can produce microcycle conidia in submerged fermentation. However, *M. anisopliae* s. l. does not seem to be able to do so. While solid substrate fermentation is the most direct process for production of aerial conidia, submerged fermentation for microcycle conidia is included here for completeness. Two difficulties exist for the inexperienced entomologist wishing to obtain microcycle conidia: efficient separation of these propagules from mycelium and their processing into a dry form for formulation and use.

#### *1. Fermentation media and conditions*

*a.* Beauveria bassiana

The key nutrient condition for optimal microcycle conidia production by *B. bassiana* seems to be the absence of organic nitrogen. Instead an inorganic source, namely $KNO_3$ is substituted. The basic medium (Thomas *et al.*, 1987) is presented in the Appendix. The medium should be inoculated with a conidial suspension to produce an initial spore concentration of $0.5–1.0 \times 10^4$ spores/ml. Shake flask fermentation is typically at 150 r.p.m. and 24–27°C and uses aerial conidia from agar media as an inoculum, although preliminary liquid can be also used. Initially blastospores and mycelium are formed, but by 4 days the preponderance of propagules are microcycle conidia and hyphal fragments; microcycle conidiation plateaus by day 6. We have to note here that the ability to produce microcycle conidia is a strain characteristic and some strains of *B. bassiana* produce very few microcycle conidia in the medium described (Jaronski, unpublished data).

*b.* Metarhizium acridum

The published literature indicates that only *M. acridum* produces microcycle conidia; isolates of the typical *M. anisopliae* sensu lato do not (Jenkins & Prior, 1993; Leland *et al.*, 2005). Two media, found in the literature, are presented in the Appendix. Inoculate medium with conidia from agar medium to achieve $1.2 \times 10^5$ conidia/ml of medium. Typical shake flask fermentation conditions are as described earlier. Microcycle conidia concentration in the medium usually plateaus by Day 6.

*2. Harvesting*

The key task is to separate the microcycle conidia from mycelium to create what will ultimately be a material that can be applied through standard hydraulic nozzles. This task can be most easily accomplished by filtering the whole culture through a single layer of Miracloth® (produced by Merck KGaA and distributed by CalBiochem, Fisher Scientific or EMDChemicals) or loosely packed glass wool in a column. One is left with a conidial suspension that needs to be concentrated for immediate use or completely dried. The former task can be achieved by centrifugation, although if the volumes are considerable this process can be laborious. Filtration as described for *I. fumosorosea* blastospores is possible and addition of a DE filtration aid at 5–10% w/v is essential. The resulting filter cake must be dried rapidly, e.g., by exposure of thin (< 1 cm) layers of filter cake to dry air flow. The methods described earlier for *I. fumosorosea* blastospores should be sufficient, but one can expect viabilities < 80%. Lyophilization of the filter cake is more desirable, but may require one or another protectant excipients. The interested reader is referred to Toegel *et al.* (2010) for more details.

*3. Quantification and viability*

The methods described for aerial conidia and blastospores are applicable to microcycle conidia and the reader should refer to the appropriate sections in our chapter and Chapter VII.

## ACKNOWLEDGMENTS

The senior author would like to thank Julie Grace, formerly Mycotech Corporation and USDA ARS, now with United Tribes Technical College, Bismarck ND, USA for her contributions in developing many of the solid substrate techniques described here, and the original training manual, *The Art of Fermentation or the Tao of Fungi*, from which these methods have been adapted.

## REFERENCES

Bartlett, M. C., & Jaronski, S. T. (1984). Mass production of entomogenous fungi for biological control of insects. In M. N. Burge (Ed.), *Fungi in Biological Control Systems* (pp. 61–85). Manchester, UK: Manchester University Press.

Bidochka, M. J., Pfeifer, T. A., & Khachatourians, G. G. (1987). Development of the entomopathogenic fungus *Beauveria bassiana* in liquid cultures. *Mycopathologia, 99*, 77–83.

Bosch, A., & Yantorno, O. (1999). Microcycle conidiation in the entomopathogenic fungus *Beauveria bassiana* (Vuill.). *Process Biochem., 34*, 707–716.

Faria, M., & de & Wraight, S. A. (2007). Mycoinsecticides and Mycoacaricides: A comprehensive list with worldwide coverage and international classification of formulation types. *Biol. Control, 43*, 237–256.

Hallsworth, J. E., & Magan, N. (1995). Manipulation of intracellular glycerol and erythritol enhances germination of conidia at low water availability. *Microbiology, 141*, 1109–1115.

Inch, J. M. M., Humphreys, A. M., Trinci, A. P. J., & Gillespie, A. T. (1986). Growth and blastospore formation by *Paecilomyces fumosoroseus*, a pathogen of brown planthopper (*Nilaparvata lugens*). *Trans. Brit. Mycol. Soc., 87*, 215–222.

Jackson, M. A. (1997). Optimizing nutritional conditions for the liquid culture production of effective fungal biological control agents. *J. Ind. Micro. Biotech., 19*, 180–187.

Jackson, M. A., & Bothast, R. J. (1990). Carbon concentration and carbon-to-nitrogen ratio influence submerged-culture conidiation by the potential bioherbicide *Colletotrichum truncatum* NRRL 13737. *Appl. Environ. Micro., 56*, 3435–3438.

Jackson, M. A., & Jaronski, S. T. (2009). Production of microsclerotia of the fungal entomopathogen *Metarhizium anisopliae* and their potential for use as a biocontrol agent for soil-inhabiting insects. *Mycol. Res., 113*, 842–850.

Jackson, M. A., & Payne, A. R. (2007). Evaluation of desiccation tolerance of blastospores of *Paecilomyces fumosoroseus* (Dueteromycotina: Hyphomycetes) using a lab-scale, air-drying chamber with controlled humidity. *Biocontrol Sci. Tech, 17*, 709–719.

Jackson, M. A., & Schisler, D. A. (1992). The composition and attributes of *Colletotrichum truncatum* spores are altered by the nutritional environment. *Appl. Environ. Microbiol., 58*, 2260–2265.

Jackson, M. A., & Schisler, D. A. (1994). Liquid culture production of microsclerotia of *Colletotrichum truncatum* for use as bioherbicidal propagules. *Mycol. Res., 99*, 879–884.

Jackson, M. A., Cliquet, S., & Iten, L. B. (2003). Media and fermentation processes for the rapid production of high concentrations of stable blastospores of the bioinsecticidal fungus. *Paecilomyces fumosoroseus. Biocontrol Sci. Tech, 13*, 23–33.

Jackson, M. A., McGuire, M. R., Lacey, L. A., & Wraight, S. P. (1997). Liquid culture production of desiccation tolerant blastospores of the bioinsecticidal fungus. *Paecilomyces fumosoroseus. Mycol. Res., 101*, 35–41.

Jaronski, S. T., & Jackson, M. A. (2008). Efficacy of *Metarhizium anisopliae* microsclerotial granules. *Biocontrol Sci. Technol., 18*, 849–863.

Jenkins, N. E. & Goettel, M. S. (1997) Methods for mass production of microbial control agents of grasshoppers and locusts. In M.S. Goettel & D.L. Johnson, *Microbial Control of Grasshoppers and Locusts. Mem. Ent. Soc. Canada, 171*, 37–48.

Jenkins, N. E., & Prior, C. E. (1993). Growth and formation of true conidia by *Metarhizium flavoviride* in a simple liquid medium. *Mycol. Res., 97,* 1489–1494.

Kassa, A., Stephan, D., Vidal, S., & Zimmermann, G. (2004). Production and processing of *Metarhizium anisopliae* var. *acridum* submerged conidia for locust and grasshopper control. *Mycol. Res., 108*, 93–100.

Kirsop, B. E., & Snell, J. J. S. (1984). *Maintenance of Microorganisms*. London, UK: Academic Press, Inc.

Leland, J. E., McGuire, M. R., Jaronski, S. T., Grace, J. A., Ulloa, M., Park, Y.-H., & Plattner, R. D. (2005). Strain selection of a fungal entomopathogen, *Beauveria bassiana*, for control of plant bugs (*Lygus* spp.) (Heteroptera: Miridae). *Biol. Control, 35*, 104–114.

Sahayaraj, K., & Namachivayam, S. K. R. (2008). Mass production of entomopathogenic fungi using agricultural products and by products. *African J. Biotechnol., 7*, 1907–1910.

Sakamoto, M., Inoue, Y., & Aoki, J. (1985). Effect of light on the conidiation of *Paecilomyces fumosoroseus. Trans. Mycolog. Soc. Japan, 26*, 499–509.

Shearer, J. F., & Jackson, M. A. (2006). Liquid culture production of microsclerotia of *Mycoleptodiscus terrestris*: a potential biological control agent for the management of hydrilla. *Biol. Control, 38*, 298–306.

Stephan, D., & Zimmerman, G. (1997). Mass production of *Metarhizium flavoviride* in submerged culture using waste products. In S. Krall, R. Peveling, & D. Ba Diallo (Eds.), *New Strategies in Locust Control* (pp. 227–229). Basel, Germany: Birkhauser Verlag.

Thomas, K. C., Khachatourians, G. G., & Ingledew, W. M. (1987). Production and properties of *Beauveria bassiana* conidia cultivated in submerged culture. *Can. J. Microbiol., 33*, 12–20.

Toegel, S., Salar-Behzadi, S., Horaczek-Clausen, A., & Viernstein, H. (2010). Preservation of aerial conidia and biomasses from entomopathogenic fungi *Beauveria brongniartii* and *Metarhizium anisopliae* during lyophilization. *J. Invertebr. Pathol., 105*, 16–23.

Vega, F. E., Jackson, M. A., Mercadier, G., & Poprawski, T. J. (2003). The impact of nutrition on spore yields for various fungal entomopathogens in liquid culture. *World J. Microbiol. Biotechnol., 19*, 363–368.

Vidal, C., Fargues, J., Lacey, L. A., & Jackson, M. A. (1998). Effect of various liquid culture media on morphology, growth, propagule production, and pathogenic activity to *Bemisia argentifolii* of the entomopathogenic Hyphomycete. *Paecilomyces fumosoroseus. Mycopathologia, 143*, 33–46.

Wikipedia. (2011). *Water Activity.* downloaded from. http://en.wikipedia.org/wiki/Water_activity July 31, 2011.

Zabriskie, D. W., Armiger, W. B., Philips, D. H., & Albano, P. A. (1980). *Trader's Guide to Media Formulation*. Memphis, TN: Trader's Protein.

Zhang, S., & Xia, Y. (2008). Identification of genes preferentially expressed during microcycle conidiation of *Metarhizium anisopliae* using suppression subtractive hybridization. *FEMS Microbiol. Lett., 286*, 71–77.

## APPENDIX

**1.** Medium for the production of liquid inoculum for solid substrate production of conidia

Add to 1 l of reverse osmosis or deionized water,

20–30 g glucose or sucrose
15–20 g yeast extract
4 g $K_2HPO_4$
15 ml (17 g) Corn Steep Liquor (Sigma-Aldrich)
(Optional) 10 mg gentamycin (our choice for antibiotic because it is autoclavable, but streptomycin (10 mg/l) or chloramphenicol (25 mg/l) may be substituted. Neither is autoclavable and they must be prepared as sterile solutions and added aseptically to the autoclaved and cooled medium.

Heat water, but not to boil. Add ingredients, except antibiotic. Bring medium to a gentle boil for 1–2 min, then lower heat and add antibiotic. Pour into flasks; cap or place foam plugs on each flask and cover stopper with aluminum foil; autoclave for 30 min at 121° C. when individual quantities are less than 1 l; 45–55 min if individual volumes are 1–2 l; cool before use. May be stored several weeks at room temperature if capped tightly to prevent water loss.

**2.** Nutritional composition of liquid medium for evaluating culture growth and propagule formation by fungal entomopathogens.

3. The carbon and nitrogen content of media formulations used to evaluate the potential of fungal entomopathogens for liquid culture fermentation. Cultures grown in a basal salts medium supplemented with trace metals and vitamins at 28°C and 300 r.p.m. in a rotary shaker incubator. The carbon source is glucose and the nitrogen source is acid hydrolyzed casein.

| **Medium component** | **Concentration (per l)** |
|---|---|
| Nitrogen Source—Acid Hydrolyzed Casein | Variable* |
| Carbon Source—Glucose | Variable* |
| *Minerals* | |
| $KH_2PO_4$ | 4.0 g |
| $CaCl_2$ $2H_2O$ | 0.8 g |
| $MgSO_4$ $7H_2O$ | 0.6 g |
| $FeSO_4$ $7H_2O$ | 0.1 g |
| *Trace metals* | |
| $CoCl_2$ $6H_2O$ | 37 mg |
| $MnSO_4$ $H_2O$ | 16 mg |
| $ZnSO_4$ $7H_2O$ | 14 mg |
| *Vitamins* | |
| Thiamin, Riboflavin, Pantothenate, Niacin, Pyridoxamine, Thiotic acid | 500 μg each |
| Folic acid, Biotin, Vitamin $B_{12}$ | 50 μg each |

*Dependant on the carbon concentration and carbon-to-nitrogen (C : N) ratio of the medium.

4. Basic *Isaria* spp. blastospore production medium, per liter of medium (Jackson *et al.*, 2003)

Prepare in 800 ml deionized water,

| *Carbon concentration (g/l)* | *Carbon-to-nitrogen (C : N) Ratio** | *Glucose (g/l)* | *Casamino acids (g/l)* |
|---|---|---|---|
| 8 | 10 : 1 | 10.0 | 10.0 |
| 8 | 30 : 1 | 16.6 | 3.4 |
| 8 | 50 : 1 | 18.0 | 2.0 |
| 36 | 10 : 1 | 45.0 | 45.0 |
| 36 | 30 : 1 | 75.0 | 15.0 |
| 36 | 50 : 1 | 81.0 | 9.0 |

*Calculations based on 40% carbon in glucose and 8% nitrogen, 50% carbon in Casamino acids.

Glucose 50 g
Nitrogen source (~ 8% N, e.g., acid hydrolyzed casein or yeast extract) 25 g
Minerals:
- $KH_2PO_4$ 2.0 g
- $CaCl_2 \cdot 2H_2O$ 0.4 g
- $MgSO_4 \cdot 7H_2O$ 0.3 g

Trace metals:
- $CoCl_2 \cdot 6H_2O$ 37 mg
- $FeSO_4 \cdot 7H_2O$ 50 mg
- $MnSO_4 \cdot H_2O$ 16 mg
- $ZnSO_4 \cdot 7H_2O$ 14 mg

Vitamins:
500 μg each
- Thiamin
- Riboflavin
- Pantothenate
- Niacin
- Pyridoxamine
- Thiotic acid

50 μg each
- Folic acid
- Biotin
- Vitamin B12

Add 200 ml water containing $1 \times 10^7$ conidia/ml

5. Medium for microsclerotial production (36 g [C]): 30 : 1 or 50 : 1 C : N)

For 100 ml:

20% Glucose solution[1] 37.5 or 40.5 ml[2]
2 × Basal medium[3] 50 ml
Casamino acids 1.5 or 0.9 g**
Distilled $H_2O$ 2.5 or 0**
Inoculum 10 ml
Antifoam 204 (Sigma Aldrich) 0.01 ml

[1]20% (w/v) Glucose solution 160 g in 800 ml water (autoclave separately)
[2]Use 37.5 , 1.5, 2.5, resp. for 30 : 1 C : N; use 40.5, 0.9, 0, resp. for 50 : 1 C : N
[3]2 × Basal medium:
50 × Vitamin mix[4] 40 ml

$KH_2PO_4$ 4.0 g
$MgSO_4 \cdot 7H_2O$ 0.6 g
CaC12 · $2H_2O$ 0.8 g
$FeSO_4 \cdot 7H_2O$ 0.1 g
$ZnSO_4 \cdot 7H_2O$ 14 mg (20 ml of Stock)[5]

$MnSO_4 \cdot 7H_2O$ 16 mg (20 ml Stock)[6]
$CoCl_2 \cdot 6H_2O$ 37 mg (20 ml Stock)[7]
Distilled $H_2O$ 900 ml

[5]ZnSO4·7H2O 100 × Stock solution 0.07 g in 100 ml $dH_2O$
[6]MnSO4·7H2O 100 × Stock solution 0.08 g in 100 ml $dH_2O$
[7]CoCl2·6H2O 100 × Stock solution 0.18 g in 100 ml $dH_2O$
[4]50 × Vitamin mix:

Thiamine 2.50 g (500 μg/l final)
Riboflavin 2.50 g (500 μg/l final)
Ca Pantothenate 2.50 g (500 μg/l final)
Niacin 2.50 g (500 μg/l final)
Pyridoxamine 2.50 g (500 μg/l final)
Thiotic acid 2.50 g (500 μg/l final)
Folic acid 2.50 g (500 μg/l final)
Biotin 0.25 g (50 μg/l final)
Vitamin B1 20.25 g (50 μg/l final)
Distilled $H_2O$ to 100 ml
Sterile filter; store at 4°C protected from light

Inoculum sufficient to achieve ~1–5 × $10^6$ blastospores/conidia per ml final medium volume

Use baffled flasks, incubate at 300–400 r.p.m., 28°C.

**6.** Media for production of microcycle conidia

a. *Beauveria bassiana* (Thomas *et al.*, 1987)

Per liter of medium,

Glucose, or fructose 50.0 g
$KNO_3$ 10.0 g
$KH_2PO_4$ 5.0 g
$MgSO_4$ 2.0 g
Trace salts:
$CaCl_2 \cdot 2H_2O$ 50.0 mg
$FeC1_3 \cdot 6H_2O$ 12.0 mg
$MnSO_4 \cdot H_2O$ 2.5 mg
$CO(NO_3)_2 \cdot 6H_2O$ 0.25 mg
$Na_2MoO_4 \cdot 2H_2O$ 0.2 mg
$ZnSO_4 \cdot 7H_2O$ 2.5 mg
$CuSO_4 \cdot 5H_2O$ 0.5 mg

Adjust pH of medium to 5.0 with 0.1 N NaOH.

Autoclave at 121°C for a duration appropriate for the volume.

b. *Metarhizium acridum* (Leland *et al.*, 2005)

Per liter of medium,

Brewer's yeast 40 g
Fructose or sucrose 40 g
Lecithin (L-α-phosphatidylcholine from soybean, Sigma Aldrich) 50 ml
Polyethylene glycol (PEG200), 50 ml

Autoclave at 121°C for a duration appropriate for the volume.

CHAPTER IX

# Methods for study of the Entomophthorales

ANN E. HAJEK,* BERNARD PAPIEROK[†] & JØRGEN EILENBERG[‡]

*Cornell University, Department of Entomology, Ithaca, NY 14853–2601, USA
[†]Institut Pasteur, Collection des Champignons, 25–28, rue du Dr Roux, 75724 Paris Cedex 15, France
[‡]University of Copenhagen, Department of Agriculture and Ecology, Thorvaldsensvej 40,
1871 Frederiksberg C., Denmark

## 1. INTRODUCTION

The order Entomophthorales (Zygomycotina) includes more than 200 species that are pathogenic to insects and mites. These entomopathogens are notable for the epizootics they induce in populations of many insects, including Homoptera, Lepidoptera, Orthoptera and Diptera. Their most noteworthy characteristic is that asexual spores, conidia, are forcibly discharged from the conidiophores, which develop at the host surface from hyphal bodies penetrating and elongating through the cuticle. This is the case for most entomopathogenic Entomophthorales, but not for species of the genus *Massospora,* for which the conidia detach passively, or for one member of the *Entomophaga grylli* species complex which produces cryptoconidia inside of cadavers (Humber & Ramoska, 1986). Moreover, conidia in some *Neozygites* species infecting mealybugs are not as forcibly discharged as for other species (Le Rü *et al.*, 1985). For most species, if conidia produced from insects do not land on a suitable place on the host, i.e. a site from which an infection may take place, a ‘primary conidium’ may form a ‘secondary conidium’. Secondary conidia can be produced on insect cuticle (e.g., on appendages of the cadaver itself), on plant surfaces, on soil, etc. Secondary conidia may themselves produce tertiary conidia, and even higher orders of conidia may thereafter be produced.

Most entomopathogenic Entomophthorales possess two further biological characteristics: (1) the fungus multiplies as protoplasts and/or hyphal bodies (having a cell wall) within hosts, after having invaded the host; and (2) hyphal bodies can develop into thick-walled environmentally resistant resting spores inside, or more rarely, outside the cadaver. Resting spores, which are azygospores or zygospores according to the species, allow the fungus to survive under adverse conditions. In most cases, the resting spore is a dormant stage and requires environmental stimulation for germination. At least two species are not known to make resting spores: *Pandora neoaphidis* (known previously as *Entomophthora aphidis* and then *Erynia neoaphidis*) (Remaudière & Hennebert, 1980) and *Entomophthora schizophorae* (Eilenberg & Philipsen, 1988; Nielsen *et al.*, 2003). By contrast, only resting spores are known for some species (e.g., *Tarichium* species) and for other species only resting spores are produced from some cadavers (e.g., Hajek & Shimazu, 1996).

These characteristics should be kept in mind when working with entomopathogenic Entomophthorales. Of primary importance for studies requiring conidia, these spores must be discharged from cadavers or cultures by the fungus and the timing of conidial discharge cannot always be manipulated or predicted fully. When working with resting spores, providing the conditions under which these spores will come out of dormancy and germinate can be challenging. For isolation, specialized media are usually needed. These conditions influence the design of methods and techniques for isolating these fungi, keeping them in culture, and conducting bioassays to study pathogenicity and virulence. A major challenge for all

MANUAL OF TECHNIQUES IN INVERTEBRATE PATHOLOGY
ISBN 9780123868992

scientists studying Entomophthorales is thus to get acquainted with the fact that these fungi do not behave like insect pathogenic fungi from the Hypocreales, such as *Beauveria bassiana* or *Metarhizium anisopliae*.

In this chapter, we will describe methods for identification, isolation, *in vivo* and *in vitro* growth, and bioassays. We provide citations from the literature for studies in which different methods were used so that readers with interests in using specific methods can access more detailed descriptions. As arthropod-pathogenic Entomophthorales can infect both insects and mites, we use the terms 'arthropods' and 'insects' interchangeably within this chapter.

## 2. COLLECTING AND PREPARING SPECIMENS

Once an arthropod has been located that has been killed by an entomophthoralean pathogen or is suspected of being infected, the first step is to collect and then to prepare it in order to (1) illustrate and preserve the morphological structures which will be used for identification, (2) attempt to isolate the etiological agent, and (3) to possibly initiate infection experiments.

### A. Collecting cadavers

Insects dying from entomophthoralean infections can frequently be recognized by conspicuous positioning on the substrate (e.g., Orthoptera killed by *E. grylli* systematically adhere to the tops of grasses), a peculiar color, the presence of sporulating structures, or possibly the presence of rhizoids, which allow the cadaver to be fixed to the substrate.

According to the pathogen species as well as the host species, cadavers can be found in various characteristic locations:

- on grasses and small plants, in open fields and in forest areas;
- on trees, on the underside of leaves or on bark, especially at the border of forests;
- on substrates adjacent to aquatic ecosystems (banks of lakes, ponds, streams, and rivers), often very close to the water level on mosses, stones, and plants;
- on inner walls of structures, e.g., houses, cellars, cow sheds, and henhouses;
- on the soil surface.

Once cadavers are discovered, the collecting procedures are as follows:

1. Carefully remove cadavers from the substrate, using fine forceps. If possible, the cadaver should be removed with a piece of the adhering substrate.
2. Place each specimen in a separate clear plastic box or a glass tube. The container should not be completely closed in order to prevent the atmosphere inside from becoming saturated, which generally allows saprophytic contaminants to develop rapidly. For this reason, it is advisable to use ventilated boxes or glass tubes with the opening covered by a piece of gauze or with a plastic lid riddled with small holes. Often a small piece of a leaf added to the tube will ensure that cadavers neither dry out nor become too wet.
3. Transport boxes or tubes to the laboratory, keeping them cool during transport to slow down, if not arrest, fungal development.

### B. Collecting living infected specimens

Collecting living infected specimens is suitable for species sporulating on living insects or which are difficult to find in the field after host death. This is also appropriate when one needs to collect living infected insects that are expected to die and short-lived structures produced shortly after host death are needed. An advantage of sampling living, infected specimens is that they will die in a relatively clean environment (a tube or dish) without contaminating material present. Insects are sampled using appropriate devices and procedures.

### C. Preparing field-collected specimens in the laboratory

For dead infected arthropods, specimens should be handled as soon as possible after returning to the laboratory, according to the procedures described briefly below and in more detail in Section 4. Living infected hosts should be maintained in incubators at appropriate temperatures (usually not $> 25°C$) for several days, until host death. Individuals should be monitored daily, and living or dead insects exhibiting symptoms of infection should be removed for fungal isolation. For those species requiring isolation from protoplasts, hemolymph samples can be taken from surface-sterilized, living insects that are suspected of being infected (see Section 4C).

Two procedures can be used to collect conidia produced from a cadaver of an insect bearing conidiophores: (1) the 'descending conidia' method, as described by Papierok (1989), Silvie & Papierok (1991), and Eilenberg *et al.* (1992); and (2) the method of 'ascending conidia' (Keller, 1987, 1994). We briefly describe these methods here, but more details are provided in Section 4A.

For the 'descending conidia' showering method, the cadaver is placed on a dab of Vaseline or directly on top of a moistened piece of tissue, paper towel or filter paper that is attached to the inside of a small sterile Petri dish lid (e.g., 35 × 10 mm). For larger cadavers (more than 10−15 mm long) larger Petri dishes can be used but the risk of contamination is then greater than with smaller dishes (contamination is an issue if isolation is the goal). To prevent contamination, large cadavers should be cut into several sections (e.g., the whole abdomen, pieces of a few abdominal segments, thorax), and then prepared as smaller specimens. Open wounds from the cutting can be sealed by Vaseline. The Petri dish lid plus cadaver is inverted over a microscope slide (Figure 9.1) under high humidity, e.g., within a plastic box containing a moistened paper towel. Conidial production begins within a few hours to a day, depending on whether conidiophores were already formed.

In the case of the method of 'ascending conidia' showering, either insects bearing conidiophores already or insects with no emerging conidiophores that have been surface sterilized (as described in Section 4A1) are placed in the bases of small Petri dishes (e.g., 35 × 10 mm) on water agar or on a moistened piece of tissue or filter paper. As with the 'descending conidia' method, a humid environment is necessary for optimizing conidial discharge. A microscope slide is placed above the dish to collect discharged conidia (Figure 9.2).

In both cases the slide has to be replaced by a new slide after enough conidia have been discharged onto it. We recommend preparing several slides with conidia per specimen. Slides can be kept under laboratory conditions until they are observed under a microscope. The cadaver producing conidia can then be used for isolation attempts or infection experiments (see Sections 4 and 7). After these operations, cadavers can be stored in 65−70% ethanol or dry, in which case sometimes the fungus is still alive for some period of time (see Section 6C).

## 3. SPECIES IDENTIFICATION

The different structures of taxonomic importance for entomophthoralean genera are described in Chapter VI. Keller (2007a) has also described the structures used for identification of Entomophthorales and Papierok & Bałazy (2007) have provided the technical advice necessary for observing them. Respective keys to families of Entomophthorales containing entomopathogenic species, to subfamilies and to genera of

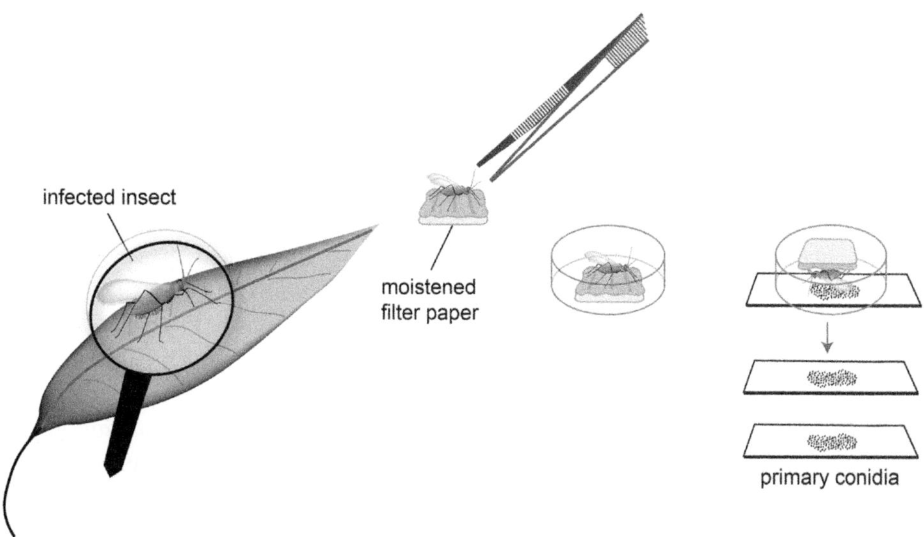


**Figure 9.1** Procedure for collecting 'descending' primary conidia from an insect infected by an entomophthoralean fungus.

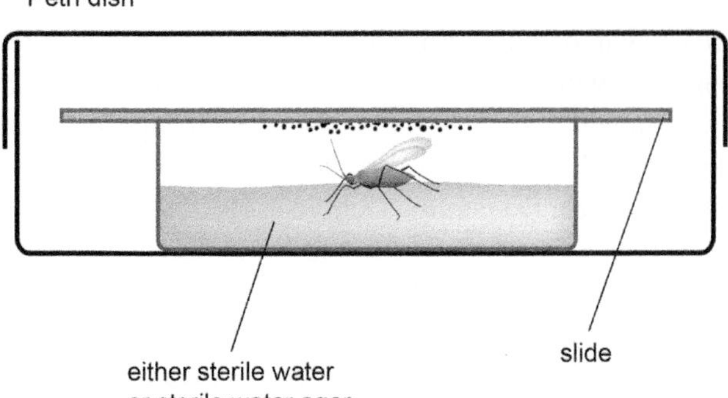


**Figure 9.2** Procedure for collecting 'ascending' primary conidia from an insect infected by an entomophthoralean fungus.

Entomophthoraceae have been provided by Keller (2007b). Along with identification of all entomophthoralean species, the species of host should also be determined, if at all possible.

For identification of Entomophthorales at the species level, the following characteristics are used: size and shape of primary conidia; size, shape and mode of formation of secondary conidia; morphology of hyphal bodies, cystidia, rhizoids; size and shape of resting spores; number and size of nuclei; host. There is currently no key published for all entomophthoralean species. Partial keys are available for some groups, however [e.g., keys to the species of *Entomophthora* or *Neozygites* (Keller, 2007b)].

Entomophthoralean species cannot always be identified or diagnosed using morphological features. Observing conidial morphology, frequently an important character for identification, is not always possible because conidia are not always formed or do not survive long. During collection of fragile specimens, important structures for morphological identification can be lost or damaged. While resting spores are more persistent structures, they often do not have many of the features needed for distinguishing species. For some species the structures necessary for identification might not be present in all specimens. Some more cryptic species can not be identified easily by morphology alone, regardless of the structures present (e.g., some members of the *Entomophthora muscae* complex). Therefore, in addition to the use of morphological features for identification, molecular techniques are now often used for diagnosis and identification of entomophthoralean species as well as for distinguishing strains. While numerous methods can possibly be used (see Castrillo & Humber, 2009), here we will present only those methods for identification and diagnosis with more frequent and relatively recent applications to Entomophthorales.

Generally, molecular methods for identification require extraction of DNA from samples followed by application of some variations of the polymerase chain reaction (PCR). Standard kits for DNA extraction from samples can be used. If the sample of interest contains only dormant, thick-walled resting spores, these must be broken open in order to extract DNA from within but care must be taken so that DNA is not damaged in the process. A bead beater with large, dense beads has been used to break *E. maimaiga* resting spores (Castrillo *et al.*, 2007) while a tissue homogenizer has been used to break resting spores of *E. grylli* (Bidochka *et al.*, 1996).

After DNA extraction, sequences of DNA called primers are bound to variable regions of the entomophthoralean genome and then amplified using PCR. Different approaches for identification using molecular methods are possible based on whether a clean culture is available or samples are potentially contaminated with DNA from other organisms, e.g., field-collected cadavers. More approaches are possible when the fungus has been isolated because general (= universal) primers can be used and there is no danger that these primers will bind to contaminating DNA.

Among techniques that require clean cultures, randomly amplified polymorphic DNA (RAPD), which uses random, short primers, has been used to investigate associations of fungi with hosts or to investigate geographic origins of species. This fast and simple method is used less frequently today due to lack of reproducibility among laboratories (see Enkerli & Widmer, 2010). Universally primed PCR (UP-PCR) and repetitive element PCR (rep-PCR) are also simple but are more reliable because primers are longer. Probably the most widely used universal primers for the Entomophthorales are based on the internal transcribed spacer (ITS) region of the ribosomal DNA (rDNA). Size of ITS and RAPD patterns helped to distinguish *P. neoaphidis* and closely related species (Rohel *et al.*, 1997). Use of the universal ITS primer successfully differentiated among two species of *Pandora* and *Zoophthora radicans* (Tymon *et al.*, 2004).

If an entomophthoralean sample could contain contaminants, e.g., DNA from the host's body or from bacteria potentially contaminating host cuticle, the choice of molecular method is critical. For potentially contaminated samples, only primers that are specific to the group of interest can be used as non-specific primers could bind to the contaminant and provide an incorrect result. Such an approach is also useful when monitoring the release of a given strain for biological control purposes. For example, specific primer pairs were developed (Agboton *et al.*, 2009) and used to evaluate samples of fungal-killed mites from areas where entomophthoralean biological control introductions had been made, when a very closely related fungal strain was native to the area of introduction (Agboton *et al.*, 2011).

For entomophthoralean samples that are potentially contaminated, researchers must use more specific methods. A commonly used method is PCR-RFLP with entomophthoralean specific primers (often ITS or LSU) and subsequent use of restriction enzymes that cut the PCR product at specific locations; resulting banding patterns on gels can help with identification (see Enkerli & Widmer, 2010 for review of examples). As an alternative, general primers can be used to amplify loci for which there are sequences available in GenBank. In this approach, after PCR with primers, fragments from the samples of interest are sequenced and sequences are compared with known sequences, e.g., in GenBank. With this approach, genes of the ribosomal subunit (SSU or LSU) or ITS are recommended as they are currently available for the highest number of species.

While we have focused on uses of molecular techniques for identification and diagnosis, molecular techniques are also being used increasingly for evaluation of environmental samples containing entomophthoraleans, e.g., to study genetic variability within a fungal species, to quantify entomophthoralean spores in soil samples and to evaluate the results of biological control releases (see review by Enkerli & Widmer, 2010).

## 4. ISOLATION

In the context of this chapter, isolation means establishment of *in vitro* cultures. As a general rule of thumb, entomophthoralean fungi are mostly relatively fastidious so complex nutrient media are required for their isolation and growth. Further, many species germinate and grow slowly (or have not yet been isolated *in vitro* at all, e.g., *Neozygites fresenii*). These characteristics, accompanied with the potential for contaminating microorganisms on or inside the infected insect, make isolation of Entomophthorales a challenge.

Entomophthoralean fungi can be isolated using conidia, resting spores or the vegetative stages within hosts. Entomophthorales can also be isolated from environmental samples, especially soil samples, given that these fungi are likely to be able to survive in the soil as resting spores for as long as several years and as mycelia or conidia for at most several months (e.g., over the winter for some species). Perhaps most commonly, entomophthoralean fungi are isolated by directly inoculating various media with conidia discharged from cadavers. Hosts are usually killed when the fungus has exhausted the body as a source for nutrients and the body is filled with fungal cells, after which conidiophores grow out through the cuticle and produce conidia that are actively discharged. There are a few exceptions to most general rules about Entomophthorales regarding conidial production: some species make conidia before the host is dead, some species do not actively discharge conidia or do not even produce conidia externally and some species do not make conidia at all. Therefore, in several instances, direct isolation using conidia is not possible and other methods have been designed that allow fungal isolation using other fungal stages (i.e. hyphal bodies and protoplasts or resting spores). In general, when using growing fungal stages from within living hosts (i.e. hyphal bodies or protoplasts) or short-lived conidia, isolation must often be performed as quickly as possible after collection. Cadavers filled with fungal cells (= mummies) can often be stored in the cold before isolation (see Section 6C).

### A. Isolation using conidia

Isolation of entomopathogenic Entomophthorales using conidia can be undertaken using two basically similar showering methods: (1) the method of 'descending conidia', as described by Papierok (1989), Silvie & Papierok (1991), and Eilenberg *et al.* (1992), or (2) the method of 'ascending conidia' (Keller, 1987, 1994; Jensen *et al.*, 2006). Given that conidial discharge is required, these isolation methods are not suitable for species that do not discharge conidia.

#### *1. The 'descending conidia' showering method (Figure 9.1)*

This method is based on conidia being discharged downwards (see also Section 2C).

1. A cadaver of an arthropod bearing conidiophores is placed on top of a moistened piece of sterile tissue, paper towel or filter paper. Cadavers that are filled with hyphal bodies but from which conidiophores have not yet emerged should be surface sterilized by dipping them successively in 65–70% ethanol (10–15 s), 2% sodium hypochlorite solution (2–3 min), and sterile water (a few seconds in two successive baths).
2. After preparing the infected specimen, the Petri dish lid is inverted over the base of a sterile Petri dish, which can contain culture medium. Conidial production begins within a few hours to a day, depending on the developmental stage of the fungus when sporulation conditions are initiated.
3. Conidia are collected for periods of varying lengths (from a few minutes to a few hours), depending on the intensity of sporulation. After sporulation, the lid with the specimen attached is replaced by a sterile lid.
4. If a Petri dish with an empty base is used, conidia that land on the bottom of the Petri dish can be transferred to media by rubbing a piece of solid medium across the conidia and subsequently using this spore-bearing medium to inoculate a culture tube (Figure 9.3). Alternatively, after showering, liquid medium (see Section 4F3) can be poured into the Petri dish bottom. An advantage of adding liquid medium afterwards is that conidia will be 'within' and not 'on top of' the liquid medium, which can improve germination and vegetative growth.

### *2. The 'ascending conidia' showering method (Figure 9.2)*

This method is based on collection of conidia being discharged upwards. A main reason for using the 'ascending' method is that by allowing conidia to be projected upwards, one will avoid the risk that contaminants will drop downwards from the cadaver into the medium.

1. Prepare specimens as described in Section 2C and place on a sterile, moist substrate, e.g., sterile water, water agar or moistened paper towels.
2. A sterile slide or coverslip is placed above the dish to collect discharged conidia.
3. This device is kept in a sterile chamber (e.g., a Petri dish of larger size) until sufficient conidia have been collected. The conidia are then removed by rubbing a piece of solid medium across the showered spores and subsequently transferring these spores to solid medium in a culture tube.
4. Alternatively, invert a small Petri dish. Conidia are projected from a cadaver attached to the inside of the lid upwards into the bottom of the dish. After projection of conidia, the dish is turned over, liquid medium is poured on to the conidia, and a new lid is added. This works best with > approximately 100 conidia per dish (Delilabera Jr & Eilenberg, unpublished observations).

## B. Isolation using resting spores

Isolation from resting spores formed inside the host's body has been successful in some cases (see Tyrrell & MacLeod, 1975).

1. Cadavers filled with resting spores are soaked in sterile water, then gently broken up and sonicated. To reduce the risk of contamination, after soaking and breaking apart cadavers, resting spores can be washed through a sieve (e.g., 63 μm) to separate them from insect body parts.
2. Surface sterilize resting spores using a 2–3% sodium hypochlorite solution (2–3 min) followed by sterile washing or 5 min exposure to 30 ppm mercuric chloride followed by five washes in sterile water (Perry & Latgé, 1982).
3. Resting spores are then spread onto media.

In the case of resting spores formed outside the cadaver, remove clumps of spores with fine forceps and place them in sterile distilled water, then break up spore clumps using a glass rod or fine forceps. Follow with sonication and washing and then spread resting spores onto media as described above.

It is frequently very difficult to separate resting spores that have already been released from cadavers from other micro-organisms in the environment, especially if they are collected from the soil or plants. In these cases, attempting to start cultures on media or even agar can lead to overgrowth of the slowly germinating entomophthoralean resting spores by fast-growing saprophytes. As an alternative, when susceptible healthy insects are available, exposing them to environmental samples containing germinating resting spores can yield successful isolation (Hajek *et al.*, 2000).

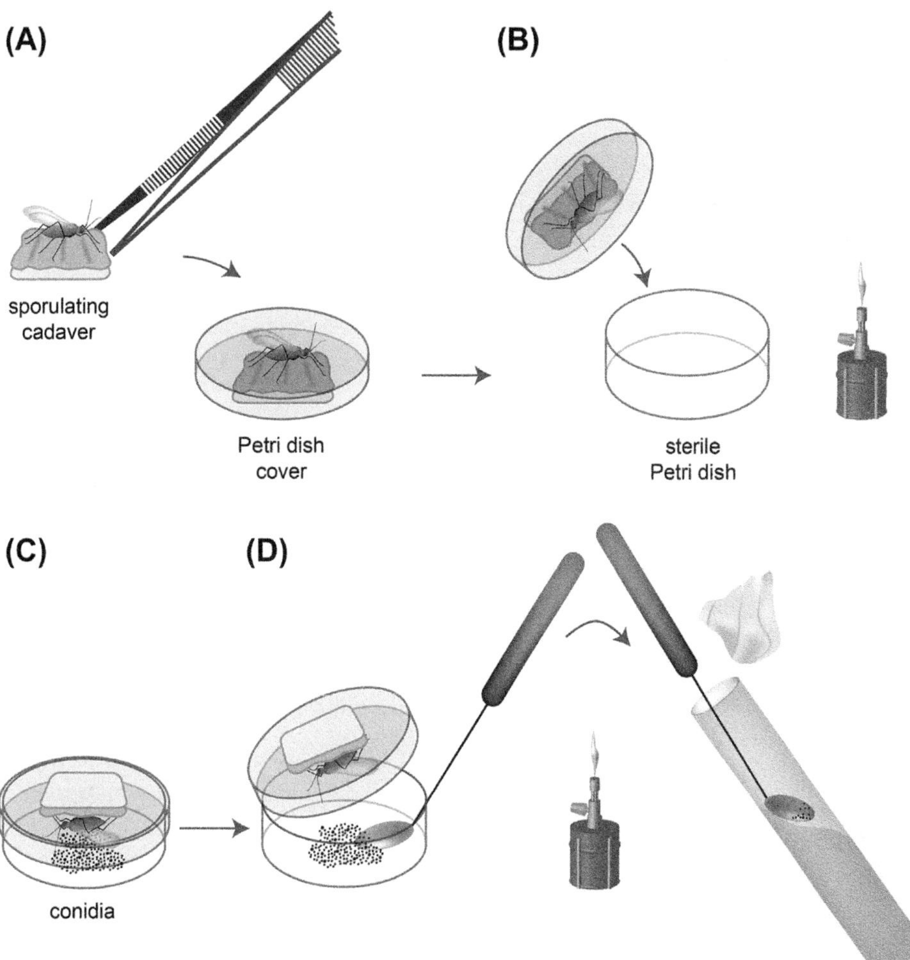


**Figure 9.3** Successive steps for isolating Entomophthorales from infected hosts using the 'descending conidia' showering method. (A) Attach a sporulating cadaver to the underside of the lid of a Petri dish. (B) Invert the lid with the sporulating cadaver over a sterile Petri dish base (under sterile conditions). (C) Let the cadaver produce conidia for a certain time, depending on the intensity of sporulation. (D) Rub a piece of solid medium across the spores which landed on the bottom of the Petri dish, and use that inoculated piece of medium to inoculate a culture tube (or plate) under sterile conditions.

1. Cage test insects over the environmental sample containing germinating resting spores on a moist substrate for varying lengths of time. For high densities of resting spores, exposures that are too long can result in 100% death from septicemia, so shorter exposures are better, if germination rate is high. The rate of resting spore germination can be checked by placing a coverslip over the resting spores.
2. Rear exposed insects for several days after exposure to resting spores, but before death isolate the fungus using vegetative stages in the hemolymph (Section 4C). In particular, this method for isolation has been very successfully used for entomophthoralean species with conidia that will not initiate cultures in liquid media (Hajek *et al.*, 2000).
3. Alternatively, after arthropods infected from germinating resting spores die, the fungus can be isolated if conidia are produced externally on cadavers (see Section 4A).

## C. Isolation using vegetative stages: hyphal bodies and protoplasts

It is occasionally useful to attempt isolation of entomophthoralean species from infected, living insects in which the pathogen is developing as protoplasts and/or hyphal bodies (Hajek *et al.*, 2000). The procedure to be used for large specimens (e.g., caterpillars) consists of:

1. Superficially cleaning the insect cuticle with 65−70% alcohol or 2−3% sodium hypochlorite.
2. Collecting a small amount of hemolymph either directly from the hemocoel with a syringe and an appropriate needle, or as it is released after removal of a proleg.
3. Inoculating the medium.

For very hairy caterpillars, bacterial contamination has been a problem in isolating fungal pathogens. In these cases, infection of a permissive species from a clean laboratory colony is achieved by conidial showers from field-collected infected hosts or by exposure to germinating resting spores. It is important to repeat this procedure, varying the inoculation concentration, because excessive inoculation can result in septicemia and not mycosis. Just before inoculated larvae die, hemolymph samples are collected (as above). The hemolymph, which contains fungal cells, is then introduced into Grace's insect cell culture medium plus 0.5% fetal bovine serum plus gentamicin (see Section 4F3). Fungi isolated using this procedure will grow as hyphal bodies or protoplasts according to the species. Initiation of growth can be very slow (frequently one to several weeks), and cultures initially growing with cell walls can eventually change to growth as protoplasts, although this transition can be slow (Hajek *et al.*, 2000).

## D. Isolation using whole specimens (living or dead)

This method can be useful for small arthropods like aphids, thrips and mites.

1. Surface sterilize specimens with ethanol and/or sodium hypochlorite and then rinse with sterile water. Because surface sterilization can be harmful to the pathogen when the host's cuticle is broken, this method should be used only with insects without emergent fungal structures.
2. If needed, individual surface-sterilized cadavers can then be crushed using sterile forceps or a sterile toothpick in a very small volume of liquid medium (50−150 μl) in 24-well tissue culture plates.
3. When cell division is observed, add fresh medium several times over a few days until the culture reaches 500−1000 μl.

Using this method, mite-pathogenic *Neozygites* spp. and *Entomophthora thripidum* were isolated from living infected mites and thrips, respectively (Grundschober *et al.*, 1998; Freimoser *et al.*, 2000; Delalibera *et al.*, 2003). Furthermore, this method has proven successful in cases where other methods have been difficult, e.g., for the aphid pathogen *Entomophthora planchoniana* (Freimoser *et al.*, 2001). When used for large scale isolation of aphid-infecting Entomophthorales, Remaudière *et al.* (1981) simply placed cadavers on solid media.

## E. Isolation from soil samples

Detection of entomopathogenic Entomophthorales in the soil is based on the use of living insects confined over soil collected from locations where entomophthoralean infections were previously noted, as first experienced with aphid pathogens by Latteur (1977). The procedure for isolation of aphid pathogens is as follows:

1. Collect soil samples, preferably not > 1.5 cm deep, without disturbing the soil surface and place these samples in sealed Petri dishes (and store at 4°C, if necessary).
2. Confine batches of laboratory-produced aphids for several hours on moist (but not saturated) soil samples in Petri dishes at room temperature, beneath gauze covered by moistened paper towelling.
3. Remove aphids and hold them on plants for approximately 24 h in a saturated atmosphere.
4. Return the aphids to normal rearing conditions and check for diseased specimens (up to 5−6 days following exposure at 20°C) and isolate the fungus as described earlier.

Isolation from soil using aphids has been used to quantify the inoculum of aphid pathogenic fungi in soil by Nielsen *et al.* (2003). A similar method also works for other insects, e.g., caging *Lymantria dispar* larvae over soil containing resting spores of *E. maimaiga* has repeatedly been successfully used for isolation (e.g., Nielsen *et al.*, 2005).

The *'Galleria bait method'* (Zimmermann, 1986), originally described for trapping entomoparasitic nematodes in soil, has proven efficient for detection of hypocrealean entomopathogenic fungi. In a few cases this method was also used to isolate Entomophthorales, although mostly *Conidiobolus coronatus* (Meyling & Eilenberg, 2006).

## F. Media for isolation

Classical mycological media, such as potato dextrose agar or malt extract agar, are unsuitable for the great majority of Entomophthorales; these species simply will not germinate or will grow poorly on these more simple types of media. Entomophthoralean species generally need richer nutrients to grow and develop. Commonly, solid media for successfully growing most Entomophthorales contains egg yolk. However, requirements of some species are more elaborate and, consequently, media, at times, have to be supplemented, e.g., with vitamins and amino acids. In addition, some species can only be isolated and grown using insect cell culture media or other liquid media.

### *1. Media containing egg yolk*

Numerous recipes for media for growing Entomophthorales that contain egg yolk are mentioned in the literature. These media are generally quite rich and care must be taken to avoid contamination. We provide recipes for two variations routinely used in the Institut Pasteur (for routine laboratory culture, Sabouraud dextrose agar (SDA) supplemented with egg yolk and milk (Papierok, 1978) and for field isolation, coagulated egg yolk with milk (Papierok & Charpentié, 1982) and the recipe used for all purposes by the USDA Agricultural Research Service Collection of Entomopathogenic Fungi in Ithaca, NY [egg yolk/Sabouraud maltose agar (SMA): EYSMA] (see Appendix). Many other basically similar media, providing variations in ingredients and preparation, both of which can be critical, and some selected references are provided below: coagulated egg yolk (Müller-Kögler, 1959; Gustafsson, 1965), egg yolk/SMA (Soper *et al.,* 1975), egg yolk with milk/SDA (Milner & Soper, 1981), egg yolk/SDA (Keller, 1987), coagulated egg yolk with milk (Keller, 1987), egg yolk/peptone broth agar (Bałazy, 1993), and Nemoto agar (Shimazu, 1993).

### *2. Complex media*

Due to poor growth of several Entomophthorales species on media containing egg yolk, more complex media supplemented with additional components (vitamins, salts, lipids, proteins, etc.) have been developed. For instance, Ben-Ze'ev (1980) (in Keller, 1994) developed '*Entomophthora* complete medium', containing a carbon source (e.g., dextrose, maltose), a solution of 11 salts, tryptophan, yeast extract, casein hydrolysate, a solution of nine vitamins, and agar. This complex medium has the advantage of being transparent, which allows microscopic observation of cultures. Furthermore, it also allows isolation of a few species which are difficult or impossible to isolate using solid media containing egg yolk, e.g., *E. grylli* and *E. muscae* (Keller, 1987, 1994).

### *3. Insect cell culture-type media and substitutes*

Tyrrell & MacLeod (1972) first observed the formation and proliferation of protoplasts originating from conidia of *Entomophaga aulicae* in Grace's medium (Grace, 1962). Since then, this insect cell culture medium and modifications of it have been used not only for growing entomophthoralean protoplasts, but also for isolating species of this fungal group that exhibit complex nutritional requirements. Other complex liquid media have also been tested, as well as a simpler medium, the GLEN medium (G = glucose, LE = levure extrait which is yeast extract and N = NaCl) (Latgé & Beauvais, 1987; Beauvais & Latgé, 1988).

The insect cell culture medium developed by Grace is available commercially (Invitrogen). Fungal isolations in Grace's medium are often attempted in 35-mm diameter Petri dishes or 25-$cm^2$ canted neck flasks containing a layer of this medium supplemented with 5% or 10% (v/v) heat-inactivated fetal bovine serum (Dunphy *et al.,* 1978; Latgé & Beauvais, 1987; Soper *et al.,* 1988). In the case of the canted neck flasks, 10 ml of media are usually used and the flask is laid on its side to maximize the surface area of the media. In the peculiar case of *E. grylli,* MacLeod *et al.* (1980) found that a grasshopper thorax and aqueous abdominal extract (without digestive tract, wings and other appendages) added to the Grace's medium plus 5% fetal bovine serum was necessary for successful isolation.

Modifications of Grace's medium for growth of *E. aulicae* have been investigated extensively by Dunphy & Nolan. *E. aulicae* could be grown in synthetic eastern hemlock looper hemolymph (Dunphy & Nolan, 1989) which has been patented (US Patent number 5,728,572, 17 March 1998) and in a simplified defined medium with only eight amino acids (Nolan, 1988). *E. aulicae* protoplasts did not need vitamins, protein, or organic acids but required Grace's medium, hematin, and oleic acid, with or without shaking (Nolan, 1988). Leite *et al.* (2003) reported that the addition of salts to a basic medium of sugars (especially including glucose rather than sucrose) plus lactalbumin hydrolysate and yeastolate resulted in increased biomass of *Batkoa* spp., *Furia* spp. and *Neozygites floridana*, while addition of vitamins and amino acids had less effect. Other commercially available tissue culture media, e.g., Mitsuhashi and Maramorosch's (Promocell), MGM-443 (Mitsuhashi, 1982), have also been successfully used for growing Entomophthorales (Soper *et al.,* 1988; Eilenberg *et al.*, 1992). Use of the simple GLEN medium designed by Beauvais & Latgé (1988) resulted in successful growth of *E. aulicae* protoplasts. GLEN medium has, without supplements, also been used for isolation of *Strongwellsea castrans* and *Entomophthora* species (Eilenberg *et al.*, 1992) (see Appendix).

Breakthroughs have been made for some fastidious species that had never been grown *in vitro* by supplementing cell culture media with additional materials and occasionally including aeration (Latgé *et al.*, 1988; Leite *et al.*, 2000). Hyphal bodies of the thrips pathogen *Neozygites parvispora* grew best in Grace's medium plus lactalbumin hydrolysate, yeastolate, 10% *Manduca sexta* or *Spodoptera littoralis* hemolymph (adjusted to 10 mM glutathione to prevent melanization) and 20% fetal bovine serum (Grundschober *et al.*, 1998). *Entomophthora thripidum* and *E. planchoniana* cells multiplied in Grace's medium supplemented with yeastolate, lactalbumin hydrolysate and 10% fetal bovine serum (Freimoser *et al.*, 2000, 2001) while *E. planchoniana* also grew well on GLEN plus 10% fetal bovine serum. In a comparison of growth of two mite-pathogenic *Neozygites* species in 11 different types of cell culture media, the more fastidious *Neozygites tanajoae* grew best in IPL-41 insect medium (Invitrogen) plus 5% fetal bovine serum, 0.3% lactalbumin hydrolysate, and 0.3% yeastolate but both *N. tanajoae* and *N. floridana* also grew well in SF900 II SFM media supplemented with 5% fetal bovine serum (Delalibera *et al.*, 2003).

## G. Tricks of the trade

As is the case with any biological material, experience is invaluable when attempting to isolate Entomophthorales.

### *1. Choice of isolation media*

Choosing an isolation medium depends mainly on what one knows about the fungus expected to be isolated. Species of Entomophthorales differ widely in the ease with which they can be isolated and grown *in vitro*. Some species of Entomophthorales are easier to isolate, e.g., many species of *Conidiobolus* can be grown on more simple media with only glucose and yeast extract. In contrast, some species of other genera have never been cultured (e.g., *Neozygites fresenii*; Vingaard *et al.*, 2003). In between these extremes, genera like *Entomophthora* contain species that can readily be isolated, such as *Entomophthora culicis*, while other species like *E. schizophorae* and *E. muscae* are fastidious (= having complex nutritional needs) and have seldom been successfully cultured (Eilenberg *et al.*, 1990). Whether one is attempting to isolate a fungus that may be a new species, that is not well known, or that is suspected to be a well-known species, it is recommended to try several types of solid and liquid media and several different fungal propagules (conidia, hyphal bodies) to optimize the chances for successful isolation. Isolation from conidia might be preferable because this allows for both collection of conidia on slides to aid in identification and retention of other fungal structures (e.g., conidiophores, hyphal bodies, etc.) as well as saving cadavers for records and subsequent taxonomic study. If possible, however, it is best to try both solid and liquid media, using different fungal stages because different stages of entomophthoralean species can differ in growth requirements. In particular, while hyphal bodies or protoplasts may grow in cell culture media, conidia of the same species can fail to differentiate in cell culture media but they might differentiate and grow on an egg yolk-based medium (Humber, 1994).

### *2. Specific considerations in preparing media*

When preparing and pouring media, it is advised:

- To autoclave the different ingredients separately.
- To avoid running egg yolk-based medium (which requires strictly sterile conditions) down along the inner wall of the tubes or Petri dishes. This can disturb observations of initial fungal growth, and can even favor contamination if the medium is in contact with the cotton plug or opening of the Petri dish.
- When adding serum additives do not heat but use a sterile filter.

*3. Antibiotics*

Because of entomophthoralean sensitivity to many antibiotics, they are not always added to isolation media for entomopathogenic Entomophthorales. The media presented in this chapter do not normally include antibiotics. However, for isolation of *E. maimaiga* from gypsy moth (*L. dispar*) larvae (large hairy lepidopteran larvae), flasks of Grace's medium (9.5 ml) plus fetal bovine serum (0.5 ml) are regularly amended with 45 μl gentamicin (A. E. Hajek, unpublished data). This is possible because gentamicin has no effect on protoplasts of this species. Similarly, streptomycin and penicillin were demonstrated to have no effect on growth of *Z. radicans* isolates, unlike tetracycline, chloramphenicol and actidione (Ben-Ze'ev, 1980).

*4. How to avoid contamination*

*a. Maintaining sterility*

As far as possible, preparation of specimens and fungal isolation should be carried out in a laminar flow hood. When such equipment is not available, a sterilized portable Plexiglas box (e.g., 25 × 40 × 40 cm) with holes for access on one side and a hinged lid has been used successfully. Under field conditions, fungal isolation can be conducted on any surface, preferably covered with a plastic tablecloth. It is recommended that the surface be disinfected with 65–80% alcohol and drafts be reduced before attempting isolation. All supplies and equipment used should be sterile. A portable Bunsen burner, e.g., Labogaz® (Camping Gaz International), is suitable for sterilizing supplies and equipment in the field. Special materials such as filter paper or paper towels, water for moistening the substrate for cadavers, and cotton plugs for small isolation tubes (see Section 2), should be sterilized in appropriate packaging. If necessary, bottled water can be used.

*b. Working rapidly*

Isolations from conidia should be attempted as soon after insect death as possible, optimally within a few hours. This is especially important for cadavers that already bear conidiophores and have to be prepared as specified in Sections 2C and 4A. Sporulating structures of most Entomophthorales are usually functional for less than 24 h at temperatures of 15–20°C and high humidity. Moreover these structures are rapidly contaminated, leading to contamination of isolates by bacteria or saprophytic fungi. Therefore, the period of time during which conidia are collected should be as short as possible. The intensity of spore discharge should be regularly monitored. If necessary, development of conidiophores can be slowed by cooling (e.g., 4°C), although this should be avoided, if possible. If cooling must be used, cadavers should not be stored in a saturated atmosphere. Relative success in being able to slow sporulation by cooling appears to be species dependent.

Cadavers with conidiophores not yet emerging through the cuticle can be stored for 24 h or more at 4°C or below, provided they are kept under dry conditions. Such cadavers are used for long term storage of mite-pathogenic *Neozygites tanajoae* mummies which then sporulate when returned to warmer and humid conditions (see Section 6C). However, the longer the cadaver is kept in cool conditions, the longer the time required for subsequent appearance of conidiophores once cadavers are returned to higher temperatures and high humidity.

If conidiophores have not formed within 24–48 h of insect death, cadavers should be dissected and observed under the light microscope. Lack of conidiation can result from poorly growing hyphal bodies and/or internal production of resting spores instead of conidia.

*5. Incubation conditions*

As a general rule, tubes or Petri dishes inoculated with conidia or vegetative cells should be incubated at an average temperature of 20°C (18–23°C). When using resting spores in soil for isolation of *E. maimaiga*, 15°C is regularly used (Hajek *et al.*, 2000; Nielsen *et al.*, 2005).

*6. Checking isolation containers*

*a. Monitoring fungal development*

Once incubated, isolation tubes or Petri dishes can be observed daily in order to monitor germination of the inoculated spores. Opaque media (e.g., SDA supplemented with egg yolk and milk, EYSMA, and coagulated egg yolk and milk) should be observed with a dissecting microscope. Details of fungal development are often difficult to see until mycelium is nearing sporulation. Transparent media (e.g., GLEN and insect tissue culture-based liquid media) can be examined with an inverted microscope. The fungal developmental rate varies according to species. On solid media, discharge of conidia onto inner walls of tubes or Petri dishes can be a sign of successful isolation. When resting spores are used as inoculum, the absence of development can simply result from spore dormancy.

*b. Checking the identity of the isolated entomophthoralean species*

The form and size of conidia produced on solid isolation media should be compared to those of conidia from the original infected specimen. One must keep in mind that, as a general rule, conidia produced *in vitro* are slightly larger than conidia from cadavers and may also contain a higher number of nuclei (Eilenberg *et al.*, 1990). In practice, conidia are obtained from solid media in a similar way as from cadavers (see Section 4A). A piece of mycelial mat is removed aseptically from the isolation container and placed on a moistened piece of filter paper (or paper towel) which is then attached to the inside of a Petri dish lid for isolation using 'descending' or 'ascending' methods for capturing conidia when they are discharged.

Conidia are sometimes not produced on the rich media often necessary to grow Entomophthorales. For cultures on egg yolk media, for many species conidia will not be produced until the fungus is starved (e.g., mycelium is transferred to water agar or moist paper towels) (e.g., *Furia gastropachae*; Filotas *et al.*, 2003). In tissue culture media, conidia can be produced in limited quantities only on the surface and to induce sporulation, the fungal mycelium must be transferred to solid media (e.g., *E. schizophorae* and *E. muscae;* Eilenberg *et al.*, 1990). Alternatively, host insects can be infected by injecting protoplasts from cultures, which will lead to development of mycosis and subsequent production of conidia on cadavers after host death.

*c. Overcoming contamination*

Fungal contaminants (e.g., *Penicillium, Mucor, Cladosporium,* etc.) can grow rapidly in entomophthoralean cultures. Often, very little can be done to overcome this situation once it has started and the best course of action is simply to discard that culture and try again if possible. A few species of Entomophthorales (mostly in the genus *Conidiobolus*) which can be fully or partly saprotrophs are an exception. These species can be contaminants on arthropods infected by other species of pathogenic Entomophthorales (Remaudière *et al.,* 1976; Papierok, 1985). In cases like this, isolation of *Conidiobolus* species does not demonstrate that they are primary pathogens.

There are some possible methods to try if microbial contamination is detected and reisolation is not possible. Bacterial contamination can occasionally be overcome because, initially, bacteria often grow slowly and growth is localized. Once a bacterial colony or other contaminant is first detected, a piece of the mycelial mat should be removed sufficiently far from the contaminant and transferred to fresh medium. In some cases, the surface of the culture can be contaminated, while entomophthoralean conidia are stuck to the inner wall of the container. Attempts to save the culture can be made by scraping conidia from the walls of the tube with a piece of sterile solid medium and transferring them to fresh medium. Finally, a projection of conidia upwards to the inverted bottom of a Petri dish and then pouring liquid medium into the dish can also assist in saving the culture.

## 5. *IN VITRO* CULTURE AND PRODUCTION

### A. Routine culture maintenance

Once an isolated entomophthoralean species is successfully growing on media, the culture has to be maintained and eventually stored frozen (see Section 6C and Chapter X). Many species of Entomophthorales can be maintained on solid media in Petri dishes or in tubes in the laboratory but some can only be grown *in vitro* in liquid media. Entomopathogenic fungal species unable to grow on media can be kept in culture using the host (see Section 6).

*1. Solid media*

*a. The most convenient media*

For keeping Entomophthorales in culture on solid media, the following two recipes have proven successful in many cases: SDA supplemented with egg yolk and milk, and EYSMA (see Appendix). Coagulated egg yolk (see Appendix) and *'Entomophthora complete medium'* can be tried for species growing poorly on the previous two media. A simple type of medium consisting of SDA supplemented with sesame oil and sucrose fatty acid esters was found to be appropriate for maintenance and preservation of several species of Entomophthorales (Feng & Xu, 2001), although adding egg yolk resulted in faster growth.

*b. Techniques for transfer*
Techniques for fungal transfer are as follows:

1. carefully remove a small piece of the mycelial mat (3–4 × 3–4 mm) from the medium;
2. transfer it to a new tube or Petri dish by first making a small depression in the middle of the surface of the medium and then placing the mat on the medium so that conidiophores will be produced upwards.

The growth of the fungus will occur partly from direct development of mycelium, and partly from germination and subsequent formation of hyphal bodies from conidia that have been discharged from the piece of culture and have landed on the new medium. We suggest keeping inoculated tubes in a slightly tilted position for a few days after inoculation. This favors landing of discharged conidia on the medium rather than on the tube walls (Gustafsson, 1965).

*c. Maintenance conditions*
Freshly inoculated media should be kept at 18–20°C for 2 weeks in order to secure consistent growth of the initial fungal inoculum. Afterwards, cultures can be stored at lower temperatures, for instance 12°C and even 8°C, so that the fungus grows more slowly and transfers are needed less frequently. Cultures should preferably be kept in the dark.

*d. Frequency of subculture*
Transfers should be made every 2 or 3 months at most, when maintaining the fungus at 12 or 8°C, respectively. However, some very fast-growing Entomophthorales may need more frequent transfers, e.g., every 4–6 weeks. Nevertheless, it is highly recommended that cultures are checked weekly or biweekly once the fungus has invaded the medium surface. Subculturing should be undertaken before fresh conidiophores are no longer detectable and/or the mycelial mat starts to liquefy.

*e. Possible loss of culture capability*
Entomophthoralean cultures maintained on solid media can deteriorate (with a diminution of growth rate and/or loss of conidial production) if successively transferred too many times. However, this phenomenon is species- or even strain-specific. Such a decline has been observed with cultures of *P. neoaphidis* (Rockwood, 1950), *Zoophthora* spp., *Erynia* spp., and strains of *E. culicis* (B. Papierok, unpublished data). Conversely, cultures of almost all strains of *Conidiobolus obscurus* maintained for years have not changed (B. Papierok, unpublished data).

*2. Liquid media*

Liquid media are used for maintenance of several species of Entomophthorales, especially those growing in the protoplast stage. Media suitable for hyphal bodies should include dextrose and hydrolysates of proteins as basic components. Two commonly used media for protoplasts are Grace's medium supplemented with fetal bovine serum (see Section 4F3) and GLEN medium, with or without serum additions. Protoplast cultures are usually maintained in 25-cm$^2$ canted neck tissue culture flasks at 18–20°C in total darkness. In this size flask, 9.5 ml Grace's medium and 0.5 ml fetal bovine serum is often inoculated with 1 ml of protoplast culture. Cultures must be transferred to fresh media based on the density and speed of growth of the inoculum. *Entomophaga maimaiga* is transferred into Grace's medium supplemented with fetal bovine serum every 3–5 days. Transfer is necessary once protoplasts become swollen, spherical, and begin to appear granular internally. However, great care must be taken to limit the number of subcultures to prevent loss of virulence (Hajek *et al.,* 1990b). To guard against using cultures that have lost virulence, *E. maimaiga* was frozen as abundant individual samples of protoplasts from the same original culture and samples were regularly thawed so that individual fungal lines were never subcultured more than 10 times. *Entomophthora schizophorae* was transferred into GLEN medium without serum additions every 14 days (Eilenberg *et al.*, 1990). Approximately 0.5–1 ml of cultured cells should be transferred into 10 ml fresh medium to ensure sufficient fungal biomass since protoplasts are fragile and some cells may not survive the transfer.

As an alternative, liquid cultures can be maintained at 4°C and then subculturing is necessary only once a month. In order to keep isolates for longer, freezing in −80°C or under liquid nitrogen is necessary (see Chapter X).

## B. Laboratory-scale production

By laboratory-scale production, we are referring to growing more of the entomophthoralean species than necessary for culture maintenance, usually for conducting bioassays. Laboratory-scale production can be done using solid media, following methods described above but with the goal of producing specified quantities of conidia or biomass. The biomass produced by changing medium composition can differ significantly (Urbanczyk *et al.*, 1992). The classical method of measuring biomass as radial growth on solid media is still a valid tool for studies of many species of Entomophthorales (Gúzman-Franco *et al.*, 2008) although it can be difficult to see the edges of growth when using opaque egg yolk-based solid media. For all laboratory-scale production, some form of stock culture must be maintained in a way that prevents degradation due to extended subculture (see above).

### *1. Hyphal bodies and protoplasts*

Protoplasts and hyphal bodies represent basic vegetative developmental stages of Entomophthorales that occur within infected hosts. They are produced in solid and liquid media but cannot easily be separated from solid medium ingredients, especially the fatty ones (e.g., egg yolk and milk) so they are best produced in liquid culture. Protoplasts, the precursors for hyphal bodies, are also best produced in liquid medium although they are often fragile and special care is needed.

Dextrose (as a carbon source) and hydrolysates of proteins (as sources of other nutrients, basically nitrogen) are especially convenient substances for growth of various entomophthoralean species (Latgé, 1975, 1981; Latgé & Remaudière, 1975). GLEN medium, which also includes these two basic components, is used for growth of entomophthoralean protoplasts and hyphal bodies (see Appendix).

Once prepared, the liquid medium is distributed into Erlenmeyer flasks, the volume per flask depending on the flask size. For instance, 150-ml flasks are filled with 50 ml medium as less medium per flask creates better aeration. Flasks containing GLEN or another suitable liquid medium are autoclaved for 30 min at 120°C. Flasks are initially inoculated with a few pieces of mycelial mat removed from cultures on solid medium or can be inoculated with protoplasts from liquid medium. Mycelium that is used should be in a fast-growing stage, e.g., taken from the periphery of fungal growth of 2- to 5-day-old cultures on solid medium incubated at 20°C, while protoplasts should be in the exponential growth phase in liquid medium. Flasks should be placed on a reciprocating shaker (100 oscillations/min) or a rotary shaker with rather moderate rotation speed (150 r.p.m., or, for protoplasts, even down to 50 r.p.m.) in darkness at 20–25°C. Depending on the fungal species and growth conditions, optimal production of fungal cells should take 2–5 days, but can take more time (e.g., in the case of *E. thripidum* 10–20 days were required for optimal production *in vitro* of protoplasts and then hyphae; Freimoser *et al.*, 2003).

For collecting hyphal bodies, liquid cultures at the appropriate stage should be filtered, possibly using a vacuum pump. Hyphal bodies collected on the sterile filter paper should then be washed with sterile distilled water. Subcultures can be made by inoculating flasks with a 5% (v/v) suspension of hyphal bodies from optimally growing liquid cultures.

When cultivated in liquid media, hyphal bodies of many species aggregate. In order to prevent or at least reduce this phenomenon it is useful to add magnetic rods to flasks, and, immediately after inoculation, to place flasks on a magnetic stirrer for a few minutes.

Laboratory-scale production methods have also been developed using shelled grains of broomcorn millet or millet powder, *Panicum miliaceum*, for growing *P. neoaphidis*, *Z. radicans* and *Pandora nouryi* as granular cultures (Hua & Feng, 2003, 2005; Zhou & Feng, 2009, 2010). Studies were extended to report good sporulation of *P. neoaphidis* grown on millet granules after storage at 5°C for 200 days (Feng & Hua, 2005). Mycelium of *P. nouryi* entrapped with millet powder and alginate still sporulated after storage at 6°C for 120 days (Zhou & Feng, 2010).

### *2. Conidia*

Entomophthoralean conidia are produced in the aerial environment. Methods for obtaining conidia from cadavers have been described earlier (Section 4A). With cultures on solid media, spontaneous conidiation is observed at the surface for some species, e.g., many species from the genera *Conidiobolus* and *Zoophthora*, although for other species, there are practically no spontaneously formed conidia in cultures, e.g., *Conidiobolus osmodes* or *E. culicis*. Conidia can also be produced spontaneously from liquid cultures grown in flasks on a shaker. Under these conditions, conidiophores appear on mycelial aggregates, either floating on the medium surface or stuck to the inner walls of the flasks.

The best way to obtain conidia from *in vitro*-grown hyphal bodies is as follows:

1. Detach the culture from the medium (solid medium) or filter the culture.
2. Place the mycelium, or the masses of hyphal bodies, on very moist filter paper (or directly on a new solid medium) or attach the filter paper bearing the fungus to the inside of a Petri dish lid.
3. Invert the lid above the base.

For most species, a few hours delay will occur before conidial discharge starts. A good trick is to carry out these preparations in the late afternoon and hold the plates overnight at 12–14°C. This chilling generally results in an adequate cover of conidiophores on the fungal mass by the following morning. However, timing can be species specific as *E. muscae* and *E. schizophorae* have taken up to 2 weeks to initiate conidial production under these conditions (Eilenberg *et al.*, 1990). Once conidiation begins, conidia are produced over several days, so patience is required. Laboratory-scale production in liquid medium has proven successful for numerous species, including *N. parvispora* (Grundschober *et al.*, 1998, 2001) and *E. grylli* (Sanchez Peña, 2005).

As a special consideration, for some fungal species, such as some members of the genus *Zoophthora* (B. Papierok, unpublished observation), a completely saturated atmosphere can be unfavorable to conidiophore formation and subsequent conidial discharge. In these cases, Petri dishes with mycelium should not be tightly closed.

### 3. Resting spores

Some entomophthoralean species are known to produce resting spores *in vitro*. For those species, production can occur in solid media, especially media including egg yolk. However, as with hyphal bodies, it is most practical to obtain this developmental stage in liquid culture (see Section 5A2). Methods for laboratory-scale production of resting spores of *Conidiobolus* spp. and *E. maimaiga* have been published (Perry & Latgé, 1980; Latgé & Sanglier, 1985; Kogan & Hajek, 2000).

Induction of resting spore production occurs when carbon or nitrogen sources are lacking and the resulting numbers of resting spores produced are in direct proportion to the quantity of nutrients. This is also affected by the type of nitrogen source; yeast extract is optimal although results can vary between different batches of this product. In a simple basic liquid medium containing only 40 g/l dextrose and 10 g/l yeast extract, dextrose can be replaced by a vegetable oil, e.g., sunflower oil (30 ml/l). In order to maximize yields, respective amounts of carbon and nitrogen sources should most likely be slightly adapted to species and even to fungal strains.

To remove variability due to using yeast extract, attempts have been made to develop chemically defined media. A simple medium containing dextrose, L-arginine, L-leucine, glycine and mineral salts allowed growth and sporulation of *Conidiobolus thromboides* (Perry & Latgé, 1980). However, higher sporulation rates were consistently obtained on a medium containing yeast extract (Perry & Latgé, 1980). According to Latgé & Sanglier (1985), the optimal defined medium for sporulation of C. *obscurus* contains dextrose, 11 amino acids, four vitamins (thiamin, biotin, folic acid and pantothenic acid), and four salts (phosphates, magnesium and zinc sulphates, manganese chloride).

Basically, culture conditions for production of resting spores are similar to those for production of hyphal bodies. However, formation of resting spores can be more sensitive to physical conditions during growth. A suitable combination of factors includes an initial pH of 6.5, a temperature of 20°C, and complete darkness (Latgé *et al.*, 1978; Latgé & Sanglier, 1985). Some compounds (trehalose, glycerol, some amino acids) inhibited formation of *E. maimaiga* resting spores *in vitro* (Kogan & Hajek, 2000). Time for complete resting spore formation can take 7–21 days for *E. maimaiga* and resting spore formation differed significantly by isolate.

Once completely formed, resting spores can be recovered by filtration and washed with sterile distilled water. The sporulation rate can be estimated by determining the number of spores per milliliter with a hemocytometer. Use of the hemocytometer is described in Chapter VII, although we suggest counting four to five 1-mm$^2$ squares [i.e. there are nine 1-mm$^2$ areas on each side (or chamber) of a standard hemocytometer] in four to five hemocytometer chambers for increased accuracy (A. E. Hajek, unpublished data).

## C. Large-scale production

For large-scale production of Entomophthorales (mycelium, hyphal bodies or resting spores), the fungus has been produced in industrial-type liquid media in fermenters, or in larger carboys. Using fermenters involves successive inoculations of increasingly larger fermentation chambers. Production in fermenters was first achieved for resting spores of C. *thromboides* and C. *obscurus* (Gröner, 1975; Soper *et al.*, 1975; Latgé *et al.,* 1977; Latgé & Perry, 1980). Nolan (1990, 1993) successfully developed a methodology for mass production of hyphal bodies of *E. aulicae* in liquid media. *Pandora nouryi* (Zhou & Feng, 2010), *Batkoa* spp., *Furia* spp., *N. floridana* and *Z. radicans* (Leite *et al.*, 2005) have also been successfully mass produced.

*1. Media*

Suitable industrial-type media for mass production could contain the following ingredients: for *C. thromboides* resting spores: 80 g/l dextrose (industrial grade product), and 20 g/l soybean flour (and/or cottonseed flour) (industrial grade product); for *C. obscurus* resting spores: 60 ml/l unrefined corn oil, and 40 ml/l corn steep liquid; for *P. neoaphidis* hyphal bodies: 60 g/l dextrose, and 20 g/l yeast extract; for *Z. radicans* hyphal bodies: 40 ml/l corn syrup, 10 g/l yeast extract, and 10 g/l peptone (pH 6.8) (Latgé *et al.*, 1977; Papierok, 2007; B. Papierok, unpublished data). For *E. aulicae,* medium for mass production of hyphal bodies includes a highly modified Grace's medium containing 13 amino acids supplemented with 0.8% tryptic soy broth and 0.4% calcium caseinate (Nolan, 1993). These media should be supplemented with up to 1% of an antifoam agent.

*2. Culture conditions*

In general, the fungus is first grown in flasks on a shaker, following the procedure described in Section 5A2. Flasks should contain at most 30% (v/v) of medium. After 24–96 h of growth, according to species and flask size, the mycelia can be used for inoculating a 2-l fermenter. The volume of inoculum should amount to 8–10% of the volume of the medium. After a second 24–96 h, the mycelia thus produced will serve as inoculum for a larger fermenter (i.e. 6, 10 or up to 25 l).

Physical conditions for fermentation of three entomophthoralean species studied by Leite *et al.*, (2005) include complete darkness, 20–23°C, agitation of 200–700 r.p.m., pH of 6.2–6.5, and an aeration of 1 (or 0.5) volume of air per volume of medium per minute (v/v/m). Alternatively, Z. *radicans* grown in 25-l carboys is agitated by constantly bubbling air through it. Such conditions allow a maximum yield of *P. neoaphidis* hyphal bodies or Z. *radicans* mycelium within 3 days. Production of mature resting spores of *C. obscurus* required 6–8 days (Latgé, 1980) and, depending on the medium, production of hyphal bodies of *N. floridana* can require more time (Leite *et al.*, 2005).

*3. Fermentation, harvesting and storing*

Fermentation of Entomophthorales has been successful in containers of up to 25 l and results with larger containers (e.g., 100 l and more) have mostly been unsuccessful due to contamination. Contamination is a major concern due to the richness of the nutrient media but also the length of time necessary for production. In order to circumvent this problem for resting spore production, one could apply the process developed for laboratory-scale production by Latgé & Perry (1980). These authors took into consideration two facts: (i) resting spore formation is induced within 4 days; and (ii) after this time, the pre-resting spores (prespores) need no further nutrients. Accordingly, fermentation was stopped after 4 days to gather the prespores, wash them with water, formulate in humid clay, and then maintain them at 20°C. Under these conditions, prespores matured within 4–5 additional days and contamination of culture media was prevented.

For *Z. radicans,* after growth the fungus is dried using a procedure named the 'marcescence process' (Wraight *et al.*, 2003). First, the mycelium is collected on cheesecloth and the liquid is removed under vacuum. Next, the mycelium is washed several times with deionized water and mats are placed on wire racks to dry and incubate for 2 h. Then, mycelial mats are sprayed with sterile 10% maltose until saturated, and dried for 4–5 h. Mats are incubated at 4°C overnight then dried until 'crunchy' under continuous air flow (e.g., in a fume hood). Mats are then ground into a powder and flash frozen by immersion of containers holding mycelial powder into liquid nitrogen for storage (McCabe & Soper, 1982; Wraight *et al.*, 2003). Unfortunately, standard methods used for milling and freezing *Z. radicans* and *P. neoaphidis* produced in this way destroyed the cultures (Li *et al.*, 1993) so these methods require further investigation. In a project working toward mass-production, *P. neoaphidis* mycelium was grown in liquid culture and then mixed with a calcium alginate matrix (Shah *et al.*, 2000). For formulated or unformulated mycelium, survival was improved when *P. neoaphidis* was dried slowly and stored at 10°C, rather than at higher temperatures.

## 6. *IN VIVO* CULTURE AND PRODUCTION

Culture and production of Entomophthorales using living hosts can be desirable for several reasons. First, it can be useful in cases where isolation and growth *in vitro* proved unsuccessful (some species of Entomophthorales have never been cultured *in vitro*). Second, when maintaining the fungus *in vitro* the risk of changes in the fungus toward attenuation is reduced when cultures are broken with passages in hosts. Third, culture using hosts could be useful as a means for storing entomophthoralean material. Numerous authors have kept small-scale entomophthoralean cultures using hosts in order to conduct laboratory infection experiments. Initiation and maintenance of such cultures need detailed knowledge about hosts and pathogens, e.g., the time from infection to sporulation, the duration of sporulation and sporulation intensity. It is generally difficult to mass produce the fungi in such conditions, so this method has principally been limited to smaller scale production for laboratory use.

### A. *In vivo* culture

For culture using hosts, these steps should generally be followed:

1. Use the smallest containers or cups possible for showering conidia, since this ensures a high density of inoculum close to the insects becoming infected.
2. The culture is usually initiated using infected insects that recently died, which serve as the source of inoculum to infect healthy hosts.
3. Humid conditions are usually necessary to promote conidial discharge but humidity that is too high for too long can be detrimental to the healthy insects being showered, so care must be taken.
4. Conidial discharge must be monitored (see Section 7) and showered insects must be removed after the correct concentration of conidia has been achieved and placed into new containers or cups.
5. Showered arthropods are often maintained at higher humidities in the new containers for up to 24 h (depending on host insect and fungus) after showering to make sure that conidia on the surfaces of potential hosts successfully infect. After this time, insects that were showered are reared under normal rearing conditions.
6. If primary conidia have been projected onto the walls or bottom of inoculation cups, additional insects can be placed in these cups and these will eventually become infected by secondary conidia. This is especially useful if few of the initially showered insects became infected.

### B. *In vivo* production

Large-scale production using a colony of hosts in the laboratory has been developed for *E. muscae* infecting *Musca domestica* (Mullens, 1986) and this method proved sufficient for biological control experiments (Six & Mullens, 1996).

As an alternative, harvesting cadavers produced during natural or induced epizootics in the field can be another source of *in vivo* inoculum. Steinkraus & Boys (2005) successfully harvested more than 30,000 cotton aphids (*Aphis gossypii*) infected with *N. fresenii* and then dried and froze the cadavers. The sporulation success of the stored cadavers was 70.4%.

### C. Storage within infected hosts

Special consideration should be taken if fungal cells produced in hosts must be stored. By incubating at lower temperatures, the development of the fungus within cadavers can be slowed down for some days, while longer storage needs drying and/or freezing.

To investigate storage of hyphal stages, mummified aphids killed by *P. neoaphidis* have been successfully stored for 8 months at 0°C at humidities of 20 or 50% (Wilding, 1973), and mummified cassava green mites killed by *N. tanajoae* have been successfully stored at 4°C for 2 years, with no differences in subsequent sporulation at humidities during storage between 5 and 37% RH (Wekesa & Delalibera, 2008). For somewhat larger hosts, *E. aulicae* or *Z. radicans* remained viable for at least 1 year within cadavers of spruce budworm larvae maintained at 0% RH and 4 or 20°C (Tyrrell, 1988) and for at least 8 months within cadavers of diamondback moth larvae stored at 20% RH and 4°C (Pell & Wilding, 1992).

Entomophthoraleans within small hosts (i.e. aphids and mites) have been successfully stored for much longer at freezing temperatures. Cadavers of cotton aphids were kept for up to 68 months at −14°C, although virulence declined over that time (Vingaard *et al.*, 2003). Mummies of cassava green mites, placed with silica gel before storage, have been successfully stored at −10°C for 10 years (Wekesa & Delalibera, 2008).

Few studies of resting spore storage in hosts have been conducted, perhaps due to the difficulties of determining whether resting spores are alive but dormant or dead; dormancy of resting spores remains a poorly understood phenomenon so methods for release from dormancy are seldom known. Mature resting spores that are alive but dormant will not become stained (see Section 7C2) and will not germinate. Some *E. maimaiga* resting spores stored at 4°C and 0–76% RH for 9.5 months successfully germinated (Hajek *et al.*, 2001). One study with *E. maimaiga* demonstrated that resting spores from cadavers maintained at 15 and 20°C on 1.5% water agar for 7 weeks after host death did not enter dormancy (as they normally would) and the resting spores subsequently continued germinating although stored at 4°C for up to 8 months (Hajek *et al.*, 2008).

## 7. BIOASSAYS

Bioassays are defined as a measure of the response to various stimuli that is produced in living organisms (Onstad *et al.*, 2006). As such, we will discuss separately bioassays that evaluate interactions between pathogens and hosts and then bioassays that evaluate various aspects regarding entomophthoralean spores.

### A. Studying interactions between Entomophthorales and hosts

When studying any pathogen, it is essential to experimentally infect healthy insects. This represents one of Koch's postulates (see Chapter I). Furthermore, such a procedure is the basis of possible culture of the pathogen using hosts (see Section 6). Experimental infections allow study of the mode of action of the pathogen, adding to our fundamental knowledge of pathogenesis and virulence. From the applied point of view, bioassays with entomopathogenic fungi help to determine their potential in biological control of pests.

Given the normal route of arthropod infection by fungi, conidia must come into contact with the host cuticle. It is using this basic information that the first attempts to infect insects with Entomophthorales were successfully undertaken. In fact, most infection experiments are still carried out in this way. As an alternative, for some experiments fungal material is injected into the insect hemocoel.

Bioassays with an entomophthoralean species usually begin with tests to evaluate pathogenicity (i.e. the potential ability of a pathogen to produce disease; Onstad *et al.*, 2006). Examples of pathogenicity studies would be bioassays used to test the host range of a pathogen (Papierok & Wilding, 1981; Hajek *et al.*, 1995; Delalibera & Hajek, 2004; Jensen *et al.*, 2006; Ribeiro *et al.*, 2009) or the susceptibility of hosts of different life stages or under different conditions (Dromph *et al.*, 2002; B. Papierok, unpublished data). Such bioassays are generally followed by tests of virulence (i.e. the quantifiable disease-producing power of a microorganism; Onstad *et al.*, 2006), which can include host response to pathogen dose or concentration or which compare different isolates. Some examples are studies comparing time to host death and sporulation according to fungal isolates, as in the case of *C. obscurus* and different aphid species (Papierok & Wilding, 1981) and *E. maimaiga* and hosts from different locations (Nielsen *et al.*, 2005). Bioassays can also include tests of the activity of different stages of pathogens, especially in response to a diversity of conditions, or the effects of fungal infection on factors such as host behavior (Watson *et al.*, 1993) or reproduction (e.g., Xu & Feng, 2002).

#### *1. Infection methodology*

*a. Conidial inoculation*

There are three ways of exposing insects to conidia discharged either from cadavers or cultures:

- place test insects in a shower of conidia; or
- bring test insects and a surface covered with conidia into contact; or
- expose test insects to suspensions of known concentrations of conidia.

Once conidia are on the insect cuticle, test insects must be maintained for a limited time in a saturated atmosphere (100% RH) to ensure germination of conidia and subsequent penetration by the fungus.

*Conidial showers.* Conidia can be obtained from cadavers filled with hyphal bodies or from mycelia produced in either solid or liquid media. These sources of inoculum should be initially prepared as specified for cadavers (see Sections 2C or 4D) or from cultures (see Section 5A). These sources of conidia should be inverted over hosts, so that only primary conidia will be showered. Coverslips or slides placed beneath showers allow quantification as conidia/mm$^2$. Germinating resting spores can also be used as sources of conidia, as first explained by Valovage *et al.* (1984), although, at least for *E. maimaiga,* infections initiated by germ conidia produced by resting spores are fundamentally different from infections initiated by conidia (i.e. infections initiated by germ conidia only yield conidia while infections initiated by conidia from cadavers produce either conidia or resting spores; Hajek, 1997).

Groups of insects can be exposed to conidial inoculum in tubes, with the size of the tubes depending on the size of insect. For instance, 55 mm high × 25 mm diameter tubes are suitable for adult stages of aphids (Papierok & Wilding, 1979). Moist substrates bearing the sources of inoculum can be cut to the diameter of the tube and inverted across the opening. A fine gauze or grid can be placed between the 'discharging' source of the inoculum and the insects being showered to prevent the insects from making direct contact with the culture but allow the conidia to pass through. Groups of insects can also be exposed in Petri dishes with cadavers placed above or mycelial mats inverted over the hosts under moist conditions.

Mycelial mats rarely produce conidia synchronously over the entire mat. Therefore, depending on the size of target arthropods and Petri dishes, the mats should be rotated regularly or the dish containing the arthropods should be placed on a revolving turntable so that all insects are showered with equivalent numbers of conidia. As one example, individual 29-ml cups containing forest tent caterpillars (*Malacosoma disstria*) were placed on a 150-mm-diameter platform rotating at 0.5 r.p.m., 10 cm below sporulating mycelium within a humid chamber (Filotas *et al.*, 2003). All cups were lined with Sigmacote (Sigma) so that the larvae could not escape and one cup contained only water agar so that the conidial density could be quantified.

*Surface contamination.* Insects can be infected with Entomophthorales by being brought into contact with a surface covered with conidia. Such a procedure is especially suitable for larvae and apterous adult forms, which would move on such a surface.

The main approach is to favor contact of target insects with a surface covered with showered conidia. As early as 1871, Brefeld infected caterpillars of *Pieris brassicae* with *Z. radicans* by exposing test insects to cabbage leaves dusted with conidia of the fungus. However, substrates besides leaves can be used for exposures, e.g., moist filter paper, plaster of Paris, plastic or glass Petri dishes. In practice, surfaces are covered with conidia by placement above or below sporulating cadavers or cultures (as described earlier in this chapter). This latter procedure is the only method suitable for testing passively detached secondary conidia, such as the capilliconidia of *Zoophthora phalloides* (Glare *et al.,* 1985) and *Neozygites* spp. (Delalibera & Hajek, 2004).

An alternative approach is to ensure direct contact of test organisms with sporulating cultures. This can be used for cultures that spontaneously produce numerous conidia across the culture surface, e.g., C. *coronatus.*

*Conidial suspensions.* Conidia, ejected from cadavers or mycelium cultured on solid or in liquid media, are collected in sterile distilled water supplemented with a few drops of wetting agent (e.g., Atmos 300, Tween 40 or 80). Suspensions are brought into contact with test insects by means of: (1) direct spraying (a hand sprayer or a glass chromatography sprayer can be convenient); (2) topically applying a microvolume of a highly concentrated suspension; or (3) dipping insects into the suspension.

As an example of the steps involved in dipping insects into a conidial suspension, *E. maimaiga* conidia were showered into 0.25% Atmos 300 and 0.10% Tween 80 (Shimazu & Soper, 1986). Conidia were then collected by centrifugation at approximately 448 g every hour and stored at 4°C so that none germinated before bioassays were begun; the conidial suspension could not be kept overnight before use in bioassays as conidia do not survive well in the detergent solution. Once an adequate number of conidia have been collected, the suspension is centrifuged and the supernatant is replaced with 0.025% Atmos 300 and 0.01 % Tween 80 to minimize effects of the detergent on test insects. The conidial concentration of the suspension is then estimated using a hemocytometer. Individual insects are sequentially briefly dipped into the conidial suspension but, as conidia settle with time, the suspension should regularly be agitated.

*Post-inoculation incubation.* Immediately after being inoculated with conidia, insects should be kept in a saturated atmosphere and possibly allowed to feed before being replaced in normal rearing conditions. The length of the period of very high humidity especially depends on each entomophthoralean species/test insect system and the temperature. At an average temperature of 18–20°C for instance, the humid period needed can range from 12 to 24 h.

A saturated atmosphere can be produced by keeping containers with test insects in a humid chamber. With winged insects, e.g. aphid alates or mosquitoes, it is convenient to simply apply small pieces of moistened filter paper to the interior of containers (Dumas & Papierok, 1989). Alternatively, the presence of water-soaked paper towels, cotton balls or free water in containers can be used to increase humidity and thus promote infection of non-winged insects, such as caterpillars (Krejzová, 1971). For insects that can be cannibalistic, individuals are placed separately in plastic Petri dishes containing slightly moistened filter paper. Petri dishes are then sealed for the required amount of time.

*Special considerations.* Throughout all types of bioassays, conidial age and state are critically important to bioassay success. In a humid environment, entomophthoralean conidia will begin to germinate in several hours. These conidia frequently die when dried for very long. Therefore, conidial showering or collection must be conducted rapidly. For conidial collection, if showering is slow, conidia can be collected on an hourly basis and refrigerated for at least several hours (but with reduced survival if kept overnight). Cold temperatures arrest development so that conidia used for bioassays are at approximately the same stage.

Directly exposing insects to conidial showers is the best method for small (winged or apterous) insects like aphids or mosquitoes. Such insects can be confined in small containers for several hours (up to 24 h) without damage. This method is also the most similar to natural conditions. However, it is very difficult in practice to expose insects to the same dosage during replicated experiments.

In the case of larger insects, especially caterpillars, infection using surfaces covered with conidia can be appropriate. However, it is important to try to expose insects to conidia of the same range of ages (and not yet germinating) in replicated experiments. Mobile insects can be maintained in showers of conidia by immobilizing them using thread and modeling clay (Gabriel, 1968).

Although convenient, the method involving application of an aqueous suspension of conidia can give inconsistent results. Care must be taken to maintain accurate homogeneous suspensions of conidia. Moreover, contact with free water may alter the adherent coat of the conidia, making them more sensitive to external conditions.

Length of exposure of test insects to conidial showers or surfaces covered with conidia depends on the intensity of showering or on the number of spores on the surface, respectively. In these cases, the use of cover slips or slides for quantifying conidial density over time or at least the use of a dissecting microscope to look for presence of conidia directly on hosts is highly recommended. Using any of the conidial inoculation methods, exposure of insects to excessive doses can result in death due to bacteremia or septicemia instead of mycosis; this can be a common problem when initiating studies, exposing insects to showers of conidia for too long with subsequent quick death due to septicemia. We caution that exposure to quantified, lower densities of conidia is a better way to start studies.

*b. Injection of fungal material*

Experimental infection can also be achieved by injection of hyphal bodies or protoplasts into the insect hemocoel. Because of the size of these fungal cells, this procedure is most frequently used only for larger insects that can be injected using a larger gauge needle. Microinjection has been regularly used with *L. dispar* larvae, as small as second instars (A. E. Hajek, unpublished data) and for grasshoppers (Ramoska *et al.*, 1988).

Hyphal bodies and protoplasts should be cultivated in liquid media (see Section 5A2). The fungal material to be injected should come from 1- or 2-day-old cultures in the log phase of growth. The culture can be injected without special preparation (e.g., as for protoplasts of *E. maimaiga* in Grace's medium; Nielsen *et al.*, 2005) or after being gently centrifuge-washed (e.g., for *E. aulicae* protoplasts, 4000 × *g*, 4°C, 5 min; cf. Dunphy & Chadwick, 1985). With hyphal bodies, cultures can be filtered and the mycelium washed with sterile distilled water before being placed into a final volume of sterile distilled water.

Suspensions of hyphal bodies or protoplasts should be introduced into insects by intra-hemocoelic injection (in 10–50-μl amounts, according to the size of the insects) via the posterior end or, for caterpillars, the base of a proleg. It is most important to insert cells into the hemocoel, with special care not to puncture the gut during this process. Microsyringes can be fitted with a drawn-glass needle or with a standard metal needle, e.g., 23 gauge for *E. maimaiga* protoplasts. The needle size used must be large enough so that it does not clog and protoplasts must be agitated regularly within the syringe (e.g., by shaking the syringe) because they settle. After injection, insects can be maintained under normal growth conditions; when injection is used, there is of course no need for maintenance in a saturated atmosphere after infection.

## B. Designing assays and analyzing data

### *1. Principles of bioassays*

- Bioassays should be reproducible, and in practice should be reproduced several times (i.e. with different batches of inoculum, beginning on different days) to document the extent to which results are consistent.
- Bioassays should be conducted using test insects that are of standardized condition (e.g., of the same instar and not infected by another pathogen, unless this is part of the study).
- Bioassays should be conducted using inoculum that is also of standard characteristics across the different repetitions of the study.
- Methods, conditions and processing used during the bioassay must be consistent for all replications.
- To document that the treatment caused resulting effects, negative control insects (not treated with the pathogen but receiving similar conditions) must be included for each replicate of the study.
- In some cases, as when testing the survival of the pathogen under different conditions, positive controls should also be included (test insects treated with a pathogen and held under conditions known to result in successful infection).

### *2. General steps for infection bioassays*

The successive steps for all host/entomophthoralean bioassays include:

1. Obtain inoculum, either *in vivo* or *in vitro* (as described earlier in this chapter).
2. Expose groups of insects to various estimated quantities of inoculum.
3. Establish the most favorable conditions for the infection and disease incubation process.
4. Regularly monitor infected specimens to record time of death and subsequent sporulation.
5. Analyze data using appropriate statistical tests, based on objectives of the research.

### *3. Infection bioassays using conidia*

Because of the trans-cuticular route of fungal infection in insects, bioassays with fungi can be based on spraying test insects with suspensions of infective spores at given concentrations. However, such a procedure does not always work well with entomophthoralean conidia, due to the relatively brief conidial survival time and occasional alteration of conidia when suspended in water. Consequently, many classical bioassays using conidia involve the exposure of insects to showers of conidia. Therefore, the actual numbers of conidia landing on individual insects cannot be exactly standardized before application. However, the densities of conidia can be measured during conidial showers, often as conidia/mm$^2$.

The procedures for estimating the infectivity of entomophthoralean conidia are based on the general methodology for infecting insects with these fungi. For conidial showers, the number of conidia landing on an insect is regulated by the amount of time the insect is exposed to a conidial shower. Groups of test insects can be immobilized under conidial showers, if necessary, using a dedicated device, $CO_2$, or by cooling. Immobilization allows more accurate determination of the concentrations received by the insects. If it is difficult to immobilize the test insects, at least they can be prevented from climbing out of containers by applying a film of silicone (e.g., Sigmacote) or Vaseline to vertical surfaces. Conidial concentrations are estimated by counting the conidia (number/mm$^2$) in sample areas (e.g., on glass coverslips) on the same level as insects under the conidial shower. This procedure was first used by Wilding (1976), who immobilized apterous adults of the pea aphid, *Acyrthosiphon pisum*, during conidial showers.

In most cases, however, it is definitely more practical not to immobilize insects under conidial showers (Figure 9.4). Test insects are submitted to conidial showers either individually or in groups. Under these conditions, the actual conidial concentrations received by insects cannot be accurately estimated, but can be estimated for a fixed time period by counting the number of conidia landing on the

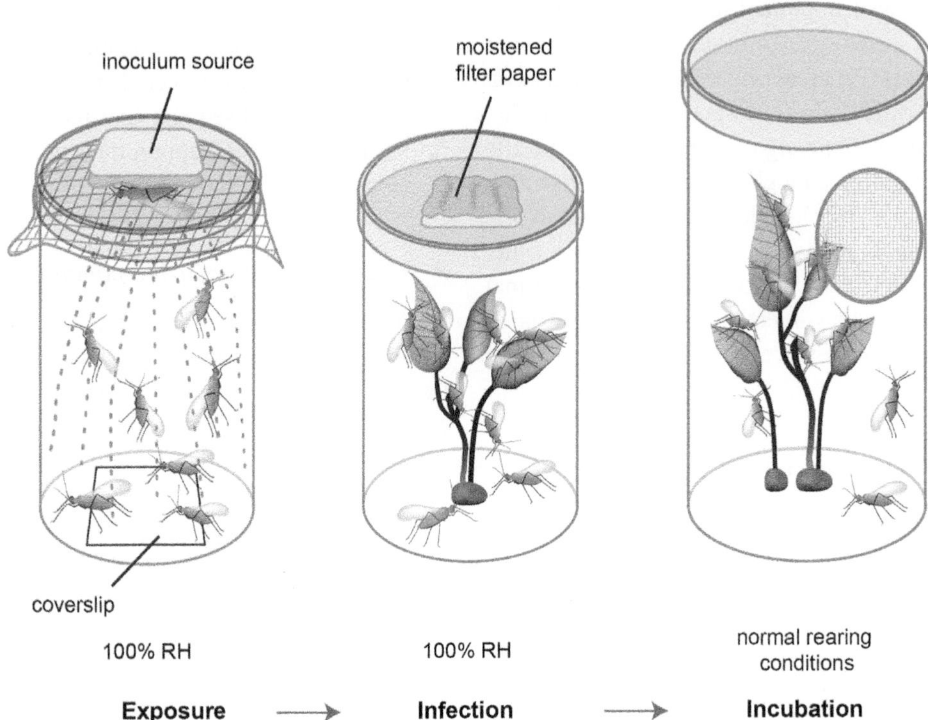


**Figure 9.4** Successive steps for assaying virulence of Entomophthorales against aphids (after Papierok & Wilding, 1979).

bottom of the container during the period of exposure. Therefore, concentration is estimated by the number of conidia/mm$^2$ on the sample areas. This method was shown to be suitable for insects of small size, such as aphids exposed to various entomophthoralean species (Papierok & Wilding, 1981; Papierok *et al.*, 1984; Shah *et al.*, 2003) and for mosquitoes (Dumas & Papierok, 1989).

Alternatively, the production of conidia can be considered uniform during the experiment, which is acceptable for short exposures (up to 60 min). The quantity of inoculum is then estimated by averaging the numbers of conidia projected in equal amounts of time before and after the insect exposure. This method is suitable for any insects, but especially for larger ones, e.g., caterpillars, which can prevent accurate collection of conidia on the bottom of containers because the larvae walk on top of the showered conidia (Vandenberg & Soper, 1979).

In most of these studies test insects have been exposed to a quantity of inoculum in groups of five to 10 individuals per container. For example, a bioassay designed to accurately estimate fungal infectivity could involve four exposure periods with five groups of insects exposed per period, which would test up to 20 different concentration levels.

Another approach was used when concentration/response bioassays were conducted with *F. gastropachae* (Filotas *et al.*, 2006). Conidial concentrations were calculated by averaging conidial densities in two individual control cups for each exposure group of lepidopteran larvae within separate cups on a turntable, for a total of 20 different concentrations tested during this study.

### *4. Infection bioassays using vegetative stages*

As an alternative, for host/pathogen systems where this is possible, bioassays using injection of protoplasts and/or hyphal bodies do not face the problems of conidial showers, as they can be conducted using a range of exact doses of cells for injection. Bioassays with vegetative stages growing in cell culture medium, protoplasts and/or hyphal bodies, basically involve injection of suspensions of fungal cells of specific concentrations into test insects.

### *5. Types of infection bioassays and data analysis*

The general goal of bioassays with Entomophthorales is to characterize or quantify the responses of arthropods exposed to inoculum. Bioassay data that can be recorded include such factors as mortality (because entomophthoralean species cause acute disease), time to death, type and extent of sporulation, presence of spores on surfaces of insects, and presence of fungal cells within hosts (e.g., Ribeiro *et al.*, 2009).

As a special type of bioassay evaluating mortality, researchers are interested in testing whether a pathogen can infect a host and subsequently in determining the virulence of the pathogen (i.e. the relative dose or concentration required to kill and how quickly the host is killed). To quantify these relationships, as with other pathogens, test insects are exposed to various doses or concentrations of infective units. If there is any mortality in the controls, treatment data should be altered using Abbott's correction (see Chapter VII). For studies where $LD_{50}$ (dose killing 50% of the test insects) or $LC_{50}$ (pathogen concentration killing 50% of the test insects) will be determined and probit analysis will be used, it is important to use at least three concentrations that yield mortality between 25 and 75%. Then, the inoculum concentrations applied in subsequent bioassays should be kept within the range of the corresponding values. Results are conventionally expressed in terms of regression of probit-transformed proportion mortality on log dose and the $LD_{50}$ or $LC_{50}$ is estimated (Finney, 1962, 1971; e.g., Shah *et al.*, 2004). Computer programs for probit analysis test for significance of the regression and calculate the slope of the regression line, its standard error, the heterogeneity (chi-squared), the $LC_{50}$, and 95% fiducial limits.

The time from infection to death, which contributes to the comparison of the virulence, can be calculated as mean values for time to death, or as $LT_{50}$ (time from pathogen exposure until 50% of inoculated insects die). One must not use probit analysis to calculate $LT_{50}$ values when there is lack of independence between data points; thus, you can not use probit for analysis when monitoring the same group of insects daily, although probit analysis could be used if a different set of hosts is monitored each day. A most commonly used method for calculation of $LT_{50}$ at present is by statistical programs conducting survival analysis (e.g., SAS). Survival analysis is also useful for analysis of data from bioassays where monitoring hosts ended before all insects died so that the data are censored.

### *6. Tricks of the trade*

As with any bioassay, test insects should be as homogeneous as possible (see Section 7A1). The developmental stage, the age, and the physiological state of hosts should be standardized. Moreover, the laboratory colony should be well established and healthy. In order to reduce possible heterogeneity in results when the purpose of the experiment is to compare fungal lines, it is advisable to use only the offspring from two insect parents, when possible.

The goal for many bioassays with Entomophthorales has been to evaluate pathogenicity or to test host/pathogen interactions under specific conditions. The fact that there are relatively few papers presenting results from bioassays of Entomophthorales where $LD_{50}$ or $LC_{50}$ were calculated emphasizes the difficulties in applying this approach, typical of toxicological studies, to studies of entomophthoralean fungi. However, the dose/response bioassays yielding $LD_{50}$ or $LC_{50}$ values have proven accurate for several fungal species (*Conidiobolus* spp., *P. neoaphidis, Z. radicans, F. gastropachae*) and different insect hosts (aphids, caterpillars, mosquitoes) (Wilding, 1976; Papierok *et al.*, 1984; Dumas & Papierok, 1989; Filotas *et al.*, 2006). Practical details (e.g., sizes of assay containers, times of exposure, etc.) have to be adapted to every entomophthoralean species/host association.

The most frequently published values for the slopes of regression lines for dose response of entomophthoralean pathogens versus arthropod hosts range from 1 to 2. These values appear relatively low compared with many chemical insecticides. Such values could result from rather high variability among the groups of arthropods submitted to given amounts of inoculum. As mentioned above, the requirement for very homogeneous test insects is therefore essential. On the other hand, variation in results can be observed between various values of $LD_{50}$ or $LC_{50}$, estimated using one strain with one test insect species. A mean log $LD_{50}$ or $LC_{50}$ and a mean slope can be calculated when slope estimates are not significantly different from each other. These problems have been discussed by Vandenberg & Soper (1979) and Oger & Latteur (1985). The latter authors demonstrated that the following experimental conditions [five doses, 10 *A. pisum* apterous adults per trial, and eight repetitions (i.e. 400 aphids total)] provided the most efficient method for statistically distinguishing $LC_{50}$ values of two different fungal strains.

## C. Laboratory studies of conidia and resting spores

Conidia are the infective spores of Entomophthorales. Resting spores are not directly infective but they germinate to produce and actively discharge germ conidia which are able to penetrate the host cuticle. When using bioassays to explore virulence of fungal strains, the survival, activity and number of infective spores are important to quantify. Furthermore, beside the virulence expressed in terms of $LD_{50}$ or $LC_{50}$, the numbers of spores produced from cadavers is a parameter of interest when considering the virulence of a strain. We present below some of the most common experiments conducted when studying entomophthoralean conidia and resting spores.

### *1. Quantifying conidial discharge*

As regards the number of conidia discharged from cadavers of insects killed by Entomophthorales, this parameter is most readily estimated by collection of conidia. Conidia per cadaver can be directly counted or quantified using a spectrophotometric method. Estimation by direct counting can be accomplished according to methods developed for species infecting aphids, flies and caterpillars. Living, infected specimens or cadavers should be prepared as specified in Sections 2A, 2B. Shimazu & Soper (1986) suspended cadavers of caterpillars over beakers containing a known volume of 1% Triton-X plus 0.2% maleic acid. The conidia that were collected were counted to estimate total conidial discharge per cadaver; this procedure made sure that discharged conidia were killed because if they had germinated, it would have become very difficult to count individual conidia. Milner (1981) and Papierok & Wilding (1981) allowed dead, infected insects to discharge conidia onto microscope slides, which were replaced at regular intervals. Spores were counted within sample areas on each slide and the total numbers were then estimated (e.g., Li *et al.*, 2006).

Wilding (1971) and Mullens & Rodriguez (1985) have used 'spore trains', that move a spore-collecting surface past a sporulating cadaver, and thus allow calculation of the approximate number of conidia produced per hour, and consequently, the total number of conidia discharged per cadaver. A somewhat similar method to evaluate the number of spores discharged over time involved use of a 'sporometer' (Hajek & Soper, 1992). This instrument consisted of a cylinder (originally from a hygrothermograph) that rotated once every 24 h around which a strip of Melinex tape (high-clarity polyester film) was attached. A cadaver just beginning to sporulate was attached over a specific location on the tape and the tape was replaced 24 h later. The numbers of spores on the tape were associated with temperature and relative humidity conditions recorded adjacent to the sporometer. Using a similar concept, a sporulation monitor was developed to collect conidia ejected from cadavers or *in vitro*-produced cultures over 24- or 100-h periods (Pell *et al.*, 1998). This monitor was later enhanced to be semi-automated with image analysis so that the sizes and shapes of conidia could be compared as well as the rate of conidial production (Bonner *et al.*, 2003).

An original photometric technique for estimating conidial production from infected aphids was developed by Courtois *et al.* (1983). The principle of this technique, utilizing a precision luxmeter and a dark room, is based on the partial absorption by conidia of a luminous flux shining through. This method involves the preliminary calculation of the linear relation between observed photometric values and given numbers of conidia (estimated by direct counts). A similar technique, but using a spectrophotometer, was developed by Dumas (1986) for quantification of conidia produced from cultures.

Another question that has been asked about conidial production regards the distance of conidial discharge. Conidial discharge by two species of *Entomophthora* from *M. domestica* cadavers was studied by suspending a cadaver from the top of an aquarium and collecting conidia on microscope slides placed at varying distances below (Six & Mullens, 1996).

### *2. Evaluating conidial viability*

Whether conidia are alive or dead can be determined using fluorescent stains (Butt, 1997). With fluorescein diacetate and propidium iodide, conidia of *Z. radicans* and *E. maimaiga* stained green if alive and red if dead (Firstencel *et al.,* 1990). This technique cannot be used for resting spores because dormant resting spores (the normal condition for many species) autofluoresce and, in addition, live/dead stains (as well as other stains) cannot penetrate through intact double walls of dormant resting spores (A. E. Hajek, unpublished data).

### *3. Evaluating conidial germination*

Another way to study spore viability is to conduct germination tests. Because conidia are not long-lived, germination tests can be used to evaluate whether they are alive or dead or to study the impact of abiotic or biotic factors on germination. For example, *P. neoaphidis* conidia were used to evaluate the impact of soil pollution with six heavy metal ions on germination (Tkaczuk, 2005) and *P. nouryi* conidia were used to evaluate the effect of eight insecticides on germination (Li *et al.*, 2006).

Conidial germination tests frequently involve showering conidia onto water agar, covering the conidia with a coverslip and incubating at an appropriate temperature. Entomophthoralean conidia do not need exogenous nutrients to germinate. Under a coverslip, entomophthoralean conidia will grow as germ tubes. A germ tube longer than the spore diameter is generally considered as an indication of germination. If a coverslip is not used, care must be taken not to count the secondary conidia produced by germinating primary conidia as ungerminated primaries. This problem can be partly avoided by inverting plates with conidia so that secondaries produced are not deposited on plates.

It is critical to determine the optimal length of time after showering at which to make germination counts. For *E. maimaiga,* 50% germination in constant dark is reached 3.4 h after showering onto 1.5% water agar at 20°C; in this system, results in constant light were dramatically different (9.0 h for 50% germination) so lighting conditions are an important variable to control (Hajek *et al.,* 1990a). Trying to make germination counts after germ tubes have grown extensively can potentially be very difficult or impossible due to interweaving of long germ tubes so care must be taken to make counts before extensive growth has occurred.

For adequate replication, studies of germination should not use the same showering culture on the same day for several replicates. Germination tests should be conducted on several different days using different source cultures of the same isolate for conidial production, and at least 100 conidia should be scored per replicate.

### 4. *Evaluating resting spore germination*

Resting spore germination tests are actually not indications of whether resting spores are alive or dead because these spores can remain dormant for several years before germinating. Therefore, germination tests demonstrate activity of resting spores during the period of investigation. Resting spore germination can be evaluated by placing resting spores on water agar or in a film of water and incubating at an appropriate temperature. As with conidia, nutrient agars are not required. Resting spores are generally obtained from cadavers and samples are frequently contaminated with bacteria and saprophytic fungi. To avoid contamination during germination trials, 30 ppm mercuric chloride was added to spores for 5 min, spores were then washed five times after which germination was tested (Perry & Fleming, 1989). Caution must be taken when using this mercuric chloride washing because it can kill resting spores if exposure is too long (A. E. Hajek, unpublished data).

Two main stages of resting spores can be distinguished in the Entomophthorales after they have developed: quiescent spores that are able to germinate immediately after being produced and dormant spores that are unable to germinate without cessation of dormancy. Germinability can be evaluated by keeping batches of resting spores in a layer of sterile water in a humid chamber at 15°C; contact with free water seems necessary for germination. Germination occurs slowly and asynchronously and can continue for a sample for a few weeks. If germination does not occur, spores should be considered dormant, and batches should then be kept for several months at 4°C and 100% RH, to satisfy dormancy requirements as demonstrated with several species, including *C. obscurus* (Latgé *et al.*, 1978; Perry & Latgé, 1982), *Z. radicans* (Perry *et al.*, 1982), and *F. gastropachae* (= *Furia crustosa*) (Perry & Fleming, 1989). However, providing these conditions to simulate winterization does not always result in germination in all species.

When germinating, the interior of resting spores becomes granular, the double wall thins to a single wall, and a relatively thick germ tube emerges (see Hajek *et al.*, 2008). Resting spore germination occurs much more slowly than conidial germination. After several months at 4°C to break dormancy, 55% of *C. obscurus* resting spores germinated at the optimum temperature of 15°C over a period of 3 weeks (Latteur *et al.*, 1982). Resting spores of *F. gastropachae* stored for 6 months at 4°C and then brought to 24°C began germinating within 3 days, with a maximum of germination of 80% by 30 days (Perry & Fleming, 1989). Observations of resting spore germination must therefore be made on an appropriate schedule.

## ACKNOWLEDGMENTS

We acknowledge the scientists who have developed the methods we report, although we have not been able to cite all of the papers where these methods have been mentioned. We are grateful to Richard Humber, Italo Delalibera Jr., Siegfried Keller, Charlotte Nielsen, and Stephen Wraight for sharing information about techniques and Annette Bruun Jensen for constructive criticism on parts of the chapter. We acknowledge the fine technical assistance of Béatrice de Cougny, from the Service Image et Reprographie of the Institut Pasteur for preparation of figures and Keith Ciccaglione for assistance with references. We would like to thank Cornell University, the Institut Pasteur and the University of Copenhagen for constant support of research on entomopathogenic fungi.

## REFERENCES

Agboton, B. V., Delalibera, I., Jr. Hanna, R., & von Tiedemann, A. (2009). Molecular detection and differentiation of Brazilian and African isolates of the entomopathogen *Neozygites tanajoae* (Entomophthorales: Neozygitaceae) with PCR using specific primers. *Biocontrol Sci. Technol., 19*, 67–79.

Agboton, B. V., Hanna, R., & von Tiedemann, A. (2011). Molecular detection of establishment and geographical distribution of Brazilian isolates of *Neozygites tanajoae*, a fungus pathogenic to cassava green mite, in Benin (West Africa). *Exp. Appl. Acarol., 53*, 235–244.

Bałazy, S. (1993). *Fungi (Mycota). vol. XXIV. Entomophthorales*. Szafera, Krakow: Polska Akademia Nauk, Instytut Botaniki im. W, pp. 356.

Beauvais, A., & Latgé, J.-P. (1988). A simple medium for growing entomophthoralean protoplasts. *J. Invertebr. Pathol., 51*, 175–178.

Ben-Ze'ev, I. (1980). *Systematics of Entomopathogenic Fungi of the* "Sphaerosperma" *Group (Zygomycetes: Entomophthoraceae) and their Prospects for use in Biological Pest Control*. Ph.D. Thesis, Hebrew University of Jerusalem, pp. 100.

Bidochka, M. J., Walsh, S. R. A., Ramos, M. E., St. Leger, R. J., Silver, J. C., & Roberts, D. W. (1996). Fate of biological control introductions: Monitoring an Australian fungal pathogen of grasshoppers in North America. *Proc. Natl. Acad. Sci., USA, 93*, 918–921.

Bonner, T. J., Pell, J. K., & Gray, S. N. (2003). A novel computerised image analysis method for the measurement of production of conidia from the aphid pathogenic fungus *Erynia neoaphidis*. *FEMS Microbiol. Lett., 220*, 75–80.

Brefeld, O. (1871). Untersuchungen über die Entwicklung der *Empusa muscae* und *Empusa radicans* und die durch sie verursachten Epidemien der Stubenfliegen und Raupen. *Abh. Naturforsch. Ges. Halle, 12*, 1–50.

Butt, T. M. (1997). Complementary techniques: Fluorescence microscopy. In L. A. Lacey (Ed.), *Manual of Techniques in Insect Pathology* (pp. 355–365). San Diego: Academic Press.

Castrillo, L. A., & Humber, R. A. (2009). Molecular methods for identification and diagnosis of fungi. In S. P. Stock, J. Vandenberg, N. Boemare, & I. Glazer (Eds.), *Insect Pathogens: Molecular Approaches and Techniques* (pp. 50–70). Wallingford, UK: CABI.

Castrillo, L. A., Thomsen, L., Juneja, P., & Hajek, A. E. (2007). Detection and quantification of *Entomophaga maimaiga* resting spores in forest soil using real-time PCR. *Mycol. Res., 111*, 324–331.

Courtois, P., Latteur, G., & Oger, R. (1983). Méthode rapide d'estimation de la quantité de conidies primaires de *Erynia neoaphidis* (Remaudière and Hennebert) (Zygomycètes, Entomophthorales) produites par des momies de *Acyrthosiphon pisum* (Harris) (Hom.: Aphididae). *Parasitica, 39*, 173–182.

Delalibera, I., Jr., & Hajek, A. E. (2004). Pathogenicity and specificity of *Neozygites tanajoae* and *Neozygites floridana* (Zygomycetes: Entomophthorales) isolates pathogenic to the cassava green mite. *Biol. Control, 30*, 608–616.

Delalibera, I., Jr., Hajek, A. E., & Humber, R. A. (2003). Use of cell culture media for cultivation of the mite pathogenic fungi *Neozygites tanajoae* and *Neozygites floridana*. *J. Invertebr. Pathol., 84*, 119–127.

Dromph, K. M., Pell, J. K., & Eilenberg, J. (2002). Influence of flight and colour morph on susceptibility of *Sitobion avenae* to infection by *Erynia neoaphidis*. *Biocontrol Sci. Technol., 12*, 753–756.

Dumas, J.-L. (1986). Utilisation de la spectrophotométrie pour l'estimation des concentrations de conidies d'Entomophthorales dans les essais d'infection d'insectes. *Parasitica, 42*, 127–136.

Dumas, J.-L., & Papierok, B. (1989). Virulence de l'Entomophthorale *Zoophthora radicans* (Zygomycètes) à l'égard des adultes de *Aedes aegypti* (Dipt.: Culicidae). *Entomophaga, 34*, 321–330.

Dunphy, G. B., & Chadwick, J. M. (1985). Strains of protoplasts of *Entomophthora egressa* in spruce budworm larvae. *J. Invertebr. Pathol., 45*, 255–259.

Dunphy, G. B., & Nolan, R. A. (1989). Development of *Entomophaga aulicae* protoplasts in synthetic eastern hemlock looper hemolymph. *Can. J. Microbiol., 35*, 304–308.

Dunphy, G. B., Nolan, R. A., & MacLeod, D. M. (1978). Comparative growth and development of two protoplast isolates of *Entomophthora egressa*. *J. Invertebr. Pathol., 31*, 267–269.

Eilenberg, J., & Philipsen, H. (1988). The occurrence of Entomophthorales on the carrot fly (*Psila rosae* F.) in the field during two successive seasons. *Entomophaga, 33*, 135–144.

Eilenberg, J., Bresciani, J., & Latgé, J.-P. (1990). Primary spore and resting spore formation *in vitro* of *Entomophthora schizophorae* and *E. muscae*, both members of the *E. muscae*-complex (Zygomycetes). *Cryptogam. Bot., 1*, 365–371.

Eilenberg, J., Wilding, N., & Bresciani, J. (1992). Isolation *in vitro* of *Strongwellsea castrans* (Fungi: Entomophthorales) a pathogen of adult cabbage root flies, *Delia radicum* (Dipt.: Anthomyiidae). *Entomophaga, 37*, 65–77.

Enkerli, J., & Widmer, F. (2010). Molecular ecology of fungal entomopathogens: molecular genetic tools and their applications in population and fate studies. *BioControl, 55*, 17–37.

Feng, M.-G., & Hua, L. (2005). Factors affecting the sporulation capacity during long-term storage of the aphid-pathogenic fungus *Pandora neoaphidis* grown on broomcorn millet. *FEMS Microbiol. Lett., 245*, 205–211.

Feng, M.-G., & Xu, Q. (2001). A simple method for routine maintenance and preservation of entomophthoraceous cultures. *J. Invertebr. Pathol., 77*, 141–143.

Filotas, M. J., Hajek, A. E., & Humber, R. A. (2003). Prevalence and biology of *Furia gastropachae* (Zygomycetes: Entomophthorales) in populations of the forest tent caterpillar (Lepidoptera: Lymantriidae). *Can. Entomol., 135*, 359–378.

Filotas, M. J., Vandenberg, J. E., & Hajek, A. E. (2006). Concentration–response and temperature-related susceptibility of the forest tent caterpillar (Lepidoptera: Lasiocampidae) to the entomopathogenic fungus *Furia gastropachae* (Zygomycetes: Entomophthorales). *Biol. Control, 39*, 218–224.

Finney, D. J. (1962). *An Introduction to Statistical Science in Agriculture* (2nd ed.). Copenhagen, Denmark: Munksgaard, pp. 216.

Finney, D. J. (1971). *Probit Analysis* (3rd ed.). Cambridge, UK: Cambridge Univ. Press, pp. 333.

Firstencel, H., Butt, T. M., & Carruthers, R. I. (1990). A fluorescence microscopy method for determining the viability of entomophthoralean fungal spores. *J. Invertebr. Pathol., 55*, 258–264.

Freimoser, F. M., Grundschober, A., Aebi, M., & Tuor, U. (2000). *In vitro* cultivation of the entomopathogenic fungus *Entomophthora thripidum*: isolation, growth requirements, and sporulation. *Mycologia, 92*, 208–215.

Freimoser, F. M., Grundschober, A., Tuor, U., & Aebi, M. (2003). Regulation of hyphal growth and sporulation of the insect pathogenic fungus *Entomophthora thripidum in vitro*. *FEMS Microbiol. Lett., 222*, 281–287.

Freimoser, F. M., Jensen, A. B., Tuor, U., Aebi, M., & Eilenberg, J. (2001). Isolation and *in vitro* cultivation of the aphid pathogenic fungus *Entomophthora planchoniana*. *Can. J. Microbiol., 47*, 1082–1087.

Gabriel, B. P. (1968). Histochemical study of the insect cuticle infected by the fungus *Entomophthora coronata*. *J. Invertebr. Pathol., 11*, 82–89.

Glare, T. R., Chilvers, G. A., & Milner, R. J. (1985). A simple method for inoculating aphids with capilliconidia. *Trans. Br. Mycol. Soc., 85*, 353–354.

Grace, T. D. C. (1962). Establishment of four strains of cells from insect tissues grown *in vitro*. *Nature (London), 195*, 788–789.

Gröner, A. (1975). Production of resting spores of *Entomophthora thaxteriana*. *J. Invertebr. Pathol., 26*, 393–394.

Grundschober, A., Tuor, U., & Aebi, M. (1998). *In vitro* cultivation and sporulation of *Neozygites parvispora* (Zygomycetes: Entomophthorales). *System. Appl. Microbiol., 21*, 461–469.

Grundschober, A., Freimoser, F. M., Tuor, U., & Aebi, M. (2001). *In vitro* spore formation and completion of the asexual life cycle of *Neozygites parvispora*, an obligate biotrophic pathogen of thrips. *Microbiol. Res., 156*, 247–257.

Gustafsson, M. (1965). On species of the genus *Entomophthora* Fres. in Sweden. II. Cultivation and physiology. *Lantbrukshögskolans Annaler, 31*, 405–457.

Gúzman-Franco, A. W., Clarke, S. J., Alderson, P. G., & Pell, J. K. (2008). Effect of temperature on the *in vitro* radial growth of *Zoophthora radicans* and *Pandora blunckii*, two co-occurring fungal pathogens of the diamondback moth *Plutella xylostella*. *BioControl, 53*, 501–516.

Hajek, A. E. (1997). *Entomophaga maimaiga* reproductive output is determined by spore type initiating an infection. *Mycol. Res., 101*, 971–974.

Hajek, A. E., & Shimazu, M. (1996). Types of spores produced by *Entomophaga maimaiga* infecting the gypsy moth *Lymantria dispar*. *Can. J. Bot., 74*, 708–715.

Hajek, A. E., & Soper, R. S. (1992). Temporal dynamics of *Entomophaga maimaiga* after death of gypsy moth (Lepidoptera: Lymantriidae) larval hosts. *Environ. Entomol., 21*, 129–135.

Hajek, A. E., Butler, L., & Wheeler, M. M. (1995). Laboratory bioassays testing the host range of the gypsy moth fungal pathogen *Entomophaga maimaiga*. *Biol. Control, 5*, 530–544.

Hajek, A. E., Carruthers, R. I., & Soper, R. S. (1990a). Temperature and moisture relations of sporulation and germination by *Entomophaga maimaiga* (Zygomycetes: Entomophthoraceae), a fungal pathogen of *Lymantria dispar* (Lepidoptera: Lymantriidae). *Environ. Entomol., 19*, 85–90.

Hajek, A. E., Humber, R. A., & Griggs, M. H. (1990b). Decline in virulence of *Entomophaga maimaiga* (Zygomycetes: Entomophthorales) with repeated *in vitro* subculture. *J. Invertebr. Pathol., 56*, 91–97.

Hajek, A. E., Shimazu, M., & Knoblauch, B. (2000). Isolating *Entomophaga maimaiga* using resting spore-bearing soil. *J. Invertebr. Pathol., 75*, 298–300.

Hajek, A. E., Wheeler, M. M., Eastburn, C. C., & Bauer, L. S. (2001). Storage of resting spores of the gypsy moth fungal pathogen, *Entomophaga maimaiga*. *Biocontrol Sci. Technol., 11*, 637–647.

Hajek, A. E., Burke, A., Nielsen, C., Hannam, J. J., & Bauer, L. S. (2008). Effects of cold storage and isolate on germination of *Entomophaga maimaiga* azygospores without dormancy. *Mycologia, 100*, 833–842.

Hua, L., & Feng, M.-G. (2003). New use of broomcorn millets for production of granular cultures of aphid-pathogenic fungus *Pandora neoaphidis* for high sporulation potential and infectivity to *Myzus persicae*. *FEMS Microbiol. Lett., 227*, 311–317.

Hua, L., & Feng, M.-G. (2005). Broomcorn millet grain cultures of the entomophthoralean fungus *Zoophthora radicans*: Sporulation capacity and infectivity to *Plutella xylostella*. *Mycol. Res., 109*, 319–325.

Humber, R. A. (1994). Special considerations for operating a culture collection of fastidious fungal pathogens. *J. Industr. Microbiol., 13*, 195–196.

Humber, R. A., & Ramoska, W. A. (1986). Variations in entomophthoralean life cycles: Practical implications. In R. A. Samson, J. M. Vlak, & D. Peters (Eds.), *Fundamental and Applied Aspects of Invertebrate Pathology* (pp. 190–193). Wageningen: Found. 4th International Colloquium Invertebrate Pathology.

Jensen, A. B., Thomsen, L., & Eilenberg, J. (2006). Value of host range, morphological, and genetic characteristics within the *Entomophthora muscae* species complex. *Mycol. Res., 110*, 941–950.

Keller, S. (1987). Arthropod-pathogenic Entomophthorales of Switzerland. I. *Conidiobolus, Entomophaga* and *Entomophthora*. *Sydowia, 40*, 122–167.

Keller, S. (1994). Working with arthropod-pathogenic Entomophthorales. *Bull. OILB/SROP, 17*, 287–307.

Keller, S. (2007a). Fungal structures and biology. In S. Keller (Ed.), *Arthropod-pathogenic Entomophthorales: Biology, Ecology, Identification* (pp. 27–54). Luxembourg: COST Action 842, EUR 22829, COST Office.

Keller, S. (2007b). Systematics, taxonomy and identification. In S. Keller (Ed.), *Arthropod-pathogenic Entomophthorales: Biology, Ecology, Identification* (pp. 111–133). Luxembourg: COST Action 842, EUR 22829, COST Office.

Kogan, P. H., & Hajek, A. E. (2000). Formation of azygospores by the insect pathogenic fungus *Entomophaga maimaiga* in cell culture. *J. Invertebr. Pathol., 75*, 193–201.

Krejzová, R. (1971). Versuchsinfektionen der Raupen von *Galleria mellonella* L. und *Antheraea pernyi* L. durch Vertreter der *Entomophthora*-Gattung. I. *Vest. Cs. Spol. Zool., 35*, 107–113.

Latgé, J.-P. (1975). Croissance et sporulation de 6 espèces d'Entomophthorales. 1. Influence de la nutrition carbonée. *Entomophaga, 20*, 201–207.

Latgé, J.-P. (1980). Sporulation de *Entomophthora obscura* Hall & Dunn en culture liquide. *J. Microbiol., 26*, 1038–1048.

Latgé, J.-P. (1981). Comparaison des exigences nutritionnelles des Entomophthorales. *Ann. Microbiol. (Inst. Pasteur), 132B*, 299–306.

Latgé, J.-P., & Beauvais, A. (1987). Wall composition of the protoplastic Entomophthorales. *J. Invertebr. Pathol., 50*, 53–57.

Latgé, J.-P., Eilenberg, J., Beauvais, A., & Prévost, M.-C. (1988). Morphology of *Entomophthora muscae* protoplasts grown *in vitro*. *Protoplasma, 146*, 166–173.

Latgé, J.-P., & Perry, D. (1980). The utilization of an *Entomophthora obscura* resting spore preparation in biological control experiments against cereal aphids. *Bull. OILB/SROP III, 4*, 19–25.

Latgé, J.-P., & Remaudière, G. (1975). Croissance et sporulation de 6 espèces d'Entomophthorales. III. Influence des concentrations de carbone et d'azote et du rapport C/N. *Rev. Mycol., 39*, 239–250.

Latgé, J.-P., & Sanglier, J. J. (1985). Optimisation de la croissance et de la sporulation de *Conidiobolus obscurus* en milieu défini. *Can. J. Bot., 63*, 68–85.

Latgé, J.-P., Soper, R. S., & Madore, C. D. (1977). Media suitable for industrial production of *Entomophthora virulenta* zygospores. *Biotechnol. Bioengineer., 19*, 1269–1284.

Latteur, G. (1977). Sur la possibilité d'infection directe d'aphides par *Entomophthora* à partir de sols hébergeant un inoculum naturel. *C.R. Acad. Sci. Paris, série D, 284*, 2253–2256.

Latteur, G., Destain, J., Oger, R., & Godefroid, J. (1982). Étude de la production de conidies par les spores durables de *Conidiobolus obscurus* (Hall et Dunn) Remaud. et Kell., une Entomophthorale pathogène de pucerons. *Parasitica, 38*, 139–161.

Le Rü, B., Silvie, P., & Papierok, B. (1985). L'Entomophthorale *Neozygites fumosa* pathogène de la Cochenille du manioc, *Phenacoccus manihoti* (Hom.: *Pseudococcidae*) en République populaire du Congo. *Entomophaga, 30*, 23–29.

Leite, L. G., Alves, S. B., Batista Filho, A., & Roberts, D. W. (2003). Effect of salts, vitamins, sugars and nitrogen sources on the growth of 3 genera of Entomophthorales: *Batkoa, Furia*, and *Neozygites*. *Mycol. Res., 107*, 872–878.

Leite, L. G., Alves, S. B., Batista Filho, A., & Roberts, D. W. (2005). Simple, inexpensive media for mass production of three entomophthoralean fungi. *Mycol. Res., 109*, 326–334.

Leite, L. G., Smith, L., Moraes, G. J., & Roberts, D. W. (2000). *In vitro* production of hyphal bodies of the mite pathogenic fungus *Neozygites floridana*. *Mycologia, 92*, 201–207.

Li, W., Xu, W.-A., Sheng, C.-F., Wang, H.-T., & Xuan, W.-J. (2006). Factors affecting sporulation and germination of *Pandora nouryi* (Entomophthorales: Entomophthoraceae), a pathogen of *Myzus persicae* (Homoptera: Aphididae). *Biocontrol Sci. Technol., 16*, 647–652.

Li, Z., Butt, T. M., Beckett, A., & Wilding, N. (1993). The structure of dry mycelia of the entomophthoralean fungi *Zoophthora radicans* and *Erynia neoaphidis* following different preparatory treatments. *Mycol. Res., 97*, 1315–1323.

McCabe, D. E., & Soper, R. S. (1982) Preparation of an entomopathogenic fungal insect control agent. U.S. Patent Pending #06/419637.

MacLeod, D. M., Tyrrell, D., & Welton, M. A. (1980). Isolation and growth of the grasshopper pathogen *Entomophthora grylli*. *J. Invertebr. Pathol., 36*, 85–89.

Meyling, N. V., & Eilenberg, J. (2006). Occurrence and distribution of soil borne entomopathogenic fungi within a single organic ecosystem. *Agric. Ecosys. Environ., 113*, 336–341.

Milner, R. J. (1981). Patterns of primary spore discharge of *Entomophthora* spp. from the blue green aphid, *Acyrthosiphon kondoi*. *J. Invertebr. Pathol., 38*, 419–425.

Milner, R. J., & Soper, R. S. (1981). Bioassay of *Entomophthora* against the spotted alfalfa aphid *Therioaphis trifolii* f. *maculata*. *J. Invertebr. Pathol., 37*, 168–173.

Mitsuhashi, J. (1982). Media for insect cell cultures. In K. Maramorosch (Ed.), *Advances in Cell Cultures. Vol. 2* (pp. 133–196). New York: Academic Press.

Mullens, B. A. (1986). A method for infecting large numbers of *Musca domestica* (Diptera: Muscidae) with *Entomophthora muscae* (Entomophthorales: Entomophthoraceae). *J. Med. Entomol., 23*, 457–458.

Mullens, B. A., & Rodriguez, J. L. (1985). Dynamics of *Entomophthora muscae* (Entomophthorales: Entomophthoraceae) conidial discharge from *Musca domestica* (Diptera: Muscidae) cadavers. *Environ. Entomol., 14*, 317–322.

Müller-Kögler, E. (1959). Zur Isolierung und Kultur insektenpathogener Entomophthoraceen. *Entomophaga, 4*, 261–267.

Nielsen, C., Hajek, A. E., Humber, R. A., Bresciani, J., & Eilenberg, J. (2003). Soil as an environment for winter survival of aphid-pathogenic Entomophthorales. *Biol. Control, 28*, 92–100.

Nielsen, C., Keena, M., & Hajek, A. E. (2005). Virulence and fitness of the fungal pathogen *Entomophaga maimaiga* in its host *Lymantria dispar*, for pathogen and host strains originating from Asia, Europe and North America. *J. Invertebr. Pathol., 89*, 232–242.

Nolan, R. A. (1988). A simplified, defined medium for growth of *Entomophaga aulicae* protoplasts. *Can. J. Microbiol., 34*, 45–51.

Nolan, R. A. (1990). Enhanced hyphal body production by *Entomophaga aulicae* protoplasts in the presence of a neutral and a positively charged surface under mass fermentation conditions. *Can. J. Bot., 68*, 2708–2713.

Nolan, R. A. (1993). An inexpensive medium for mass fermentation production of *Entomophaga aulicae* hyphal bodies competent to form conidia. *Can. J. Microbiol., 39*, 588–593.

Oger, R., & Latteur, G. (1985). Description et précision d'une nouvelle méthode d'estimation de la virulence d'une Entomophthorale pathogène de pucerons. *Parasitica, 41*, 135–150.

Onstad, D. W., Fuxa, J. R., Humber, R. A., Oestergaard, J., Shapiro-Ilan, D. I., Gouli, V. V., Anderson, R. S., Andreadis, T. G., & Lacey, L. A. (2006). *An abridged glossary of terms used in invertebrate pathology* (3rd ed.). http://www.sipweb.org/glossary/. accessed 23 March 2011.

Papierok, B. (1978). Obtention *in vivo* des azygospores d'*Entomophthora thaxteriana* Petch, champignon pathogène de pucerons (Homoptères Aphididae). *C.R. Acad. Sci. Paris, sèrie D, 286*, 1503–1506.

Papierok, B. (1985). Données écologiques et expérimentales sur les potentialitiés entomopathogènes de l'Entomophthorale *Conidiobolus coronatus* (Costantin) Batko. *Entomophaga, 30*, 303–312.

Papierok, B. (1989). On the occurrence of Entomophthorales in Finland. I. Species attacking aphids (Homoptera, Aphididae). *Ann. Entomol. Fennici, 55*, 63–69.

Papierok, B. (2007). Isolating, growing and storing arthropod-pathogenic Entomophthorales. In S. Keller (Ed.), *Arthropod-pathogenic Entomophthorales: Biology, Ecology, Identification* (pp. 66–81). Luxembourg: COST Action 842, EUR 22829, COST Office.

Papierok, B., & Balazy, S. (2007). Collecting arthropod-pathogenic Entomophthorales: how to find them in the field and how to collect, prepare and store the appropriate material for classical identification purposes? In S. Keller (Ed.), *Arthropod-pathogenic Entomophthorales: Biology, Ecology, Identification* (pp. 56–65) Luxembourg: COST Action 842, EUR 22829, COST Office.

Papierok, B., & Charpentié, M. J. (1982). Les champignons se développant en Côte-d'Ivoire sur la fourmi *Paltothyreus tarsatus* F. Relation entre l'Hyphomycète *Tilachlidiopsis catenulata* sp. nov. et l'Ascomycète *Cordyceps myrmecophila* Cesati 1846. *Mycotaxon, 14*, 351–368.

Papierok, B., & Wilding, N. (1979). Mise en évidence d'une différence de sensibilité entre 2 clones du puceron du pois *Acyrthosiphon pisum* Harr. (Homoptères: Aphididae), exposés à 2 souches du champignon Phycomycète *Entomophthora obscura* Hall & Dunn. *C.R. Séances Acad. Sci. Paris, série D, 288*, 93–95.

Papierok, B., & Wilding, N. (1981). Étude du comportement de plusieurs souches de *Conidiobolus obscurus* (Zygomycètes Entomophthoraceae) vis-à-vis des pucerons *Acyrthosiphon pisum* et *Sitobion avenae* (Hom. Aphididae). *Entomophaga, 26*, 241–249.

Papierok, B., Valadão, L., Tôrres, B., & Arnault, M. (1984). Contribution à l'étude de la spécificité parasitaire du champignon entomopathogène *Zoophthora radicans* (Zygomycètes, Entomophthorales). *Entomophaga, 29*, 109–119.

Pell, J. K., & Wilding, N. (1992). The survival of *Zoophthora radicans* (Zygomycetes: Entomophthorales) isolates as hyphal bodies in mummified larvae of *Plutella xylostella* (Lep.: Yponomeutidae). *Entomophaga, 37*, 649–654.

Pell, J. K., Barker, A. D., Clark, S. J., Wilding, N., & Alderson, P. G. (1998). Use of a novel sporulation monitor to quantify the effects of formulation and storage on conidiation by dried mycelia of the entomopathogenic fungus *Zoophthora radicans*. *Biocontrol Sci. Technol., 8*, 13–21.

Perry, D. F., & Fleming, R. A. (1989). *Erynia crustosa* zygospore germination. *Mycologia, 81*, 154–158.

Perry, D. F., & Latgé, J.-P. (1980). Chemically defined media for growth and sporulation of *Entomophthora virulenta*. *J. Invertebr. Pathol., 35*, 43–48.

Perry, D. F., & Latgé, J.-P. (1982). Dormancy and germination of *Conidiobolus obscurus* azygospores. *Trans. Br. Mycol. Soc., 78*, 221–225.

Perry, D. F., Tyrrell, D., & DeLyzer, A. J. (1982). The mode of germination of *Zoophthora radicans* zygospores. *Mycologia, 74*, 549–555.

Ramoska, W. A., Hajek, A. E., Ramos, M. E., & Soper, R. S. (1988). Infection of grasshoppers (Orthoptera: Acrididae) by members of the *Entomophaga grylli* species complex (Zygomycetes: Entomophthorales). *J. Invertebr. Pathol., 52*, 309–313.

Remaudière, G., & Hennebert, G. L. (1980). Révision systématique de *Entomophthora aphidis* Hoffm. in Fres. Description de deux nouveaux pathogènes d'aphides. *Mycotaxon, 11*, 269–321.

Remaudière, G., Latgé, J.-P., Papierok, B., & Coremans-Pelseneer, J. (1976). Sur le pouvoir pathogène de quatre espèces d'Entomophthorales occasionnellement isolées d'aphides en France. *C.R. Acad. Sci. Paris, série D, 283*, 1065–1068.

Remaudière, G., Latgé, J.-P., & Michel, M.-F. (1981). Écologie comparée des Entomophthoracées pathogènes de Pucerons en France littorale et continentale. *Entomophaga, 26*, 157–178.

Ribeiro, A. E. L., Gondim, M. G. C., Jr., Calderan, E., & Delalibera, I., Jr. (2009). Host range of *Neozygites floridana* isolates (Zygomycetes: Entomophthorales) to spider mites. *J. Invertebr. Pathol., 102*, 196–202.

Rockwood, L. P. (1950). Entomogenous fungi of the family Entomophthoraceae in the Pacific Northwest. *J. Econ. Entomol., 43*, 704–707.

Rohel, É, Couteaudier, Y., Papierok, B., Cavelier, N., & Dedryver, C. A. (1997). Ribosomal internal transcribed spacer size variation correlated with RAPD-PCR pattern polymorphisms in the entomopathogenic fungus *Erynia neoaphidis* and some closely related species. *Mycol. Res., 101*, 573–579.

Sanchez Peña, S. R. (2005). *In vitro* production of hyphae of the grasshopper pathogen *Entomophaga grylli* (Zygomycota: Entomophthorales): Potential for production of conidia. *Fl. Entomol., 88*, 332–334.

Shah, P. A., Aebi, M., & Tuor, U. (2000). Drying and storage procedures for formulated and unformulated mycelia of the aphid-pathogenic fungus *Erynia neoaphidis*. *Mycol. Res., 104*, 440–446.

Shah, P. A., Clark, S. J., & Pell, J. K. (2003). Direct and indirect estimates of *Pandora neoaphidis* conidia in laboratory bioassays with aphids. *J. Invertebr. Pathol., 84*, 145–147.

Shah, P. A., Clark, S. J., & Pell, J. K. (2004). Assessment of aphid host range and isolate variability in *Pandora neoaphidis* (Zygomycetes: Entomophthorales). *Biol. Control, 29*, 90–99.

Shimazu, M. (1993). How to isolate, culture and preserve insect pathogens: Entomopathogenic fungi. In H. Iwahana, M. Okada, Y. Kunimi, & M. Shimazu (Eds.), *Research Manual on Insect Pathogens. Special Issue of Monthly Plant Protection Journal No.2* (pp. 24–42), Tokyo, Japan: Japan Plant Protection Association, (in Japanese).

Shimazu, M., & Soper, R. S. (1986). Pathogenicity and sporulation of *Entomophaga maimaiga* Humber, Shimazu, Soper and Hajek (Entomophthorales: Entomophthoraceae) on larvae of the gypsy moth, *Lymantria dispar* L. (Lepidoptera: Lymantriidae). *Appl. Ent. Zool., 21*, 589–596.

Silvie, P., & Papierok, B. (1991). Les ennemis naturels d'insectes du cotonnier au Tchad. Premières données sur les champignons de l'ordre des Entomophthorales. *Coton Fibres Trop., 46*, 293–308.

Six, D. L., & Mullens, B. A. (1996). Distance of conidial discharge of *Entomophthora muscae* and *Entomophthora schizophorae* (Zygomycotina: Entomophthorales). *J. Invertebr. Pathol., 67*, 253–258.

Soper, R. S., Holbrook, F. R., Majchrowicz, I., & Gordon, C. C. (1975). Production of *Entomophthora* resting spores for biological control. *Maine Life Sci. Agric. Exp. Sta., Tech. Bull., 76*, 1–15.

Soper, R. S., Shimazu, M., Humber, R. A., Ramos, M. E., & Hajek, A. E. (1988). Isolation and characterization of *Entomophaga maimaiga* sp. nov., a fungal pathogen of gypsy moth, *Lymantria dispar*, from Japan. *J. Invertebr. Pathol., 51*, 229–241.

Steinkraus, D. C., & Boys, G. O. (2005). Mass harvesting of the entomopathogenic fungus, *Neozygites fresenii*, from natural field epizootics in the cotton aphid, *Aphis gossypii*. *J. Invertebr. Pathol., 88*, 212–217.

Tkaczuk, C. (2005). The effect of selected heavy metal ions on the growth and conidial germination of the aphid pathogenic fungus *Pandora neoaphidis* (Remaudière et Hennebert) Humber. *Polish J. Environ. Stud., 14*, 897–902.

Tymon, A. M., Shah, P. A., & Pell, J. K. (2004). PCR-based molecular discrimination of *Pandora neoaphidis* isolates from related entomopathogenic fungi and development of species-specific diagnostic primers. *Mycol. Res., 108*, 419–433.

Tyrrell, D. (1988). Survival of *Entomophaga aulicae* in dried insect larvae. *J. Invertebr. Pathol., 52*, 187–188.

Tyrrell, D., & MacLeod, D. M. (1972). Spontaneous formation of protoplasts by a species of *Entomophthora*. *J. Invertebr. Pathol., 19*, 354–360.

Tyrrell, D., & MacLeod, D. M. (1975). *In vitro* germination of *Entomophthora aphidis* resting spores. *Can. J. Bot., 53*, 1188–1191.

Urbanczyk, M. J., Zabza, A., Bałazy, S., & Peczynskaczoch, W. (1992). Laboratory culture media and enzyme-activity of some entomopathogenic fungi of *Zoophthora* (Zygomycetes, Entomophthoraceae). *J. Invertebr. Pathol., 59*, 250–257.

Valovage, W. D., Nelson, D. R., & Frye, R. D. (1984). Infection of grasshoppers with *Entomophaga grylli* by exposure to resting spores and germ conidia. *J. Invertebr. Pathol., 43*, 274–275.

Vandenberg, J. S., & Soper, R. S. (1979). A bioassay technique for *Entomophthora sphaerosperma* on the spruce budworm, *Choristoneura fumiferana*. *J. Invertebr. Pathol., 33*, 148–154.

Vingaard, M. G., Steinkraus, D. C., Boys, G. O., & Eilenberg, J. (2003). Effects of long-term storage at −14°C on the survival of *Neozygites fresenii* (Entomophthorales: Neozygitaceae) in cotton aphids (Homoptera: Aphididae). *J. Invertebr. Pathol., 82*, 97–102.

Watson, D. W., Mullens, B. A., & Petersen, J. J. (1993). Behavioral fever response of *Musca domestica* (Diptera: Muscidae) to infection by *Entomophthora muscae* (Zygomycetes: Entomophthorales). *J. Invertebr. Pathol., 61*, 10–16.

Wekesa, V. W., & Delalibera, I., Jr. (2008). Long-term preservation of *Neozygites tanajoae* (Entomophthorales: Neozygitaceae) in cadavers of *Mononychellus tanajoae* (Acari: Tetranychidae). *Biocontrol Sci. Technol., 18*, 621–627.

Wilding, N. (1971). Discharge of conidia of *Entomophthora thaxteriana* Petch from the pea aphid *Acyrthosiphon pisum* Harris. *J. Gen. Microbiol., 69*, 417–422.

Wilding, N. (1973). The survival of *Entomophthora* spp. in mummified aphids at different temperatures and humidities. *J. Invertebr. Pathol., 21*, 309–311.

Wilding, N. (1976). Determination of the infectivity of *Entomophthora* spp. *Proc. 1st. Int. Coll. Invertebr. Pathol. Kingston.*, 296–300.

Wraight, S. P., Galaini-Wraight, S., Carruthers, R. I., & Roberts, D. W. (2003). *Zoophthora radicans* (Zygomycetes:

Entomophthorales) conidia production from naturally infected *Empoasca kraemeri* and dry-formulated mycelium under laboratory and field conditions. *Biol. Control, 28*, 60–77.

Xu, J.-H., & Feng, M.-G. (2002). *Pandora delphacis* (Entomophthorales: Entomophthoraceae) infection affects the fecundity and population dynamics of *Myzus persicae* (Homoptera: Aphididae) at varying regimes of temperature and relative humidity in the laboratory. *Biol. Control, 25*, 85–91.

Zhou, X., & Feng, M.-G. (2009). Sporulation, storage and infectivity of obligate aphid pathogen *Pandora nouryi* grown on novel granules of broomcorn millet and polymer gel. *J. Appl. Microbiol., 107*, 1847–1856.

Zhou, X., & Feng, M.-G. (2010). Improved sporulation of alginate pellets entrapping *Pandora nouryi* and millet powder and their potential to induce an aphid epizootic in field cages after release. *Biol. Control, 54*, 153–158.

Zimmermann, G. (1986). The '*Galleria* bait method' for detection of entomopathogenic fungi in soil. *J. Appl. Entomol., 102*, 213–215.

## APPENDIX: MEDIA

**1.** Sabouraud dextrose agar supplemented with egg yolk and milk

55% Sabouraud dextrose agar
25% dextrose yeast extract agar
20% of a mixture of egg yolk (60%) and milk (40%)

The respective amounts of each ingredient are calculated according to the final volume of medium wanted. The volume of a medium-sized egg yolk is about 15–18 ml. The different ingredients are prepared separately as follows:

- for Sabouraud agar: dextrose (20 g/l), peptone (10 g/l), and agar (20 g/l) are placed in a flask with distilled water and sterilized at 121°C for 15 min, and then maintained at 50–60°C;
- for dextrose–yeast extract-agar: dextrose (20 g/l), yeast extract (10 g/l), agar (20 g/l), and distilled water are placed in a flask and sterilized at 121°C for 15 min, and then kept at 50–60°C. It is most efficient to sterilize the above two flasks simultaneously;
- for milk: sterilize whole milk at 110°C for 15 min, and keep at room temperature;
- for egg yolk: place fresh eggs in a mixture of 90% alcohol (200 ml) and 2% sodium hypochlorite (800 ml) for surface sterilization. Break egg shells with forceps near the flame of a Bunsen burner. Gently remove the egg yolk, and pour it into a sterile graduated cylinder;
- add milk to egg yolks in the graduated cylinder, and stir the contents with a sterile glass rod, until the mixture is homogeneous;
- pour the egg yolk–milk mixture into the flask containing dextrose–yeast extract-agar, and shake;
- pour the accumulated mixture into the flask containing Sabouraud agar, and shake;
- immediately pour the medium into sterile Petri dishes or tubes. An automatic dispenser, such as a multiple pipetter (e.g., Multipette, Eppendorf) fitted with a 50-ml sterile reservoir (e.g., Combitips, Eppendorf), is efficient;
- if tubes are being used, immediately push a cotton plug in after filling, and lightly tilt until the medium hardens.

**2.** Coagulated egg yolk and milk

60% egg yolk
40% whole milk

Because this medium is harder than the previous one, it is quite useful for isolation by rubbing showered conidia off of the surfaces of Petri dishes (see Section 4A). For this reason, and for work away from a laboratory, this medium is frequently prepared in small tubes (60 × 12 mm). The procedure is as follows:

- prepare the two ingredients separately, and then mix them together as explained above for the preparation of Sabouraud-dextrose-agar supplemented with egg yolk and milk;
- distribute the medium in small tubes, each tube being plugged with cotton and then lightly tilted immediately after filling;
- sterilize in the oven at 100°C for 30 min;
- let rest at room temperature for 24 h;
- replace cotton plugs with rubber ones, taking extreme care to plug securely.

These tubes can be stored for several months or even a few years, as long as they are kept in the dark. After inoculation, tubes must be plugged with cotton.

**3.** Egg yolk/Sabouraud maltose agar (EYSMA)

24 g maltose
6 g peptone
6 g yeast extract
9 g agar
12 eggs
600 ml distilled water

- To each of six 500-ml Fleakers® (18-cm tall beakers that are convenient to use; Corning Inc.) add 100 ml distilled water.
- To each of three of the Fleakers, add 8 g maltose and to each of the remaining three Fleakers, add 2 g peptone, 2 g yeast extract, and 3 g agar.
- Cover and autoclave all Fleakers for 15 min.
- Soak 12 chicken eggs in 50% ethanol for approximately 30 min.
- In a sterile hood, add two broken egg yolks to each Fleaker. After egg yolks are added, procedures must be undertaken quickly due to rapid coagulation as ingredients cool. Therefore, it is easiest to add egg yolks to only one pair of Fleakers at a time.
- Thoroughly mix the contents of pairs of Fleakers containing different ingredients by pouring back and forth, and dispense into Petri dishes.

**4.** Beauvais and Latgé protoplast medium (GLEN)

0.4% dextrose
0.5% yeast extract (Difco)
0.65% lactalbumin hydrolysate (Difco)
0.77% NaCl
10% fetal bovine serum

Technical considerations are as follows: osmotic pressure $400 \pm 20$ mOsm, pH 6.5, autoclaved 30 min at 115°C, distributed into 35-mm diameter Petri dishes. Lactalbumin hydrolysate can be replaced v/v with yeast extract.

CHAPTER X

# Preservation of entomopathogenic fungal cultures

RICHARD A. HUMBER

USDA-ARS Biological Integrated Pest Management Research Unit, Robert W. Holley Center for Agriculture and Health, 538 Tower Road, Ithaca, NY 14853-2901, USA

## 1. INTRODUCTION

All research or applied studies using live organisms requires a constant supply of them in a suitable condition. Work with fungi usually requires keeping cultures, a task that is both easier and facilitates more possible research approaches than dealing, for example, with migratory birds, marine mammals, mountain gorillas, mature redwood trees, or even many insects.

The isolation and growth of microbial cultures are dealt with elsewhere in this book. While this chapter focuses on fungi, the techniques described here apply equally for nearly all other types of entomopathogens.

No matter why or how one may store cultures, all preservation techniques increase the time between transfers to periods ranging from several weeks or months to many years with a minimal loss of viability or other key properties of the organism during storage. Each of these preservation techniques has strengths and weaknesses (Table 10.1; also see Ryan *et al.*, 2000). Once one's needs to store cultures go beyond the most casual level, it is very important to choose the preservation technology that best fits one's needs with a convenient, affordable level of technological sophistication. Much time, anguish, and money can be saved by carefully weighing the real purposes and needs to preserve cultures *before* committing one's effort and financial and physical resources to any specific storage technique. The demands for space, materials, record keeping, and labor are much lower for researchers maintaining a few cultures being used in current research or teaching than for laboratories keeping small archival collections with dozens to several hundred cultures, or for general service culture collections that are actively acquiring, storing, and distributing large numbers of cultures.

Formulations of microbial biocontrol agents are means to assure the short-term preservation of a pathogen in a viable, quiescent state during shelf storage to be activated and infective upon application. Most entomopathogenic fungal formulations (granular, liquid, emulsifiable concentrates, encapsulation in starch or alginate, etc.) incorporate conidia harvested after a suitable means of mass production. One notably different approach to fungal biocontrol generates flaked or granular fragments of dried mycelium that sporulate when rehydrated (McCabe & Soper, 1985; Rombach *et al.*, 1989; Verkleij *et al.*, 1992) or (dried) cultures of fungi on whole grains (e.g., Feng *et al.*, 2000; Hua & Feng, 2005; Gindin *et al.*, 2006). Techniques to produce and to formulate entomopathogens are not covered further in this manual, but Burges (1998) provides detailed discussion about formulations of microbial biocontrol agents as does a series of books entitled *Pesticide Formulations and Applications Systems* (with varying editors and publishers).

There is a rich literature on the preservation of fungal cultures, and this chapter cannot review all possible techniques. It does, however, present a wide range of

MANUAL OF TECHNIQUES IN INVERTEBRATE PATHOLOGY
ISBN 9780123868992

2012 Published by Elsevier Ltd.

**Table 10.1** Advantages and disadvantages of various preservation techniques.

| *Preservation method* | *Advantages* | *Disadvantages* |
|---|---|---|
| *Storage temperature ≥ 0°C (room temperature or refrigerated)* | | |
| Serial transfer (stored at 4°C) | Technologically simple | Phenotypic characters may change |
| | Allows continuous monitoring of phenotype | Continuous need for materials, labor |
| Mineral oil | Inexpensive, technologically simple | Space intensive; tubes must be upright |
| | | Must monitor for contaminants |
| Distilled water stasis | Inexpensive, technologically simple | Must check and maintain water levels |
| | Needs little space if using small vials | Must monitor for contaminants |
| Lyophilized (freeze-dried) | Standard methods for many fungi | Initial expense for equipment is large |
| | Dried cultures are easily mailed | Ampoules should be refrigerated |
| | Can store at room temperature (but better at 4°C) | Not suitable for some fungi |
| *Storage at < 0°C in standard or ultracold electric (mechanical) freezers* | | |
| Silica gel (at −20°C) | Inexpensive and technologically simple | Long-term success depends on security of screw camp closure |
| | Uses standard appliance-type freezer | Subject to losses during power failures |
| | No requirement to store > 1 tube per culture | |
| | Can store at −20°C or room temperature | |
| Standard freezer (−20°C) | Inexpensive and readily available | Not recommended (except for silica gel or cultures on filter paper) |
| | | Subject to losses during power failures |
| Ultracold freezer (−80 or −120°C) | Convenient, widely available in many laboratories | Ice crystals *may* form and grow depending on handling of cryovials |
| | Long-term storage without liquid N2 | Subject to losses during power failures |
| | Probably suitable for all cultured (and noncultured) fungi | Liquid nitrogen back-up is strongly recommended |
| *Storage at ≤−120°C in liquid nitrogen ($LN_2$) freezers (dewars)* | | |
| | Unsurpassed duration of storage | High initial and continuing costs |
| | Long-term viability is maximized | Continuous access to affordable $LN_2$ is required |
| | All types of cultured (or noncultured) fungi can be stored | |

**Table 10.1** Advantages and disadvantages of various preservation techniques—Cont'd

| *Preservation method* | *Advantages* | *Disadvantages* |
|---|---|---|
| Vapor phase (−120 to −196°C) | $LN_2$ supply can not contaminate vial contents | Temperature stability depends on dewar design and storage racking system |
| | No risk of vials exploding when thawed | Must continuously monitor and maintain $LN_2$ levels |
| | | Automatic $LN_2$ level control and replenishment depends on uninterrupted electric service |
| Immersed in $LN_2$ (−196°C) | Temperature is stable while immersed | No guarantee against leakage of $LN_2$ into cryovials |
| | Ice crystal formation during storage is unlikely | Vial contents may be contaminated by $LN_2$ leakage |
| | Monitoring of nitrogen levels is minimized | Vials with leaked $LN_2$ may explode when thawed |
| | Unaffected by electric power failures as long as $LN_2$ remains in dewar | |

Also see Ryan *et al.* (2000) for another approach to circumstances-based 'best' technique for preserving fungal cultures.

techniques used for preserving cultures of fungal entomopathogens. More detailed discussions of the most used of these techniques can be found in more comprehensive compendia on fungal preservation (Onions, 1983; Simione & Brown, 1991; Smith & Onions, 1994).

All protocols given here can be adapted to meet specific needs and limitations. The goal here is not to encourage the exact reproduction of the given protocols, but to preserve viable cultures by the most successful, practical, and reasonable means compatible with a laboratory's capacities and constraints.

## 2. PRESERVATION AT TEMPERATURES ABOVE FREEZING

Culture storage in ambient room temperatures rather than under refrigeration may require fewer resources than any other preservation strategy, but also seems to be the least dependable and shortest-term approach and is not advised. However, storage at such an elevated temperature is possible with serial transfers or cultures submitted to mineral oil or water stasis, silica gel crystals, or even freeze-drying. It is important to remember that the phenotypic and genotypic stabilities of cultures are affected by the storage technique used and by the storage temperature (and its stability) throughout the period of storage.

### A. Serial transfer

This most obvious way to store cultures is best suited for relatively short-term studies (over a few weeks or months), but it can serve well to maintain small numbers of cultures for many years. Repeated subculturing, however, may also have such deleterious outcomes as losses of pathogenicity, virulence, and/or sporulation but there is no way to predict whether or when any of these vital characters might be lost during subculturing. Cultures of entomopathogenic fungi must be checked carefully for such morphological changes and, if possible, bioassayed periodically for losses of essential pathological characters. The USDA-ARS Collection of Entomopathogenic Fungal Cultures (ARSEF; Ithaca NY, USA) inoculates cultures with both hyphae and spores (if both are available) for nearly all its entomopathogens. For some fungal genera, however, Fennell (1960) and Onions (1971) discuss which fungal structures (hyphae, spores, etc.) constitute appropriate inoculum for serial transfers of fungi in general; the ARSEF staff use both hyphae and, if available, spores as the preferred inoculum.

## B. Mineral oil

Storing culture slants under a layer of sterile mineral oil is one of the oldest, simplest, and least expensive methods for long-term culture preservation. This approach is still widely used, especially for fungi that do not tolerate freeze-drying or where cryogenic storage is too costly.

The oil both prevents desiccation and diminishes gas exchange, thus reducing fungal metabolism to a very low level. If space is available to store racks of tubes, this is a common alternative for the storage of a very diverse range of fungi. Onions (1971) warns that McCartney bottles or other screwcap bottles with rubber gaskets should be avoided unless the gaskets are removed since oil-soluble components in the rubber may be toxic to cultures.

Cultures under mineral oil may remain viable for decades (Cavalcanti, 1991; Silva *et al.*, 1994). Pathogenicity of entomopathogens may remain after months of storage (Balardin & Loch, 1988), but the pathogenicity or virulence of some taxa might decline after many years of storage under oil.

### *1. Set up for storage*

1. Use vigorously growing culture slants in glass culture tubes.
2. Autoclave a supply of heavy mineral oil and re-autoclave 24–48 h later to kill any bacterial spores activated by the first autoclaving.
3. Aseptically cover the culture slant with sterile oil to a depth of 1 cm over the upper edge of the medium in an upright tube.
4. Cover tubes with tight caps or plugs and apply a couple of layers of paraffin film as a further vapor barrier and to exclude mites or other contaminants.
5. Store tubes upright in racks (at 4°C if possible for longest viability).

### *2. Culture recovery*

1. With a sterile scalpel, loop, or needle, recover an explant from the submerged culture.
2. Drain excess oil from the explant and place on fresh medium.
3. Reseal and return the tube to long-term storage.
4. Monitor the culture for viability and/or contamination.

## C. Distilled water stasis

Storing metabolically inactive fungi in sterile distilled water may be the simplest preservation technique discussed here and can succeed for many diverse fungi including human or plant pathogens (Castellani, 1967; Figueiredo & Pimentel, 1975), but may be especially useful for organisms that cannot be freeze-dried such as oomycetes (no longer regarded as true fungi; Clark & Dick, 1974) and Entomophthorales (López Lastra *et al.*, 2002). Some fungi may survive for up to 20 years in sterile water (Hartung de Capriles *et al.*, 1989) but most fungi lose viability much sooner. López Lastra *et al.* (2002) confirmed the utility of water stasis for a selection of diverse entomopathogens; this technique can probably be used to store nearly any culturable fungal pathogens of invertebrates.

The set-up for water stasis requires nothing more than sterile screwcap tubes or vials and sterilized water; tap water can be used if distilled or deionized water is unavailable. Cultures in water stasis can be kept at room temperature but may survive longer if refrigerated.

Some basic problems for this technique are easily avoided: too much inoculum for the volume of water may jeopardize the ability of the fungus to withstand long-term storage; the volume of water should be at least 40-times greater than of the inoculum blocks. Bringing too much medium to the storage tube may result in excessive residual nutrient levels that shorten the longevity of the stored fungus. Periodically add sterile water to units whose levels are visibly lower than when first set up (due to evaporative loss from inadequately sealed tubes).

### *1. Set up for storage*

1. Use vigorously growing, relatively young cultures for inoculum.
2. Dispense sterile water into sterile, preferably screw-capped storage tubes or vials.
3. Inoculate tubes with small (*ca.* 1 mm$^3$) blocks of a culture and/or an aqueous suspension of spores. The volume of water should be $\geq$40-times the total volume of the culture inoculum preserved in it.
4. Cover tubes with tight-fitting caps. Seal with paraffin film or by dipping tube tops in melted paraffin for

greater security against evaporation, mites, or contamination during storage.

5. Store tubes upright at room temperature or in a refrigerator.

*2. Culture recovery*

1. Recover a block from the tube with a sterile scalpel, loop, or needle, and place (fungus side down) on to fresh medium.
2. Monitor the culture for viability and/or contamination.
3. Reseal the tube and return to long-term storage.

## 3. PRESERVATION BY FREEZE-DRYING (LYOPHILIZATION)

Lyophilization is the most commonly used among the more technologically demanding means to preserve fungal germplasm and is the primary technique used by most culture collections. Cultures can be lyophilized relatively quickly, occupy relatively little storage space, and are easily sent elsewhere.

Freeze-drying is effective for nearly all conidial fungi, ascomycetes, basidiomycetes. It is not usually useful for fungi with very 'watery' cells with large vacuolar volumes (e.g., oomycete 'fungi' and Entomophthorales) but may succeed if the spores are not heavily vacuolated (e.g., the sporangiospores of many zygomycete fungi). It is generally understood that spores can tolerate lyophilization better than vegetative hyphae.

Most small- to medium-scale users use either manifold or centrifugal freeze-dryers and preserve cultures in commercial lyophil ampoules or glass tubes that are individually flame-sealed. Large-scale operations usually use serum bottles dried on shelves in a large vacuum chamber and sealed under vacuum by the lowering of a pressure plate onto the partially seated rubber stoppers.

The protocol for lyophilizing fungi is only a summary; detailed instructions such as those in Simione & Brown (1991), Smith & Onions (1994) or Day & McLellan (1995) are needed only by laboratories having a lyophilizer. However, because anybody studying fungi may receive and need to revive lyophilized cultures, detailed instructions for doing so are given below.

### A. Set up for storage using manifold (and shelf-freezing) lyophilizer

1. Use sporulating cultures for inoculum.
2. Close ampoules with small cotton plugs, then autoclave and dry the plugged ampoules. (With shelf-freezing units, serum bottles are partially plugged with slotted rubber stoppers before autoclaving.)
3. Cover sporulating cultures with sterilized skim milk solution and suspend spores and hyphae, aliquot small quantities of this suspension to sterile ampoules or serum bottles, and 'cure' the contents for several h in a refrigerator. **Note**: If refrigerated overnight, some fungi such as *Metarhizium* species may clear this opaque milk carrier but this clearance does not destroy the ultimate viability of the processed culture.
4. *Rapidly* freeze the preparations in a mix of dry ice and either ethanol or propylene glycol or by placing ampoules in an ultracold freezer or in the vapor phase in a liquid nitrogen dewar. (This freezing is the first step completed by shelf-freezing lyophilizers.)
5. Attach frozen ampoules to a strong vacuum on the lyophilizer. Preparations must remain frozen during the initial stages of vacuum desiccation. (These steps are automatically accomplished in shelf-freezing units.)
6. After desiccation, ampoules are flame-sealed while still under vacuum. (The last step of a run in shelf freezing units is to expand a bladder in the vacuum chamber to push the shelves and partially sealed serum bottles to the top of the chamber, thereby seating the rubber stoppers fully into the bottles.)
7. Store freeze-dried units at *ca.* 4°C. Viability will probably be lost sooner if kept at room temperature rather than refrigerated.

### B. Culture recovery

1. Shippers of freeze-dried cultures usually provide detailed directions on how to open and reconstitute such material. See http://www.ars.usda.gov/SP2UserFiles/Place/19070510/lyophil_recovery.pdf for recovery of ARSEF isolates from either flame-sealed ampoules or rubber-capped serum bottles.
2. If an ampoule is not pre-scored, score the neck with a file or diamond pencil. (For capped serum bottles, remove the pre-cut aluminum disk to expose the rubber cap's diaphragm.)

3. Surface sterilize in 70% ethanol or sodium hypochlorite solution (e.g., a 1 : 1 dilution of commercial bleach), wrap the scored ampoule in a sterile paper wipe moistened (but not soaking!) with ethanol and break at the scoring. (For serum bottles, disinfect the exposed rubber seal.)
4. Add sterile water or liquid medium (the type and quantity of liquid will usually be indicated by the culture's sender) to the freeze-dried contents to reconstitute the culture. (For serum bottles, it is easiest to use a sterile, disposable syringe to add water, to resuspend the reconstituted culture, and to transfer the contents to fresh medium.)
5. Rest the material in a sterile hood for 1–30 min to soften and to rehydrate the dried pellet. Resuspend and transfer the reconstituted mixture on to fresh culture medium.
6. Monitor the culture for viability and/or contamination.

## 4. PRESERVATION IN A FROZEN STATE

Standard home-type freezers (operating at *ca.* −20°C) might seem to be an ideal and inexpensive way to keep fungal cultures but can not be recommended for such purposes for entomopathogens even though Carmichael (1962) successfully stored cultures of a wide range of fungi at −20°C. If one wants to use such a freezer, it is important to note that frost-free units should be avoided since ice build-up in them is prevented by periodic heating cycles. The successful long-term preservation of frozen cultures depends strongly on temperature *stability* during storage, but it is also true that the colder the temperature during storage is, the greater the likelihood will be to retain long-term viability.

### A. Silica gel

Storage of spores on sterile, anhydrous silica gel crystals is the only technique for aerobic microbes routinely using −20°C freezers; while such preparations may be stored at room temperature, viability will probably decline sooner and faster. This method is inexpensive, simple, and reliable for many fungi (Smith, 1993) including entomopathogens (Bell & Hamalle, 1974) although fungi with high vacuolar volumes in their cells (e.g., oomycete 'fungi' and entomophthoraleans) are not suitable for storage on silica gel. Fungi may remain viable on silica gel for as long as 25 years (Sharma & Smith, 1999).

A major concern with silica gel storage is to guard against the loosening of caps on storage tubes (a particular risk in frost-free freezers), thus exposing stored fungi to over desiccation and loss of viability.

The use of anhydrous silica gel crystals as a carrier for culture propagules is limited to aerobic bacteria and fungi that grow on solid culture media. Fungi with high ratios of vacuolar to cytoplasmic volumes (e.g., oomycete 'fungi' and Entomophthorales), viruses, microsporidia, and anaerobes are unsuited for storage on silica gel.

*1. Set up for storage*

1. Use sporulating cultures for inoculum.
2. Fill 25 × 200-mm screw-cap glass tubes one-third full with 6–12-mesh, grade 40, uncolored anhydrous silica gel crystals. **Note:** The blue hydration indicator dye in some silica gel products may be toxic to fungi (Perkins, 1962).
3. Sterilize the tubes and silica gel in an oven at 160–180°C for 1–6 h to ensure that the silica gel is both sterile and fully anhydrous.
4. Dispense 1–5 ml sterile water—or an autoclaved solution of 5–7% (v/v) skim milk—into a sporulating culture. Cap and agitate the tube or rub the surface of the plate with a sterile glass rod to suspend the spores.
5. Water uptake by anhydrous silica gel is *strongly exothermic*, and prepared tubes must be well chilled in an ice bath or freezer until inoculation (to avoid killing the culture to be stored).
6. Dispense 1 ml of the suspension by drops on to cold silica gel crystals. Tilt tubes during inoculation to expose the greatest possible surface area and rotate or agitate tubes while adding the inoculum suspension.
7. Hold inoculated tubes at room temperature for a few days, rotating or agitating them periodically until all water has been absorbed and the crystals separated.
8. Store at −20°C (tubes may be kept at room temperature, but this is not preferred).
9. Check viability and sterility of the stored preparation after *ca.* 1–2 weeks.

*2. Culture recovery*

1. Sprinkle a few granules of inoculated silica gel from a tube on to fresh culture medium.
2. *Tightly* reseal tube and return to long-term storage.

## B. Cryogenic preservation in mechanical or liquid nitrogen freezers: general considerations and protocols

The wide presence of ultra-cold freezers in biological laboratories makes cryogenic storage of essential cultures available to many biologists. Culture storage in liquid nitrogen dewars allows both more secure and longer-term storage than in (electric) ultracold freezers but the hardware needed is expensive and depends on a continuous supply of liquid nitrogen that is either unavailable or prohibitively expensive for most laboratories. Material stored in ultracold freezers, however, is vulnerable to power outages in a way that nitrogen dewars are not.

Virtually any culturable fungus that can be stored by any technique noted here can be preserved cryogenically. Further, many fungi that do not tolerate preservation by freeze-drying or some other techniques can be stored cryogenically. However, despite the versatility of cryogenic techniques, there are no universally applicable protocols. Many fungi require specialized handling for cryopreservations, and all cryogenic storage techniques for germplasm have flaws and weaknesses (see Table 10.1). The viability of frozen cultures may be strongly affected by the choices of medium from which preservation is attempted, cryoprotectant, the rate of freezing, and temperature stability during storage. These same factors may also affect the viability of freeze-dried cultures (Tan *et al.*, 1995). The physical factors affecting the freezing, storage, and recovery of living cells demand an awareness of some basic chemical and physical principles that are lucidly discussed by Calcott (1978).

The fate of water in and around cells during freezing, storage, and thawing may be the most critical factor to understand: water freezes in either an amorphous (glassy or vitreous) state or crystalline (icy) state. Cryoprotectants act to favor intracellular water freezing as the glassy rather than icy state (where ice crystal growth can destroy membranes and, thereby, kill cells). Unfortunately, water can convert from a glassy to crystalline state (with an initiation and growth of ice crystals) even while still remaining frozen. This 'devitrification' is favored by fluctuating temperatures (discussed below) and, troublingly, at a few temperatures that are close to those in ultracold mechanical freezers: one of these glass (to ice) transition temperatures for frozen protein solutions is −80°C (Chang & Randall, 1992). Ice nucleation (without much crystal growth) may occur in water–glycerol mixtures at temperatures from −123 to −93°C (Vigier & Vassoille, 1987).

*1. Cryoprotectant*

Many diverse cryoprotectants can help prevent ice crystal formation during freezing, storage, and thawing of cultures. The most widely used cryoprotectants for fungi include glycerol, dimethylsulfoxide (DMSO), polyethylene glycol, and propylene glycol. Most mycological laboratories that freeze fungi use only 10% glycerol because of its simplicity and convenience and reliable effectiveness (see Sanskär & Magalhães, 1994). The ARSEF collection uses 10% glycerol as its sole cryoprotectant for all of its nearly 700 fungal taxa. A few ARSEF cultures of entomophthoraleans were frozen first in DMSO or polydextrose solutions but all later (successful) freezes of these isolates were in glycerol. Failed freezes in glycerol showing no post-preservation viability are attempted again (in glycerol) until younger—or even somewhat older—cultures grown on the same (or a different medium) freeze successfully. Isolates that resist multiple cryopreservation attempts also usually grow poorly in normal cultural conditions and die after very few serial transfers; it is worth noting that all culture collections inevitably lose some cultures no matter how much time or effort is spent in trying to maintain them.

It is the comparatively large volume of cryoprotectant in a cryovial that fixes the temperature at which the vial contents freeze with a strongly exothermic release of the 'heat of fusion'. Electronically controlled cell freezers rapidly reduce the freezing chamber temperature to dissipate this short-term release of heat and then warm the chamber so that the temperature of samples in cryovials continue to drop at the most uniform possible rate (usually *ca.* −1°C per min) throughout a freeze process.

Cryoprotectant solutions should be made and sterilized in small batches (e.g., 100 ml) to minimize contamination risks to every vial filled from that batch.

*2. Cryogenic storage units*

Plastic cryogenic vials, plastic straws, or glass ampoules are used to contain material to be frozen. The choice among these storage units is driven by the type of racking system in which frozen units are stored in a freezer or dewar, by convenience issues in handling the components of the system, by cost, and by safety considerations. Although they are not the least expensive alternative, plastic cryovials (of assorted types and sizes) are probably the most common storage units; polypropylene drinking straws are much less expensive are much less used. Glass ampoules or tubes are rarely used since they are more problematic during the freezing process and can be especially hazardous if they shatter during recovery of stored cultures.

The exclusion of liquid nitrogen from the interior of cryostorage units is critical if units are immersed in liquid nitrogen since the infiltration of liquid nitrogen—which is not sterile—can contaminate the contents with bacteria or fungal spores, but the possible explosion of a vial containing liquid nitrogen being thawed rapidly is a very real physical hazard. Sections of polypropylene drinking straws with heat-sealed ends are an inexpensive alternative to screwcap cryovials (Challen & Elliott, 1986; Stalpers *et al.*, 1987) and may be immersed safely in liquid nitrogen. Cryovials can be protected from nitrogen infiltration during immersion in nitrogen by sheathing entire sets of vials or individual vials in plastic tubing heat-shrunk on to the vials and canes, and sealed by heat-crimping the ends before freezing the material; this approach, however, is expensive, does not allow removal of a single vial from a set, and cultures delicate enough to benefit from such sheathing may not survive being heating while being sealed into the sheaths before freezing.

*3. Auxiliary equipment for culture freezing*

An inexpensive plastic freezing container—Nalgene Cryo 1°C Freezing Container, also known as 'Mr. Frosty'—is a simple device that is partly filled with isopropanol and placed in a −80°C freezer to obtain semi-controlled freezing rates of *ca.* −1°C per min for up to 18 cryovials at a time. This device can successfully preserve entomophthoralean protoplasts (which are among the most difficult forms of all entomopathogenic fungi to freeze in a viable condition). Styrofoam boxes with sides 2.5 cm thick and tops and bottoms 1.25 cm thick can also be used as freezing containers, but these may have much faster freezing rates than with an isopropanol-mediated freeze container or those in electronically controlled (and vastly more expensive) cell freezing devices.

Electronically controlled, programmable cell freezers couple high processing capacities with precisely controlled freezing rates (achieved by balancing cold nitrogen vapor and brief heating inputs) to achieve specified freezing curves. These expensive freezers are practical only for laboratories that frequently process volumes of relatively delicate germplasm whose viability demands precisely controlled freeze rates. McLaughlin *et al.* (1990) found that human sperm frozen in an uncontrolled manner in nitrogen dewars or in an electronically controlled cell freezer retained similar viabilities. The ARSEF collection finds that cryovials of most entomopathogens—but especially conidial or sexual ascomycete taxa—do not need to undergo controlled freezing but can be moved directly from refrigerator into cryostorage and retain high viability when thawed, even many years later.

*4. Freezing and thawing protocols*

*a. Set up for storage*

1. *For cultures on solid media:* dispense sterile cryoprotectant into storage units (vials, straws, etc.) and inoculate with two to four blocks of the culture (1−3 mm on a side) but assure a large excess of cryoprotectant volume to that of the inoculum. *For cultures in liquid media:* add an appropriate volume of *undiluted*, sterile cryoprotectant (e.g., 1 ml glycerol) to 9 ml of culture, and disperse quickly by gentle agitation to minimize osmotic damage to the culture. Aliquot into sterile cryovials.
2. 'Cure' fungus in cryoprotectant at 4°C for 2−48 h to allow uptake of the cryoprotectant. Most ARSEF cultures are cured overnight (*ca.* 12 h) before freezing.
3. *For uncontrolled freezing:* transfer cryovials from refrigerator directly into the storage freezer or dewar. *For controlled (or semi-controlled) freezes:* follow manufacturer instructions; cultures are usually frozen to between −40 and −50°C and then retained at that temperature for 30−120 min before being moved to the storage freezer or dewar.
4. As discussed below, check one unit of all frozen cultures 1−7 days after freezing for viability, growth rate, morphology, etc.

*b. Culture recovery*

1. ***Important safety warning***: when thawing frozen cultures ***always*** use proper personal protective gear—laboratory coat, cryogenic gloves, and a full face mask—and keep unprotected persons 2–3 m away from thawing cultures.
2. ***Important safety Warning***: if vials, ampoules, or straws are immersed in liquid nitrogen, check carefully for any visible fluid (nitrogen) content immediately upon removal from storage. *If motion is detected inside any vial, keep the vial for a few minutes at room temperature* ***under a metal cover*** *(until no liquid nitrogen is detected). Glass or plastic vials holding liquid nitrogen can explode violently when placed in warm water; the resulting debris can cause serious bodily injury and the frozen culture will contaminate the laboratory space.*
3. Thaw all cryogenically stored material directly in 37°C water. Leave units in the warm water *only* until the ice is completely melted.
4. Transfer thawed culture promptly to fresh culture medium by removing agar blocks from cryoprotectant, or pipetting thawed liquid cultures into fresh medium.

*c. Checking viability*

No matter what storage method is used, *always* check the viability and purity of a preserved sample a few days after its preparation. If the test sample is viable and uncontaminated upon recovery, the remaining samples can be kept for long-term storage. If the sample is contaminated, inviable, or fails to meet any expectations, the rest of its lot is also unacceptable and should be discarded *after* a new lot is frozen and confirmed to be acceptable. Alternatively, if a viability check is contaminated or has lost essential phenotypic properties, check the original culture (if it is still available) and pull a second unit from storage; if this second unit is in any way unacceptable, do a wholly new freeze preparation and, when the isolate has been preserved in an acceptable state, discard all unsuccessful storage units.

*d. Stability of temperature during storage*

The stability of temperature during cryostorage may be more important than the actual temperature at which storage occurs. The fewer and smaller the temperature fluctuations during frozen storage, the less likely it will be that ice crystals can form, grow, and damage the preserved cells.

MacFarlane *et al.* (1992) suggest that optimal storage conditions require both the coldest-possible temperature and greatest-possible temperature stability. This suggests that, for optimal cryostorage conditions, vials with living material should be immersed in liquid nitrogen at −196°C; in such a state the issues of nitrogen leakage into cryounits become critical. Storage in a dewar in nitrogen vapor avoids this contamination hazard but, depending on the design and construction of individual dewars, the temperature in the vapor phase ranges from −196°C at the liquid nitrogen surface interface up to *ca.* −120°C at the top of the dewar; some large-capacity dewars designed for vapor-phase storage, however, maintain a temperatures of *ca.* −196°C throughout the entire storage space.

Apart from the storage temperature, the physical racking system in which storage units are arranged within a freezer can also significantly affect the overall stability of temperature for individual stored units. Vials or straws attached to vertical canes stored in open-topped boxes immersed in liquid nitrogen (or in vapor phase) need never be removed from liquid nitrogen except to retrieve the topmost unit on a cane. Storage systems with vertical stacks of covered cryoboxes containing single layered arrays of cryovials are usually shown in cryotank sales literature and are not suitable for immersed storage; nonetheless, such a stacked racking system is not advisable for storing live cells because the necessary way of handling the stacks submits their contents to significant temperature fluxes every time material is added or removed from the stack. Unless the desired box is near the top of a stack, the entire stack may have to be removed from its normal position, placed on top of other stacks to be able to withdraw the desired box and either left there in what may be a comparatively hot location while removing the box, opening and removing material from a box (which needs to be kept in the dewar during this process), the box replaced in the stack, and then the stack returned to its normal position. In stacked box systems, the most temperature-stable positions are those at the top where a stack may be lifted a minimal distance in order to withdraw a box, place it on top of the other stacks, remove its top and the desired storage units from it, and then to close and return the box and stack to their normal storage position. In a worst-case scenario (for boxes at the bottom of the stack), *all* material in a stack might be exposed, however briefly, to temperatures from 70°C up to potentially more than 200°C hotter than the long-term storage temperature *each and every time an individual stack of boxes is retrieved from storage*. Even though each such warming period may only last a few seconds, this may be enough time and temperature change to damage the least robust fungi. If such large temperature fluctuations may seem trivial for such cold material, consider that plunging your hand into boiling water (a temperature change of only *ca.* 70−80°C) even for only a few seconds would not be a trivial or desirable experience, but material in cryostorage can be submitted to these kinds of temperature changes *many* times over years of storage. Despite this rather lurid and imperfect analogy about the effects of brief but large temperature fluctuations, practical experience with cryostorage of living cultures suggests that this sort of handling-induced temperature fluctuation does not routinely (or, at least, obviously) damage the overall recoverability of most cultures.

*5. Comments about specific cryopreservation facilities*

*a. Mechanical freezers*

*Standard freezers (−20°C).* Except for cultures preserved on silica gel (see 4.A above), standard upright or chest freezers or the freezing units of two-door refrigerator−freezer combination units are not recommended for preserving fungal cultures. The freezing units inside single-door refrigerator−freezer appliances may only achieve temperatures of *ca.* −5°C. All of these freezers are too comparatively warm and too unstable to be reliable for germplasm storage. Frost-free freezer units (whose periodic heating cycles maintain the frost-free state) may be particularly unsuitable for culture storage.

*Ultracold freezers (−80 or −120°C).* Many laboratories store cultures in ultracold mechanical freezers. Chest freezers, because they hold their cold air when opened, are more temperature-stable and preferable to upright designs. While ultra-cold freezers may serve well to store many diverse fungi, their manufacturers do recommend attaching liquid nitrogen back-up dewars to maintain the temperature during electrical outages. Further, as was noted above, the temperatures in these freezers are close to those that can pose some risks for long-term viability of frozen cells.

*b. Liquid nitrogen freezers*

Storage in or over liquid nitrogen is the most expensive but the most nearly optimal of all available preservation technologies. To depend on liquid nitrogen-based preservation for cultures requires a serious time and resource commitment and presupposes uninterruptable and affordable access to the required liquid nitrogen. Any laboratory relying on nitrogen storage has made a long-term, expensive commitment to maintain (usually large numbers of) cultures perceived to have a very high intrinsic value.

## 5. RECORD KEEPING FOR STORAGE FACILITIES

Knowing what is stored and where to locate it is of paramount importance as the amount of material being preserved increases. Record keeping often begins with either a card file (for small collections) and then moves to a flat spreadsheet type of file (with information in rows and columns) as collections grow. The use of a relational database to record information allows one-time entry of many types of essential data (e.g., names of organisms, locations from which they came, preferred culture conditions, etc.) that can then be entered automatically in many ways in multiple files and changed globally throughout a database with a single keystroke. The ARSEF collection uses a relational database application that was custom designed nearly 30 years ago, and serves to manage accession information, to track the physical locations of every stored unit in the collection, to track requests for isolates, and to generate catalogs and many types of reports about collection activities.

As collections grow in size, the need to locate a specific item of stored material rapidly increases, and the spatial tracking of items needs to be incorporated when organizing a storage facility. The ARSEF collection's database, for example, tracks the physical location in three dimensional space of each and every cryovial (among the many tens of thousands) by its position on a specific cane in a specific box with a known location in each of several cryogenic dewars.

## 6. OTHER LESS COMMON PRESERVATION TECHNIQUES

Many techniques useful in preserving live fungi are either not widely known or have not been tested with entomopathogenic fungi but might be entirely suitable for entomopathogenic fungi. Some of these as possible alternatives to the techniques discussed above deserve being noted here.

A protocol developed for basidiomycete fungi (Homolka *et al.*, 2010) freezes cultures grown on sterile perlite (a volcanic aluminum-silicate mineral) moistened with a nutrient medium such as wort broth and supplemented with 5% glycerol as a cryoprotectant. The fungus-overgrown perlite/broth/glycerol in cryovials is incubated until the perlite is overgrown by the fungus, and then tubes are frozen in an electronically controlled cell freezer. Such cultures remain viable for at least 10 years and without obvious alteration of major phenotypic or genotypic characters. This technique may not have been tried yet with any of the more fastidious fungal entomopathogens such as *Cordyceps* species, but might work very well for such fungi.

Growing fungal cultures on pieces of sterile filter paper on the surface of an agar plate (Fong *et al.*, 2000) is another simple and inexpensive method that is effective with conidial entomopathogens. After the culture grows to the point of sporulation, the filter paper and culture are removed from the agar medium, allowed to dry naturally in another Petri plate (or in a desiccator), and then cut up, placed into sterilized envelopes, and stored in a plastic bag or box (such as those intended for food storage) in a standard freezer at −20°C. Recovery of the fungus from such storage is by placing a small piece of the dried, frozen culture on to fresh culture medium.

Encapsulation of entomopathogenic fungi in beads using starch and/or alginate is an established formulation procedure to prepare fungi for temporary storage and application, but it has also been suggested that starch encapsulation may be a suitable method for long-term preservation of some fungi (Amsellem *et al.*, 1999). Another sort of technological 're-tasking' is the use of plastic beads manufactured for use in ion exchange filtration procedures as suitable carriers for the adsorption of fungal spores and cryostorage in ultracold freezers (Chandler, 1994; Belkacemi *et al.*, 1997).

## REFERENCES


Amsellem, Z., Zidack, N. K., Quimby, P. C., Jr., & Gressel, J. (1999). Long-term dry preservation of viable mycelial of two mycoherbicidal organisms. *Crop Prot., 18*, 643−649.

Balardin, R. S., & Loch, L. C. (1988). Methods for inoculum production and preservation of *Nomuraea rileyi* (Farlow) Samson. *Summa Phytopathol., 14*, 144−151.

Belkacemi, L., Barton, R. C., & Evans, E. G. V. (1997). Cryopreservation of *Aspergillus fumigatus* stock cultures with a commercial bead system. *Mycoses, 40*, 103−104.

Bell, J. V., & Hamalle, R. J. (1974). Viability and pathogenicity of entomogenous fungi after prolonged storage on silica gel at −20°C. *Can. J. Microbiol., 20*, 639−642.

Burges, H. D. (Ed.). (1998). *Formulation of Microbial Biopesticides: Beneficial Microorganisms, Nematodes and Seed Treatments*. Dordrecht, The Netherlands: Kluwer Academic Publishers, pp. 399.

Calcott, P. H. (1978). *Freezing and Thawing Microbes*. Durham, UK: Meadowfield Press, Ltd, pp. 68.

Carmichael, J. W. (1962). Viability of mold cultures stored at −20°C. *Mycologia, 54*, 432−436.

Castellani, A. (1967). Maintenance and cultivation of common pathogenic fungi of man in sterile distilled water. Further researched. *J. Trop. Med. Hyg., 70*, 181−184.

Cavalcanti, M. A. D. Q. (1991). Viability of Basidiomycotina cultures preserved in mineral oil. *Rev. Latinoam. Microbiol., 32*, 265−268.

Challen, M. P., & Elliott, T. J. (1986). Polypropylene straw ampoules for the storage of microorganisms in liquid nitrogen. *J. Microbiol. Methods, 5*, 11−22.

Chandler, D. (1994). Cryoperservation of fungal spores using porous beads. *Mycol. Res., 98*, 525−526.

Chang, B. S., & Randall, C. S. (1992). Use of subambient thermal analysis to optimize protein lyophilization. *Cryobiology, 29*, 632−656.

Clark, G., & Dick, M. W. (1974). Long-term storage and viability of aquatic Oomycetes. *Trans. Br. Mycol. Soc., 63*, 611−612.

Day, J. G., & McLellan, M. R. (1995). *Cryopreservation and Freeze-Drying Protocols*. Totowa, NJ: Humana Press, pp. 254.

Feng, K. C., Liu, B. L., & Tzeng, Y. M. (2000). *Verticillium lecanii* spore production in solid-state and liquid-state fermentation. *Bioprocess Eng., 23*, 25−29.

Fennell, D. I. (1960). Conservation of fungus cultures. *Bot. Rev., 26*, 80−141.

Figueiredo, M. B., & Pimentel, C. P. V. (1975). Métodos utilizados para conservação de fungos na micoteca da seção de micologia fitopatológica do Instituto Biológico. *Summa Phytopathol., 1*, 299−302.

Fong, Y. K., Anuar, S., Lim, H. P., Tham, F. Y., & Sanderson, F. R. (2000). A modified filter paper technique for long-term preservation of some fungal cultures. *Mycologist, 14*, 127−130.

Gindin, G., Levski, S., Glazer, I., & Soroker, V. (2006). Evaluation of the entomopathogenic fungi *Metarhizium anisopliae* and *Beauveria bassiana* against the red palm weevil *Rhynchophorus ferrugineus*. *Phytoparasitica, 34*, 370−379.

Hartung de Capriles, C., Mata, S., & Middelveen, M. (1989). Preservation of fungi in water (Castellani): 20 years. *Mycopathologia, 106*, 73−80.

Homolka, L., Lisá, L., Eichlerová, I., Valášková, V., & Baldrian, P. (2010). Effect of long-term preservation of basidiomycetes on perlite in liquid nitrogen on their growth, morphological, enzymatic and genetic characteristics. *Fung. Biol., 114*, 929–935.

Hua, L., & Feng, M. G. (2005). Broomcorn millet grain cultures of the entomophthoraean fungus *Zoophthoa radicans*: sporulation capacity and infectivity to *Plutella xylostella*. *Mycol. Res., 109*, 319–325.

López Lastra, C. C., Hajek, A. E., & Humber, R. A. (2002). Comparing methods of preservation for cultures of entomopathogenic fungi. *Can. J. Bot., 80*, 1126–1130.

MacFarlane, D. R., Forsyth, M., & Barton, C. A. (1992). Vitrification and devitrification in cryopreservation. In P. L. Steponkus (Ed.), *Advances in Low-Temperature Biology, Vol. 1* (pp. 221–278). London, UK: JAI Press.

McCabe, D.E., & Soper, R.S. (1985). Preparation of an entomopathogenic fungal insect control agent. U.S. Patent No. 4,530,834 (23 July 1985).

McLaughlin, E. A., Ford, W. C. L., & Hull, M. G. R. (1990). A comparison of the freezing of human semen in uncirculated vapor above liquid nitrogen and in a commercial semi-programmable freezer. *Hum. Reprod., 5*, 724–728.

Onions, A. H. S. (1971). Preservation of fungi. In C. Booth (Ed.), *Methods in Microbiology, Vol. 4* (pp. 113–151). London, UK: Academic Press.

Onions, A. H. S. (1983). Preservation of fungi. In J. E. Smith, D. R. Berry, & B. Kristiansen (Eds.), *The Filamentous Fungi. Fungal Technology, Vol. 4* (pp. 373–390). London, UK: Edward Arnold.

Perkins, D. D. (1962). Preservation of *Neurospora* stock cultures with anhydrous silica gel. *Can. J. Microbiol., 8*, 591–594.

Rombach, M. C., Rombach, G. M., & Roberts, D. W. (1989). Growth *in vitro* of the entomogenous fungi *Hirsutella citriformis* and *Hirsutella versicolor*. *J. Pl. Prot. Trop., 6*, 193–204.

Ryan, M. J., Smith, C., & Jeffries, P. (2000). A decision-based key to determine the most appropriate protocol for the preservation of fungi. *World J. Microbiol. Biotechnol., 16*, 183–186.

Sanskär, B., & Magalhães, B. (1994). Cryopreservation of *Zoophthora radicans* (Zygomycetes, Entomophthorales) in liquid nitrogen. *Cryobiology, 31*, 206–213.

Sharma, B., & Smith, D. (1999). Recovery of fungi after storage for over a quarter of a century. *World J. Microbiol. Biotechnol., 15*, 517–519.

Silva, A. M. M. D., Borba, C. M., & Oliveira, P. C. D. (1994). Viability and morphological alterations of *Paracoccidioides brasiliensis* strains preserved under mineral oil for long periods of time. *Mycoses, 37*, 165–169.

Simione, F. P., & Brown, E. M. (1991). *ATCC Preservation Methods: Freezing and Freeze-Drying*. Rockville, MD: American Type Culture Collection, pp. 42.

Smith, C. (1993). Long-term preservation of test strains (fungus). *Int. Biodeterior. Biodegrad., 31*, 227–230.

Smith, D., & Onions, A. H. S. (1994). The preservation and maintenance of living fungi. In *IMI Technical Handbooks, No. 2. International Mycological Institute* (2nd ed.). Wallingford, UK: CABI Publishing, pp. 122.

Stalpers, J. A., Hoog, G., de & Vlug, Ij. (1987). Improvement of the straw technique for the preservation of fungi in liquid nitrogen. *Mycologia, 79*, 82–89.

Tan, C. S., van Ingen, C. W., Talsma, H., van Miltenburg, J. C., Steffensen, C. L., Vlug, Ij. A., & Stalpers, J. A. (1995). Freeze-drying of fungi: influence of composition and glass transition temperature of the cryoprotectant. *Cryobiology, 32*, 60–67.

Verkleij, F. M., van Amelsvoort, P. A. M., & Smits, P. H. (1992). Control of the pea weevil *Sitona lineatus* L. (Col., Curculionidae) by the entomopathogenic fungus *Metarhizium anisopliae* in field beans. *J. Appl. Entomol., 113*, 183–193.

Vigier, G., & Vassoille, R. (1987). Ice nucleation and crystallization in water–glycerol mixtures. *Cryobiology, 24*, 345–354.

CHAPTER XI

# Research methods for entomopathogenic microsporidia and other protists

LEELLEN F. SOLTER*, JAMES J. BECNEL[†] & JIRI VÁVRA[‡]

*Illinois Natural History Survey, Prairie Research Institute, University of Illinois, 1816 S. Oak Street, Champaign, IL 61820, USA
[†]Center for Medical, Agricultural, and Veterinary Entomology, United States Department of Agriculture, Agricultural Research Service, 1600 S.W. 23rd Drive, Gainesville, Florida, FL 32608, USA
[‡]Institute of Parasitology, Academy of Sciences of the Czech Republic and Department of Parasitology, Faculty of Science, University of South Bohemia, Ceske Budejovice, Czech Republic, and Department of Parasitology, Faculty of Science, Charles University, Prague, Czech Republic

## 1. INTRODUCTION

The insect parasites treated in this chapter are a broadly diverse group of unicellular eukaryotic organisms belonging to the previous conventional taxon Protozoa, believed to represent a stem taxon at the base of multicellular organisms (Whittaker, 1969). In recent classifications of organisms, the existence of Protozoa is either not recognized (Adl *et al.*, 2005, Walker *et al.*, 2011) or exists in a modified form that includes only a fraction of organisms formerly recognized in the taxon (Cavalier-Smith, 2010). Protists representing most of the former Protozoa are now distributed in six kingdoms (or supergroups) to which all eukaryotic organisms on Earth belong. These groups are relatively well supported by molecular phylogeny.

Protist–insect symbiotic relationships reflect the full range of possible interactions, from commensalism, whereby the host simply provides a spatial niche for the life-cycle of the protist, to mutualistic relationships where the protist and host are each essential for the survival of the other, to parasitism, interactions in which the protist is truly pathogenic to the insect host. Although any tissue or organ of insects may be infected, the various groups of protists characteristically infect specific tissues in insect larval stages and adults. Amoebae and flagellates generally target parts of the digestive tract (pharynx, salivary glands, proventriculus, midgut, Malpighian tubules). Apicomplexa, are intracellular and epicellular, usually infecting gut tissues, Malpighian tubules and fat bodies, but also have forms that develop in the hemocoel. Microsporidia are strictly intracellular and their representatives can be tissue specific or systemic, occurring in nearly all body tissues. Insect pathogenic ciliates appear to be restricted to the hemocoel, gut lumen and body surface.

The focus in this chapter will be on those groups of protists that are pathogenic to their insect hosts (Table 11.1), although some basic data necessary for the identification of non-pathogenic taxa are provided. Protists that are vectored by blood feeding insects, especially the flagellates (trypanosomes and their relatives) and Apicomplexa (*Plasmodium* and other blood parasites), although extremely important medical and veterinary pathogens, are beyond the scope of this chapter and are mentioned only briefly. Because the Microsporidia are by far the most widely studied of the entomopathogenic protists, most of the methods and techniques described here were drawn from research on this group.

MANUAL OF TECHNIQUES IN INVERTEBRATE PATHOLOGY
ISBN 9780123868992

**Table 11.1** Orientation table showing simplified evolutionary relationships of insect infecting protists (partly based on data in Adl *et al.*, 2005 and Walker *et al.*, 2011).

| *Kingdom* | *Group* | *Subgroup* | *Lower-group 1* | *Lower-group 2* | *Representative* | *Former taxonomy* |
|---|---|---|---|---|---|---|
| Excavata | Discoba | Euglenozoa | Kinetoplastids | Trypanosomatids | *Trypanosoma* | Sarcomastigophora |
| | Metamonads | Fornicata | Retortomonads | | *Retortomonas* | |
| | | Parabasalids | Trichomonads | | *Pseudotrypanosoma* | |
| | | | Hypermastigids | | *Trichonympha* | |
| | | Preaxostyla | Oxymonads | | *Oxymonas* | |
| Amoebozoa | Amoebae | | | | *Maplhigamoeba* | |
| Chromalveolata | Alveolata | Apicomplexa | Gregarines | Eugregarines | *Gregarina* | Apicomplexa |
| | | | | Neogregarines | *Mattesia* | |
| | | | Coccidia | Adeleid coccidia | *Adelina* | |
| | | | | Haemosporina | *Plasmodium* | |
| | | Ciliophora | | | *Tetrahymena* | Ciliophora |
| Archaeplastida | Viridiplantae | Chlorophyta | Trebouxiophytes | | *Helicosporidium* | *Unknown* |
| Opisthokonta | Fungi-related | Nephridiophagids | | | *Nephridiophaga* | *Unknown* |
| | | Microsporidia | | | *Nosema* | Microspora |

## 2. IDENTIFICATION

### A. Description of entomopathogenic protistan groups

This section provides a brief account of the protists that are commonly found in insects and the hosts in which they occur. Modern classification treatises, based on relatedness of organisms established on molecular and ultrastructural characters, do not use the formal high-order taxon designations, phylum, class, order, suborder, and family, but simply group the organisms in a hierarchical order (Adl *et al.*, 2005, Walker *et al.*, 2011). The reason for this is the recognition of the enormous diversity of protists and the constant influx of new data, making any rigid classification system immediately obsolete. In this chapter we will respect this approach. We treat groups of insect-pathogenic protists supported by molecular phylogeny and structural evidence as independent units, but inform the reader on their evolutionary affinity.

The major protistan entomopathogens and insect symbionts are flagellate organisms (kinetoplastids, metamonads) belonging to the kingdom Excavata; amoebic organisms belonging to the kingdom Amoebozoa; microsporidia, now affiliated with the kingdom Opisthokonta; and organisms of the kingdom Chromalveolata, that move by gliding (Alveolata/Apicomplexa) or by cilia (Alveolata/Ciliophora). Additional sources that characterize the various protistan organisms, although most use traditional taxonomies, include 'Protozoology' (Kudo, 1966), the brief but very informative booklet *How to Know Protozoa* (Jahn & Jahn, 1949), *An Illustrated Guide to the Protozoa* (Lee *et al.*, 2000), *Atlas of Insect Diseases* (Weiser, 1977), *Biological Control of Vectors* (Weiser, 1991), *Insect Pathology* (Tanada & Kaya, 1993; Solter *et al.*, 2012), and *Protistology* (Hausmann *et al.*, 2003). The variety of protistan organisms to be identified preclude a detailed account here of life cycles, some of which are quite complex. Many of these pathogens are host specific, thus the host groups can be a useful starting point for identifying a pathogen. Table 11.2 provides a short key to entomopathogenic protists that enables the researcher to quickly distinguish the major groups. Methods for observing and identifying these organisms are found in Chapter I.

**Table 11.2** Key to primary entomopathogenic protists.

| | | |
|---|---|---|
| A | Motile stages are the prevailing life-cycle form, cysts may be also present | |
| A1 | Motile stages equipped with a flagellum or flagella | Trypanosomastid flagellates<br>Metamonad flagellates |
| A2 | Motile stages with rows of cilia | Ciliates |
| A3 | Motile stages move by cytoplasmic pseudopodia | Amoebae |
| A4 | Motile stages move by gliding (no conspicuous movement organelles present) | Gregarines |
| B | Most or all stages non-motile, developmental stages and spores or cysts are located both inside and outside the cells | |
| B1 | Most conspicuous form is a thick-walled relatively large ($\geq$ 10 μm) gametocyst that contains sporocysts enclosing sporozoites | Gregarines, Coccidia |
| B2 | Ovoid, pyriform, rod-like spores, usually 2–10 μm; vacuole-like area at one (broader) spore pole. A long evaginable filament-like structure can be extruded from the opposite spore pole | Microsporidia |
| B3 | Spores similar in size to B2, but without vacuole or evaginable filament, but with a mid-longitudinal concavity | *Nephridiophaga* |
| B4 | Spores (5–10 μm) are discoidal, contain three cells and a cell in the form of a long spiral | *Helicosporidium* |

### 1. *Kingdom Excavata*

The kingdom Excavata contains flagellates, which are, or ancestrally were, equipped with a ventral feeding groove supported by a microtubular cytoskeleton. In older taxonomic schemes, these organisms were placed in the Phylum Sarcomastigophora with some of the amoebae. Numerous insect symbionts belong to two clades, *Kinetoplastids* and *Metamonads* (see Walker *et al.*, 2011 for definitions). Both groups are further divided into several subgroups: trypanosomatids, representing insect inhabiting kinetoplastids and retortamonads, and parabasalids and oxymonads representing insect dwelling metamonads.

#### *a. Kinetoplastids—Trypanosomatids*

*Diagnostic features.* Kinetoplastids are flagellates characterized by the presence of a kinetoplast, an organelle that stains like the nucleus, but is actually a specific part of the mitochondrion in which large amounts of extranuclear DNA are accumulated. Trypanosomatids are kinetoplastids with a single flagellum and with the kinetoplast situated in the proximity of the basal body (kinetosome) of the flagellum. Size range: 4—15 μm.

*Occurrence.* Trypanosomatids are the most commonly occurring flagellates in insects and are found primarily in the orders Hemiptera and Diptera. They inhabit the gut and Malphigian tubules lumen or attach to epithelial cells of the esophagus. They are more rarely observed in salivary glands or the insect mouthparts, especially in the cases when the flagellate is digenetic (has a second host) and the insect serves as its vector (e.g., *Leishmania, Trypanosoma, Phytomonas*).

*Methods.* Trypanosomatids are observed in fresh microscopic smears of the gut contents, hemolymph, or the salivary glands in insect saline. Giemsa staining of dry or wet smears reveals the nucleus and kinetoplast. Many trypanosomatids can be cultivated, and cultivation is often used for multiplication of material from an insect gut as prelude to molecular biology identification methods (Lee & Soldo, 1992; Westenberger *et al.*, 2004).

*Identification.* Rod-shaped or undulating bodies move rapidly in fresh preparations, also round non-motile stages might be present. Staining reveals a relatively large nucleus and the kinetoplast stained as a dense body located close to the basal body of the single flagellum. Trypanosomatids exist in several structural and developmental morphs and genera are identified by the spatial relationship between the kinetoplast and the nucleus, presence or apparent absence of the flagellum, flagellar form, and the position of its emergence from the cell (Lee, *et al.*, 2000) (Figure 11.1). Some genera have a single morph, but usually there are several morphs in each genus representing different life-cycle stages. Precise determination of many forms requires the use of molecular methods involving RNA and protein coding genes (Podlipaev *et al.*, 2004; Svobodova *et al.*, 2007).

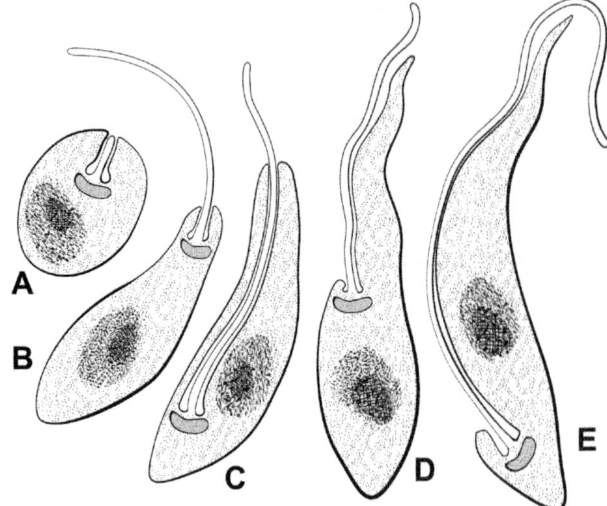


**Figure 11.1** Developmental forms (morphs) of kinetoplastid flagellates. A. amastigote; B. promastigote; C. opisthomastigote; D. epimastigote; E. trypomastigote. (Drawing by K. Helms.)

*Pathogenicity.* Species occurring in the gut are seldom pathogenic and usually act as subpathogenic stressors, unless the multiplication of the flagellates is so extensive that the digestive tract is blocked. Species that penetrate to the salivary glands via hemolymph tend to be more pathogenic to the insect host.

*Major trypanosomatid genera in insects.*

- *Leptomonas*, characterized by the dominant presence of promastigote stages that sometimes occur together with cyst-like stages. Hosts: Hemiptera, Diptera, Hymenoptera, Blattoidea, Lepidoptera, Siphonaptera,

Anoplura, in the digestive tract lumen. Nonpathogenic. Example species: *Leptomonas ctenocephali* from the dog flea *Ctenocephalides canis*; *Leptomonas pyrrhocoris*, a ubiquitous parasite of the hemipteran *Pyrrhocoris apterus*.

- *Herpetomonas*, characterized by the dominant occurrence of promastigotes, occurring together with opisthomastigotes. Hosts: Diptera, in the digestive tube lumen. Nonpathogenic. For example: *Herpetomonas muscarum* from *Musca domestica* and many other flies.
- *Crithidia* have choanomastigote stages in the lumen of the digestive tracts of insects in the orders Diptera, Hemiptera, Trichoptera, and Hymenoptera. Although thought to be generally nonpathogenic, some species, e.g., *Crithidia fasciculata* from various culicine mosquitoes and *Crithidia bombi* in bumble bees, cause sublethal effects (Shykoff & Schmid-Hempel, 1991; Brown *et al.*, 2000). *Crithidia mellificae* is a honey bee gut inhabitant.
- *Walleceina* flagellates have promastigote and choanomastigote stages. Typical is the occurrence of choanomastigote-like stages, but without a free flagellum (endomastigotes). Small genus. *Wallaceina inconstans* typically occurs in the gut of the capsid or mirid bug *Calocoris sexguttatus*.
- *Blastocrithidia* has epimastigote stages and cyst-like stages. Hosts: Diptera, Hemiptera, Siphonaptera, typically located in the digestive tube lumen. They are pathogenic in some species. For example, *Blastocrithidia triatomae* in kissing bugs, *Triatoma* spp.
- *Phytomonas* has promastigote stages. These are plant pathogens, living in plant vascular systems, with insects serving as vectors. Transmission occurs when the flagellate invades the gut lumen, then the salivary glands of the insect hosts. This migration suggests some degree of pathogenicity to the insect hosts (Hemiptera and some Diptera), e.g. *Phytomonas davidi* from various Hemiptera feeding on plants in the family Euphorbiaceae.
- *Leishmania* has promastigote stages in the digestive tract of sandfly adults and amastigote stages in vertebrate macrophages, causing human and animal leishmaniasis. Although they apparently live as nonpathogenic commensals in their insect hosts, their attachment to the stomodeal valve may hinder the feeding of the insect, e.g., *Leishmania* spp. in the gut of *Phlebotomus* spp. Phlebotomine Diptera are hosts of several other trypanosomatids and serve as vectors to leishmania-related organisms including the genera *Sauroleishmania* and *Endotrypanum*.
- *Rhynchoidomonas* has mostly trypomastigote stages; the flagellum, however does not form a prominent undulating membrane as in *Trypanosoma* spp. Epimastigotes and amastigotes also occur, e.g., *Rhynchoidomonas lucilliae* in Malpighian tubules of calliphorid and muscoid flies.
- *Trypanosoma* occurs as trypomastigotes in vertebrate tissues (typically blood). Trypomastigote, epimastigote and amastigote stages occur in the digestive tract, salivary glands and mouthparts of insects. Diptera, Hemiptera and Siphonaptera serve as vectors of some species to man and animals, causing trypanosomiasis. Pathogenicity to insects varies and depends on the lifecycle of the flagellate in the host. Some of these flagellates live as non-pathogenic commensals in the digestive tract of their insect hosts (e.g., *Trypanosoma cruzi* in the gut lumen as epimastigotes and in the rectum as trypomastigotes of reduviid bugs), while some penetrate from the gut to the hemocoel and the salivary glands (e.g., *Trypanosoma rangeli* in reduviid bugs) and are thus pathogenic to the insect vector. A number of insects in dipteran families (e.g., Culicidae, Simuliidae, Tabanidae, Muscidae, Glossinidae, Hippoboscidae) serve as insect hosts and vectors of the genus *Trypanosoma*. Development in the insect host may be either very complex, involving different parts of the digestive system and salivary glands (e.g., *T. brucei* s.l. in *Glossina* spp.), or limited to survival in the proboscis of the insect vector, enabling mechanical transmission (e.g., *T. evansi* in horse flies and *Stomoxys* flies). No reliable data are available concerning the pathogenicity of these flagellates for their insect hosts.

Molecular phylogeny shows that several of these genera are polyphyletic and group with unrelated organisms. Several new genera (e.g., *Sergeia, Angomonas, Strigomonas*) have been erected, based on RNA, protein coding genes and the presence of bacterial symbionts in the cell to rectify this situation (see Podlipaev *et al.,* 2004; Svobodova *et al.*, 2007; Teixeira *et al.*, 2011).

### *b. Metamonads*

The main unifying character of metamonad flagellates, in addition to some structural characters, is the lack of typical mitochondria. These organisms live in microxic or anoxic insect gut environments. Several phylogenetically supported subgroups occur in insects, the retortamonads, parabasalids and oxymonads, differing mostly in structural details of their cytoskeleton and flagellar apparatus.

- *Retortamonads* are representative of the Metamonada/Fornicates subgroup of Excavates

*Diagnostic features.* Small, uninucleate flagellates, 6–14 μm, with two or four flagella. One flagellum is directed backwards and moves in a groove-like depression situated in the anterior part of the body representing a cytostome. The cytostome depression is bordered by two characteristic fibrils. Cysts are broadly pyriform (pear-like) in shape.

*Occurrence.* Hindgut lumen of insects such as *Tipula* spp. larvae, *Phyllophaga* spp., *Gryllotalpa* spp., *Blatta orientalis* and *Amitermes* spp..

*Methods.* Fresh preparation of the gut contents, dry smear stained with Giemsa, wet smear stained with Heidenhain's iron hematoxylin, or Protargol silver staining (Nie, 1950; Lee & Soldo, 1992).

*Identification.* Two flagella are present in those species inhabiting the insect gut. The wet smear staining mentioned above will reveal the flagella and fibrils bordering the cytostome. The nucleus, both flagella and cytostomal fibrils, are stained in cysts.

*Pathogenicity.* Nonpathogenic to their hosts.

- *Parabasalids*, a large and important subgroup of the Metamonada

Parabasalids are structurally variable protists occurring in insect gut (mostly lower termites and roaches); all possess a parabasal apparatus represented by an enormously developed, sometimes multiplied Golgi apparatus, connected by fibers to flagellar kinetosomes. Some forms possess a non-contractile microtubular rod (axostyle). The parabasal apparatus, basal bodies, nucleus and axostyle form a unit called karyomastigont. Canonical mitochondria are missing (these flagellates live under anoxic conditions) and are replaced by a derived organelle, the hydrogenosome. Parabasalian flagellates are most simply divided into two lower groups: Trichomonads and hypermastigids, although detailed phylogeny investigations suggest the existence of several other phylogenetic units (Cepicka *et al.*, 2010).

- Trichomonads

*Diagnostic features.* Medium-size protists, usually with a single nucleus and four to six flagella (insect forms), one of which is a trailing flagellum attached to the cell surface that may form an undulating membrane. The cell contains an axostyle. Some insect-inhabiting members of this order have cells with several karyomastigonts: many nuclei, many flagella and many individual axostyles joining into an intracellular stem-like structure.

*Occurrence.* Many representatives of this order are parasites of vertebrates, but a number of genera occur also in the gut of termites.

*Methods.* Fresh preparation of gut contents, wet smear of gut contents stained with Heidenhain's iron hematoxylin or Protargol silver staining.

*Identification.* See Lee *et al.* (2000).

*Pathogenicity.* Harmless commensals.

*Representatives. Pseudotrypanosoma*, *Hexamastix*, *Tricercomitus*, *Calonympha*.

- Hypermastigids

*Diagnostic features.* Highly organized, multiflagellated but uninucleate. Flagella are distributed as either an anterior tuft, one or more plates or in spiral rows. Staining often reveal highly developed and organized parabasal bodies (actually Golgi apparatus) arranged in proximity of basal bodies of flagella. Axostyle usually present, sometimes in the form of axostylar fibers; missing in several genera.

*Occurrence.* Mutualistic symbionts living in the guts of lower termites, and roaches (particularly woodroaches).

*Methods.* Fresh preparation of gut contents, wet smear of gut contents stained with Heidenhain's iron hematoxylin or Protargol silver staining.

*Identification.* See Lee *et al.* (2000).

*Pathogenicity.* None. These flagellates are mutualistic symbionts, facilitating host digestion of cellulose; some hosts can not survive without their presence.

*Representatives. Trichonympha, Spirotrichonympha, Lophomonas* in colon of omnivorous roaches in the genera *Blatta, Periplaneta* and *Blatella*.

- Oxymonads

*Diagnostic features.* Small to large in size (10−200 μm), uninucleate to exceptionally multinucleate, flagellates, usually with four flagella arranged in two pairs. Either three flagella are oriented forward, the fourth one serving as trailing flagellum and partly adhering to the cell, or all four flagella are directed backwards adhering for some length to the cell surface. There are no mitochondria or Golgi. A microtubular rod, the axostyle, is situated lengthwise through the body of the flagellate. Many forms have developed adaptations for existence inside the host: a prominent extensible rostellum, which is the holdfast organelle attaching the flagellate to the host gut, and a contractile axostyle facilitating the movement in dense gut contents.

*Occurrence.* Most oxymonads occur in the gut of wood-eating insects (termites and woodroaches) and also in guts of xylophagous insect larvae (e.g., *Melolontha* spp., *Cetonia* spp.) and in *Tipula* spp. larvae.

*Methods.* Fresh preparation of gut contents, wet smear stained with Heidenhain's iron hematoxylin or Protargol silver staining.

*Identification.* See Lee *et al.* (2000).

*Pathogenicity.* Harmless mutualists and endocommensals.

*Representatives. Oxymonas* spp. in termites and the wood roach *Cryptocercus, Monocercomonoides* in *Tipula* larvae.

### *2. Kingdom Amoebozoa*

*Diagnostic features.* Movement by tubular pseudopodia. Amoebozoa are probably monophyletic in origin, their division into lower groups is problematic because some have only amoebic forms, others may develop resistant cysts, and some possess flagellated forms during parts of their life-cycles.

*Occurrence.* Most amoebae are free-living, many are parasitic in various invertebrate and vertebrate hosts, a few live in the gut or Malpighian tubules of insects.

*Methods.* Fresh preparation of gut contents, tissue smears of Malpighian tubules in insect saline. Dry smears stained with Giemsa, wet smears stained with hematoxylin stains.

*Identification.* Very difficult, involving the type of pseudopodia formation, the structure of the nucleus after staining of wet smear, ability to form cysts and flagellate stages, and the infected host. See Lee *et al.* (2000).

*Pathogenicity and representatives.* Some amoebae are harmless commensals in the gut of insects (e.g., *Endamoeba blattae* in cockroaches and several species of the same genus in gut of termites). Only three genera, *Malamoeba* from Orthoptera, *Malpighamoeba* from *Apis mellifera* (Hymenoptera), and *Malpighiella* from the dog flea (*Ctenocephalides canis*) are known to be pathogenic in insects. They have amoebic forms and cysts in the lumen of the gut and Malpighian tubules; e.g., *Malamoeba locustae* from the Malpighian tubules of grasshoppers of several genera. Massive numbers of cysts in the Malpighian tubules may cause stunted growth and death of the host.

### *3. Kingdom Chromalveolata*

Chromalveolata is a vast supergroup of morphologically variable protists united phylogenetically and evaluated recently using several slightly differing concepts (Chromista—Cavalier-Smith 2010; part of the SAR clade—Walker *et al.*, 2011). Numerous insect pathogens belong to the Alveolata, a large chromalveolata clade, and are characterized ultrastructurally by a plasmalemma subtended by flattened vesicles called cortical alveoli. Insect pathogens occur in two monophyletic subgroups of Alveolata, the Apicomplexa and Ciliophora.

#### *a. Apicomplexa*

All members of the Apicomplexa clade are parasitic. The diagnostic feature of these protists is the 'apical complex', a complex of organelles present in motile, infective stages, the sporozoite. The apical complex consists of a conoid (a cone of spirally wound microtubules), rhoptries and micronemes (tear-shaped secretory gland-like formations) that aid entry into host cells. All Apicomplexa are totally or partly intracellular or epicellular during at least a part of their life-cycle. Extracellular stages move by gliding. Apicomplexan parasites are most often and most easily identified, however, by their resistant oocysts that contain a specific number of sporozoites. The oocysts are oval or lemon-shaped refractile bodies with thick walls, most often containing another cyst-like body (or bodies), the sporocyst. Filiform or banana shaped, uninucleate sporozoites are present in different, often diagnostic, numbers inside the oocyst. A new host is infected when an oocyst is ingested and sporozoites escape into the digestive tract.

Many representatives of the Apicomplexa clade occur in insects, the gregarines and coccidians being the most important. The distinguishing characters between these two groups are the size of gamonts (sexual stages)—large in gregarines, small in coccidians; and the modification of the apical complex to form a conspicuous feeding and holdfast organelle (epimerite or mucron) in the trophic stages in gregarines. The apical complex of coccidians is not modified or only slightly modified in stages where it exists.

- Eugregarines

*Diagnostic features.* Most eugregarines are epicellular parasites, with the anterior epimerite (in segmented eugregarines) or mucron (in non-segmented eugregarines) embedded in a host cell (Figure 11.2A). Some gregarines are fully intracellular, at least for part of the life-cycle. The epimerites and mucrons have a wide range of forms from a simple, button-like type to very complicated structures with protrusions in the form of hairs, hooks etc., interdigitating with the plasmalemma of the host cell. The form of the epimerite/mucron, the body size of the eugregarine, the location within the host, and the host species are major taxonomic characters used in eugregarine classification. Eugregarines are often observed at the mature trophozoite stage having detached from the host gut epithelium and holdfast resorbed, and swimming freely in the gut lumen. The trophozoite stage is a long oval or filiform cell, usually in the range of several tens or even hundreds of microns, moving by slow, gliding movement without any change in body shape. The trophozoite of some species is divided by a transverse septum (septate gregarines) with an anterior part (protomerite) and the nucleus is situated in the posterior part (the deutomerite). Eugregarines are frequently observed in syzygy (Figure 11.2B), a prelude to gamete formation in which two individuals are attached either tailwise or sidewise. Another characteristic stage of eugregarines is the gametocyst, a large spherical formation in which two gregarines in syzygy encyst and later transform into numerous gametes. After gamete formation the mature gametocyst contains several hundreds of lemon shaped oocysts, the shape of which is also diagnostic. Oocysts are released from the gametocyst by rupture of the cyst wall (Figure 11.2C) or are evacuated by means of evaginable tubes, the sporoducts.

*Occurrence.* Eugregarines are parasites of invertebrates and their presence is expected in all insect orders. They are usually restricted to the gut although some species live in Malpighian tubules or in the coelomic cavity.

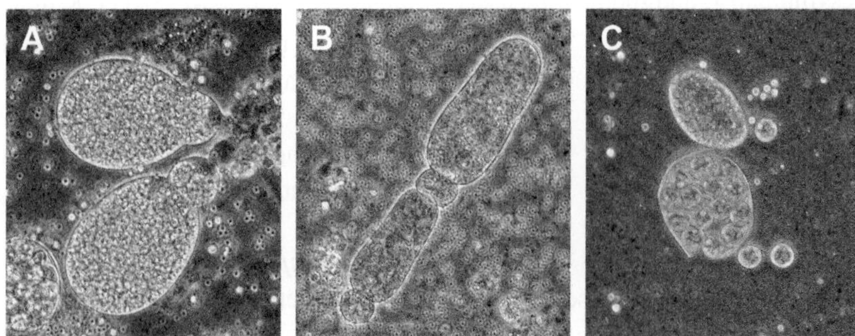


**Figure 11.2** Eugregarines isolated from an embiopteran. A. Trophozoites with epimerite embedded in host tissues. B. Trophozoites in syzygy, in preparation for forming gametocyst. C. Oocytes released from gametocyst. (Photos by LFS.)

**Table 11.3** Key to the families of eugregarines.

| | | |
|---|---|---|
| 1a | Round gametocysts in body cavity | Diplocystidae |
| 1b | Trophozoites and gametocysts in the gut | 2 |
| 2a | Epimerite conspicuous; sometimes structurally modified | 3 |
| 2b | Epimerite is structurally simple | 4 |
| 3a | Gametocysts with one or several sporoducts | Gregarinidae |
| 3b | Gametocyst without sporoducts, open by rupture | Hirmocystidae |
| 4a | Gametocyst with a two-layer envelope, with large, cup-shaped residual body. Oocysts evacuated first as a cyst like structure, later being dispersed in chains | Stylocephalidae |
| 4b | Gametocyst with a one-layer envelope, open by rupture, oocysts not in chains | Actinocephalidae |

(Modified from Weiser, 1966.)

*Methods.* Fresh preparation of gut contents and body cavity fluids. Wet smear stained with hematoxylin stains will reveal single nucleus in the posterior segment of the gregarine body and the septum, when present. Epimerites or mucrons are rarely seen in permanent preparations because they frequently remain embedded in host tissues during preparation. Staining sometimes reveals fine longitudinal ribs on the eugregarine pellicle. These are epicytar folds thought to be involved in movement or feeding. Sporozoites within the oocyst can be revealed by Giemsa staining after acid hydrolysis (Weiser, 1976) or by staining with hematoxylin stains (e.g., Heidenhain's, see Chapter XV).

*Identification.* See Lee *et al.* (2000). Table 11.3 is a simple key for determination of the primary eugregarine families from insects.

*Pathogenicity.* Because eugregarines do not divide as trophozoites—the only divisions occur during formation of oocysts (which often takes place outside of the host)—their pathogenic effect on the host is minimal, although extremely heavy infections may cause gut blockage. However, species that live in Malpighian tubules and in the coelomic cavity may cause some pathogenic effects (Lantova *et al.*, 2010). There are also reports of harmful metabolic effects of eugregarines on their hosts (Schilder & Marden, 2007).

*Representatives. Gregarina* spp. in the gut of the mealworm, *Tenebrio*; *Ascogregarina* spp. in gut and Malpighian tubules of treehole and other mosquito species, e.g., *Ascogregarina taiwanensis* in *Aedes albopictus*.

- Neogregarines

*Diagnostic features.* Morphologically dissimilar to eugregarines, usually smaller and having, with the exception of a single genus, a non-segmented body. Reproduction is more extensive than in the eugregarines, involving merogony (= schizogony) during the vegetative part of the life-cycle (Figure 11.3) and another merogony during gamogony and sporogony.

*Occurrence.* Neogregarines occur as epicellular, and frequently as intracellular parasites in various organs in the hemocoel, Malpighian tubules, intestines, and the fat body of insects, their most typical hosts.

*Methods.* Dry smears stained with Giemsa. Wet smears stained with hematoxylin stains. These reveal uninucleate trophonts, multinucleate meronts, merozoites with single nuclei, gametocysts and oocysts. Sporozoites within the oocyst can be revealed by Giemsa staining after an acid hydrolysis (see Chapter XV) or by staining with hematoxylin stains (e.g., Heidenhain's).

*Identification.* Common genera of neogregarines are listed in Table 11.4.

*Pathogenicity.* Because there is extensive multiplication of the parasite in the host, neogregarines are more

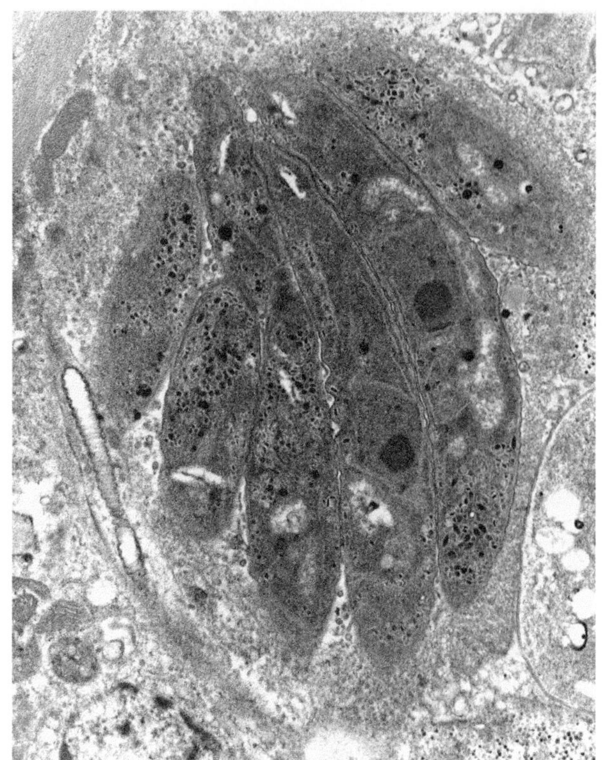

**Figure 11.3** Developing merozoites of *Mattesia oryzaephili, a grain beetle parasite.* (TEM micrograph courtesy of Jeffrey Lord.)

pathogenic than the eugregarines and the infection is often lethal.

*Representatives. Mattesia dispora* in the fat body of the flour moth *Ephestia kuehniella; Farinocystis tribolii* in the fat body of the mealworm *Tribolium molitor.*

*b. Coccidia*

Coccidia are all intracellular parasites and multiply extensively during both the vegetative and sporogonic phases of the life-cycle. Most species are host specific. Almost any tissue can be infected, but the infection sites of individual species can be quite specific. Insect-infecting coccidia belong to two groups, adeleid coccidia and Haemosporina.

- *Adeleid coccidia*

*Diagnostic features.* Intracellular parasites multiply in host cells by merogony, forming multinucleate meronts and uninucleate merozoites. Sexual stages (gamonts) are formed in relatively small numbers and may occur in long-lasting syzygy, the initial stage of zygote formation, the male gamont being much smaller and adhering to the much larger female gamont. Round or oval oocysts with a specific number of sporocysts and sporozoites are formed as a final stage of development.

*Occurrence.* Various invertebrates; most representatives are found in insects. Most genera of *Adelina*-like monoxenous coccidians infect only insect hosts (Table 11.5); several genera are dixenous parasites of vertebrates with an invertebrate (arthropods, insects in some cases) as a primary host for the sexual development and formation of sporozoites (*Hepatozoon/Haemogregarina*-like coccidia).

*Methods.* Dry smears stained with Giemsa, wet smears stained with hematoxylin stains. Sporozoites within the oocysts are revealed by Giemsa staining after acid hydrolysis (see Methods) or by staining with hematoxylin stains (e.g., Heidenhain's).

*Identification.* The following genera are partly or entirely entomogenous: *Adelina*, *Chagasella*, *Legerella*, *Ithania,* and *Rasajeyna*. The diagnostic stage is the oocyst, a spherical or sub-spherical thick walled body, approximately 10–50 μm in size, containing spherical or ellipsoidal sporocysts, which, in turn, contain sporozoites. One genus has oocysts with no sporocysts and with naked sporozoites inside. The shape and size of the oocyst, the number of sporocysts it contains, and the number of sporozoites within each sporocyst are diagnostic markers (Table 11.5).

*Pathogenicity.* Pathogenic effects are more pronounced as more multiplication cycles take place in the insect host, resulting in increased mortality of the host or its greater sensitivity to insecticides. No reliable data exist on pathogenic effects on insect hosts of the *Hepatozoon/ Haemogregarina*-like coccidia.

- *Haemosporina*

*Diagnostic features.* Coccidian parasites using insects as vectors. In the insect host only the sexual portion of life-cycle takes place and, in contrast to the adeleid coccidia, does not involve the syzygy of gamonts. Oocysts formed in the gut tissues are thin walled and have no sporocysts. When mature, the oocyst breaks open, releasing filiform

**Table 11.4** Key to the most common genera of neogregarines infecting insects.

| | | |
|---|---|---|
| 1a | Forms having two merogonies in the life-cycle with different size of nuclei ('micronuclear' and 'macronuclear' merogony). Ophryocystidae | 2 |
| 1b | One merogony, with relatively large nuclei present, trophozoites band-like and wide. Schizocystidae | 7 |
| 2a | Meronts of the first and second merogonies are morphologically different. Gametocytes at the early syzygy are uninucleate. Ophryocystinae | 3 |
| 2b | Meronts of the first and second merogony are morphologically similar, but are multinucleate. Gametocytes at early syzygy multinucleate. Machadoellinae | 6 |
| 3a | Trophozoites and merozoites with root-like extensions attaching the parasite to host cells, gametocyst with single oocyst | *Ophryocystis* |
| 3b | Trophozoites and merozoites without root-like extensions, gametocyst with several oocysts | 4 |
| 4a | One or two oocysts in a gametocyst | *Mattesia* |
| 4b | Gametocyst with more than two oocysts | 5 |
| 5a | Maximum 40 oocysts per gametocyst | *Menzbieria* |
| 5b | >40, but usually 200–300 oocysts per gametocyst | *Lipocystis* |
| 6a | Trophozoites worm-like, gametocytes in syzygy usually with four nuclei, gametocyst with four oocysts | *Machadoella* |
| 6b | Trophozoites round in shape, gametocytes in syzygy with eight nuclei, ~30 spores in a gametocyst | *Farinocystis* |
| 7a | Oocysts egg-like. Caulleryellinae | 8 |
| 7b | Oocysts naviculate | 9 |
| 8a | One oocyst per gametocyst | *Tipulocystis* |
| 8b | Eight oocysts per gametocyst | *Caulleryella* |
| 9a | Trophozoites round or oval in shape | 10 |
| 9b | Trophozoites long and band-like, disintegrating into groups of merozoites | *Schizocystis* |
| 10a | Oocysts naviculate, with four spines at each pole, more than 100 oocysts per gametocyst | *Syncystis* |
| 10b | Oocysts naviculate to egg-shaped, without spines, 16 oocysts per gametocyst | *Lipotropha* |

(Modified from Weiser, 1966.)

sporozoites that migrate to the salivary glands of the insect.

*Occurrence.* In gut tissues, haemolymph and salivary glands of blood sucking insects of the order Diptera.

*Methods.* Dry smears stained with Giemsa, wet smears stained with Giemsa or hematoxylin stains. Oocysts can be seen on the surface of gut upon dissection (Lee & Soldo, 1992).

*Pathogenicity.* Shortened lifespan of the female host.

*Representatives.* Haemosporina are extremely important pathogens, causing serious human and animal diseases. Mosquitoes transmit human and animal malaria caused

**Table 11.5** Genera of known important monoxenous coccidia infecting insects.

| *Genus* | *Oocyst shape* | *No. sporocysts per oocyst* | *No. sporozoites per sporocyst* |
|---|---|---|---|
| *Adelina* | Subspherical–spherical | 3–30 | 2 |
| *Chagasella* | Ovoid | 3 | 4–6 or more |
| *Legerella* | Spherical | 0 | 15–40 |
| *Ithania* | Spherical | 1–4 | 9–33 |
| *Rasajeyna* | Double-layer wall | Up to 18 | 1 |
| *Barrouxia* | Spherical | Many | 1 |

by the genus *Plasmodium*. Hippoboscids, midges and *Chrysops* are vectors of *Haemoproteus* bird parasites, and black flies and ceratopogonid midges transmit the bird parasite *Leucocytozoon*.

### *c. Ciliophora*

*Diagnostic features.* Ciliates are characterized as organisms propelled by rows of cilia and possessing two types of nuclei: a large macronucleus involved in vegetative functions of the organism, and a small micronucleus involved in sexuality. Beneath the plasmalemma, the cell of a ciliate contains flattened, membrane-bound alveoli and a complex system of microtubules and fibers called the infraciliature, a major taxonomic character of ciliates (Lee *et al.*, 2000).

*Occurrence.* Most ciliates are free-living organisms, but many parasitic forms exist. Only a few genera infect insects as endoparasites. *Lambornella clarki* and *Tetrahymena* spp. occur in the body cavity of Diptera (Culicidae, Chironomidae and Simuliidae). Parasitism is identified by the presence of actively swimming organisms in the hemolymph of the host. These parasitic ciliates also have free-living infectious stages that actively seek hosts in the larval habitat. *Nyctotherus* and *Clevelandella* occur as intestinal parasites of roaches. Both these ciliate genera are characterized by a buccal region with a row of membranelles, this region being prominent and anteriorly located in the cells of *Nyctotherus* and less prominent and posteriorly located in the cells of *Clevelandella*. Many more ciliate genera and species occur as epibionts on insect larvae and adults in aquatic environments.

*Methods.* Ciliate nuclei are revealed with a wide variety of stains (hematoxylin, Giemsa, etc.). Infraciliature is revealed by silver impregnation techniques reviewed by Foissner (1991). Useful information on ciliate surface structures may be obtained by simply mixing 2–4% aqueous nigrosine or 10% aqueous opal blue in 1 : 1 proportion with thick suspension of ciliates, smearing and drying ('relief staining').

*Identification.* Classification of ciliates is based on infraciliature organization, revealed to some extent by silver staining, primarily in the area around the cytosome and in most detail by transmission electron microscopy (TEM).

Representatives of several genera of ciliates live attached to the surface of water-dwelling insects or their larvae. Most of these epibiotic ciliates belong to the Peritrichida/Sessilina group, are bell or goblet shaped with a ciliary belt at the apical end of body, and are attached to the substratum at the posterior end of body, usually by a stalk. These ciliates use the insect body merely as an attachment site as they feed on bacteria and small particulate food brought to the cytosome by whirling of the cilia. Although many of these ciliates are specifically adapted to live on certain parts of insect bodies, they are not pathogenic. However, in certain situations some peritrichous ciliates can cover aquatic insects (e.g., moribund mosquito larvae) with a dense mat of cells growing on the cuticle and effectively hinder movement. This is an abnormal situation indicating poor-quality water (excess bacteria).

*Pathogenicity.* *Lambornella* causes decreased survivorship of mosquito adults and parasitic castration of its female hosts. *Tetrahymena* spp. are pathogenic when ciliate numbers in hemolymph are high (Egerter & Anderson, 1985). Occasionally, a dead insect larva is observed to be full of actively moving ciliates. Usually, these ciliates are not the pathogens that caused the death of the host, but are feeding as scavengers on decaying host tissues and bacteria. Histiophagous ciliates of the genera *Tetrahymena* and *Ophryoglena* are often found in dead insect larvae.

*Representatives.* *Lambornella clarki* in the tree-hole mosquito *Aedes sierrensis*, *Tetrahymena* spp. (e.g., *Tetrahymena chironomi* in midge larvae), *Nyctotherus ovalis* in the hindgut of roaches *Periplaneta* and *Blatta*, *Epistylis* spp. attached to the surface of mosquito larvae.

### *4. Kingdom Archaeplastida*

Organisms in this kingdom, including both microscopic green algae and terrestrial plants, are better known as Plantae. By rule, these organisms contain plastids, however in many parasitic forms the plastids are reduced and do not contain photosynthetic pigments. There is only one insect-infecting representative of this kingdom.

- *Helicosporidium* spp.: pathogenic green algae

*Helicosporidium* belongs to the Trebouxiophytes group of algae, which includes many common green algae known from aquatic or terrestrial habitats (e.g., *Chlorella*). Despite their 'green affinity', *Helicosporidium* spp. lack photosynthetic pigments but possess genes of a reduced plastid. This genus forms infective spores consisting of more than one cell.

*Diagnostic features.* Discoidal spores are composed of three centrally stacked round cells (Figure 11.4A,C) surrounded by three to four coils of another cell in the form of a long filament (Figure 11.4B). All four cells are packaged in a smooth, somewhat transparent shell (Figure 11.4C). Spores sizes range from 5–9 μm in diameter, with a thickness of approximately 1/2 to 3/4 the diameter. Upon ingestion, the spore dehisces (Figure 11.4B), the coiled spring-like cell straightens and appears to break the outer shell (Kellen & Lindegren, 1974). The spring cell is equipped with barbs at the penetrating tip, which allow it to penetrate through the gut into the hemocoel and tissues of the host where reproduction and formation of new spores takes place (Bläske-Lietze *et al.*, 2006).

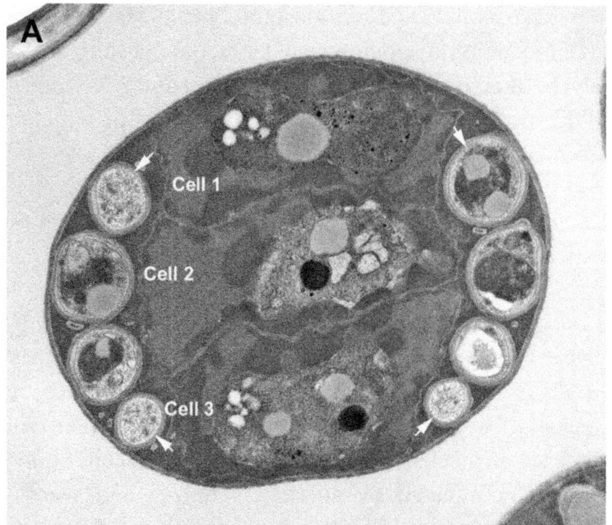


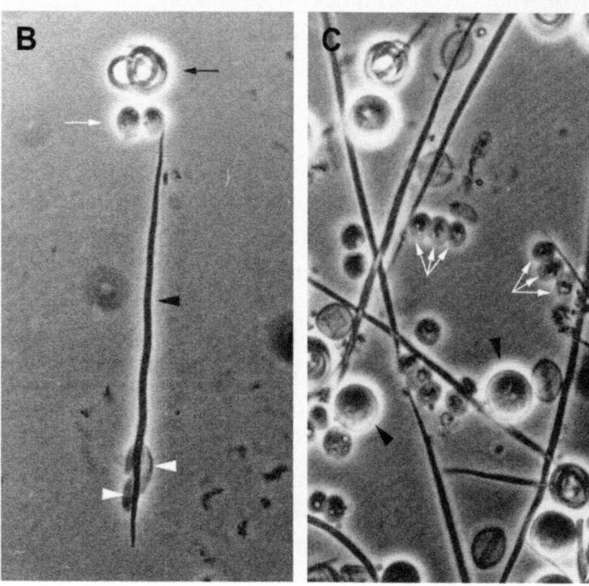


**Figure 11.4** *Helicosporidium* sp. A. Transmission electron micrograph showing three stacked cells and coils of the infective, spear-like cell (arrows). B. Germinating *Helicosporidium* sp. Dehiscing spore (black arrow) consists of the forms observed below, including two of the three cells of unknown function (white arrow) and an infective spear-like spring cell (black arrowhead). Remains of the spore coat cling to the infective cell (white arrowhead). C. Intact spores (black arrowheads) and germinated spring cells with three round cells (white arrows) released per spore. (Photos by JJB & LFS.)

*Helicosporidium* from mosquito and black fly larvae are infectious to lepidopteran larvae, suggesting a low degree of physiological host specificity (Avery & Undeen 1987; Boucias *et al.*, 2001). Common sites of infection are muscle, fat body cells and hemolymph.

*Occurrence.* Culicidae, Simuliidae, Lepidoptera, Collembola, and Coleoptera.

*Methods. Helicosporidium* spp. are readily identified by phase contrast microscopy of squash preparations from living or dead hosts. Giemsa staining differentiates the nuclei and cytoplasm of developmental stages and of the three stacked cells but not of the filamentous cell. Spores can be germinated by alternately drying and wetting them. The inner cells quickly lyse in water but the coiled cell persists, looking much like a small roundworm. The empty shell is oval and the site of breakage has inward turning edges.

*Identification.* The spores are characteristic and are not easily mistaken for any other protists, particularly after they have been germinated or crushed under a coverslip, revealing the coiled filamentous cell.

*Pathogenicity.* The degree of pathogenicity appears to be variable. *Helicosporidium* readily kills mosquito and lepidopteran larvae, even though these might be atypical hosts (Avery & Undeen, 1987).

*Representative. Helicosporidium parasiticum* in *Dasyhelea obscura* (Ceratopogonidae).

### 5. Kingdom Opisthokonta

Kingdom Opisthokonta includes animals, fungi, and phylogenetically related protists. Opisthokonts may be flagellated or non-flagellated; some phylogenetically younger groups apparently lost the flagellum. Flagellated forms often have a single, posterior flagellum, which may be the ancestral type of flagellation. Two non-flagellated groups of Opisthokonts are believed to be related to Fungi, which have also lost their flagella.

#### a. Nephridiophagids

*Nephridiophaga* spp. are non-flagellated protists of uncertain phylogeny but are believed to be related (with weak support) to the zygomycetous fungi (Wylezich *et al.*, 2004). They live in the Malpighian tubules of insects, where multinucleate sporogonic plasmodia give rise endogenously to thick-walled chitinous spores.

*Diagnostic features.* Organisms occurring as small multinuclear plasmodia, later undergoing differentiation of nuclei into generative nuclei and somatic nuclei, the latter remaining unchanged in the plasmodium. A thick spore wall without a polar filament or spore lid is formed around each generative nucleus. The spores are 5–8 μm in size and have a mid-longitudinal concavity.

*Occurrence.* Roaches, Coleoptera, Dermaptera, Hymenoptera (*Apis mellifera*).

*Methods.* Examination of fresh smear preparation of Malpighian tubules with a phase contrast microscope will reveal the spores. Giemsa stained smears of the same material will reveal small multinucleate plasmodia.

*Identification.* See Lange (1993) and Radek *et al.* (2011).

*Pathogenicity.* None observed.

*Representatives. Nephridiophaga periplanetae* from *Periplaneta americana*.

#### b. Microsporidia

Microsporidia are non-flagellated protists characterized by production of spores equipped with a unique infection mechanism that enables spore contents to be injected by means of an evaginable tube directly into the cytoplasm of cells of their hosts. Microsporidia are the most species-rich and important protistan pathogens of insects and are also pathogens of nearly all animal phyla and some protists. Molecular evidence suggests that microsporidia are related to Fungi but the exact nature of this relationship (ancestors to Fungi or derived from extant Fungi?) is debated (Keeling, 2009). Because they are intracellular parasites, dependent on the host cell for all resources, microsporidia are extremely reduced organisms both morphologically and genetically (Keeling & Fast, 2002). In extreme cases, the genome of some species is smaller than that of bacteria (Peyretaillade *et al.*, 2011). The structural and genomic reduction complicates the search for the evolutionary history of microsporidia. Microsporidian classification is based on life-cycle, ultrastructure and molecular data (Wittner & Weiss, 1999). Insects are hosts to type species of more than 90 genera.

*Diagnostic features.* All microsporidia are obligatory intracellular organisms, multiplying in the host cell in the form of small paucinucleate meronts or plasmodia. They generally have sublethal deleterious effects on their hosts; excessive mortality, reduced longevity, and/or reduced fecundity may be an indication of microsporidian infection in laboratory-reared insects. Intense fat body infections in lightly pigmented hosts can sometimes be observed as pale or white coloration. Occasionally, soft-bodied insects will appear to be somewhat deformed, swollen by masses of developing spores in hypertrophied cells. Patent infections are rarely observed in live insects with opaque and heavily sclerotized cuticles, and spores can occur in numbers that appear to fill the host entirely without producing obvious signs. Host immune response to microsporidia is sometimes observed as dark spots of melanized parasites in tissues that can be observed beneath the cuticle. Infected tissues are frequently identifiable during dissection by swelling and altered color, often a 'milky' or 'puffy' appearance, due to masses of spores or spore-filled cysts in the cells.

Researchers usually encounter the environmentally resistant, infective spores ('environmental spores') of microsporidia, which are relatively easily indentified. The spores are generally small; most entomopathogenic microsporidia are approximately 2–6 μm in size. Spore shapes may be round oval, pyriform, less frequently also reniform, long oval to nearly tubular and are refringent under phase-contrast microscopy (Figure 11.5A). A vacuolar space can be observed at the posterior (broader) pole of thin-walled aquatic spores ('posterior vacuole') and internally infective 'primary' spores (Figure 11.5B). The spores contain an eversible polar tube through which the spore contents, including nuclei and cytoplasm (sporoplasm), are inoculated into a host cell when the spore germinates. Environmental spores, the most obvious and frequently encountered stage, are released into the habitat via frass, and possibly silk and regurgitated material, and when dead infected hosts decompose. Some microsporidia form a single type of spore, some others form spores of several types, either in the same or in different hosts. The more resistant environmental spores serve for inter-host transmission and the spontaneously germinating primary spores for spread of infection within the host. This second type of spore is usually encountered in direct examination of smeared insect tissues. Giemsa stains show a characteristic light blue staining of the spore which appears as a blue outline and a 'thumbprint' or 'horseshoe' of light blue in the center, possibly due to

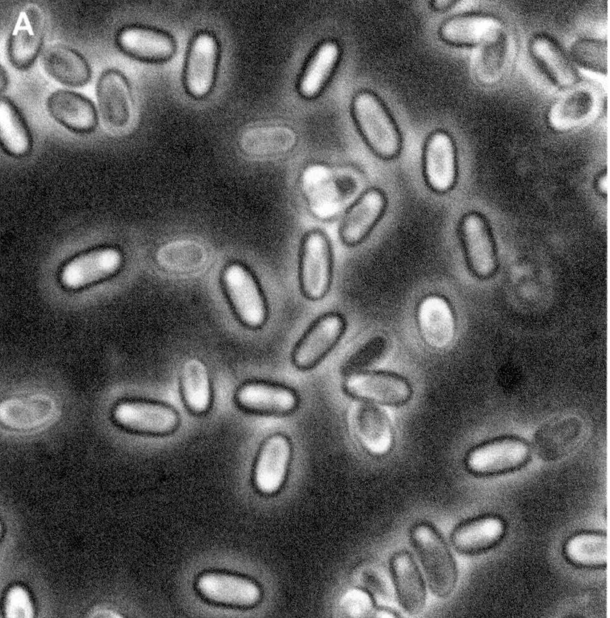


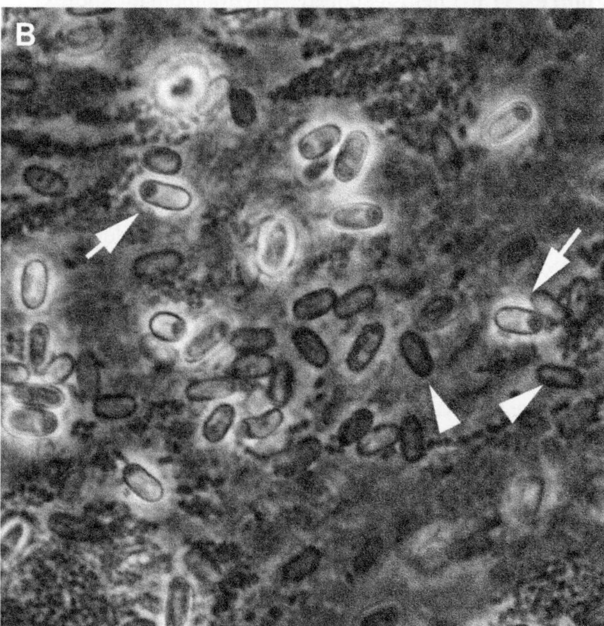


**Figure 11.5** Microsporidian spores isolated from *Otiorhynchus sulcatus.* A. Mature 'environmental' spores are refringent under phase contrast microscopy. B. Internally infective 'primary' spores with visible posterior vacuole (arrows) and germinated primary spores (arrowheads) that appear as 'shells' in midgut tissues of the host. (Photos by LFS.)

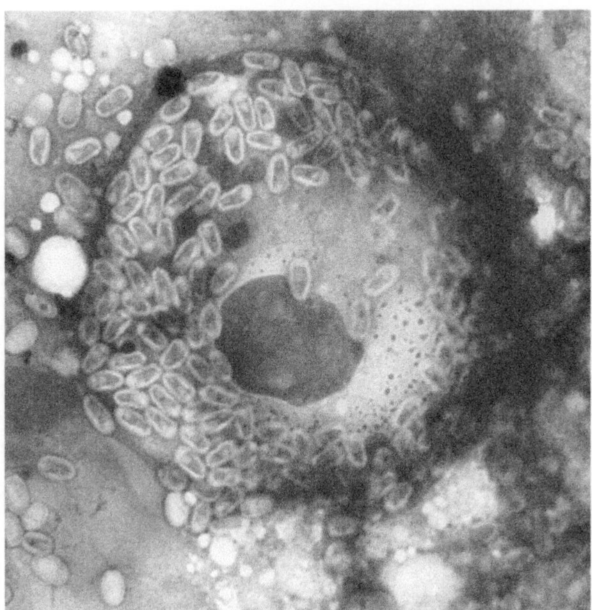

**Figure 11.6** Giemsa-stained environmental spores of a microsporidium infecting a host midgut cell. Note the blue 'thumbprint' or smear that is often diagnostic for mature microsporidian spores. (Photo by LFS.) Please see the color plate section at the back of the book.

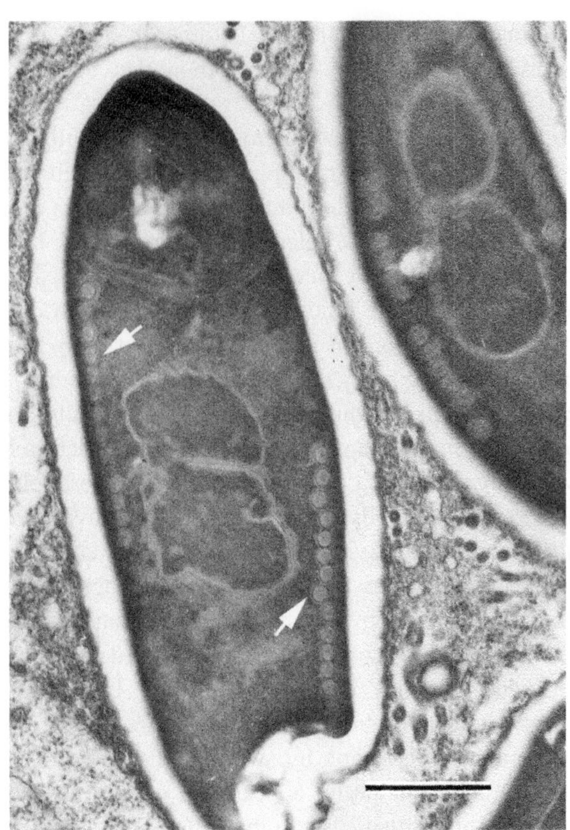

**Figure 11.7** Diplokaryotic microsporidian spore *(Vairimorpha disparis)* with cross-sections of the polar filament coils (arrows). Scale bar = 0.5 μm. (Photo by JV with permission of Blackwell Publishing.)

indentation of the spore wall during fixation (Figure 11.6). The nuclear configuration, either a single nucleus or two tightly adhering nuclei in the form of a 'coffee bean' (diplokaryon) is revealed in spores only by staining preceded by acid hydrolysis (see below).

TEM reveals the very complex internal organization of the spore (Vávra & Larsson, 1999). The internally coiled polar tube (with the appearance of a filament) is the most diagnostic feature of a microsporidium (Figure 11.7). The presence of polar filaments (Figure 11.8) can be confirmed under light microscopy by inducing the spores to germinate mechanically by crushing them slightly under the coverslip, or by chemical means (Keohane & Weiss, 1999). Spores are quite resistant and can be recovered from insects that have been pinned, alcohol-preserved or badly decayed.

Vegetative forms appear under light microscopy variously as very small cells with smooth cytoplasm and either single or diplokaryotic nuclei, or as larger plasmodia with multiple nuclei. Giemsa stains show a characteristic smooth blue cytoplasm and bright pink nuclei.

*Occurrence.* Microsporidia are parasites of animals and other protists, but especially arthropods. They probably occur in all insect orders. Development is entirely

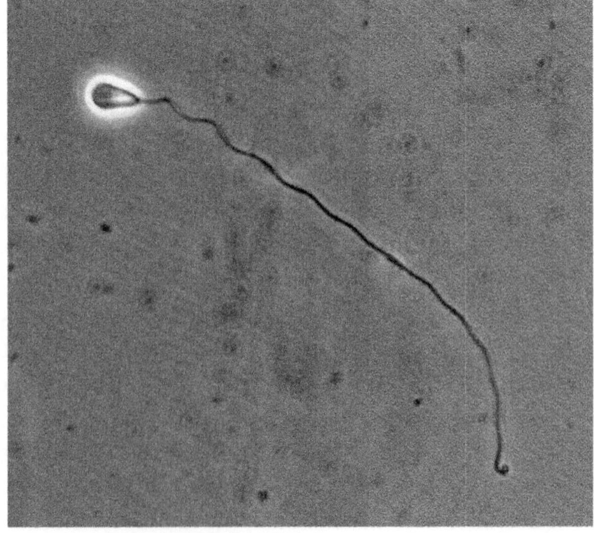

**Figure 11.8** Germinating microsporidian spore with polar tube extruded. (Photo by JJB.)

intracellular with the target tissues depending on species. Gut epithelial cells, fat body cells, silk, and salivary glands, Malpighian tubules and muscles are the most commonly parasitized tissues, but systemic infections are also common. Examination of insects in laboratory colonies, especially those unable to pupate and that remain in the larvae stage after a population peak, often yields a microsporidian infection.

*Methods.* Thin dry smears on glass slides or coverslips are stained with Giemsa or Gram stain (or a modification). The Robinow—Piekarski method—Giemsa stain following acid hydrolysis—is used for determining the number of nuclei, either one (monokaryotic), or two (diplokaryotic) in microsporidian spores on dry smears. Wet smears are stained with Heidenhain's hematoxylin. Lacto-aceto orcine (Hazard & Brookbank, 1984) is used to reveal chromosomes in wet smears. Heidenhain's hematoxylin, Giemsa-colophonium and a modified Gormori triple stain can all be used for revealing spores in tissue sections (Hazard & Brookbank, 1984). Erytrosine or eosine, 1% in distilled water, added to a drop of spores is absorbed by the surface of *Vairimorpha* spp. octospores to distinguish them from *Nosema*-type (diplokaryotic) spores (Figure 11.9A). Methods for preparing these stains are found in Chapter XV. For additional special techniques, see Vávra & Maddox (1976) and Hazard *et al.* (1981).

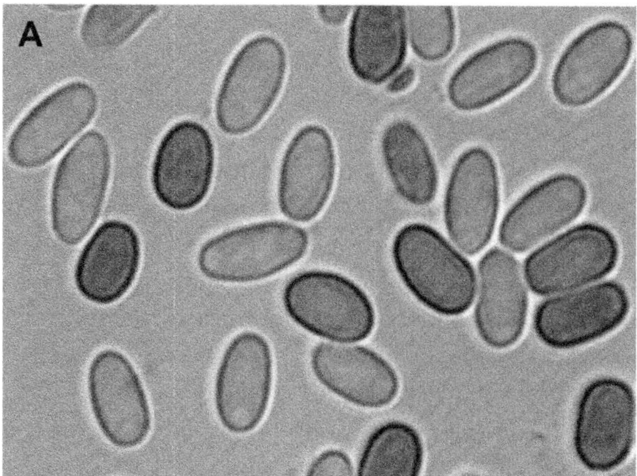


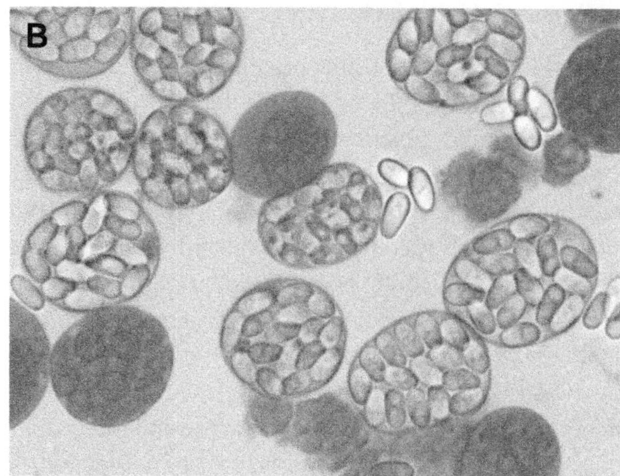


**Figure 11.9** Special stains for microsporidia. A. The spore wall of *Vairimorpha* octospores is thicker than that of the diplokaryotic spores, allowing octospores to take up 1% eosine stain. B. The sporophorous vesicle of *Vavraia culicis* is stained with 1% Congo Red; the spores are not. (Photos by JV.). Please see the color plate section at the back of the book.

*Identification.* Microsporidian classification requires an understanding of the developmental cycles of the organism and of structural characters, including the fine structure of developmental stages and spores. Finding of spores alone makes the identification possible only in exceptional cases. Classification is based on several major characters:

1. Number of nuclei and their configuration during the initial multiplicative cycle of the organism (merogony), in the presporal stages (sporogony) and in spores. The nuclei can be either single (monokaryotic) or associated as diplokarya, and some species form plasmodia, vegetative forms with multiple nuclei. Spores originate by internal differentiation of a single presporal cell. Spores have either a single nucleus or one diplokaryon.
2. Spores produced by the parasite may be in direct contact with the host cell cytoplasm or enclosed in a membrane-like formation that can be one of two types depending on species, a sporophorous vesicle (SPOV) or a parasitophorous vacuole (PV). SPOVs have been determined ultrastructurally to be formed by the microsporidium; PVs appear to be formed by host cells. Within the SPOVs, microsporidian spores are usually arranged in a characteristic number, one to many depending on species. The SPOV is a very fragile structure in some microsporidian species, and sometimes encloses only a single spore, therefore, its definite presence or absence may need to be verified by TEM. An aqueous solution of 1% Congo Red penetrates the SPOV and stains its volume, the spores remain unstained (Figure 11.9B).
3. Because microsporidia are usually host specific, knowledge of the pathogen complex of a host is an

excellent aid in species identification; however, some microsporidian species have complex life-cycles with more than one type of spore formed either in the same or different hosts.

4. Molecular data have resolved many species identification problems by elucidating species relationships and genetic variation within species, as well as verifying diagnoses. The small subunit rRNA gene is currently the most frequently used for identification and phylogenetic information but is highly conserved and may not distinguish closely related species and strains which, however, may be revealed by sequencing of the internal transcribed spacer (ITS) (Weiss & Vossbrinck, 1999).

Table 11.6 provides some orientation for identifying microsporidia in insects. More detailed information about individual genera can be found in Sprague *et al.* (1992), Larsson (1999), and Canning & Vávra (2000). Because microsporidia are usually relatively host specific and reproduce in specific tissues, this information is also provided. Classification based on spore morphology can be difficult because some microsporidia have complicated life-cycles and form several types of spores. In some cases, different sporulation cycles occur in different stages of the host, for example, in larvae and imagos, but some species can also form different spore types in the same host, sometimes in the same tissues.

*Pathogenicity.* Microsporidia are well adapted to the parasitic way of life and are generally slow to kill the infected host; many species take advantage of the entire host life span to maximize spore production. Some species produce spores in massive numbers in the late stages of infection, which may kill the host. Many microsporidian species are transmitted vertically from infected female to offspring in or on the surface of the eggs, called transovum transmission. If embryos are infected, this special type of transovum infection is called transovarial infection. Microsporidia that can be acquired orally but are then transovarially transmitted by female hosts tend to produce less intense (and often systemic) infections, allowing the host to survive long enough to eclose, mate and oviposit. Nevertheless, effects such as a shortened adult lifespan and lower fecundity are often reported for these species, as well as high mortality in transovarially infected larvae.

*Representatives. Nosema* spp. in various Lepidoptera (e.g., *Nosema bombycis* in the silkworm *Bombyx mori*, *Vairimorpha necatrix* in the gypsy moth *Lymantria*); *Amblyospora* spp., *Vavraia culicis* in mosquitoes, *Nosema apis* and *N. ceranae* in the honey bee.

## B. Describing species

Even if the researcher does not intend to describe a new species, the details presented below offer guidelines for examining and studying the pathogen. Some information is easy to obtain; collecting data on other characteristics may require rearing the host, experimental infections and timed examinations—tasks that may not be possible for some hosts or pathogens, especially if hosts are only available from field collections.

### *1. Information required for species descriptions*

- Host species, including any obligate intermediate hosts
- Tissues infected
- Transmission mechanisms
- Interface (vesicle membranes surrounding the pathogens, separating them from host cell cytoplasm, or none)
- Any other pathogen–host relationships
- Development of the pathogen in the host tissues; include the entire life-cycle if possible
- Spore type(s) including:
  - Number of nuclei
  - Shape and size (preferably fresh and stained)
- Spore wall structure—endospore and exospore; surface ornamentations or projections
- Internal structures as appropriate and diagnostic within each group
- Locality
  - Habitat type
  - Geographic location—GPS coordinates (preferred) or proximity to a durable well known landmark.
- Location of deposition of specimens
- Nucleic acid sequences; SSU rDNA is standard
- Other available information, such as:
  - Experimental host range
  - Comparison of morphological, biological and molecular characteristics with similar species

**Table 11.6** Genera of microsporidia and associated insect host orders.

| *Microsporidia with type species in insect hosts* | | | |
|---|---|---|---|
| Insect order | Microsporidian genus | Insect order | Microsporidian genus |
| Collembola | *Auraspora* | Diptera | *Simuliospora* |
| Coleoptera | *Anncaliia* | | *Spherospora* |
| | *Cannngia* | | *Spiroglugea* |
| | *Chytridiopsis* | | *Striatospora* |
| | *Endoreticulatus* | | *Systenostrema* |
| | *Ovavesicula* | | *Tabanispora* |
| Diptera | *Aedispora* | | *Toxoglugea* |
| | *Amblyospora* | | *Toxospora* |
| | *Andreanna* | | *Trichoctosporea* |
| | *Anisofilariata* | | *Tricornia* |
| | *Bohuslavia* | | *Tubilinosema* |
| | *Campanulospora* | | *Vavraia* |
| | *Caudospora* | | *Weiseria* |
| | *Chapmanium* | | *Ringueletium* |
| | *Coccospora* | | *Scipionospora* |
| | *Crepidulospora* | | *Semenovaia* |
| | *Crispospora* | | *Senoma* |
| | *Cristulospora* | Ephemeroptera | *Mitoplistophora* |
| | *Culicospora* | | *Pankovaia* |
| | *Culicosporella* | | *Stempellia* |
| | *Cylindrospora* | | *Telomyxa* |
| | *Dimeiospora* | | *Trichoduboscqia* |
| | *Edhazardia* | Hemiptera | *Becnelia* |
| | *Evlachovaia* | Hymenoptera | *Antonospora* |
| | *Flabelliforma* | | *Burenella* |
| | *Golbergia* | | *Kneallhazia* |
| | *Hazardia* | Isoptera | *Duboscqia* |
| | *Helmichia* | Lepidoptera | *Cystosporogenes* |
| | *Hessea* | | *Larssoniella* |
| | *Hirsutusporos* | | *Nosema* |
| | *Hyalinocysta* | | *Orthosomella* |

(*Continued*)

**Table 11.6** Genera of microsporidia and associated insect host orders.—Cont'd

| | | | |
|---|---|---|---|
| Diptera | *Intrapredatorus* | Lepidoptera | *Vairimorpha* |
| | *Janacekia* | Odonata | *Nudispora* |
| | *Krishtalia* | | *Resiomeria* |
| | *Merocinta* | Orthoptera | *Heterovesicula* |
| | *Napamichum* | | *Johenrea* |
| | *Neoperezia* | | *Liebermannia* |
| | *Octosporea* | | *Paranosema* |
| | *Octotetraspora* | Psocoptera | *Mockfordia* |
| | *Parapleistophora* | Siphonaptera | *Nolleria* |
| | *Parastempellia* | | *Pulicispora* |
| | *Parathelohania* | Thysanura | *Buxtehudea* |
| | *Pegmatheca* | Trichoptera | *Episeptum* |
| | *Pernicivesicula* | | *Issia* |
| | *Pilosporella* | | *Paraepisetum* |
| | *Polydispyrenia* | | *Tardivesicula* |

*Microsporidia in insects with type species in other animal classes*

| Microsporidian genus | Insect hosts | Microsporidian genus | Insect hosts |
|---|---|---|---|
| *Cougourdella* | Trichoptera | *Pyrotheca* | Trichoptera |
| *Encephalitozoon* | Orthoptera | *Thelohania* | Many insect orders |
| *Gurleya* | Diptera, Lepidoptera Ephemeroptera, Odonata, Isoptera | *Tuzetia* | Ephemeroptera, Coleoptera, Odonata |

*2. Deposition of reference slides and live type material*

When a new species is described, a type specimen should be deposited with the International Protozoan Type Collection at the National Museum of Natural History of the Smithsonian Institution or other repository. Correspondence should be addressed to: Museum Specialist, Department of Invertebrate Zoology, Smithsonian Institution Museum Support Center, MRC 534, 4210 Silver Hill Road, Suitland, MD 20746-2863, USA. A deed of gift form must be submitted with the type slides (information can be found at http://invertebrates.si.edu/donation.htm).

The material must be permanently mounted on glass microscope slides and affixed with labels listing genus, species, author, and year; material type (paratype, lectotype, etc.); collection locality; genus, species, and class of the host. Slides must be arranged in order and be accompanied by a listing of this information: stain, collector, number of specimens, original field number, identifier, pertinent remarks, and higher classification. A short letter should be sent with the specimens stating that they are being donated. Wet specimens (specimens in alcohol or formalin) must contain labels with the same information (except stain) or be accompanied by research papers containing this information.

Type material (specimens from which a new species was described) will form a separate collection and must be accompanied by separate data sheets. The descriptive paper of a type material must be accepted for publication or be in press before type specimens are submitted to the museum. At least one copy of each publication should be supplied for the museum's library.

The museum affixes an accession number to each slide and all pertinent papers can be located by this number at any time. All material is assigned a catalog number in addition to the accession number. Slides are stored numerically by catalog number. Wet specimens have a special section in the collection area. Slides requiring catalog numbers for forthcoming publications are rapidly accommodated. Collections are loaned in full or in part to investigators or institutions; the duration of the loan being arrived at by negotiation, but not to exceed one year. Loans are renewable annually upon request with justification. Utmost care of materials loaned is assumed.

Viable spore samples may be of interest to the American Type Culture Collection (ATCC), Protistology Collection, 10801 University Boulevard, Manassas, Virginia, VA 20110-2209, USA. Contact a collections specialist to determine the requirements to archive live cultures (http://www.atcc.org/EmailaCollectionScientist/tabid/722/Default.aspx).

## C. General techniques useful for identification of protists

The variety of protists occurring in insects precludes a detailed overview of techniques used in their identification here; however, the reader can find many useful and detailed techniques of how to observe, preserve, stain, and cultivate protists (including those from insects) in Lee & Soldo (1992).

### *1. Light microscopy*

#### *a. Fresh smears*

Live infected insects are best for examination. Pathologies can be noted without confounding changes associated with decomposition. Secondary microbial activity, particularly fungi and bacteria in decomposing insects, can obscure the cause of death and may degrade the pathogen of interest beyond recognition. Presporal stages of protists are not easily visible in fresh smears and deteriorate quickly; environmental spores of most species are the most easily observed stages. Larger insects may need to be macerated or homogenized separately and a small part of the fluid, without pieces of cuticle, placed on a slide for examination. Small insects can be crushed whole under a coverslip. It is often best to remove the gut contents before macerating the insect for a cleaner preparation. To determine the site of infection, the insect should be dissected and tissues carefully removed and examined individually. Phase-contrast is usually the best optical system to use for examining fresh smears.

#### *b. Preserved material*

Preserved material can also be examined for infection, although decomposition of tissues can confound diagnosis. Environmental spores in ethanol, methanol, preserved tissues, or even in dried, pinned insects can be observed and identified; however, many identification features of the pathogens will be lost in material that has been preserved in these ways. Formalin is a better preservative for infected insects (if a fume hood is available—formaldehyde is a known carcinogen). Infected insects or tissues should be preserved by the method most appropriate for the technique to be used later for examining and studying the parasite.

#### *c. Measuring pathogen life stages*

The size and shape of the spore or cyst is a fairly consistent feature of most protistan species, although these characters may vary between hosts or may be influenced by temperature during development. Spore size is measured in length and width at the maximum plane of each. Published measurements are preferably from live spores; measurements of stained spores and developmental stages can also be informative, but may vary considerably depending on osmotic conditions at the time the smear was made or the response to fixation. Measurements have greater precision as the number of spores measured increase; 30 spores are usually adequate for accuracy.

*Ocular micrometer.* The ocular micrometer is the most frequently available device for measuring objects in a microscopic field. This is simply a ruled grid placed in one of the oculars of the microscope and calibrated with a stage micrometer. This device can provide an estimate of size but lacks precision at the small sizes of most protistan spores. It is not adequate for a species description.

*Digital imaging.* Recent advances in digital photography and projection have provided accurate methods for measurement of microbes. Images are measured with software programs such as Meta Imaging Series by Metaview (1998), or QuickPHOTO MICRO by Promicra for Olympus®.

*Immobilization.* Immobilization of spores is a necessity in precisely measuring microbes. Spores move with the flow of the suspending medium under the coverslip and are subjected to Brownian motion. Immobilization methods include use of different suspending media including agar, polylysine and oil.

Agar immobilization method (Vávra, 1964)

1. Cover a large area of a microscope slide with molten 1.5% agar. Draw off excess agar using the edge of a coverslip to create a thin, even, lightly convex layer of the agar, then allow to cool.
2. Place a very small drop of concentrated spores, $< 5\ \mu l$, in the center of a coverslip and invert on top of the solidified agar. The spores will be immobilized between the coverslip and the agar.
3. Spores that are in clear focus at both ends should be measured.

Polylysine

Coating the slide with a film of polylysine will immobilize spores (Mazia *et al.*, 1975).

Oil

Paraffin (mineral) oil can also serve as a trap to immobilize spores (Vávra, 1964). A drop of the oil is placed on a slide and a coverslip with a small drop of dense spore suspension applied on top of the oil. Water, having a better affinity for glass, spreads out on the surface on the coverslip, leaving spores individually trapped in 'holes' in the oil. Phase-contrast microscopy cannot be used to examine water–oil slides (Vávra, 1976). Desiccation of any of these preparations can be slowed by sealing the edges of the coverslip with soft petroleum jelly.

*2. Ultrastructure (transmission electron microscopy)*

Arrangement and type of internal organelles are critical for identification of microsporidia and other protists. Use of TEM to evaluate the ultrastructure of internal organelles and the interface with the host cell became the best option in the 1960s, allowing researchers to describe in detail the nuclear configuration, number and type of polar filament coils, type of polaroplast, location of posterior vacuole and anchoring disk among other features. Chapter XV provides detailed methods for TEM, necessary to observe these characters.

*3. Molecular characters*

The spores of many protists are morphologically similar within each major group, with few external characters that are useful for taxonomic evaluation. The microsporidia pose a particular problem because of the large number of species, nearly all of which possess similar internal and external characters that overlap within and among taxonomic clades. Molecular data have shown that many of these characters are pleisiomorphic or have been lost in multiple clades, suggesting that characters previously relied upon for species descriptions and systematics are frequently not phylogenetically meaningful within the larger group. Section 4 covers detailed methods for obtaining genetic information needed for species descriptions.

## 3. TECHNIQUES FOR MANAGING SPORES

### A. Extraction from the host

Intact infected organisms or excised infected tissues can be triturated by a number of methods to extract environmental spores using simple equipment such as a tissue grinder (Figure 11.10), mortar and pestle, or a countertop blender, depending on the particular organism and the volume of material being processed. For example, while a blender is a good choice for large batches of soft-bodied terrestrial insects infected with microsporidia that produce small, tough spores, the large fragile aquatic spores of *Edhazardia aedis* are damaged by such harsh disruption methods. Large, heavily sclerotized insects are best dissected and infected tissues removed for homogenizing individually or in small batches (Figure 11.11). Tissue extraction also facilitates cleaning of the spores because cuticle, hairs and gut contents are removed. Some species of microsporidia infect only the fat body tissues and, because the organ is almost entirely replaced by spores, little further purification of spores is necessary.

Because the quality of tap water varies by time and locale and may contain microorganisms that can reproduce in insect tissue homogenate and compromise the spores, it is best to use sterile deionized or distilled water for spore extraction procedures; however, filtered and autoclaved tap water may also be a reasonable choice.

Some spores are easily germinated by mechanical agitation or stimulation by solutes from the triturated host

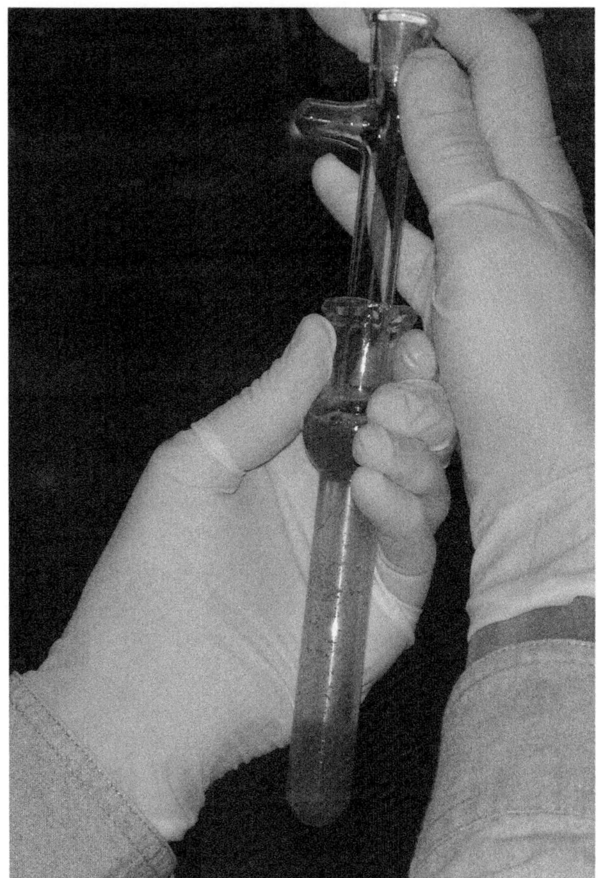

**Figure 11.10** Insect tissues are homogenized in a glass tissue grinder. The resulting suspension is then filtered to remove large pieces of insect tissue and centrifuged to pellet microsporidian spores. (Photo by LFS.)

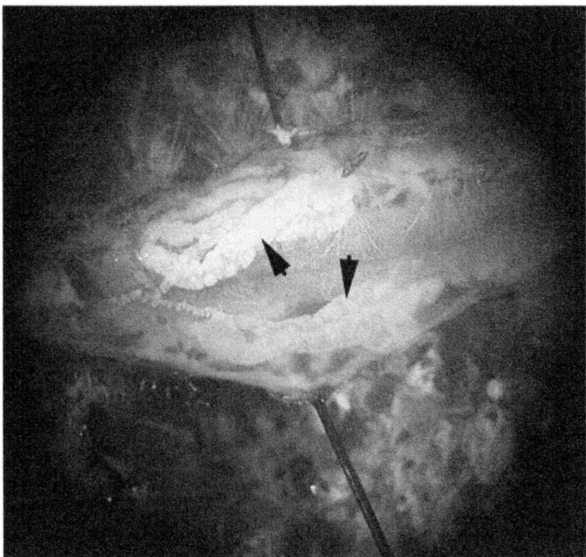

**Figure 11.11** Silk glands of gypsy moth larva infected with *Nosema lymantriae* can easily be removed and homogenized for isolation of spores with little host tissue contamination. (Photo by G. Hoch.)

tissues and may be lost during the extraction process. Keeping the volume of water high in relation to the amount of tissue to be triturated helps avoid this problem. Germination of microsporidian spores can be prevented by extracting spores in a pH 9.0 buffered, 0.001–0.05 M ammonium chloride solution. Maintaining cool temperatures throughout the process also limits adventitious germination.

## B. Purification

Purity of the spore suspension facilitates counting, measuring, and storage, and is also needed for biochemical, molecular, and proteomic studies. Bacterial and fungal growth quickly destroys stored aliquots of spores, therefore, the time that spores remain in insect material or other media that supports microbial growth should be minimized. Although purification may not be particularly important if material is frozen or dried, there is opportunity for microbial growth when the material is thawed.

### *1. Filtration and centrifugation*

Purification is usually accomplished by a sequence of filtrations and centrifugations. Small quantities of triturated host material can be filtered through about 2–5 mm of wet cotton packed in a syringe (Undeen & Becnel, 1992). Larger quantities can be vacuum-filtered through a cotton pad in a Buchner funnel. Laboratory tissues (Kimwipes, etc.), cheesecloth, ultrafine hardware cloth or other fabric folded in the manner of a coffee filter in a small funnel may also serve as filtration beds for removal of large pieces of host material. Some microsporidian spores, such as those of *Caudospora* spp., have external ornamentation that increases loss during filtration. Until experienced with a particular organism, microscopically examine all filtrates, supernatant, and residues for spores before discarding them. An extra wash of the filter material with sterile water may increase yield.

Because of their high density, a series of washes and centrifugations will free most microsporidian spores of dissolved and particulate contaminants. The density of spores is usually higher than that of most host tissues and centrifugation concentrates them at or near the bottom of the residue where, if sufficient spores are present, they appear as a white pellet. The supernatant and the detritus layers above the pellet are carefully mixed without disturbing the pellet and then decanted or removed with a pipette. The pellet with spores can then be re-suspended in sterile water and the process repeated if necessary. Another reason to extract infected tissues from the host before processing is the density of insect cuticle, which is similar to that of spores and is difficult to separate from the spores. If a considerable loss of spores is acceptable, a fairly clean suspension can be obtained in one or two centrifugation cycles. The process of 'triangulation' expands on this process and, while time consuming, yields a fair harvest of clean spores using only low-speed centrifugation (Cole, 1970).

### *2. Density gradient centrifugation*

Density gradient centrifugation is fast and the yield is high. Sucrose gradients can be used for terrestrial-host microsporidia that tolerate desiccation, but some microsporidia, especially those of aquatic hosts, may be desiccated and killed by sucrose concentrations high enough to suspend them. Colloidal silica does not have this disadvantage. Two silica colloids, Ludox® HS40 (DuPont), with a density of 1.303 (Undeen & Avery, 1983) and Percoll®, a Sigma product (Jouvenaz, 1981), with a density of 1.130 g/ml, are commonly used for this purpose. Both are alkaline (pH 9.7) and Ludox is unstable at pH 7.0. High pH causes some spores to germinate, a problem that can be solved by the addition of *ca.* 0.01 M ammonium chloride to the gradient components before mixing (Undeen & Avery, 1983). Percoll can be neutralized but its density is too low to suspend most microsporidian spores, which have density values in the range of 1.180–2.200 (Undeen & Solter, 1996); spores will be pelleted in the bottom of the centrifuge tube.

#### *a. Self-forming continuous Ludox gradients (modifications by Neil Sanscrainte)*

(http://web-mcb.agr.ehime-u.ac.jp/english/methods/gradient.htm; Abe & Davies, 1986.)

As an alternative to using density gradient mixers, which can be time consuming and do not always produce uniform gradients, the simple procedure below will prepare multiple, reproducible, self-forming Ludox or Percoll density gradients by diffusion without any specialized equipment (Abe & Davies, 1986).

*Materials.*

- Centrifuge capable of at least 10,000 × *g*
- Centrifuge tubes
- Ludox® HS-40 (or Percoll®)
- 1.0 M $NH_4Cl$ (optional)

*Procedure.*

1. Place $^1/_3$ of the centrifuge tube volume (i.e. 10 ml for 30-ml centrifuge tubes or 5 ml for 15-ml centrifuge tubes) of the higher density solution (Ludox or Percoll) in the centrifuge tube. Adding 0.01–0.05 M $NH_4Cl$ to the higher density solution reduces the risk of spores germinating in the gradient.
2. Layer the same amount of the lighter-density solution (sterile DI or distilled water) on top, being careful not to disturb the interface. If $NH_4Cl$ was added to the higher-density solution in the previous step, add 0.01–0.05 M $NH_4Cl$ to the lower-density solution to reduce the risk of spores germinating in the gradient.
3. Close the centrifuge tubes with cork or rubber stoppers. Gently lay the tubes on their sides and the two layers will diffuse together. Allow gradients to form for 1–2 h for 10–15-ml tubes and 2–3 h for 30–50-ml tubes.
4. Slowly straighten the tubes and let them 'rest' for 5 min. The gradients will then be ready to use.
5. An aliquot of the filtered crude spore extract is layered on top of the gradient. For high purity do not overload the gradient (e.g., 1 ml suspension or less for a 35-ml Ludox gradient in a 50-ml tube). If the spore suspension is too concentrated, considerable detritus that otherwise might remain above the spore band, will be carried down with the spores.
6. The gradients have commonly been centrifuged at about 16,000 × *g* for 30 min but optimum centrifugation has never been experimentally determined. Allow the centrifuge to decelerate slowly; braking will create a vortex, mixing the upper region of the gradient.
7. After centrifugation, the mature spores are concentrated in a white band, 2–3 mm wide, 60–70% down the gradient, below most of the contaminants (Figure 11.12). Immature spores, soft insect tissues and bacteria will be layered above the spores. Insect

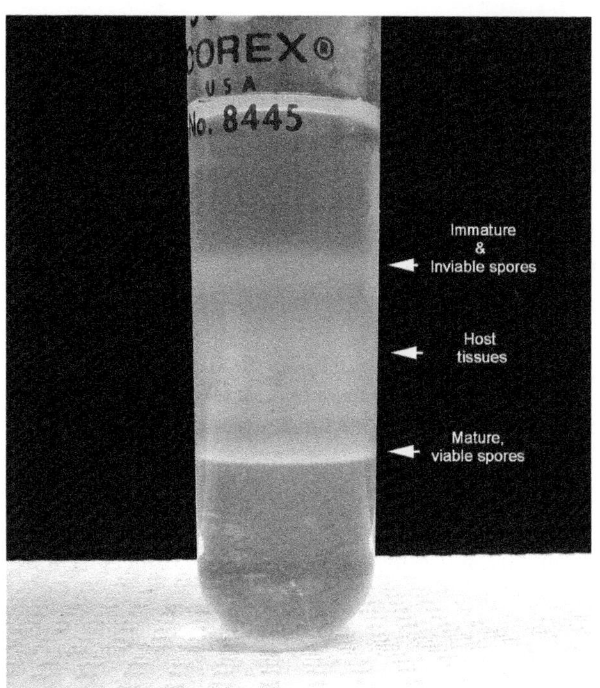


**Figure 11.12** Ludox HS-40 continuous gradient. Viable spores form the bottom layer. (Photo by N. Sanscrainte.)

viral polyhedra and some fungal spores, but little else, are found with or below the spores. Although mature spores of most microsporidian species will be in this zone, there is no guarantee because spore densities differ somewhat. *E. aedis*, for example, has a lower and more variable density, demonstrating the need to determine the density of each species.

8. If there are too few spores to form a visible band, the gradient can be fractionated and each fraction examined for spores. Even if spores are present in small numbers they can be found by diluting the fractions with water (so that the spores do not remain suspended) and centrifuged to concentrate any spores that might be present into the bottom of the tube. Alternatively, a pipette can be carefully inserted into the gradient and small samples taken at various levels to be microscopically examined for spores.
9. As soon as centrifugation is complete and the spores are located, they should be removed immediately from the gradient and washed two to three times in water to remove the silica. Spores can be damaged by leaving them for several hours in Ludox.

Percoll can be autoclaved to provide sterility and neutralized to enhance the survival of cells. In a procedure used by Iwano & Kurtti (1995), spores were layered on the top of neutralized 100% Percoll. Centrifuged at 39,000 $\times g$ for 40 min, a 'shelf' of silica upon which the spores layered, formed near the bottom of the gradient.

*b. Discontinuous gradients*

Another method that does not require a density gradient mixer is a discontinuous gradient. These gradients can be constructed by carefully layering a concentration series of the gradient material, starting with most concentrated at the bottom. With a little experimentation, a discontinuous gradient of only two phases can used to affect purification of spores. The lower concentration (top layer) is made sufficiently dense to just allow passage of the spores, and the higher density (lower layer) just sufficiently dense to suspend the spores (Undeen *et al.*, 1993).

*c. Continuous flow density gradient centrifugation*

Liter quantities of spore suspension can be purified by feeding the suspensions slowly into density gradients in a continuous-flow centrifuge (Undeen & Avery, 1983). Because the diluted crude spore suspension is fed slowly onto the gradient, the same quantity of gradient material will purify many times more spores than could be accommodated in a single batch.

*Procedure.*

1. Filter the suspension to prevent blockage of the passageways in the distribution head.
2. If necessary, a small amount of detergent (0.1% vv, sodium dodecyl sulfate) can be added to the filtered spore suspension to prevent formation of a layer of fat that tends to trap spores near the inlet tube.
3. Make density gradients as described above in specialized tubes for the continuous-flow centrifuge.
4. Stir the crude spore suspension continuously to prevent settling of the spores and feed slowly into the centrifuge, which is set at the relative centrifugal force described above for density gradients. Save the outflow to check for presence of spores before discarding.
5. After all the spore suspension has passed through the centrifuge, water is fed through the system to clear spores and debris from the tubing.
6. The centrifuge is stopped without braking, the tubes are removed and the bands containing the spores are withdrawn. If the tube is opaque, the contents of the

gradient can be suctioned from it through small-diameter tubing (*ca.* 1–2 mm). The suction tube is placed at the bottom of the gradient; fluid passing through the tube first will be clear high-concentration Ludox and then become cloudy as the spores begin to pass through. At this point begin collecting the fluid, stopping when the fluid once again becomes clear.

7. Rinse the spores free of the Ludox as described above.

*3. Aseptic spores*

Aseptic spores are needed to inoculate cell cultures or inject into an insect host (Undeen & Alger, 1975; Undeen & Maddox, 1973; Pilley *et al.*, 1978).

*a. Density gradient centrifugation*

Spores that are of sufficiently high density to settle below the bacterial contaminants can be cleaned with a Ludox density gradient as described above. Spores such as *E. aedis* that are much lower density cannot be sufficiently cleaned in this way. To limit the numbers of bacteria present at the outset, the spores must be extracted from living hosts and cleaned immediately thereafter. The spores must be well dispersed and the gradients must not be overloaded or bacteria can be carried down and layer with the spores. The band of spores is extracted from the gradient, cleaned by several rinses in sterile water and may be treated with antibiotics. This method does not provide absolute sterility, therefore, the spores must be used immediately, before the bacterial levels increase. Fungal contamination can also be a problem because some fungal spores are similar in density to microsporidian spores.

*b. Sterile dissection*

Sterile spores can be obtained from live hosts by sterile dissection of infected tissues (Weiser, 1978). The host is surface sterilized with 3–5% household bleach [bleach = 3% sodium hypochlorite (NaClO)] or ethanol, pinned, and the integument cut longitudinally on the ventral side, taking care not to puncture the gut. Infected tissues are removed with sterile implements, transferred to sterile water, and rinsed two to three times in sterile water to remove host contaminants. Spores obtained from the host asceptically can be inoculated into cell cultures or injected into insect hosts without concerns about septicemia.

*c. Antibiotics*

Tetracycline, streptomycin, kanamycin, and amphotericin B (e.g., Fungizone®) are antibiotics used to suppress growth of contaminating microbes in microsporidian spore suspensions. *V. necatrix* spores harvested from lepidopteran larvae were stored at 4°C for as long as 3 years in 100 mg/ml tetracycline hydrochloride and 500 mg/ml neomycin sulfate with no ill effects (Pilley *et al.*, 1978).

*4. Obtaining spores from the habitat*

Samples of water from an aquatic habitat can be bioassayed or examined microscopically for the presence of protistan spores. Water is first filtered through a thin layer of cotton or coarse filter paper to remove small organisms and other larger particles. Concentration of particles by centrifugation increases the odds of detecting spores. A continuous-flow centrifuge is used to concentrate particulates from several liters of water; microsporidia and *Helicosporidium* spp. were both isolated by this method (Avery & Undeen, 1987). Microsporidia from terrestrial hosts occupying the vegetation around or above the aquatic sample site also have been obtained by this technique. PCR and sequencing can be used to determine relationships with described species.

## D. Storage

Optimal storage conditions need to be experimentally determined for each species. The environmental conditions that are typical for the host and that released environmental spores normally experience are useful guidelines. Generally speaking, microsporidia, and perhaps other protists from terrestrial hosts, tolerate desiccation and freezing; those from aquatic hosts do not. Microsporidia from most aquatic hosts must be stored in distilled or deionized water. When in doubt, the safest course of action is to purify the spores and hold them in a sterile aqueous suspension in a refrigerator; however, experience with *E. aedis* spores has shown that not all spores are tolerant of refrigerator temperatures. Brooks (1988) presents an excellent review of spore storage. A sodium azide water solution (0.01–0.02%) prevents bacterial and fungal growth in samples of water-stored spores and is a suitable storage medium for some microsporidia for periods of up to several weeks.

*1. Refrigeration*

Highly purified spores of most species survive well in cold (*ca.* 5°C), deionized or distilled water. Refrigerated spores should be held in tightly capped vials and checked frequently for evaporation. Longevity varies considerably among species with some remaining viable longer than 2 years (Brooks, 1988). *Anncaliia algerae* spores have retained their viability after 10 years' storage in the refrigerator (Undeen & Vávra, 1997). Microbial activity in the suspension medium is deleterious to microsporidian spores. Contamination by fungal spores is difficult to avoid because many have densities similar to those of microsporidian spores, making them impossible to remove by even the best of purification schemes. Massive fungal overgrowth is often responsible for destroying vials of spores in the refrigerator, even when the spores appear to be quite clean otherwise. A mixture of 100 mg penicillin, 100 units streptomycin and 0.25 mg Fungizone per milliliter of suspension medium is a combination that is routinely used to retard microbial growth. Purity of the spore suspension is, however, the critical factor for optimizing refrigerated storage; antibiotics provide some protection but will not substitute for purification.

*E. aedis* is the only microsporidium known to be adversely affected by temperatures just above freezing (0−5°C), a trait that might be shared by other microsporidia of tropical origin. It survives for approximately 1 month at temperatures between 10 and 30°C but less than 24 h in the refrigerator (Undeen *et al.*, 1993). The developmental stages within *Aedes aegypti* eggs and larvae are more tolerant of chilling than the spores.

*2. Freezing*

*a. Household-type freezer*

*Spores.* Most terrestrial microsporidia species can be stored frozen at −20 to −30°C in a household-type freezer. The inclusion of 50% glycerol in the suspension as a cryoprotectant is frequently required. Even under the best of conditions, repeated freezing and thawing causes spores to lose viability (Maddox & Solter, 1996), therefore, spores should be stored in small aliquots so that they need to be thawed only once.

*Carcasses. Paranosema locustae* spores are routinely stored in frozen grasshopper carcasses (Henry & Oma, 1981). This method appears to be as efficacious as storage in water, and does not require cleaning the spores before storage. *P. locustae* spores formulated on a bran bait for application have a relatively short shelf-life; therefore, the spores are immediately extracted in water from carcasses removed from frozen storage and are sprayed on the bran shortly before field application is anticipated.

*b. Ultra cold (−60 to −100°C) and liquid nitrogen*

Ultra cold temperatures are probably the most reliable long-term storage method for all microsporidia except those from aquatic hosts; however, a few terrestrial species, e.g., *Nosema ceranae* from honey bees, appear to be somewhat cold sensitive and lose viability more quickly than most (Gisder *et al.*, 2010).

*Spores.* Spores that are tolerant of 0°C freezing conditions can be stored for long periods of time in ultra-cold freezers and in liquid nitrogen, although time periods for protists have not been tested for freezers between −60 and −100°C. Cryoprotectants such as glycerol, dimethylsulfoxide or sucrose are frequently required (Maddox & Solter, 1996). The high density of sucrose and the slight toxicity of dimethylsulfoxide leave glycerol as the preferred cryoprotectant. Experimentation with cryoprotectant concentration might be necessary. Fifty per cent sucrose or glycerol, or about 10% dimethylsulfoxide, are good starting points. Procedures for preparing spores for storage in liquid nitrogen are found in Sections 3B1 and 2. To avoid freezing and thawing spores a number of times, spores should be stored in several small aliquots. In one routine procedure, 0.5 ml each of microsporidian spore suspension and glycerol are placed in a cryopreservation vial and then plunged directly into liquid nitrogen, without pre-cooling. Crude homogenates of spores can be frozen directly but antimicrobial agents should be added for protection of the spores after the vials are thawed, or the homogenate should be cleaned by filtering and centrifugation in sterile water immediately after thawing.

*Insect carcasses.* Preservation of spores can sometimes be accomplished by freezing a whole intact host at ultra cold temperatures or in a cryovial in liquid nitrogen, either dry or suspended in a 50 : 50 v : v solution of glycerol and water.

*Cell cultures.* Stocks of cultured cells are commonly stored in liquid nitrogen. Whenever tested, the microsporidian infecting the cells also survived under cryopreservation. In one study (Sohi & Wilson, 1976), infected cells were mixed with 10% dimethylsulfoxide and cooled at 1°C/min from room temperature to −40°C in an ethanol–solid $CO_2$ bath, then plunged into liquid nitrogen. Spores were viable after rapid thawing in a 30°C water bath. Developmental stages of *A. algerae* and *Nosema eurytremae* also survived freezing in the host cell (Jaronski, 1984).

### *3. Dry storage*

#### *a. Desiccation*

The spores of some microsporidia such as *Tubiliosema* (= *Nosema*) *whitei* naturally survive for extended periods in the dead dried host and then germinate as soon as they come into contact with water. Most terrestrial host microsporidia are to some degree tolerant to desiccation but they usually survive longer in refrigerated aqueous suspensions or frozen.

### *4.* In situ *storage*

Some species, such as *E. aedis*, are best 'stored' within the living host. A pathogen of the mosquito *A. aegypti*, *E. aedis* is vertically transmitted within the egg and will retain viability as long as the host eggs remain viable, a period of several months.

## E. Field collection and transport of live or fixed material

Field collections of insects infected with microsporidia and other protists require the same basic procedures that are used for hosts infected with other pathogens and parasites. General collection protocols are covered in Chapter I; specific issues concerning transport and handling of microsporidia-infected invertebrates are covered here.

### *1. Shipping live infected hosts*

**Note:** A permit may be required to ship or receive live organisms across international boundaries or interstate in the USA.

It is often necessary to ship microsporidia to another laboratory for an expert opinion on its identity or for cooperative research. Shipment time for unfixed material should be kept to a minimum to limit microbial activity that can destroy the sample. Live host material is always preferred because presporal developmental stages of protists deteriorate along with the host cells immediately after host death. Shipping live infected hosts ensures that all life stages are available for examination. A living host is particularly important for preparation of specimens for electron microscopy. Live hosts can be shipped by domestic or international courier in labeled sealable boxes or vials supplied with host plant material (if allowed), paper strips or crushed paper toweling to reduce trauma.

### *2. Shipping host carcasses*

Environmental spores usually remain identifiable in dry dead insect bodies. Protistan spores and cysts in terrestrial hosts usually withstand desiccation of the host without immediate loss of viability and can be shipped in dried carcasses. Ensure that vials have air circulation (air holes or cotton plugs) to reduce the possibility of microbial decomposition. Microsporidian spores from aquatic insects do not appear to tolerate desiccation and putrefaction can occur during shipping; however, if insect bodies or tissues are shipped *via* express mail in sterile water, the spores may retain viability.

### *3. Shipping live spores*

In order to avoid the possibility of desiccation or putrefaction, spores can be purified and shipped in sterile (autoclaved) deionized or distilled water. If the suspensions still contain host material or other contaminants, the addition of streptomycin and Fungizone will help to protect the spores from microbial activity. Live spores allow the collaborator to infect susceptible hosts to produce life-cycle stages for species descriptions and other research that can be performed if the host is available and can be reared.

### *4. Shipping fixed host tissues and dried smears*

Small pieces of infected tissues excised from live hosts can be fixed in buffered 1–2% glutaraldehyde and shipped for later preparation for electron microscopy (see Chapter XV). (Care should be taken to either process the tissues or remove to buffer within approximately 1 week and store at 4°C.) Tissues that have been fixed in formalin, ethanol, or other common histological fixatives are useful for diagnosis of microsporidioses but provide limited information concerning the identity of the microsporidium. Ethanol is not a satisfactory fixative for morphological study of microsporidian spores. Ethanol-fixed microsporidia processed for electron microscopy produce poor results; however, fixation in 95–100% ethanol appears to be an adequate means for preserving nucleic acids for sequencing (Frampton *et al.*, 2008; Bourgeois *et al.*, 2011).

Dried tissue smears on microscope slides can easily be mailed. These samples can be shipped unfixed, fixed with absolute methanol, or stained before mailing. Dried smears, especially unfixed slides, need protection from household insects and humidity.

## 4. MOLECULAR AND PROTEOMIC TECHNIQUES

### A. Extraction of DNA and other substances from microsporidian spores

DNA extraction for general PCR and sequencing for single genes or partial genes has improved significantly in recent years. Methods are well documented in a growing literature, but some modifications are particularly useful for extracting DNA from small, tough microsporidian spores. Microsporidia infecting terrestrial hosts, for example, possess a much thicker endospore than species from aquatic hosts, and small spores (> 3 μm) from terrestrial hosts may be even more difficult to fracture. Methods must therefore be adapted to individual microsporidian species.

#### *1. Extraction of nucleic acids*

*a. Germination of spores*

Ribosomal DNA is easily extracted by agitation of the spores with glass beads, but this procedure can cause heating and excessive shearing of nuclear DNA. A less abrasive method for obtaining the spore contents is to germinate the spores *in vitro*. In this procedure, the primary issue of concern is protecting the nucleic acids from nucleases. A reasonably successful procedure is described in Undeen & Cockburn (1989). When germinating high concentrations of spores, the polar tubes entangle, forming a solid mass of spores, a problem that can be ameliorated by the addition of 2-mercaptoethanol, which dissolves the polar tubes.

*b. Agitation with glass beads*

For species producing spores that can not be germinated *in vitro*, spores are commonly disrupted by agitating them vigorously with glass beads. Cell disrupters such as the French press, Parr Bomb, X-press and sonicators can not break most microsporidian spores. The method used to break spores is dictated by the substance to be recovered. Nuclear DNA must be treated gently to prevent excessive shearing of the long molecules.

*Procedure.*

1. Cleaned spores are suspended in 0.2 ml Bead Beater Solution (BBS), 0.4 ml Tris-saturated phenol and combined with 0.4 g glass beads (0.5 mm diameter; smaller beads, e.g., 0.1 mm may be needed for very small spores such as *Endoreticulatus*) in a 1.5-ml Eppendorf tube.
2. The tube is capped, covered with parafilm and shaken with a bead beater for one min at high speed.
3. The tube is centrifuged in a microcentrifuge for 1 min at 10,000 r.p.m.
4. The aqueous phase (top) is withdrawn and transferred to another tube.
5. A 0.2-ml aliquot of BBS is added to the phenolic phase; this mixture is vortexed, centrifuged, and the aqueous phase is again extracted and combined with the first aqueous supernatant.
6. At 4°C, the aqueous supernatant is extracted with Tris-saturated phenol, centrifuged for 5 min, then:
7. Extracted with an equal volume of a 1 : 1 solution of phenol and chloroform.
8. A final extraction is made with an equal volume of chloroform.
9. The nucleic acids remaining in the aqueous phase are precipitated by adding one part sodium acetate (3 M) to nine parts final DNA extract and 2.5 parts cold absolute ethanol followed by chilling at −80°C for 15 min.
10. The nucleic acids are pelleted by centrifugation, dried to remove the alcohol, re-suspended in 20–30 μl of 10 mM, pH 8.5 Tris buffer, and stored at −80°C.

A less abrasive agitation can be used to extract nuclear DNA from spores using procedures similar to those described for the bead beater (Undeen & Cockburn, 1989). Small volumes (0.1–0.5 ml) of spore suspension ($10^7$–$10^9$ spores/ml) are combined with equal volumes of 0.5 μl glass beads in 10 × 75 mm glass culture tubes and shaken at high speed on a vortex mixer for 30–60 s. Approximately 60–80% of the spores are disrupted and the DNA is not severely sheared.

Spores can be mechanically disrupted but this step is no longer necessary for DNA extraction from most species; it has been replaced by heat and chemical disruption that does not destroy the DNA. Two efficient methods are 'HotSHOT' using EDTA as a lysis buffer, and methods using chelating materials such as Chelex 100® produced by BioRad.

*c. HotSHOT method (Truett* et al.*, 2000) Pre-heat PCR block to 95°C*

1. Into clean 200-µl PCR tubes, add 25 µl of lysis buffer (EDTA and NaOH).
2. Add homogenized spores or tissues to the tube.
3. Incubate samples in a PCR machine for 30 min at 95°C.
4. Cool samples to 4°C in the block.
5. Add 25 µl neutralization buffer (Tris-HCl).
6. Vortex samples and incubate at 4°C for 10–15 min.
7. Centrifuge samples in microfuge to eliminate debris (bottom of the tube).
8. Use 1.5 µl* of this lysate in 25-µl PCR reactions.
9. Samples can be stored at −20°C. Multiple freeze/thaws may cause PCR to fail.

* (EDTA and Tris may interfere with PCR, use as little as possible or eliminate all salt and EDTA by alcohol precipitation, or by DNA binding column before PCR and storage.)

*d. Chelex® method (Walsh* et al.*, 1991; modified by W.-F. Huang)*

1. Prepare Chelex buffer: 90% molecular water, 5% Chelex 100 and 5% Tween 20, in a 100-µl batch. If tissue/spore sample is not pure, add 2–2.5 µl proteinase K (1 ng/µl final concentration) to a 200-µl PCR tube.
2. Preheat PCR block to 95°C.
3. Mix spore suspension with buffer, 10 : 90 v : v and snap lids on tightly.
4. Vortex samples in Chelex slurry for 15 s.
5. Spin samples briefly (10–15 s) in a microcentrifuge. This step is to ensure that all liquid goes to the bottom and air bubbles are eliminated.
6. Incubate samples for 20–30 min at 95°C (add 2 h at 56°C prior to this step if proteinase K is used). Cool to 4°C.
7. Spin tubes at highest speed in microcentrifuge to eliminate debris and Chelex resin (Chelex resin will interfere with PCR.)
8. Extract supernatant for preparation for PCR.

*e. gDNA extraction*

The following protocol has been used on multiple species of microsporidia to obtain sufficient quantities of high-quality genomic DNA (gDNA) suitable for 454 Sequencing of complete genomes. (N. Sanscrainte, personal communication).

For gDNA extraction, the Omniprep™ Genomic DNA Extraction kit (G-Biosciences) is modified (based on recommendations from the Broad Institute's Genome Sequencing Sample Repository) and used as follows. Spores are purified using a continuous Ludox gradient, repeated if necessary (Section 3B2a) and are quantified on a hemacytometer. A pellet of $1 \times 10^8$ to $7 \times 10^8$ spores is re-suspended in 500 µl of Genomic Lysis Buffer (G-Biosciences). A 500-µl volume of glass beads (425–600 µm), siliconized using Sigmacote® (Sigma-Aldrich) per manufacture's protocol, is added to the spores/Genomic Lysis Buffer mixture. Spores are then mechanically disrupted using the Mini-Beadbeater-1 (BioSpec) at setting 48 (4800 r.p.m.) for 30 s. Hemacytometer counts show > 70% of spores are disrupted. Disruption time may vary depending on the species of microsporidia but excessive time may damage the DNA. The mixture is incubated with 50 µg/ml LonglifeTMRNase (G-Biosciences) for 30 min at 42°C, after which incubation with 200 µg/ml Longlife™ Proteinase K (G-Biosciences) is performed at 60°C for 3 h. After returning to room temperature, a 500-µl chloroform extraction is performed, and the aqueous layer transferred to 50 µl DNA Stripping Solution (G-Biosciences), mixed by inversion, and incubated at 60°C for 10 min. An initial 100 µl of Precipitation Solution (G-Biosciences) is added followed by additional 50-µl aliquots until a white precipitate forms, followed by centrifugation at $21{,}000 \times g$ for 7 min (or longer if precipitate is not all pelleted). To precipitate gDNA, the supernatant is then added to 500 µl of ice-cold isopropanol and held on ice for 30–60 min before centrifuging at $21{,}000 \times g$ for 10 min at 4°C. The gDNA pellet is then washed in 70% ethanol, air dried, and re-suspended in 30–50 µl of TE buffer. As 8–12 µg of gDNA is optimal for 454 Sequencing, several preps as described above may be required to obtain these quantities, depending on the microsporidian species. Visualization on SDS−PAGE should show RNA-free gDNA, appearing as a smear due to the bead beating of the spores.

*f. DNA extraction from preserved material on glass slides*

Hylis *et al.* (2005) isolated DNA from microsporidia archived on glass microscope slides for up to 50 years. Over half of the material, from both terrestrial and aquatic hosts on slides that were fixed with methanol and stained with Giemsa, produced partial sequences, up to 594 base pairs, of large- and small-subunit rDNA. The fragments were sufficient, along with host data, to determine the species of the microsporidia.

*g. Purity assessment*

A quality-control quantitative PCR (qPCR) check can be performed to ensure that the microsporidian gDNA contains < 10% contaminating host DNA (E. Troemel & C. Cuomo, personal communication). Primer sets should be designed for two microsporidian genes and two host genes and show ≥ 90% efficiency. When qPCR is run on a microsporidia gDNA sample with both the microsporidia and the host primers, the microsporidian primers should produce an amplicon at least three cycles before the host specific primers amplify the host gDNA. By performing this quality control step with two different primer sets, the microsporidian gDNA purity is confirmed.

*2. Extraction of proteins*

Foaming and heating must be avoided to prevent denaturing of the proteins that will inactivate enzymes. Three methods of spore disruption are suggested.

*a. Agitation with glass beads*

One aliquot (1 ml) high concentration of spores and 400 mg 0.5-mm glass beads are added to 2-ml microcentrifuge tubes and are ruptured using a mini beadbeater at 4600 r.p.m. for 5 min total shaking time, stopping every 30 s to chill the vial in an ice bath for 30 s (modified by P. Solter from Prigneau *et al.*, 2000). This method allows processing of two samples at once, with one being chilled while the other is in the beadbeater. Ruptured cell homogenates are centrifuged and supernatant is drawn off into new tubes. The supernatant is dried in an evaporative centrifuge and re-suspended in 1 ml solubilization buffer (Bio-Rad ReadyPrep™ sequential extraction reagent 3, containing 5 M urea, 2 M thiourea, 2% CHAPS, 2% SB 3−10, 2 mM tributylphosphine, 40 mM Tris, and 0.2% pH 3/10 ampholytes). Homogenates are incubated for 30 min at room temperature with intermittent vortexing in preparation for gel electrophoresis.

*b. Freeze-grinding*

Spores are frozen in a mortar and then ground with a pestle (Conner, 1970). The material is allowed to thaw, pool in the bottom of the mortar, and then refrozen. Several cycles of freezing can disrupt about 95% of the spores. The materials are not subjected to any heating by this procedure and are usable for immunological studies without further extraction. This procedure has also been followed with spores that were frozen at −196°C in liquid nitrogen (Strick, 1993).

Protein extraction was optimized by a combination of treatment with glass beads, freezing-sonication and lysis buffers for studies of proteins involved in microsporidian host invasion (Wang *et al.*, 2007).

*3. Extraction of carbohydrates*

Spores can be disrupted for extraction of sugars by grinding with glass beads for ~ 1 min using either a bead beater or a vortex mixer without need for cooling.

*Procedure.*

1. A 200-μl sample of spores suspended in deionized water is combined with an equal volume of 0.45-mm glass beads in a 10 × 75 mm borosilicate culture tube.
2. The tube is shaken for 1 min at the highest speed on a vortex mixer or in a mini bead beater for 50 s.
3. The homogenate (~ 100 μl) is withdrawn from the beads; an additional 100-μl aliquot of deionized water is added to the tube, the tube shaken, and the homogenate withdrawn and added to the first homogenate in a 1.5-ml Eppendorf tube.
4. The homogenates are immediately placed in boiling water for 5 min to stop enzymatic activity, then centrifuged at high speed for 5 min.
5. The supernatant is retained for carbohydrate assays.

## B. Sequencing

Details regarding sequencing protocols and primers for the microsporidian ribosomal DNA genes used for identification are found in Weiss & Vossbrinck (1999), and recent primary literature addresses details for sequencing of particular species, including sequences for specific primers (see GenBank http://www.ncbi.nlm.nih.gov). An approximately 1200-bp sequence of the small subunit (SSU) rRNA gene is most frequently sequenced for species identification; however, to identify microsporidian strains or closely related species, the more genetically variable internal transcribed spacer (ITS) region is often sequenced, requiring flanking sequences from the SSU and large subunit (LSU) on either side. Typically, the SSU (1.2–1.3 kb) is located at the 5′-end of the rDNA repeat unit, the LSU (2.8–3.3 kb) at the 3′-end, and the ITS (29–60 bp) between them. Unlike the rDNA in other eukaryotes, there are exceptions to this configuration, particularly in the 'true' *Nosema* group, those species most closely related to *N. bombycis*. In this group, and possibly in other phylogenetic clades, the sequence is reversed, with the LSU at the 5′-end and the SSU at the 3′-end. In this case, the ITS region is much larger, 160–180 bp, and without homology to the shorter, typical ITS region. The same primers are used for polymerase chain reaction and sequencing, but the sequence orientation and number of base pairs sequenced will be different from more typical species (Huang *et al.*, 2004).

Complicating the use of rDNA sequencing is the presence of polymorphic rDNA. Polymorphisms may be found in the ITS region of many eukarotes but, in microsporidia, polymorphisms may also occur in the SSU and LSU genes. The haplotypes may vary by single base pair substitutions or by single or multiple base pair insertions or deletions at any site in the rDNA. The variation may range from 96–99% similarity among haplotypes, higher than for species distinctions in some cases. Polymorphisms may primarily occur in one gene or region, allowing sequencing of another; for example, the ITS region may be much less polymorphic than the SSU, or *vice versa*; however, sequencing these species may require the use of cloning to isolate haplotypes.

## 5. PROPAGATION AND PRODUCTION

Large numbers of protistan spores can sometimes be obtained from field-collected insects; however, to assure that the spores of interest are one species, production in the laboratory is preferred as a source of material for experimental work. Laboratory propagation of a protist is generally limited by two factors, the amenability of the host to laboratory culture and the ability to infect the host with the pathogen. The latter is a problem for many aquatic microsporidian species that produce a spore type in one host that infects an unknown or difficult-to-rear intermediate host. A more convenient alternate (laboratory) host can often be used for production of microsporidia that do not have a high degree of physiological host specificity.

Spores can be fed to the hosts on artificial diet or on their natural food. Experimentation is necessary to determine the optimum dosage for best spore production. Feeding too many spores can kill the host too quickly, reducing spore production, too few spores may result in a low rate of infection. The infectious and lethal dosage ranges are generally quite broad, however, up to 10-times the 100% infectious dosage is often tolerated without excessive mortality. Outlined below are methods that have been routinely used for the production of some commonly studied microsporidia. A more thorough review of mass production methodologies can be found in Brooks (1980, 1988) and specifics for particular species are found in the primary literature.

### A. *In vivo* production of microsporidia

#### *1.* Edhazardia aedis

*E. aedis* spores are produced in vertically infected larvae from infected female *A. aegypti*. Spore production thus requires two host generations. Second instar larvae are infected by placing them in deionized water inoculated with $10^3$–$10^4$ spores/ml with larval density at about 1 larva/ml. A small amount of larval food is added to ensure normal feeding. After 24 h, the larvae are transferred to a larger volume of water suitable for optimal larval development. Pupae are collected and transferred to cages. The emerging adults are supplied with cotton soaked with sugar solution and provided a blood meal so that eggs will be produced. Eggs are collected and hatched and the larvae are reared for 5–7 days until pupation begins. Most of the infected larvae are delayed in development and fail to pupate; spores are harvested from fourth instar larvae.

#### *2.* Paranosema locustae

*Paranosema* (= *Nosema*) *locustae* spores have been produced in quantities sufficient to treat thousands of acres of pasture and rangeland in efforts to suppress grasshopper populations (Henry & Oma, 1981). To infect grasshoppers, lettuce is sprayed with *P. locustae*, $10^6$ spores per 2000 5th instar nymphs. Nymphs are fed for two consecutive days and then again on day 4. Infections develop slowly until 32 days post-inoculation; there are nearly $5 \times 10^9$ spores per male and about twice that in females (Henry & Oma, 1981). For this microsporidium and others that have a wide host range, there is considerable latitude among the Orthoptera in the choice of a production host.

*3.* Vairimorpha necatrix *and* Vairimorpha disparis

*Vairimorpha necatrix* infects noctuid (Lepidoptera) larvae and has a moderately broad host range within the host order. *Vairimorpha disparis* is ecologically host specific to the gypsy moth. Optimal spore production is accomplished by feeding late 3rd or early 4th instar larvae approximately $10^3$ spores each. Spores can be spread on the surface of meridic diet or, for bioassay purposes, pipetted on to a piece of diet small enough to ensure that it is entirely eaten within 24 h. Spores develop in the fat body tissues and, at a rearing temperature of ~24°C, are ready for harvest in 12–18 days. Pupation of infected larvae is either delayed or does not occur at all. On the order of $10^9$ spores will be produced in each larva, a mixture of the elongated, diplokaryotic type and octospores (meiospores). Octospores mature later, therefore, to obtain more octospores, harvest after 14 days.

*4.* Nosema *and* Endoreticulatus *spp.*

*Nosema* spp. and the *Endoreticulatus shubergi* species complex infecting Lepidoptera develop more slowly and are less virulent than *V. necatrix*, therefore infections should be initiated sooner—in late 2nd or early 3rd instar larvae depending on timing of the larval development period. Larval development is often slowed by disease and spores can be harvested 14–20 days post infection. *E. schubergi* infects the midgut and is easily harvested directly from these tissues; most lepidopteran *Nosema* spp. are systemic and are harvested from all internal tissues. *N. apis* and *N. ceranae* from honey bees are both restricted to the midgut tissues.

*5. Producing microsporidia with broad host ranges*

*A. algerae*, *Vavraia culicis*, and other microsporidia that infect small insects and have broad physiological host ranges, can be produced more efficiently in a larger alternate host. For these two mosquito pathogens, *Helicoverpa zea* larvae are reared individually (they are cannibalistic) to the 3rd or 4th instar and then starved overnight. A small drop of spore suspension (~ 10 μl) is then added to each container (a small piece of artificial diet can also be inoculated) and the larvae are held again for several hours or overnight. The larvae are then returned to individual containers of diet and reared to the adult stage. Spores are harvested when the adults begin to die from the disease. Approximately $10^9$ spores are produced in each *H. zea*, 1000 × the yield from a mosquito. Many *Nosema*-type and *Vairimorpha* species can be reared in *Spodptera exigua* or *H. zea* if hosts are not available.

*6. Injection of spores*

Some microsporidian species can be propagated in alternate hosts by intra-haemocoelomic injection of an aseptic spore suspension. Larger insects can be injected with a disposable 'tuberculin syringe' (1 ml, 25 gauge, 3/8 inch needle). A smaller needle and a microapplicator produce more consistent results because of better control over dosage and less tissue damage and wounding (Undeen & Maddox, 1973; Pilley *et al.*, 1978; Weiser, 1978). Lepidopteran larvae are best injected through the planta (base) of a proleg, with the needle entry at a shallow angle, sliding beneath the epidermis to avoid puncturing the gut; microbes released from the gut cause fatal septicemia. Injection through the proleg also reduces bleeding, as does anesthesia with $CO_2$, or immobilizing larvae and decreasing the turgor of the hemolymph. Chilling makes the insect easier to handle but inoculated insects must be warmed immediately because cold temperatures can seriously reduce spore germination. The stage of the host injected and the stage from which spores are harvested depend upon the development time of the microsporidium in the host, as well as the minimum size and age of the organism that can be injected (Pilley *et al.*, 1978). Spore production of *A. algerae* in *H. zea* is maximized by injecting 3rd instar larvae and harvesting spores from the adult moths a few days after emergence. It is unlikely that all species of microsporidia will infect insects by injection as easily as *A. algerae* and *V. necatrix*. The spores must be able to germinate in the blood in order to infect the susceptible tissues. Microsporidia that are the least host and tissue specific have the best chance of infecting laboratory hosts by this route, although the natural host also may be susceptible to injection by a host specific pathogen, for example, *V. disparis* injected into the gypsy moth larvae.

## B. *In vitro* propagation

### *1. Cell-free media*

Only pathogens that can reproduce outside living cells can be grown in cell-free media. The mosquito parasitic ciliates, *Tetrahymena pyriformis* and *Lambornella clarki* replicate both in the habitat and in the host. These pathogens can be cultured *in vitro* in a vitamin-supplemented, septic cerophyl (powdered wheat leaf) extract (Washburn *et al.*, 1988). The medium must be inoculated with a bacterium before use to provide a food source for the ciliates. The algal pathogen *Helicosporidium* spp., isolated from the black fly *Simulium jonesi,* replicated and encysted in cell-free TC100 + 10% fetal calf serum media (Boucias *et al.*, 2001). A small percentage of the vegetative forms encysted, suggesting that parasitism provides the enriched environment needed for optimal development, but also that this algal entomopathogen may have a free-living stage or cycle. In addition, the bumble bee trypanosome *Crithidia bombi* was propagated in brain–heart infusion media (Popp & Lattroff, 2010). Although the authors did not state whether development was complete, two growth-phase dependent forms of the mosquito pathogen *Crithidia fasciculata* were produced in similar media (Scolaro *et al.*, 2005). Other pathogenic protists are more fastidious, requiring the host or at least cultures of host cells for development.

### *2. Cell culture*

Mass production of microsporidian spores in cell culture is not yet practical. Some entomopathogenic microsporidia are easily inoculated into cell culture (Jaronski, 1984; Kurtti *et al.*, 1990; Hayasaka, 1993), and some are capable of infecting cells derived from organisms that are distantly related to the natural host (Ishihara, 1968; Undeen, 1975; Monaghan *et al.*, 2011) and in tissues that do not reflect natural tropism (Monaghan *et al.*, 2009). However, only a few species (notably some of the pathogens of vertebrates and *A. algerae*) have been maintained in culture for more than a few generations. Often, spores produced in cell culture are under-developed or malformed (Tsai *et al.*, 2009), similar to the aberrations found when microsporidia are inoculated into non-target but partially permissible live hosts (Solter & Maddox, 1998). Kurtti *et al.* (1994) achieved 70 generations of *N. furnacalis* in insect cell culture, but noted that after several generations the internally infective primary spore type predominated and, after 40 generations, the isolates lost infectivity due to production of aberrant spores. Nevertheless, cell culture can be used for studies of microsporidia as enumerated by Monaghan *et al.* (2009): (1) produce pathogens in sufficient numbers for study, particularly for DNA analysis; (2) study the early stages of development; (3) study cell biology and kinetics of the various microsporidian stages; (4) characterize the responses of host cells to infection; and (5) study the effects of potential treatments for microsporidian infections.

Cell culture can be inoculated with a microsporidium by the addition of explanted tissues obtained by sterile dissection from an infected host (Sohi, 1971; Sohi & Wilson, 1976). More often, established cell lines are inoculated with aseptic spores. Initiating germination of the spores in order to infect the cells is frequently necessary.

### *3. Modification of the culture medium*

Infection of cultured cells occurs when the spores actively germinate, a process that is usually complete within a few min. Inoculation of cultured cells with microsporidia that germinate in simple salt solutions (e.g., *A. algerae*) is straightforward, but microsporidia and other entomopathogenic protists are adapted to germinate in the host's gut, so the culture medium may need to be altered temporarily for spore germination and infection to occur. Germination requirements of the spores must be met without damaging the cells, or the spores must be pretreated to germinate in media that is normally not stimulatory. The amount of time that cells are immersed in germination medium is minimized by replacing the culture medium as soon as germination is complete. During the germination period, the temperature should be held near the upper limits for the survival of the microsporidium or the cell cultures so that the maximum percentage germination is obtained.

### *4. Conditioning the spores*

Spores of some species of microsporidia require pretreatment ('priming'), and then will germinate in a second solution, the culture medium in this instance. There is no one method that is always successful. Some techniques are described below.

*a. High pH-neutralization*

Many microsporidia germinate in a near-neutral solution after priming for 10–30 min in an alkaline solution (pH 11.0 or above), usually a low concentration of NaOH or KOH (approximately 0.01 M). In addition to an alkaline pretreatment, some spores require chelation of bivalent metal ions for optimum germination. The following method was used by Kurtti *et al.* (1990) to inoculate cell cultures with spores of *V. necatrix*. Spores in an aseptic suspension, at sufficient concentration for 5–10 spores/cell, were suspended in 5 ml of 5 mM EDTA in 0.5 mM Tris-HCl, pH 7.5 for 30 min at room temperature. The spores were pelleted by centrifugation at 260 × *g*, re-suspended and held for 30 min in a priming solution of 0.01 M KOH in 0.17 M KCl. Cells were suspended in culture medium and both the cells and the spores were centrifuged at 260 × *g* for 5 min at room temperature. The cell pellet was re-suspended in 1 ml of 0.17 M KCl in 1 mM Tris-HCl with 10 mM EDTA at pH 8 (the germination solution) and immediately mixed with the spore pellet. After 3 min, the suspension of cells and spores was poured into 30 ml of fresh culture medium plus 50 mg/ml gentamycin to prevent bacterial growth and placed in the appropriate culture vessels. It should be noted that the cells were suspended in the germination medium first and the spores added last. Germination proceeds so rapidly after stimulation that if the spores are added first, many germinate before they come into contact with the cells.

*b. Other priming systems used*

Spores of *Nosema michaelis* from the blue crab were primed for 90–120 min in Michaelis veronal-acetate buffer (9.7 g sodium acetate and 2.9% (wv) sodium barbiturate in 500 ml $CO_2$-free distilled water); they discharged in cell culture medium (199 GIBCO (R)) with glutamine and Hank's salts (Weidner, 1976). Gisder *et al.* (2011) dried *N. ceranae* and *N. apis* spores for 30 min under vacuum for immediate use and added 100 µl cell suspension and 400 µl 0.1 M sucrose in PBS-buffer to the dried spores to initiate germination. After 5 min incubation, the spore-cell-suspension was transferred to tissue culture flasks containing 4.5 ml TC-100 medium + 11% FCS, 250 µg/ml penicillin/streptomycin and 125 µl antimycotic/antibiotic solution.

*c. Timing*

Although germination usually occurs quickly after stimulation, with careful timing spores can be stimulated, even in solutions that are unfavorable to cells, then transferred to the cell cultures in a volume of germination solution too small to affect the culture medium. The amount of time between stimulation and germination (eversion of the polar tube) can be extended by stimulating the spores in a high concentration of sucrose (about 1.7 M for *A. algerae*) or polyethylene glycol. The process of germination will continue after dilution in the culture medium (Undeen & Frixione, 1990).

## 6. CONTROLLING INFECTIONS IN INSECT COLONIES

Protists, especially microsporidia, can cause serious problems in insect colonies. An organism later determined to be the microsporidium *N. bombycis*, was determined by Louis Pasteur to be the causative agent for 'pebrine disease' in silkworms and is transmitted from infected females to progeny in the eggs. Using this knowledge, Pasteur initiated new uninfected colonies from offspring of moths that were isolated for oviposition and determined by microscopy to be uninfected after oviposition was complete (Steinhaus, 1975). Today, the "Pasteur Technique" is still the most reliable method to rid a laboratory colony of unwanted infection by microsporidia.

Selection of uninfected individuals is even less complicated when the contaminating microsporidium is not transmitted transovarially. Removal of dead adults from oviposition containers and rinsing the eggs with water is sufficient to keep colonies of anopheline mosquitoes free of infection by *A. algerae* (Alger & Undeen, 1970). A 5% household bleach solution is effective in destroying microsporidian spores on the surface of insect eggs and on laboratory equipment and surfaces. Some drugs, e.g., fumagillin, are fed to insects for control of microsporidia but none are known to eliminate the disease.

### A. Selecting uninfected progeny

In order to eliminate microsporidia from an insect colony, the following procedure should be used.

1. Isolate gravid females individually, or one unmated female and one or two males (to ensure mating), in sterilized containers.
2. After oviposition is completed, dissect the female (and males if present) and examine for signs of infection.
3. Rear only the offspring of uninfected parent insects and spot check the cohort for infection.
4. Destroy all cohorts produced by infected adults or in which infection is found.
5. If the resulting new colony is uninfected, destroy the infected parental colony and sterilize the rearing rooms and all equipment used for rearing the new colony.

### B. Sanitation

If a contaminating microsporidium is not transovarially transmitted, disease can be controlled by sterilization of all equipment used to rear the animal. All dead adults and other insect material must be separated from the eggs, and eggs rinsed in distilled water. If spores are deposited along with the eggs and adhere to them (transovum transmission), 0.25–5.0% bleach can be used for sterilization, the concentration depending on the ability of host eggs to tolerate the bleach. In most cases, a concentration of bleach and treatment time can be determined that will kill the microsporidian spores without damaging the eggs. Sanitation measures used in bacteriology labs are good guidelines for avoiding contamination problems with microsporidian spores. All counter tops, glassware, dissection instruments, and even pens, pencils, chemical jars, spray bottles—anything that is used in proximity to the microsporidia and the insects—must be routinely sterilized by heat or wiped down with 5% household bleach (the least expensive effective antimicrobial) or another antimicrobial. Some spores, particularly those that tolerate desiccation, are difficult to kill and some agents might not be completely effective; for example, ethanol can evaporate before all the spores are destroyed. These measures should remain a part of routine colony maintenance.

## 7. BIOASSAYS AND EXPERIMENTAL INFECTIONS

Bioassays are used to determine the viability, virulence and/or relative infectivity of a protist pathogen.

### A. Conducting bioassays

To conduct a quantitative bioassay, spores must be counted and a dilution series made so that a known number or concentration of spores is fed to the host. A control group, fed only the suspension medium but otherwise handled in the same manner as the test groups, must always be included as a check for mortality unrelated to the pathogen. The test insects must be healthy and of uniform age and size. All other environmental factors subject to control must be as consistent as possible within and between trials. The trials should be repeated to assure consistency and to provide sufficient numbers to demonstrate statistical significance. Details of statistical analyses of bioassay data are found in Chapter VII and in Marcus & Eaves (2000).

#### *1. Quantification of infective units*

To calculate a dosage or concentration for any bioassay, the spores or other infectious stages must be accurately counted. The most frequently used tool to count microsporidian spores is the hemocytometer, designed to count blood cells. The counting grid is divided into 25 large squares, each divided into 16 smaller squares. The procedures and calculations for different types of hemocytometers, e.g., Brite-line® phase contrast hemocytometer, Neubauer and Petroff–Hausser, are presented in Chapters II and VII and included in the manufacturer's instructions.

#### *2. Dosing insects in bioassays*

Bioassays differ primarily by the method of feeding inoculum to the host. A dosage is the known number of spores or other infectious stages fed to the host. (A dose is the number of infectious stages per body weight, and is rarely used in insect studies.) A concentration is a known number of infectious stages per volume of suspension or per unit surface area made available to the host, either on the surface of the host's food or suspended in water for filter feeders, but the exact number of spores ingested by individual test organisms cannot be determined.

*a. Methods for oral inoculation*

Dosages of infectious agents are usually fed to the test host on a small amount of its natural food or artificial diet contaminated with a known number of spores. The amount of food used is standardized within the test and should be sufficiently small that the host can consume all of it within a short period. Any test organism failing to consume the entire amount within a specified time period is eliminated from the test. Another method is to offer a starved host a small (1–2 µl) droplet of spore suspension on a calibrated bacteriological inoculating loop. The loop is dipped into a suspension of spores by a standard technique to ensure that an equal volume is picked up each time, and then the droplet is touched to the mouthparts of the test insect. Some lepidopoteran larvae will imbibe the spore suspension as soon as the loop touches the mouthparts. Only those hosts that consume the entire quantity should be included in the assay. Droplet feeding can be used for neonate larvae for which the average volume of the first ingestion of a liquid is known (previously calculated) and for species that possess a sufficiently translucent epidermis to observe whether a full volume of dyed spore suspension has been ingested. This method has been used extensively in bioassays of *Nosema pyrausta* infecting the European corn borer (Hughes & Wood, 1981).

Bioassay results are usually presented as an LD (lethal dosage) or ID (infective dosage) followed by a subscript percentage, without the % sign, most often 10, 50, or 90. For example, $LD_{50}$ is the dosage that, on average, kills 50% of the larvae; $ID_{90}$ is the dosage expected to infect 90% of the test insects. For further explanation, see Chapter VII.

Lethal or infective concentrations are used when the exact numbers of spores ingested by the test organisms can not be determined. A concentration series is applied either per volume or per unit area and forms the basis of an assay in which results are reported as lethal or infective concentrations instead of dosages, e.g., $LC_{50}$ or $IC_{50}$. Uniformity in application is the key to consistency for these techniques. Spore concentrations are used for filter feeding organisms, such as mosquito larvae; spores are mixed with water in a known concentration. The volume of water in the exposure arena and the time that larvae remain in it are important. The chronic nature of many protistan diseases is frequently (although not always) expressed in reduced longevity and fecundity, as well as high mortality in transovarially infected offspring, and for many interactions, the effects are better assessed in terms of loss of reproductive potential than outright mortality.

*3. Bioassay procedures*

The following bioassay procedure for aquatic hosts is typical but many variations are possible and even necessary. A reasonable range of dosages or concentrations is established in a preliminary 'range-finding' assay. A 10-fold dilution series is generally used for the preliminary assay.

*a. Dilution series*

A dilution series of pathogen infectious stages is made within the dosage range of the preliminary assay, generally between ~5 and 95% infection rates. A double dilution series is usually satisfactory but at least five dosage groups within the targeted infection range should be included in the assay. Dilutions are preferable to measuring material directly from the stock factors. With the same concentration of spores in the water, larvae in a large volume might receive a higher dosage than larvae in a small volume with the same concentration. Likewise, larger numbers of larvae may receive a smaller dosage than fewer larvae in an equal volume of water. Standardization of variables such as temperature, volume of water, duration of exposure, age/stage of larvae and food availability must, therefore, be rigidly controlled. Results are more accurate when larger volumes are measured and each test group receives the same volume.

Infection is often a more useful parameter than lethality for protists. Mortality due to infection by protists is extended over such a long time period that mortality by natural causes can be excessively high.

*b. Selection of test organisms*

All test organisms must be healthy and of the same age and size. As few as 10 individuals per dosage group can be used, but higher numbers provide more robust results. Staging of colony insects is used to choose insects that have completed a molt to the target stage within a short time period.

*c. Replicates and controls*

To guard against the loss of a dosage group and to assess the variability within the assay, each dosage group should be set up in triplicate. Always include a control group that is treated in an identical manner except that no spores are included in the inoculum. Occasionally, positive controls are needed, for example, when studying interactions of pathogens in mixed infections. Further details can be found in Chapters II, IV, and VII, and references in Navon & Ascher (2000).

*d. Expose test organisms*

Feed infectious spores to the test organisms by appropriate means (host and pathogen species specific) using the same volume for each dosage group. Include 'sham controls' which subject identical groups of insects to the same handling and dosage procedure except that no pathogens are included in the inoculum.

*e. Development time*

Allow sufficient time for the infections to develop, depending on the lifecycle of the pathogen, or for mortality to occur, maintaining groups under uniform conditions optimum for development of the host and the pathogen.

*4. Scoring bioassays*

*a. Mortality*

Count the number of dead and live test organisms in each group, including the control (uninfected) group. As dead insects are sometimes missing (decomposed or eaten by others), the usual procedure is to count the live insects. The number of dead per treatment = the number in the initial group minus the live insects.

1. Combine the numbers from all repetitions of each dosage group.
2. Calculate the percentage of dead insects in each group: 100 × (number dead/total number of insects).
3. Using Abbott's Formula, correct the percentage mortality: corrected % mortality = 100 × (T%−C%/100%−%C), where T% = the percentage of dead test organisms and C% = the percentage of dead control organisms.

*b. Infectivity*

1. Count the number of infected and uninfected organisms in each dosage group and in the control group. Examine and score both dead and live animals.
2. Calculate % infection: 100 × (number infected/total number scored). In this case the missing organisms are omitted from consideration.
3. Analyze data using probit analysis with a variety of software statistical packages such as SPSS® (IBM®), SAS (SAS Institute, Inc.), R (R Project: http://www.R-project.org/), S (Alcatel-Lucent/Bell Labs) or LdP Line (Bakr, http://www.ehabsoft.com/ldpline/).

**B. Viability tests**

Viability can be assessed without infecting host animals. Viable microsporidian spores are brightly refringent when inspected by phase-contrast microscope. They must be also capable of germination, a rapid process that is easily observed by closing the iris diaphragm when using bright-field or, better, with a phase-contrast microscope. Within a few minutes of stimulation, spores usually lose their refringency and a discharged polar tube can be seen (Figure 11.7). Spores treated with gamma radiation or ultraviolet light are the only demonstrated exceptions to this generality; irradiated spores may be capable of germination for several days after they are no longer capable of infecting the host (Undeen & Vander Meer, 1990). Viability is probably indicated by *in vitro* germination even when the germination stimuli are obviously abnormal, that is, different from those prevailing in the gut of the host.

*1. Viability stains*

Inviable microsporidian spores appear less refractive under phase-contrast lighting but there have been no reliable visual cues or established staining methods to differentiate live from dead spores until recently. Sytox Green and Calcoflour White MR2 were shown to differentially stain viable and inviable spores of the human pathogen *Encephalitozoon cuniculi* in cell culture (Green *et al.*, 2000), and infectious spores of *Encephalitozoon intestinalis* stained poorly while smaller non-infectious spores were highly permeable to the dyes (Hoffman *et al.*, 2003). A modification of the Green *et al.* (2000) method using DAPI stain was used to determine viability of *N. ceranae* (Fenoy *et al.*, 2009), however, no corresponding bioassays were reported to validate the assessment. Sytox Green was also used to distinguish dead spores of *Thelohania solenopsae* from fire ants (*Solenopsis invictae*) (Hale, 2006). *T. solenopsae* can not be assayed by feeding spores.

*2. Spore germination*

The spores of many microsporidia germinate in alkaline (pH 9.0−11.0) solutions of monovalent salts (KCl, CsCl,

NaCl, RbCl, KI, KBr and many others) (Ohishima, 1964; Weidner *et al.*, 1984; Keohane & Weiss, 1999). The best ions and pH conditions for a particular species require experimentation. Many species germinate when rehydrated in a buffered salt solution after they have been subjected to partial dehydration (Undeen & Avery, 1984), observed by placing a thin drop of spore suspension on a slide and permitting it to dry until only a small wet spot remains in the center. The area is then flooded with a drop of the putative germination solution and a cover slip applied. A ring of germinated spores, marking the zone where the proper degree of drying has occurred, is seen a few min later. Partial desiccation can be obtained in a more quantitative manner by placing the spores in a 1.6–1.8 M sucrose solution for about 15 min, then flooding with the test solution. Although sometimes producing high percentages of germination, drying–rehydration is not necessarily the physiologically normal stimulus.

Spores of many species will germinate in the presence of 2–5% hydrogen peroxide ($H_2O_2$), although sometimes at very low percentages. Contrary to other germination stimuli, germination by $H_2O_2$ is often better at low temperatures. The effect of calcium in the germination medium is variable with species (see Ishihara, 1967; Undeen, 1978, 1983; Malone, 1984; Weidner & Halonen, 1993).

Germinated spores appear dark with an obviously empty spore case. If examined within a few min after germination, the sporoplasm can be seen at the end of the thin polar tube in the form of a minute drop of cytoplasmic material.

### *3. Spore density*

Viable microsporidian spores contain high concentrations of sugars that are gradually lost from inviable spores (Undeen & Solter, 1996) and can be measured. A Ludox density gradient can sometimes be used when no other method is possible; sugar-depleted spores (and immature spores, as well) are less dense than viable spores and usually form a band above the viable spore band in density gradients.

## ACKNOWLEDGMENTS

We thank Wei-Fone Huang for sharing his expert procedural modifications of molecular techniques; Neil Sanscrainte for sharing methods for Ludox gradients; and Mirek Hylis for comments on the manuscript.

## REFERENCES

Abe, S., & Davies, E. (1986). Quantitative analysis of polysomes using a baseline from uncentrifuged blank gradients. *Mem. College Agricul., Ehime University, 31*, 187–199.

Adl, S. M., Simpson, A. G., Farmer, M. A., Andersen, R. A., Anderson, O. R., Barta, J. R., Bowser, S. S., Brugerolle, G., Fensome, R. A., Fredericq, S., James, T. Y., Karpov, S., Kugrens, P., Krug, J., Lane, C. E., Lewis, L. A., Lodge, J., Lynn, D. H., Mann, D. G., McCourt, R. M., Mendoza, L., Moestrup, O., Mozley-Standridge, S. E., Nerad, T. A., Shearer, C. A., Smirnov, A. V., Spiegel, F. W., & Taylor, M. F. (2005). The new higher level classification of eukaryotes with emphasis on the taxonomy of protists. *J. Eukaryot. Microbiol, 52*, 399–451.

Alger, N. E., & Undeen, A. H. (1970). The control of a microsporidian, *Nosema* sp. in an anopheline colony by an egg-rinsing technique. *J. Invertebr. Pathol., 15*, 321–337.

Avery, S. W., & Undeen, A. H. (1987). The isolation of microsporidia and other pathogens from concentrated ditch water. *J. Am. Mosq. Control Assoc., 3*, 54–58.

Bläske-Lietze, V.-U., Shapiro, A. M., Denton, J. S., Botts, M., Becnel, J. J., & Boucias, D. G. (2006). Development of the insect pathogenic alga *Helicosporidium*. *J. Eukaryot. Microbiol., 53*, 165–176.

Boucias, D., Becnel, J. J., White, S. E., & Bott, M. (2001). *In vitro* development of the protist *Helicosporidium* sp. *J. Eukaryot. Microbiol., 48*, 460–470.

Bourgeois, L., Beaman, L. D., & Rinderer, T. E. (2011). Preservation and processing methods for molecular genetic detection and quantification of *Nosema ceranae*. *Sci. Bee Cult., 3*, 1–5.

Brooks, W. M. (1980). Production and efficacy of protozoa. *Biotech. Bioeng., 22*, 1415–1440.

Brooks, W. M. (1988). Entomogenous protozoa. In C. M. Ignoffo (Ed.), *CRC Handbook of Natural Pesticides. Volume V Microbial Insecticides Part A. Entomogenous Protozoa and Fungi* (pp. 1–150). Boca Raton, FL: CRC Press, Inc.

Brown, M. J. F., Loosli, R., & Schmid-Hempel, P. (2000). Condition-dependent expression of virulence in a trypanosome infecting bumblebees. *Oikos, 91*, 421–427.

Canning, E. U., & Vávra, J. (2000). Phylum Microsporida. In J. J. Lee, G. F. Leeddale, & P. Bradbury (Eds.), *An Illustrated Guide to the Protozoa* (pp. 39–126). Lawrence, KS: Society of Protozoologists.

Cavalier-Smith, T. (2010). Kingdoms Protozoa and Chromista and the eozoan root of the eukaryotic tree. *Biol. Lett., 6*, 342–345.

Cepicka, I., Hampl, V., & Kulda, J. (2010). Critical taxonomic revision of parabasalids with description of one new genus and three new species. *Protist, 161*, 400–433.

Cole, R. J. (1970). The application of the 'triangulation' method to the purification of *Nosema* spores from insect tissues. *J. Invertebr. Pathol., 15*, 193–195.

Conner, R. M. (1970). Disruption of microsporidian spores for serological studies. *J. Invertebr. Pathol, 15*, 138.

Egerter, D. E., & Anderson, J. R. (1985). Infection of the western treehole mosquito, *Aedes sierrensis* (Diptera: Culicidae), with *Lambornella Clark* (Ciliophora: Tetrahymenidae). *J. Invertebr. Pathol., 46*, 296–304.

Fenoy, S., Rueda, C., Higes, M., Martín-Hernández, R., & del Aguila, C. (2009). High-level resistance of *Nosema ceranae*, a parasite of the honeybee, to temperature and desiccation. *Appl. Environ. Microbiol., 75*, 6886–6889.

Foissner, W. (1991). Basic light and scanning electron microscopic methods for taxonomic studies of ciliated protozoa. *Eur. J. Protistol., 27*, 313–330.

Frampton, M., Droege, S., Conrad, T., Prager, S., & Richards, M. H. (2008). Evaluation of specimen preservatives for DNA analyses of bees. *J. Hym. Res., 17*, 195–200.

Gisder, S., Hedtke, K., Möckel, N., Frielitz, M. C., Linde, A., & Genersch, E. (2010). Five-year cohort study of *Nosema* spp. in Germany: Does climate shape virulence and assertiveness of *Nosema ceranae*? Appl. Environ. *Microbiol., 76*, 3032–3038.

Gisder, S., Möckell, N., Linde, A., & Genersch, E. (2011). A cell culture model for *Nosema ceranae* and *Nosema apis* allows new insights into the life-cycle of these important honey bee-pathogenic microsporidia. *Environ. Microbiol., 13*, 404–413.

Green, L. C., LeBlanc, P. J., & Didier, E. S. (2000). Discrimination between viable and dead *Encephalitozoon cuniculi* (microsporidian) spores by dual staining with SYTOX Green and Calcofluor White M2R. *J. Clin. Microbiol., 38*, 3811–3814.

Hale, M. W. (2006). *Host/parasite Interactions Between* Solenopsis invicta *(Hymenoptera: Formicidae) and* Thelohania solenopsae *(Microsporida: Thelohaniidae)*. MS Thesis: Texas A & M University.

Hausmann, K., Hulsmann, N., & Radek, R. (2003). *Protistology* (3rd ed.). Stuttgart, Germany: Schweizerbart'sche Verlagsbuchhandlung. pp. 379.

Hayasaka, S., Sato, T., & Inoue, H. (1993). Infection and proliferation of microsporidians pathogenic to the silkworm *Bombyx mori* L. and the Chinese oak silkworm *Antheraea pernyi* in lepidopteran cell lines. *Bull. Natl. Inst. Seric. Entomol. Sci. O, 7*, 47–63.

Hazard, E. I., & Brookbank, J. W. (1984). Karyogamy and meiosis in an *Amblyospora* sp. (Microspora) in the mosquito *Culex salinarius*. *J. Invertebr. Pathol., 44*, 3–11.

Hazard, E. I., Ellis, E. A., & Joslyn, D. J. (1981). Identification of microsporidia. In H. D. Burges (Ed.), *Microbial Control of Plant Pests and Plant Diseases* (pp. 163–182). New York, NY: Academic Press.

Henry, J. E., & Oma, E. A. (1981). Pest control by *Nosema locustae*, a pathogen of grasshoppers and crickets. In H. D. Burges (Ed.), *Microbial Control of Pests and Plant Diseases 1970–1980* (pp. 573–586). New York, NY: Academic Press.

Hoffman, R. M., Marshall, M. M., Polchert, D. M., & Jost, B. H. (2003). Identification of two subpopulations of *Encephalitozoon intestinalis*. *Appl. Environ. Microbiol., 69*, 4966–4970.

Huang, W.-F., Tsai, S.-J., Lo, C.-F., Soichi, Y., & Wang, C.-H. (2004). The novel organization and complete sequence of the ribosomal RNA gene of *Nosema bymbycis*. *Fungal Genet. Biol., 41*, 473–481.

Hughes, P. R., & Wood, H. A. (1981). A synchronous peroral technique for the bioassay of insect viruses. *J. Invertebr. Pathol., 37*, 154–159.

Hylis, M., Weiser, J., Oborník, M., & Vávra, J. (2005). DNA isolation from museum and type collection slides of microsporidia. *J. Invertebr. Pathol., 88*, 257–260.

Ishihara, R. (1967). Stimuli causing extrusion of polar filaments of *Glugea fumiferanae* spores. *Can. J. Microbiol., 13*, 1321–1332.

Ishihara, R. (1968). Growth of *Nosema bombycis* in primary cell cultures of mammalian and chicken embryos. *J. Invertebr. Pathol., 11*, 328–329.

Iwano, H., & Kurtti, T. J. (1995). Identification and isolation of dimorphic spores from *Nosema furnacalis* (Microspora: Nosematidae). *J. Invertebr. Pathol., 65*, 230–236.

Jahn, T. L., & Jahn, F. J. (1949). *How to Know the Protozoa*. Wm. C. Brown Co. pp. 234.

Jaronski, S. T. (1984). Microsporida in cell culture. *Adv. Cell Culture, 3*, 183–299.

Jouvenaz, D. P. (1981). Percoll: An effective medium for cleaning microsporidian spores. *J. Invertebr. Pathol, 37*, 319.

Keeling, P. J. (2009). Five questions about Microsporidia. *PloS Pathogens, 5*, 1–3.

Keeling, P. J., & Fast, N. M. (2002). Microsporidia: Biology of highly reduced intracellular parasites. *Ann. Rev. Microbiol., 56*, 93–116.

Kellen, W. R., & Lindegren, J. E. (1974). Life-cycle of *Helicosporidium parasiticum* in the naval orangeworm, *Paramyelois transitella*. *J. Invertebr. Pathol., 23*, 202–208.

Keohane, E. M., & Weiss, L. M. (1999). The structure, function and composition of the microsporidian polar tube. In M. Wittner, & L. M. Weiss (Eds.), *The Microsporidia and Microsporidiosis* (pp. 196–224). Washington, DC: ASM Press.

Kudo, R. (1966). *Protozoology*. Springfield, IL: C. C. Thomas. pp. 1174.

Kurtti, T. J., Munderloh, U. G., & Noda, H. (1990). *Vairimorpha necatrix*: Infectivity for and development in a lepidopteran cell line. *J. Invertebr. Pathol., 55*, 61–68.

Kurtti, T. J., Ross, S. E., Liu, Y., & Munderloh, U. G. (1994). *In vitro* developmental biology and spore production in *Nosema furnacalis* (*Microspora: Nosematidae*). *J. Invertebr. Pathol., 63*, 188–196.

Lange, C. E. (1993). Unclassified protists of arthropods: The ultrastructure of *Nephridiophaga periplanetae* (Lutz & Splendore, 1903) n. comb., and the affinities of the Nephridiophagidae to other protists. *J. Eukaryot. Microbiol., 40*, 689–700.

Lantova, L., Ghosh, K., Svobodova, M., Braig, H. R., Rowton, E., Weina, P., Volf, P., & Votypka, J. (2010). The life-cycle and host specificity of *Psychodiella sergenti n.* sp. and *Ps. tobbi* n. sp. (Protozoa: Apicomplexa) in sand flies *Phlebotomus sergenti* and *Ph. tobbi* (Diptera: Psychodidae). *J. Invertebr. Pathol., 105*, 182–189.

Larsson, J. I. R. (1999). Identification of microsporidia. *Acta Protozool., 38*, 161–197.

Lee, J. J., & Soldo, S. T. (1992). *Protocols in Protozoology.* Lawrence, KS: Society of Protozoologists at Allen Press. pp. 240.

Lee, J. J., Leedale, G. F., & Bradbury, P. (2000). *An Illustrated Guide to the Protozoa.* Lawrence, KS: Society of Protozoologists. pp. 1432.

Maddox, J. V., & Solter, L. F. (1996). Long term storage of viable microsporidian spores in liquid nitrogen. *J. Eukaryot. Microbiol., 43*, 221–225.

Malone, L. A. (1984). Factors controlling *in vitro* hatching of *Vairimorpha plodiae* (Microspora) spores and their infectivity to *Plodia interpunctella*, *Heliothis virescens*, and *Pieris brassicae*. *J. Invertebr. Pathol., 44*, 192–197.

Marcus, R., & Eaves, D. M. (2000). Statistical and computational analysis of bioassay data. In A. Navon, & K. R. S. Ascher (Eds.), *Bioassays of Entomopathogenic Microbes and Nematodes* (pp. 249–293). Wallingford, UK: CABI Publishing.

Mazia, D., Schatten, G., & Sale, W. (1975). Adhesion of cells to surfaces coated with polylysine. *J. Cell Biol., 66*, 198–200.

MetaView. (1998). *Meta Imaging Series 4.5. West.* Chester, PA: Universal Imaging Corporation.

Monaghan, S. R., Kent, M. L., Watral, V. G., Kaufman, R. J., Lee, L. E. J., & Bols, N. C. (2009). Animal cell cultures in microsporidial research: their general roles and their specific use for fish microsporidia. *In Vitro Cell. Dev. Biol., 45*, 135–147.

Monaghan, S. R., Rumney, R. L., Nguyen, T. K. V., Bols, N. C., & Lee, L. E. J. (2011). *In vitro* growth of microsporidia *Anncaliia algerae* in cell lines from warm water fish. *In Vitro Cell. Dev. Biol., 47*, 104–113.

Navon, A., & Ascher, K. R. S. (Eds.). (2000). *Bioassays of Entomopathogenic Microbes and Nematodes*. Wallingford, UK: CABI Publishing.

Nie, D. (1950). Morphology and taxonomy of the intestinal protozoa of the Guinea-pig. *Cavia porcella. J. Morphol., 86*, 381–493.

Ohishima, K. (1964). Stimulative or inhibitive substances to evaginate the filament of *Nosema bombycis* as studied by the neutralization method. *Annot. Zool. Jpn, 37*, 102.

Peyretaillade, E., El Alaoui, H., Diogon, M., Polonais, V., Parisot, N., Biron, D. G., Peyret, P., & Delbac, F. (2011). Extreme reduction and compaction of microsporidian genomes. *Res. Microbiol., 162*, 598–606.

Pilley, B. M., Canning, E. U., & Hammond, J. C. (1978). The use of a microinjection procedure for large-scale production of the microsporidian *Nosema eurytremae* in *Pieris brassicae*. *J. Invertebr. Pathol., 32*, 355–358.

Podlipaev, S. A., Sturm, N. R., Fiala, I., Fernandes, O., Westenberger, S. J., Dollet, M., Campbell, D. A., & Lukes, J. (2004). Diversity of insect trypanosomatids assessed from the spliced leader RNA and 5S rRNA genes and intergenic regions. *J. Eukaryot. Microbiol., 51*, 283–290.

Popp, M., & Lattorff, M. G. (2010). A quantitative *in vitro* cultivation technique to determine cell number and growth rates in strains of *Crithidia bombi* (Trypanosomatidae), a parasite of bumblebees. *J. Eukaryot. Microbiol., 58*, 7–10.

Prigneau, O., Achbarou, A., Bouladoux, N., Mazier, D., & Desportes-Livage, I. (2000). Identification of proteins in *Encephalitozoon intestinalis*, a microsporidian pathogen of immunocompromised humans: An immunoblotting and immunocytochemical study. *J. Eukaryot. Microbiol., 47*, 48–56.

Radek, R., Wellmanns, D., & Wolf, A. (2011). Two new species of *Nephridiophaga* (Zygomycota) in the Malphigian tubules of cockroaches. *Parasitol. Res., 109*, 473–482.

Schilder, R. J., & Marden, J. H. (2007). Parasites, proteomics and performance: Effects of gregarine gut parasites on dragonfly flight muscle composition and function. *J. Exper. Biol., 210*, 4298–4306.

Scolaro, E. J., Ames, R. P., & Brittingham, A. (2005). Growth-phase dependent substrate adhesion in *Crithidia fasciculata*. *J. Eukaryot. Microbiol., 52*, 17–22.

Shykoff, J. A., & Schmid-Hempel, P. (1991). Incidence and effects of four parasites in natural populations of bumble bees in Switzerland. *Apidologie, 22*, 117–125.

Sohi, S. S. (1971). *In vitro* cultivation of hemocytes of *Malacosoma disstria* Hübner (Lepidoptera, Lasiocampidae). *Can. J. Zool., 49*, 1355–1358.

Sohi, S. S., & Wilson, G. G. (1976). Persistent infection of *Malacosoma disstria* (Lepidoptera, Lasiocampidae) cell culture with *Nosema disstriae (Microsporida, Nosematidae). Can. J. Zool., 54*, 336–342.

Solter, L. F., & Maddox, J. V. (1998). Physiological host specificity of microsporidia as an indicator of ecological host specificity. *J. Invertebr. Pathol., 71*, 207–216.

Solter, L. F., Becnel, J. J. & Oi, D. H. (2012). Microsporidian entomopathogens. In Insect Pathology and Microbial Pest Control. F. E. Vega & H. K. Kaya (eds.), San Diego, Elsevier.

Sprague, V., Becnel, J. J., & Hazard, E. I. (1992). Taxonomy of Phylum Microspora. *Crit. Rev. Microbiol., 18*, 285–293.

Steinhaus, E. A. (1975). *Disease in a Minor Chord.* Ohio State University Press. pp. 488.

Strick, H. (1993). Disruption of microsporidian spores for biochemical analysis. *Zeit. Angew. Zool., 77*, 3–4.

Svobodova, M., Zidkova, L., Cepicka, I., Obornik, M., Lukes, J., & Votypka, J. (2007). *Sergeia podlipaevi* gen. nov., sp. nov. (Typanosomatidae, Kinetoplastida), a parasite of biting midges (Ceratopogonidae, Diptera). *Int J. Syst. Evol. Micr., 57*, 423–432.

Tanada, Y., & Kaya, H. K. (1993). *Insect Pathology.* San Diego, CA: Academic Press, Inc. pp. 666.

Teixeira, M. M. G., Borghesan, T. C., Ferreira, R. C., Santos, M. A., Takata, C. S. A., Campaner, M., Nunes, V. L. B., Milder, R. V., de Souza, W., & Camargo, E. P. (2011). Phylogenetic validation of the genera *Angomonas* and *Strigomonas* of trypanosomatids harboring bacterial endosymbionts with the description of new species of trypanosomatids and of proteobacterial symbionts. *Protist, 162*, 503–524.

Truett, G. E., Heeger, P., Mynatt, R. L., Truett, A. A., Walker, J. A., & Warman, M. L. (2000). Preparation of PCR-quality mouse genomic DNA with hot sodium hydroxide and tris (HotSHOT). *Biotechniques, 29*, 52–54.

Tsai, Y. C., Solter, L. F., Wang, C. Y., Fan, H. S., Chang, C. C., & Wang, C. H. (2009). Morphological and molecular studies of a microsporidium (*Nosema* sp.) isolated from the three spot grass yellow butterfly. *Eurema blanda arsakia (Lepidoptera: Pieridae) J. Invertebr. Pathol., 100*, 85–93.

Undeen, A. H. (1975). Growth of *Nosema algerae* in pig kidney cell cultures. *J. Protozool., 22*, 107–110.

Undeen, A. H. (1978). Spore hatching processes in some *Nosema* species with particular reference to *Nosema algerae* Vávra & Undeen. *Misc. Publ. Entomol. Soc. Am., 11*, 29–50.

Undeen, A. H. (1983). The germination of *Vavraia culicis* spores. *J. Protozool., 30*, 274–277.

Undeen, A. H., & Alger, N. E. (1975). The effect of the microsporidian, *Nosema algerae*, on *Anopheles stephensi*. *J. Invertebr. Pathol., 25*, 19–24.

Undeen, A. H., & Avery, S. W. (1983). Continuous flow-density gradient centrifugation for purification of microsporidia spores. *J. Invertebr. Pathol., 42*, 405–406.

Undeen, A. H., & Avery, S. W. (1984). Germination of experimentally nontransmissible microsporidia. *J. Invertebr. Pathol., 43*, 299–301.

Undeen, A. H., & Becnel, J. J. (1992). Longevity and germination of *Edhazardia aedis* (Microspora: Amblyosporidae). *Biocontrol Sci. Technol., 2*. 257–256.

Undeen, A. H., & Cockburn, A. F. (1989). The extraction of DNA from microsporidia spores. *J. Invertebr. Pathol., 54*, 132–133.

Undeen, A. H., & Frixione, E. (1990). The role of osmotic pressure in the germination of *Nosema algerae* spores. *J. Protozool., 37*, 561–567.

Undeen, A. H., & Maddox, J. V. (1973). The infection of non-mosquito hosts by injection with spores of the microsporidian. *Nosema algerae. J. Invertebr. Pathol, 22*, 258–265.

Undeen, A. H., & Solter, L. F. (1996). The sugar content and density of living and dead microsporidian (Protozoa: Microspora) spores. *J. Invertebr. Pathol., 67*, 80–91.

Undeen, A. H., & Vander Meer, R. K. (1990). The effect of ultraviolet radiation on the germination of *Nosema algerae* Vávra & Undeen (Microsporida: Nosematidae) spores. *J. Protozool., 37*, 194–199.

Undeen, A. H., & Vander Meer, R. K. (1994). Conversion of intrasporal trehalose into reducing sugars during germination of *Nosema algerae* (Protista: Microspora) spores: A quantitative study. *J. Eukaryot, Microbiol., 41*, 129–132.

Undeen, A. H., & Vávra, J. (1997). Research methods for entomopathogenic Protozoa. In L. A. Lacey (Ed.), *Manual of Techniques in Insect Pathology* (pp. 117–151). New York, NY: Academic Press.

Undeen, A. H., Johnson, M. A., & Becnel, J. J. (1993). The effects of temperature on the survival of *Edhazardia aedis* (Microspora: Amblyosporidae), a pathogen of *Aedes aegypti*. *J. Invertebr. Pathol., 61*, 303–307.

Vávra, J. (1964). Recording microsporidian spores. *J. Insect Pathol., 6*, 258–260.

Vávra, J. (1976). Structure of the microsporidia. In L. A. Bulla, & T. C. Cheng (Eds.), *Comparative Pathobiology Vol. 1: The Biology of the Microsporidia* (pp. 2–74). London, UK: Plenum Press.

Vávra, J., & Larsson, J. I. R. (1999). Structure of the Microsporidia. In M. Wittner, & L. M. Weiss (Eds.), *The Microsporidia and Microsporidiosis* (pp. 7–84). Washington, DC: ASM Press.

Vávra, J., & Maddox, J. V. (1976). Methods in microsporidiology. In L. A. Bulla, & T. C. Cheng (Eds.), *Comparative Pathobiology Vol. 1. The Biology of the Microsporidia* (pp. 298–313). London, UK: Plenum Press.

Walker, G., Dorrell, R. G., Schlacht, A., & Dacks, J. B. (2011). Eukaryotic systematics: A user's guide for cell biologists and parasitologists. *Parasitology*. (in press).

Walsh, P. S., Metzger, D. A., & Higuchi, R. (1991). Chelex 100 as a medium for simple extraction of DNA for PCR-based typing from forensic material. *Biotechniques, 10*, 506–513.

Wang, J-Y., Chambon, C., Lu Huang, C-D., Vivarès, K-W., C.P., & Texier, C. (2007). A proteomic-based approach for the characterization of some major structural proteins involved in host–parasite relationships from the silkworm parasite. *Nosema bombycis (Microsporidia) Proteomics, 7*, 1461–1472.

Washburn, J. O., Gross, M. E., Mercer, D. R., & Anderson, J. E. (1988). Predator-induced trophic shift of a free-living ciliate: Parasitism of a mosquito larva by their prey. *Science, 240*, 1193–1195.

Weidner, E. (1976). The microsporidian spore invasion tube: The ultrastructure, isolation, and characterization of the protein comprising the tube. *J. Cell Biol., 71*, 23–34.

Weidner, E., Byrd, W., Scarborough, A., Pleshinger, J., & Sibley, D. (1984). Microsporidian spore discharge and the transfer of polaroplast organelle membrane into plasma membrane. *J. Protozool., 31*, 195–198.

Weidner, E., & Halonen, S. K. (1993). Microsporidian spore envelope keratins phosphorylate and disassemble during spore activation. *J. Eukaryot. Microbiol., 40*, 783–788.

Weiser, J. (1976). Staining of the nuclei of microsporidian spores. *J. Invertebr. Pathol., 28*, 147–149.

Weiser, J. (1977). *An Atlas of Insect Diseases*. Prague, Czech Republic: Academia. pp. 240.

Weiser, J. (1978). Transmission of microsporidia to insects via injection. *Vest. Cs. Spol. Zool, 42*, 311–317.

Weiser, J. (1991). *Biological control of vectors (Manual for Collecting, Field Determination and Handling of Biofactors for Control of Vectors)*. New York, NY: John Wiley and Sons. pp. 189.

Weiss, L. M., & Vossbrinck, C. R. (1999). Molecular biology, molecular phylogeny, and molecular diagnostic approaches to the microsporidia. In M. Wittner, & L. M. Weiss (Eds.), *The*

*Microsporidia and Microsporidiosis* (pp. 129−171). Washington, DC: ASM Press.

Whittaker, R. H. (1969). New concepts of kingdoms or organisms. Evolutionary relations are better represented by new classifications than by the traditional two kingdoms. *Science, 163*, 150−160.

Wittner, M., & Weiss, L. M. (Eds.). (1999). *The Microsporidia and Microsporidiosis*. Washington DC.: ASM Press. pp. 553.

Weiser, J. (1966). *Nemoci hmyzu (Insect Diseases)*. Praha: Academia. pp. 554 (In CZech).

Westenberger, S. J., Sturm, N. R., Yanega, D., Podlipaev, S. A., Zeledón, R., Campbell, D. A., & Maslov, D. A. (2004). Trypanosomatid biodiversity in Costa Rica: Genotyping of parasites from Heteroptera using the spliced leader RNA gene. *Parasitology, 129*, 537−547.

Wylezich, C., Radek, R., & Schlegel, M. (2004). Phylogenetic analysis of the 18S rRNA identifies the parasitic Protist *Nephridiophaga blattellae* (Nephridiophagidae) as a representative of the Zygomycota (Fungi). *Denisia, 13*, 435−442.

CHAPTER XII

# Nemetode parasites, pathogens and associates of insects and invertebrates of economic importance

S. PATRICIA STOCK* & HEIDI GOODRICH-BLAIR[†]

*Department of Entomology, University of Arizona, Forbes Building, Room 410, 1140 E. South Campus Drive, Tucson, AZ 85721-0036, USA
[†]Department of Bacteriology, University of Wisconsin, Microbial Sciences Building, Room 4550, 1550 Linden Drive, Madison, WI 53706, USA

## 1. INTRODUCTION

Nematodes are unsegmented roundworms that belong to the phylum Nematoda, within the super-phylum Ecdysozoa, along with arthropods and other organisms that build and shed cuticle, in a process called ecdysis (Aguinaldo *et al.*, 1997). Nematodes are not only highly diverse, but also complex and biologically specialized metazoans. They have a pseudocoelom (a false cavity lined by mesoderm along the epidermis), and have complete excretory, nervous, digestive, reproductive, and muscular systems but lack circulatory and respiratory systems. The alimentary tract consists of a mouth followed by the buccal cavity or stoma, esophagus, intestine, rectum, and anus. The sexes are usually separate. The male's reproductive system opens ventrally into the rectum forming a cloaca, has one or two testes, and has spicules that are used as copulatory structures. The one or two sclerotized spicules may or may not be guided by another sclerotized structure called the gubernaculum. The adult female has one or two ovaries with the vulva located ventrally and near mid-body or more posteriorly.

Approximately 25,000 species have been described, and 500,000−100,000,000 species are estimated to exist worldwide (Lambshead *et al.*, 2003). Nematodes are ubiquitous, occupying numerous niches and existing in nearly every available habitat on every continent, utilizing many trophic strategies and lifestyles. Many species are associated with invertebrates, and these relationships may span from fortuitous to antagonistic (Bedding *et al.*, 1983; Poinar, 1983; Petersen, 1985; Gaugler & Kaya, 1990; Wilson *et al.*, 1993, 1994).

Commensalism is one of the most common associations between nematodes and insects and can be categorized as phoretic or necromenic (Kiontke & Sudhaus, 2006). Phoretic nematodes may be found externally on various areas of the host surface (i.e. elytra in insects, on or around the mantle in molluscs). They usually use the host for transport but not as food. Necromenic nematodes may be found in the body cavity of the host where they do very little, if any, damage to their host. Usually, these necromenic nematodes have an 'invasive stage' which enters an insect, waits for the death of the host, and then feeds on bacteria and other microorganisms that proliferate on the insect carcass. Generally, necromenic

MANUAL OF TECHNIQUES IN INVERTEBRATE PATHOLOGY
ISBN 9780123868992

associations are more specific than phoretic associations. It has been suggested that necromeny represents a pre-adaptation for the evolution of true parasitism because the nematode is exposed to low oxygen levels, high temperatures, and toxic host enzymes (Weischer & Brown, 2000).

Parasitism is another type of association between invertebrates and nematodes. In this case, nematodes may cause variable deleterious effects on their insect hosts including sterility, reduced fecundity, reduced longevity, reduced flight activity, delayed development, or other behavioral, physiological and morphological changes (Poinar, 1979). Some nematodes (i.e. steinernematids and heterorhabditids) have symbiotic relationships with bacteria that kill their hosts quickly (within 48 h). Because of this rapid kill, the term 'entomopathogenic' is used to describe these nematodes.

This chapter's focus is on the techniques used for identifying, isolating, propagating, bioassaying, and preserving nematodes that are parasitic in or pathogenic to insects and other invertebrates of economic importance. In many cases, the techniques for obligate parasites (e.g., steinernematids, heterorhabditids, mermithids, tetradonematids, sphaerulariids, and allantonematids) involve rearing the insect hosts, but we will not discuss insect rearing techniques in this chapter. Rather, we will concentrate on the nematodes with particular emphasis placed on the steinernematids and heterorhabditids and their bacterial symbionts, because they are the most widely studied group at this time. Finally, many techniques in this chapter use chemicals and/or equipment that require safety measures. Be sure that personal protection devices (eyeglasses, gloves, laboratory coats and aprons, fume hoods, etc.) are used and instructions from the manufacturers for handling chemicals and equipment are followed.

## 2. MAJOR GROUPS OF NEMATODES ASSOCIATED WITH INSECTS AND OTHER INVERTEBRATES OF ECONOMIC IMPORTANCE

### A. Current classification scheme

Among invertebrate parasites, there are 30 nematode families that are associated with insects and other invertebrates (Stock & Hunt, 2005). Seven of these families have the potential for being considered as biological control agents: Mermithidae, Allantonematidae, Neotylenchidae, Sphaerularidae, Rhabditidae, Steinernematidae and Heterorhabditidae. Of these, the most widely studied group are the so-called "entomopathogenic nematodes", also known as EPN. In addition to these taxa, a member of the family Rhabditidae, *Phasmarhabditis hermaphrodita* (Schneider), is known to suppress several slug species, and has recently been developed as a biological molluscicide (Wilson *et al.*, 1993; Glen & Wilson, 1997; Wilson, 2007).

In this chapter, we have adopted the new classification scheme suggested by De Ley & Blaxter (2002) to list those groups with biological control potential. This classification is based on the 18S ribosomal DNA phylogenetic framework proposed by Blaxter *et al.* (1998). This molecular framework recognizes the presence of three basal clades: dorylaimids, enoplids and chromadorids. Relationships between these clades are yet not fully resolved, but existing data supports sister taxon status of dorylaimids and enoplids (De Ley & Blaxter, 2002). In this new taxonomic system, dorylaimids and enoplids are encompassed within the class Enoplea Inglis, 1983 (Figure 12.1). The Chromadorea Inglis, 1983 comprise the majority of taxa within Nematoda, including all the former Secernentea. Seven out of eleven nematode families currently considered in biological control of invertebrate pests are grouped within the Chromadorea, whereas the Mermithidae, are members of the Enoplea (Table 12.1).

### B. Diagnostic characters of most relevant invertebrate parasites

#### *1. Family Steinernematidae Chitwood & Chitwood, 1937*

Steinernematids are obligate insect pathogens and are characterized by their mutualistic association with bacteria of the genus *Xenorhabdus.* Steinernematidae currently comprises two genera, *Steinernema* Travassos, 1927, with more than 70 species and *Neosteinernema* Nguyen & Smart, 1994, with only one species *N. longicurvicauda.* Diagnostic traits include: adults with truncated to slightly round head. Six fused lips, with distinct tips, and with one labial papilla each. Four cephalic papillae present. Amphids small. Stoma reduced, short and wide, with inconspicuous sclerotized walls (Figure 12.2B). Esophagus rhabditoid, set off from intestine (Figure 12.3C).

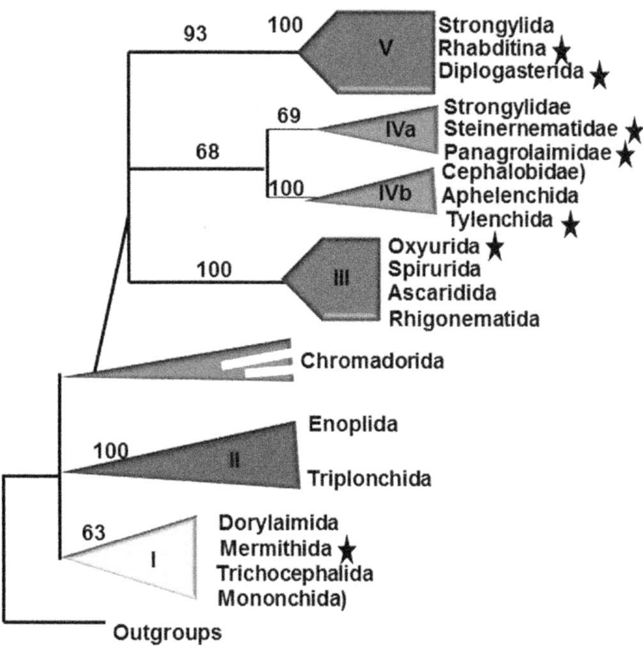


**Figure 12.1** Nematoda molecular phylogenetic framework (modified from Blaxter *et al.*, 1998). (★) Groups with species that are parasites and/or pathogens of invertebrates.

Nerve-ring usually surrounding isthmus or anterior part of basal bulb. Females have paired opposed ovaries. Vagina short, muscular. Vulva located near middle of body, with or without protruding lips. Epiptygma present or absent. Male with single reflexed testis. Spicules paired, symmetrical. Gubernaculum present. One single mid-ventral and 10–14 pairs of genital papillae present of which seven to 10 pairs pre-cloacal. Tail rounded, digitated or mucronated, without bursa (Figure 12.4C). Third-stage infective juvenile with collapsed stoma. Cuticle annulated, lateral field with six to eight ridges in middle of body. Esophagus and intestine collapsed. Specialized bacterial pouch located at beginning of intestine and of variable shape. Excretory pore distinct, anterior to nerve ring. Tail conoid or filiform, with variable hyaline portion. Phasmids present, prominent or inconspicuous.

### *2. Family Allantonematidae Pereira, 1931*

Allantonematids have a single heterosexual cycle. Adult females are parasites of the hemocoel of mites and insects (Siddiqi, 2000). Within this family, members of *Thripinema* Siddiqi, 1986, are known to parasitize thrips (Thysanoptera: Thripidae). A free-living stage occurs in flowers, buds and leaf galls of plants that attack thrips. Key diagnostic characters include: pre-parasitic female and free-living male with small stylet (less than 15 μm long) with or without knobs. Esophageal glands elongated, lobe-like; sub-ventral glands extending past dorsal lobe. Tail conoid or sub-cylindrical. Pre-parasitic female with small vulva and short vagina. Post-vulval sac short or absent. Uterus elongated. Parasitic female obese, sac-like, elongate or spindle-shaped. Reproductive organs filling body cavity. Uterus not everted. Vulva a small transverse slit or indistinct. Male with outstretched testis. Spicules arcuate, pointed, usually less than 25 μm long. Gubernaculum usually present. Bursa present or absent.

### *3. Family Neotylenchidae Thorne, 1941*

Members of this family have a free-living generation alternating with an insect parasitic generation. *Beddingia* Thorne, 1941, currently comprises 17 nominal species with *B. siricidicola* Bedding, 1968, a parasite of the wood wasp *Sirex noctilio*, being the only taxon currently used in biological control. Free-living stages have either a smooth or finely striated cuticle. Stylet well developed, less than 20 μm long, basal knobs may be bifid (Figure 12.2D). Esophagus fusiform, basal bulb absent. Esophageal glands free in body cavity, extending over intestine. Orifice of dorsal gland close to stylet base. Nerve ring generally around the esophageal isthmus,

**Table 12.1** List of invertebrate parasitic and/or pathogenic taxonomic groups of nematodes.

| *Class* | *Subclass* | *Order* | *Suborder* | *Infraorder* | *Super Family* | *Family* |
|---|---|---|---|---|---|---|
| ENOPLEA Inglis, 1983 | DORYLAIMIA Inglis, 1983 | Mermithida Hyman, 1951 | Mermithina Andrássy, 1974 | | Mermithoidea Braun, 1883 | Mermithidae Braun, 1883 |
| CHROMADOREA Inglis, 1983 | CHROMADORIA Pearse, 1942 | Rhabditida Chitwood, 1933 | Tylenchina Thorne, 1949 | Panagrolaimomorpha De Ley & Blaxter, 2002 | Strongyloidoidea Chitwood & McIntosh, 1934 | Steinernematidae Travassos, 1927 |
| | | | | Tylenchomorpha De Ley & Blaxter, 2002 | Sphaerularoidea Lubbock, 1861 | Allantonematidae Pereira, 1931 |
| | | | | | | Neotylenchidae Thorne, 1941 |
| | | | Rhabditina Chitwood, 1933 | Rhabditomorpha De Ley & Blaxter, 2002 | Rhabditoidea Örley, 1880 | Rhabditidae Örley, 1880 |
| | | | | | Strongyloidea Baird, 1853 | Heterorhabditidae Poinar, 1975 |
| | | | | Diplogasteromorpha De Ley & Blaxter, 2002 | Diplogasteroidea Micoletzky, 1922 | Diplogasteridae Micoletzky, 1922 |

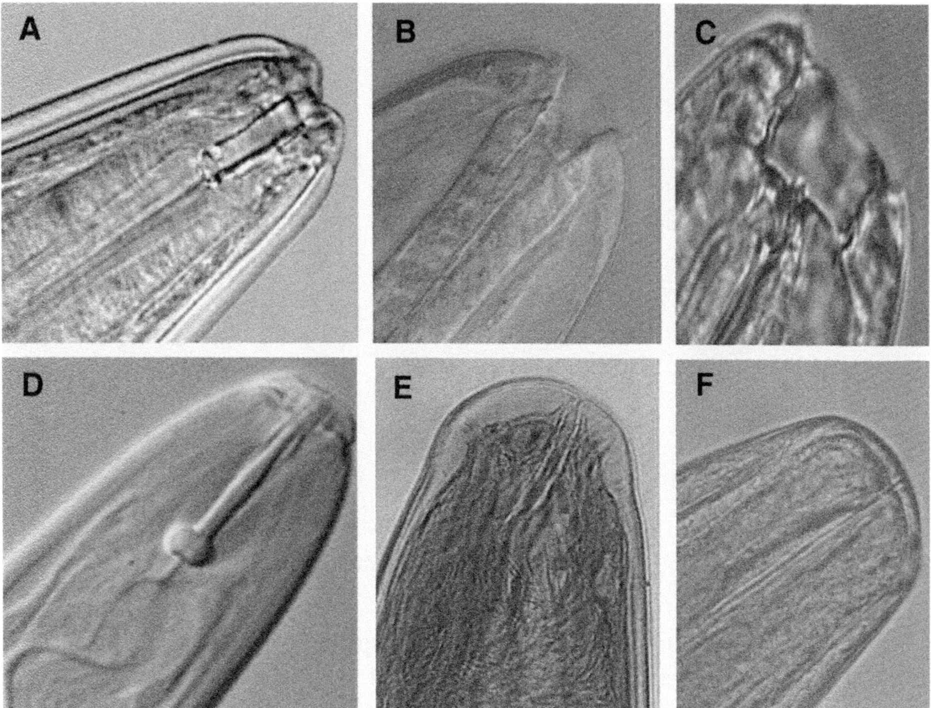


**Figure 12.2** Types of stoma. A. Rhabditid; B. Steinernematid/Heterorhabditid; C. Diplogasterid; D. Tylenchid; D, E. Mermithid (E: adult female; F: second-stage juvenile).

posterior to, or at level of, esophago–intestinal junction. Excretory pore anterior or posterior to nerve ring. Female monodelphic prodelphic. Vulva in posterior region, post-vulval sac present or absent. Tail conoid, sub-cylindroid or cylindroid. Male monorchic, testis outstretched. Bursa present or absent. Spicules paired, small, cephalated or arcuate, distally pointed. Gubernaculum present or absent. Pre-adult female (free-living) with hypertrophied

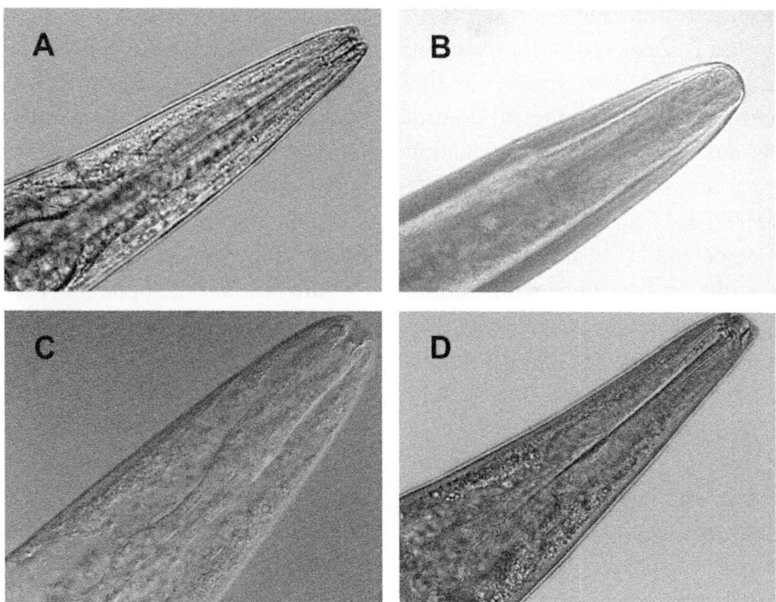


**Figure 12.3** Types of esophagus. A. Rhabditid; B. Mermithid; C. Steinernematid/Heterorhabditid; D. Diplogasterid.

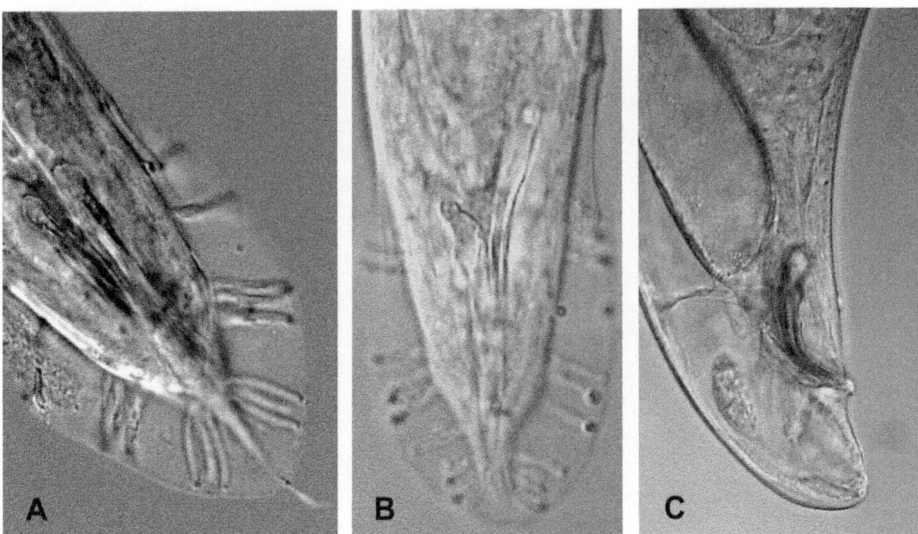


**Figure 12.4** Tail types. A, B. With bursa: A. leptoderan; B. peloderan. C. Without bursa.

stylet and esophagus. Ovary immature. Uterus long. Mature parasitic female, obese, sausage-shaped or elongate tuboid. Stylet and esophagus non-functional. Uterus hypertrophied but not everted.

### *4. Family Rhabditidae Örley, 1880*

Most members of this family are free-living microbivores. Species such as *Phasmarhabditis*, *P. hermaphrodita* (Schneider, 1859), *P. papillosa* (Schneider, 1866) and *P. neopapillosa* (Mengert in Osche, 1952) have parasitic associations with terrestrial slugs and snails. In particular, *P. hermaphrodita* is capable of killing several slug and snail pests and it is the only species in this nematode family currently used as a biological control agent. The nematode is amenable for mass production and it is commercialized as a molluscicide (Wilson *et al.*, 1994; Glen & Wilson, 1997; see also Chapter XIII). The mode of action of this nematode is yet not fully understood, but it is known that association with certain bacteria such as *Moraxella osloensis, Providencia rettgeri* and *Pseudomonas fluorescens* strongly influences its pathogenic action (Wilson *et al.*, 1995). Key diagnostic traits are: stoma commonly cylindrical without distinct separation of cheilo-, gymno- and stegostom. Stoma two or more times as long as wide (Figure 12.2A). Usually, with six distinct lips, each with one cephalic papilla. Amphids pore-like. Esophagus clearly divided into corpus (procorpus and metacorpus) and post-corpus (isthmus and valvated muscular portion) (Figure 12.3A). Male spicules separate or fused distally. Gubernaculum present. Bursa mostly well developed, peloderan or leptoderan, occasionally small or rudimentary (Figure 12.4A,B). Nine or 10 pairs of genital papillae (bursal rays). Female with one or two ovaries.

### *5. Family Heterorhabditidae Poinar, 1976*

Heterorhabditidae consists of one genus, *Heterorhabditis* Poinar, 1976, with *Heterorhabditis bacteriophora* as the type and 17 other species described. Heterorhabditids have a similar life-cycle to steinernematids, but adults resulting from infective juveniles (IJs) are hermaphroditic. Eggs laid by the hermaphrodites produce juveniles that develop into males and females or IJs. The males and females mate and produce eggs that develop to IJs. Key diagnostic traits include: adults with six distinct protruding pointed lips surrounding oral aperture. Each lip bearing one labial papilla. Stoma short and wide (Figure 12.2D). Esophagus rhabditoid. Corpus cylindrical, metacorpus not differentiated. Isthmus short. Basal bulb pyriform with reduced valve (Figure 12.3B). Excretory pore usually located at level of basal bulb. Hermaphrodite (first generation) with an ovotestis. Vulva located near middle of body. Post-anal swelling present or absent. Tail terminus blunt, with or without a mucro. Female (second generation) amphidelphic, ovaries with reflexed portions often extending past vulva opening. Vulva located near middle of body, with or without protruding lips. Tail conoid; post-anal swelling present or absent. Bursa peloderan or leptoderan (Figure 12.4A, B). Male (second generation) monorchic. Spicules paired, symmetrical, straight or arcuate, with pointed tips. Gubernaculum slender, about half length of spicules. Bursa open, peloderan, attended by a complement of nine pairs of bursal rays (papillae). Third-stage IJ ensheathed in cuticle of second-stage juvenile. Cuticle of J2 with longitudinal ridges throughout most of body length, and a tessellate pattern in anteriormost region. Lateral field with two ridges. Prominent cuticular dorsal tooth present. Excretory pore located posterior to basal bulb. Tail short, conoid, tapering to a small spike-like tip.

### 6. *Family Diplogasteridae Micoletzky, 1922*

Diplogasterids are usually predators or omnivores but can also be bacterial feeders. Only a few genera (i.e. *Butlerius, Fictor* and *Mononchoides*) have been studied as biological control agents of plant parasitic nematodes (Stock & Hunt, 2005). Diagnostic characters include: lip region never set off by a constriction usually composed of six distinct lips or six fused lips. Amphids pore-like. Stylet absent. Stoma variable, usually broad and short with stegostom containing denticles, warts or teeth (Figure 12.2C). Esophagus with a median valvated bulb and a basal valveless bulb (Figure 12.3D). Female gonad usually paired. Male with paired spicules and gubernaculum. Bursa usually small or absent. Male tail often with nine pairs of genital papillae and a pair of phasmids. Three pairs of genital papillae located preanal.

### 7. *Family Mermithidae Braun, 1883*

This family comprises numerous genera, many of which are poorly characterized by contemporary taxonomic standards. The group is in critical need of revision before a workable key can be constructed (Stock & Hunt, 2005). All known species are obligate parasites of terrestrial and aquatic arthropods and other invertebrates. Mermithids parasitize many different insect groups, including Orthoptera, Dermaptera, Hemiptera, Lepidoptera, Diptera, Coleoptera, and Hymenoptera. Mermithids with significant biological control potential include *Romanomermis culicivorax*, a parasite of mosquito larvae (Petersen, 1985), *Oesophagomermis* (= *Filipjevimermis*) *leipsandra* Poinar & Welch, 1968, a parasite of larval banded cucumber beetle *Diabrotica balteata* (Creighton & Fassuliotis, 1981), *Mermis nigrescens*, a parasite of grasshoppers (Webster & Thong, 1984), and *Agamermis unka*, a parasite of white and brown planthoppers (Choo *et al.*, 1989, Choo & Kaya, 1994). Long slender nematodes sometimes reaching a length of 50 cm, but usually between 1 and 10 cm. Cuticle smooth or with criss-crossed fibers. Anterior end containing two, four, or six cephalic papillae and rarely a pair of lateral mouth papillae. Amphids tube-like or modified pouch like. Esophagus modified into a slender tube surrounded posteriorly by stichosomal tissue (Figures 12.2E, 12.3B). Intestine modified into a trophosome or food storage organ forming a blind sac soon after the nematodes enter a host. Pre-parasitic juveniles (J2) with a functional stylet and a pair of penetration glands that degenerate after host invasion (Figure 12.2F). Ovaries paired; muscular vagina straight or curved. Males with a single fused or paired spicules. Gubernaculum and bursa absent. Several rows of genital papillae usually present.

## C. Generalized life-cycle

Most nematodes have a simple life-cycle (Figure 12.5). The mated female lays eggs and usually the nematodes develop to the second juvenile (J2) stage before hatching. We refer to the immature stages as juveniles rather than larvae to avoid confusion with the immature stages of insects. Each subsequent juvenile stage feeds and molts to the next stage. The J2 molts to J3 and then to J4 that molts to the adult stage. Most nematodes are amphimictic (requiring males and females), but some groups are hermaphroditic or parthenogenetic (requiring females

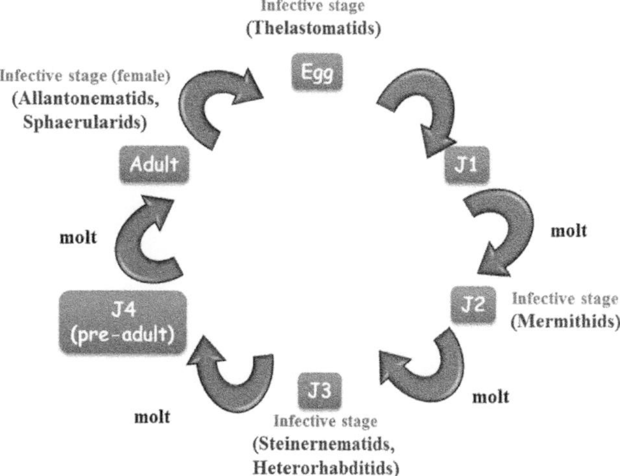


**Figure 12.5** Nematodes generalized life-cycle indicating insect infective stages.

only) or have alternate gamogenetic and parthenogenetic life-cycles.

The stage of entomogenous and EPNs that is infective varies depending on the group. In some cases, the infective stage is the egg (e.g., thelastomatids), the J2 (e.g., mermithids) or the J3 that is often referred to as the dauer or IJ (e.g., steinernematids and heterorhabditids, diplogasterids) or the mated female (e.g., allantonematids, sphaerulariids, and phaenopsitylenchids). In the last group, the adult males are non-infective and die after mating.

Different juvenile stages can be recognized by the presence/absence of key structures and/or organs, especially those of the digestive and reproductive systems. Morphology of immature stages varies significantly among different nematode groups. Below we describe the morphology of immature stages (J1–J4) of EPNs to help readers recognize each of these stages (Figure 12.6).

### *Eggs*

EPN eggs are oval (average 50 × 30 μm) and the developing juveniles can be observed within those eggs that are mature.

### *First juvenile stage (J1)*

J1 usually hatch within the first 2–10 h after egg laying. This stage is characterized by its translucent appearance and small size, usually between 10 and 20 μm long. Using DIC microscopy only rudimentary structures can be seen inside (Figure 12.6A) and pumping of the esophagus can be observed almost immediately upon hatching. J1 have an overall length of 100–200 μm and a body diameter of 15–20 μm.

### *Second juvenile stage (J2)*

J2 body is partially transparent and has a body length of 250–350 μm and a diameter of 25–30 μm. In this stage the digestive system is more or less completely formed. In particular the stoma (mouth part) and esophagus are well differentiated. Observing the mouth part at this stage can aid in discriminating rhabditids from EPN taxa. In the former group the stoma is usually twice as longer as wide, whereas in EPN (both steinernematids and heterorhabditids) the stoma is short (Figure 12.6B).

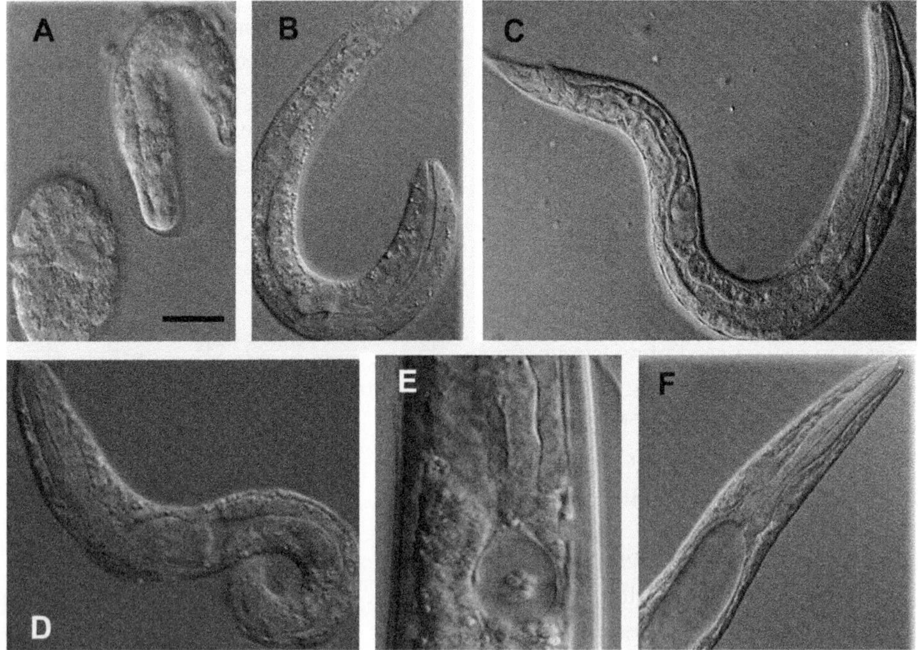


**Figure 12.6** Immature stages of entomopathogenic nematodes. A. Eggs: morulated (bottom left) and with J1 inside (top right). B. Second-stage juvenile (J2). C, D. Early third-stage juvenile (J3). E. Anterior intestinal portion of infective juvenile (IJ of *Steinernema*, notice bacterial receptacle). F. Fourth-stage juvenile (J4) or predadult, notice mature esophagus and opened anterior intestine portion. Scale bar as in A. A, B = 30 μm; C, D = 50 μm; E = 10 μm; F = 25 μm.

*Third juvenile stage (J3)*

J3 are about 1.5- to 3-times larger than J2. In this stage the digestive system is fully developed and the intestine is very dense due to the presence of storage material (Figure 12.6C, D). J3 are characterized by a transparent area initially near mid-body, which is the genital primordium (Figure 12.6C). Sex of the nematodes can not be determined in this stage. In this stage J3 of steinernematids and heterorhabditids can be differentiated from each other. For steinernematids the excretory pore is usually located above the nerve ring, whereas for heterorhabditids the excretory pore is located below the nerve ring. Steinernematid J3 infective stages have a receptacle in the anterior portion of the intestine where they harbor symbiotic bacteria (Figure 12.6E).

*Fourth juvenile stage (pre-adult)*

J4 are characterized by a relatively long body in comparison with J3, ranging from 100 μm or more depending on the EPN species. Development of reproductive organs is noticeable and sexes can be discerned. Usually gonads look darker than other organs (i.e. digestive system). The anterior portion of the intestine is wide and the bacterial receptacle is not present (Figure 12.6F). Pre-adult female ovaries and uterus can be differentiated from male testis. In the males the gonad (testis) is straight and in the females the gonads have a sigmoid shape particularly noticed in the mid-body region.

**Note:** size of different juvenile stages will vary for different nematode species.

## 3. SAMPLING TECHNIQUES

The purpose of sampling will determine where and how many samples will be taken. Samples can be taken from the hosts' habitat (soil, manure, under tree bark, insect galleries, etc.), or natural hosts may serve as the sampling unit. Familiarity with the life-cycle of the nematode and/or the insect will make selection of the proper time for sampling easier. If the prevalence of nematode parasitism of the larval, pupal or adult insects is to be determined, sampling should commence when these stages are present in the field and should be conducted over a period of time.

### A. Soil sampling

*1. Collection of samples*

Stratified random soil sampling should be taken from an area for intensive study (Campbell *et al.*, 1995) or over a geographical area for extensive study (Akhurst & Bedding, 1986; Hominick & Briscoe, 1990a, b; Griffin *et al.*, 1991; Hara *et al.*, 1991). In the former case, soil samples from a transect need to be taken from the study area over time. In the latter case, samples may be taken over a range of elevations, soil textures, and habitats (e.g., cultivated fields, forests, pastures, parks, seashores, riparian areas). At a given site, soil samples should be taken to a depth of at least 15 cm using a soil corer, trowel, post-hole digger, auger, sampling tube, hand shovel, etc. The sample volume should contain a minimum of 1 kg. At each site, collect at least three random samples over an area of 2–4 m$^2$. Within a sample, three subsamples (e.g., 10 × 10 × 15 cm) should be taken whenever possible. Depending on the sampling objective, the samples may be combined or kept separate. Subsamples should be placed in a plastic bag (use of two bags is recommended to avoid leaking of samples), mixed, and kept in a cooler (8–15°C) during transport to the laboratory. Between samples, the collecting tool needs to be thoroughly washed with water and carefully disinfected by wiping it with 70% alcohol or 0.5% bleach solution to prevent contamination of the next sampling unit.

At a minimum, the following information from each sampling site should include: (1) site location including locality/area/site name and GPS coordinates information; (2) date; (3) habitat; (4) associated vegetation; (5) temperature; (6) elevation; (7) recording and, if possible, identification of insect or other invertebrate host present in the site where the sample was collected.

Additionally, a portion of the soil sample should be sent to a soil analysis laboratory to characterize it for sand, silt, and clay content and organic matter, pH, and electrical conductivity and other desired soil parameters.

Sampling for insects can be done according to procedures detailed by Pedigo & Buntin (1994) and Leather (2005). Insects in soil, litter, manure, or plant material can be collected by dry or wet sieving, flotation, centrifugation, sedimentation, or Berlesse funnel or through combinations of one or more of the listed methods. Insects in aquatic habitats can be sampled using a mosquito dipper, sampling submerged vegetation, or sampling with various types of nets and traps. For aerial insects, the following traps can be employed: sticky, water, light, host plant, sound, and pheromone traps. Soil insects can be captured with various traps such as pitfall; food, carbon dioxide; carrion or dung; and animal traps, or an insect net or an aspirator can be also be used.

Many of the above sampling methods have been used to collect nematode-infected insects or to study nematode infections in insect populations. For example, pheromone traps or white sheets illuminated with ultraviolet light have been employed for noctuid moths parasitized by *Noctuidema* spp. (Rogers *et al.*, 1993), sticky traps, sweep nets, bloody boards or dung traps for face flies infected with *Paraiotonchium autumnale* (= *Heterotylenchus autumnalis*) (Jones & Perdue, 1967; Kaya & Moon, 1978; Kaya *et al.*, 1979), infested logs in emergence cages for bark beetles (Choo *et al.*, 1987), and aspirators for leafhoppers parasitized by *Agamermis unka* (Choo *et al.*, 1995). In addition, pitfall traps have been used to assess the impact of applied nematodes on non-target insects (Georgis *et al.*, 1991). However, the researcher must be careful in interpreting the data from a single sampling technique because nematode-infected insects may have aberrant behaviors that may bias the estimate of nematode prevalence (Wheeler, 1928; Poinar & van der Laan, 1972; Poinar *et al.*, 1976; Kaya *et al.*, 1979).

### 2. *Extraction techniques*

Usually, methods considered for extraction of insect and/or invertebrate nematodes are derived from techniques developed with plant-parasitic nematodes (Kaya & Stock, 1997). Most of these methods are based on the fact that nematodes have positive hydrotropism and once exposed to water, they would be attracted and move towards it. The most common methods are: Baermann funnel, sieving (gravity-screening), elutriation (Byrd *et al.*, 1976), and centrifugal flotation (Jenkins, 1964). Southey (1986) and Hooper & Evans (1993) provide examples of modifications of the above methods as well as other methods that are not covered in this chapter. One method that is unique for EPNs is the insect-bait method developed by Bedding & Akhurst (1975). Below we provide a brief description of these methods. Readers should refer to original sources for expanding information on this topic.

#### *a. Baermann funnel/mist chamber*

This technique is simple and inexpensive, and it mostly useful for recovering motile nematodes from a small sample. It is not recommended for collection of inactive nematodes or ones with poor movement. Users should be aware that mortality of nematodes due to lack of oxygen in the water, is common. Therefore we recommend samples to be checked every 12 h and to consider a small amount of sample to be processed. Below we describe the construction of a Baermann funnel (Figure 12.7).

1. Insert a piece of rubber tubing to the funnel stem and close with a pinch clamp.
2. Place the funnel in a suitable rack, place a stainless steel wire screen inside the funnel that serves as a basket and support.
3. Wrap the material that contains the nematodes with a piece of muslin or tissue paper.
4. Place the wrapped material into the wire basket.
5. Add water to the funnel until the wrapped material is touching the water. The motile nematodes will emerge from the material and settle into the rubber tubing.

**Figure 12.7** Baermann funnel set up.

6. Loosen the clamps after a few hours and collect the water containing the nematodes into a beaker or test tube.

A modification of this technique is the mist chamber, which basically considers the Baermann funnel system but in this case funnels are kept in a chamber where a fine tepid mist is sprayed intermittently (e.g., 1 min in every 10 min) over the funnels containing the material with the nematodes. The funnels' stems in this case have no tubing and are placed directly into a receptacle (usually a test tube or a beaker) that collects the water that drips from the funnel (Figure 12.8).

Aeration in this modified system is not limiting, the downward movement of water allows for all nematodes to be collected in the receptacle, and it does not require continual monitoring by an individual to collect nematodes. However water should be collected every 24 h. The disadvantage is that a more elaborate system is required and therefore, it is more costly to set up and maintain.

*b. Sieving*

This technique and its variations use the difference in size and specific gravity between nematodes and soil components. It is not dependent upon nematode movement and large samples can be processed relatively quickly. The nematodes can be obtained within 30 min, but the initial investment in equipment is greater and an experienced worker is needed.

The method was described in detail by Hooper & Evans (1993) and Ayoub (1980). Various sieves (2 mm down to 38 μm), two beakers (250 and 600 ml), Syracuse watch glass or small Petri dishes, two stainless steel pans or plastic buckets, 4 l or greater in capacity, and a rubber hose attached to a cold water faucet are needed (Figure 12.9). The idea is to capture nematodes on the sieves so that large nematodes (> 250 μm), average size nematodes (> 90 μm), small adults and larger juvenile nematodes (> 45 μm), and very small juveniles (> 38 μm) are trapped on the 250-, 90-, 45- and 38-μm aperture sieves, respectively. Therefore, the size of the nematodes being sampled determines which series of sieves are used. In addition, Peloquin & Platzer (1993) obtained *Tetradonema plicans* eggs by rearing infected sciarid hosts in the laboratory. The nematode eggs were extracted from the compost by passing them through 149-, 53-, and 43-μm sieves and collecting the eggs on a 25-μm sieve.

1. Place a soil sample (*ca.* 200–400 $cm^3$) that has been thoroughly mixed into a 600-ml beaker.
2. Place the soil sample into one of the stainless steel pans or plastic buckets (Pan A).
3. Add 2 or 3 l of cold water and mix the soil and water by hand.
4. Thoroughly agitate, and pour the soil–water mixture through the largest sieve (2 mm) into the second pan or bucket (Pan B).
5. Wash the sieve with water from the rubber hose over Pan B.
6. Discard any sediment in Pan A and wash the pan for the next step.
7. Agitate the material in Pan B, let it set for 10–60 s.
8. Pour the material into the next smaller sieve (710 μm) into clean Pan A.

**Figure 12.8** Mist chamber.

**Figure 12.9** Equipment for sieving technique.

9. Wash the sieve over Pan A. Heavy sediments in Pan B are discarded.
10. Repeat the process with smaller-aperture sieves (250, 90, 63, and 38 μm).
11. Depending upon the size of the nematode, place the final sieved material into the 250 ml beaker and allow settling for 1–2 h.
12. Remove the supernatant carefully by decanting or siphoning.
13. Leave about 40 ml of the material for examination. However, the sample can be further processed using the centrifugal flotation technique.

*c. Centrifugal flotation*

This technique developed by Jenkins (1964) is very useful after large samples have been processed with the sieving (Kung *et al.*, 1990) and/or elutriation (Hooper & Evans, 1993) techniques. The concept is to separate the nematodes from the soil particles and organic debris by floating them out in a solution of specific gravity greater than their own. The technique can provide nematodes for examination in a few minutes, is generally more efficient than other extraction techniques, and can extract living as well as dead nematodes. It is not very useful for large nematodes and requires expensive equipment, and the extraction solution can distort or kill the nematodes because of osmotic stress.

1. Place the extract containing the nematodes into a 50-ml centrifuge tube.
2. Add sufficient water until the centrifuge tube is filled up to 10 cm from the top.
3. Spin in a centrifuge at 2900 × *g* for 5 min.
4. Pour off and discard supernatant.
5. Add a solution of sucrose at a specific gravity of about 1.18 (484 g of sucrose dissolved in water and made up to 1 l) to the centrifuge tube containing the pellet until it is filled up to 10 cm from the top.
6. Mix the pellet and sugar solution thoroughly with a stirring rod or vortex mixer.
7. Centrifuge at 2900 × *g* for 1 min.
8. Pour the supernatant through a fine sieve (5–38 μm), depending on the size of the nematodes).
9. Rinse the nematodes from the sieve into a beaker. (They should be removed from the sucrose solution as quickly as possible to avoid killing them.) Other than sucrose, solutions of NaCl, $MgSO_4$, or $ZnSO_4$ can be used (Curran & Heng, 1992; Hooper & Evans, 1993).

*d. Insect baiting*

The most commonly used method for collecting EPNs from soil involves the use of 'trap' insects. The technique was originally developed by Bedding & Akhurst (1975) and used last instar larvae of the greater wax moth (*Galleria mellonella*). However, other insects such as crickets, mealworms, and housefly larvae may also be used (Kaya & Stock, 1997). These insects are easily reared or are readily obtained from commercial companies selling live fishing baits and/or reptile food in many countries. Briefly, the technique is as follows:

1. Remove any debris collected with your samples. Rocks, pieces of wood or bark, leaves, or any other organic residue should be removed to avoid contamination with saprophagous microorganisms.
2. Add water as necessary. The soil to be baited should be moistened (but not too damp) to facilitate the movement of nematodes. Usually a spray bottle is used to add water to slowly moisturize the soil.
3. Place approximately 250–500 $cm^3$ of moist soil in a clean container.
4. Add five to 10 last instar larvae of *G. mellonella* (or other insect species and/or suitable stage) on the soil surface of each sample.
5. Cover the containers with lids and invert containers (Figure 12.10).
6. Place containers at room temperature (usually 22–24°C).

**Note:** Other temperatures should be used if possible, especially depending on the origin of the samples. Usually a cooler temperature in temperate (e.g., 15°C) and a warmer one in tropical (e.g., 30°C areas) are desired.

7. Remove dead insects from the soil every 2–3 days. Depending upon the purpose of the study, healthy larvae can be added to the soil sample to extract more nematodes from the soil.
8. Rinse cadavers in sterile water and place them separately on filter paper in a Petri dish (60 × 15 mm).
9. After 2–3 days in the Petri dish (5–7 days after initial soil exposure), place dead larvae with signs of infection by EPNs (Poinar, 1979) on a modified White trap (Kaya & Stock, 1997) for collecting IJs (see procedure below).

This method is simple, inexpensive and selective for EPNs (but if the insect dies in the soil, free-living

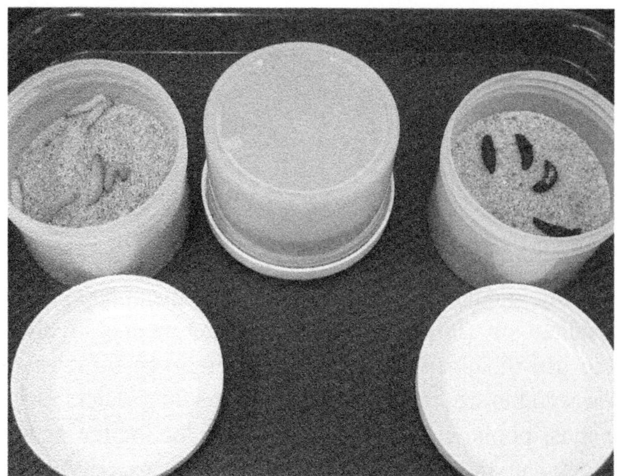

**Figure 12.10** Insect baiting technique. Left: container with live *G. mellonella* larvae; middle: container is turned upside down; right: container after 5 days showing EPN-infected *G. mellonella* larvae.

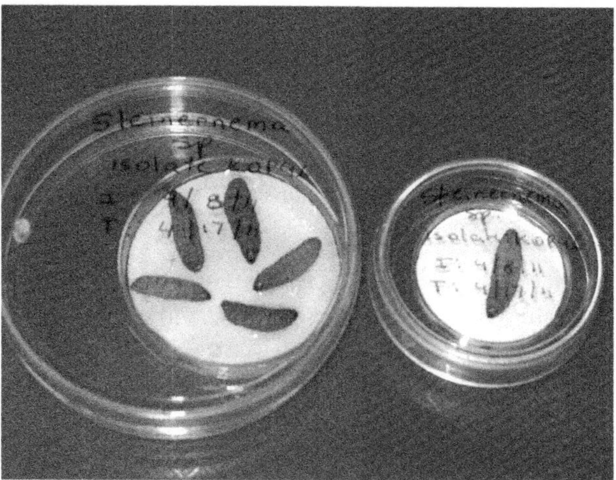

**Figure 12.11** Modified White trap.

microbivore nematodes may colonize the cadaver and give the impression of an EPN-infection). The disadvantages are that it is only useful for EPNs and can be time consuming, usually requiring exposure of the insects for 5−7 days. Additionally, if the nematodes are host specific, they may not infect the trap insect.

## B. Nematode recovery from infected cadavers

### *1. Modified White trap*

The modified White trap was described by Kaya & Stock (1997). This trap is a modification of the one developed by White (1927). Briefly, the setup of the modified White trap is a follows (Figure 12.11):

1. Place the top of a 50−60-mm diameter Petri dish inside a larger dish (100 mm).
2. Set one single circular filter paper (Whatman #1) inside the smaller dish.
3. Fill the outer (larger) Petri dish with *ca.* 20 ml of sterile distilled water.
4. Select only cadavers with good signs of infection: *Steinernema*-infected cadavers have an ochre, brown (but black for *S. siamkayai*) coloration, and brick-red to dark purple (but greenish for *H. zealandica*) coloration usually for *Heterorhabditis*- infected cadavers.
5. Place the *G. mellonella* cadavers on the filter paper of the smaller dish making sure cadavers do not touch each other. Do not place water into the dish that holds the cadavers.
6. Cover the dish with a lid and place the dishes on a bench or tray at room temperature until emergence occurs.
7. Label the dishes accordingly. We recommend adding the following information: Nematode name (species/isolate), infection date (date infection was set), and trap date (this is the date you set up the modified White trap) (Figure 12.11).

Certain experiments require the user to consider the recording of nematode emergence from a single insect cadaver. For this purpose we recommend using a 60-mm Petri dish as the larger dish, and use the top of a smaller dish (35 mm) to be placed inside the larger dish. As described above, place one single piece of filter paper (usually 32 mm) in the smaller dish and place the cadaver on top of it, then follow the procedures described above.

Some species, especially *S. glaseri*, emerge from the cadavers as ‘pre-IJs.’ If the pre-IJs emerge directly into the water, they will not successfully develop into IJs or may have a reduced number of symbiont load. Good results have been obtained by forcing pre-IJs to travel some distance. One method is to place the cadavers on a specially prepared Petri dish with Plaster of Paris.

### *2. Whitehead and Hemming tray*

If a large number of cadavers need to be processed for collection of emerging nematodes, the Whitehead and Hemming tray may be used (Whitehead & Hemming,

1965). Briefly, this trapping technique consists of a stainless steel sieve to support the infected insects. Alternatively, the sieve is made by removing the bottom of a container and replacing it with a plastic mesh (90 μm) that has large enough openings for nematodes to pass through. For set up of the Whitehead and Hemming tray, proceed as follows:

1. Place a double layer of paper tissue or fine mesh nylon cloth (Nitex® nylon mesh 20–40 μm) on the sieve.
2. Place the cadavers on the tissue or cloth.
3. Place the sieve on a collecting tray/pan.
4. Add water to the pan so that the cloth or paper tissue is moist. The insects should not be covered with water as the nematodes within will die from anoxia.
5. Hold cadavers on the sieve for 1 week for IJs to migrate into the water. (For further information on harvesting nematodes, see Section 9B). Emergence of IJs varies according to the nematode group: for steinernematids, 8–12 days after infection and for heterorhabditids, 15–21 days after infection.
6. Expose the collected IJs to new insect larvae to confirm pathogenicity and complete Koch 'pre-IJs.' tissues to find the nematodes.

### 3. Enzymatic digestion method

To facilitate counting the number of nematodes, Mauleón *et al.* (1993) developed a pepsin enzyme technique that has made dissection of nematode infected cadavers an easier task (see Appendix II).

1. Place the cadaver into a Petri dish (100 x 15 mm).
2. Pour the pepsin solution (ca. 15–20 ml) to cover the insect.
3. Dissect the insect.
4. Place the Petri dish in an incubator at 37°C on a shaker (120 rpm/min) for 2 hr.
   **Note:** Do not let the digestion go beyond 2 h without close monitoring because the nematodes may also become digested. The pepsin will digest away most of the insect tissue but will keep the nematodes intact. After digestion, the dishes can be held from 4 to 15°C for 2–3 days until counting can be done. In this case, dilute with an equal volume of water before storage.
5. Count and sex the nematodes using a dissecting stereomicroscope.

The advantages of this method are that a large number of insects can be processed quickly and counted immediately or at a later time, and it is not necessary to spend time searching through the insects' tissues to find the nematodes.

## C. Aquatic samples

### 1. Collection of samples

Aquatic nematodes can be sampled quantitatively or qualitatively (Poinar, 1991). The former method involves the use of auger or probe to remove core samples of measurable size from the bottom of standing water (lakes, ponds, pools, etc.). Techniques used for marine nematodes can be applied for freshwater forms (Holme & McIntyre, 1971; Downing & Rigler, 1984). Qualitative methods vary and depend on the water flow and the zone to be sampled. In sampling nematodes from beds of fast-flowing streams, a net, a set of sieves and instruments to turn over rocks are needed. Filipjev & Schuurmans-Stekhoven (1959) described a number of devices to collect fresh water nematodes. These include dredges, bottom catchers, plankton nets, sledge trawls, mud-suckers and worm nets.

### 2. Extraction techniques

Although the majority of freshwater nematodes are too small to view without magnification, the free-living stages of many aquatic mermithids are large enough to see and can be picked from the samples with an 'L-shaped needle' (Figure 12.12). Other small mermithids or insect-parasitic

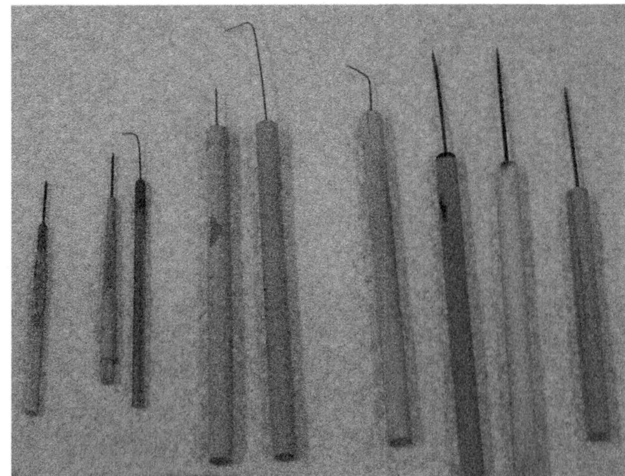

**Figure 12.12** Different types of needles for picking nematodes.

nematodes that escape visual detection can be collected with various extraction procedures.

The extraction methods used for soil and plant-parasitic nematodes (i.e. Baermann funnel, mist chamber, sieving, centrifugal flotation methods, etc.) are suitable for freshwater forms.

### D. Extraction efficiency

Extraction efficiency depends on a number of factors including the method utilized, soil type, nematode species and size, and skills of personnel involved in this procedure (Barker, 1985; Barker *et al.*, 1985). Relative extraction efficiency for various plant-parasitic nematode species have been determined (Barker, 1985; Barker *et al.*, 1985), but little information is available for nematodes associated with insects. Saunders & All (1982) compared the efficiency of recovering *Steinernema carpocapsae* IJs from soil using the Baermann funnel, centrifugation flotation, and flotation-sieving methods. They found that the Baermann funnel technique was superior to centrifugation flotation and flotation sieving methods. Kung *et al.* (1990) used the centrifugal flotation technique and obtained an extraction efficiency of 52% for *S. carpocapsae* with a 45% sucrose solution and 65% for *S. glaseri* with a 75% sucrose solution. Extraction efficiencies were used for studying persistence (survival) of the nematodes over time in soil (addition of a known number of nematodes) as follows:

$$\text{Extraction efficiency} = \frac{\text{No. of live nematodes extracted}}{\text{No. of nematodes inoculated} \times \text{extraction efficiency}}$$

Several studies have considered the percentage of trap insects infected in a given soil sample as an indication of presence or absence of EPNs. Generally, a high percentage indicated that many nematodes were present and a low percentage indicated that very few nematodes were present. This information was qualitative and not quantitative.

Fan & Hominick (1991) developed an efficiency assay for determining the numbers of steinernematid and heterorhabditid IJs capable of infecting a host insect. This assay is described below.

1. Place known numbers of IJs of a given species into soil or use field soil known to have IJs.
2. Add a trap insect to the soil to determine the actual numbers of nematodes capable of infecting the host insect.
3. After a specified time (24−72 h depending upon temperature), remove the trap insect and rinse in water.
4. Dissect the insect immediately in Ringer's solution or M9 buffer solution or hold on moistened filter paper.

The idea is to allow the nematodes sufficient time to kill the host and develop to adults (which facilitate counting) but before progeny production (the presence of progeny makes counting difficult and timing the dissection before this happens is critical). Another trap insect can be added to the soil and the process repeated until no more nematodes are trapped. They found that it took nearly 36 days of consecutive sampling before infection ceased. Two or three consecutive assays over 12 days recovered 75% or more of the nematodes. The Fan and Hominick method is widely used to provide quantitative data of the number of nematodes that could penetrate and establish in the hosts and is a useful parameter for comparing the efficacy of different nematode species under specific conditions.

## 4. DIAGNOSTIC METHODS

### A. Nematodes

#### *1. Classic (morphological) methods and techniques*

*a. Collection and initial preparation*

A good stereomicroscope is essential for nematode identification and should have a range of magnifications between 10 and 100 ×, a fairly flat field, and good resolution. Illumination by transmitted light should be as even as possible. Handling most nematodes in distilled water presents no problem; however, the use of a saline solution (i.e. Ringer or M9 buffer; see Appendix II) is highly recommended to avoid osmotic shock. For living material, the nematodes can be mounted in a saline solution or water on a glass slide with supports (i.e. placing small pieces of glass fiber or Pliobond® cement (Lee *et al.*, 2009) between the slide and the coverslip and a cover-slide sealed with paraffin, Pliobond® cement or nail polish.

1. Perform dissections in a Petri dish with Ringer's or M9 buffer solution and under a stereomicroscope.

2. Lift individual specimens out of the solution with the help of a handling needle (Figure 12.12) of which there are many types: nylon toothbrush bristle, steel needle, tungsten needle, bamboo splinter, etc.

**Note:** For smaller nematodes, an eyebrow hair glued on to the end of a mounted needle is very useful and does not damage the nematodes. See Appendix II for making of tungsten needles.

3. Use the lowest microscope magnification (to give good depth of focus and working distance) and start with nematodes that are near the center of the dish.
4. Place the needle underneath the nematode and lift up quickly so that the nematode is pulled through the meniscus.

**Note:** The timeframe for doing dissections may vary according to the different species/isolates and the nematode stage(s) the user may be interested in recovering. If the life-cycle of a nematode is not known, we recommend users to set up infections so that multiple cadavers can be dissected every day to follow up on the nematode development/life-cycle and recover the various stages. Additionally, careful dissections of different organs may need to be considered separately to help locate nematodes inside the host's body.

**Note:** For steinernematids, dissections should be carried out 2–6 days after infection to recover first generation adults (males and females), and 5–9 days after infection for second generation adults. For heterorhabditids, dissections should be done 3–5 days after infection to recover first generation hermaphrodites, and 7–9 days after infection for second generation adults (males and amphimictic females). Infective juveniles of both steinernematids and heterorhabditids should be recovered during the first 3 days after initial emergence.

*b. Killing and fixing*

*(1) Heating.* Prior to fixation, it is critical that nematodes (especially those of larger size) are killed properly to make sure their bodies are relaxed and do not coil. One of the methods considered for killing/relaxing of nematodes is by applying heat. Briefly, the procedure is as follows:

1. Rinse collected nematodes three times in saline solution and place them in a vial (Nalgene®, 3 ml with screw caps) containing 1–1.5 ml of saline solution.
2. Close vial and place it in a water bath at 60°C for 2 min.
3. Concomitantly, use another tube containing an equal volume of fixative (see Appendix II) without nematodes and place it in the water bath at the same temperature as the saline solution containing the nematodes is being heated.
   **Note:** This is a critical step that should be followed carefully to avoid fixation problems such as swelling of the nematodes or detachment of the nematodes' cuticle.
4. Once the nematodes are killed, add the fixative to the vial.
5. Maintain the nematodes in this 1 : 1 saline solution–fixative mix for 12–24 h.
6. Replace the 1 : 1 saline solution–fixative mix with 100% fixative.

**Note:** Nematodes can remain stored in the glass vials for longer periods of time, without being permanently mounted. We recommend users to seal the vials with parafilm and preferably store them in a cabinet to avoid evaporation of formaldehyde fumes.

An alternative method that kills and fixes nematodes in one single step is that of Seinhorst (1966). Briefly, specimens are collected in a very small drop (*ca.* 1 ml) of water in a Syracuse watch glass or similar deep concave vessel. The fixative (i.e. TAF) is heated to 100°C and an excess (3–4 ml) is then added to the vessel containing the nematodes. Nematodes remain in this solution for 12 h, and then the solution is replaced with double-strength fixative (see Appendix II).

*(2) Cooling.* Nematodes can also be killed by placing them at cold temperatures. For this purpose, follow the following steps:

1. Place sample in a small vial containing a minimal amount of water/saline solution and place it in a refrigerator (*ca.* 4°C).
2. This process may usually take a few minutes only (10–60 min).
3. Once the nematodes are relaxed, add fixative previously heated at 65°C.
4. Allow the fixative to infiltrate for at least 24 h.
5. Transfer the nematodes to a Syracuse watch glass with as little fixative as possible.

*c. Preparation of permanent mounts*

Several nematode structures and/or organs may be obscured by the granular and darker appearance of the intestine. Specimens can be cleared by processing them to lactophenol or glycerin. Although both solutions are good mounting media, nematodes will keep almost indefinitely if they are processed to glycerin. The procedure was originally described by Seinhorst (1959). Below is a modification of this procedure:

1. Transfer fixed nematodes (i.e. in TAF, FAA, etc.) to a Syracuse watch glass containing 0.5 ml of Solution I (20 parts 95% ethanol, 1 part glycerin, 79 parts distilled water; Figure 12.13).
2. Place the Syracuse watch glass in a desiccator and add 95% ethanol to the desiccator so that the space below the holding shelf is half full.
3. Place the desiccator in an oven for at least 12 h at 35°C to allow slow evaporation of the ethanol from Solution I in the watch glass.
4. Remove the watch glass from the desiccator.
5. Fill the watch glass with Solution II (5 parts of glycerin, 95 parts of 95% ethanol), place in a Petri dish that should be partially opened to allow slow evaporation of the ethanol, and put it back in an oven for 3 h at 40°C. After processing the nematodes in pure glycerin, they are ready for mounting.

Specimens can be mounted on quality glass slides (76 × 25 × 1 mm). However, we recommend use of aluminum double-coverslip slides (Cobb, 1917) because they allow viewing of specimens from either side, are durable, and are less likely to crack during shipping or rough handling. See Figure 12.14 for different supplies and slides available. When using aluminum holders, a square coverslip must be slipped in the aluminum carrier between two pieces of cardboard. The edges of the carrier must then be pressed to hold the cardboard and the coverslip in place. Below we describe a procedure developed by Lee *et al.* (2009) used to make a permanent slide.

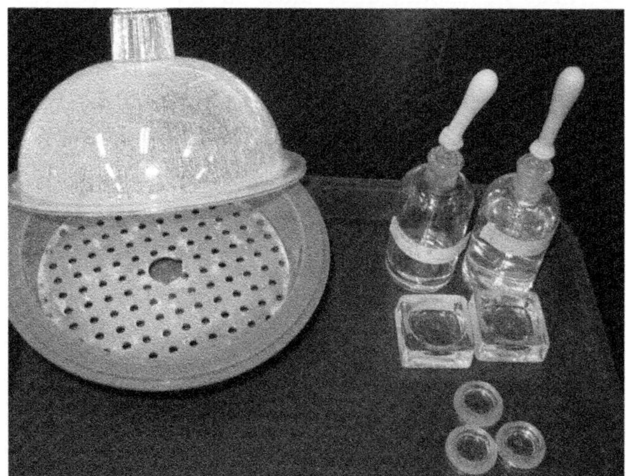

**Figure 12.13** Supplies needed for Seinhorst fixation method. Left: Desiccator filled with 95% ethanol; right: Seinhorst I and II solutions and watch glasses.

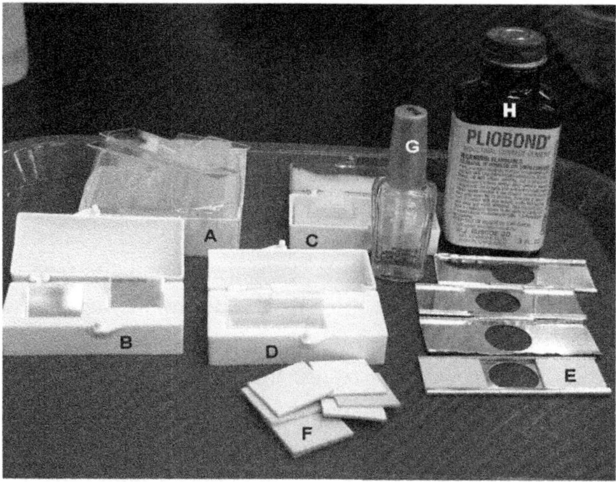


**Figure 12.14** Supplies needed for making permanent slides. A. Glass slides; B,C,D. different types of coverslips; E. Aluminum frames (Cobb slides); F. cardboards; G,H. sealants (nail polish, Pliobond cement).

1. Place a small drop of glycerin in the center of a glass slide.
2. Position your sample(s) in the glycerin
3. Make three small dots of Pliobond® contact cement around the glycerin.

**Note:** Pliobond® is fast-drying so you should aliquot a few drops into a Petri or glass vial each time you use it. The cement drops will serve as spacers such that specimens will not be crushed when the coverslip is in place. When the contact cement is wet, it may be runny so you may have to daub the same spot until it become slightly tacky and builds up to the desired thickness.

4. Carefully place a circular coverslip on top of everything. Glycerin should now spread out and fill the area beneath the coverslip.
5. Apply Pliobond® with a brush all around the edge of the coverslip. Don't worry if it makes contact with some of the glycerin—it will dry.
6. Seal with clear nail polish or more contact cement if the glycerin is leaking. (Check after a day or two.)

*d. Preparation of temporary mounts*

For various reasons users may need to make temporary mounts with live specimens. Making agar pads on glass slides is one of the best options to avoid drying of the samples that eventually will either stress or kill the nematodes under observations.

*(1) Agar pads.* Agar pads were developed for *Caenorhabditis elegans* (Brenner, 1974). Below we summarize the procedure involved in the preparation of these agar pads. If observations require the immobilization of the specimens, we recommend adding anesthetics such as Levamizol or Na azide. Sodium azide is a poison and must be handled carefully. Avoid contact with dust and contact with skin (including with agar containing azide).

Steps to follow:

1. Clean surface of slides to be used by wiping with ethanol
2. Put two layers of labeling tape along the outer short axis of the slide. Leave sufficient space between the tape for the coverslip (we recommend 18 × 18 coverslips).
3. Coat another slide with Rain-X™ on the alcohol cleaned side.
4. Melt a 4% agar suspension in M9 buffer using a microwave oven. Make sure you do not boil the agar out of the flask.
5. Cool the melted agar in a 50°C water bath and dispense about 450 μl of the 50°C agar on to the slide.
6. Be ready to quickly put the Rain-X™-treated slide on top of the agar
7. Allow the slides to cool. Remove the top slide by pulling in the direction of the tape.
8. Store the slides in a closed container containing water saturated paper towels to prevent the agar from drying.

*(2) Glycerin jelly.* Certain structures such as head and labial papillae and the number of longitudinal chords are diagnostic characters for generic or specific determination of certain groups of nematodes (i.e., rhabditids and mermithids). These structures are best seen by making cross-sections. Below we describe the technique used to make cross-sections and mount them.

1. Place nematodes that have already been processed with glycerin in a drop of glycerin and cut into small sections (anterior end or mid-body) with a razor blade or a small oculist's scalpel.
2. Transfer the section, with the help of a needle, to a drop of melted glycerin jelly (see Appendix II) to the center of a coverslip and orient in the proper position.
3. Invert the coverslip and gently place on a coverslip on the aluminum slide or directly on a glass slide with pieces of glass fiber to prevent the jelly from touching the slide.
4. Seal the mount with paraffin or nail polish.

*e. Electron microscopy techniques and applications*

*(1) Scanning electron microscopy.* Scanning electron microscopy (SEM) is a very useful tool for the visualization and interpretation of certain nematode features that can not be appreciated with a light microscope but are important for taxonomic identification. Below we describe a protocol by Lee *et al.* (2009).

1. Relax and heat kill nematodes by placing them in a glass vial filled with M9 or phosphate buffer (pH 7.4) inside a water bath at 60°C for 2 min.
2. Once cooled to room temperature, rinse nematodes three times with M9 buffer or phosphate buffer (5 min each change).
3. Pre-fix in 8% glutaraldehyde buffered in cacodylate at pH 7.3, and leave nematodes overnight in this solution.
4. Rinse nematodes three times in M9 buffer solution (5 min each change) and once in distilled water (5 min).
5. Post-fix in 1% osmium tetroxide ($OsO_4$) for 1 h.
6. Rinse nematodes three times in water (5 min each wash) and dehydrate them using a series of ethanol concentrations (30, 50, 70, 90, 95 and 100%).
7. Dry the nematodes to critical point with liquid $CO_2$.
8. Mount on SEM stubs, and coat twice with gold for 1 h.

*(2) Transmission electron microscopy.* Transmission electron microscopy (TEM) is a powerful tool for examination of the microanatomy of biological tissues, cells and organisms including nematodes. Specimens in the TEM are examined by passing an electron beam through them, disclosing detailed information of the internal structure of specimens. The greatest obstacle to examining biological material with the electron microscope is proper preparation of the specimens so their ultrastructure is preserved as intact as possible. Additionally, the limited penetrating power of electrons,

requires samples to be thin or sliced into thin sections (50–100 nm) to allow electrons to pass through. Thin sections need to be stained with salts of heavy metals (i.e. uranium, lead, osmium) that are electron opaque. Below we describe the procedure.

1. Rinse nematodes in sterile distilled water three times and fix overnight in 3% glutaraldehyde buffered with 0.05 M sodium cacodylate, pH 7.2 in individual chambers.
2. Post-fix specimens in 2% aqueous osmium tetroxide for 1 h.
3. Rinse three times (10 min each wash) with distilled water.
4. Dehydrate nematodes in a graded series of ethanol transfers (10, 30, 50, 70, 90 and 100% [vol/vol]) for 10 min each.
5. Clear nematodes with propylene oxide twice (30 min each).
6. Infiltrate them with a mixture of 50% Eponate resin (Ted Pella, Inc., Redding, CA, USA) and 50% propylene oxide for 24 h. Consider repeating this step two additional times.
7. Do a final embedding in 100% Eponate 12 resin in a mold.
8. Proceed to sectioning of the specimens with an ultramicrotome. There are many manufacturers and models of ultramicrotomes and blades, and users will consider using the one that is available to them. Depending on the purpose of the study, thickness of the sectioning may vary. Usually, 70- to 90-nm sections are used.
9. Collect sections on uncoated 200-mesh copper grids for electron microscopy (Ted Pella, Inc., Redding, CA, USA).
10. Stain grids with saturated aqueous uranyl acetate at room temperature for 20 min followed by lead citrate at room temperature for 2–4 min.
11. Grid-mounted sections are now ready for examination.

*f. Type and voucher specimens*

*(1) Conditioning of specimens.* Specimens used for the diagnosis of a new species/genus must be deposited in a public collection. Accordingly, holotype and paratypes should be prepared as permanent slides (see Section 4), with the following information accompanying the specimens.

**a.** Slide number or code (usually given by the collection's curator once description of the new specimen has been accepted for publication).
**b.** Nematode species name.
**c.** Type designation (holotype, paratype). Also indicate number of specimens per slide.
**d.** Type host (if known).
**e.** Type locality. May include GS information if possible
**f.** Author's name.
**g.** Collector's name.

*(2) Deposition of type and voucher specimens.* Type and voucher specimens should be deposited prior to the publication of the description of the new taxon in a national depository. In the United States, the US Department of Agriculture Nematode Collection at Beltsville, Maryland and the Nematode Collection (UCDNC) at the Department of Nematology, University of California Davis, are the sites where the largest nematode collections are housed, whereas in Europe the collection at the Museum National d'Histoire Naturelle, Paris, France is a major site for type specimens. If sufficient material is available, specimens should be placed in established collections on different continents.

Usually primary types (i.e. holotypes) are unavailable for loan, but guidelines may vary with each institution. Other type specimens (i.e. paratypes) are loaned to recognized specialists for up to 1 year. Researchers are not permitted to change labels or remount specimens. When returning slides, the following recommendations should be followed:

1. Alert the curator that you are mailing the slides before the package is mailed.
2. Place slides into the mailer and put a small piece of cotton on the top of each label to prevent the slides from moving.
3. Place the mailer inside a larger box, with at least 5–8 cm of packing material between the mailer and the box.
4. Send the package by registered mail.

*2. Molecular diagnostic methods*

Molecular techniques are widely used for quick diagnosis or to supplement morphological diagnosis of nematodes including those that are invertebrate parasites and associates. Depending on the goal of the study, different molecular markers are available (Stock *et al.*, 2001; Nguyen *et al.*, 2001; Perlman *et al.*, 2003; Spiridonov *et al.*, 2004). Readers should be aware that certain markers are only useful for diagnostic and/or taxonomic purposes (i.e. aid in the description of a taxon) and should not be considered for systematic studies such as inference of evolutionary histories of taxa. A detailed explanation of markers and step-by-step techniques and methods used for diagnostics and phylogenetic studies of insect parasitic nematodes including conditioning of samples, nucleic acid extraction, PCR and sequencing can be found in Stock (2009). Table 12.2 provides a summary of the most commonly used markers for diagnosis of insect associates, parasitic and pathogenic nematodes

*3. Cross-hybridization assays*

Cross-hybridization assays have been used as a supplemental tool for diagnosis of EPNs. The technique provides evidence to support the 'biological species concept', which defines species in terms of interbreeding as proposed by Mayr (1942). According to this concept, "species are groups of interbreeding natural populations that are reproductively isolated from other such groups". If the hybridization assays show there is no progeny as a result of the crosses between two isolates, then the isolate which species identity is being tested is confirmed as a new species.

We recommend considering two to four other species [usually morphologically similar or closely related (based on phylogenetic molecular evidence)] for comparisons.

*(1) Hanging blood drop.* This technique was originally developed by Poinar & Thomas (1966) and later modified by Kaya & Stock (1997). Below we describe the latter method:

1. Surface-sterilize *G. mellonella* larvae in 70% ethanol.
2. Bleed larva by puncturing between the head and the first prothoracic segment.
3. Carefully place a drop of hemolymph on a coverslip. Make sure hemolymph and not fat is collected.
4. To prevent hemolymph from drying, add 10 μl of serum-free medium for insect tissue culture (SF-900 II SFM Gibco®).
5. Place 30–50 surface sterilized IJs (see procedure in Section 10.B.1.e.) in the drop.
6. Turn the coverslip upside down, and gently place on deep concave slide.
7. Place the slide in a Petri dish (100 × 15 mm) containing one disc of filter paper saturated with water. Moist filter paper will prevent sample from drying. Cover dish with the top.
8. Place the dish inside a plastic bag (to prevent water loss) and incubate at 25–27°C.
9. Observe daily and remove pre-adult (J4) males and females.
10. Place males and females of the isolates to be tested in new hanging drop slides with adults of the opposite sex of the other isolates (ratio of five males to five females is recommended).

**Note:** Evaluation of the mating should be done over a 10-day period. Controls consist of crosses of the same isolates. The presence of progeny is considered positive, indicating the mated pair is of the same species, whereas absence of progeny is a negative result indicating the mated pair is different species. Crosses should have sufficient replications to validate the results.

*(2) Lipid agar plates.*

1. Prepare lipid agar plates as described in Appendix II. Consider using 5-cm Petri dishes for this purpose.
2. Inoculate plates with Phase I bacteria isolated from the nematode strain or species from which the female partner in the cross is derived.
3. Incubate the plates at 25°C overnight.
4. For each cross, place 10 virgin females and 10 males of the appropriate strain or species on each of six lipid agar plates.
5. Check plates under a microscope on a daily basis for the appearance of outcross progeny and record observations.
6. After 3 days and if males have died and/or no egg development is observed in the females, add 10 additional males on each plate to ensure that viable males are always available to fertilize the females.
7. Continue to monitor the plates under a microscope daily for three additional days.

**Note:** In successful crosses, progeny will be visible after 2–3 days.

8. Collect IJs from these crosses after 2 weeks.
9. Transfer to fresh lipid agar plates so that a hybrid line can be established.

**Table 12.2** A selection of primers considered for sequencing of insect parasitic and pathogenic nematodes (from Stock, 2009).

| *Primer* | *Orientation (R = reverse; F = forward)* | *Amplified gene* | *Sequence* | *Comments* | *Reference* |
|---|---|---|---|---|---|
| D2aF | F | 28S rDNA | 5′-ACAAGTACCGTGAGG GAAAGT | Nematoda | Nunn (1992) |
| D3bR | R | 28S rDNA | 5′- TGCGAAGGAACCAG CTACTA | Nematoda | Nunn (1992) |
| D2bF | F | 28S rDNA | 5′-GACCCGTCTTGAAAC ACGGA | Nematoda | Nunn (1992) |
| D3aR | R | 28S rDNA | 5′-TCCGTGTTTCAAGAC GGGTC | Nematoda | Nunn (1992) |
| 391 | F | 28S rDNA | 5′-AGCGGAGGAAAAGAA ACTAA | *Steinernema* spp. | Stock *et al.*, 2001 |
| 501 | R | 28S rDNA | 5′-TCGGAAGGAACCGC TACTA | *Steinernema* spp. | Stock *et al.*, 2001 |
| Ferg-ID2B | F | 28S rDNA | 5′- AGTAACCTCTTGCAC CAAAC | *Fergusobia* spp. | Ye *et al.*, 2007 |
| D2F1-Ferg | R | 28S rDNA | 5′-AGTACCGTGAGGGAAA GTTGAA | *Fergusobia* spp. | Ye *et al.*, 2007 |
| D2F2-Ferg | F | 28S rDNA | 5′-GGAAAGTTGAAAAGC ACTTTG | *Fergusobia* spp. | Ye *et al.*, 2007 |
| D2R-Ferg | R | 28S rDNA | 5′-GATAGTTCGATTAGTCTT TCGCCC | *Fergusobia* spp. | Ye *et al.*, 2007 |
| Ferg5 | F | 28S rDNA | 5′-GAAGAGAGAGTTAAAG AGCACG | *Fergusobia* spp. | Ye *et al.*, 2007 |
| Ferg3 | R | 28S rDNA | 5′-GATAGTTCGATTAGTC TTTC | *Fergusobia* spp. | Ye *et al.*, 2007 |
| 93 | F | ITS rDNA | 5′-TTGAACCGGGTAAAAGTCG | *Steinernema, Heterorhabditis* spp. | Stock *et al.*, 2001 |
| 94 | R | ITS rDNA | 5′-TTAGTTTCTTTTCCTCCGCT | *Steinernema, Heterorhabditis* spp. | Stock *et al.*, 2001 |
| AB28 | F | ITS rDNA | 5′-ATATGCTTAAGTTCAGCGGGT | *Steinernema, Heterorhabditis* spp. | Curran & Robinson, 1993 |
| TW81 | R | ITS rDNA | 5′-GTTTCCGTAGGTGAACCTGC | *Steinernema, Heterorhabditis* spp. | Curran & Robinson, 1993 |
| 18s | F | ITS rDNA | 5′-TTGATTACGTCCCTGCCCTTT | *Steinernema* spp. | Vrain *et al.*, 1992 |

(*Continued*)

**Table 12.2** A selection of primers considered for sequencing of insect parasitic and pathogenic nematodes (from Stock, 2009).—Cont'd

| *Primer* | *Orientation (R = reverse; F = forward)* | *Amplified gene* | *Sequence* | *Comments* | *Reference* |
|---|---|---|---|---|---|
| 28s | R | ITS rDNA | 5′-TTTCACTCGCCGTTACTAAGG | *Steinernema* spp. | Vrain *et al.*, 1992 |
| 505 | F | Mitochondrial 12S rDNA | 5′-GTTCCAGAATAATCGGCTAGAC | *Steinernema* spp. | Nadler *et al.*, 2006a |
| 506 | R | Mitochondrial 12S rDNA | 5′-TCTACTTTACTACAACTTA CTCCCC | *Steinernema* spp. | Nadler *et al.*, 2006a |
| 507 | F | Mitochondrial *cox*1 | 5′-AGTTCTAATCATAA(A/G)GA TAT(C/T)GG | *Steinernema* spp. | Nadler *et al.*, 2006a |
| 588 | R | Mitochondrial *cox* 1 | 5′-TAAACTTCAGGGTGACCA AAAAATCA | *Steinernema* spp. | Nadler *et al.*, 2006a |
| 527 | F | 18S rDNA | 5′-CTAAGGAGTGTGTAACA ACTCACC | Cephalobina including *Steinernema* sp. | Nadler *et al.*, 2006b |
| 532 | R | 18S rDNA | 5′-AATGACGAGGCATTTGGC TACCTT | Cephalobina including *Steinernema* sp. | Nadler *et al.*, 2006b |
| 18S-G18S4 | F | 18S rDNA | 5′-GCTTGTCTCAAAGATTA AGCC | Nematoda | De Ley & Blaxter, 2002 |
| 18S-18P | F | 18S rDNA | 5′-TGATCCWKCYGCAGGTTCAC | Nematoda | De Ley & Blaxter, 2002 |
| COI-F1 | F | Mitochondrial *cox*1 | 5′- CCTACTATGATTGGTGGTTTTGGT AATTG | Tylenchina | Kanzaki & Futai, 2002 |
| COI-R2 | R | Mitochondrial *cox*1 | 5′-GTAGCAGCAGTAAAATAA GCACG | Tylenchina | Kanzaki & Futai, 2002 |
| 537 | F | 18S rDNA | 5′-GATCCGTAACTTCGGGAAAAG GAT | Cephalobina including *Steinernema* sp. | Nadler *et al.*, 2006b |
| 531 | R | 18S rDNA | 5′-CTTCGCAATGATAGGAAGAGCC | Cephalobina including *Steinernema* sp. | Nadler *et al.*, 2006b |
| 5F | F | 18S rDNA | 5′-GCGAAAGCATTTGCCAAGAA | Mermithidae | Vandergast & Roderick, 2003 |
| 18S-9R | R | 18S rDNA | 5′-GATCCTTCCGCAGGTTCACCT | Mermithidae | Vandergast & Roderick, 2003 |

10. Include the following controls for each cross:
    a. Virginity/self-fertility test: 10 virgin females are placed without males on to each of six lipid agar plates.
    b. Mating test/self-cross: 10 virgin females and 10 males of the same strain or species are placed on lipid agar plates, and 10 additional males are added 3 days later.

**Note:** The result of a cross between different isolates is taken as valid if: (1) there is no progeny in the virginity test, and (2) there is progeny in the self-cross.

Because IJs of *Heterorhabditis* species always develop into hermaphroditic females in the first adult generation, the second-generation amphimictic adults are used for the cross-mating studies (Dix, 1994). For this purpose the technique to collect second-generation adults is as follows:

1. Set up infections with IJs of the appropriate strain or species as described in Section 6. It is advisable to infect a series of *G. mellonella* larvae at 2-day intervals to ensure that female nematodes at a suitable stage for cross-breeding are available.
2. Obtain second generation adults from cadavers between 6 and 8 days post-infection by dissecting the *G. mellonella* cadavers in saline solution (see Appendix I).
3. Select female nematodes with immature gonads using a stereomicroscope at 50 × magnification.
4. Collect immature females using a needle or by aspiration, using a microcapillary pipette that is drawn out to give an external diameter of 150–250 μm.

**Note:** As an additional precaution, collect virgin females only from those cadavers in which no more than 5–10% of the second-generation females have started oogenesis. Virgin females can be used immediately for cross breeding or can be placed on a lipid agar plate (see Appendix II) for 1–2 days to confirm that they are producing only unfertilized eggs.

5. Collect males in the same manner as females; however, their age or state of development is not critical.

## B. Entomopathogenic bacteria

### *1. Classic (morphological) methods and techniques*

#### *a. General growth characteristics*

*Photorhabdus* and *Xenorhabdus* are sensitive to oxidants that build up in media exposed to light (Xu and Hurlbert, 1990). Therefore, grow cultures in media that has either been stored in the dark or to which pyruvate has been added to 0.1% (w/v) final concentration. For the latter, add 1 g/l pyruvate to media before autoclaving or add from sterile 20% stock after autoclaving. Unlike other enteric bacteria, including *Photorhabdus*, *Xenorhabdus* is catalase negative. This allows it to be distinguished from possible *Escherichia coli* contaminants by dropping hydrogen peroxide on colonies. If they 'bubble' vigorously (a result of catalase-catalyzed conversion of $H_2O_2$ to water and oxygen) then they are not *Xenorhabdus*.

Optimal temperature for growth will vary depending on the bacterial isolate. Generally, *Photorhabdus* and *Xenorhabdus* grow best at 28–30°C and do not thrive at higher temperatures (e.g., 37°C). Also, generally strains do not survive cold storage (e.g., 4°C).

*Xenorhabdus* and *Photorhabdus* can grow in Luria broth (LB), nutrient broth (NB), minimal medium, Grace's insect medium (see recipes in Appendices), or insect hemolymph. *X. nematophila* and *X. bovienii* both require nictonic acid supplementation for growth in minimal medium, as they lack the *nad* genes necessary for synthesis of this cofactor (Goodrich-Blair, unpublished). *Photorhabdus* do not require any vitamin or growth factor supplements, and can grow in standard M9 minimal medium with $NH_4+$ as a nitrogen source and glucose, mannose, or N-acetylglucosamine as a carbon source (D. Clarke, personal communication). Hemolymph for growth is collected from either *Manduca sexta* or *G. mellonella* insects as follows:

1. Surface sterilize insect larvae by dipping in 95% ethanol.
2. Soak up extra ethanol using a lab tissue.
3. Place larvae in a Petri dish cooled on ice.
4. Use one hand to bend the larvae at the middle, holding the head and tail.
5. Use a needle to penetrate one or two legs in the middle of the body and press gently to squeeze out the hemolymph into a Falcon tube cooled on ice.
6. Filter the hemolymph using a 0.2-μm filter.
7. Add glutathione ($C_{10}H_{17}N_3O_6S$ fw 307.3 reduced form, Sigma catalog number G-6529) to a final concentration of 5 mM.
8. This mixture can be frozen in aliquots at −80°C.
9. For growing bacteria, wash and re-suspend an overnight culture in phosphate buffered saline, pH 7.4.

10. Inoculate into hemolymph at a 1:100 ratio, wrap the tube in foil (to help preserve the hemolymph) and shake (220 r.p.m. on a floor shaker or 56 r.p.m. on a rotating drum) at 30°C.
11. Monitor growth using a spectrophotometer at $OD_{600}$ or by sampling over time and dilution plating to count colony forming units.

*b. Phenotypic assays and virulence modulation*

*Xenorhabdus* and *Photorhabdus* undergo phenotypic variation between primary and secondary form. Typically, the form that is isolated from nematodes is the primary form. During laboratory culture and during recovery growth from the freezer some cells in the population will often convert to secondary form, which has altered gene expression and enzyme activities (Cowles *et al.*, 2007). Another phenomenon, virulence modulation, has been observed in two species of *Xenorhabdus* (Park *et al.* 2007; Goodrich-Blair, unpublished). In these species, variants occur that have attenuated virulence, but no other observed phenotypic differences from their virulent counterparts. The cause of virulence modulation is not known, but has been observed in frozen stocks. Investigators studying *Xenorhabdus* pathogenesis should closely monitor the virulence of their strains and if a reduction is noted, isolate and test single colonies to recreate a virulent stock. The assays described below are used to characterize standard phenotypes of *Xenorhabdus* and *Photorhabdus* strains.

*c. Microscopy and cell morphology*

*Xenorhabdus* and *Photorhabdus* are rod-shaped bacteria (~ 1 × 2 μm), although cell shape can become more spherical during stationary phase. To observe cells microscopically the following protocol can be used:

1. Grow the strain of interest overnight in Luria–Bertani broth (See Appendix II.7) at 30°C on a shaker (220 r.p.m. on a floor shaker or 56 r.p.m. on a rotating drum).
2. Apply a 2% agarose pad to a glass microscope slide (Figure 12.15).
   a. Place two pieces of tape on top of one another on two microscope slides, and place a third slide between the taped slides.
   b. Apply 30 μl of 2% molten agarose to the middle slide.
   c. Press a fourth slide down on top of all three slides to create a flat surface.

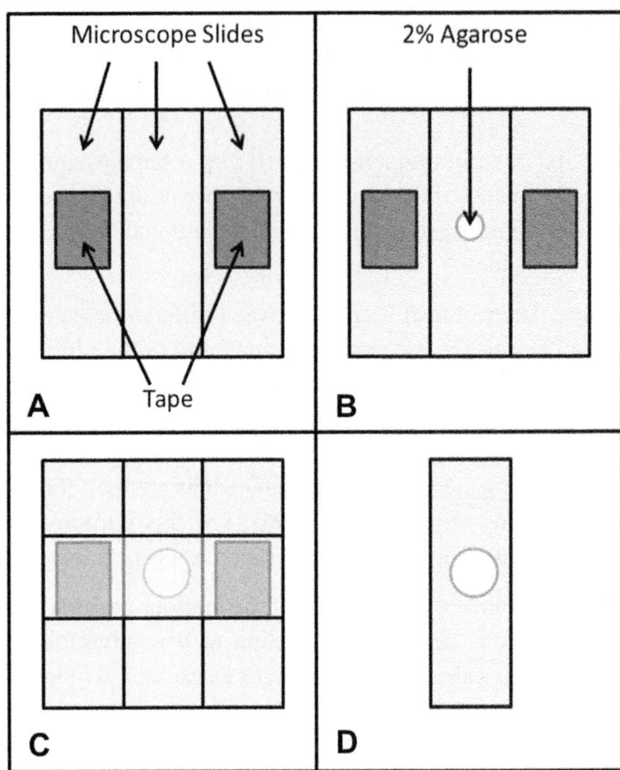


**Figure 12.15** Agarose slides for bacterial cell microscopy.

   d. The middle slide containing the agar pad can now be used for microscopy.
3. Place 20 μl of the overnight culture to the agarose pad and cover with a coverslip.
4. Visualize cells using differential interference contrast microscopy at 100 ×.

*d. Dye binding (NBTA)*

For most strains of *Xenorhabdus* phenotypic switching can be reliably monitored using colony color (indicative of dye binding) on NBTA medium (see Appendix II for recipe). After several days of growth on this medium, primary form cells will appear blue and secondary form cells red. Although *Xenorhabdus* grow slowly on NBTA it is the best indicator medium, as substituting LB for NB leads to false positive calls for primary form.

*e. Antibiotic resistance*

Most strains of *Xenorhabdus* spp. appear to be resistant to ampicillin. Newly isolated strains should be tested for their resistance profile. Generally antibiotic concentrations used are: ampicillin (Amp), 100–150 μg/ml; chloramphenicol (Cm), 15 μg/ml; erythromycin (Erm),

200 μg/ml; streptomycin (Str), 12.5 μg/ml; kanamycin (Kan), 20–25 μg/ml.

*(1) Assay to monitor antibiotic production. Xenorhabdus* and *Photorhabdus* are well known for production of antimicrobial compounds effective against fungi and gram-negative and gram-positive bacteria. The following is a standard assay for monitoring antibiotic activity (Jarosz, 1996; Volgyi, *et al.* 1998).

1. Prepare tryptic soy agar plates. Per liter: 30 g tryptic soy, 20 g agar and autoclave.
2. Place ~ 10–50 ml of an overnight culture of the test strain in the center of the tryptic soy agar plate and grow overnight.
3. Kill the test strain by exposing the plate, in a hood, to chloroform for 30 min. This can be accomplished by placing the chloroform in a small, short container, and inverting the agar plate over it.
4. Using 3-ml soft agar overlay, place the indicator strain (e.g., *Bacillus subtilis, Micrococcus luteus,* or *E. coli*) over the killed test strain. Incubate for 24–72 h at 30°C. Absence of bacteria growth indicates production of antimicrobial factors.

*f. Assay to monitor protease activity*

Boemare *et al.* (1997) describe techniques to assess several phenotypic characteristics of *Xenorhabdus* and *Photorhabdus* including protease activity detected by clearing of milk substrate.

1. Prepare milk agar plates. Per liter: 5 g yeast extract, 10 g tryptone, 5 g NaCl, 20 g agar mixed in 700 ml water and in separate flask 30 g milk powder in 300 ml water. Autoclave flasks separately, and when cool to the touch combine the two liquids, pour into Petri plates.
2. Spot 1–2 μl of overnight cultures to milk plates. Several spots/strains can be applied to a single plate.
3. Observe clearing zone around culture after several days incubation at 30°C.

*g. Assay for hemolytic activity*

1. Prepare blood agar plates. Per liter: 15 g agar, 10 g beef extract, 10 g peptone, 5 g NaCl, and 950 ml $H_2O$. Autoclave, then add 50 ml defibrinated blood (e.g., sheep, rabbit, horse) (Colorado Serum Company, Denver, CO, USA).
2. Add 1–2 μl of overnight culture of test strain to blood agar plates. Several spots/strains can be applied to a single plate.
3. Observe clearing zone around culture after a few days incubation at 30°C.

*h. Assay for lipolytic activity*

1. Prepare Tween agar plates. Per liter: 10 g bacto-peptone, 5 g NaCl, 0.11 g $CaCl_2$–$2H_2O$, 15 g agar, 900 ml $H_2O$. In separate flask mix 90 ml of $H_2O$ with 10 ml Tween (Sigma). Autoclave both flasks, cool to ~ 55°C, mix, and pour into Petri plates. *Xenorhabdus* strains have been shown to have activity against Tween 20, 40, 60, and/or 80.
2. Spot 5 μl of an overnight culture of the test strain on to Tween agar plates. Several spots/strains can be applied to a single plate.
3. Observe clearing zone around culture after a few days' incubation at 30°C.

*i. Assay for bacterial motility*

1. Prepare soft agar swim plates. Per liter: 10 g tryptone, 5 g yeast extract, 5 g NaCl, 2.5 g agar.
2. Add 5 l of overnight culture to center of soft agar plates. Apply only one spot per plate.
3. Observe diameter and pattern of culture after ± 16 h incubation at 30°C.

*j. Luminescence*

*Photorhabdus* isolates are bioluminescent. Luminescence can be monitored either visually in the dark, or quantitatively using default settings on a luminometer (e.g., a multi-mode plate reader).

*2. Deposit type strains*

Like for other bacteria and microorganisms, entomopathogenic bacteria, can be deposited at the ATCC collection. Users can find instructions for deposition of type specimens or any novel strains at: http://www.atcc.org/. Typically, both a primary and a secondary form isolate should be deposited.

*3. Molecular diagnostic methods*

*a. Recommended reference strain name for molecular data*

Two *Xenorhabdus* genomes, *X. bovienii* SS-2004 (NCBI accession NC_013892) and *X. nematophila* ATCC 19061

(NCBI accession: NC_014228) have been sequenced. The following nomenclature has been established for their gene names: XBJ1_XXXX and XNC1_XXXX, referring to Genus (X) species ('b' for *bovienii* and 'n' for *nematophila*, respectively), nematode host (*S. jollieti* and *S. carpocapsae,* respectively), and strain sequenced (each is thus far the first among their species to be sequenced and are therefore designated '1'). Future molecular sequence depositions should follow this nomenclature.

*b.* Xenorhabdus *and* Photorhabdus *species diagnostics using* 16S *and other genes (Table 12.3)*

*16S rDNA gene.* This gene region has been extensively surveyed both for inferring phylogenetic relationships and for identification and characterization of bacterial species including *Xenorhabdus* and *Photorhabdus* (Liu *et al.*, 1997; Tailliez *et al.* 2006, 2010; Lee & Stock, 2010). Moreover, sequences of this gene have also been used with other molecular approaches such as ERIC and RAPD

**Table 12.3** List of primers considered for diagnostics and phylogenetics of *Xenorhabdus* and *Photorhabdus*.

| *Primer* | *Region* | *Orientation* | *5–3 Sequence* |
|---|---|---|---|
| 16SP1* | 16s rDNA | Fwd | GAAGAGTTGATCATGGCTC |
| 16SP2* | 16s rDNA | Rev | AAGGAGGTGATCCAGCCGCA |
| SP1* | 16s rDNA | Rev | ACCGCGGCTGCTGGCACG |
| SP2* | 16s rDNA | Rev | CTCGTTGCGGGACTTAAC |
| recA[†] | *recA* | Fwd | CCAATGGGCCGTATTGTTGA |
| recA-R[†] | *recA* | Rev | TCATACGGATCTGGTTGATGAA |
| *serC*[†] | serC | Fwd | CCACCAGCAACTTTGTCCTTTC |
| *serC*-R[†] | serC | Rev | AAAGAAGCAGAAAAATATTGCAC |
| serC-F1[‡] | serC | Fwd | CGTTTGCTGATTTCYTGTAA |
| *serC*-F2[‡] | serC | Fwd | CAACRCGGTTGATATACA |
| serC-F3[‡] | serC | Fwd | CCCTGCTCTTTCARCCA |
| *serC*-R1[‡] | serC | Rev | CCKGATTTTGGYGAYGATAA |
| *serC*-R2[‡] | serC | Rev | TATTGYCCTAATGAAAC |
| gyrB* | gyrB | Fwd | ATTGGCACTGTATGGTATCAC |
| gyrB* | gyrBrev | Rev | TACTCATCC ATTGCTTCATCATCT |
| gyrB* | gyrBSP1 | Fwd | ATAACTCTTATAAAGTTCCG |
| gyrB* | gyrBSP2 | Fwd | CGGGTTGTATTCGTCACGGCC |
| gyrB* | gyrBSP4 | Rev | GCAGTAAATATTTTCCTGG |
| gltX* | gltX1 | Fwd | GCACCAAGTCCTACTGGCTA |
| gltX* | gltX2(R) 5′- | Rev | GGCATRCCSACTTTACCCATA |
| gltX* | gltX3 | Rev | TCCATATCCCAGTCATC |
| dnaN* | dnaN1 | Fwd | GAAATTYATCATTGAACGWG |
| dnaN* | dnaN2 | Rev | CGCATWGGCATMACRAC |
| dnaN* | dnaN6 | Rev | GTTRTTRCTGCCAATCTG |

*From Tailliez *et al.* 2006;
[†]from Sergeant *et al.* 2006;
[‡]from Lee & Stock (2010).

profiling, RFLP pattern analysis, and also phenotypic traits, to infer phylogenetic relationships of entomopathogenic bacteria. Despite the widely accepted use of this gene for diagnostic purposes and phylogenetic inference, many studies have shown lack of variation between 16S rDNA sequences (Dauga, 2002; Wertz *et al.*, 2003). Tailliez *et al.* (2006) noted 16s rDNA sequence divergence between any two *Xenorhabdus* species was low (3–5%). Moreover, lateral gene transfer (LGT), especially among enteric bacteria, has become a confounding factor in determining phylogenies for prokaryotes. The exact level of gene transfer is not known for *Xenorhabdus* bacteria, but is thought to occur even at the deepest branches of the Enterobacteriaceae (Doolittle, 1999).

*Housekeeping genes.* Core housekeeping genes are known to be resistant to high levels of LGT because there is little selective advantage to acquiring a novel function—unlike the high selective pressure for gaining genes such as pathogenicity islands and those that encode toxins, which may play an important role in adaptation to local environments. Housekeeping genes are being increasingly used to infer phylogenies of many organisms including entomopathogenic bacteria, because they are highly expressed and highly conserved, yet evolve at a rate faster than 16s rDNA, therefore making them useful for species level comparisons (Lawrence *et al.*, 1991).

So far, five protein coding genes: *recA* (recombinase protein), *serC* (phosphoserine aminotransferase), *gyrB* (DNA gyrase subunit B), *dnaN* (beta subunit of DNA polymerase III holoenzyme), *rplB*, *gtlX* (encodes glutamyl-tRNA synthetase) have been considered to both clarify the taxonomic status and assess the evolutionary relationships between *Photorhabdus* and *Xenorhabdus* species and strains (Akhurst, *et al.*, 2004; Lee & Stock, 2010; Tailliez *et al.*, 2010).

## 5. STAINING METHODS

### A. Live nematodes

Stained IJs can be used for ecological studies and may be a means to separate them from naturally occurring IJs. For *in vitro*-reared nematodes, stains such as Sudan (Sudan II, Sudan III) or oil soluble blue can be added into the medium. Nematodes will ingest stains and the color will be present in the intestine and fat body (Yang & Jian, 1988). The stain in the nematodes should last for several months and it will be lost when they infect a host. No adverse effect on nematode fecundity or infectivity has been observed.

Ogura (1993), using a different artificial diet supplemented with 0.5% Sudan III, could not obtain stained *S. kushidai*, but the incorporation of 0.4% neutral red in the medium produced stained IJs. The red pigment was discernible in the intestinal tract of the IJs. Survival of the stained *S. kushidai* IJs appeared to be less than that of the unstained ones.

### B. Nematodes in insects

There are occasions when the location of nematodes within a host is needed. Dissection may reveal the tentative location of the nematodes, but a more exact method is to stain the nematode *in situ*. This method is useful for those insects with soft or transparent cuticle.

#### *1. Staining*

**a.** Fix the parasitized larva (or adult) in 70% ethanol for at least 24 h. If a beetle larva is being processed, puncture it to allow stains to penetrate efficiently.
**b.** Transfer the larval specimen to Grenacher's borax carmine (see Appendix II).
**c.** Leave for at least 12 h.
**d.** After this time, remove specimens from the stain and rinse excess stain with several changes of 70% ethanol.

#### *2. De-staining*

**a.** De-stain in acid alcohol (70% ethanol + 2% HCl) until the nematodes can be seen inside the insect's body. For insects with very thin integument de-staining requires a few seconds, whereas for those with thicker cuticles it may take 15–30 min.
**b.** Stop de-staining by placing the insect in plain 70% ethanol; if more staining is needed, reinitiate by placing into the acid alcohol.

#### *3. Dehydration*

**a.** Place the insect into 70% ethanol for 2 h to remove acid from the tissues.
**b.** Transfer the insect to 80% ethanol for 2 h and then to 95% ethanol for 2 h.

*4. Counterstaining*

**a.** Counterstain by adding one or two drops of the stock solution of 1% Fast green in 95% ethanol (see Appendix II) in a Syracuse watch glass.
**b.** Dip specimen into the diluted stain for 1–5 s while observing under a dissecting microscope.
**c.** Remove immediately to 95% ethanol and examine. Repeat if necessary. The cuticle of the insect should have a green blush.
**d.** Place specimen into absolute ethanol for 1–2 h. Do not leave longer than 2 h as the green will discolor and fade.

*5. Clearing and mounting*

**a.** Transfer specimen directly into methyl salicylate for *ca.* 5 min.
**b.** Remove for mounting as soon as the specimen has cleared.
**c.** Mount specimen directly from methyl salicylate into permount, damar balsam, or neutral Canada balsam.

## 6. CULTURING METHODS

### A. Culturing of entomopathogenic nematodes

*1.* In vivo *culturing*

Most steinernematids and heterorhabditids can be produced in *G. mellonella* larvae or other suitable insect hosts. Species such as *Steinernema kushidai, S. scarabaei* and *S. scapterisci* are not very virulent to lepidopteran larvae, therefore other insect hosts such as white grubs, mealworms, mole crickets or house crickets, should be used. The set-up of an 'infection trap' is as follows (Figure 12.16):

1. Adjust the IJ suspension to a concentration of 200 IJs/ml.
2. Evenly distribute 1 ml of the IJ suspension on a 10-cm-diameter piece of filter paper (Whatman #1) in the lid of a 100 × 15 mm plastic Petri dish.
3. Add 10 insect larvae to the dish. The goal is to have about 20 IJs/larva; too many IJs per larva produce few progeny because of competition and/or contamination with foreign bacteria.
4. Cover the lid with the inverted Petri bottom.

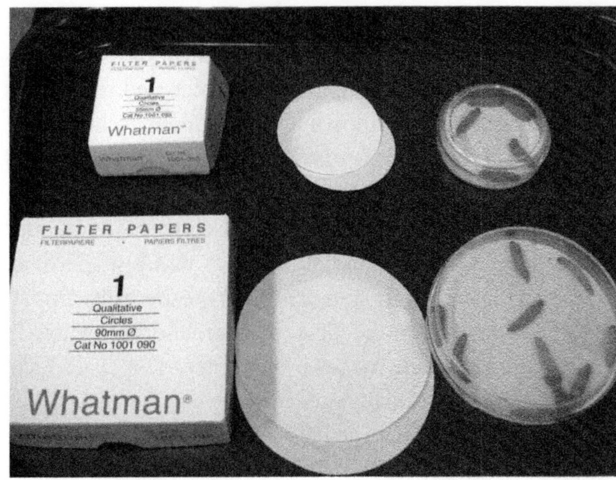


**Figure 12.16** Supplies needed to set up an infection chamber.

5. Label the Petri dish and store it in a plastic bag (to conserve moisture) at room temperature (in the dark). For the labeling of the dish, include the following information:
   **a.** Nematode species name (if known).
   **b.** Isolate code/designation.
   **c.** Date infection trap was set up.
6. Place Petri dish inside a plastic Ziploc® bag and incubate in the dark at either room temperature or in an incubator at approximately 20–25°C.

**Note:** When working with more than one species, make sure that you wash your hands and use sterilized tools to avoid cross contamination of the cultures. Alternatively, wear gloves and spray them with 70% ethanol before switching to a new culture.

7. After 3–5 days, place the cadavers in a modified White trap (as described in Section 3).

**Note:** If the cadavers smell putrid, the culturing was not successful.

*2.* In vitro *culturing*

EPNs can be successfully cultured *in vitro*. Different agar media have been developed depending on the purpose of the study. Below we describe three of the most commonly used methods. The ingredients for nematode culture include a source of nutrients for the symbiotic bacterium and a sterol source for the nematodes. Our focus will be on monoxenic culture in the laboratory rather than on commercial production and on solid rather liquid culture.

*a. Dog-food agar method*

One of the artificial media for steinernematid and heterorhabditid production uses dry dog or cat food as the main ingredient (Hara *et al.*, 1981; Ogura & Haraguchi, 1993) (see Appendix II).

1. After sterilization and cooling of the medium, seed with Phase I bacterium from the species to be cultured (see Section 4).
2. Let the bacterium grow for 1–2 days and inoculate with surface-sterilized IJs or eggs or from another monoxenic culture (see Section 10.B.1.a.).

**Note:** Alternatively, surface-sterilized IJs can be placed on the medium without the Phase I bacterium (see Section 4); the IJs will release the bacterial cells from their intestines but development will be delayed by 2–3 days.

3. After 2–3 weeks, harvest the IJs or, if in test tubes, store at cooler temperatures until needed.

To harvest from Petri dishes, proceed as follows:

1. Use a modified version of the White trap.
2. Remove the cover from the Petri dish and place the bottom of the Petri dish into a larger dish (10–15 mm diameter).
3. Add water to the larger dish.
4. Place four discs of filter paper (5 or 10 cm diameter, depending on the size of the dish being used) on the medium and allow a part of each filter paper to touch the water. The IJs will migrate into the water using the filter paper as a bridge.
5. Collect IJs from the water as previously mentioned.

*b. Liver–kidney agar method*

The liver–kidney agar allows EPN to grow without an insect host. For this method the only precaution to exercise is to sterilize nematodes prior to their inoculation on the dish. There is no need to streak the bacterial symbionts prior to the inoculation with nematodes.

This method has been used to grow successfully aposymbiotic (i.e. without symbiont) EPN.

The recipe and instructions for preparing this agar are described in Appendix II (see also Figure 12.17).

Because of its rich nature, this medium is prone to contamination. Therefore, we suggest that users wait until the dish is full with nematodes and IJs are observed crawling up the dish to place it on a modified White trap.

*c. Lipid agar method*

This method has been widely used to grow both EPN and their symbiotic bacteria. It can be used in a number of bioassays including growth of aposymbiotic nematodes and nematode–bacterium specificity assays. The method has also been used to perform cross-hybridization assays required to validate new species status based on the biological species concept. Briefly the technique is as follows:

1. Prepare and pour lipid agar according to instructions provided in Appendix II on 5- or 6-cm Petri dishes.
2. Streak agar with desired bacterial lawn.

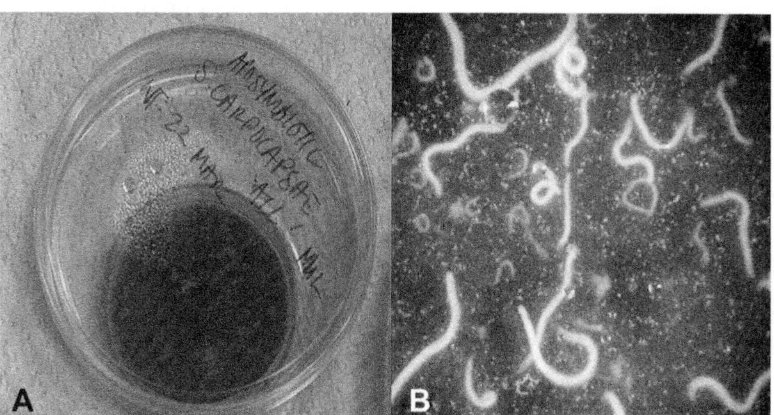


**Figure 12.17** Liver–kidney agar. A. White trap with emerging IJs; B. close-up showing nematodes in various stages.

3. Incubate at 25°C for 24–48 h.
4. Add either aposymbiotic eggs on liver–kidney agar, or J1s of the desired nematode species/isolate.
5. Cultures should produce IJs in about 12 days.
6. After about 9 days, transfer the plates into a larger Petri dish (10 cm) containing water (i.e. Modified White trap) to allow IJs to emerge.
7. Collect IJs according to procedures described in Section 3.

*d. Chicken-offal medium method*

Bedding (1981, 1984) developed a technique whereby large numbers of nematodes may be produced using chicken offal medium on a porous foam substrate such as polyether polyurethane. This substrate provides the largest surface-to-volume ratio while providing adequate interstitial space. Glass flasks serve as rearing containers (see Appendix I), but autoclavable bags have also been used.

1. After autoclaving, allow the medium to cool and inoculate with Phase I bacterium of the nematode species to be cultured.
2. Inoculate with 5 ml of nutrient broth with 2- to 3-day-old bacterium.
3. One day after bacterial inoculation, inoculate with *ca.* 1000 surface-sterilized IJs or nematodes from another monoxenic culture. Care should be taken to maintain monoxenicity during the transfer.
4. Harvest IJs from the flasks in 2–4 weeks. Harvesting IJs can be accomplished as follows:
   **a.** Pile foam 5 cm deep on a 20-mesh (833 μm) sieve.
   **b.** Place the sieve in a pan of water with the water level adjusted so that the foam is just submerged. If a mist chamber is available, the pan and sieve may be placed in it for 2–24 h.
   **Note:** Do not pour water over the foam as this washes particles of homogenate into the nematode suspension. Within 2 h, 95% of the IJs will migrate into the water.
   **c.** Sediment IJs and rinse to remove the particulate matter by passing through a sieve (90–38 μm depending on nematode size).
   **d.** Rinse with water several times until it is clear.

A drop in production can be due to a number of problems. These include, but are not limited to, contamination, and reversion of Phase I bacterium to Phase II, unsuitable incubation temperatures, improper moisture content, and poor genetic stock. Contamination will sometimes be visually evident in the form of fungal or bacterial colonies or 'unusual' coloration or exudates. Often this will be associated with putrid or 'unusual' odors. If contamination is suspected, purity can be routinely verified using NBTA or MacConkey agar (see Appendix II).

Some nematode species do better at lower temperatures and placing the flask at temperatures below 25°C will increase production. If the foam is too dry or too wet, adjust the amount of medium placed on the foam. If the medium starts to dry out during culture, holding the flask at higher humidity may help.

Production levels can remain adequate while quality, (measured in terms of infectivity), may drop in activity. *In vivo* passage is recommended after several generations of *in vitro* culture. Another method is to store IJs in liquid nitrogen and use this source to initiate new cultures (see Section 8).

It is difficult to foresee every potential problem. With some knowledge of nematode and bacterium biology, solutions for most problems should not be too difficult to pinpoint. However, *in vitro* production is not easy, especially for the novice, and trial and error will improve the system.

## B. *In vivo* rearing of *Romanomermis culicivorax*

Mermithids are obligate parasites and, therefore, rearing of these nematodes requires a suitable host. Species such as *Romanomermis culicivorax, Heleidomermis magnapapula, Tetradonema plicans,* among others have been cultured *in vivo* in their respective insect host. Poinar (1979) summarized culturing techniques for many species of nematodes that show biological control potential. In addition, Petersen (1984) and Petersen & Willis (1972) discussed mass production of *R. culicivorax* in mosquitoes. Mullens & Velten (1994) cover the laboratory culture of *H. magnapapula* in midges. Peloquin & Platzer (1993) provide detailed methods for rearing *T. plicans* in sciarid flies. Below we describe a technique used for rearing *R. culicivorax* (Kaya & Stock 1997).

1. Collect J2 nematodes at the desired rate to newly hatched mosquito larvae in the rearing pans (*Culex pipiens* and *Culex quinquefasciatus* can be reared in the laboratory and both are susceptible to the mermithid).

**Note:** The ratio of J2s to mosquito larvae can vary from 2.5 : 1 to 12 : 1. A ratio of 5 : 1 can produce 95% parasitization of mosquitoes in small containers, but for mass production a ratio of 12 : 1 is needed to achieve the same level of parasitism. A higher ratio will result in a sex ratio skewed towards males because males predominate when several individuals occur in the same host.

2. Incubate the rearing pans at 27°C.
3. Feed mosquito larvae finely ground rabbit chow or other suitable diet as needed. Do not overfeed because the rearing containers will become cloudy and foamy. This situation can be minimized by aerating the rearing pan.
4. Transfer the larval mosquitoes to a modified Whitehead and Hemming tray (see Section 3) 6 days after exposure to nematodes.

**Note:** Post-parasitic nematodes (J4s) will settle to the bottom container and the uninfected mosquitoes and cadavers will be retained in the upper container. The J4s, which will begin emerging on the seventh day, can be collected from the bottom container.

5. Concentrate the J4s in a small container.
6. Allow the J4s to settle and discard the supernatant.
7. Repeat the procedure several times to eliminate most of the debris.
8. Transfer the J4s to a pan containing clean coarse sand (1.5 cm deep) covered to a depth of 1 cm with dechlorinated water.

**Note:** Fine sand is not recommended because it becomes tightly compacted and inhibits nematode movement.

9. Cover the pan with a loose-fitting lid and store at room temperature.
10. Remove any visible dead nematodes and decant the water 3 weeks after setting the mermithids in the coarse sand.
11. Absorb the excess water in the pan with paper towels.
12. Store the pan for an additional 4–15 weeks before use. Eggs will be laid in the sand, and J2s can be obtained by flooding the pan. The best hatching of J2s occurs after 6 weeks of storage and declines after 20 weeks (Petersen & Willis, 1972).

## C. Culturing of insect associates and facultative parasites

### *1. Microbivore nematodes*

As described in Section 2, many free-living nematodes may sporadically associate with insects and contingent on the insect health and/or association of these nematodes with pathogenic microbes, they may kill an insect or invertebrate host. Because these nematodes are more likely to be facultative parasites rather than obligate (requiring a specific host association) they can be propagated *in vitro*. Stock *et al.* (2001) developed an agar medium, baby food (BF) agar (see Appendix 1) that has been successful in allowing growth of microbivore nematodes. Briefly the procedure is as follows

1. Recover nematodes from an insect host (dead or alive) by rinsing the insect with saline solution.
2. Collect solution with nematodes in a beaker and rinse it three times to remove any contaminant.
3. If possible separate mature and immature nematode stages.
4. Aliquote a 100-μl suspension in a 10-cm BF agar dish.
5. Follow nematode development daily and maintain cultures by sub-culturing them periodically.

**Note:** Time-frame for this procedure will vary depending on the nematodes considered. Sub-culturing is usually done by cutting a 1-cm$^2$ piece of agar containing nematodes and transferring it to a new agar dish.

Certain necromenic nematodes such as *Pristionchus* spp. have a more specific association with an insect host. For example *P. pacificus* and *P. entomophagus* have a preference for scarab beetles (Hermann *et al.*, 2007; Rae *et al.*, 2008). Therefore, sampling of beetles and extraction of nematodes from them to establish laboratory cultures is necessary. Below we summarize the procedure described by Rae *et al.*(2008).

1. Cut beetles in half transversely with scissors.
2. Place half sections on 6-cm Nematode Growth Medium (NGM) agar plates (see Appendix II).
3. Store dishes at room temperature and inspect them daily for 7–14 days for nematode movement and/or reproduction in the cadavers.
4. Remove nematodes individually with the help of a needle.
5. Wash nematodes in M9 buffer for 1–2 min.
6. Transfer nematodes to LB plates and incubate at room temperature.
7. Bacteria growing on LB plates are assumed to be 'beetle-derived' bacteria and will eventually support nematode growth.

*2. Slug nematodes*—Phasmarahbditis hermaphrodita

*P. hermaphrodita* can be reared both *in vivo* (considering various slug species as its host) and *in vitro*. Below we summarize techniques for both rearing alternatives. See also Chapter XIII for additional methods.

*a.* In vivo *rearing*

Several slug species have been used successfully to rear *P. hermaphrodita* such as *Deroceras reticulatum, M. gagates* and *L. pseudoflavus*. The technique was described by Rae *et al.* (2010) as follows:

1. Make 'slug sandwich' by placing individual slugs between two pads of absorbent cotton wool.
2. Place slugs in 30- or 50-ml centrifuge tubes.
3. Inoculate slugs with approximately with 100 *P. hermaphrodita* infective stages.
4. After 48 h transfer slugs to Petri dishes and check periodically for mortality.
5. Transfer dead slugs to modified to White traps.
6. Watch for nematode emergence and collect emerging nematodes from the water of the White trap.

*b.* In vitro *rearing*

As previously described, *P. hermaphrodita* is a bacterial-feeding nematode. Unlike EPNs, no specific symbiotically associated bacterium has been isolated from this nematode to date. When grown *in vitro,* different bacterial species can influence both yields and virulence of the nematodes. For commercial production, *P. hermaphrodita* is always reared in a monoxenic culture with the bacterium *Moraxella osloensis* (Wilson *et al.*, 1993). See Chapter XIII for details on this technique.

## D. Culturing methods for symbiotic bacteria

*1. From insect cadavers (Lee & Stock, 2010)*

Bacteria can be isolated from EPN-infected cadavers. For this purpose users should set up infection as described in Section 3. Extraction of symbiotic bacteria should be done 2–4 days after initial infection. Extraction of symbiotic bacteria, should be done under sterile conditions in a laminar flow bench. The procedure is as follows:

1. Surface sterilize cadavers by submerging them first in 1% bleach solution followed by 70% ethanol. Rinse and blot dry on sterile filter paper.
2. With a sterile needle or thin-point forceps, rupture the cadaver's cuticle.

**Note**: Insect hemolymph at this time will no longer be liquid. Macerated insect tissues with the consistency of a paste should be the normal appearance.

3. Use a sterile microbiological loop and gently introduce it into the cadaver's body to collect a sample.
4. Streak loop on a suitable agar for bacteria growth (see Appendix II). NBTA is recommended.
5. Place agar plates in an incubator at 25°C.
6. To test for positive identity of bacterial colonies the catalase test described in Section 4b should be used.

*2. Isolation of bacteria from infective juvenile nematodes using a hand-held grinder*

1. Transfer 1 ml of IJ suspension to a sterile 1.5-ml microfuge tube (a minimum of 300 IJs).
2. Pellet IJs by spinning for 3 min at maximum speed (at least 13,000 r.p.m.) in a microfuge.
3. Remove water and add 1 ml 1.0% bleach.
4. Incubate for 1 min at room temperature.
5. Pellet nematodes as in step 2.
6. Remove bleach and rinse with 1 ml sterile water.
7. Repeat wash steps 5 and 6 for a total of two washes.
8. Remove water from last wash and re-suspend the nematode pellet in 0.5–1.0 ml sterile distilled water.
9. Quantify IJ concentration by counting at least three 2- to 5-μl drops and averaging.
10. Dilute IJs to 4 IJs/μl.
11. Move 50 μl of the IJ suspension into a sterile microcentrifuge tube containing 200 μl LB-broth.
12. Using a hand-held motor driven grinder and polypropylene pestle (Kontes), grind nematodes for 2 min.
13. Plate immediately onto two LB-pyruvate plates using a flame sterilized spreader.
    a. Plate 50 μl of the homogenate onto the first plate.
    b. Dilute homogenate 1 : 3 in LB-broth and plate 50 μl on to a second plate.
14. Incubate plates overnight at 30°C (or optimal temperature for symbiont).
15. Inspect plates for colonies.
    a. One of the two plates should contain countable colonies
    b. Plates may need to be incubated for an additional day to produce colonies large enough for counting.

**Note**: When doing multiple samples, steps 11−13 should be done in an immediate sequence. Several batches of IJs may be sterilized at one time and left in water while other samples are processed.

## 7. NEMATODE QUANTIFICATION METHODS

Nematodes can be quantified in one of 3 ways: (1) direct count; (2) dilutions; and (3) volumetric estimation (Woodring & Kaya, 1988).

### A. Direct counts

This method is useful only for a small number of nematodes, usually not higher than 50 individuals. With this method, nematodes are counted individually under a dissecting microscope. Proceed as follows:

1. Transfer the nematodes to the desired place with the aid of a micropipette or microdispenser.
2. Check the micropipette or microdispenser to verify that all nematodes were expelled.

For 1−50 nematodes, direct counting is much more accurate and is preferred over dilution. However, for numbers above 50, the time saved using a dilution is justified over the small loss in accuracy.

### B. Dilution

Users should prepare a dilution of about100 nematodes/ml to facilitate counting. Procedure for make a dilution is described below

1. Mix the initial suspension as homogeneously as possible (a stir bar in a beaker suspension works well).
2. Remove 1 ml of the initial suspension and add to X ml water, where X is estimated to yield a new suspension of approximately100 nematodes/ml. This is a 1 : X + 1 dilution.
3. Count the nematodes in the dilution under a dissecting microscope.
4. Take samples from a thoroughly mixed, diluted suspension.
5. Count on a counting slide that holds a known volume and provides grids to aid counting. An 'eelworm counter' in which 1 ml of nematode suspension occurs on a grid is commercially available (Figure 12.18).

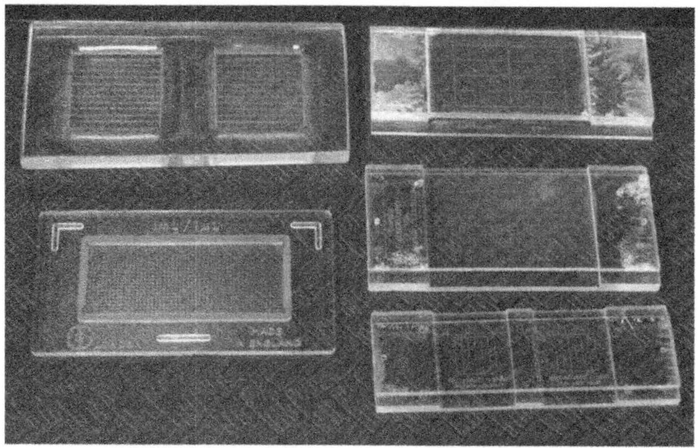

**Figure 12.18** Types of counting devices.

Conventionally, nematodes touching or crossing the top and right-hand sides of a grid are counted whereas those touching or crossing the bottom or left-handed side are not counted. In this manner, the same nematode will not be counted twice (Figure 12.19).

Alternatively, 1 ml of suspension may be spread on a microscope slide or Petri dish and counted. If the dilution has too few (< 50/ml) or too many nematodes to easily count, make a new dilution.

Statistically, an acceptable sample size is 10 counts; however, most workers probably count three to five samples from two to three dilutions. To determine the

**Figure 12.19** Counting grid showing suggested directionality for counting samples.

concentration of nematodes in the original suspension, the following formula is used:

$$\frac{N \times 1(X+1)}{M} = S$$

where N = average number of nematodes per counted sample, M = number of ml per counted sample, S = concentration (nematodes/ml) in initial suspension, X + 1 = dilution.

To prepare dilutions with a given number of nematodes per ml, dilute the initial suspension or the counted suspension. Statistically, it would be preferable to use the counted suspension, but practically it may contain far too few nematodes/ml. The following formula is used:

$$A = \frac{D \times C}{B}$$

where A = ml of suspension of known concentrations, the suspension to be diluted; B = number of nematodes/ml in suspension; C = final volume, in ml, of new dilution; D = desired concentration in new dilution.

Then: C−A = ml of water to be added to make a new dilution.

### C. Volumetric counts

Large numbers of nematodes can be estimated on the settled volume. This technique is useful when the nematodes are similar in size, such as IJs of EPNs. Calibrate the system in the following way:

1. Add 10 ml of a known (high) concentration of nematodes to a graduated centrifuge tube.
2. Centrifuge at a given r.p.m. for a set period of time (e.g., 300 r.p.m. for 1 min).
3. Divide the number of nematodes in the tube by the volume (in ml) that the nematodes now fill to obtain the number of nematodes/ml (V).

'V' should be estimated several times from different samples to verify that the system is relatively consistent for the given species and procedure. Once the system is calibrated, the number of nematodes may be estimated by multiplying the volume taken up by the appropriate V. However, this method is the least precise and repeatability may be a problem, particularly involving different workers.

## 8. PRESERVATION METHODS

### A. Nematodes

#### *1. Short-term preservation*

EPNs can be stored (with no aeration) in distilled water in tissue culture flasks. Generally, steinernematids can be stored at 4–15°C for 6–9 months and heterorhabditids for 3–4 months before sub-culturing is needed. Below we describe the protocol adopted in the P. Stock laboratory.

1. After collecting IJs from a modified White trap, examine their activity. Most species are active at room temperature. However, IJs of certain *Steinernema* spp. may remain still and will need to be probed for activity. Another good indication that these IJs are not dead is to determine if they have a characteristic 'J' shape. IJs that are straight without the 'J' shape are probably dead.
2. If host tissues or large numbers of dead IJs are present in the water of the trap consider a 'separatory (sieving) step' before storage.
3. Separation of active IJs can be accomplished by using a 400-mesh sieve.
4. If other nematode stages besides IJs are present, these 'non-IJ stages' can be removed by rinsing the nematodes in 0.4% methylbenzethonium chloride solution for 15 min. Non-IJ stages will die and IJs can now be collected prior to rinsing them using the technique described in the harvesting section.
5. Optimize IJ concentration. Usually, 1500–3000 IJs/ml is a good number to store in a flask (Figure 12.20). Lower concentrations should be used for larger nematodes such as *S. glaseri* (500–1,000 IJs/ml).

**Note:** We recommend storing no more than 50 ml of IJ suspension in a 250-ml tissue culture flask. For smaller flasks the IJ suspension volume should be reduced accordingly. For example for 50 ml tissue culture flasks, store no more than 15 ml of IJ suspension.

6. A drop of Triton X-100 (wetting agent) may be added to prevent the IJs from sticking to the surface of the containers.
7. To prevent the formation of rosettes (clumps) commonly formed by heterorhabditids, a few drops of sodium bicarbonate solution (1 g $NaH_2CO_3$/50 ml of H2O) may be added (from Woodring & Kaya, 1988).

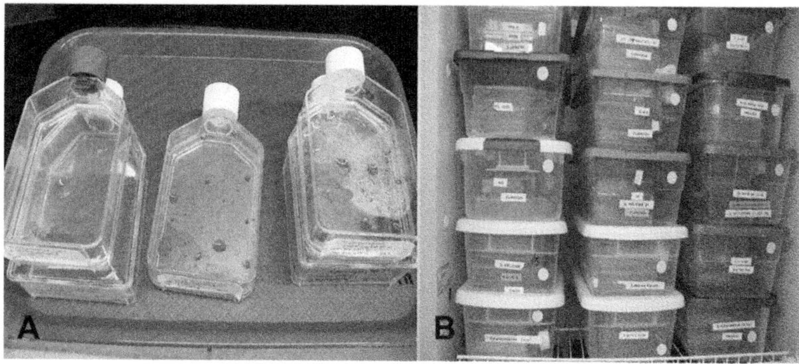


**Figure 12.20** EPN short-term storage. Right: 250-ml tissue-culture flasks. Left: flasks are stored in plastic boxes in an incubator at temperatures between 10 and 15°C.

8. Tissue culture flasks should be stored flat so that there is good exchange of air.
9. The amount of water in a tissue culture flask should be no more than 1 cm deep when the tissue culture flask is stored horizontally (Figure 12.21).

*2. Long-term preservation*

*a. Glycerol stocks*

Temperatures below −130°C (the recrystallization point of ice) can assure long-term, and possibly indefinite preservation of certain biological specimens (White & Wharton, 1984). Successful cryopreservation, however, almost always requires (1) an appropriate pretreatment of the specimens with substances (cryoprotectants) that minimize intracellular and/or intercellular crystal formation, (2) a precisely controlled rate of cooling of the specimens, at least during the early stages of freezing, and (3) a controlled rate of thawing (Triantaphyllou & McCabe, 1989).

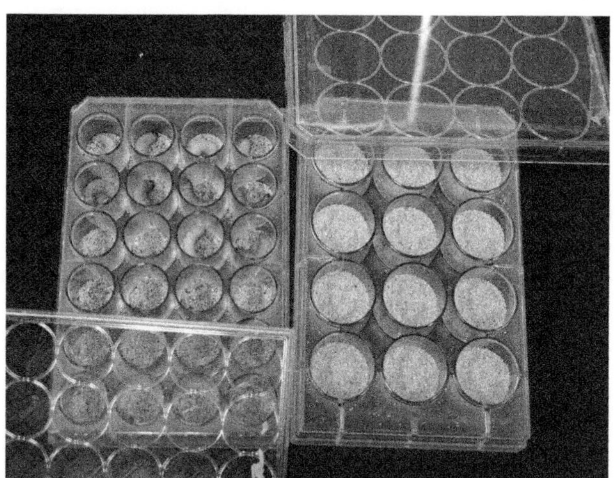

**Figure 12.21** One-on-one assay set up with a 24-well plate (left) and a 12-well plate (right).

Popiel & Vasquez (1991) developed a reliable method for the cryopreservation of IJs of *S. carpocapsae* and *H. bacteriophora*. Curran & Heng (1992) introduced some modifications.

1. Place specimens for cryopreservation in prepared glycerol solution at 2 ×final concentration, and mix equal weights (final weights from 10 to 100 g) of nematode suspension and glycerol solution by stirring on a magnetic stirrer.
2. Pour the suspension into a Petri dish and incubate for 24, 48 or 72 h at 23°C.
3. After incubation, remove excess glycerol solution by suction filtration of the IJs on to a Whatman #42 filter paper supported in a Buchner funnel.
4. Apply suction until IJs no longer appear glossy.
5. Rinse IJs under suction with *ca.* 15 ml 70% methanol (23°C).
6. Wash IJs off the filter paper with cool 70% methanol (5°C) into a tapered 15-ml centrifuge tube on ice.
7. Incubate IJs for 10 min on ice.
8. Re-suspend in cool 70% methanol, agitate after 5 min, and allow to sediment.
9. Transfer the IJ pellet to cool 2-ml round-bottomed polypropylene cryotubes with silicone rubber seals.
10. Plunge immediately into liquid nitrogen.
11. Thaw IJs by immersing the opened cryotube in 15–30 ml of Ringer's solution at 23°C for 24 h.

**Note:** Incubation time and glycerol concentration varies among nematode species and strains. Therefore users should make sure a testing time is set to monitor and optimize this technique.

*b. Other substrates*

Nematodes can also be stored on various substrates such as charcoal, alginate, or clay. Georgis & Manweiler (1994) cite the original references, and Woodring & Kaya (1988) detail the alginate process. The approach is primarily commercial, and the reader is referred to the above publications for further details.

### B. Bacterial symbionts

*1. Short-term storage*

Bacterial colonies can be stored at 12–25°C and routinely sub-cultured monthly (12°C) or twice weekly (25°C) on NBTA or MacConkey agar (see Appendix II) so that colonies reverting to Phase II can be recognized and not used.

It is not recommended to use this method of bacterial maintenance for any strain for which genetic or molecular approaches will be used, since maintenance on plates will select for mutants with a selective advantage for laboratory growth and potentially against mutualism and pathogenicity traits. For this reason, permanent frozen storage is recommended.

*2. Long-term storage*

1. Grow bacteria in LB medium between 28 and 30°C for 24–48 h based on observable growth for each species and/or strain following procedures described in Section 4.
2. Transfer 900 μl of LB culture to a 2-ml sterile cryogenic storage tube.
3. Add 600 μl of 50 : 50 LB : glycerol solution.
4. Freeze at −80°C.
5. Prepare several tubes for each sample.

**Note:** We advise recording times of retrieval of samples to avoid contamination. Samples can be retrieved from a cryogenic vial using a sterile stick or swab. We recommend a maximum of 10 extractions per sample.

Frozen stocks are used to inoculate cultures by removing a visible piece of frozen material from the tube (do not allow tube to thaw) using a sterile stick or a flamed hot loop, and immediately placing this material in liquid or on solid medium. Although variable, this inoculum should result in turbid growth within 24 h if no antibiotics are added. Older stocks of *Xenorhabdus* may be difficult to revive from frozen stocks. In this event, a larger piece of frozen material streaked on to a plate can be successful. Alternatively, the entire stock can be thawed and plated, although this is not ideal for the reasons stated above for short-term storage methods.

## 9. SHIPPING AND HANDLING OF SAMPLES

### A. Nematodes

*1. Soil samples*

Nematodes can be shipped embedded in sterile soil or sand. For this purpose small containers of 200–300 cm$^3$ with a screw cap can be used. Make sure caps are sealed with tape or parafilm to prevent water loss and leaking of soil during transit. Samples containing soil and/or infected host material must be packed inside two nested and sealed heavy duty plastic bags (Ziploc™ or other similar brand with zipper option are recommended) for transport. Plastic bags must be sealed inside a sturdy, sealed leak-proof cardboard, plastic, or metal box for shipping.

*2. Infected cadavers*

Another way to ship insect parasites is to consider infected hosts. For this purpose it is recommended that cadavers are sent individually in either 15-ml plastic centrifuge tubes or micro-centrifuge tubes (1.7 ml), depending of the insect size. Tubes should be conditioned as follows:

1. Cut a filter paper (Whatman # 1) into four pieces (four triangles).
2. Place one piece of filter paper inside the micro-centrifuge tube.
3. Add approximately 100 μl of nematode suspension to the filter paper.
4. Add one *G. mellonella* larva or other suitable insect and/or stage.
5. Seal the tube cap with parafilm.
6. Label tube accordingly including date you set up the infection and nematode species/isolate name.

With this method, nematodes will infect the insect during transit. Individually packed cadavers will be better preserved and crushing of specimens will be reduced with this method.

We do not recommend sending infection traps with multiple infected cadavers as shaking of samples through transit will crush cadavers and contaminate samples.

*3. Aqueous nematode samples*

A third option available to submit nematode cultures is by shipping IJ aqueous suspensions. For this purpose samples should be processed as follows:

1. Prepare a nematode suspension following procedures described in Section 7 (short storage). Use either a 15-ml or a 50-ml centrifuge tube with screw cap.
2. Fill half of the tube with hygrophobic cotton.
3. Fill tube with IJ suspension until cotton is soaked
4. Make sure the whole volume of the suspension is soaked by the cotton.
5. Close tube and add parafilm to seal it.
6. Label tube with pertinent information.
7. Place tubes in plastic bags (Ziploc type).
8. Follow procedures described in the soil shipping section.

### B. Symbiotic bacteria

*1. Shipping of bacterial strains on LB agar plates*

1. Prepare a 10-cm LB agar Petri plate (with antibiotics/pyruvate if required).
2. Streak bacteria from frozen stock (see Section 8) on to LB plate and incubate at 30°C overnight.
3. Once growth has occurred the following day, wrap plate with parafilm to prevent dehydration.
4. To prepare for shipment, wrap plate(s) in padding (e.g., paper towels or bubble wrap) to provide protection against cracking and place in additional bubble mailer for protection.

*2. Shipping of frozen bacterial strains in cryogenic vials*

1. Prepare an insulated Styrofoam box by adding ~250 g of dry ice to box.
2. From freezer, place frozen tube/culture in a small plastic bag to limit movement within Styrofoam box.
3. Place sample in box and add additional paper/plastic to box (to limit movement of dry ice and frozen bacterial tube).
4. Tape and seal the Styrofoam box with packing tape.
5. Place safety hazard/dry ice sticker on outside of box to declare the amount of dry ice within, and ship.

## 10. BIOASSAYS WITH ENTOMOPATHOGENIC NEMATODES AND SYMBIOTIC BACTERIA

### A. Nematodes

*1. One-on-one assay*

This method has mostly been considered to assess interactions of a single host with a given EPN inoculum. Normally the one-on-one assays are conducted in 24 well plates, however users may modify the method by employing smaller or larger well capacity plates (Figure 12.21). To set up the basic 24-well plate assay proceed as follows:

1. Use a 24-well tissue culture plate with lid (8 × 12.2 cm) with each well having an area of *ca.* 2 cm$^2$.
2. Place a piece of filter paper inside each well. A small amount of sand (approximately 1 g) can also be used instead of filter paper.

**Note**: You may need to use a hole-punch (Crafts store have a wide selection available) to cut circles of filter paper to a diameter suitable to the diameter of the well.

3. Place an IJ in each well. This can be done using a 25-μl micropipette.
4. Place one last instar *G. mellonella* larva into each well.
5. Cover with the lid and incubate at the desired temperature.

**Note:** You may want to consider placing the well plate in a plastic bag to avoid moisture loss.

6. Record mortality at 24-h intervals.

*2. Sand barrier assay*

This technique has mostly been used to assess EPN host foraging behavior. Briefly assembly of the chamber is as follows:

1. Use sieved sand (or soil). Particle size is dependent upon purpose.

2. Use 5 cm × 5 cm (4.4 cm inner diameter) polyvinyl-chloride (PVC) cylinders (tubes) (Figure 12.22). Covers of 5-cm plastic Petri dishes make excellent bottoms and lids for the cylinders (Figure 12.23).
3. Put a dish bottom on a tube and fill it with moist sand (−15 kPa or 8% moisture).
4. Place one insect host at the bottom of the tube.

**Note:** To avoid crushing the insect, first make a small depression in the sand. Replace bottom.

5. Apply the required number of IJs in 0.5 ml of water at the top of the tube (end opposite to insect).
6. Follow procedures as in the Petri dish assay from Section 6.A.1. It may be difficult to determine whether the host is dead and its condition may only be recorded during removal of the insect host from the tube.

**Figure 12.22** Set up for and barrier assay.

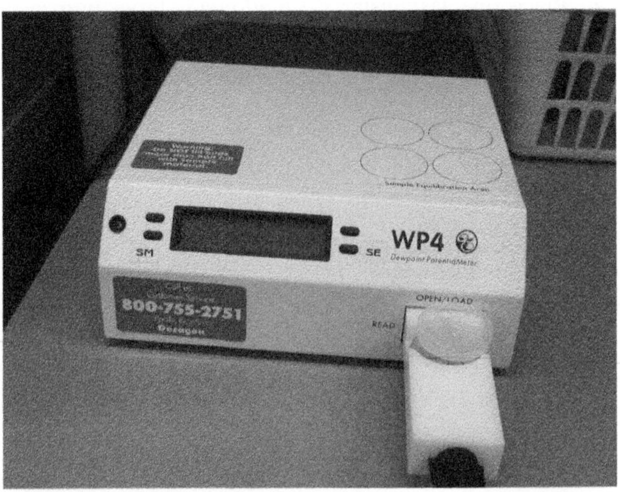


**Figure 12.23** Soil water-potential meter apparatus.

*3. Handling soil for bioassays*

*a. Conditioning of substrates*

Most nematodes associated with insects spend part of their life-cycle in the soil, therefore it is important to have some understanding of soil science. A good general textbook on soils such as Brady & Weil (2008) should be consulted. As stated earlier, the soil texture and other physical and chemical parameters should be analyzed.

Potting soil with varying amounts of organic matter can be used and it is available from nurseries and other suppliers. Field soil can also be used. It should be screened before use to eliminate unwanted materials (rocks, debris, etc.). The soil can be used raw or can be pasteurized or sterilized. If raw soil is used, there may be unwanted antagonists that can affect the nematodes (Kaya & Koppenhöfer, 1996). Soil may be pasteurized at 62°C for 2 h. This will eliminate the microfauna and many fungi and bacteria but may not kill all microorganisms. If soil is sterilized, its properties will change (Skipper & Westerman, 1973; Lotrario *et al.*, 1995). Autoclaved soil should not be used immediately because toxic components from organic matter and microorganisms may occur. Although pasteurization is less drastic than autoclaving, both soils should be air dried and kept at room temperature for a few weeks before using.

Many laboratories use sand as a test medium. There are different types of sand available and users should make decisions on the type of sand, based on the type of experiments they will perform and the needs they may have. We recommend using fine, coarse sand such as the one used for children's sandboxes, or silica sand used for construction. Both sand types are commercially available in many countries. Sand should be washed and sterilized prior to its use.

*b. Soil moisture*

Soil moisture has important effects on nematode activity. Much of the earlier nematode studies evaluating soil moisture effects on EPNs can not be interpreted because soil moisture content was usually expressed as percentage of water holding capacity or percentage of soil dry weight. These soil moisture measurements are not related to the availability of water to the nematodes. Water potential is a more meaningful measure biologically because two soils with the same water potential make water equally available to nematodes, even if their water content, measured as a percentage of soil dry weight differ. Water potential is defined as the chemical potential of water per unit volume and has the dimensions of pressure. The units of measurements are usually expressed in bars, megapascals (Mpa), or kilopascals (kPa), and are negative values. Saturated soil has a water potential at nearly 0 kPa. As the soil becomes drier, the water potential becomes increasingly negative. Soil water potential has two components, the matric potential which is the capillary and absorption forces associated with the soil matrix, and the solute potential which is the osmotic forces caused by the salt in the soil solution. The matric potential is essentially equal to the water potential as long as the soil is low in salts. For most soils, the matric potential is the important component.

Water potential of soils can be determined using a pressure plate apparatus (Studdert *et al.*, 1990), but this method takes several days to conduct and requires specialized equipment. Fawcett & Collis-George (1967) and Hamblin (1981) developed a rapid method for determining water potential by using Whatman No. 42 5.5-cm diameter filter paper. The filter paper only absorbs the water that is available and not bound to the soil. The filter paper protocol is as follows:

1. Obtain the dry weight of the filter paper before placing into the soil because the weight of the filter paper varies by 10%.
2. Sandwich the weighed filter paper between two other pieces of filter paper and place it in the soil.
3. Allow filter paper to equilibrate with the water in soil. Equilibration time of the paper in the soil varies with wet soils equilibrating within a few minutes and dry soils taking up to 36 h.
4. Retrieve the weighed filter paper after 2–4 h.
5. Reweigh the paper and calculate the % moisture. It is obtained as follows:

$$\frac{\text{wet weight} - \text{dry weight}}{\text{dry weight}} = \%\text{ moisture of filter paper}$$

6. Determine the water potential (kPa) of any soil type by reading off the graph presented by Hamblin (1981).
7. There are currently available apparatuses that have been developed to measure water potential that allow measurements ranging from 0 to −1000 kPa (Figure 12.24).

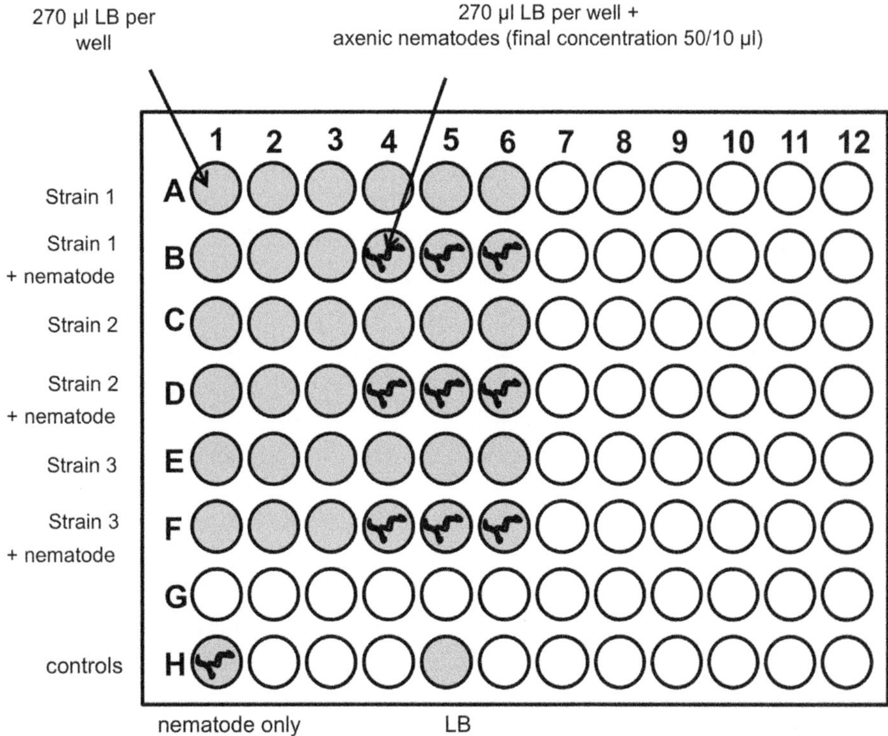


**Figure 12.24** Schematic representation of a dilution series plate.

## B. Symbiotic bacteria

### *1. Bioassays to study nematode–bacteria mutualisms*

The ability of symbiotic bacteria to support the growth and development of the nematode host, as well as to colonize the infective stage can be assessed by both *in vivo* (i.e. within insects) and *in vitro* (i.e. on plates) methods. The methods described below allow the investigator to assess the symbiotic traits of different bacterial strains or species on various nematode hosts. To do this, it is necessary to eliminate the natural symbiont from the nematode host.

#### *a. Preparing aposymbiotic nematodes*

Aposymbiotic nematodes also can be obtained by isolating nematode eggs from gravid females. These eggs can then be used directly, or to inoculate liver–kidney agar (see Section 6a and Appendix II) for development of aposymbiotic IJs.

1. Inoculate each of 5 × 10-cm lipid agar (LA) plates (Appendix II) with 800 µl of overnight bacterial culture. Rotate the plates gently until the liquid is distributed. Incubate the plates overnight at 30°C (or the preferred temperature for the symbiont strain).
2. To each plate add ~ 5000 IJ nematodes and continue incubation in the dark at room temperature for an additional 4–5 days.
3. When adults have developed, dislodge all nematodes using sterile distilled $H_2O$ and remove them into a 50-ml capped conical tube using a Pasteur pipette.
4. When nematodes have settled to the bottom of the tube, remove and discard the water and re-suspend the nematodes in 45 ml fresh $H_2O$.
5. Repeat the settle–wash cycle two to three additional times, until the water appears clear.
6. Add 50 ml of axenizing solution [2.4% (v/v) NaOCl, 0.25 N KOH] and incubate for 10 min with shaking.
7. Centrifuge the tubes in a tabletop centrifuge 7–10 min at ~ 2000 × *g*.
8. Discard the supernatant and re-suspend the pellet in axenizing solution as above, then repeat the procedure.
9. Re-suspend the pelleted material (nematode eggs) in LB and wash two to three times.
10. Re-suspend the eggs in 5–10 ml of LB and split in two small, sterile Petri dishes.
11. Aposymbiotic eggs can now be used or applied to chosen agar.

**Note:** The nematodes will remain viable for 2–3 days at room temperature or 4–5 days at 4°C. At room temperature, the eggs will hatch into J1 juveniles after overnight incubation.

#### *b.* In vivo *cultivation of nematodes and bacteria*

1. Prepare aposymbiotic nematodes and set up insects for injections.
2. Subculture bacterial strains from overnight cultures and grow for 18 h.
3. Select insects (e.g., *G. mellonella* or *M. sexta*) of approximately the same size for injection.
4. Set up dilution plates for injections. In specific wells, nematodes and bacteria will be combined at the appropriate concentrations so that a single injection will deliver both.

**Note:** The approximate dilution level that will yield the desired number of bacterial cells must be empirically determined before the experiment by serially diluting test cultures, plating ~ 10-µl drops of each dilution on LB agar and counting the number of colony forming units (cfu). Injection of approximately 200 bacterial cells in 10 µl is a recommended target, and the flanking dilution levels may also be injected to ensure the desired levels are obtained. For those wells that will be used for injections of both nematodes and bacteria, consider the final volume to be 300 µl and re-suspend nematodes in LB to a concentration of approximately 1500 nematodes per 270 µl. Inoculate the appropriate wells with this suspension.

5. Plate 10 µl of each control on LB-pyruvate to assess contamination.
6. Inject each insect with 10 µl of suspension.
   - **a.** Rinse syringe thoroughly in ethanol and then water.
   - **b.** To inject, anesthetize the insects on ice and then inject between prolegs just under the integument.
7. Inject first with the nematode only and LB-pyruvate controls.
8. Dilute each bacterial strain immediately before injection. Serially dilute with LB-pyruvate 1 : 10 log dilutions, switching tips between each dilution step (Figure 12.25).
   - **a.** In the example, transfer 30 µl of culture into the first well containing 270 µl of LB. Mix

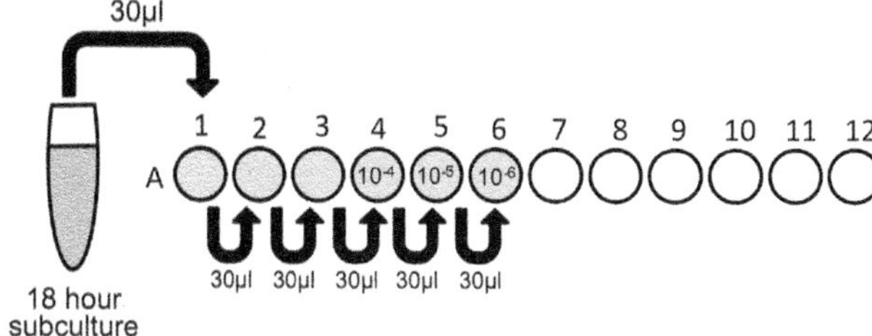


**Figure 12.25** Schematic representation showing serial dilution procedure.

thoroughly. Change tips, then transfer 30 µl from well 1 to well 2, and repeat the process.

   **b.** Keep the plate covered (e.g., with foil) as much as possible to prevent light damage.

**9.** Once a bacterial strain is serially diluted, proceed with injections and plating for that strain.

   **a.** Rinse the syringe thoroughly in ethanol and then water.

   **b.** Plate 10 µl from one dilution (e.g., well #6) on an LB-pyruvate plate to determine 'before injection' counts.

   **c.** Inject 12 insects with 10 µl from that dilution.

   **d.** Plate 10 µl from the same dilution on an LB-pyruvate plate to determine 'after injection' cfu counts.

   **e.** Repeat steps a–d for other wells (e.g., #5 and #4) to be tested for the same dilution.

**10.** Repeat steps 8 and 9 for each injection treatment.

**11.** Place insects back into containers with food, or in Petri dishes lined with filter paper.

**12.** Insects can be monitored to determine time until death, and replicates can be dissected to observe nematode development. Cadavers can be placed in White traps (see Section 3; Figure 12.11) to monitor IJ production (both timing and number).

*c.* In vitro *cultivation of nematodes and bacteria*

1. Inoculate lipid agar (LA) plates (Appendix II) with 800 µl each of bacterial overnight culture. Rotate the plates gently until the liquid is distributed. Incubate the plates from 1–4 days at 30°C (or the preferred temperature for the symbiont strain).
2. Inoculate plates with aposymbiotic nematodes (either ~40 IJs or ~1000 hatched J1 juveniles).
3. Nematode development can be monitored easily using a dissecting microscope.
4. When IJs begin to form, agar plates can be placed in a modified White trap.

*d. Monitoring colonization of the infective juvenile stage of nematodes*

Numerous methods have been developed to visualize or quantify the bacterial population within infective juvenile nematodes. These include approaches that examine either individuals or populations of nematodes. In many *Xenorhabdus* and *Photorhabdus* species, genetic manipulation is possible, and can be used to introduce the gene encoding green fluorescent protein. Such strains are easily visualized within nematodes, and allow measurement of colonization frequency within a population. The assays described below do not require specific bacterial strains.

*e. Monitoring colonization of individual nematodes using a pestle (Goetsch* et al., *2006)*

1. Transfer 1 ml of IJ aqueous suspension to a sterile 1.5-ml microfuge tube (a minimum of 300 IJs).
2. Pellet IJs by spinning twice for 10 s each at maximum speed in a microfuge.
3. Remove water and add 1.0 ml 1.0% bleach.
4. Incubate for 2 min at room temperature.
5. Pellet nematodes as in step #2.
6. Remove bleach and rinse with 1 ml sterile water.
7. Repeat wash steps 5 and 6 for a total of five washes.
8. Remove water from last wash and re-suspend the nematode pellet in 0.5–1.0 ml sterile distilled water.
9. Pipette a single live nematode into 100 µl LB broth in a sterile microfuge tube.
10. Homogenize for 70 s with a sterile motor-driven polypropylene pestle (Kontes®).
11. Plate 50 µl of homogenate on LB-pyruvate plates and count resulting colonies.

*f. Monitoring colonization of individual nematodes using the guillotine assay (Vivas & Goodrich-Blair, 2001)*

1. Pipette individual nematodes into ~2-µl drops of distilled water on a microscope slide.
2. Using a dissecting microscope and working as quickly possible to avoid evaporation, place a sharp razor blade with the blade directly perpendicular to the microscope slide, and near the head.
3. Gently rock the razor blade back and forth over the head, and it should break off, extruding the bacteria.

Not all attempts will be successful, due to surface tension.

4. Pass the microscope slide through a flame to heat-fix the sample.
5. Stain with 0.1% (wt/vol) crystal violet.
6. Examine with bright-field microscopy under 40–100 × magnification.

**Note:** An alternative approach is to spread multiple nematodes on a single slide and chop randomly with the razor blade then continue from step 4.

*g. Monitoring average colonization levels in a nematode population using sonication (Heungens et al., 2002)*

The absolute levels of colonization determined by this assay can fluctuate between experiments, ranging from 30–200 cfu/IJ. It is therefore critical to include appropriate wild-type controls in every experiment. Colonization phenotypes can be divided into five categories: full colonization (> cfu/IJ); partial colonization (~10–25 cfu/IJ); low colonization (~1–10 cfu/IJ); very low colonization (between 0 and 1 cfu/IJ); and no colonization (< 0.005 cfu/IJ, below our limit of detection). Microscopic analysis indicates there are likely to be hundreds of cells within each infective juvenile. While some of these may be non-culturable (e.g., non-viable) it is likely that sonication assay is an under-representation of actual levels of bacterial colonization. The protocol provided below was optimized for *S. carpocapsae* and *X. nematophila*, and uses 10,000 IJs. However, the optimum number of IJs, as well as the sonication time should be optimized for each IJ-bacterium pair by the investigator.

The following protocol is based on analyzing 20 samples.

1. Prepare in advance the following:
   - **a.** Sterile distilled $H_2O$ (3 l total).
   - **b.** Standards for IJs number in 13-mm test tubes (5,000, 10,000, and 20,000 IJs/ml in 1 ml each).
   - **c.** Twenty pieces of filter paper (Whatman #1). If using 60-mm filter paper, use a whole piece; if using 100-mm filter paper, cut into quarters, for a total of 20 quarter pieces.
   - **d.** Cut nylon filter paper into squares large enough to sit on the glass filter column; make enough for one per sample.
   - **e.** Millipore nylon net filters; filter type: 11um NY11; 30 cm × 3 m; CAT NO: NY1100010. This comes as a roll (30 cm × 3 m). Cut into squares about the size of the column stand.
   - **f.** Stopper with inserted column stand.
   - **g.** Set up filter flask apparatus (Figure 12.26—surface sterilization set up).
   - **h.** Wash all components (there is no need to sterilize).
   - **i.** Set up wash/rinse buckets (one bucket with soapy water, one bucket with rinse water).
   - **j.** Collect 40 × 13-ml test tubes.
   - **k.** Collect 20 × 20-ml test tubes.
2. Protocol
   - **a.** Collect IJ samples from White traps in 18-ml glass test tube.
   - **b.** Rinse all samples before proceeding.
     - **(i)** Rinse once in $H_2O$, let IJs settle (~ 15 min) and remove supernatant to ~ 0.5–1 in. above the settled IJs.
     - **(ii)** Fill to top of tube with $H_2O$.
     - **(iii)** Repeat step ii.
   - **c.** Make 0.5% bleach sterilization solution (42 ml 12% bleach in 1 l sterile $H_2O$).
   - **d.** Start sterilizations on a 3-min cycle (do this in front of a timer so that all samples are sterilized and rinsed for a similar amount of time).
     - **(i)** At 0:00 min add bleach solution to the first sample: add a small (~ 1/4 full) amount of bleach to the test tube, briefly vortex to mix the IJs and then fill the rest of the way with the bleach solution.
     - **(ii)** Add bleach solution to the Millipore column on the filter apparatus until the column is full; some will drip out by gravity afterwards.
     - **(iii)** At 2:35 min add a small (~ 1/4 full) amount of bleach to the test tube for sample #2, vortex and fill the rest of the way with the bleach solution.
     - **(iv)** Refill Millipore filter column with bleach (some has drained out by gravity).
     - **(v)** At 3:00 min turn on the suction and suck down the bleach solution you just added, then add sample #1 (which has been in bleach for 3 min).

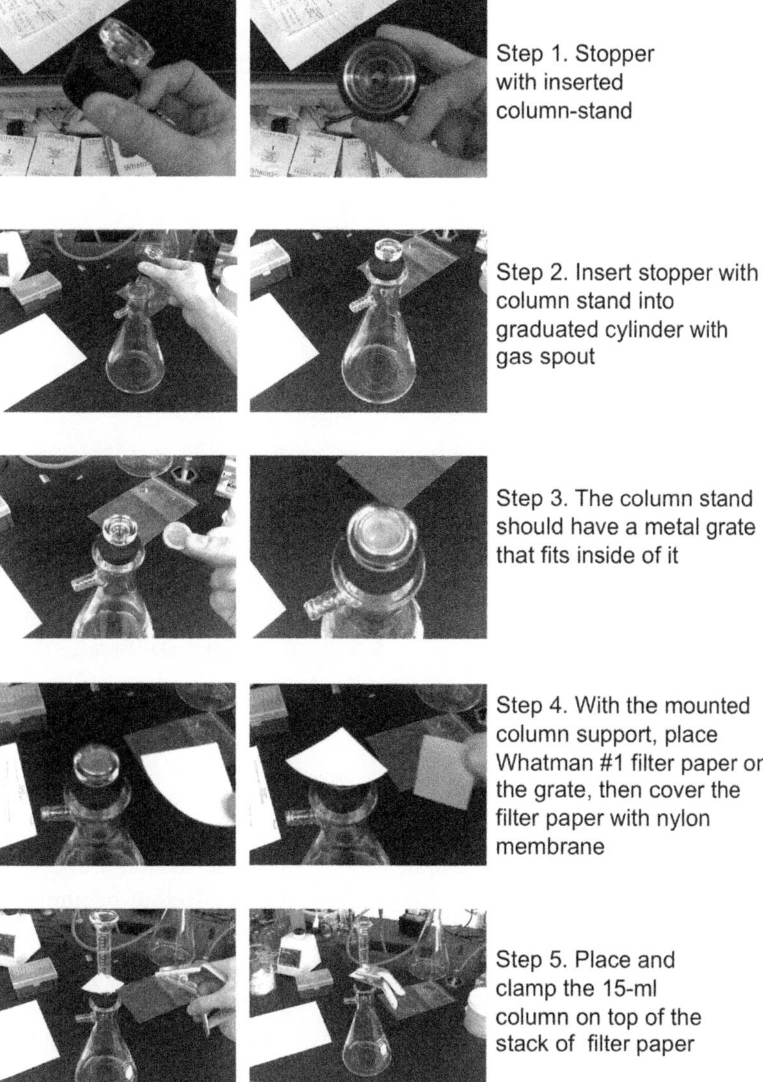


**Figure 12.26** Set-up for filter-flask apparatus.

**(vi)** Add clean sterile double distilled $H_2O$ to the column just before the sample has been sucked all the way to the bottom (keeping the filter at the bottom always submerged, but allowing it to run mostly empty so that each rinse is a fairly complete rinse rather than a dilution of the previous rinse).

**(vii)** Repeat rinse two more times (three times in total). After adding the third rinse, pull the suction tube off of the flask when ~5 ml of solution are left in the bottom of the column on the filter.

**(viii)** Using a pipette collect greater than 10,000 IJs in a 13-mm test tube (optimally in ~1 ml, but collect multiple-milliliter volumes if necessary).

**(ix)** Wash the Millipore filter column in the soap bucket with a scrubber wand (~5 s is sufficient).

**(x)** Rinse the Millipore filter column in the rinse water (~5 s is sufficient).

**(xi)** Set the Millipore column up on the glass filter, replacing the filter paper and nylon membrane (Figure 12.27).

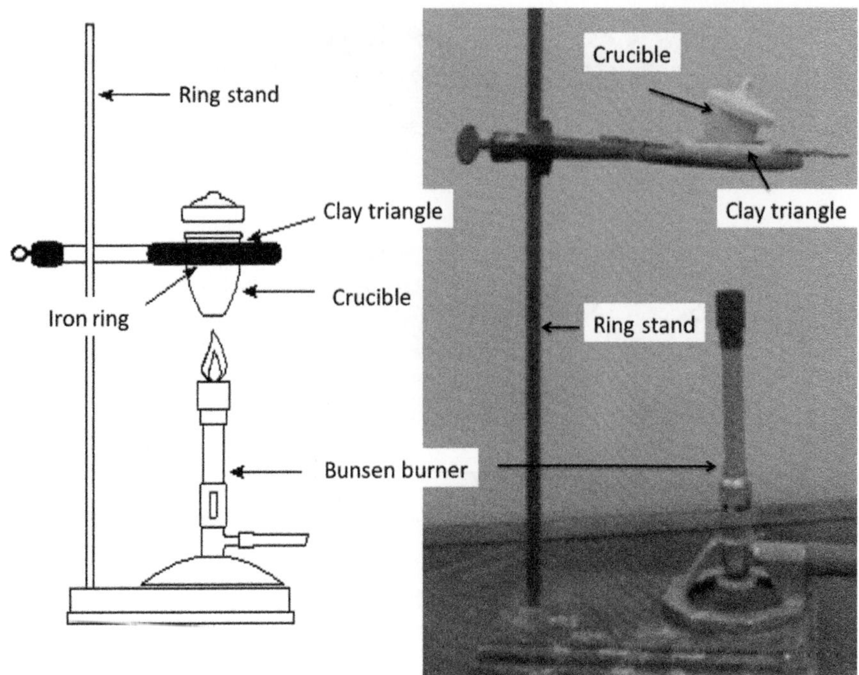


**Figure 12.27** Set-up for making tungsten needles.

**(xii)** At 5:35 min repeat above for (2:35) sample #3.

**(xiii)** At 6:00 min repeat above (3:00) for sample #2.

**(xiv)** Continue on this cycle of adding bleach at each 2:35 time point, and rinsing the bleach away at each 3:00 interval until all samples are surface sterilized and rinsed.

**e.** Equilibrate samples so each has ~10,000 IJs in 1 ml sterile $H_2O$.

**(i)** Visually compare each sample to a standard series.

**(ii)** For concentrated samples, use a 1-ml pipetter with 1 ml sterile $H_2O$ and insert the tip into the liquid in the tube. Slowly add water to the tube to bring the IJs to the desired density (based on visual comparison). Without removing the tip from the liquid, remove the same amount of liquid just added, leaving diluted IJs in the initial 1-ml volume.

**(iii)** If the sample concentration is less than 10,000 IJs/ml return to the White trap and sterilize additional nematodes.

**f.** Add 850 μl sterile LB broth to each sample so that each sample has 1.85 ml of liquid.

**g.** Sonicate samples for 1 min each in a water bath sonicator (e.g., Branson). Place the tubes in a single location, and only sonicate one tube at a time. Before beginning, test the water bath to identify the optimum location for the tube, by listening to the pitch and intensity of the sound generated and observing the IJ behavior. A location that gives quiet sound with visibly moving nematodes appears to be optimal.

**(i)** Turn on sonicator.

**(ii)** Place 13-mm test tube in holder.

**(iii)** Push the test tube until the region containing IJs is completely submerged in the water in the sonicator (but keep the cap of the tube above the level of the water).

**h.** Plate the sonicate to determine cfu.

**(i)** Perform a 1 : 200 dilution with each sample.

**(ii)** Plate and spread replicate 40 μl of the 1 : 200 dilution as well as (on a separate plate) the undiluted sonicate. The number of colonies derived from the former is equivalent to the cfu/IJ. If no colonies are seen in either plating, the colonization level is given as the limit of detection: $< 0.005$ cfu/IJ.

*2. Bioassays to assess entomopathogenic bacteria pathogenesis and virulence*

*a. Bacterial injection into insects*

Insects are commercially available from various companies. Users should locate insect companies in their respective countries or decide to rear the preferred insects in their facilities. The protocols below are optimized for injections of *M. sexta. M. sexta* eggs arrive in small plastic vials, and should be visually inspected upon delivery. A yellow color indicates the eggs are likely to hatch by the next day and should be processed immediately. Although immediate processing is recommended, it can be delayed if the insects are not yellow. Be sure to have insect diet prepared for the day of insect delivery. *M. sexta* eggs take ~1–3 days from arrival to hatch, then these insects take ~7 days to reach 4th instar (10 days total).

*(1) Injections.*

**a.** When *M. sexta* larvae are approximately 8 days old, users should streak bacterial strains to be tested on agar plates. This is 2 days before doing the injections.

**b.** At approximately day 9 (one day before injections) bacterial cultures should be started to allow overnight growth in LB from single colonies (Certain labs grow bacteria statically in microtiter dishes to limit the amount of growth prior to injection), unless injecting colonies directly from plates (in this case, wait another day).

**c.** On the injection day (day 10), make sure insects have the characteristics of $4^{th}$ instar larvae (dark foot pads, prominent dorsal stripes, weight of 0.2 g or more) and are not molting (do not have bulb or 'cap' on tip of head).

**d.** Make labels that list the strain you will be injecting, the dilution, and the number (1–10). Place labels on individual insect containers.

*(2) Dilutions.*

**a.** Use multichannel pipette to add 270 μl of 1 × PBS (phosphate-buffered solution) to rows of a microtiter plate (one row per strain to be injected) (Figure 12.24)

**b.** Monitor the $OD_{600}$ of each strain to ensure they are similar among the treatments.

**c.** Prepare strains.

1. Pellet 500 μl of each strain to be tested.
2. Re-suspend in 1 ml PBS and pellet again.
3. Re-suspend in 500 μl of PBS.

**Note:** Alternatively, colonies can be injected, in which case each colony is re-suspended in ~200 μl of PBS.

**d.** Dilute one strain at a time, 1 : 10, down a row. Wait until you have completed injections for this strain before diluting the other strains.

1. To dilute, add 30 μl to the top well (or the amount calculated based on the $OD_{600}$ measured as above, supplemented to a total of 30 ml with PBS).
2. Mix several times, change tips, transfer 30 μl from well A to well B, mix.
3. Change tips, transfer from B to C. Continue down to the bottom of the row changing tips each time.
4. Plate 10 μl of each dilution of the strain on an LB-pyruvate 10-cm Petri dish.
5. Plate dilutions both before and after injections for an average cfu/ml count.

*3. Injections*

**a.** Put 1–2 ml ethanol (95%) and distilled sterile water in separate yellow cap tubes.

**b.** Group insects by strain/inoculum. Place a few insects in their containers on ice.

**c.** Rinse the syringe six times with ethanol (three times each in two separate tubes) and three times with water.

**d.** From the appropriate dilution of the microtiter plate, suck 10 μl of bacteria into the syringe.

**e.** Inject behind 1st proleg. Be careful to slide the needle just under the integument (not poking directly into the insect) and also not to slide the needle in and back out.

**f.** Return insect to container, remove from ice.

**g.** Continue injections.

**h.** Wash out syringe after each strain (not after each animal, unless the syringe gets contaminated).

**i.** Repeat injections with other dilutions of that strain (if applicable—this is not always necessary).

**j.** Make dilutions for the next strain.

**Note:** For beginners, it is wise to practice injecting insects with PBS to develop the skill. Regardless of experience, it is important to include a PBS control in every experiment, as it also checks for needle contamination, which may indicate improper needle sterilization between injections.

*4. Monitoring injected insects*

**a.** Record how many insects are dead (do not respond to probing/ shaking, and changing color at intervals.

**b.** Count the colonies on the plates to determine inoculum level.

*5. Monitoring orally active toxicity*

The protocol below is designed to monitor insecticidal toxins secreted by bacteria into the extracellular milieu. However, toxicity of the bacteria themselves can also be monitored by simply adding them directly to the food.

1. Grow the bacterial culture for 24 h at 30°C in LB (or other medium of choice).
2. Normalize cultures to the same $OD_{600}$.
3. Spin at 10,000 r.p.m. for 15 min to pellet cells.
4. Place supernatant on an Amicon Ultra-15 Centrifugal Filter Unit with a 10-kDa filter (Millipore Corporation).
5. Spin at $2345 \times g$ until the supernatant is $10 \times$ concentrated. When concentrating 15 ml, start spin time at 10 min, then check and spin in 2-min increments until $10 \times$ concentrated.
6. Place 1 $cm^3$ cubes of insect diet on a piece of foil and pipette 200 ml of the concentrated supernatant (or LB control) on to each cube. Prepare two cubes for each treatment. Let them dry in a hood just until all the liquid has evaporated. It is important not to have any free standing liquid on the diet, however, the insects will not eat dried-out food. Wrap one replicate cube in foil and store at 4°C.
7. Monitor insect molting and choose insects that have molted within several hours of each other. Weigh 2nd instar larvae and place those with ~25 mg starting weight (can range from ~15 to 50 mg, but ensure that all insects in a cohort are approximately the same weight) on treated diet (one insect per cube).

Weigh daily until ~96 h. At 48 h replace any remaining food with the second prepared cube.

## ACKNOWLEDGMENTS

We would like to thank lab members in both the P. Stock and H. Goodrich-Blair laboratories for their numerous contributions and assistance in the development and/or improvement of many techniques and methods described in this chapter. Specifically, we would like to acknowledge the following lab members: Ming-Min Lee, Victoria Miranda, Sam-Kyu Kim, Yolanda Flores-Lara, Rousel Orozco, Katherine Plichta, John McMullen II (P. Stock Lab), John Chaston, Elizabeth Hussa, Darby Renneckar-Sugar, Kristen Murfin, Aaron Andersen, Jennifer Knack, Xiaojun Lu (H. Goodrich-Blair lab). We also would like to extend our appreciation to our colleagues David Clarke (Corke University) and Steven Forst (University of Wisconsin Milwaukee) for sharing lab protocols and other techniques with us. We also extend our appreciation to the National Science Foundation for their continued financial support (IOS-0840932 (S.P.S., H.G.B.); IOS-0920631(H.G.B.); IOS-0919565 (S. P. S): DEB-0640899 (S. P. S).

## REFERENCES

Aguinaldo, A. M. A., Linford, L. S., Rivera, M. C., Garey, J. R., Raff, R. A., & Lake, J. A. (1997). Evidence for a clade of nematodes, arthropods and other moulting animals. *Nature, 387*, 489–493.

Akhurst, R. J., & Bedding, R. A. (1986). Natural occurrence of insect pathogenic nematodes (Heterorhabditidae and Steinernematidae) in soil in Australia. *J. Aust. Entomol. Soc., 25*, 241–244.

Akhurst, R. J., Boemare, N. E., Janssen, P. H., Peel, M. M., Alfredson, D. A., & Beard, C. E. (2004). Taxonomy of Australian clinical isolates of the genus *Photorhabdus* and proposal of *Photorhabdus asymbiotica* subsp. *asymbiotica* subs. nov. and *P. asymbiotica* subsp *australis* subsp. nov. *Int. J.Syst.Evol. Microb., 54*, 1301–1310.

Ayoub, S. M. (1980). *Plant Nematology an Agricultural Training Aid.* Sacramento, CA: NemaAid Publications.

Barker, K. R. (1985). Nematode extraction and bioassays. In K. R. Barker, C. C. Carter, & J. N. Sasser (Eds.), *An Advanced treatise on Meloidogyne Volume II Methodology* (pp. 20–35). Raleigh, NC: North Carolina State University Graphics.

Barker, K. R., Carter, C. C., & Sasser, J. N. (1985). *An Advanced Treatise on Meloidogyne Volume II Methodology.* Raleigh, NC: North Carolina State University Graphics.

Bedding, R. A. (1981). Low cost *in vitro* mass production of *Neoaplectana* and *Heterorhabditis* species (Nematoda) for field control of insect pests. *Nematologica, 27*, 109–114.

Bedding, R. A. (1984). Large-scale production, storage and transport of the insect-parasitic nematodes *Neoaplectana spp.* and *Heterorhabditis*. *Ann. Appl. Biol., 101*, 117–120.

Bedding, R. A., & Akhurst, R. J. (1975). A simple technique for the detection of insect parasitic rhabditid nematodes in soil. *Nematologica, 21*, 109–110.

Bedding, R. A., Molyneux, A. S., & Akhurst, R. J. (1983). *Heterorhabditis* spp., *Neoaplectana* spp., and *Steinernema kraussei*: Interspecific and intraspecific differences in infectivity for insects. *Exp. Parasitol., 55*, 249–257.

Blaxter, M. L., De Ley, P., Garey, J. R., Liu, L. X., Scheldeman, P., Vierstraete, A., Vanfleteren, J. R., Mackey, L. Y., Dorris, M., Frisse, L. M., Vida, J. T., & Thomas, W. K. (1998). A molecular evolutionary framework for the phylum Nematoda. *Nature, 392*, 71–75.

Boemare, N., Thaler, J. O., & Lanois, A. (1997). Simple bacteriological tests for phenotypic characterization of *Xenorhabdus* and *Photorhabdus* phase variants. *Symbiosis, 22*, 167–175.

Brady, N. C., & Weil, R. C. (2008). *The Nature and Properties of Soils*. Prentice Hall, p. 965.

Brenner, S. (1974). The genetics of *Caenorhabditis elegans*. *Genetics, 77*, 71–94.

Byrd, D. W., Jr., Barker, K. R., Ferris, H., Nusbaum, C. J., Griffin, W. E., Small, R. H., & Stone, C. A. (1976). Two semi-automatic elutriators for extracting nematodes and certain fungi from soil. *J. Nematol., 8*, 206–212.

Campbell, J. F., Lewis, E., Yoder, F., & Gaugler, R. (1995). Entomopathogenic nematode (Heterorhabditidae and Steinernematidae) seasonal population dynamics and impact on insect populations in turfgrass. *Biol. Control, 5*, 598–606.

Choo, H. Y., & Kaya, H. K. (1994). Biological control of the brown planthopper by a mermithid nematode. *Korean J. Appl. Entomol., 33*, 207–215.

Choo, H. Y., Kaya, H. K., Shea, P., & Noffsinger, E. M. (1987). Ecological study of nematode parasitism in *Ips* beetles from California and Idaho. *J. Nematol., 19*, 495–502.

Choo, H. Y., Kaya, H. K., & Kim, J. B. (1989). *Agamermis unka* (Mermithidae) parasitism of *Nilaparvata lugens* in rice fields in Korea. *J. Nematol., 21*, 254–259.

Choo, H. Y., Kaya, H. K., & Kim, H. H. (1995). Biological studies on *Agamermis unka* (Nematoda: Mermithidae), a parasite of the brown planthopper *Nilaparvata lugens*. *Biocontrol Sci. Technol., 5*, 209–223.

Cobb, N. A. (1917). *Notes on nemas*. Contribution Science Nematology No. 5. pp. 117–128.

Cowles, K. N., Cowes, C. E., Richards, G. P., Martens, E. C., & Goodrich-Blair, H. (2007). The global regulator Lrp contributes to mutualism, pathogenesis and phenotypic variation in the bacterium *Xenorhabdus nematophila*. *Cell Microbiol., 9*, 1311–1323.

Creighton, C. S., & Fassuliotis, G. (1981). A laboratory technique for culturing *Filipjevimermis leipsandra*, a nematode parasite of *Diabrotica balteata* (Insecta: Coleoptera). *J. Nematol., 13*, 226–227.

Curran, J., & Heng, J. (1992). Comparison of three methods for estimating the number of entomopathogenic nematodes present in soil samples. *J. Nematol., 24*, 170–176.

Curran, J., & Robinson, M. P. (1993). Molecular aids to nematode diagnosis. In K. Evans, D. L. Trudgill, & J. M. Webster (Eds.), *Plant Parasitic Nematodes in Temperate Agriculture* (pp. 545–564). Wallingford, UK: CABI Publishing.

Dauga, C. (2002). Evolution of the *gyrB* gene and the molecular phylogeny of Enterobacteriaceae: a model molecule for molecular studies. *J. Syst. Evol. Microbiol., 52*, 531–547.

De Ley, P., & Blaxter, M. (2002). Systematic position and phylogeny. In D. L. Lee (Ed.), *The Biology of Nematodes* (pp. 1–30). New York, NY: Taylor and Francis.

Doolittle, W. M. (1999). Lateral Genomics. *Trends in Biochemistry Science, 24*, M5–M8.

Downing, J. A., & Rigler, F. H. (1984). *A Manual on the Methods for the Assessment of Secondary Production in Freshwaters*. Oxford, UK: Blackwell.

Fan, X., & Hominick, W. M. (1991). Efficiency of the *Galleria* (wax moth) baiting technique for recovering infective stages of entomopathogenic rhabditids (Steinernematidae and Heterorhabditidae) from sand and soil. *Rev. Nematol., 14*, 381–387.

Fawcett, R. G., & Collis-George, N. (1967). A filter-paper method for determining the moisture characteristics of soil. *Aust. J. Exp. Agric. Anim. Husb., 7*, 162–167.

Filipjev, I. N., & Schuurmans-Stekhoven, J. H., Jr. (1959). *Agricultural Helminthology*. Leiden, The Netherlands: Brill.

Gaugler, R., & Kaya, H. K. (1990). *Entomopathogenic Nematodes in Biological Control*. Boca Raton, FL: CRC Press, Inc. pp. 365.

Georgis, R., & Manweiler, S. A. (1994). Entomopathogenic nematodes: a developing biological control technology. In K. Evans (Ed.), *Agricultural Zoology Reviews, Vol. 6* (pp. 63–94). Andover, UK: Intercept.

Georgis, R., Kaya, H. K., & Gaugler, R. (1991). Effect of steinernematid and heterorhabditid nematodes (Rhabditida: Steinernematidae and Heterorhabditidae) on nontarget arthropods. *Environ. Entomol., 20*, 815–822.

Glen, D. M., & Wilson, M. (1997). Slug-parasitic nematodes as biocontrol agents for slugs. *Agro Food Ind. Hi Tech, 8*, 23–27.

Goetsch, M., Owen, H., Goldman, B., & Forst, S. (2006). Analysis of the PixA inclusion body protein of *Xenorhabdus nematophila*. *J Bacteriol., 188*, 2706–2710.

Götz, P., Boman, A., & Boman, H. G. (1981). Interaction between insect immunity and an insect-pathogenic nematode with symbiotic bacteria. *Proc. R. Soc. Lond. B, 212*, 333–350.

Griffin, C. T., Moore, J. F., & Downes, M. J. (1991). Occurrence of insect-parasitic nematodes (Steinernematidae, Heterorhabditidae) in the Republic of Ireland. *Nematologica, 37*, 92–100.

Hamblin, A. P. (1981). Filter-paper method for routine measurement of field water potential. *J. Hydrol., 53*, 355–360.

Hara, A. H., Gaugler, R., Kaya, H. K., & LeBeck, L. M. (1991). Natural populations of entomopathogenic nematodes (Rhabditida: Steinernematidae and Heterorhabditidae) from the Hawaiian Islands. Environ. *Entomol., 20*, 211–216.

Hara, A. H., Lindegren, J. E., & Kaya, H. K. (1981). *Monoxenic mass production of the entomogenous nematode,* Neoaplectana carpocapsae *Weiser on dog food/agar medium*. Science & Education Administration, Advances in Agricultural Technology, Western Series No. 16.

Herrmann, M., Mayer, W. E., Hong, R. L., Kienle, S., Minasaki, R., & Sommer, R. J. (2007). The nematode

*Pristionchus pacificus* (Nematoda: Diplogastridae) is associated with the oriental beetle *Exomala orientalis* (Coleoptera: Scarabaeidae) in Japan. *Zool. Science, 24*, 883–889.

Heugens, K., Cowes, C. E., & Godrich-Blair, H. (2002). Identification of *Xenorhabdus nematophila* genes required for mutualistic colonization of *Steinernema carpocapsae* nematodes. *Mol. Microbiol., 45*, 1337–1353.

Holme, N. A., & McIntyre, A. O. (1971). *Methods for the Study of Marine Benthos*. Oxford, UK: IBP Handbook No. 16. Blackwell.

Hominick, W. M., & Briscoe, B. R. (1990a). Survey of 15 sites over 28 months for entomopathogenic nematodes (Rhabditida: Steinernematidae). *Parasitology, 100*, 289–294.

Hominick, W. M., & Briscoe, B. R. (1990b). Occurrence of entomopathogenic nematodes (Rhabditida: Steinernematidae and Heterorhabditidae) in British soils. *Parasitology, 100*, 295–302.

Hooper, D. J., & Evans, K. (1993). Extraction, identification and control of plant parasitic nematodes. In K. Evans, D. L. Trudgill, & J. M. Webster (Eds.), *Plant Parasitic Nematodes in Temperate Agriculture* (pp. 1–59). Wallingford, UK: CABI Publishing.

Inglis, W. G. (1983). An outline classification of the phylum Nematoda. *Aust. J. Zool., 31*, 243–255.

Jarosz, J. (1996). Ecology of anti-microbials produced by bacterial associates of *Steinernema carpocapsae* and *Heterorhabditis bacteriophora*. *Parasitology, 112*, 545–552.

Jenkins, W. R. (1964). A rapid centrifugal-flotation technique for separating nematodes from soil. *Plant Dis. Rep, 48*, 692.

Jones, C. M., & Perdue, J. M. (1967). *Heterotylenchus autumnalis*, a parasite of the face fly. *J. Econ. Entomol., 60*, 1393–1395.

Kanzaki, K., & Futai, K. (2002). A PCR primer set for determination of phylogenetic relationships of *Bursaphelenchus* species within the *xylophilus* group. *Nematology, 4*, 35–41.

Kaya, H. K., & Koppenhöfer, A. M. (1996). Effects of microbial and other antagonistic organisms and competition on entomopathogenic nematodes. *Biocontrol Sci. Technol., 6*, 421–434.

Kaya, H. K., & Moon, R. D. (1978). The nematode *Heterotylenchus autumnalis* and face fly *Musca autumnalis*: a field study in northern California. *J. Nematol., 10*, 333–341.

Kaya, H. K., & Stock, S. P. (1997). Techniques in insect nematology. In L. A. Lacey (Ed.), *Manual of Techniques in Insect Pathology* (pp. 281–324). London, UK: Academic Press.

Kaya, H. K., Moon, R. D., & Witt, P. L. (1979). Influence of the nematode, *Heterotylenchus autumnalis*, on the behavior of the face fly, *Musca autumnalis*. *Environ. Entomol., 8*, 537–540.

Kiontke, K., & Sudhaus, W. (2006). Ecology of *Caenorhabditis* species. In The *C. elegans* Research Community (Ed.), *WormBook*. doi/10.1895/wormbook.1.37.1. http://www.wormbook.org.

Kung, S. P., Gaugler, R., & Kaya, H. K. (1990). Soil type and entomopathogenic nematode persistence. *J. Invertebr. Pathol., 55*, 401–406.

Lambshead, P., John, D., & Boucher, G. (2003). Marine nematode deep-sea biodiversity: Hyperdiverse or hype? *J. Biogeography, 30*, 475–485.

Lawrence, J. G., Ochman, H., & Hartl, D. L. (1991). Molecular and evolutionary relationships among enteric bacteria. *J. Gen. Microbiol., 137*, 1911–1921.

Leather, S. R. (2005). *Insect Sampling in Forest Ecosystems. Ecological Methods and Concepts*. Oxford, UK: Blackwell Publishing. pp. 320.

Lee, M. M., & Stock, S. P. (2010). A multigene approach for assessing evolutionary relationships of *Xenorhabdus* spp. (Gamma-Proteobacteria), the bacterial symbionts of entomopathogenic *Steinernema* nematodes. *J. Invertebr. Pathol., 104*, 67–74.

Lee, M. M., Sicard, M., Skeie, M., & Stock, S. P. (2009). *Steinernema boemarei* n. sp. (Nematoda: Steinernematidae), a new entomopathogenic nematode from southern France. *Syst. Parasitol., 72*, 127–141.

Liu, J., Berry, R., Poinar, G., & Moldenke, A. (1997). Phylogeny of *Photorhabdus* and *Xenorhabdus* species and strains as determined by comparison of partial 16S rRNA gene sequences. *Int. J. Syst. Bacteriol., 47*, 948–951.

Lotrario, J. B., Stuart, B. J., Lam, T., Arands, R. R., O'Connor, O. A., & Kosson, D. S. (1995). Effects of soil: implications for sorption isotherm analyses. *Bull. Environ. Contam. Toxicol., 54*, 668–675.

Mauleon, H., Briand, S., Laumond, C., & Bonifassi, E. (1993). Utilisation d'enzyme digestives pour l'etude du parasitisme des *Steinernema* et *Heterorhabditis* envers les larves d'insectes. *Fundam. Appl. Nematol., 16*, 185–191.

Mayr, E. (1942). *Systematics and the Origin of Species*. Harvard University Press. pp. 334.

Mullens, B. A., & Velten, R. K. (1994). Laboratory culture and life history of *Heleidomermis magnapapula* in its host, *Culicoides variipennis* (Diptera: Ceratopogonidae). *J. Nematol., 26*, 1–10.

Nadler, S. A., Bolotin, E., & Stock, S. P. (2006a). Phylogenetic relationships of *Steinernema* (Cephalobina, Steinernematidae) based on nuclear, mitochondrial, and morphological data. *System. Parasitol, 63*, 159–179.

Nadler, S. A., De Ley, P., Mundo-Ocampo, M., Smythe, A. B., Stock, S. P., Bumbarger, D., Adams, B. J., De Ley, I. T., Holovachov, O., & Baldwin, J. G. (2006b). Phylogeny of Cephalobina (Nematoda): molecular evidence for substantial phenotypic homoplasy and incongruence with traditional classifications. *Mol. Phylogen. Evol., 63*, 161–181.

Nguyen, K. B., Maruniak, J., & Adams, J. B. (2001). Diagnostic and phylogenetic utility of the rDNA internal transcribed spacer sequences of *Steinernema*. *J. Nematol., 33*, 73–82.

Nickle, W. R., & MacGowan, J. B. (1992). Grenacher's borax carmine for staining nematodes inside insects. *J. Helminthol. Soc. Wash, 59*, 231–233.

Nunn, G. B. (1992). *Nematode Molecular Evolution*. Ph.D. thesis. UIC: University of Nottingham.

Ogura, N. (1993). A method to produce neutral-red-labelled infective juveniles of *Steinernema kushidai*. *Jpn. J. Nematol., 23*, 37–38.

Ogura, N., & Haraguchi, N. (1993). Xenic culture of *Steinernema kushidai* (Nematoda: Steinernematidae) on artificial media. *Nematologica, 39*, 266–273.

Park, Y., Herbert, E. E., Cowles, C. E., Cowles, K. N., Menard, M. L., Orchard, S., & Goodrich-Blair. (2007). Clonal variation in *Xenorhabdus nematophila* virulence and suppression of *Manduca sexta* immunity. *Cell Microbiol., 9*, 645–656.

Pedigo, L. P., & Buntin, G. D. (1994). *Handbook of Sampling Methods for Arthropods in Agriculture*. Boca Raton, FL: CRC Press. pp. 794.

Peloquin, J. J., & Platzer, E. G. (1993). Control of root gnats (Sciaridae: Diptera) by *Tetradonema plicans* Hungerford (Tetradonematidae: Nematoda) produced by a novel culture method. *J. Invertebr. Pathol., 62*, 79–86.

Perlman, S. J., Spicer, G. S., Shoemaker, D., & Jaenike, J. (2003). Associations between mycophagous *Drosophila* and their *Howardula* nematode parasites: a worldwide phylogenetic shuffle. *Mol. Biol., 12*, 237–249.

Petersen, J. J. (1984). Nematode parasitism of mosquitoes. In W. R. Nickle (Ed.), *Plant and Insect Nematodes* (pp. 797–820). New York, NY: Marcel Dekker.

Petersen, J. J. (1985). Nematodes as biological control agents: Part I. *Mermithidae. Adv. Parasitol, 24*, 307–344.

Petersen, J. J., & Willis, O. R. (1972). Procedures for the mass rearing of a mermithid parasite of mosquitoes. *Mosq. News., 32*, 226–230.

Poinar, G. O., Jr. (1979). *Nematodes for Biological Control of Insects*. Boca Raton, FL: CRC Press.

Poinar, G. O., Jr. (1983). *The Natural History of Nematodes*. Englewood Cliffs, NJ: Prentice–Hall Inc., pp. 323.

Poinar, G. O., Jr. (1991). Nematoda and Nematomorpha. In J. N. Thorp, & A. P. Covich (Eds.), *Ecology and Classification of North American Freshwater Invertebrates* (pp. 249–283). San Diego, CA: Academic Press.

Poinar, G. O., Jr., & van der Laan, P. A. (1972). Morphology and life history of *Sphaerularia bombi*. *Nematologica, 18*, 239–252.

Poinar, G. O., Jr., & Thomas, G. M. (1966). Significance of *Achromobacter nematophilus* sp. nov. (Achromobacteriaceae: Eubacteriales) associated with a nematode. *Int. Bull. Bacteriol. Nomencl. Taxon., 15*, 249–252.

Poinar, G. O., Jr., Lane, R. S., & Thomas, G. M. (1976). Biology and redescription of *Pheromermis pachysoma* (V. Linstow) n. gen. comb. (Nematoda: Mermithidae), a parasite of yellowjackets (Hymenoptera: Vespidae). *Nematologica, 22*, 360–370.

Popiel, I., & Vasquez, E. M. (1991). Cryopreservation of *Steinernema carpocapsae* and *Heterorhabditis bacteriophora*. *J. Nematol., 23*, 432–437.

Rae, R. G., Riebesell, M., Dinkelacker, I., Wang, Q., Herrmann, M., Weller, A. M., Dietrich, C., & Sommer, R. J. (2008). Isolation of naturally associated bacteria of necromenic *Pristionchus* nematodes and fitness consequences. *J. Exp. Biol., 211*, 1927–1936.

Rae, R. G., Tourna, M., & Wilson, M. J. (2010). The slug parasitic nematode *Phasmarhabditis hermaphrodita* associates with complex and variable bacterial assemblages that do not affect its virulence. *J. Invertebr. Pathol., 104*, 222–226.

Rogers, C. E., Marti, O. G., Jr., & Simmons, A. M. (1993). *Noctuidonema guyanense* (Nematoda: Aphelenchoididae): host range and pathogenicity to the fall armyworm, *Spodoptera frugiperda* (Lepidoptera; Noctuidae). In R. Bedding, R. Akhurst, & H. K. Kaya (Eds.), *Nematodes and the Biological Control of Insect Pests* (pp. 27–32). East Melbourne, Victoria: CSIRO Publications.

Sambrook, J., Fritsch, E. F., & Maniatis, T. (1989). *Molecular Cloning: a Laboratory Manual*. Cold Spring Harbor, NY: Cold Spring Harbor Laboratory Press.

Saunders, M. C., & All, J. N. (1982). Laboratory extraction methods and field detection of entomophilic rhabditoid nematodes from soil. *Environ. Entomol, 11*, 1164–1165.

Seinhorst, J. W. (1959). A rapid method for the transfer of nematodes from fixative to anhydrous glycerin. *Nematologica, 4*, 117–128.

Seinhorst, J. W. (1966). Killing nematodes for taxonomic study with hot F. A. 4:1. *Nematologica, 12*, 178.

Sergeant, M., Baxter, L., Jarret, P., Shaw, E., Ousley, M., Winstanley, C., & Morgan, A. U. (2006). Identification, typing, and insecticidal activity of *Xenorhabdus* isolates from entomopathogenic nematodes in United Kingdom soil and characterization of the *xpt* toxin loci. *Appl. Microbiol., 72*, 5895–5907.

Siddiqi, M. R. (2000). *Tylenchida Parasites of Plants and Insects*. Wallingford, UK: CABI Publishing. pp. 633–751.

Skipper, H. D., & Westerman, D. L. (1973). Comparative effects of propylene oxide, sodium azide, and autoclaving on selected soil properties. *Soil Biol. Biochem., 5*, 409–414.

Southey, J. F. (1986). *Laboratory Methods for Work with Plant and Soil Nematodes* (6th ed.). London, UK: Ministry of Agriculture, Fisheries and Food. Her Majesty's Stationary Office.

Spiridonov, S. E., Reid, A. P., Podrucka, K., Sergei, A., & Moens, M. (2004). Phylogenetic relationships within the genus *Steinernema* (Nematoda: Rhabditida) as inferred from analyses of sequences of the ITS1-5.8S-ITS2 region of rDNA and morphological features. *Nematology, 6*, 547–566.

Stock, S. P. (2009). Molecular approaches and taxonomy of insect parasitic and pathogenic nematodes. In S. P. Stock, J. Vandenberg, N. Boemare, & I. Glazer (Eds.), *Insect Pathogens: Molecular Approaches and Techniques* (pp. 70–98). Wallingford, UK: CABI Publishing.

Stock, S. P., & Hunt, D. (2005). J. In P. S. Grewal, R. U. Ehlers, & D. Shapiro-Ilan (Eds.), *Nematode Morphology and Taxonomy. In: Nematodes as Biological Control Agents* (pp. 3–31). Wallingford, UK: CABI Publishing.

Stock, S. P., Campbell, J. F., & Nadler, S. A. (2001). Phylogeny of *Steinernema* Travassos, 1927 (Cephalobina: Steinernematidae) inferred from ribosomal DNA sequences and morphological characters. *J. Parasitol., 87*, 877–889.

Studdert, J. P., Kaya, H. K., & Duniway, J. M. (1990). Effect of water potential, temperature, and clay-coating on survival of *Beauveria bassiana* conidia in a loam and peat soil. *J. Invertebr. Pathol., 55*, 417–427.

Tailliez, P., Pages, S., Ginibre, N., & Boemare, N. (2006). New insights into diversity in the genus *Xenorhabdus*, including the description of ten novel species. *Int. J. Syst. Evol. Microb., 56*, 2805–2818.

Tailliez, P., Laroui, C., Ginibre, N., Paule, A., Pages, S., & Boemare, N. (2010). Phylogeny of *Photorhabdus* and *Xenorhabdus* based on universally conserved protein-coding sequences and implications for the taxonomy of these two genera. *Int. J. Syst. Evol. Microbiol., 60*, 1921–1937.

Triantaphyllou, A. C., & McCabe, E. (1989). Efficient preservation of root-knot and cyst nematode in liquid nitrogen. *J. Nematol., 21*, 423–426.

Vandergast, A. G., & Roderick, G. K. (2003). Mermithid parasitism of Hawaiian *Tetragnatha* spiders in a fragmented landscape. *J. Invertebr. Pathol., 84*, 128–136.

Vivas, E. I., & Goodrich-Blair, H. (2001). *Xenorhabdus nematophilus* as a model for host–bacterium interactions: *rpoS* is necessary for mutualism with nematodes. *J. Bacteriol., 183*, 4687–4693.

Volgyi, A., Fodor, A., Szentirmai, A., & Forst, S. (1998). Phase variation in *Xenorhabdus nematophilus*. *Appl. Environ. Microbiol., 64*, 1188–1193.

Vrain, T. C., Wakarchuk, D. A., Levesque, A. C., & Hamilton, R. A. (1992). Intraspecific rDNA restriction fragment length polymorphism in the *Xiphinema americanum* group. *Fund. Appl Nematol., 15*, 563–573.

Webster, J. M., & Thong, C. H. S. (1984). Nematode parasites of Orthopterans. In W. R. Nickle (Ed.), *Plant and Insect Nematodes* (pp. 697–726). New York, NY: Marcel Dekker.

Weischer, B., & Brown, D. J. F. (2000). *An Introduction to Nematodes*. Moscow, Russia: Pensoft.

Wertz, J. E., Goldstone, C., Gordon, D. M., & Riley, M. A. (2003). A molecular phylogeny of enteric bacteria and implications for bacterial species concept. *J. Evol. Biol., 16*, 1236–1248.

Wheeler, W. M. (1928). *Mermis* parasitism and intercastes among ants. *J. Exp. Zool., 50*, 165–237.

White, G. F. (1927). A method for obtaining infective nematode larvae from cultures. *Science, 66*, 302–303.

White, W., & Wharton, K. L. (1984). Development of a cryogenic preservation system. *Am. Lab.*, 65–76.

Whitehead, A. G., & Hemming, J. R. (1965). A comparison of some quantitative methods of extracting small veriform nematodes from soil. *Ann. Appl. Biol., 55*, 25–38.

Wilson, M. J. (2007). Terrestrial mollusk pests. In L. A. Lacey, & H. K. Kaya (Eds.), *Field Manual of Techniques in Invertebrate Pathology* (pp. 751–765). Dordrecht, The Netherlands: Springer.

Wilson, M. J., Glen, D. M., & George, S. K. (1993). The rhabditid nematode *Phasmarhabditis hermaphrodita* as a potential biological control agent for slugs. *Biocontrol Sci. Technol., 3*, 503–511.

Wilson, M. J., Glen, D. M., George, S. K., Pearce, J. D., & Wiltshire, C. W. (1994). Biological control of slugs in winter wheat using the rhabditid nematode *Phasmarhabditis hermaphrodita. Ann. Appl. Biol., 125*, 377–390.

Wilson, M. J., Glen, D. M., George, S. K., & Hughes, L. A. (1995). Biocontrol of slugs in protected lettuce using the rhabditid nematode *Phasmarhabditis hermaphrodita. Biocontrol Sci. Technol., 5*, 233–242.

Woodring, J. L., & Kaya, H. K. (1988). Steinernematid and heterorhabditid nematodes: a handbook of techniques. *Southern Cooperative Series Bulletin. 331*. Fayetteville, AK: Arkansas Agricultural Experiment Station. pp. 30.

Xu, J., & Hurlbert, R. E. (1990). Toxicity of irradiated media for *Xenorhabdus* spp. *Appl. Environ. Microbiol., 56*, 815–818.

Yang, Y. H., & Jian, H. (1988). Labelling living entomopathogenic nematodes with stains. *Chin. J. Biol. Control, 4*, 59–61, (In Chinese, English Abstract).

Ye, W., Giblin-Davis, R. M., Davies, K. A., Purcell, M. F., Scheffer, S. J., Taylor, G. S., Center, T. S., Morris, K., & Thomas, W. K. (2007). Molecular phylogenetics and the evolution of host plant associations in the nematode genus *Fergusobia* (Tylenchida: Fergusobiinae). *Mol. Phylogen. Evol., 45*, 123–141.

## APPENDIX I

### 1. Manufacturing of Tungsten wire needles (McClure, personal communication)

*Supplies*

Tungsten wire
Sodium nitrite ($NaNO_2$)
The wooden dowel from the handle of a dissecting needle
Ceramic crucible
Bunsen burner
Ring stand
Pipe-clay triangle or wire mesh for supporting crucible
Goggles, gloves, and a lab coat are recommended. See Figure 12.27 for setup.

*Procedures*

1. In the fumehood, set up a Bunsen burner similar to the example in Figure 12.28.
2. Place crucible on mesh/clay triangle and fill 3/4 full with $NaNO_2$.
3. Light the Bunsen burner and begin melting the $NaNO_2$ with a low blue flame.

**Note:** You MUST melt with an open flame. $NaNO_2$ will not melt satisfactorily with a hot plate.

4. Stab a length of tungsten wire into the wooden dowel so you can easily handle it in the following steps.

**4a.** Make the wire only slightly longer than you want your needle to be.

**4b.** If you desire a bend in your needle, now is a good time to put it in. Tungsten is brittle, and will become even more so after treatment with $NaNO_2$.

5. When the $NaNO_2$ is melted, dip the end of the wire into the hot liquid and it will gradually melt down. Don't hold the wire in the liquid. Instead, keep dipping and removing it and you will gradually 'pull' out a needle-sharp tip.
6. When you have your desired tip, let cool and wash your needles in water before use. Affix to any handle you like.

### 2. Making Na–Azide slides

*Supplies*

Microscope slides
Labeling tape
5 × M9 buffer
4% agar
Na azide (1 M)
100–95% ethanol (for cleaning slides)
Rain-X (available at auto and discount stores)
Microwave
Water bath (set to 50°C)
Pipettor (P-1000) and tips
5 × M9 buffer

Make a suspension of 4% (w/v) agar in 1X M9 buffer. For each 100ml of suspension, add 1 ml of 1 M Na-azide.

---

## APPENDIX II

### 1. Grenacher's borax carmine stain (Nickle & MacGowan, 1992)

1 g carmine
2 g borax
50 ml $H_2O$

**a.** Boil the ingredients in a covered vessel for 30 min or until the carmine dissolves.
**b.** Add 50 ml of 70% ethanol.
**c.** Allow the solution to stand 1–2 days and filter through filter paper.
**d.** Use the filtrate for staining.

Counterstain

1 g Fast green in 99 ml 95% ethanol (1% Fast green)

### 2. Pepsin solution

8 g pepsin
23 g NaCl
20 ml HCl
940 ml distilled water

**a.** Stir the mixture until the ingredients go into solution.
**b.** Keep the solution at 4°C until used, but prolonged storage can inactivate the enzyme. If there is any doubt about the freshness of the solution, prepare a new batch or prepare only enough solution needed for the situation.

### 3. Saline Solutions

*Ringer's solution (Woodring & Kaya, 1988)*

9 g NaCl
0.4 g KCl
0.4 g $CaCl_2$
0.2 g $NaH_2CO_3$
1 l dd $H_2O$

*M9 buffer solution (Brenner, 1974)*

15 g $KH_2PO_4$
30 g $Na_2HPO_4$
25 g NaCl
5 ml $MgSO_4$ (1 M)
dd$H_2O$ to volume

### 4. Fixatives (Seinhorst, 1959, 1966; Southey, 1986; Woodring & Kaya, 1988)

Numerous fixatives have been recommended for studying different features of nematode anatomy. Double-strength fixatives are prepared using half the amount of water indicated below. The commonly used fixatives are:

*TAF*

7 ml formalin (40% formaldehyde)
2 ml triethanolamine
91 ml dd $H_2O$

*F.A. 4:1 or F.A. 4:10*

10 ml formalin
1 or 10 ml glacial acetic acid
up to 100 ml dd $H_2O$

*F.A. A*

20 ml 95% ethanol
6 ml formalin
1 ml glacial acetic acid
40 ml dd $H_2O$

*Buffered formalin*

1000 ml 40% formaldehyde
4 g $NaH_2PO_4.H_2O$
6.5 g $Na_2HPO_4$

### 5. Hard glycerin jelly (Southey, 1986)

**a.** Soak 20 g of gelatin in 40 ml of distilled water for 2 h.
**b.** Add 50 ml of glycerin and 1 ml of phenol.
**c.** Place in water bath (70–80°C) for 10–15 min and stir until the mixture is homogenous.

### 6. Agar media for *in vitro* cultivation of nematodes

**Note:** For solid media, add a stir bar before autoclaving to media containing agar, cover tops with aluminum foil, and mix agar well before autoclaving. Autoclave on liquid cycle (slow exhaust) for 20 min.

*a. Baby Food (BF) Agar (Stock* et al.*, 2001)*

10 g agar
6 g Beechnut™ baby food (or any other brand available with three-cereal mix).
1 l distilled $H_2O$. The baby food seems to have some heat resistant spores in it. To eliminate these spores, we use a double autoclave technique called tyndallization. Briefly, proceed as follows:

**1.** Autoclave agar for 30–45 min (liquid setting).
**2.** Allow agar to solidify overnight.
**3.** Autoclave again for at least 30 min.
**4.** Allow agar to cool and pour into sterile Petri dishes.

*b. Liver–kidney agar*

50 g ground beef kidney
50 g ground beef liver
2.5 g NaCl—**Note:** Final NaCl concentration should be 0.5% NaCl.
500 ml water
7.5 g agar

Standard agar concentration for most media recipes is 1.5–2.0% (7.5–10.0 g/500 ml).
**Note:** The liver and kidney should be blended well enough so that the two meats are essentially a liquid pulp or thick paste. Doing this seems to lessen coagulation of the meat when it is autoclaved. The meat also tends to remain more evenly distributed in the molten agar; although, it never really stays homogenously dispersed without frequent mixing when pouring the plates. We grind an entire piece of meat and use only the required amount and freeze the remaining liver and kidney. Below we provide a modified procedure from that originally described by Poinar & Thomas (1966):

**1.** Add 500 ml of water to an Erlenmeyer flask.
**2.** Add agar and NaCl.
**3.** Add beef liver and kidney.
**4.** Autoclave for 15 min at 121°C.
**Note:** You will need to continuously hand-stir agar before pouring into plates to make sure chunks of meat are not poured.
**5.** Use 5-cm Petri dishes.

*c. Lipid agar*

8 g nutrient broth
15 g agar
5 g yeast extract
890 ml $ddH_2O$
10 ml 0.2 g/ml $MgCl_2 \cdot 6H_2O$
4 ml corn oil
96 ml corn syrup mix (7 ml corn syrup in 89 ml $H_2O$)

**1.** Mix together nutrient broth, agar and yeast extract.
**2.** Pour mix in a 2-l Erlenmeyer flask.
**3.** Add double-distilled $H_2O$.
**4.** Add $MgCl_2 \cdot 6H_2O$.
**Note:** This can be added after autoclaving.
**5.** Autoclave for 15 min at 121°C.
**6.** Add sterile mix of corn oil and corn syrup mix.
**7.** Stir/swirl vigorously making sure oil is dispersed evenly.

**Note:** The oil will not dissolve into the liquid, but ensure it is in tiny droplets. You can add a stir bar in the beginning and continuously stir your liquid while plating if you want to 'pour' your plates using a glass pipette. This is done to ensure an even distribution of oil and volume to each plate since variation in agar thickness or oil distribution can lead to inconsistent nematode development between plates. If this is not an issue, simply ensure the oil is dispersed evenly while pouring from the flask/bottle.

## 7. Media for bacterial growth

*1. Liquid media*

*1. Luria Bertani Broth (for 1 l)*
5 g yeast extract
10 g tryptone
5 g NaCl

1. Place all solid ingredients into a 2-l Erlenmeyer flask.
2. Add 1 l $ddH_2O$.
3. Autoclave at 120°C for 3 min on using liquid setting.

**Note:** For dark LB keep flask and medium in dark or add pyruvate to 0.1% final. One gram of powder can be added before autoclaving. Alternatively, pyruvate can be added after autoclaving from a 20% sterile stock.

*2. Grace's Media (for 500 ml)*
22.85 g powdered insect cell culture medium (Gibco®)
0.175 g $NaHCO_3$
450 ml $H_2O$

1. Add insect cell culture medium to the water, stir until dissolved. (Do not heat the water).
2. Add 0.175 g (0.35 g/l) sodium bicarbonate to the water, stir until dissolved.
3. Adjust pH of medium to 6.1 with 1 N NaOH.
4. Add additional water to bring the solution to final volume (500 ml).
5. Filter and sterilize.

*3. M9 minimal medium (for* Photorhabdus*) (Sambrook* et al. *1989)*
6 g $Na_2HPO_4$
3 g $KH_2PO_4$
0.5 g NaCl
1 g $NH_4Cl$

Adjust pH to 7.4, autoclave then add sterile solutions of:

2 ml 1 M $MgSO_4$
10 ml 20% glucose (or alternative carbon source, sterilize by filtration)
0.1 ml 1 M $CaCl_2$

*a. Luria–Bertani Agar*

5 g yeast extract
10 g tryptone
5 g NaCl
20 g agar
1 l $H_2O$

1. Place in a 2-l flask with stir bar.
2. Add 1 l $ddH_2O$.
3. Autoclave.
4. Stir while cooling.

**Note:** When cool enough to touch with hand add then any desired supplements (e.g. antibiotics) and pour into Petri dishes.

*b. NBTA agar (for 1 l)*

8 g nutrient broth
15 g agar
0.25 g bromothymol blue
1 l $H_2O$

1. Add water into a 2-l Erlenmeyer flask.
2. Pour all solid ingredients into the water.
3. Autoclave.
4. Stir while cooling.
5. Add 0.04 g triphenyltetrazolium chloride (TTC).

*4. Minimal medium for* Xenorhabdus nematophila *and* X. bovienii

*Bottle 1: salts*
100 mg nicotinic acid (both *X. nematophila* and *X. bovienii* are auxotrophs for this vitamin)
3 g $KH_2PO_4$
7 g $K_2HPO_4$
2 g $(NH_4)_2SO_4$
1–2 pellets NaOH
500 ml $H_2O$

**Note:** pH of this solution should be 7.0.

*Bottle 2: agar*
15 g agar (alternatively use Noble agar if stringent conditions are required)
1 g sodium pyruvate
450 ml $dH_2O$

*Supplements*
10 ml of SL4 salts
9.1 g glucose

$0.7$ g $MgCl_2\cdot(H_2O)_6$
Bring to 50 ml with $ddH_2O$

*Salts*

*SL-6 stock (used to make SL-4)*
500 mg $MnCl_2\cdot 4H_2O$
300 mg $H_3BO_3$
200 mg $CoCl_2\cdot 6H_2O$
100 mg $ZnSO_4\cdot 7H_2O$
30 mg $Na_2MoO_4\cdot 2H_2O$
20 mg $NiCl_2\cdot 6H_2O$
10 mg $CuCl_2\cdot 2H_2O$
1 l $H_2O$

Mix all ingredients and autoclave.

*SL-4 trace elements stock*
0.5 g EDTA
0.2 g $FeSO_4\cdot 7H_2)$
100 ml SL-6
900 ml $H_2O$

Mix all ingredients and autoclave.

*Procedure*

1. Autoclave Bottles 1 and 2 separately, cool, and combine while stirring, making sure they are quite cool before mixing.
2. Mix together supplements, heat to dissolve and filter through a 0.2-μm filter into mixed bottle 1 + 2 media.

**Note:** Before pouring can also add 0.1 % casamine acids (*Xenorhabdus* grow much better with amino acids). Also you can add glycerol rather than glucose as the carbon source.

CHAPTER XIII

# Pathogens and parasites of terrestrial molluscs

MICHAEL J. WILSON

AgResearch, Ruakura Research Centre, East Street, Private Bag 3123, Hamilton 3240, New Zealand

## 1. INTRODUCTION

Terrestrial molluscs, (slugs and snails) are an extremely diverse group of animals that have colonized all the inhabited continents and some have become pests of a wide range of crops. As a group, they are not particularly well-studied in comparison with the insects, probably because few people receive any dedicated training in malacology. There is no formal biological distinction between slugs and snails, with animals being considered slugs when they have no external shell, or when the external shell is small in comparison to the size of the body. All slug forms have evolved from snail forms, and this is considered to have happened on several separate occasions (Solem, 1978). Furthermore, there are many intermediate forms, with small external shells.

While there are approximately 35,000 species of terrestrial molluscs, the fauna of cultivated land is dominated by a few slug forms and it is usually slug forms that damage crops. One European slug in particular, *Deroceras reticulatum* (Figure 13.1), is a pest throughout its home range, and is also a serious invasive pest in North and South America, and Australasia. Some snails are pests of citrus trees and small-scale horticultural crops. A comprehensive review of molluscs as crop pests throughout the world was published by Barker (2002).

In the vast majority of cases, pest molluscs are controlled using molluscicidal bait pellets containing either metaldehyde, carbamate compounds or iron phosphate (for review of chemical control, see Bailey, 2002). However, the chemical baits tend to have poor efficacy, and there are environmental concerns regarding non-target effects (particularly with carbamates) and contamination of ground/drinking water (particularly with metaldehyde).

There is one commercially available biological molluscicide, the nematode *Phasmarhabditis hermaphrodita* that is sold in several European countries (see review by Rae *et al.*, 2007). The nematode agent is environmentally benign, but is expensive to produce and is only sold for use in high-value crops. Thus, there is a clear need for a low-cost biological molluscicide.

Compared with insect pathology, pathology of terrestrial molluscs is in its infancy and there is much to be learned. However, many of the general techniques developed for other invertebrate pathogens and parasites described within this volume (particularly Chapters II, III, VII, XI, and XII) should be broadly transferable for work with terrestrial molluscs. The current chapter gives an overview of techniques needed for working with

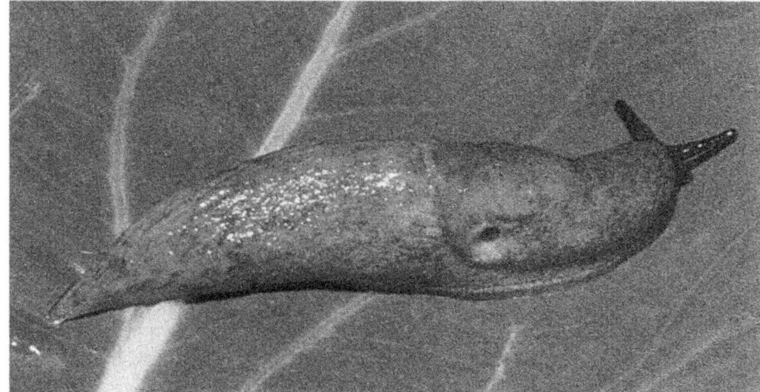

**Figure 13.1** *Deroceras reticulatum*, the most widely distributed species of pest mollusc in the world, is found in Europe, North and South America, Australasia and parts of Asia. It is particularly susceptible to infection by the nematode parasite *Phasmarhabditis hermaphrodita*.

MANUAL OF TECHNIQUES IN INVERTEBRATE PATHOLOGY
ISBN 9780123868992

terrestrial molluscs, and then briefly summarizes current knowledge of the major groups of pathogens and parasites that infect molluscs, describing techniques specifically developed for their study.

## 2. GENERAL TECHNIQUES FOR WORKING WITH TERRESTRIAL MOLLUCS

### A. Collection and maintenance of slugs and snails

It is possible to rear slugs and snails in the laboratory but it is labor intensive, so most workers rely on field collection. Snails tend to rest above ground and can be hand collected. This is best done early in the morning or evening when the animals are most active. Slugs often rest in the soil or under stones during the day and can be collected by placing refuges in the field.

#### *1. Collection of slugs and snails*

1. Place roof tiles or wooden boards out in an area of grassland or cereal field, having first cleared any vegetation below (Figure 13.2).
2. Put a small amount (approx. 5 g) of wheat or oat bran under each tile (small amounts can be obtained as supermarket breakfast cereals, large sacks can be purchased cheaply from animal feed suppliers).
3. Leave the tiles in the field for at least 2 days then collect slugs from underneath.

**Figure 13.2** Slug refuges, such as this one made from a pot saucer, are a convenient way of collecting slugs from the field.

4. Traps can be left out indefinitely, and bran added when necessary. If bran becomes moldy, discard and replace with fresh bran.
5. Under dry conditions refuges can be placed in irrigated patches of grass.
6. Once collected, slugs can be kept in non-airtight plastic boxes lined with moistened absorbent cotton wool, or damp paper towels. For long-term storage, keep at 4°C and feed with cabbage, carrots, or bran. Clean the boxes and change food at least fortnightly.

Slugs can be identified to species using one of several keys, e.g., Chichester & Getz, 1973; Cameron *et al.*, 1983.

### B. Dissection of slugs

If looking for living parasites (e.g., nematodes or ciliates) it is best to dissect slugs alive so moving parasites can be seen. Slugs can be anesthetized prior to dissection by being kept in an atmosphere of $CO_2$ for 15 min. Generally, parasites in such anesthetized slugs are still active. For snails, the animal has to be removed from the shell first, which can be achieved by holding the head in forceps and tugging the shell. The protocol below relates to slugs, but dissection of snails once removed from the shell is very similar.

#### *1. Methods for dissection of slugs (see Figure 13.3)*

1. Place slugs on a strip of Blu-Tac (or similar material) within a water-filled Petri dish base.
2. Slugs should then be pinned at the head and tail end using entomological pins.
3. Remove the mantle and expose the animal's vestigial shell. *Phasmarhabidits,* other rhabditid nematodes and *Cosmocercoides dukae* are usually found in the mantle cavity in close association with the shell. *Angiostoma asperse* is found in the equivalent site in the snail *Cantareus asperses.*
4. The slug's kidney can be seen immediately below the shell, separated by some thin membranous tissues.
5. Using sharp forceps take a small section of kidney and place into the dissecting water. The kidney is the typical site of reproduction for the ciliate parasite *Tetrahymena rostrata.* If present, numerous ciliates will swim out. If so, the kidney can be removed and

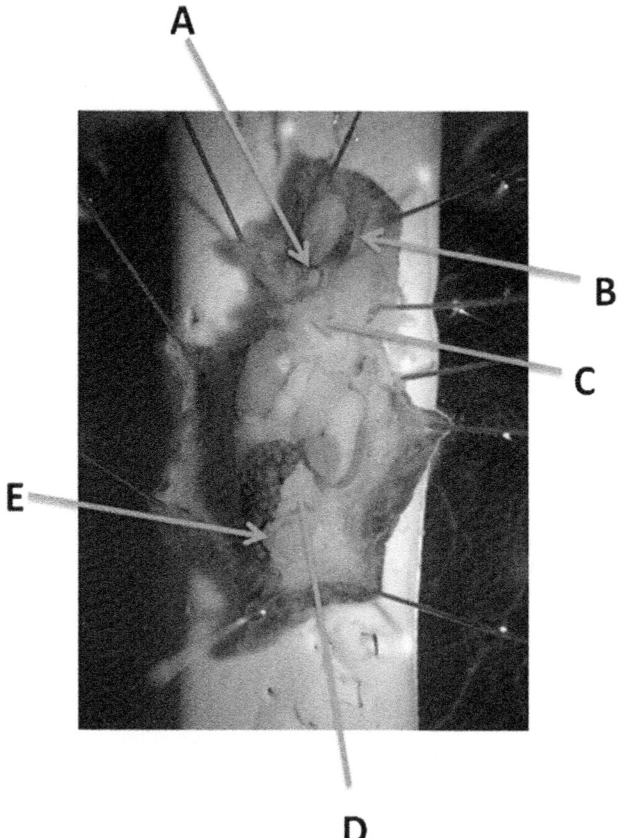


**Figure 13.3** Dissected *Deroceras reticulatum* showing sites of infection for common ciliate and nematode parasites. A. Esophagus, directly behind the buccal mass is where most *Angiostoma* nematodes are found. B. Proximal reproductive structures and salivary glands where *Agfa* nematodes are found. C. Kidney where ciliate protozoan *Tetrahymena rostrata* is found. D. Digestive gland (hepatopancreas) where *Angiostoma glanicola* nematodes and *Tetrahymena limacis* ciliates are found. E. lower reaches of intestinal tract where Alaninematidae nematodes are found. Please see the color plate section at the back of the book.

used as a source of ciliates to establish cultures (see Section 3E below).

6. Gradually cut a long slit from the tail to the rear of the mantle cavity. Peel back the skin on both sides and pin into the Blu-Tac.
7. Repeat cutting the skin from the front of the mantle to the head, and again peel back.
8. At the front of the animal are the salivary glands and reproductive tract where the nematodes *Agfa* spp., *Nemhelix bakeri* and *Hugotdiplogaster neozelandia* are found.
9. The intestine is intertwined around the hepatopancreas (liver) and is difficult to remove intact. Most *Angiostoma* spp. are found in the anterior part of the intestine that can be removed separately. *Alaninema* spp. tend to be found towards the rear of the intestinal tract.
10. The large hepatopancreas at the rear is the site of infection of the ciliate *Tetrahymena limacis* and also site of infection of the nematode *Anigostoma glandicola.*
11. Once all internal organs have been examined, they can be removed and the foot muscles inspected for presence of *Alloionema appendiculatum* nematodes.

## 3. MAJOR GROUPS OF PATHOGENS AND PARASITES ASSOCIATED WITH TERRESTRIAL MOLLUSCS

In the following sections, current knowledge of pathogens and parasites is reviewed and key references are cited. Techniques developed specifically for working with molluscan pathogens and parasites, where available, are described.

### A. Viruses

There are no confirmed reports of viruses infecting slugs. David *et al.* (1977) reported virus-like particles in *D. reticulatum*, but these were later found to be particles of galactogen and glycogen (Kassanis *et al.,* 1984). It is highly likely that viruses occur and those interested in molluscs viruses are directed to Chapter II of this volume for general methods.

### B. Fungi

There are no recorded cases of fungi parasitizing slugs but there are reports of fungi attacking slug eggs. *Pachonia chlamydosporium* has been shown to cause 97% mortality of *D. reticulatum* eggs (Trevet & Esslemont, 1938) and *Arthrobotrys* sp. is known to parasitize eggs of *Arion circumscriptus* (Arias & Crowell, 1963). There have not been any studies on the feasibility of using either of these fungi as biological control agents, possibly because Trevet

& Esslemont (1938) considered that they would be unsuitable.

## C. Microsporidia

The most recent review of microsporidia associated with slugs and snails is that of Selman & Jones (2004). Apart from reports of a *Steinhausia* sp. from captive *Partula turgida* snails (Cunningham & Daszak, 1998), all work has concentrated on *Microsporidium novacastriensis* infecting *D. reticulatum.* Jones & Selman (1984) studied the parasitic action and biological control potential of *M. novocastriensis* and showed that it caused chronic infection inhibiting growth, longevity, fecundity, and feeding activity. The infection is characterized by the presence of numerous small refractile spores infecting the gut cells. These can, under certain circumstances, destroy the gut completely, allowing bacteria into the hemocoel which may result in death. Its potential as a biological control agent may be limited because it only infects *D. reticulatum* and like all microsporidia, can not be grown *in vitro.* However, evidence from New Zealand suggests that *M. novacastriensis* can regulate populations of *D. reticulatum* (Selman & Jones, 2004) and thus, it may be possible to utilize this parasite in conservation biological control approaches. To do this, it is necessary to monitor infection levels.

### *1. Visualising* Microsporidium novocastriensis *in slugs*

When viewed under phase-contrast microscopy, the microsporida appear as highly refractile spores.

They can be viewed in small samples of fresh material by simply smearing the samples onto a microscope slide. Alternatively they can be viewed in whole preserved and sectioned slugs.

1. Fix whole slugs by immersion in alcoholic Bouin's fluid (see Chapter XV), then dehydrate in absolute ethanol.
2. Embed slugs in ParaPlast® wax (melting point of approx. 56°C, available from most laboratory suppliers) under a vacuum.
3. Using a microtome, cut sections at approx. 6 µm.
4. Stain with Giemsa or Heidnenhain's hematoxylin and eosin solutions (see Chapter XV).

Photographs of the various life stages of *M. novocastriensis* are included in Jones & Selman (1985).

## D. Bacteria

Bacterial diseases (and non-microbial diseases) of terrestrial molluscs have been reviewed by Raut (2004). It should be noted that, while Raut (2004) described symptoms of several diseases of slugs and snails for which no microbial agents have yet been isolated, this does not necessarily mean the diseases are non-microbial in nature.

Like all animals, slugs and snails have an associated bacterial flora, and there have been many studies investigating the nature and development of this flora, with emphasis on how intestinal bacteria may be involved in digestion (Charrier, 1990; Charrier *et al.*, 1998; Walker *et al.*, 1999). There have also been several studies investigating the potential of terrestrial molluscs to carry bacteria that are pathogenic to humans (Shrewsbury & Barson, 1947; Yamada *et al.,* 1960; Elliot, 1969; Sproston *et al.,* 2006, 2010). There have been many fewer studies in which the effects of bacteria on the mollusc host have been studied in relation to biological control. Furthermore, the majority of these studies are inconclusive, or other authors have failed to repeat early promising results.

The symptoms of eight disease types of Indian pest molluscs have been described (Raut & Panigrahi, 1989), seven of these diseases being found in slugs. The etiology of these diseases has not yet been investigated but one disease, a leucodermia-type disease in the pest slug *Laevicaulis alte*, has identical symptoms to a disease of the giant African snail, *Achatina fulica*, thought to be caused by the bacterium *Aeromonas hydrophila* (Mead, 1961). Field observations suggest that this disease is capable of reducing populations of both slugs (Raut & Mandal, 1986) and snails (Raut & Ghose, 1977). However, while Dean *et al.* (1970) showed a statistically significant association of *A. hydrophila* with this disease, the bacterium has never been subjected to Koch's postulates to demonstrate a causal relationship. The mode of transmission of this disease is not understood and the potential of *Aeromonas hydrophila* as a biological control agent for terrestrial molluscs has not been investigated.

The aquatic snail *Biomphalaria glabrata* has been investigated as a possible target for biological control because it is a vector of schistosomiasis in man. Pan (1956) reported an acid-fast pathogen in this snail but stated that it had low virulence and had little potential as a control agent. Another bacterium, *Bacillus pinottii* has been successfully used in field trials against this snail in Egypt (Dias & Dawood, 1955), but Trip (1961) and Wright (1968) did not achieve control of the snail using the same bacterium. *Vibrio parahaemolyticus* has been shown to be pathogenic to *B. glabrata* (Ducklow *et al.*, 1980) and Singer *et al.* (1997) demonstrated that some isolates of *Bacillus brevis* are also pathogenic towards this species. However, none of these bacteria has been assayed against terrestrial molluscs, and none has been developed further for control of *B. glabrata*.

It has been reported that certain commercially available insecticidal strains of *Bacillus thuringiensis* are highly pathogenic to *D. reticulatum* (Terytze & Hofmann, 1986), but Hommay and Wilson (unpublished data), using the same commercially available strains, found no activity and suggested that the high mortality of slugs recorded by Terytze & Hofmann (1986) might be a result of the high temperature (22°C) at which they kept their slugs.

Thus, there are some promising leads and the potentially low production costs of bacteria make them attractive candidates for microbial pesticides. To this end, Howlett *et al.* (2009) described a simple method for mass screening of bacterial isolates for pathogenicity to slugs. Slugs are exposed to the bacteria incorporated into a bran based diet. The assay uses bacteria removed from the surface of Petri dish cultures rather than pelleted cells from liquid cultures, as this will incorporate both living cells and any extracellular toxins produced during growth. The following method is an improvement of that described by Howlett *et al.* (2009).

### *1. Bioassay of bacteria for molluscicidal activity (Figure 13.4)*

1. Inoculate the test bacterium as a lawn over the entire surface of plates of tryptic soy agar (TSA) and incubate for 48 h at 24°C.
2. Using a sterile rounded spatula, remove all bacterial growth from the plates and suspend in 12 ml sterile distilled water (SDW).
3. Mix 11 ml of the resulting bacterial suspension with 11 g dry weight oat bran and stir thoroughly.
4. Dispense 0.5 g of the bran/bacterium mix into each of 10 small weighing boats (3 × 3 × 0.5 cm).
5. Add each of the 10 weighing boats to a test arena consisting of a 9-cm Petri dish lined with filter paper moistened with 2 ml water and then add one *Deroceras* sp. slug.

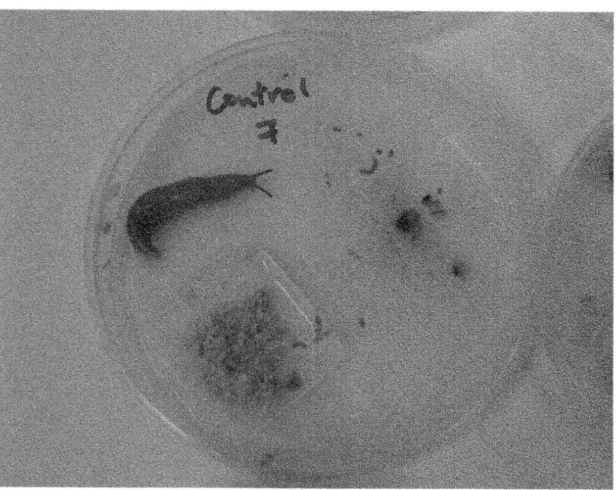


**Figure 13.4** Simple bioassay system for screening bacteria for pathogenicity to slugs, modified from Howlett *et al.*, 2009. Individual slugs are kept in a Petri dish with a diet of oat bran soaked in the test bacterium.

6. Prepare controls as above using bran and SDW.
7. Incubate the test arenas at 15°C with a 16 : 8 h light dark cycle for 4 days, but remove the bran/bacterium mix after 24 h.
8. Record slug mortality on days 3 and 4.

Suggested modifications could include substituting SDW for a buffer suitable for testing soil bacteria, e.g., quarter strength Ringer's solution.

## E. Protozoa

Reviews of protozoa infecting terrestrial gastropods can be found in Stephenson & Knutson (1966) and Van As & Basson (2004). In most cases, previous studies concentrated on the parasites rather than on their effects on host slugs and it is not possible to assess which, if any of the protozoa would be of use as biological control agents. However, a notable exception is the work of Brooks (1968) who studied the biology and parasitic action of two species of holotrichous cilliates, *T. rostrata* and *T. limacis*. *T. rostrata* (Figure 13.5) was found to be highly pathogenic, reducing the longevity and fecundity of infected *D. reticulatum*. The parasite can be grown in artificial medium, is capable of surviving in the soil as a free-living organism and can be transmitted trans-ovum. This ciliate has been found in nine different slug species, and Brooks (1968) considered it to have considerable potential as a biological control agent. Brooks (1968) incubated most of his bioassays at room temperature, and there is some evidence that the pathogenicity of this ciliate is reduced at lower temperatures at which slugs are typically damaging (Brooks, 1968; Wilson *et al.*, 1998; Selman, personal communication). *T. limacis* was found to have low pathogenicity, poor survival in the field and was more difficult than *T. rostrata* to cultivate *in vitro*.

### *1. Isolation and cultivation of ciliates*

Because mixed infections of *T. rostrata* and *T. limacis* can occur in slugs, laboratory cultures should be established using individual ciliates.

1. Dissect slugs in quarter strength Ringer's (QSR) solution and using mounted needles, tease apart the kidney and hepatopancreas (*T. rostrata* is typically found in the slug's kidney below the mantle, whereas *T. limacis* is typically found in the hepatopancreas).
2. If present, numerous ciliates will start to swim in the QSR.
3. Individual ciliates can be handled using capillary pipettes.
4. Transfer individual ciliates to 10-ml tubes of autoclaved 1% proteose peptone containing 1000 units/ml of penicillin G and streptomycin sulphate to inhibit growth of bacterial contaminants (antibiotics should be filter sterilized and added to the medium after autoclaving).

### *2. Bioassays of ciliates against slugs (after Brooks, 1968)*

1. Grow ciliates as above for approx. 4 days at 25°C.
2. Centrifuge ciliates at 500 × *g* for 10 s, then re-suspend in sterile QSR.
3. Concentration of ciliates can be determined in a subsample using a hemocytometer (see Chapter XI).
4. Prepare Petri dishes with a thin layer (1−2 mm) of 2% (w/w) agar.
5. Add known numbers of ciliates (Brooks used 5000 per dish) in 1 ml QSR and spread evenly over the surface.
6. Add one slug (Brooks used *D. reticulatum*) to each and leave for 3 days' exposure period.
7. Transfer slugs to 250-ml cups half-filled with autoclaved soil and seal with a non-airtight lid, e.g., a Petri dish lid. Feed slugs with carrot or celery.

Brooks (1968) conducted all experiments at room temperature and showed good efficacy of *T. rostrata* under these conditions. However, slugs tend to be more active under cooler conditions. Wilson *et al.*, 1996 confirmed the findings of Brooks that *T. rostrata* was pathogenic towards *D. reticulatum*, but showed that this effect was lost at 10°C.

## F. Nematodes

More is known about nematodes associated with terrestrial molluscs than any other group of pathogens or parasites and the nematode *Phasmarhabditis hermaphrodita* is the only commercially available bio-molluscicide. Reviews by Mengert (1953), Grewal *et al.* (2003) and Morand *et al.* (2004) describe the main taxa and features of nematodes associated with terrestrial slugs and snails. Surveys of nematodes parasitizing molluscs have been done in Europe (Mengert, 1953; Morand, 1988; Ross *et al.*, 2010a), North America (Gleich *et al.*, 1977; Ross *et al.*, 2010a), South-East Asia (Pham Van Luc *et al.*, 2005), Australia (Charwat & Davies, 1999), and parts of South Africa (Ross *et al.*, 2012). These studies have shown that there are eight families of nematodes that associate with terrestrial molluscs: the Agfidae, Alaninematidae, Alloionematidae, Angiostomatidae, Cosmocercidae, Diplogasteridae, Mermithidae, and Rhabditidae. Work by Ross *et al.* (2010b) has shown that the slug-parasitic Angiostomatidae, Agfidae, and Rhabditidae (represented by the genus *Phasmarhabditis*) form a tight monophyletic clade, suggesting that all species within these groups are derived from a common ancestral slug-colonizing nematode.

These different families tend to exhibit different lifestyles, and with few exceptions are found within distinct regions of the host.

The key distinguishing features of these families are shown in Figure 13.6. In most cases it is easy to identify nematodes to family based on the stages of nematodes found in the slug (adults or juveniles) and the site of infection as revealed by dissection (see Section 2 above).

The Agfidae (Figure 13.6G), represented by three species, are very characteristic in appearance with an extremely thin neck region. They are found predominantly as adults parasitizing the salivary glands and reproductive tract of slugs. The Angiostamatidae (Figure 13.6D) are predominantly found as adults or fourth stage larvae in the

**Figure 13.5** Silver stained *Tetrahymena rostrata* found in the kidney of *Deroceras reticulatum*.

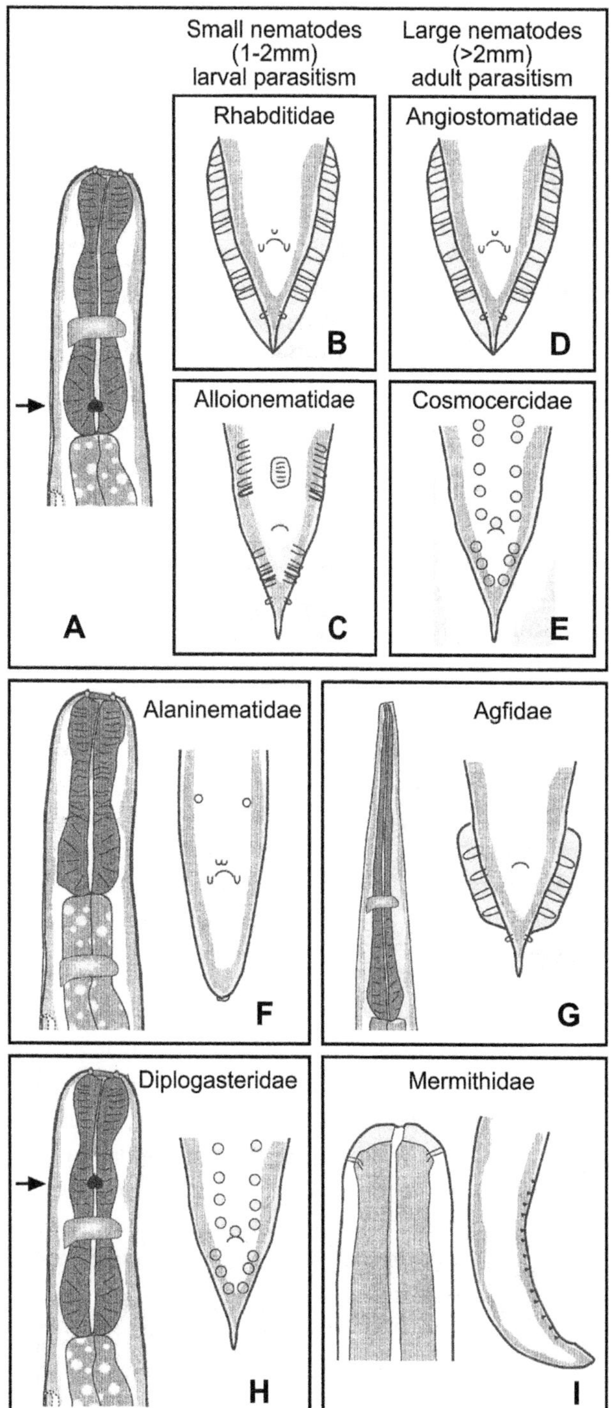


**Figure 13.6** Simplified diagnoses of nematode families associated with terrestrial molluscs based on male tail morphology and position of the valvated bulb within the pharynx. Reprinted with permission from Morand *et al.* (2004).

anterior intestinal tract of molluscs. The Alaninematidae (Figure 13.6F) have only been found in the southern hemisphere (Indonesia, Kenya, and New Zealand) and are found in the digestive tract (lower portions) or pallial cavity. The Alloionematidae (Figure 13.6C), is represented by the single species of mollusc parasite, *Alloionema appendiculatum* found as 3rd and 4th stage larvae parasitizing the foot muscles. This nematode appears to be widespread and has been found in North America, Europe, South Africa and Australasia. The Cosmocercidae (Figure 13.6E) are mostly parasites of reptiles and amphibians, but two genera are found associated with molluscs: *Cosmocercoides dukae* is found as adults in the pallial cavity of North American molluscs, and three *Nemhelix* spp. are found as adults in the reproductive organs of European snails. Only one species of the Diplogateridae (Figure 13.6H), *Hugodiplogaster neozelandia,* has been found as adults parasitizing the reproductive tract of *Athorcophorus bitentaculatus* slugs in New Zealand. However, it is not uncommon to find juvenile diplogasterids within slugs and snails. The Mermithidae (Figure 13.6I) are found as large (1–10 cm) preadult larvae living in the general body cavity.

The most problematic group in terms of identification and distinguishing true parasites from non-trophic associates are the rhabditidae. While two species of *Phasmarhabditis* (*P. hermaphrodita* and *P. neopapillosa*) are clearly true parasites of some slug species, there are many rhabditid nematodes associated with terrestrial molluscs either phoretically (i.e. for transport) or necromenically (an association in which the dauer larvae of the nematode enter the mollusc, and remain dormant until the mollusc dies, after which the nematodes feed, develop, and reproduce on the cadaver). Rhabditids known to associate with molluscs non-parasitically include several *Caenorhabditis* spp. (*C. elegans, C. brigssae, C. remanei, C. fruticicolae*) (Kiontke & Sudhaus, 2006) and *Pelodera teres* (Mengert, 1953) and it is likely that other rhabditids also exhibit this behavior. While *Phasmarhabditis hermaphrodita* can parasitize and develop in some slugs, particularly *D. reticulatum,* it lives necromenically in some larger slugs, e.g., *Arion ater* (Rae *et al.*, 2009). Thus, if dissection reveals an adult rhabditid, it is likely to be to be *Phasmarhabditis* spp., whereas presence of dauer juveniles may or may not represent true parasitism. Since it is not possible to identify all rhabditids from dauer larvae, it is recommended to put the dauer on to a freeze-killed slug or plate of kidney growth medium. These can then be incubated at 15°C until the juvenile becomes adult and can be identified. Alternatively DNA can be extracted directly from the dauer juvenile and sequenced to give a preliminary identification.

Apart from *Phasmarhabditis* spp. most parasitic nematodes seem well adapted to life within their hosts without causing much harm, and thus have little potential for use as inundative biological control agents. However, it may well be that their presence does check population growth and thus, it may be possible to manipulate parasitism in conservation biological control programs. Ross *et al.* (2010a) suggested that parasite release may be a factor in the success of European slug species when they invade new territories. Thus, the ability to determine parasite prevalence in natural slug communities is of value. Methods to do so typically involve dissection of slugs and identification of nematodes by either morphological or molecular (sequencing) methods.

In addition to the *Phasmarhabditis* spp., there are probably other rhabditid nematodes with potential for biocontrol of slugs and snails. Charwat *et al.* (2000) demonstrated that an un-named rhabditid species was capable of reducing the fecundity and lifespan of the snail *Cernuella virgata.* Workers who are interested in studying nematodes associated with slugs and snails will need to be able to identify nematodes to family following dissection of the host. To assess the biocontrol potential of rhabditid nematodes, workers will need to be able to mass rear the nematodes for use in bioassays, know how to monoxenize the nematodes to investigate the relationships between nematodes and bacteria, and be able to bioassay nematodes against slug hosts. These specific techniques are described in the following sections, whereas general nematological methods for preserving nematodes and preparing permanent mounts, *etc.*, are described in Chapter XII.

### *1. Method for DNA extraction of and PCR amplification of nematode rRNA genes for identification*

Field-collected slugs can have mixed nematode infections (Caberet & Morand, 1990; Ivanova & Wilson, 2009) so any sequence-based identifications should be done using individual nematodes. In the case of rhabditid nematodes, individual juveniles, or gravid females can be used to start cultures on kidney agar (see Appendix), or on killed slug cadavers. Then, large numbers of nematodes from the resulting cultures can be used.

There are numerous published methods for extracting DNA from individual nematodes. However, the author's personal experience has shown the following method based on Chelex® to be very efficient, even when used with small rhabditid larvae.

1. Pick individual nematodes using a bristle and transfer to 25 µl of 5% Chelex and 5 µl proteinase K solution (600 µl/ml) stored in 0.2-ml tubes.
2. Incubate the tubes at 60°C for 30 min.
3. Heat the lysate to 94°C for 10 min.
4. Transfer lysate to −20°C for 20 min.
5. Centrifuge at 8000 r.p.m. for 5 min.
6. Remove supernatant which can be stored frozen.
7. Use 2 µl of supernatant as DNA template in PCR reaction.

### *2. PCR amplification*

Most work on molecular diagnostics/phylogenetics for nematodes have used ribosomal RNA genes, with the most common sequence being the 18S subunit which has been used extensively in phylogenetic studies (Blaxter *et al.*, 1998; van Megen *et al.*, 2009) and the majority of slug nematode sequences are available for this gene. However, as time goes on it is likely sequences of other ribosomal genes including ITS and 28S will become more widely available, and these are likely to be of more use for identifying nematodes to species (see Table 13.1

primers for amplifying nematode, rRNA genes). A protocol used for the sequencing of 18S rRNA genes for nematodes from slugs is as follows:

1. Primary denaturation at 94°C for 5 min.
2. Run 35 cycles at 94°C for 60 s, 55°C for 90 s, and 72°C for 2 min, followed by a final 72°C for 10 min.

The PCR reaction products can be visualized on 1% agarose gel and DNA from successful PCRs can be cleaned prior to sequencing using Qiagen QIAquick® PCR Purification Kits or similar proprietary products.

### 3. *Simple method for growing* P. hermaphrodita *on killed slugs*

An easy way to produce sufficient inoculum for use in bioassays against slugs, is to rear the nematodes on slug cadavers. This mimics the way in which the nematodes grow in nature. While *P. hermaphrodita* infects live slugs and starts to reproduce within living slugs, the vast majority of nematode multiplication takes place on the slug cadaver, which decomposes rapidly. Placing nematodes on slug cadavers in modified White traps, similar to those used by entomopathogenic nematode workers allows collection of dauer larvae in water (Figure 13.7). White traps as described in Chapter XII can be used, but *P. hermaphrodita* adults often migrate into the water too. The trap described below provides good clean suspensions of dauer larvae. Because field collected slugs can harbor parasites or necromenic nematodes it is essential that slugs are killed in such a way to kill such internal nematodes.

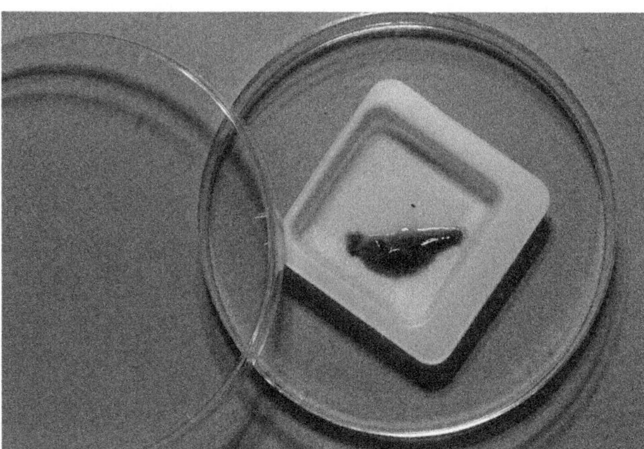

**Figure 13.7** Modified White trap for collecting dauer larvae of rhabditid nematodes reared on slug cadavers (Photo by T. James, with permission).

1. Kill *D. reticulatum* slugs by freezing at −20°C for at least 48 h or by dropping slugs into boiling water (slugs will die instantly).
2. Line a 2 × 2 cm plastic weighing boat with filter paper.
3. Secure the weigh boat in the center of a 9-cm Petri dish using a small piece of Blu-Tac.
4. Fill the Petri dish with tap water so that the bottom of the weigh boat, but not the top is immersed in water.
5. Incubate at 16°C. Dauer larvae start to collect in the water typically within 10–20 days.

### 4. *Method for monoxenization of* P. hermaphrodita

While the method described above for rearing *P. hermaphrodita* on dead slugs is satisfactory for initial pathogenicity tests of new isolates, if an isolate is to be further developed it should ideally be grown in monoxenic culture with a known bacterium. Failure to do so may allow toxic or pathogenic bacteria to enter the cultures.

1. Place gravid adult females in sterile QSR and rupture with a hypodermic needle or glass rod.
2. Transfer eggs to 10-ml sterile 10% NaClO for 5 min.
3. Concentrate eggs by centrifugation for 2 min at 800 × *g*.
4. Remove supernatant and replace with fresh sterile QSR.
5. Shake or vortex egg suspension to clean any residual NaClO.
6. Re-centrifuge eggs at 800 × *g* for a further 2 min.
7. Inoculate half of a Petri dish of kidney agar (see appendix) with the test bacterium.
8. Place cleaned eggs on the bacteria-free side of the Petri dish, at least 2 cm from the bacterial lawn.
9. Incubate at 15°C.

Viable eggs will hatch and larvae will migrate into the bacterial lawn. If eggs are not completely sterile, bacterial growth will be seen around the eggs, and along the nematode trails leading from hatched eggs to the bacterial lawn.

Commercial *P. hermaphrodita* product is reared on *Moraxella olsoensis,* but the nematode is capable of growth on a wide range of bacteria (Wilson *et al.*, 1995a). Furthermore, there is no evidence that in nature *P. hermaphrodita* associates with *M. osloensis* or any other bacterium in a way analogous to entomopathogenic nematodes (Rae *et al.*, 2010). However, different bacteria may alter virulence of the nematodes (Wilson *et al.*,

1995b). Nematodes reared in monoxenic culture should thus always be bioassayed against slugs using the simple method described below.

*5. Simple Petri dish assay for slug mortality*

Rae *et al.* (2008) described a simple method for measuring slug mortality when exposed to *P. hermaphrodita*. A slightly improved version of this assay is given below.

1. Fill replicated large Petri dishes (13.6 cm diameter) with 100 g air dried loam soil.
2. Add 13,0000 lab-reared or commercially produced *P. hermaphrodita* suspended in 30 ml tap water (equivalent to 90 dauer juveniles/cm$^2$, three-times the recommended field rate).
3. Prepare control dishes by adding 30 ml tap water without nematodes.
4. Add 10 *D. reticulatum* to each dish.
5. Seal dishes with Parafilm® and store in an incubator at 16°C.
6. Record mortality every day for 18 days.
7. Remove dead slugs and examine for presence of nematodes.

Glen *et al.* (2000) demonstrated that nematode-induced inhibition of slug feeding is a more sensitive assay than mortality. If feeding data are required:

1. Set up infection plates as above.
2. After 3 days, transfer individual slugs to 9 cm Petri dishes lined with moist filter paper.
3. Add a previously weighed 3 cm diam. disc of lettuce leaf to each dish.
4. Re-incubate slugs at 16°C.
5. Re-weigh lettuce discs and replace with fresh ones every 3 days until the end of the experiment.

**Table 13.1** Primers for amplifying ribosomal rRNA genes that have been used successfully with nematodes associated with terrestrial molluscs.

| *Primer name* | *Sequence (5′→ 3′)* | *rRNA region amplified* | *Reference* |
|---|---|---|---|
| G18S4 | GCTTGTCTCAAAGATTAAGCC | 18S | Blaxter *et al.* (1998) |
| 26R | GCTTTCGTAAACGGAAGAATG | 18S | Blaxter *et al.* (1998) |
| 22F | TCCAAGGAAGGCAGCAGGC | 18S | Blaxter *et al.* (1998) |
| 1080JR | TCCTGGTGGTGCCCTTCCGTCAATTTC | 18S | Ross *et al.* (2010b) |
| 24F | AGRGGTGAAATYCGTGGACC | 18S | Blaxter *et al.* (1998) |
| 18P | TGATCCWKCYGCAGGTTCAC | 18S | Blaxter *et al.* (1998) |
| 55F | GCCGCGAATGGCTCGGTATAAC | 18S | Ross *et al.* (2010b) |
| 920DR | CTTGGCAAATGCTTTCGCAG | 18S | Ross *et al.* (2010b) |
| 555F | AGCCGCGGTAATTCCAGCTC | 18S | Ross *et al.* (2010b) |
| 1165SR | CGTGTTGAGTCAAATTAAGCCGCAGG | 18S | Ross *et al.* (2010b) |
| 18s-5F | GCGAAAGCATTTGCCAAGAA | 18S | Vandergast & Roderick (2003) |
| 18s-9R | GATCCTTCCGCAGGTTCACCT | 18S | Vandergast & Roderick (2003) |
| h (D2A) (F) | ACAAGTACCGTGAGGGAAAGTTG | 28S | Subbotin *et al.* (2006) |
| g (D3B) (R) | TCGGAAGGAACCAGCTACTA | 28S | Subbotin *et al.* (2006) |
| AB28 | ATATGCTTAAGTTCAGCGGGT | ITS | Joyce *et al.* (1994) |
| TW81 | GTTTCCGTAGGTGAACCTGC | ITS | Joyce *et al.* (1994) |
| 18S | TTGATTACGTCCCTGCCCTTT | ITS | Vrain *et al.* (1992) |
| 26S | TTTCACTCGCCGTTACTAAGG | ITS | Vrain *et al.* (1992) |

### 6. *Protocol for determination of feeding*

1. Set up infection plates as above.
2. After 3 days, transfer individual slugs to 9-cm Petri dishes lined with moist filter paper.
3. Add a previously weighed 3-cm-diameter disc of lettuce leaf to each dish.
4. Re-incubate slugs at 16°C.
5. Re-weigh lettuce discs and replace with fresh ones every 3 days until the end of the experiment.

## ACKNOWLEDGMENTS

Funding to write this chapter was provided by AgResearch, New Zealand.

## REFERENCES

Arias, R. O., & Crowell, H. H. (1963). A contribution to the biology of the gray garden slug. *Bull. South. Calif. Acad. Sci., 62*, 83–97.

Barker, G. M. (2002). *Molluscs as Crop Pests*. Wallingford, UK: CABI Publishing, pp. 400.

Bailey, S. E. R. (2002). Molluscicidal baits for control of terrestrial gastropods. In G. M. Barker (Ed.), *Molluscs as Crop Pests* (pp. 33–54). Wallingford, UK: CABI Publishing.

Blaxter, M. L., de Ley, P., Garey, J. R., Liu, L. X., Scheldeman, P., Vierstraete, A., Vanfleteren, J. R., Mackey, L. Y., Dorris, M., Frisse, L. M., Vida, J. T., & Thomas, W. K. (1998). A molecular evolutionary framework for the phylum Nematoda. *Nature, 392*, 71–75.

Brooks, W. M. (1968). Tetrahymenid ciliates as parasites of the gray garden slug. *Hilgardia, 39*, 205–276.

Caberet, J., & Morand, S. (1990). Single and dual infections of the land snail *Helix aspersa* with *Muellerius capillaries* and *Alloionema appendiculatum* (Nematoda). *J. Parasitol., 76*, 579–580.

Cameron, R. A. D., Evesham, B., & Jackson, N. (1983). A field key to slugs of the British Isles. *Field Stud., 5*, 807–824.

Charrier, M. (1990). Evolution, during digestion, of the bacterial flora in the alimentary system of *Helix aspersa* (Gastropoda Pulmonata): a scanning electron microscope study. *J. Mollusc. Stud., 56*, 425–433.

Charrier, M., Combet-Blanc, Y., & Ollivier, B. (1998). Bacterial flora in the gut of *Helix aspersa* (Gastropoda Pulmonata): evidence for a permanent population with a dominant homolactic intestinal bacterium, *Enterococcus casseliflavus*. *Can. J. Microbiol., 44*, 20–27.

Charwat, S. M., & Davies, K. A. (1999). Laboratory screening of nematodes isolated from South Australia for potential as biocontrol agents of helicid snails. *J. Invertebr. Pathol., 74*, 55–61.

Charwat, S. M., Davies, K. A., & Hunt, C. H. (2000). Impact of a rhabditid nematode on survival and fecundity of *Cernuella virgata* (Mollusca: Helicidae). *Biocontol. Sci. Technol., 10*, 147–155.

Chichester, L. F., & Getz, L. L. (1973). The terrestrial slugs of north-eastern North America. *Sterkiana, 50*, 11–42.

Cunningham, A. A., & Daszak, P. (1998). Extinction of a species of land snail due to infection with a microsporidian parasite. *Cons. Biol., 12*, 1139–1141.

David, W. A. L., Taylor, C. E., & Atkey, P. T. (1977). Non-occluded virus-like particles in the mollusc, *Agriolimax reticulatus*. *J. Invertebr. Pathol., 29*, 242–243.

Dean, W. W., Mead, A. R., & Northey, W. T. (1970). *Aeromonas liquefaciens* in the giant African snail *Achatina fulica*. *J. Invertebr. Pathol., 16*, 346–351.

Dias, E., & Dawood, M. M. (1955). Preliminary trials on the biological snail control with *Bacillus pinottii* in Egypt. *Mem. Instit. Oswald. Cruz, 53*, 13–29.

Ducklow, H. W., Tarraza, J. R., & Mitchell, R. (1980). Experimental pathogenicity of Vibrio parahaemolyticus for the schistosome-bearing snail Biomphalaria glabrata. *Can. J. Microbiol., 26*, 503–506.

Elliot, L. P. (1969). Certain bacteria, some of medical interest, associated with the slug *Limax maximus*. *J. Invertebr. Pathol., 15*, 306–312.

Gleich, J. G., Gilbert, F. F., & Kutscha, N. P. (1977). Nematodes in terrestrial gastropods from central Maine. *J. Wildlife Dis., 13*, 43–46.

Glen, D. M., Wilson, M. J., Brain, P. B., & Stroud, G. (2000). Feeding activity and survival of slugs, *Deroceras reticulatum*, exposed to the rhabditid nematode, *Phasmarhabditis hermaphrodita*: a model of dose response. *Biol. Control, 17*, 73–81.

Grewal, P. S., Grewal, S. K., Tan, L., & Adams, B. J. (2003). Parasitism of molluscs by nematodes: types of associations and evolutionary trends. *J. Nematol., 35*, 146–156.

Howlett, S. A., Burch, G., Sarathchandra, U., & Bell, N. L. (2009). A bioassay technique to assess the molluscicidal effects of microbes. *New Zeal. Plant Prot., 62*, 7–11.

Ivanova, E. S., & Wilson, M. J. (2009). Two new species of *Angiostoma* Dujardin, 1845 (Nematoda: Angiostomatidae) from British terrestrial molluscs. *Syst. Parasitol., 74*, 113–124.

Jones, A. A., & Selman, B. J. (1984). A possible biological control agent of the grey field slug (*Deroceras reticultum*). *Proceedings of the 1984 British Crop Protection Conference, 1*, 261–266.

Jones, A. A., & Selman, B. J. (1985). *Microsporidium novocastriensis* n.sp., a microsporidian parasite of the grey field slug, (*Deroceras reticulatum*). *J. Protozool., 32*, 581–586.

Joyce, S. A., Reid, A., Driver, F., & Curren, J. (1994). Application of polymerase chain reaction (PCR) methods to the identification of entomopathogenic nematodes. In A. M. Burnell,

R. U. Ehlers, & J. P. Masson (Eds.), *Genetics of Entomopathogenic Nematode—bacterium Complexes* (pp. 178—187). Luxemborg: European Commission Directorate-General XII.

Kassanis, B., Woods, R. D., & MacFarlane, I. (1984). Galactogen, a virus-like particle from slugs. *Ann. Appl. Biol., 105*, 587—589.

Kiontke, K., & Sudhaus, W. (2006). Ecology of *Caenorhabditis* species. In The C. elegans Research Community. (Ed.), *WormBook*. WormBook, doi/10.1895/wormbook.1.37.1. http://www.wormbook.org.

Mead, A. R. (1961). *The Giant African Snail—a Problem in Economic Malacology*. Chicago, IL: The University of Chicago Press, pp. 257.

Mengert, H. (1953). Nematoden und schneken. *Zeitschrift fur Morphologie und Okologie Tiere, 41*, 311—349.

Morand, S. (1988). *Contribution a l'Etude d'un Systeme Hotes—parasites: Nematodes Associés a Quelques Mollusques Terrestres*. These doctorate. France: Universite de Rennes, pp. 265.

Morand, S., Wilson, M. J., & Glen, D. M. (2004). Nematodes (Nematoda) parasitic in terrestrial gastropods. In G. M. Barker (Ed.), *Natural Enemies of Terrestrial Molluscs* (pp. 525—557). Wallingford, UK: CABI Publishing.

Pan, C. T. (1956). Studies on the biological control of schistosome-bearing snails: A preliminary report on pathogenic micro-organisms found in *Australorbis glabratas*. *J. Parasitol, 42*(Suppl. 4), 135—140.

Pham Van Luc, P., Spiridonov, S. E., & Wilson, M. J. (2005). *Aulacnema monodelphis* n.g. n.sp. and *Angiostoma coloaense* n. sp. (Nematoda: Rhabditida: Angiostomatidae) from terrestrial molluscs of Vietnam. *Syst. Parasitol, 60*, 91—97.

Rae, R., Verdun, C., Grewal, P. S., Robertson, J. F., & Wilson, M. J. (2007). Biological control of terrestrial molluscs using *Phasmarhabditis hermaphrodita*—progress and prospects. *Pest Manag. Sci., 63*, 1153—1164.

Rae, R. G., Robertson, J. F., & Wilson, M. J. (2008). Susceptibility and immune response of *Deroceras reticulatum, Milax gagates* and *Limax pseudoflavus* exposed to the slug parasitic nematode *Phasmarhabditis hermaphrodita*. *J. Invertebr. Pathol., 97*. 61—19.

Rae, R. G., Roberston, J. F., & Wilson, M. J. (2009). Chemoattraction and host preference of *Phasmarhabditis hermaphrodita*. *J. Parasitol, 95*, 517—526.

Rae, R. G., Tourna, M., & Wilson, M. J. (2010). The slug parasitic nematode *Phasmarhabditis hermaphrodita* associates with complex and variable bacterial assemblages that do not affect its virulence. *J. Invertebr. Pathol., 104*, 222—226.

Raut, S. K. (2004). Bacterial and non-microbial diseases in terrestrial gastropods. In G. M. Barker (Ed.), *Natural Enemies of Terrestrial Gastropods* (pp. 599—611). Wallingford, UK: CABI Publishing.

Raut, S. K., & Ghose, K. C. (1977). Outbreak of leucoderima-like disease in the giant land snail, *Achatina fulica* Bowdich from West Bengal. *Ind. J. Anim. Health*, 93—94, June 1977.

Raut, S. K., & Mandal, R. N. (1986). Disease in the pestiferous slug *Laevicaulis alte* (Gastropoda: Veronicellidae). *Malacol. Rev., 19*, 106.

Raut, S. K., & Panigrahi, A. (1989). Diseases of Indian pest slugs and snails. *J. Medic. Appl. Malacol, 1*, 113—121.

Ross, J. L., Ivanova, E. S., Severns, P. M., & Wilson, M. J. (2010a). The role of parasite release in invasion of the USA by European slugs. *Biol. Invasions, 12*, 603—610.

Ross, J. L., Ivanova, E. S., Spiridonov, S. E., Waeyenberge, L., Moens, M., Nicol, G. W., & Wilson, M. J. (2010b). Molecular phylogeny of slug-parasitic nematodes inferred from 18S rRNA gene sequences. *Mol. Phylogenet. Evol., 55*, 738—743.

Ross, J. L., Ivanoval, E. S., Sirgel, W. F., Malan, A. P., & Wilson, M. J. (2011). Diversity and distribution of nematodes associated with terrestrial slugs in the Western Cape Province of South Africa. *J. Helminthol.* In Press. DOI:10.1017/S0022149X11000277.

Selman, B. J., & Jones, A. A. (2004). Microsporidia (Microspora) parasitic in terrestrial gastropods. In G. M. Barker (Ed.), *Natural Enemies of Terrestrial Gastropods* (pp. 579—597). Wallingford, UK: CABI Publishing.

Shrewsbury, J. F. D., & Barson, G. J. (1947). A contribution to the study of the bacterial flora of *Arion ater*. *Proc. Soc. Appl. Bacteriologist, 2*, 70—76.

Singer, S., Van Fleet, A. L., Viel, J. J., & Genevese, E. E. (1997). Biological control of the zebra mussel *Dreissena polymorpha* and the snail *Biomphalaria glabrata*, using gramicidin S and D and molluscicidal strains of *Bacillus*. *J. Indust. Microb. Biotechnol., 18*, 226—231.

Solem, A. (1978). Classification of land mollusca. In V. Fretter, & J. Peake (Eds.), *Pulmonates, Vol. 2A* (pp. 49—97). New York, NY and London, UK: Academic Press.

Sproston, E. L., McRae, M., Ogden, I. D., Wilson, M. J., & Strachan, N. J. C. (2006). Slugs: potential novel vectors of *Escherichia coli* O157. *Appl. Environ. Microb., 72*, 144—149.

Sproston, E. L., Ogden, I. D., MacRae, M., Forbes, K. J., Dallas, J. F., Sheppard, S. K., Cody, A., Colles, F., Wilson, M. J., & Strachan, N. J. C. (2010). Multi-locus sequence types of *Campylobacter* carried by flies and slugs acquired from local ruminant faeces. *J. Appl. Microb., 109*, 829—838.

Stephenson, J. W., & Knutson, L. V. (1966). A resume of recent studies of invertebrates associated with slugs. *J. Econ. Entomol., 59*, 356—360.

Subbotin, S. A., Sturhan, D., Chizhov, V. N., Vovlas, N., & Baldwin, J. G. (2006). Phylogenetic analysis of Tylenchida Thorne, 1949 as inferred from D2 and D3 expansion fragments of the 28S rRNA gene sequences. *Nematol., 8*, 455—474.

Terytze, K., & Hoffman, G. (1986). Die Wirkung von Bakterienpraparaten (*Bacillus thungiensis* Berliner) zur bekampfung von nactschnecken in gerbera-bestanden. *Arch. Phytopathol. Pflanenschuts Berlin, 22*, 361—363.

Trip, M. R. (1961). Is *Bacillus pinotii* pathogenic in *Australorbis glabratus*? *J. Parasitol., 47*, 464.

Trevet, I. W., & Esslemont, J. M. (1938). A fungous parasite of the eggs of the gray field slug. *J. Quekett Microcsopical Club, 4th series, 1*, 1—3.

Van As, J. G., & Basson, L. (2004). Ciliophoran (Cilophora) parasites of terrestrial gastropods. In G. M. Barker (Ed.), *Natural Enemies of Terrestrial Gastropods* (pp. 559–578). Wallingford, UK: CABI Publishing.

Vandergast, A. G., & Roderick, G. K. (2003). Mermithid parasitism of Hawaiian *Tetragnatha* spiders in a fragmented landscape. *J. Invertebr. Pathol., 84*, 128–136.

Van Megen, H., Van Den Elsen, S., Holterman, M., Karssen, G., Mooyman, P., Bongers, T., Holovachov, O., Bakker, J., & Helder, J. (2009). A phylogenetic tree of nematodes based on about 1200 full-length small subunit ribosomal DNA sequences. *Nematol., 11*, 927–950.

Vrain, T. C., Wakarchuk, D. A., Levesque, A. C., & Hamilton, R. I. (1992). Intraspecific rDNA restriction fragment length polymorphism in the *Xiphinema americanum* group. *Fund. Appl. Nematol., 15*, 563–573.

Walker, A. J., Glen, D. M., & Shewry, P. R. (1999). Bacteria associated with the digestive system of the slug *Deroceras reticulatum* are not required for protein digestion. *Soil Biol. Biochem., 31*, 1387–1394.

Wilson, M. J., Glen, D. M., George, S. K., & Pearce, J. D. (1995a). Selection of a bacterium for the mass production of *Phasmarhabditis hermaphrodita* (Nematoda Rhabditidae) as a biocontrol agent for slugs. *Fund. Appl. Nematol., 18*, 419–425.

Wilson, M. J., Glen, D. M., Pearce, J. D., & Rodgers, P. B. (1995b). Monoxenic culture of the slug parasite, *Phasmarhabditis hermaphrodita* with different bacteria in liquid and solid phase. *Fund. Appl. Nematol, 18*, 159–166.

Wilson, M. J., Coyne, C. C., & Glen, D. M. (1998). Low temperatures suppress growth of the ciliate *Tetrahymena rostrata* and pathogenicity to field slugs (*Deroceras reticulatum*). *Biocontrol Sci. Technol., 8*, 181–184.

Wright, C. A. (1968). Some views on biological control of trematode disease. *Trans. Royal Soc. Trop. Med. Hyg., 62*, 320–324.

Yamada, G., Yonemoto, S., Matsumoto, H., Toda, N., & Ibusuki, O. (1960). Studies on slugs as vectors of pathogenic microbes. *J. Osoka City Med. Cent, 2*, 707–717.

## APPENDIX

### Preparation of kidney medium for growth of *Phasmarhabditis* spp. nematodes in liquid or solid culture

1. Prepare an aqueous suspension of 3.5% homogenized pig kidney, 2.5% yeast extract and 3% corn oil (% w/v) and autoclave at 121°C.
2. During the heating, proteins congeal. Filter out the precipitated lumps through muslin.
3. Dispense the desired volume into culture flasks, plug with bungs of non-absorbent cotton wool and re-autoclave at 121°C.
4. To make kidney agar, add technical grade agar at 2% (w/v) to the muslin filtered medium and re-autoclave. Pour the medium aseptically into Petri dishes using standard methods.

Incubate cultures at 15°C. Liquid culture flasks should be rotated at 200 rpm. Fluted or baffled flasks provide better aeration and nematode yields than standard conical flasks.

CHAPTER XIV

# Testing the pathogenicity and infectivity of entomopathogens to mammals

JOEL P. SIEGEL

USDA/ARS, Commodity Protection and Quality Unit, San Joaquin Valley Agricultural Sciences Center, Parlier, CA 93648, USA

## 1. INTRODUCTION

Although it may seem a matter of common sense that known vertebrate pathogens will not be used in agriculture or in vector control, there are regulatory mechanisms in place that require that potential insect control agents be evaluated for their safety to vertebrates. Furthermore, with advances in taxonomy and reclassification, organisms that once seemed far removed from vertebrate pathogens may now fall into the same genus or share genes associated with toxin production or virulence factors. Safety testing is the mechanism designed to produce the data necessary to address these concerns. While in most countries, safety testing is the province of specialized laboratories, in the long run an understanding of the philosophy behind these tests will facilitate communication between the investigator and the testing laboratory.

Mammalian safety screens are a subset of a larger grouping of tests that assess the effects of a microbial pest control agent (MPCA) on nontarget organisms (NTO) which include plants, fish, beneficial arthropods, birds, and mammals. There is a shared philosophy behind many of these protocols and the same principles used in evaluating the potential infectivity of an MPCA to mammals can be applied to other vertebrates. Ultimately, NTO testing can be viewed as attempts to manipulate a candidate organism into doing something it would not do in nature, either by providing access to hosts outside its natural range or by varying both the dose and route of exposure in order to produce infection and/or mortality. There is currently some divergence between the testing philosophies of the European Union (EU) and United States. The EU follows the precautionary principle, in which the burden of proof requires that the individual demonstrate that the MPCA is not harmful, although each member state may differ in the application of this principle, while the United States follows a different philosophy that may take both the cost and benefit into consideration. In practice, these philosophies tend to merge, although one can argue that the precautionary principle sets a higher standard. Because there is no incentive for a company to share its data and perhaps aid competitors, the majority of safety test data are proprietary and mammalian and NTO studies are rarely published in refereed journals. This can fuel a public perception that these agents are unsafe, despite the fact that hundreds of studies have been conducted. It is quite possible that the same safety question will be revisited repeatedly, at a cost of many animal lives, as well as time and money, but that is the nature of this system.

These tests are conducted with the expectation that the infectivity and the majority of the toxicity tests will be negative, because of the specificity of entomopathogens to arthropods. In the most favorable testing scenario, all of the mammalian safety tests produce

MANUAL OF TECHNIQUES IN INVERTEBRATE PATHOLOGY
ISBN 9780123868992

negative data and a relatively straightforward decision to proceed can be made. In practice, negative results may be a problem when the officials reviewing the data are used to assessing chemical studies and there is no endpoint (mortality). While it is always possible to kill an animal if the dose is high enough, through suffocation, embolism, or blockage of the gastrointestinal tract, these results have little relevance to the real world. If mortality occurs after an invasive route of administration, its significance must be judged in the context of the experimental protocol-dose, route of administration, etc. An adverse outcome in any one screen may not be grounds for automatic rejection but further tests will probably be conducted to quantify the effect. The goal of this chapter is to briefly summarize the history and philosophy behind these tests and then present a series of single-exposure, short-term tests that can be used to assess the mammalian safety of a candidate organism. The tests presented are based on the registration requirements of the United States Environmental Protection Agency, subdivision M (Anonymous, 1988) and guidelines published by the World Health Organization (WHO) (Anonymous, 1981). These screens may serve as a starting point for the registration of an entomopathogen or a point of comparison for an existing protocol.

In 1993 the WHO reviewed the testing protocols of the United States, Canada, and European Economic Community (now the EU), with the goal of drafting guidelines for the registration of MPCAs. There were no final guidelines produced nor is there a single set of guidelines used worldwide. That does not mean that there is a shortage of published protocols. Summaries of the registration requirements for North America (Betz *et al.*, 1990), the EU (Quinlan, 1990), the former USSR and Eastern Europe (Kandybin & Smirnov, 1990), and Japan (Aizawai, 1990) have been published, as well as detailed explanations of the United States Guidelines (Briggs & Sands, 1992; Campbell & Sands, 1992; Fisher & Briggs, 1992; Kerwin, 1992; Siegel & Shadduck, 1992; Spacie, 1992). *Bacillus thuringiensis* Berliner is undoubtedly the single most tested MPCA and detailed summaries of the numerous safety tests conducted have been summarized by the International Programme on Chemical Safety Environmental Health Criteria 217, 1999; Glare & O'Callaghan, 2000; Lacey & Siegel, 2000; Siegel, 2001; Competent Authority's Report *B. thuringiensis* subsp. *israelensis* Serotype H-14 strain AM65-52 (Anonymous, 2008). These publications are an excellent starting point but are no substitute for the regulations of the country or countries in which the MPCA will be registered.

## 2. DEFINITION OF INFECTION

The major characteristic that differentiates MPCAs from chemical toxicants is their ability to multiply and cause infection. However, determining infectivity is complicated by the fact that the term 'infection' may not be defined in the testing guidelines although infection has several definitions. The definition chosen may affect the interpretation of the test data. There are two schools of thought concerning the definition of infection, and each definition has a different implication for the relationship between microorganism and host. One school defines infection as simple colonization by microorganisms that may or may not be detrimental to the host. Colonization occurs shortly after birth and the microorganisms are restricted or compartmentalized by the host to areas where they are tolerated, such as the gastrointestinal or upper respiratory tracts. Infectious disease is then a special condition that arises when host defenses are breached and microorganisms are introduced into areas where they can not be tolerated (Siegel & Shadduck, 1992). The term 'pathogenicity' denotes the intrinsic capability of a microorganism to penetrate host defenses and cause disease, and the term 'virulence' refers to the speed by which this is accomplished (Davis *et al.*, 1973). Shapiro-Ilan *et al.* (2005) define virulence as the degree of pathogenicity within a group or species. The second school of thought links infection directly to disease (disease is a host response). Living agents cause a disruption to the host, either by multiplication in tissue, toxin production, or both. In my opinion, this second definition, which links the presence of a microorganism to tissue damage, is the more useful one for safety testing.

Using this second definition, simple recovery of an MPCA is not indicative of infection. Safety tests by design introduce MPCAs into mammals by a variety of routes and transient disturbances in the normal flora should be expected. Assuming that host immune response is constant, recovery of a portion of an inoculum from host tissue, as well as redistribution of the inoculum among host tissues, may occur over a variable length of time, depending on both the route of administration and the magnitude of the dose. Under these circumstances, infection is demonstrated when there is evidence of multiplication of an MPCA (recovery of an amount greater than injected, recovery of vegetative forms when spores are injected, failure to clear over time) coupled with tissue damage. This coupling is essential because the occurrence of tissue damage alone, or even death, can be caused by the physical nature of the inoculum and/or the host response to foreign protein (Barford *et al.*, 2010). Proper controls, such as inactivated inoculum, are useful in infectivity studies in order to determine if an adverse outcome is due to the presence of foreign protein.

## 3. DEFINITION OF PERSISTENCE

In this chapter the term 'persistence' will be used to describe the ability of an MPCA to remain viable in mammalian tissue without multiplying. Persistence occurs because clearance of an MPCA from a test animal is not instantaneous. Adlersberg *et al.* (1969) reported that intravenously injected radioiodinated latex particles were recovered from mice as long as six months after administration. The particles were redistributed between the lungs, spleen, and liver during this period and the proportion of the inoculum recovered from these organs changed over time. When environmentally resistant life stages of MPCAs such as spores or oospores are injected, they are redistributed in tissue and recovered for a variable amount of time, depending on dose and the route of injection. Viable spores of *B. thuringiensis* subsp. *israelensis* have been recovered as long as 70 days after injection and oospores of *Lagenidium giganteum* have been observed in spleen tissue four weeks after intraperitoneal injection (Siegel & Shadduck, 1990; Siegel & Shadduck, 1992; Jensen *et al.*, 2002; Mancebo *et al.*, 2011). Consequently, it is essential to distinguish between infection and persistence. Simple recovery of an MPCA from a mammal cannot automatically be construed as infection, although the significance of prolonged persistence must be addressed on a case-by-case basis.

## 4. MAMMALIAN SAFETY TESTS AND PHILOSOPHY

Initially, MPCAs underwent the entire battery of tests required for chemical toxicants, including long-term carcinogenicity studies (e.g,. a 2-year rat dietary study for *Heliothis* nuclear polyhedrosis virus and *B. thuringiensis* as well as a 23-month hen feeding study for *B. thuringiensis*) (Heimpel, 1971; Ignoffo, 1973; Burges, 1981). MPCAs also underwent infectivity studies that lasted as long as 10 weeks. This approach not only placed MPCAs at a competitive disadvantage compared to chemicals because of the additional testing required, but failed to address the fundamental differences between MPCAs and chemical toxicants. Shadduck (1983) noted that four assumptions underlay standard chemical pesticide safety tests and that these assumptions were not applicable to MPCAs. First, chemical safety tests assume that a measureable biological effect can be obtained if the dose is high enough, while it may be impossible to achieve an adverse effect for an MPCA delivered by conventional means such as diet or as an aerosol. Second, chemical tests assume that the material administered is metabolized, excreted, or both, and that the toxic effects can be predicted if one knows the routes by which the primary metabolites are eliminated. In contrast there is no evidence that MPCAs are activated by metabolism or genetically altered by passage through mammals. Third, chemical tests assume that a chemical may accumulate and exert its effects over a long period of time; the amount of material stored in body tissues increases with prolonged exposure. Both acute and long-term studies are therefore necessary to assess this effect. In contrast, there is no evidence to date that MPCAs colonize mammals or have teratogenic or carcinogenic effects, making long-term dietary studies unnecessary. Fourth and finally, chemical safety testing assumes that knowledge of the chemical structure of a toxicant enables one to make educated guesses about the toxicity and/or carcinogenicity of other molecules belonging to the same family. This can not be assumed about MPCAs; despite the close genetic relationship between *B. thuringiensis* and *B. anthracis*, *B. thuringiensis* is not a mammalian pathogen.

One philosophy of safety testing known as 'maximum challenge' arose that recognized the unique characteristics of MPCAs. This approach advocated the use of invasive exposure routes (intracerebral, intraocular, intraperitoneal injection) targeting vulnerable organ systems in short-term tests (2- to 3-week observation period) in order to achieve a measurable biological effect without the use of massive doses of an MPCA. This approach is parallel to the $LD_{50}$ concept, except that the dose is held constant and the route of exposure is varied. In a maximum challenge test, the highest dose possible is given by the route that most severely compromises an animal's natural defenses (Shadduck, 1983). Selection of target sites is based in part on a literature review as well as knowledge of vulnerable mammalian sites such as the central nervous system. This approach is not used today although elements of this philosophy, such as intravenous injection, are incorporated in numerous test protocols. The use of immunocompromised animals is also part of this maximum challenge philosophy. The rationale for this approach is that the ability of certain MPCAs to persist in mammalian tissue (bacterial spores) raises concerns about their fate in immunocompromised humans. If a researcher demonstrates that multiplication of an MPCA is blocked in immunocompromised animals, it is further evidence that the MPCA is not a mammalian pathogen.

Maximum challenge tests are controversial because of the difficulty of placing mortality data in the proper context for assessing human safety. The greatest value of an intracerebral injection study is that it is clearly a 'worse case' scenario and a negative result provides strong evidence of an MPCA's safety. Conversely, if mortality occurs it is not evidence that a candidate organism is unsuitable because even nonpathogenic organisms administered by this route can kill. Likewise, although the failure of an MPCA to infect an immunocompromised animal provides convincing evidence of safety, these tests have been challenged by researchers who argued that immunocompromised humans would succumb to opportunistic infections from a wide range of microorganisms before MPCAs could cause infection (Burges, 1981). Additionally, the various methods of immune suppression combined with the numerous laboratory strains of test animal makes comparison between studies difficult. An immunosuppressant such as cyclophosphamide has a greater effect on B-lymphocytes whereas glucocorticosteroids have a greater effect on T-lymphocytes (Dumont, 1974; Parillo & Fauci, 1979). The use of immunocompromised animals is not required in the United States, Canadian, or European Union guidelines, although this question arises periodically.

In 1981, the World Health Organization (WHO) proposed a multi-tier testing strategy for evaluating MPCAs (Anonymous, 1981). The first tier evaluated both unformulated and formulated product. The unformulated MPCA was evaluated by acute oral, inhalation, and intraperitoneal administration, combined with *in vivo* mutagenicity screens, and formulated MPCAs were evaluated by dermal, ocular, and allergenicity tests. The maximum observation period was 28 days and the endpoints of interest were persistence, infection, and irritation. Infectivity was grounds for immediate rejection, and questions concerning persistence were evaluated in the second tier, with observation periods as long as 90 days. The third-tier tests included conventional 2-year feeding studies and teratogenicity testing. This multi-tier format was adopted by the United States and elements of these protocols can be found in the testing requirements of many countries. It is noteworthy that the WHO protocol tested both formulated and unformulated product; numerous countries adopted this approach. Some have questioned the use of formulated product, because it diluted the concentration of MPCA, and there is no evidence that formulated products clear differently than unformulated MPCAs. Furthermore, formulated products are not part of the chemical testing protocols. However, since in recent years much of the public opposition to the use of MPCAs has centered on the formulation ingredients, it is prudent to include formulated product in the tests.

Currently, there is no single standard for the test dose, therefore doses vary between countries and among published studies. The WHO guidelines recommended as a basis for the oral exposure, the amount that a 70-kg man would receive from a one-hectare dose, although it could not exceed 5 g/kg body weight. Other guidelines such as the one for the United States required a minimum dose of $10^8$ organisms per mouse or rat. The intraperitoneal guidelines in the WHO protocols recognized the weight difference between mice and rats and suggested $10^6$ organisms per mouse and $10^7$ organisms per rat. No doses are specified for inhalation, ocular, and dermal exposure in the WHO guidelines while the United States guidelines required minimum doses of $10^8$ units of MPCA, $10^7$ units of MPCA, and 2 g dry weight, respectively. In recently published studies of subchronic exposure, mice were given $10^6$ colony forming units (cfu) of *B. thuringiensis* and rats as many as $10^8$ cfu by intranasal instillation (Barford *et al.*, 2010; Mancebo, 2011).

## 5. ANIMAL PROCUREMENT AND HOUSING

Animals must be procured from reputable suppliers to ensure homogeneity and freedom from disease. There is some debate about whether to use outbred or inbred lines; inbred lines are preferred when homogeneity is emphasized but outbred lines may more closely reflect the variation found in nature. Within an experiment, the animals used for the treatments should be as homogeneous as possible in terms of sex, weight and age, and housed and handled in a manner that minimizes stress and in accordance with government regulations (Anonymous, 1985; McGregor, 1986). If immunocompromised animals are used, increased care in housing and handling is needed to reduce the risk of infection by other microorganisms. In summary, the use of reputable animal suppliers, proper diet and handling will help ensure the uniformity of the test animals and reduce the likelihood of aberrant results.

Housing rodents is a special concern. In general, females are less likely to fight than males, and littermates housed together are less likely to fight. Male mice require extra attention to make sure that fighting does not occur because the individuals who are at the lowest level of the social hierarchy undergo the most stress. Stressed individuals may react abnormally when exposed to the MPCA. If an animal dies, it must be removed promptly because its cagemates may eat the cadaver and destroy tissues of interest. Housing control animals poses its own challenge to ensure that they do not become exposed to the MPCA by copraphagous activity or grooming. Using separate cages and minimizing dust exposure is the simplest solution.

## 6. MODIFIED TIER SCHEME

In the remainder of the chapter, four acute (single dose, 28-day observation period) tests, designed to assess unformulated MPCAs, are presented in detail. Toxicity in these tests may be caused by the constituents of the MPCA or by its metabolic by-products. If toxicity occurs without infectivity, conventional chemical safety tests can then be used to quantify the effect. In these tests infectivity is indicated by the failure of the inoculum to clear or by tissue damage after administration. Although it may take an inoculum more than 28 days to clear, clearance curves can be generated in this four-week time period to determine whether the MPCA is multiplying. Prior to testing, a sufficient amount of the MPCA must be produced so that a single lot is used in all tests. This same lot will form the 'mother culture' for subsequent production of the agent. If an MPCA is produced *in vivo*, extra consideration must be given to the method by which batches are standardized. Extra characterization of insect fragments or other microorganisms present in the test material may be necessary in order to register the MPCA.

### A. Acute oral administration test: rat

The purpose of this test is to determine whether a single dose of $10^8$ units of the MPCA, suspended in an appropriate vehicle and administered with a curved feeding needle (volume not to exceed 2 ml/100 g body weight) produces mortality. The observation period is 21 days. Two groups of animals must be used, consisting of a group administered the MPCA and a control group administered autoclaved MPCA. A third group of rats administered the carrier may also be included. Exposure to autoclaved MPCA is necessary to identify lesions and/or mortality arising from the presence of heat-stable toxins and/or protein. Each group should include at least 10 young male and 10 young female rats. The feeding needle should then be slowly inserted into the rat's mouth, either over the tongue or to one side. The needle should then be slowly inserted until approximately 10 cm is in the esophagus. At this point the contents should be slowly expelled. If any resistance is noted during insertion, the needle should be immediately withdrawn and then reinserted. The rat must be carefully observed for any sign of respiratory distress during this procedure (Paget & Thompson, 1979).

The rats should be weighed one day before administration of the inoculum and then denied food and water 16 h before exposure. After administration of the MPCA, food and water should be withheld for an additional 3–4 h. All animals should be observed daily following administration of the MPCA and the observation should include the condition of the skin, fur, eye, and mucous membranes, as well as respiration, behavior, and somatomotor activity. Findings of interest include tremors, convulsions, diarrhea, lethargy, excessive salivation and coma. Transient effects such as diarrhea, or loss of appetite may occur and should be noted, but are not grounds for rejection. The animals must be weighed weekly and weight loss or gain determined for each rat. If practical, feces should be collected weekly and cultured to determine if viable MPCA is shed.

At the end of the experiment all rats are killed and a gross necropsy conducted (Paget & Thompson, 1979). At a minimum, the spleen, liver, mesenteric lymph node, stomach, and portions of the intestine should be examined for the presence of lesions. If lesions are observed, samples are collected and placed in buffered neutral 10% formalin (3.7% formaldehyde) for histopathology. When possible, the MPCA should be quantified from the kidney, liver, lung, and spleen by plating samples on appropriate media. Any animals that die during the experiment must be promptly examined. If substantial mortality occurs during the first three days following administration of the MPCA, half the remaining animals in each group should be killed and necropsied 1 week after administration of the candidate organism.

An MPCA that can not clear this test is not a suitable candidate. If toxin production is suspected as the cause of mortality rather than infection, the toxic components must be isolated and characterized. They can then be tested using the chemical safety testing protocols. If the MPCA is recovered from tissue three weeks after administration, additional testing is necessary to determine the rate of clearance. Regulatory agencies may require much longer studies to quantify this effect. Other outcomes of concern in this test are weight change and illness. Given the sample size of 40 animals and the likelihood that the deleterious effect of the MPCA is large, this test has sufficient power to detect a true difference between the treatment and control group 96% of the time (Cohen, 1988).

### B. Acute abraded dermal toxicity test: rabbit

The purpose of this test is to determine whether a single dermal exposure to 2 g of the MPCA over 24 h causes infection or mortality over a 14-day period. Dermal toxicity testing is a standard toxicology protocol but I suggest modifying this test by applying the MPCA to abraded skin. Since dermal toxicity is unlikely, this test instead evaluates the role played by intact skin in preventing infection by the MPCA.

Ten young rabbits, five males and five females, are used in this test. No controls are necessary unless the characteristics of the MPCA are completely unknown. Fur should be clipped, about 10% of the body surface (as close as possible to the skin without irritating it) from the dorsal and ventral sides of the rabbit 24 h before the test. The abrasion should penetrate the stratum corneum but not the dermis. The inoculum should be evenly applied over the shaved area in a liquid or cream carrier, and then held in contact with the rabbit using porous gauze and nonirritating tape. The rabbit should be held in a manner that does not allow it to eat the inoculum, but the rabbit should not be completely immobilized. At the end of the exposure period, any residual MPCA should be removed with water.

The rabbits should be weighed 1 day before administration and then weekly. All rabbits should be observed daily following administration of the MPCA and the observation should include the condition of the skin, fur, eye, and mucous membranes, as well as respiration, behavior, and somatomotor activity. Findings of interest include tremors, convulsions, diarrhea, lethargy, excessive salivation, and coma. Particular attention should be paid to skin lesions and irritancy (irritancy can be evaluated using the protocols of McCreesh & Steinberg, 1983), weight loss, behavioral abnormalities, and death. Gross necropsies must be performed on all animals that die during the course of the experiment and samples of skin ($6 \times 6$ cm) should be collected for histology. The skin is fixed in buffered neutral 10% formalin (3.7% formaldehyde). If there is irritancy or mortality, the test should be repeated using autoclaved inoculum. This test is not invasive and an MPCA that produces deleterious effects when administered by this route in not a suitable candidate.

### C. Acute pulmonary infectivity test: rat

The purpose of this test is to determine whether a single exposure to $10^8$ units of an MPCA, instilled intranasally or intratracheally (Nicholson & Kincaid, 1982), suspended in an appropriate vehicle such as sterile distilled water (not to exceed 0.3 ml/100 g body weight) clears from the lungs over 21 days. Infection in this study is demonstrated by failure to clear and/or mortality. Mortality may also occur in this study due to mechanical obstruction of the airways or the occurrence of foreign body pneumonia. A trial study should first be conducted in order to determine whether the dose and/or method of administration would cause mortality (four animals should be used). If mortality is excessive, the dose should be reduced by at least one log and the reason for this reduction documented.

Two groups of animals are used in this test, consisting of an experimental group administered the MPCA and a control group administered the autoclaved MPCA. The control group will enable the investigator to determine whether any lesions observed in lung tissue are due to the presence of foreign protein or the introduction into the lungs during instillation of the normal microbial flora present in the pharynx. The treated group should consist of at least 10 young male and 10 young female rats and the control group should consist of five young male and five young female rats. All rats should be weighed 1 day before administration of the MPCA and at weekly intervals thereafter, and weight change calculated. In many rat strains, female weight does not exceed 250 g (males may weigh 600 g) and this sex related difference in weight gain should be considered when evaluating the results.

The majority of clearance occurs during the first two weeks after administration of the inoculum; therefore, most of the sampling should occur during this period in order to accurately determine the rate of clearance. Samples should be taken 1 h, 1 day, days 2–6, day 7, day 14, and day 21 after administration. Gross necropsies should be performed on all rats. The lungs, liver and spleen should be removed, weighed, and a portion retained for enumerating the MPCA. At a minimum, the MPCA should be enumerated from the lungs by homogenizing a weighed tissue sample in sterile distilled water (1 : 9, weight to volume), serially diluting the homogenate, and then plating the suspension on to suitable growth media. The number of colony forming units (cfu) per gram of tissue can be calculated after incubation. Additional samples of tissues from the lungs, spleen and liver should be saved for histology and fixed in buffered neutral 10% formalin (3.7% formaldehyde). The control rats should be killed three weeks after administration of the MPCA. Gross necropsies should be performed on the rats and tissue from the lungs of the control group should be collected for histology. These samples are used to determine the background level of lesions in the lungs from administration of the inoculum or exposure to other irritants.

When mortality does not occur following administration of the MPCA, the endpoint of interest is clearance from the lungs. This can be calculated by linear regression. Regression can also be used to determine the clearance rate from liver and spleen. Failure to clear from the lungs should be regarded as evidence of infection and is grounds for rejection of the candidate MPCA.

### D. Acute intraperitoneal infectivity test: outbred female mice

This study evaluates the ability of an inoculum consisting of $10^7$ units of the MPCA, suspended in 0.1 ml of carrier, to clear from the spleen over a 28-day period. The inoculum is injected using a small needle (26 or 30 gauge), but in some cases a wider bore may be necessary if the inoculum clogs the needle. The mouse should be held so that the ventral surface is facing the person performing the injection. The needle should be introduced rapidly into a point slightly left or right of the midline, and halfway between the pubic syphilis and the xiphisternum, of the ventral surface of the mouse. The syringe contents are expelled by firmly depressing the plunger (Paget & Thompson, 1979). It is important to avoid injecting the mammary glands.

The MPCA is enumerated from the spleen because this organ is more efficient by weight than the liver at filtering particles and microorganisms from the blood. The spleen is collected aseptically, weighed and then homogenized (1 : 9 weight to volume) in sterile distilled water to liberate the MPCA. Homogenate is then collected, serially diluted, and the cfu/g spleen calculated. Based on past experience, most of the inoculum is collected during the first 2 weeks. Mice are followed for 4 weeks in order to quantify the fluctuations in recovery of the MPCA that may result from liberation of the MPCA from extrasplenic sites and subsequent filtration by the spleen. If mortality occurs in this test, the toxic factors should be identified when possible. It may be necessary to inject autoclaved MPCA in order to determine if the observed toxicity is due to heat-labile compounds.

In this clearance experiment, a total of 33 mice of the same age and sex are injected and three mice are killed on days 1, 2, 4, 6, 8, 10, 14, 16, 20, 24, and 28 after exposure. When applicable, heart blood should be collected from the mice killed and cultured for the MPCA. Additional mice may be injected in order to ensure that there are three spleens available for each collection time point. The spleens are then collected and homogenized as previously described. The clearance rate of the MPCA from the spleen on a unit per gram basis can be calculated using regression. It is likely that some inoculum may be recovered from the spleen 28 days after injection, but this should not necessarily be interpreted as evidence of infection, provided that recovery is decreasing with time. However, if there is no clearance, infection may have occurred and a follow-up study assessing clearance over 90 days is necessary. In this follow-up study, the same numbers of mice are used, but the collection interval is stretched over a longer period of time. If the MPCA is present in the heart blood at the end of the study, the MPCA is still circulating in the blood and multiplication may have occurred. The regulatory authority may then require follow-up studies.

Intravenous and intraperitoneal injections are the most invasive routes of administration in the first tier studies of the United States. These tests evaluate the likelihood of infection by an MPCA when the skin is bypassed as a barrier. Intravenous injection is favored in the United States guidelines and intraperitoneal injection is reserved for an MPCA that has a large particle size that may cause an embolism. In contrast, in the WHO safety guidelines intraperitoneal injection is the preferred route of exposure because it is more challenging. In the relatively anoxic environment of the abdominal cavity, an MPCA may produce toxins that would not ordinarily be expressed in the more oxygenated bloodstream. The bulk of an intravenous inoculum is rapidly filtered by the spleen and liver within 4 h (Adlersberg *et al.*, 1969) while an intraperitoneal inoculum takes longer to clear because it must first pass through the lymphatic system before it enters the bloodstream. This delay provides an additional opportunity for the MPCA to multiply. In my opinion the more conservative route of administration is preferable.

## 7. CONCLUSION

The basic challenge of safety testing is determining when an MPCA is reasonably safe. The acute tests outlined in this chapter should help determine the hazard to mammals posed by an MPCA. In theory, one can always require an ever-expanding series of invasive tests that do not address the biology of the MPCA or likely routes of human exposure, but ultimately it will be almost impossible to evaluate the significance of the data. Unrealistic standards will drive MPCAs out of the marketplace. Burges (1981) succinctly states the challenge of evaluating an MPCA when he stated that a no-risk situation does not exist, certainly not with chemical pesticides, and even with biological agents, one can not absolutely prove a negative. Registration of a chemical is essentially a statement of usage in which risks are acceptable, and the same must be applied to biological agents.

The questions then shifts from hazard, to how much risk is acceptable? The answer to this question lies with the regulatory authority, although ultimately it is decided in both the scientific and political arenas. Ironically, although there is a vast quantity of safety data in government archives, since few of these studies are publicly available, this seeming lack of data can fuel public concern. Furthermore, even when the data for an MPCA are accepted, opponents of its use focus on the ingredients present in the proprietary formulation. It is important to acknowledge that safety testing is one step or perhaps a series of steps on a road that will ultimately lead to registration and perhaps public acceptance. What I hope to have communicated in this chapter, in addition to the protocols provided, is the fact that safety testing requirements are not static and will inevitably change in the future due to inputs from science, government, and the general public.

## ACKNOWLEDGMENTS

I am indebted to John A. Shadduck and Denis Burges and thank them for the many hours of conversation that we have had on this topic. I thank the World Health Organization Special Program for Research and Training in Tropical Diseases and the International Programe for Chemical Safety for laying the foundation for our present mammalian safety testing protocols.

## REFERENCES

Adlersberg, L., Singer, J. M., & Ende, E. (1969). Redistribution and elimination of intravenously injected latex particles in mice. *J. Reticuloendothel. Soc., 6*, 536–560.

Aizawai, K. (1990). Registration requirements and safety considerations for Microbial Pest Control Agents in Japan. In M. Laird, L. Lacey, & E. Davidson (Eds.), *Safety of Microbial Insecticides* (pp. 31–42). Boca Raton, FL: CRC Press.

Anonymous. (1981). Mammalian safety or microbial control agents for vector control: a WHO Memorandum. *Bull. Wld. Hlth. Org., 59*, 857–863.

Anonymous. (1985). *Guide for the Care and Use of Laboratory Animals*. US Department of Health and Human Services, NIH publication no. 86–23, revised 1985. United States National Institute of Health Bethesda, MD, pp. 83.

Anonymous. (1988). Toxicology guidelines for microbial pest control agents, subdivision M. United States Environmental Protection Agency. *Office of Pesticide and Toxic Substances*, pp. 303.

Anonymous. (2008). *Competent Authority's Report:* Bacillus thuringiensis *subsp.* israelensis *Serotype H-14 Strain AM65–52. Dossier According to Directive 98/8/EC*. Italy: Rapporteur Member State. ec.europa.eu/food/plant/protection/evaluation/existactive/AM65-52.pdf.

Barford, K. K., Poulsen, S. S., Hammer, M., & Larsen, S. T. (2010). Sub-chronic lung inflammation after airway exposures to *Bacillus thuringiensis* biopesticides in mice. *BMC Microbiol., 10*, 233.

Betz, F. S., Forsyth, S. F., & Stuart, W. E. (1990). Registration requirements and safety considerations for Microbial Pest Control Agents in North America. In M. Laird, L. Lacey, & E. Davidson (Eds.), *Safety of Microbial Insecticides* (pp. 3–10). Boca Raton, FL: CRC Press.

Briggs, J. D., & Sands, D. C. (1992). Overview: The effects of microbial pest control agents on nontarget organisms. In M. A. Levin, R. J. Seidler, & M. Rogul (Eds.), *Microbial Ecology: Principles, Methods and Applications* (pp. 685–688). New York, NY: McGraw–Hill.

Burges, H. D. (1981). Safety, safety testing, and quality control of microbial pesticides. In H. D. Burges (Ed.), *Microbial Control of Pests and Plant Diseases, 1970–1980* (pp. 738–769). New York, NY: Academic Press.

Campbell, C. L., & Sands, D. C. (1992). Testing the effects of microbial agents on plants. In M. A. Levin, R. J. Seidler, & M. Rogul (Eds.), *Microbial Ecology: Principles, Methods and Applications* (pp. 689–705). New York, NY: McGraw–Hill.

Cohen, J. (1988). *Statistical Power Analysis for the Behavioral Sciences* (2nd edn.) Hillside, NJ: Lawrence Erlbaum Associates. pp. 567

Davis, B. D., Dulbecco, R., Eisen, H. N., Ginsberg, H. S., & Wood, W. B. (1973). *Microbiology* (2nd ed.). New York, NY: Harper & Row.

Dumont, F. (1974). Destruction and regeneration of lymphocyte populations in the mouse spleen after cyclophosphamide treatment. *Int. Arch. Allergy, 47*, 110–123.

Fisher, S. W., & Briggs, J. D. (1992). Testing of microbial pest control agents in nontarget insects and acari. In M. A. Levin, R. J. Seidler, & M. Rogul (Eds.), *Microbial Ecology: Principles, Methods and Applications* (pp. 761–777). New York, NY: McGraw–Hill.

Glare, T. R., & O'Callaghan, M. (2000). *Bacillus thuringiensis: Biology, Ecology and Safety*. New Jersey: John Wiley & Sons, Ltd.

Heimpel, A. M. (1971). Safety of insect pathogens for man and vertebrates. In H. D. Burges, & N. W. Hussey (Eds.), *Control of Insects and Mites* (pp. 469–489). New York, NY: Academic Press.

Ignoffo, C. M. (1973). Effects of entomopathogens on vertebrates. *Ann. New York Acad. Sci., 217*, 141–164.

Jensen, G. B., Larsen, P., Jacobsen, B. L., Madsen, B., Wilcks, A., Smidt, L., & Andrup, L. (2002). Isolation and characterization of *Bacillus cereus*-like bacteria from faecal samples from greenhouse workers who are using *Bacillus thuringiensis*-based insecticides. *Int. Arch. Occup. Environ. Health, 75*, 191–196.

Kandybin, N. V., & Smironov, O. V. (1990). Registration requirements and safety considerations for microbial pest control agents in the USSR and adjacent Eastern European countries. In M. Laird, L. Lacey, & E. Davidson (Eds.), *Safety of Microbial Insecticides* (pp. 19–30). Boca Raton, FL: CRC Press.

Kerwin, J. L. (1992). Testing the effects of microorganisms on birds. In M. A. Levin, R. J. Seidler, & M. Rogul (Eds.), *Microbial Ecology: Principles, Methods and Applications* (pp. 729–744). New York, NY: McGraw–Hill.

Lacey, L. A., & Siegel, J. P. (2000). Safety and ecotoxicology of entomopathogenic bacteria. In J.-F. Charles, A. Delécluse, & C. Nielsen-LeRoux (Eds.), *Entomopathogenic Bacteria: From Laboratory to Field Application* (pp. 253–273). Dordrecht, The Netherlands: Kluwer Academic Publishers.

Mancebo, A., Molier, T., Gonzalez, B., Lugo, S., Riera, L., Artega, M. E., Bada, A. M., Gonzalez, Y., Pupo, M., Hernandez, Y., Gonzalez, C., Rojas, N. M., & Rodriguez, G. (2011). Acute oral, pulmonary and intravenous toxicity/pathogenicity of a new formulation of *Bacillus thuringiensis* var. *israelensis* SH-14 in rats. *Regul. Toxicol. Pharmacol., 59*, 184–190.

McCreesh, A. H., & Steinberg, M. (1983). Skin irritation testing in animals. In F. N. Marzuli, & H. I. Maibach (Eds.), *Dermatoxicity* (2nd ed.). (pp. 147–166) New York, NY: Hemisphere Publishing.

McGregor, D. (1986). Ethics of animal experimentation. *Drug Metab. Rev., 17*, 349–361.

Nicholson, J. W., & Kincaid, E. R. (1982). A simple device for intratracheal injections in rats. *Lab. Anim. Sci., 32*, 509–510.

Paget, G. E., & Thomson, R. (1979). *Standard Operating Procedures in Pathology*. Baltimore, MD: University Park Press.

Parillo, J. E., & Fauci, A. S. (1979). Mechanisms of unformulated action on immune processes. *Annu. Rev. Pharmacol. Toxicol., 19*, 179–201.

Quinlan, R. J. (1990). Registration requirements and safety considerations for microbial pest control agents in the European Economic Community. In M. Laird, L. Lacey, & E. Davidson (Eds.), *Safety of Microbial Insecticides* (pp. 11–18). Boca Raton, FL: CRC Press.

Shadduck, J. A. (1983). Some considerations on the safety evaluation of nonviral microbial pesticides. *Bull. WHO, 61*, 117–128.

Shapiro-Ilan, D. I., Fuxa, J. R., Lacey, L. A., Onstad, D. W., & Kaya, H. K. (2005). Definitions of pathogenicity and virulence in invertebrate pathology. *J. Invertebr. Pathol., 88*, 1–7.

Siegel, J. P. (2001). The mammalian safety of *Bacillus thuringiensis*-based insecticides. *J. Invertebr. Pathol, 77*, 13–21.

Siegel, J. P., & Shadduck, J. A. (1990). Safety of microbial insecticides to vertebrates—humans. In M. Laird, L. Lacey, & E. Davidson (Eds.), *Safety of Microbial Insecticides* (pp. 102–112). Boca Raton, FL: CRC Press.

Siegel, J. P., & Shadduck, J. A. (1992). Testing the effects of microbial pest control agents on mammals. In M. A. Levin, R. J. Seidler, & M. Rogul (Eds.), *Microbial Ecology: Principles, Methods and Applications* (pp. 745–759). New York, NY: McGraw–Hill.

Spacie, A. (1992). Testing the effects of microbial agents on fish and crustaceans. In M. A. Levin, R. J. Seidler, & M. Rogul (Eds.), *Microbial Ecology: Principles, Methods and Applications* (pp. 707–728). New York, NY: McGraw–Hill.

CHAPTER XV

# Complementary techniques: preparations of entomopathogens and diseased specimens for more detailed study using microscopy

JAMES J. BECNEL

Center for Medical, Agricultural, and Veterinary Entomology, United States Department of Agriculture, Agricultural Research Service, 1600 S.W. 23rd Drive, Gainesville, FL 32608, USA

## 1. INTRODUCTION

The science of insect pathology encompasses a diverse assemblage of pathogens from a large and varied group of hosts. Microscopy techniques and protocols for these organisms are complex and varied and often require modifications and adaptations of standard procedures. The objective of this chapter is to provide the researcher with some of the basic techniques and protocols used to study insect pathogens realizing that the guidelines must be tailored for specific needs. Many specialized protocols have been developed and for an extensive review of the literature on techniques for light and electron microscopy refer to Adams & Bonami (1991). Recommended texts on general histological techniques for light microscopy are by Luna (1960) and Barbosa (1974) with specific protocols for the diagnosis of insect diseases found in Thomas (1974) and Poinar & Thomas (1984); see also Chapter I. Information on the cytology of insects can be found in Smith (1968) and King & Akai (1982), and general cytology in Celis (2006). General procedures for electron microscopy can be found in Glauert (1974), Aldrich & Todd (1986), and Hyatt (1986).

## 2. LIGHT MICROSCOPY

The first evidence for the presence of a pathogen is often observed with either a stereoscopic or compound microscope. These observations are often crucial for making the initial diagnosis which leads to the specific approach required depending upon the type of pathogen found. Specific protocols for identification and preparation of specimens for the various types of pathogens are detailed in the previous chapters. This chapter deals with some general and specialized techniques for light microscopy and detailed procedures for the preparation and analysis of diseased specimens for electron microscopy.

There are a number of general protocols for the handling of entomopathogens and diseased individuals regardless of the host or pathogen. Because many microorganisms are found associated with healthy insects, good laboratory practices are essential to prevent contamination. This can be accomplished by maintaining good sanitation practices at all times and good sterilization protocols when required. Common sense dictates that all working surfaces and instruments be kept clean.

MANUAL OF TECHNIQUES IN INVERTEBRATE PATHOLOGY
ISBN 9780123868992

2012 Published by Elsevier Ltd.

## A. General remarks

The three most common light microscopes used for the study of entomopathogens are bright-field, phase-contrast, and confocal microscopy. The type of microscope (bright-field or phase-contrast) used is determined by the preparation of the specimen examined. In general, bright-field microscopy is suited for specimens of high contrast such as Giemsa-stained preparations or stained histological sections. Phase-contrast microscopy is useful for examination of living cells in what are normally called 'fresh preparations'. Phase-contrast is also extremely useful for the examination of 1-μm-thick unstained plastic sections of material prepared for electron microscopy. Proper alignment of the phase-contrast microscope is essential for optimum performance. A specialized type of phase-contrast microscopy is differential interference contrast microscopy (commonly referred to as Nomarski-interference). Nomarski-interference provides an apparent three-dimensional quality to the image. This technique is especially useful for examining surface structure important for taxonomic purposes.

Confocal microscopy is a powerful instrument that creates sharp images of fixed or living cells and tissues and can greatly increase optical resolution and contrast over that of a conventional microscope. This is achieved by excluding most of the light outside the focal plane of the microscope. The resulting image represents a thin cross-section and allows for three-dimensional reconstruction of a specimen. Images can be generated from reflected light off of the specimens or by stimulating a fluorescent reporter. An excellent review of confocal microscopy is presented by Semwogerere & Weeks (2005).

## B. Histological methods

### *1. Dissecting fluids*

Preliminary dissection and preparation of insect tissues for further examination requires that tissues be kept moist without damage or disruption to the tissues. Typically, saline solutions have been employed for this purpose with the most simple being a 0.85% sodium chloride (NaCl) solution. A commonly used dissecting fluid is Ringer's solution recommended as a normal salt solution for insect tissues. Several specialized physiological solutions have been developed, for example, Eide & Reinecke's saline (Eide & Reinecke, 1970) for muscoid fly sperm. Other specialized saline solutions can be found elsewhere (Barbosa, 1974).

### *2. Chemical fixation*

This is the process of stabilizing the cellular integrity of tissues for detailed histological examinations. The fixative must penetrate quickly to preserve the tissues in a natural state with a minimum of artifacts due to swelling, shrinkage, leaching, or other detrimental effects. This process often requires the preservation of whole insects or dissected tissues which are then embedded, sectioned, and stained. The selection of a fixing agent depends upon the purpose for which the tissue is intended. Generally, the live specimen is dissected in one of the saline solutions given above and the tissues of interest are then removed and placed into the fixative. Alternatively, the specimen may be dissected directly in the fixative if none of the tissue is intended for other purposes that may be adversely affected by the fixative. If whole specimens are to be fixed, it is often necessary to immerse the live specimen into the fixative and carefully make additional openings in the cuticle to allow for better infiltration of the fixative. Vacuum can also be used for difficult to fix tissues of whole specimens with hard cuticles. Tissues should be placed in at least 10-times their volume of the fixative. There are many fixatives developed for specific purposes but a few general-purpose fixatives are commonly used in insect pathology. Neutral buffered formalin (10%) is a good overall fixative that acts quickly and allows for long-term storage of tissues. The tissues must be thoroughly washed in distilled or deionized water prior to further processing. Formaldehyde is dangerous and must be used under a fume hood. Two of the most commonly used fixatives are Carnoy's and Bouin's. Carnoy's is an excellent general insect fixative because it penetrates rapidly and acts quickly. Fixation is complete for normal sized tissues (<1 cm) in 3 h and whole specimens in 12–24 h. Rinsing is in 70% alcohol (commonly ETOH) and specimens can be stored in this solution for extended periods prior to additional processing. Bouin's is also a good general fixative with fixation completed in 4–12 h depending upon the size of the tissue. It is critical that the tissues are washed thoroughly in 50% alcohol for 4–6 h (preferably agitated) to remove the picric acid. Failure to do this can adversely affect the staining of the tissue. Properly washed specimens can be stored in 70% ETOH for extended periods. Several fixatives for nematodes are presented in Chapter II.

*3. Dehydration and paraffin embedding*

Following fixation, the tissue must be dehydrated, infiltrated with paraffin and embedded in paraffin prior to sectioning and staining. The general procedure given is an example of the process but an experienced technician should be consulted prior to the undertaking of a specific project.

*a. Embedding in Paraplast™ (a paraffin–plastic mixture)*

1. Fix living insect host or freshly dissected tissue sample in Carnoy's or Bouin's fixative for 2–4 h.
2. Rinse in 70% ethanol (ETOH) for 1 h, leave in ETOH overnight. At this point, tissue may be stored in 70% ETOH.
3. Dehydrate tissue in a graded series of ETOH to tertiary-butyl alcohol (xylene can be substituted but can cause hardening of the tissues) and infiltrate with paraffin:
   - **(a)** 80 % ETOH, 2 h
   - **(b)** 95% ETOH, 2 h
   - **(c)** 100% ETOH, 1 h
   - **(d)** 100% ETOH, 1 h
   - **(e)** Absolute ETOH : butanol (1 : 1), 2 h
     Steps (f)–(h) must be done at a temperature > 25.5°C, the melting point of t-butyl alcohol.
   - **(f)** 100% butanol, 2 h
   - **(g)** 100% butanol, 2 h
   - **(h)** 100% butanol, 2 h
     Steps (i)–(k) are done in a 60°C oven.
   - **(i)** Butanol : paraffin (1 : 1), 2 h
   - **(j)** 100% paraffin, 2 h
   - **(k)** 100% paraffin (under vacuum), 2 h
4. Embed in fresh paraffin, with the tissue sample near but not on the bottom of the container ('boat'). This is done by pouring a bit of the paraffin into the container, and allowing it to harden slightly before adding the tissue sample. The paraffin containing the sample should be cooled rapidly.
5. After the paraffin has hardened, remove the container, trim the block to expose the tissue which is now ready to section.
   Sectioning and transfer of the sections to slides is probably the most tedious and difficult part of the process. Training by an experienced technician is strongly suggested but detailed procedures can be found in manuals such as Luna (1960).
6. Section the faced block with a microtome to obtain the thinnest sections possible (approximately 5 μm).
7. Place ribbons of sections on clean slides warmed on a slide warming tray set at 5°C below the melting point of the paraffin to flatten and fix them to the slide. A water bath set below the melting temperature of paraffin can be used to transfer sections to slides. Float sections in the water bath and then pick them up with the warm slides, remove excess water and dry.

Before staining, sections must be de-paraffinized. The paraffin is removed from the sections with a solvent (e.g., xylene or Hemo-De™, a natural citrus by-product, can be used in place of xylene in many cases) and the tissue rehydrated. This is done by hydration through a decreasing ETOH concentration series to distilled water. Three minutes in each solution should be sufficient. The slides should not be allowed to dry.

1 xylene : 1 absolute ETOH
Absolute ETOH
95% ETOH
70% ETOH
50% ETOH
Distilled water

*4. Staining*

Hematoxylin is one of the most common and valuable histological stains used. There are many different formulas for this stain depending on the specific objectives of the study. Many variations have been developed (Barbosa, 1974; Luna, 1960). The procedure below is a common one used for study of diseases in insects (Figures 15.1–15.3).

*a. Heidenhain's hematoxylin*

This is a classic procedure that stains nuclei a blue–black to brown–black color. It is time consuming but the stain is durable, actually improving with age. This procedure can be used on either wet smears or sectioned material.

*Wet smears.*

1. Rinse insect in distilled water; blot dry.
2. Smear the insect or selected tissue on a coverslip and drop it immediately into aqueous Bouin's fixative. The coverslip should float by surface tension, smear side down on the fixative.
3. Fix for at least 2 h.

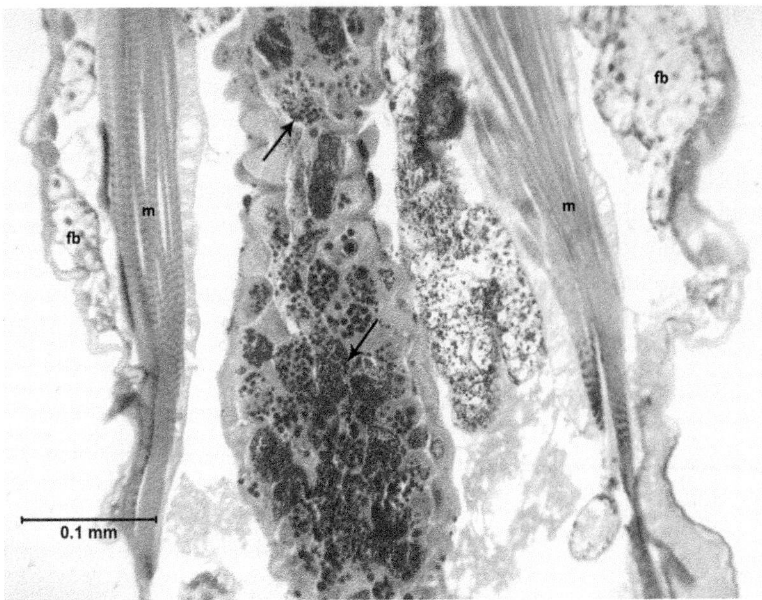


**Figure 15.1** Heidenhain's haematoxylin eosin stained longitudinal section of the larval midgut region of a midge (*Chironomus* sp.) infected with a cypovirus. Note the dense inclusion bodies of the cypovirus (arrows) in the cytoplasm of midgut cells. Muscle (m); fat body (fb). Please see the color plate section at the back of the book.

4. Rinse coverslip three times in 70% ETOH and then leave overnight in 70% ETOH.
5. After all of the yellow color (from the picric acid in the fixative) has been removed, rinse the coverslip in distilled water for 2–3 min and proceed with staining without allowing the smear to dry.

*Stain.*

1. Pre-treat in iron alum (mordant) for at least 5 h.
2. Rinse in distilled water for 3–4 min.
3. Stain in Heidenhain hematoxylin solution overnight.
4. Rinse in slowly running tap water for 5 min.

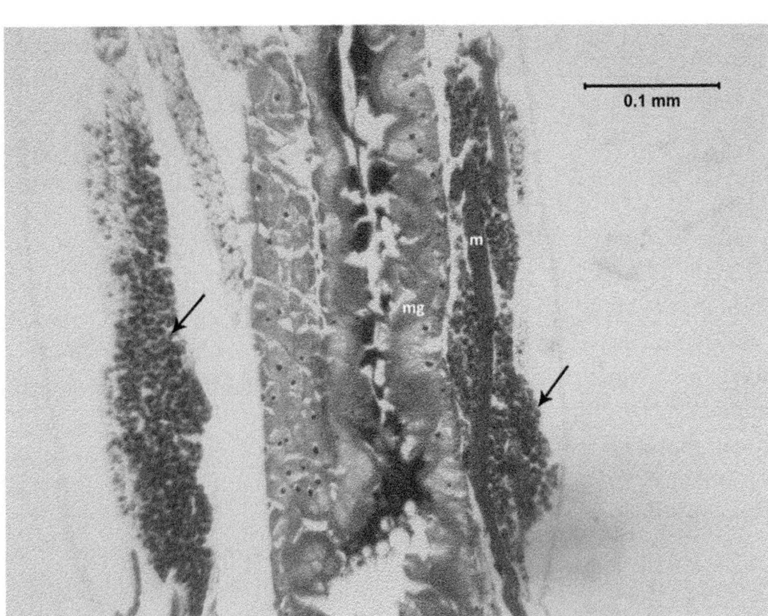


**Figure 15.2** Heidenhain's haematoxylin stained longitudinal section of the larval fat body and midgut region of a midge (*Chironomus* sp.). The fat body is filled with inclusion bodies of an entomopox virus (arrows). Muscle (m); midgut (mg).

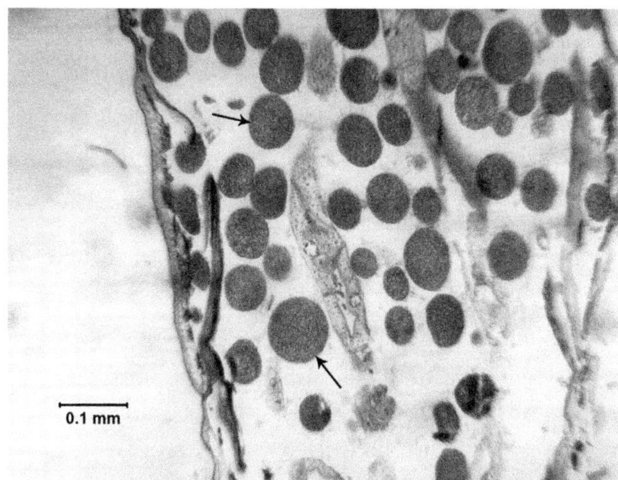


**Figure 15.3** Heidenhain's haematoxylin stained section of a black fly larval demonstrating sporangia of *Coelomycidium simulii* (arrows) in the hemocoel.

5. De-stain in iron alum until nuclei stand out sharp blue–black against a gray–tan background. Monitor the de-staining process by occasional examination under a microscope, rinsing the coverslips well in distilled water before examination.
6. After de-staining, rinse the coverslip briefly (10 s) in tap water containing a few drops (approx. 5 drops/100 ml) of concentrated ammonium hydroxide. Then rinse in slowly running tap water for 30–45 min.
7. Dehydrate in graded series of ETOH and xylene.
   (a) 70% ETOH, 0.5 min
   (b) 95% ETOH, 0.5 min
   (c) 100% ETOH, 1 min
   (d) 100% ETOH, 1 min
   (e) ETOH: xylene (1:1), 3 min
   (f) Xylene, three changes, 3 min each
8. Mount in Permount™ or other suitable mounting medium.

*Tissue sections.*

1. De-paraffinize sections (as described above).
2. Stain as described above for wet smears.
3. Rinse in slowly running tap water for 30 min.
4. Counterstain in Eosin Y for 1.5 min (this step can be omitted).
5. Dehydrate slides to xylene and mount as above.

## C. Specialized stains and protocols

Many different stains are utilized depending upon the pathogen group under study. Some of the most common stains for each group will be provided; however, other staining procedures are often required and can be found in the chapters of this manual dealing with the specific pathogens and in more specialized references (Adams & Bonami, 1991). Recipes for the stains mentioned in this section are found in the Appendix.

*1. Protists (Lee* et al.*, 1985; see also Chapter XI)*

*a. Microsporidia*

*(1). Giemsa-stain.* This stain was originally designed to examine blood for the presence of malarial parasites. It has become the most widely used stain for the identification of vegetative stages of microsporidia. The staining methods described below are methods successfully used in different laboratories and are offered here without explanation of differences in rinses and pH. Some experimentation with this stain is usually necessary to adapt it to different microsporidia, hosts and even laboratory water quality. This stain can be used for fungi (Figure 15.4) and bacteria and a procedure for utilizing Giemsa-stain for viruses can be found in Chapter II.

Air-dried tissue smears for Giemsa's stain can be made on either slides or coverslips. In either case they should be clean.

***Smearing procedure.***

1. If the host to be sampled is an aquatic one, excess water must be blotted from the surface of the organism before dissection.
2. Dissect a sample of tissue from a large host. If small, the entire organism can be crushed.

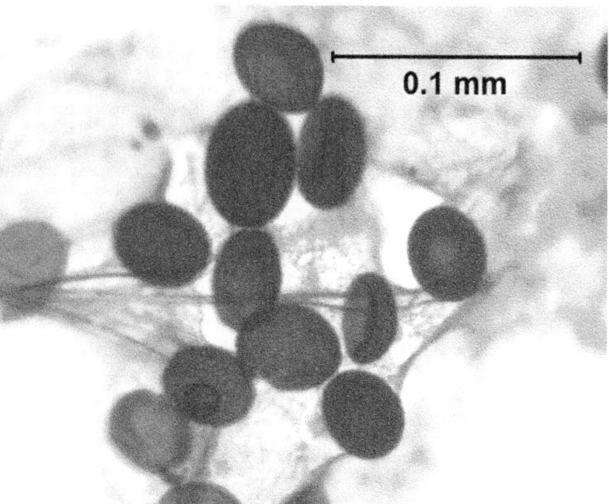


**Figure 15.4** Giemsa stained smear of sporangia of *Coelomomyces* sp. from the mosquito *Anopheles bradleyi*.

3. (a) Using a pair of forceps, press the sample against the slide with sufficient force to disrupt the host cells and release the microsporidian cells. Draw the tissue over the slide in a circular, spiral manner without passing over the same area twice. Make several small smears on a slide. Or (b), dissect a small piece of tissue from the insect and macerate it on a slide in a small drop of hemolymph or physiological saline then, with the forceps, transfer a drop of the macerate to a coverslip and spread it thinly.
4. The ideal smear is a monolayer of dispersed, disrupted cells. In reality, the cells will be sufficiently dispersed to be usable in some areas and in other regions the sample will be too thick.
5. If a frosted slide is used, label it with No. 2 pencil or India Ink, otherwise use a diamond marking pen. Ink used for the label must insoluble in both absolute methanol and water.
6. Set slide with smear side up, on slide holder and allow to air dry.
7. Float absolute methanol on the slide and fix for 5 min. After fixation, the excess methanol is poured off and the slide is stained. Alternatively, the slides can be allowed to dry and then stained, preferably within 24 h.

***Staining procedure.***

1. On a staining rack, place slides horizontally and flood with 10% Giemsa stain in pH 7.4 buffer for 10–20 min.
2. Rinse the slides in running tap water and blot dry with bibulous paper.
3. Examine after drying (usually a coverslip will be required on the dry slide) using a 16–40 × dry objective and bright-field optics. For more detailed observation immersion oil can be placed directly on stained smears for use with a higher power, oil immersion objective.
4. Nuclei stain red and cytoplasm stains blue. Acidophilic organelles in the cytoplasm will also stain red. The pH of the stain and rinse is important for proper color development. Lower pH shifts the colors toward red.
5. Apply a mounting medium (Histoclad™, Permount™ or Pro-Texx™ are appropriate; Canada Balsam cannot be used) and a coverslip to the dried slides to improve longevity of the stains.

***Alternative staining procedure.***

1. Make a smear on a coverslip, dry.
2. Place in absolute methanol for 7–10 min in a Coplin jar.
3. Remove and air dry.
4. Place in Giemsa stain solution plus one drop 0.01 M, pH 7.0 phosphate buffer, in Coplin jars for *ca.* 1.5 h.
6. Remove and rinse briefly in tap water buffered with a few drops of 6.8 pH phosphate buffer.
7. Air dry.
8. Place coverslip, stain side down, on a drop of Pro-texx™ on a clean slide.
9. Harden for 2 days.

The above procedure can also be used with smears on slides. Buffers often need adjustment because of local differences in pH of the tap water.

*(2). HCl—Giemsa (Weiser, 1976).* Acid hydrolysis prior to Giemsa's stain will reveal the number of nuclei in the spore.

1. Heat 1 N HCl to 60°C.
2. Lower the smear into the hot HCl.
3. The time in HCl will vary with the species; try 30, 60, 90 s. This can be done on one slide with a long smear by lowering the slide into the acid a bit at a time. (Alternatively, place a drop of the 1 N HCl on the smear and heat it gently over a flame, moving the slide frequently—maximum of 30 s—until the first tiny bubbles appear.)
4. Rinse for several minutes in distilled water.
5. Fix with methanol and stain with Giemsa's stain as usual.

**Note:** Hot HCl de-stains Giemsa, therefore, HCl-Giemsa can be used on slides that have been previously stained with Giemsa.

*(3). Giemsa–colophonium (Short & Copper, 1948).* This is an adaptation of Giemsa for staining paraffin sections.

1. Start with de-paraffinized sections (see Section 2.B.3.a.).
2. Fix for 5 min in absolute methanol.
3. Stain for 20–30 min in 10% Giemsa stain (staining time can vary with tissues).
4. Wash briefly in tap water.
5. De-stain (also referred to differentiation) in colophonium resin (gum rosin, 15 g in 100 ml acetone) for at least 15 s, checking occasionally under the microscope. Renew the colophonium solution if a film forms on its surface.

6. Transfer the slides to 70% acetone—30% xylene solution to remove the colophonium and stop differentiation.
7. Pass the slides through several changes of xylene until the sections clear.
8. Apply mounting medium and coverslip.

*(4). Calcofluor white (Vavra & Chalupsky, 1982).* The optical brightener, Calcofluor M2R binds to the chitinous layer of the microsporidian spore wall and makes them fluoresce in UV light. This can be a useful diagnostic technique.

One drop of water with spores is mixed with Calcofluor white (10—4 dilution) dissolved in distilled water. The slide is immediately observed under a fluorescent microscope. The spores exhibit a bright-green fluorescence. This stain also works on methanol-fixed spores and in de-paraffinized histological sections. If spores have been stored for a long time they lose the ability to bind to Calcofluor but the use of an alkaline solution (Calcofluor in 0.1 N NaOH) can restore the binding capacity.

*(5). Burri ink (Vavra & Maddox, 1976).* The simplest way to preserve the size and shape of spores on smears is to mix the spores with a solution of water soluble nigrosin stain (Burri Ink) and allow to dry. Nigrosine stain can also be applied to a smear that has already dried. The spores appear colorless on a gray background. The shape of the spores and any external appendages are revealed.

*(6). Wheatley's modified gomori trichrome (Alger, 1966).* A simple alternative to Heidenhain for staining microsporidian infected specimens.

1. Slides with de-paraffinized sections (see Section 2.B.3.a.) are dipped 10 times in 50% ETOH—HCl solution (0.1 ml concentrated hydrochloric acid per 10 ml, 50% ETOH).
2. Stain for 10 min in undiluted Wheatley's stain.
3. Dip 10 times in 90% ETOH containing 0.1 ml glacial acetic acid per 10 ml 90% ETOH.
4. Dip 10 times in 90% ETOH.
5. Dip 10 times in 100% ETOH.
6. 3 min in 100% ETOH.
7. 3 min in 1 : 1, ETOH : xylene.
8. 3 min or longer in xylene.
9. Apply a mounting medium and a coverslip.

*(7). India ink test for presence of a mucocalyx (Lom & Vavra, 1963).* Some microsporidian spores of aquatic hosts are surrounded by a mucocalyx that is thought to reduce their density, extending their time in the feeding zone of the host. This mucous layer is detected by mixing a drop of spores with a small amount of India ink on a microscope slide under a coverslip. The layer of fluid between the slide and coverslip must be thin, not much thicker than the spores. The small carbon particles will be held away from the spores by the mucocalyx, revealing it as a clear area around the spore.

*(8). Lacto-Aceto-Orcein chromosome squashes.* This is universal chromosome stain that has been adapted to examine chromosomes of microsporidia . A modification of this stain for Fungi can be found in Chapter VII but can be applied to most entomopathogens.

1. Clean slides (not siliconized) with 45% acetic acid.
2. Dissect infected tissue out into 45% acetic acid. Macerate and remove excess, hard tissues such as cuticle and head capsules, that may prevent an even spread.
3. Place silicon coverslip and flatten by applying direct pressure to the squash to prevent smearing of the cells (try both hard and soft pressure).
4. Put on dry ice, freeze, pop off coverslip, and allow to air dry.
5. Place a drop of 2% lacto-aceto-orcein in 45% acetic acid on to the squash, add coverslip, heat over alcohol lamp briefly until steam dissipates.
6. Cool briefly, place slide into ETOH, remove coverslip and add Euperol™ or Protexx™ mounting media and new coverslip.

***Alternative lacto-aceto-orcein procedure.***

1. Dissect infected tissue in small drop of water.
2. Add 1 drop of Carnoy's fixative. Fix for 1 min.
3. Remove fixative by carefully absorbing excess.
4. Add four drops of stock lacto-aceto-orcein stain; gently place coverslip.
5. After 5 min, apply direct pressure for 5 s with slide placed into folded filter paper to adsorb the excess stain (experiment with time and amount of pressure for proper staining and spread of chromosomes).

*b. Ciliates*

*(1). Fixing protozoa on a slide for permanent mounts (Farmer, 1980).* Ciliary structures and nuclei will be clearly differentiated against a gray background.

1. Place a drop of protozoa culture on a clean slide.
2. Pipette a drop of slide affixative from a height of 2–3 cm on to the sample.
3. Carefully remove excess with a pipette. Repeat steps 2 and 3, three times.
4. After approximately 15 s, move the slide through a dehydrating series of ethyl alcohols, 35–100%.
5. Clear in xylene and cover with mounting medium and coverslip.

*(2). Klein's silver stain (Farmer, 1980).* This stain reveals the tubules and other supporting structure for the cilia.

1. Place a drop of ciliates on a slide and let dry, or use the fixing method described previously.
2. Immerse for 20 min in 3% silver nitrate solution at 5–10°C.
3. Wash the slides in cold distilled water.
4. Submerge in water, expose to sunlight for 30 min (or an equivalent time under a UV lamp).
5. Dehydrate in a graded ETOH series into xylene and mount.

*2. Bacteria*

*a. Gram stain (Poinar & Thomas, 1984)*
An important bacteriological stain for diagnostic identification. Gram-positive organisms retain the violet stain and appear blue–violet; Gram-negative organisms are colored with the counterstain and appear red. A variation of this procedure can be found in Chapter III.

*Procedure.*

1. Air dry smears, lightly heat fix in flame (smear side up).
2. Flood slide with ammonium oxalate crystal violet for 1 min.
3. Rinse in tap water for 5 s.
4. Rinse with Gram's iodine then flood with this solution for 1 min..
5. Rinse in tap water for 5 s.
6. Rinse slide in three changes of n-propyl alcohol in coplin jars, 1 min each.
7. Rinse in tap water for 5 s.
8. Rinse with safranin counterstain then flood with counterstain for 1 min.
9. Rinse in tap water for 5 s, then air dry.
10. Examine under oil immersion.

*b. Flagella stain (Poinar & Thomas, 1984)*
This is used to visualize bacterial flagella.

*Culture.* Grow test organisms in 3 ml of a phosphate enriched broth medium for 16 h or less at 20°C.

*Fixation.* Add 6.0 ml of 10% formalin to the 3 ml of culture.

*Wash.*

1. Dilute the fixed culture with distilled water and centrifuge at 3000 r.p.m. for 30 min.
2. Decant and discard the supernatant, re-suspend the pellet in distilled water and centrifuge again. Repeat.
3. Suspend the pellet in distilled water until barely turbid.

*Slide preparation.*

1. Clean slides overnight in hot (70–80°C) sulfuric acid saturated with potassium dichromate.
2. Rinse slides thoroughly in tap water, then distilled water and then air dry. Slides must be kept grease free so handle only with clean forceps. Store in a clean, dry, airtight container.
3. Just prior to use, heat a slide in the flame of a Bunsen burner (the side to be used against the flame) and draw a line with a wax pencil across the slide about one third of the distance from one end. Handle slide only on the short end.
4. Place a drop of the final bacterial suspension on the distal end of the cooled slide, tilt the slide to cause the suspension to run down to the wax line. After the slide has air dried, it is ready to be stained.

*Staining procedure.*

1. Place the prepared slide on a staining rack and flood with the flagellar staining solution for 5–15 min (shorter time for new and/or warm stain, longer for old and/or cold stain).
2. Wash all stain off the slide with running tap water.
3. Air dry and examine under oil for flagella.

See Chapter III for other methods of preparing bacteria for microscopy.

*3. Fungi*

*a. Lactophenol cotton blue (Lipa, 1975)*

Hyphae and spores stain blue; fat substances stain orange-red. This is used as both a mounting media and stain for fungi. A variation of this stain can be found in Chapter VI.

1. Prepare lactophenol and add 0.5% methyl blue.
2. Place the fungal preparation into a drop of the stain on a glass slide.
3. Cover with a cover glass and heat slightly to enhance staining.
4. Cool and examine.

For additional information on staining Fungi, (see Chapter VI)

*4. Viruses*

*a. Sudan III stain for polyhedrovirus occlusion bodies (OB) (Thomas, 1974)*

This is used to differentiate virus OBs from fat droplets. Fat droplets stain red while OBs remain unstained.

1. Air dry smear.
2. Stain for 10–15 min in saturated aqueous Sudan III.
3. Rinse for 5–10 s in running tap water.
4. Air dry and examine under oil.

*b. Modified azan staining technique (Hamm, 1966)*

This is used for detection of occlusion body viruses (Nucleopolyhedrovirus, Granulovirus, Cytoplasmicpolyhedrovirus, and Entomopox virus).

*Staining procedure for paraffin sections.*

1. Rehydration from toluene via alcohols to water.
2. 50% Acetic acid, 5 min.
3. Distilled water rinse, 2 min.
4. Azocarmine (Solution 1), 15 min.
5. Distilled water rinse, 5 s.
6. Aniline, 1% in 95% alcohol, 30 s (aniline should be distilled and kept in the freezer).
7. Distilled water rinse, 5 s (change often).
8. Counterstain (Solution 2), 15 min.
9. 50% Alcohol, 10 s.
10. Absolute alcohol, two changes, 30 s each.
11. Toluene, two changes.
12. Mount in neutral, synthetic mounting medium.

***Results.*** Virus OBs: red; epicuticle: red; endocuticle: blue; muscle: light-blue to blue–green; epidermal cells: yellowish-green; fat body: yellowish-green with darker-green nuclei; nerve tissue: light-blue; silk gland: green, contents red or blue; midgut epithelium: green and blue.

***Negative staining.*** This procedure can be used to detect small non-occluded viruses with transmission electron microscopy by creating a darker background around the virus particle. This is done by using a 2% (w/v) aqueous phosphotungstic acid (PTA) adjusted to pH 7.5 with 1 N NaOH or KOH. This should be made fresh for each use. A drop of the viral suspension is placed on to a formvar-coated grid for approximately 1 min depending on the size of the particles. The excess is removed from the side with a sliver of filter paper. A drop of the PTA is then placed on the slide for 1 min and removed and the grid allowed to air dry prior to viewing. Alternatively, the viral suspension and PTA can be mixed together and then placed on to the grid. After 1 min (time will vary), remove excess and allow to dry. Modification of this procedure is usually required and references for additional information can be found in Adams & Bonami (1991).

For additional information on staining viruses see Chapter 2.

*5. Nematodes*

*a. Permanent mounts (Woodring & Kaya, 1988; Chapter XII)*

Fix the nematodes in TAF for 4–5 days. Process with glycerin via the evaporative method of Poinar (1975). Make certain specimens are free of dust and dirt. Filter solutions if necessary. Put fixed specimens in an ETOH-glycerin-water solution in a small dish. Cover all but $^{1}/_{8}$ of the surface area for 2 days and then all but $^{1}/_{4}$ for 7 days. The alcohol and water will evaporate to leave the nematodes in pure glycerin. Mount as described by Southey (1970) and in Chapter XII.

## D. Florescence microscopy

The basic principles of florescence microscopy (FM) and applications for the study of invertebrate pathogens (primarily mycopathogens) were covered in detail in the previous edition of this manual (Butt, 1997). The field continues to expand both in instrumentation advances such as confocal microscopy as well as the continued expansion of available fluorescent probes with wide application for labeling a wide range of targets in many biological systems. For example, The Molecular Probes® Handbook—A Guide to Fluorescent Probes and Labeling Technologies 11th Edition contains over 3000 technology solutions for a wide range of bio-molecular labeling and detection reagents (http://www.invitrogen.com/site/us/en/home/References/Molecular-Probes-The-Handbook.html). FM can be a powerful tool for examining the interplay of pathogens and hosts at the cellular level during growth and differentiation. Details on the application of one technique termed fluorescent *in situ* hybridization is provided below.

*1. Fluorescent* in situ *hybridization (FISH)*

FISH is a method of detecting single-stranded nucleic acid (DNA or RNA) using sequence-specific probes that can be observed directly using epifluorescence microscopy. While often used to detect specific DNA sequences in cells, it can also be used to detect mRNAs within tissue samples. There are three basic steps involved in FISH protocols: fixation of the specimens, hybridization of the fluorescent labeled probes to the homologous regions of cellular DNA or RNA, and observation of the labeled probe–target complexes. This technique has a broad range of applications but can provide valuable information when there is a need to detect pathogens in invertebrate cells and tissues, particularly early in development when a very sensitive method is required. An example of the technique is presented below and is based on a study investigating detection of a baculovirus in larval mosquitoes and induction of a programmed cell death pathway (Liu *et al.*, 2011).

**(1)** Probes are synthesized using DIG- or Fluorescein-RNA Labeling Mix (Roche) for *Culex nigriplapus* baculovirus (CuniNPV) genes *cun16*, 65, *86,* and *103 (early* expression), and *cun 24, 75, 85* (*late* expression). Detailed steps of making probes (*cun103* as an example, others use the same strategy) are:
   - **(a)** Prepare linearized plasmid DNA which will serve as a template for probe synthesis. pBlueScript vector is used because it contains T3/T7 RNA polymerase binding site.
   - **(b)** Design primers for *cun103* using online tool Primer3 (http://frodo.wi.mit.edu/primer3/). Each primer contains different restriction sites which are compatible with pBlueScript vector (e.g., for *cun103*, 5′-primer contains *Kpn1* site and 3′-primer contains *EcoR1* site). For all of these genes, the PCR product (which is also the future probe length) is ~500 bp.
   - **(c)** Perform PCR using CuniNPV genomic DNA as the template.
   - **(d)** After gel purification of the PCR product and restriction enzyme digestion (in the case of *cun103*, *Kpn1*, and *EcoR1*), ligate the fragment with *Kpn1/EcoR1* digested pBlueScript vector and transform into competent DH5a *E. coli.*
   - **(e)** Miniprep to extract the plasmid and verify by sequencing (T3 or T7 primer).
   - **(f)** Linearize the plasmid by single enzyme digestion. For instance, the order in pBlueScript plasmid is T3 - *Kpn1* - *cun106* - *EcoR1* - T7. To obtain the anti-sense probe, use *Kpn1* to linearize the template and T7 RNA polymerase to synthesize the probe; to obtain the sense probe (which serves as negative ctrl for FISH), use EcoR1 to linearize and T3 to drive the synthesis.

Synthesize probe using RNA Labeling Mix (Roche). Depending on what labeling you want for your probe, you may choose a different labeling mix, e.g., DIG-RNA labeling mix, biotin-RNA labeling mix. or FITC-RNA labeling mix, etc.

1. Assemble the *in vitro* transcription reactions on ice as below, making sure to work under RNase-free conditions:
   200 ng above linearized plasmid DNA.
   2 μl RNA-labeling mix.
   2 μl Transcription buffer.
   X μl sterile RNase-free double-distilled water to a final volume of 18 μl.
   2 μl RNA polymerase (T3 or T7).
2. Mix and centrifuge briefly.
3. Incubate at 37°C for 4 h.
4. Load 1 μl product into agarose gel to examine the quality of RNA probe.
5. For the remainder of the probe, add 0.1 volume of 4 M LiCl (~2.5 μl) and 2.5 volume of pre-chilled 100% EtOH (~75 μl), mix well.
6. Leave for 2 h at −20°C.
7. Centrifuge at 13,000 × *g* for 15 min at 4°C.

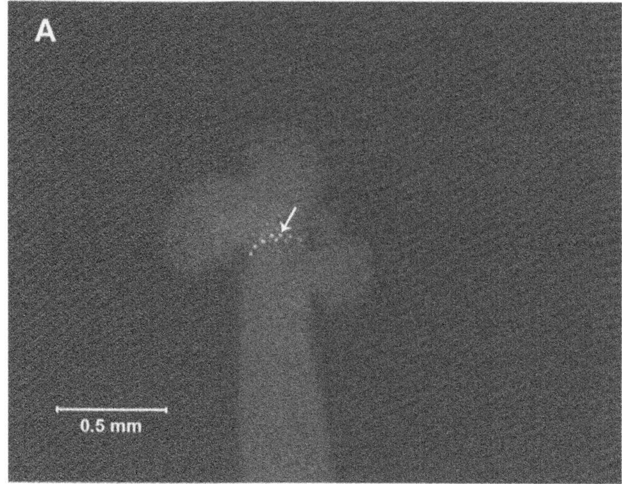


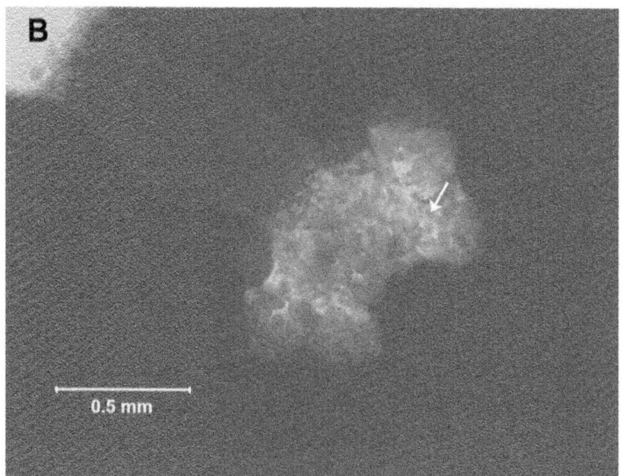


**Figure 15.5** A. Localization of *Culex nigripalpus* baculovirus (CuniNPV) replication 2 h post-infection in gastric caecum of *Culex quinquefasciatus* using a fluorescent labeled probe cocktail against early expressed genes *cun16*, *cun65*, *cun86*, and *cun103* (arrow). B. Localization of CuniNPV replication 48 h post-infection in gastric caecum of *Culex quinquefasciatus* using a fluorescent labeled probe cocktail against late expressed genes *cun24*, *cun75*, and *cun85* (arrow). Photo credit to Bo Liu and Lei Zhou. Please see the color plate section at the back of the book.

**8.** Decant the EtOH and allow the pellet to dry.
**9.** Dissolve in 50 μl hybridization buffer (50% formamide, 25% 2*SSC, 20 μg/ml yeast tRNA, 100 μg/ml ssRNA, 50 μg/ml heparin, 0.1% Tween-20).

**(2)** Tissues (in this example *Culex quinquefasicatus* mosquito larval midguts infected with the Baculovirus CuniNPV) are dissected at various times post-exposure and fixed with 4% paraformaldehyde in PBT_DEPC (0.3% Triton in PBS made with DEPC pretreated double-distilled water) for 30 min.
**(3)** The tissue is then incubated for 7 min with 50 μg/ml protease K in PBT_DEPC and the reaction is stopped by washing with 4% paraformaldehyde.
**(4)** Samples are then incubated with probes diluted in hybridization buffer (see above).
**(5)** Hybridization is performed overnight at 60°C. If necessary, HRP-conjugated anti-DIG or anti-FITC antibody (depending on what marker the probes carry) is applied after hybridization followed by signal amplification using the Tyramid Signal Amplification Kit (PerkinElmer).
Optional: It may be desirable to counter-stain the tissue with DNA dyes such as DAPI to visualize individual cell/nuclei. To stain for DAPI, the tissue is incubated for 10 min in 1 μg/ml DAPI diluted in PBT and then wash twice with PBT.
**(6)** Mount the tissue with vector shield (prevent/delay fluorescent bleaching) and observations can be made with a Leica upright fluorescent microscope using OpenLab software (Figure 15.5A, B).

## 3. ELECTRON MICROSCOPY

### A. Transmission electron microscopy (TEM)

Electron microscopy of biological materials places rather strict requirements on specimen preparation in order to obtain high-quality micrographs for detailed study. Protocols for preservation, dehydration, and embedding of tissues in a suitable medium must be carefully followed but modifications are often required depending upon the host and pathogen under investigation. Once this has been accomplished, thin sections (approximately 90−150 nm) are mounted on grids, stained, and then viewed and photographed with the electron microscope. This is a tedious process involving many steps making problem resolution a difficult task. While experience and practice are key to successful electron microscopy, the protocols and procedures given below are intended to provide a basic foundation for initiating studies utilizing the electron microscope.

*1. Tissue preparation*

This process involves the fixation of the tissue (hardening and preservation), dehydration, and infiltration with a medium that can be hardened to provide a material suitable for thin sectioning. The main goal of this process is to stabilize and preserve the fine structural details of the cells to a state near to that in the living tissues.

*a. Fixation*

This is the first step in the preparation of biological specimens for examination by electron microscopy. This process must be accomplished as soon as possible after killing the specimen so that *post mortem* changes are kept to a minimum. Aldehydes and osmium tetroxide ($OsO_4$) are the most effective fixatives for TEM. Fixatives cross-link macromolecules causing them to become immobilized and insoluble. One standard procedure involves double fixation using glutaraldehyde as the primary fixative followed by $OsO_4$. Glutaraldehyde stabilizes tissue by cross-linking proteins. $OsO_4$ reacts with lipids and certain proteins but also provides electron density to the tissue. Therefore, $OsO_4$ acts as both a post-fixative and an electron stain. Without $OsO_4$ or if the $OsO_4$ is bad, nuclear membranes and cytoplasmic membranes of the endoplasmic reticulum, Golgi and other organelles will not be preserved. Some procedures also involve a third fixative, uranyl acetate, before or during dehydration often to enhance the electron density of the material and it is therefore referred to as '*en block* stain'. It also acts as a fixative particularly for lipid components. Good fixation is usually measured by the continuity of membrane structures and the lack of obvious distortions and discontinuity in cytoplasmic details. One group of organelles to carefully examine are the mitochondria which should have clearly defined cisternae without swelling or lysis. The following is a general procedure for the fixation of insect tissues infected with a pathogen.

*General procedure.*

1. Dissect specimen in 2.5% glutaraldehyde (it is critical that the specimens be living when processed) After 5–15 min, the specimen can be cut into smaller pieces (1–2 $mm^3$).
2. Transfer pieces to fresh glutaraldehyde and fix for a total of 2.5 h at room temperature or overnight in the refrigerator.
3. Wash in 0.1 M Cacodylate buffer (pH 7.2–7.3) three times for 15 min each (for a total of 45 min). Rinses are important to prevent any reaction between the primary and post-fixative.
4. Postfix in 1.0% $OsO_4$ (pH 7.5) for 1 h 45 min to 2 h. This should be done at room temperature with the vials wrapped in foil. $OsO_4$ should be handled with great care and only under a fume hood. Gloves should be used.
5. Double-distilled water washes: three times for 15 min each (for a total of 45 min).
6. Begin dehydration or, for extended storage use, place in sucrose buffer.

An alternative to chemical fixation is freeze-substitution. This protocol was developed to avoid the many artifacts associated with conventional chemical fixation. The term freeze-substitution refers to the dissolution of ice in a frozen specimen by an organic solvent at low temperatures. The sample is quickly frozen by one of several ultra-rapid freezing techniques and then the water in the sample is substituted by an organic fluid, such as methanol, ETOH or acetone, at very low temperatures. Usually, the solvent contains a chemical fixative, such as $OsO_4$, with substitution requiring 48 h at −75 to −85°C. The sample is then brought to room temperature and infiltrated and embedded conventionally. Excellent results have been obtained with this method but the sample size is critical and is usually limited to specimens made up of individual cells and cell layers. An excellent discussion of this technique for use with fungal cells is provided by Hoch (1986).

*b. Problem tissues*

Processing certain tissues is often difficult due to either the small size of the specimens or problems with tissues that do not readily sink in the fixatives. In most cases, small specimens (cells, spores, eggs, etc.) can be embedded in agar and handled like pieces of tissue. The specimens are first fixed (at least through glutaraldehyde), and the fixative removed by centrifugation. The specimens are then washed at least twice in buffer and the specimens re-suspended in warm 2–4% ultra-low gelling agarose. After hardening, the agarose with the specimens can be cut into small pieces and handled like pieces of tissue to complete processing.

For tissues that will not sink in the fixative, small carriers can either be purchased or constructed from Beem capsules (Ted Pella, Inc. Product #130) and small wire mesh (Adams & Bonami, 1991). The tissues are placed into the holders that will sink in the fixative and can usually be removed prior to $OsO_4$ fixation. Make sure that no air bubbles are trapped around the tissues in the holder so the fixative is in contact with the tissue. A simple alternative is to overfill the vial containing the tissues with glutaraldehyde until you have a positive meniscus (the tissue will be floating on the surface). Carefully stretch a piece of parafilm over the top of the vial trapping the tissue and removing all of the air. Tighten the cap and the tissue should sink to the bottom of the vial. Process as normal.

*c. Dehydration*

After fixation, tissues must be dehydrated and embedded. Dehydration is achieved by transferring the material through an ascending alcohol or acetone series into absolute alcohol or acetone. A sample dehydrating protocol is given below but can be modified to reduce the steps by using increments of 25% (for example 25, 50, 75, 95% for 10 min each).

1. 10% ETOH, 10 min.
2. 30% ETOH, 10 min.
3. 50% ETOH, 10 min.
4. 70% ETOH, 10 min (good point for *en block* staining, wrap in foil and hold overnight).
5. 80% ETOH, 10 min.
6. 90% ETOH, 10 min.
7. 95% ETOH, 10 min.
8. 100% ETOH, 15 min.
9. 100% ETOH, 15 min.
10. 100% Acetone, 15 min.
11. 100% Acetone, 15 min.

Immediately place specimen into plastic dilutions

**Note:** Absolute alcohol and acetone must be stored over molecular sieve to insure the absence of water.

*Quick dehydration protocols using 2,2 dimethoxypropane (DMP).*

1. Add 1−2 ml DMP + 1−2 ml distilled $H_2O$ + three to four drops of 0.2 N HCL. Shake; should turn cold. Hold 5−15 min.
2. Remove solution, add 2−3 ml DMP + three to four drops HCL (No $DH_2O$). Shake; hold 5−15 min (two changes).
3. Absolute acetone three times, 15 min each.

*d. Infiltration and embedding*

The final process in tissue preparation is to infiltrate the specimens with a liquid embedding medium which is then polymerized to produce a solid block. Epoxy resins are the most commonly used media and a general protocol is given below that is easy to use and provides uniform blocks that are easy to section and stain. Other resins are available for specific purposes such as Spurr's resin which is less viscous but is more difficult to section and stain and is not as stable under the beam of the electron microscope. Embedding media should be handled with caution, paying careful attention to the safety data sheets is essential.

A combination of two epoxy resins, Araldite and Epon, is easy to prepare and has been shown to be highly reliable. After dehydration, specimens are infiltrated with the embedding medium by passing them through a series of solutions until the dehydrating agent has been completely replaced by the final embedding medium. This is done in small vials on a shaker at room temperature. Activator must be included in all dilutions. After the pure resins, tissues are transferred to capsules, filled with pure resins and polymerized in an oven.

1. 25% resin : 75% absolute acetone—4 h to overnight.
2. 50% resin : 50% absolute acetone—4 h.
3. 75% resin : 25% absolute acetone—4 h.
4. Pure resin overnight.
5. Pure resin (change vials)—all day (≈ 6 h)*.
6. Embed in Beem™ capsules which have dried at least 24 h in a 60°C oven. A small drop of fresh plastic is placed into the tip of the Beem capsule and the tissue is placed into the drop and the capsule filled with resin. Make sure to include label with block number when embedding. Leave in oven (uncovered) overnight. Be sure no air bubbles are below the tissue.
7. Remove the embedded blocks next morning and allow to cool (best for 24 h) prior to sectioning.

**Note:** For better infiltration of difficult tissues, extend the specimen in pure resin for another day (overnight) or for several days changing daily. Embed as usual.

*2. Sectioning and staining*

*a. Remarks on thick and thin sectioning*

Prior to facing and thin sectioning, thick sections (0.5–1 μm thick) can be removed from the block using a microtome and glass knife. These sections can be transferred directly to a slide and mounted with Pro-Texx™ and a coverslip. Sections can be examined directly (without staining) with phase-contrast to locate areas of interest for the final trimming (facing). Alternatively, the sections can be stained prior to covering. Once the area of interest has been determined, the block is trimmed until a 'face' of the appropriate size is obtained. The trimmed block is mounted in a holder on the ultramicrotome and automatically advanced to be sectioned by either a glass or diamond knife. Sections are floated on to water and transferred to a grid for thin sections. Thin sections are generally in the range of 90–150 nm which can be judged from the interference colors shown by the sections as they float on the water surface. Sections in this thickness range will generally appear gold with light gold sections thinner and dark gold sections thicker. Thin sections are transferred to grids or grids coated with formvar for added stability under the beam. After drying, the sections are ready for post-staining prior to viewing in the electron microscope.

*b. Post-staining*

This process serves to increase the contrast in thin sections and is usually performed immediately prior to viewing. A two-step staining protocol is commonly employed with excellent results. Grids are floated on to a drop of uranyl acetate (section side down) for 5 min. The time will vary depending on whether aqueous or methanolic uranyl acetate is used and the thickness of the sections (thicker sections take less time). The grids are passed through three rinses in deionized water (hold grid and quickly dip in water), blot and immediately submerge into a drop of lead citrate, section side up, for 5 min. Grids are rinsed three times in deionized water, blotted on filter paper and allowed to dry before viewing. For difficult to stain material, time in the uranyl acetate can be extended or various concentrations of methanolic uranyl acetate used. For extremely difficult tissues, 1% dimethyl sulfoxide (DMSO) in 100% methanolic uranyl acetate has proven useful. A possible problem when post-staining with 100% methanolic uranyl acetate is the loss of sections from the grid. This can be prevented by passing the grid under the electron beam at low intensity to adhere the sections to the grid prior to post-staining.

One of the most common problems encountered in post-staining is the presence of lead precipitate on the sections. A simple solution is to re-stain the sections in uranyl acetate which will remove the precipitate. It is then necessary to re-stain in freshly made lead citrate. Another common problem is the presence of uranyl acetate precipitate, which can be removed with oxalic acid (Avery & Ellis, 1978).

*3. Immunocytochemistry*

Localization of cellular antigens has many applications in the study of diseased invertebrates. Epoxy resins and standard fixation protocols are often not compatible with efficient and reproducible labeling of antigens and alternatives are the acrylic resins and modified protocols. LR White is an acrylic resin with very low viscosity and toxicity and can be used for both light- and electron-microscopy. The resin is available in medium and hard grades and can be cured (polymerized) by four different methods: heat, microwave, UV light, and chemical. Dehydration with ETOH is suggested as any traces of acetone can interfere with curing of the resin. Because the sections of LR White embedded tissue are hydrophilic, reagents for immunocytochemistry can easily penetrate into the tissue without the need for etching which is normally required for epoxy resins. A protocol is provided below for using LR White and modified fixation protocols for immunocytochemistry. Technical data sheets available from the supplier provide additional information and details for mixing, storage, and other protocols for conventional light- and electron-microscopy applications.

*a. Primary fixation*

Fix 1–4 h with 2.5% glutarldehyde. Alternatively, can use 4% paraformaldehyde/0.05% glutaraldehyde/0.2% picric acid in 0.1 M phosphate buffer, pH 7.3 (Somogyi & Takagi, 1982).

*b. Buffer rinse*

2 × 30 min

*c. Post-fixation*

$OsO_4$ fixation is not recommended. For improved contrast without osmium, post fixation for 1 h in 1%

tannic acid in sucrose buffer can be utilized (Berryman *et al.*, 1992) followed by 3 × 10-min buffer rinses.

*d. Dehydration*

50% ETOH, 15 min
70% ETOH, 2 × 15 min
80% ETOH, 10 min

*e. Infiltration*

2 : 1 resin : 70% ETOH 1 h
100% resin, 1 h
100% resin, overnight
100% resin, 1 h

*f. Polymerization*

Heat curing must be conducted under anaerobic conditions. One simple method uses Beem capsules. Tissue is placed into capsule and completely filled with resin, covered with a piece of parafilm and the cap is placed on top to seal. Blocks are cured overnight at 65°C. Microwave, UV and chemical curing methods can be found in the technical data sheets from the resin provider.

*g. Sectioning*

Standard sectioning techniques can be used to obtain sections for TEM and immunolocalization of the desired cellular antigens.

## B. Scanning electron microscopy (SEM)

SEM has been utilized in the study of insect pathology primarily for examining surface morphology of microsporidian spores and fungal conidia and developmental stages. Some applications have also been useful for bacteria and viruses. Usually the process involves fixation of the material, dehydration and drying, followed by mounting on to a grid or stub and applying a conductive coating. An extensive reference section on SEM is found in Adams & Bonami (1991).

### *1. Specimen preparation*

Although some specimens can be viewed without fixation, results are generally improved by fixing in both glutaraldehyde and $OsO_4$ similar to the procedures for TEM. Specimens are then dehydrated and can then be mounted and air dried. Often this results in artifacts caused by shrinkage and collapse of the specimens. Best results usually are obtained when specimens are critically point dried. Specimens must be placed into a carrier, dehydrated and critically point dried to reduce damage to the tissue. A chemical method of drying soft tissues has also been developed (Nation, 1983).

Low-temperature SEM examines samples that are rapidly frozen (frozen-hydrated) and maintained under vacuum. This is an alternative to chemical fixation and provided excellent results but requires a specialized SEM. For an excellent discussion of the procedures and protocols see Beckett & Read (1986).

### *2. Mounting and coating*

Depending on the sizes of the specimens, they can be mounted on grids or stubs. Larger specimens can be adhered to a stub with conductive silver paint after critical point drying. There are many methods for handling small specimens such as collecting them on a filter disk after fixation. The disk can then be used to carry the specimen through dehydration and critical point drying. The disk is then mounted onto a stub with conductive silver paint and coated. For SEM, a coating of a conductive metal layer (usually gold or palladium) is required. This is usually applied to the mounted specimens with a sputter coater. Experimentation is necessary to obtain a coating of suitable thickness. Specimens are then ready for examination with the SEM.

## DISCLAIMER

Mention of a commercial or proprietary product in this paper does not constitute an endorsement of this product by the United States Department of Agriculture.

## ACKNOWLEDGMENTS

The author gratefully acknowledges the assistance of Dr Lei Zhou and Bo Liu in providing the protocol and photos for the section on Fluorescent *in situ* hybridization. The services of Neil Sanscrainte in manuscript preparation are greatly appreciated.

## REFERENCES

Adams, J. R., & Bonami, J. R. (1991). *Atlas of Invertebrate Viruses* Boca Raton, FL: CRC Press. pp. 684.

Aldrich, H. C., & Todd, W. J. (1986). *Ultrastructure Techniques for Microorganisms* New York, NY: Plenum Press. p. 533.

Alger, N. E. (1966). A simple, rapid, precise stain for intestinal Protozoa. *Amer. J. Clin. Pathol., 45*, 361–362.

Avery, S. W., & Ellis, E. A. (1978). Methods for removing uranyl acetate from ultra-thin sections. *Stain Technol., 53*, 137.

Barbosa, P. (1974). *Manual of Basic Techniques in Insect Histology* Amherst: Autumn Publishers. pp. 245.

Beckett, A., & Read, N. D. (1986). Flow-temperature scanning electron microscopy. In H. C. Aldrich, & W. J. Todd (Eds.), *Ultrastructure Techniques for Microorganisms* (pp. 45–86). New York, NY: Plenum Press.

Berryman, M. A., Porter, W. R., Rodewald, R. D., & Hubbard, A. L. (1992). Effects of tannic acid on antigenicity and membrane contrast in ultrastructural immunocytochemistry. *J. Histochem. Cytochem, 40*, 845–857.

Butt, T. M. (1997). Complementary Techniques: Fluorescence microscopy. In L. A. Lacey (Ed.), *Manual of Techniques in Insect Pathology* (pp. 355–365). London, UK: Academic Press.

Celis, J. E. (Ed.). (2006) *Cell Biology Vols.* I-IV, (3rd ed.). New York, NY: Elsevier Academic Press

Eide, P. E., & Reinecke, J. P. (1970). A physiological saline solution for sperm of the house fly and the black blow fly. *J. Econ. Entomol., 63*, 1006.

Farmer, J. N. (1980). *The Protozoa: Introduction to Protozoology* St. Louis, MO: C. V. Mosby Co. pp. 732.

Glauert, A. M. (Ed.). (1974)., *Practical Methods in Electron Microscopy, Vol. II.* Amsterdam: North-Holland. pp. 353.

Hamm, J. J. (1966). A modified azan staining technique for inclusion body viruses. *J. Invertebr. Pathol., 8*, 125–126.

Hayat, M. A. (1986). *Basic Techniques for Transmission Electron Microscopy* New York, NY: Academic Press. pp. 411.

Hoch, H. C. (1986). Freeze-substitution of fungi. In H. C. Aldrich, & W. J. Todd (Eds.), *Ultrastructure Techniques for Microorganisms* (pp. 183–212). New York, NY: Plenum Press.

King, R. C., & Akai, H. (Eds.). (1982)., *Insect Ultrastructure, Vols. I-II.* New York, NY: Plenum Press.

Lipa, J. J. (1975). *An Outline of Insect Pathology.* Warszawa: PWRiL. pp. 342.

Liu, B., Becnel, J. J., Zhang, Y., & Zhou, L. (2011). Induction of reaper ortholog in mosquito midgut cells following baculovirus infection. *Cell Death Differ., 18*, 1337–1345.

Lee, J. J., Small, E. B., Lynn, D. H., & Bovee, E. C. (1985). Some techniques for collecting, cultivating and observing protozoa. In J. J. Lee, S. H. Hutner, & E. C. Bovee (Eds.), *Illustrated Guide to the Protozoa* (pp. 1–7). Lawrence: Society of Protozoologists.

Lom, J., & Vavra, J. (1963). Mucous envelopes of spores of the subphylum Cnidospora (Dolfein, 1901). *Vestn. Cesk. Spol. Zool.* 274–276.

Luna, L. G. (Ed.). (1960). *Manual of Histologic Staining Methods of the Armed Forces Institute of Pathology* (3rd ed.). New York, NY: McGraw-Hill. pp. 258.

Nation, J. L. (1983). A new method using hexamethyldisilazane for preparation of soft insect tissues for scanning electron microscopy. *Stain Technol., 58*, 347.

Poinar., G. O., Jr. (1975). *Entomogenous Nematodes.* Leiden, Netherlands: E. J. Brill. pp. 317.

Poinar, G. O., Jr., & Thomas, G. M. (1984). *Laboratory Guide to Insect Pathogens and Parasites* New York, NY: Plenum Press. pp. 392.

Semwogerere, D., & Weeks, E. R. (2005). Confocal Microscopy. In *Encyclopedia of Biomaterials and Biomedical Engineering* (pp. 705–714). New York, NY: Taylor & Francis.

Short, H. E., & Cooper, W. (1948). Staining of microscopical sections containing protozoal parasites by modification of McNamara's method. *Trans. Roy. Soc. Trop. Med. Hyg., 41*, 427–428.

Smith, D. S. (1968). *Insect Cells, Their Structure and Function* Edinburgh: Oliver & Boyd. pp. 372.

Somogyi, P., & Takagi, H. (1982). A note on the use of picric acid-paraformaldehyde fixative for correlated light and electron microscopic immunocytochemistry. *Neuroscience, 7*, 1779–1783.

Southey, J. F. (Ed.). (1970). *Laboratory Methods for Work with Plant and Soil Nematodes, Ministry of Agriculture, Fisheries and Food, Technical Bulletin 2.* London, UK: Her Majesty's Stationary Office. pp. 148.

Thomas, G. M. (1974). Diagnostic techniques. In G. E. Cantwell (Ed.), *Insect Diseases, Vol. 1* (pp. 1–48). New York, NY: Marcel Dekker.

Vavra, J., & Chalupsky, J. (1982). Fluorescence staining of microsporidian spores with the brightener 'Calcofluor White M2R'. *J. Protozool., 29*, 503 (Abs. no. 121).

Vavra, J., & Maddox, J. V. (1976). Methods in microsporidiology. In L. A. Bulla, & T. C. Cheng (Eds.), *Comparative Pathobiology, Vol. 1. The Biology of the Microsporidia* (pp. 281–319). New York, NY: Plenum Press.

Weiser, J. (1976). Staining of the nuclei of microsporidian spores. *J. Invertebr. Pathol., 28*, 147–149.

Woodring, J. L., & Kaya, H. K. (1988). *Steinernematid and Heterorhabditid Nematodes: A Handbook of Biology and Techniques* Fayetteville, AR: Southern Cooperative Series Bulletin 331. pp. 30.

## APPENDIX
### Dissecting fluids

*Ringer's solution*

| | |
|---|---|
| NaCl | 8.0 g |
| Calcium chloride ($CaCl_2$) | 0.25 g |
| Potassium chloride (KCl) | 0.25 g |
| Sodium bicarbonate ($NaHCO_3$) | 0.25 g |
| Distilled water to make | 1000 ml |

**Note:** The amount of sodium chloride can vary from 6.5 to 9.0 g depending on the organisms under study.

*Simple physiological saline*

| | |
|---|---|
| NaCl | 0.85 g |
| Distilled water to make | 100 ml |

*Eide and reinecke's physiological saline (Eide and Reinecke, 1970)*

| | |
|---|---|
| NaCl | 0.453 g |
| Magnesium chloride hexahydrate ($MgCl_2 \cdot 6H_20$) | 0.3 g |
| Sodium bicarbonate ($NaHCO_3$) | 0.035 g |
| Dextrose ($C_6H_{12}0_6$) | 1.155 g |
| Potassium chloride (KCl) | 0.107 g |
| Monosodium phosphate ($NaH_2PO_4H_2O$) | 0.04 g |
| Sodium acetate ($C_2H_3O_2Na$) | 0.025 g |
| Distilled water to make | 100 ml |

## Fixatives for light microscopy

*Buffered neutral formalin*

| | |
|---|---|
| Formalin ($CH_2O$, 37–40 %) | 100 ml |
| Distilled water | 900 ml |
| Sodium phosphate monobasic ($NaH_2PO_4 \cdot H_2O$) | 4.0 g |
| Sodium phosphate dibasic ($Na_2HPO_4$) | 6.5 g |

*Carnoy's fixative*

| | |
|---|---|
| Absolute ETOH ($C_2H_5OH$) | 60 ml |
| Chloroform ($CHCl_3$) | 30 ml |
| Glacial acetic acid ($C_2H_4O_2$) | 10 ml |

*Bouin's fixative*

| | |
|---|---|
| Saturated aqueous picric acid | 75 ml |
| Formalin ($CH_2O$, 37–40 %) | 25 ml |
| Glacial acetic acid ($C_2H_4O_2$, add just before use) | 5 ml |

*TAF (Southey, 1970)*

| | |
|---|---|
| Formalin ($CH_2O$, 37–40 %) | 7 ml |
| Triethanolamine ($C_6H_{15}NO_3$) | 2 ml |
| Distilled water | 91 ml |

See Chapter XII for additional fixatives and mounting media for nematodes.

## Stains for light microscopy

*Heidenhain's hematoxylin stain*

| | |
|---|---|
| Distilled water | 90 ml |
| Absolute ETOH | 10 ml |
| Hematoxylin (Harleco #234 - Hematoxylin stain CI #75290) | 0.5 g |

Dissolve hematoxylin in alcohol, add water and age in the dark for at least 6 weeks. Store in the dark. To use, dilute 1 : 1 with distilled water and add three drops saturated lithium carbonate ($Li_2CO_3$)/100 ml.

*Mordant*

| | |
|---|---|
| Iron alum [ferric ammonium sulfate $FeNH_4(SO_4)_2 \cdot 2H_2O$] (Iron alum crystals should have a violet-pinkish color) | 2.5 g |
| Distilled water | 100 ml |

*Wheatley's modified Gomori trichrome stain*

| | |
|---|---|
| Distilled water | 100 ml |
| Glacial acetic acid ($C_2H_4O_2$) | 1.0 ml |
| Phosphotungstic acid ($12WO_3 \cdot H_3PO_4 \cdot H_2O$) | 0.7 g |
| Chromotrope 2R ($C_{16}H_{10}N_2Na_2O_8S_2$) | 0.4 g |
| Bright green SF, certified | 0.3 g |
| Bismarck brown, certified | 0.1 g |

Combine ingredients and let stand 24–48 h before use. A sediment develops but causes no problem. The stain is stable for 1 year.

*Eosin Y stain*

| | |
|---|---|
| Eosin Y | 5 g |
| Distilled water | 100 ml |

Filter the eosin solution and add a few drops of glacial acetic acid just before use.

*Slide affixative*

| | |
|---|---|
| Saturated mercuric chloride ($HgCl_2$) | 10 ml |
| Glacial acetic acid ($C_2H_4O_2$) | 2 ml |
| Formalin ($CH_2O$, 37-40 %) | 2 ml |
| Tertiary butyl alcohol ($(CH_3)_3COH$) | 10 ml |

*Giemsa stain*

| | |
|---|---|
| Giemsa stain | 1 part |
| 0.01 M phosphate buffer, pH 7.4 | 9 parts |

Good results have been obtained with the Fisher Scientific, Baker Chemicals products and a new product from Sigma that needs only to be diluted with distilled water because it is already buffered. Phosphate uffer at pH 7.4 (premixed packets can be obtained from Fisher Scientific). The stain solution must be prepared fresh for each use.

*Gram stain*

*1. Ammonium oxalate crystal violet*
Solution A:

| | |
|---|---|
| Crystal violet (90% dye content) | 4 g |
| Dissolve in 40 ml 95% ETOH | |

Solution B:

| | |
|---|---|
| Ammonium oxalate ($C_2H_8N_2O_4 \cdot H_2O$) | 1.6 g |
| Dissolve in 160 ml distilled water | |

Mix solutions A and B 48 h before use.

*2. Gram's iodine*

| | |
|---|---|
| KCl | 2 g |
| Iodine | 1 g |

Grind in a mortar for 5–10 s. Add 1 ml distilled water and grind until all ingredients are in solution. Add 10 ml water and mix. Rinse into a reagent bottle and bring the volume to 200 ml.

*3. Counterstain*

| | |
|---|---|
| Safranin (86% dye content) | 0.5 g |
| ETOH (95%) | 20 ml |

Mix. Add to 180 ml distilled water.

*Flagella stain*

| | |
|---|---|
| A. Basic fuchsin | 1.2 g |
| Dissolve in 100 ml of 95% ETOH. | |
| B. Tannic acid | 3.0 g |
| Dissolve in 100 ml of distilled water | |
| C. NaCl | 1.5 g |
| Dissolve in 100 ml of distilled water | |

Prepare the stain by mixing equal parts of the three stock solutions. The stain solution may be stored for 1 week at room temperature, 1–2 months under refrigeration, and indefinitely if frozen.

*Lactophenol and cotton blue*

| | |
|---|---|
| Phenol crystals ($C_6H_6O_2$) | 100 g |
| Lactic acid (USP 85%) | 80 ml |
| Glycerine | 159 ml |
| Distilled Water | 100 ml |

Mix ingredients and heat until hot; add 0.5% cotton blue.

*Aqueous eosin*

| | |
|---|---|
| Eosin Y (C.I. 45380) | 1 g |
| Distilled water | 100 ml |

Mix ingredients and filter. Add several drops of glacial acetic acid to staining solution before use.

*Lacto-aceto-orcein stock (2%)*

| | |
|---|---|
| Orcein | 2 g |
| Glacial acetic acid (45%) | 50 ml |
| Lactic acid (85%) | 50 ml |

Place orcein and acids into flask and plug with cotton. Heat to near boil (do not boil!) and hold for 30 min. Filter while hot. Cool and dilute stock 1 : 3 with 45% acetic acid for the final stain.

*Modified azan staining technique (Hamm, 1966)*

*Solution 1*

| | |
|---|---|
| Azocarmine G | 0.1 g |
| Glacial acetic acid | 2 ml |
| Distilled water | 100 ml |

Dissolve azocarmine G in water and boil for 5 min. Cool and add acid. Filter before use.

*Solution 2*

| | |
|---|---|
| Phosphotungstic acid | 1.0 g |
| Aniline blue (water soluble) | 0.1 g |
| Orange G | 0.5 g |
| Fast green FCF | 0.2 g |
| Distilled water | 100 ml |

Dissolve all ingredients in water.

*ETOH-glycerine-water solution (Poinar, 1975)*

| | |
|---|---|
| 95% ETOH | 15 parts |
| Glycerine | 1 part |
| Distilled water | 5 parts |

## Fixatives, buffers and stains for electron microscopy

*Working solutions*

*0.2 M Cacodylate buffer*

| | |
|---|---|
| Cacodylate buffer stock | 50 ml |
| 0.2 M HCL | 6 ml |
| Double-distilled water to make | 100 ml |

*2.5 % Glutaraldehyde*

| | |
|---|---|
| 8% Glutaraldehyde | 10 ml |
| 0.2 M Cacodylate Buffer | 16 ml |
| Double-distilled water | 6 ml |
| Calcium chloride ($CaCl_2$) | 32 mg |

*1% $OsO_4$*

| | |
|---|---|
| 4% $OsO_4$ | 1 ml |
| 0.3 M Sucrose | 1 ml |
| 0.2 M Cacodylate buffer | 2 ml |

Wrap vial in foil during fixation.

*0.1 M Cacodylate buffer/sucrose*

| | |
|---|---|
| 0.2 M Cacodylate buffer | 5 ml |
| Double-distilled water | 5 ml |
| Sucrose | 0.1 g |

En blocs *stain*

| | |
|---|---|
| Uranyl acetate (UrAc) | 0.1 g |
| 70% ETOH | 20 ml |

Wrap in foil.

*Stocks*

*0.4 M Cacodylate buffer*

| | |
|---|---|
| 0.4 M Cacodylate buffer | 42.8 g |
| Double-distilled water to make | 500 ml |

*0.3 M Sucrose*

| | |
|---|---|
| Sucrose | 5.1 g |
| Double distilled water to make | 50 ml |

Store in refrigerator.

*4% $OsO_4$ Stock*

| | |
|---|---|
| $OsO_4$ | 1 g |
| Double distilled water | 25 ml |

Wrap in foil; dissolve at room temp, usually 24 h.; store in refrigerator.

*0.2 M HCl*

| | |
|---|---|
| Hydrochloric acid (HCl) | 1.6 ml |
| Double-distilled water to make | 100 ml |

Store in refrigerator

*Epon—Araldite*

| | |
|---|---|
| Epon 812 | 2 g |
| Araldite 502 | 1 g |
| DDSA, Hardener | 4.5 g |
| DMP-30, Activator | 4 drops |

Plastic tri-pour beakers (50 or 100 ml) are used to mix the plastics. The Epon 812, Araldite 502 and the DDSA are weighed out on a top-loading balance into the tri-pour beaker. This mixture is usually placed into a 55—60°C oven for 1—2 min to facilitate the mixing of the resins. Four drops of DMP-30 are added with a medicine dropper and mixed immediately by swirling the components. Attempt to avoid too many bubbles, but this is not crucial. The plastic will darken but should not turn orange. If the plastic turns orange then the DDSA used is probably not good. The plastics should last a long time. Only enough plastic is mixed for each use. Beem capsules are used for embedding the tissues. Do not put the lids on these during the curing process. Cure overnight in a 62—65°C oven. Larger batches can be made by multiples of the ingredients, but not more than four-times the basic formula.

*Standard reynolds lead citrate*

| | |
|---|---|
| Lead nitrate ($Pb(NO_3)_2$) | 1.33 g |
| Sodium citrate ($Na_3(C_6H_5O_4) \cdot 2H_2O$ | 1.76 g |

Freshly boiled and cooled distilled water.

1. Dissolve lead nitrate completely in 30 ml distilled $H_2O$.
2. Add sodium citrate. A heavy white precipitate will form.
3. Add 8 ml of 1 N NaOH (1 g/25 ml) and dilute to 50 ml with boiled then cooled water.
4. Mix until precipitate is dissolved.
5. pH should be 12.

Stain can be stored for several months in the refrigerator if sealed properly. Discard when precipitate forms or when contamination is found on stained grids.

*2. 5% Uranyl acetate in 50% methanol*

| | |
|---|---|
| Uranyl acetate (UrAc) | 0.5 g |
| Absolute methanol | 10 ml |
| Distilled water | 10 ml |

Wrap in foil, shake until dissolved, store in refrigerator.

This is a standard Uranyl acetate stain but can be easily modified to suit individual needs for different plastics or section thickness. For easily stained sections, aqueous or 25% methanol can be used. More difficult sections can be stained in 75 or 100% methanol. In some cases when additional staining is needed for particularly difficult material, 1% DMSO can be added to the 100% methanol.

*Formvar-coated grids*

Wash a glass microscope slide in 95% ETOH. Air dry for 1–2 min. Soak the end of the slide in dilute dish-washing detergent for 2–3 min. Wipe the slide partially dry with a Kimwipe™ but leave some of the detergent on the slide so that when it dries a detergent residue remains. When dry, wipe the slide vigorously with a dry Kimwipe™. It will feel slightly slick and waxy but will look clean. Dip the slide in 0.25% formvar dissolved in ethylene dichloride or chloroform and dry. Scrape edges of slide with a razor blade to free film from slide. Release on to water by inserting slide slowly under a water surface at a 45° angle. Place grids face down onto the floating film. Pick the film up on an index card and dry overnight in a slightly opened petri and place on the top of a 60° oven. (Contributed by Henry C. Aldrich, University of Florida)

## Safety, hazards and precautions

Many of the reagents utilized to prepare insect pathogens for study are potentially hazardous. Material safety data sheets are provided with all reagents and should be made available to all individuals who handle the material. Preventative protocols should always be followed when appropriate, such as working under a fume hood or wearing gloves and lab coats. Remediation procedures should be in place in the event of an accidental spill or exposure to toxic substances. Safety training should be a part of every laboratory's general operating procedures.

# Index

Page references followed by "f" indicate figure, by "b" indicate box, and by "t" indicate table.

# Color Plates

**Figure 1.1** Equipment and materials used for collection and initial processing of aquatic arthropods. A. Paraphernalia for collection and transport of specimens in habitat water; large debris are removed from samples at the collection site. B. Sieving field-collected water and specimens for additional removal of debris and separation of different sized arthropods. C. A one sieve subsample to facilitate closer examination of specimens. D. Funnel for separation of small specimens. E. Observation of mosquito larvae in a spot plate using a dissecting microscope. (Photographs provided by James Becnel.)

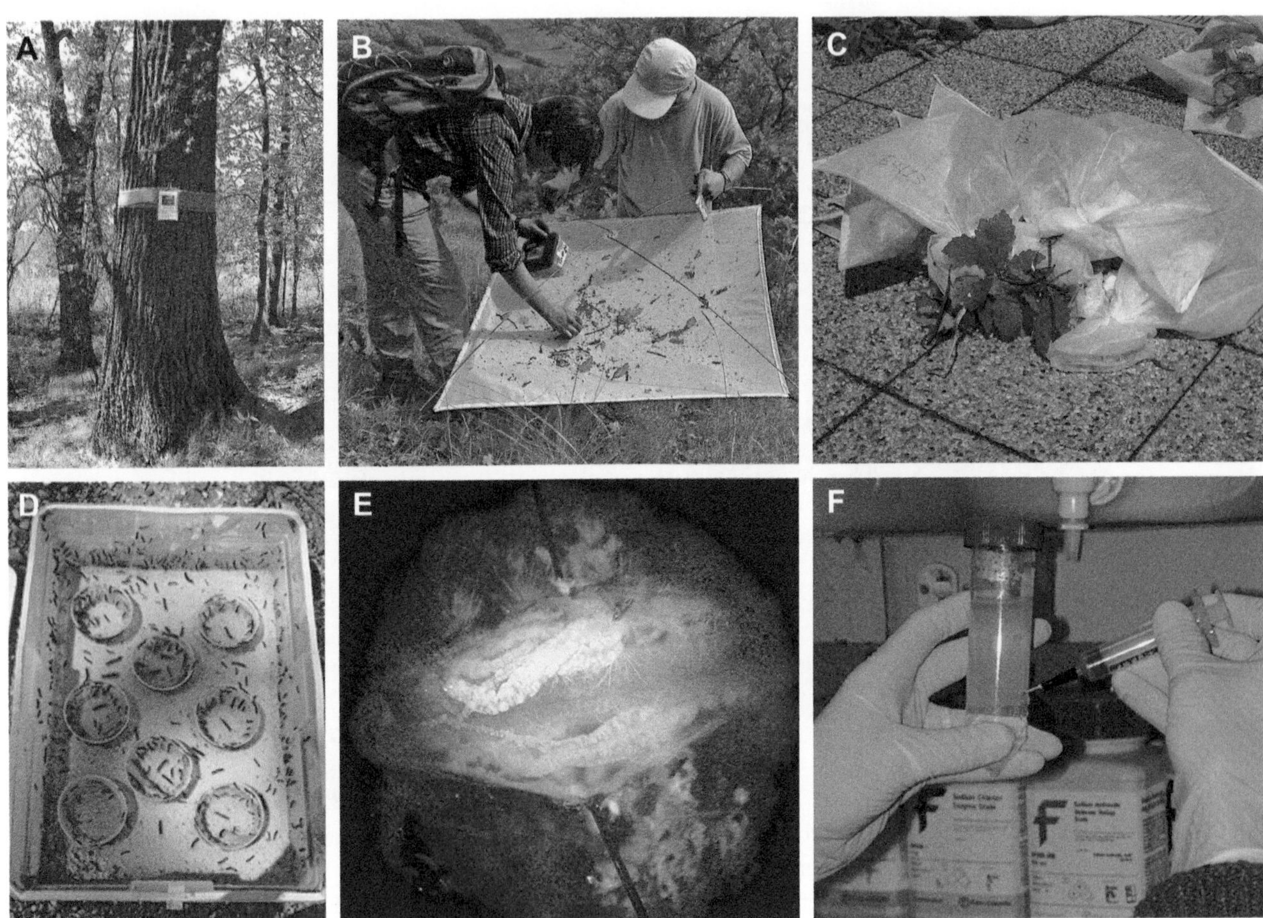


**Figure 1.2** Research methods for forest invertebrate pathology; A. Trees are banded with burlap to collect foliage feeding Lepidoptera and other tree dwelling arthropods; B. Beat sheets are useful for collecting foliage-feeding insects; C. Tree branches are infested with infected and non-infected foliage-feeding hosts and enclosed in fabric bags to study disease transmission; D. Microsporidia-infected gypsy moth larva reared for release in a classical biological control program; E. Silk glands of a lepidopteran larvae infected with a microsporidium; F. Isolation of pathogen infective units in a Ludox® HS-40 gradient. (Photographs by LFS and Gernot Hoch.)

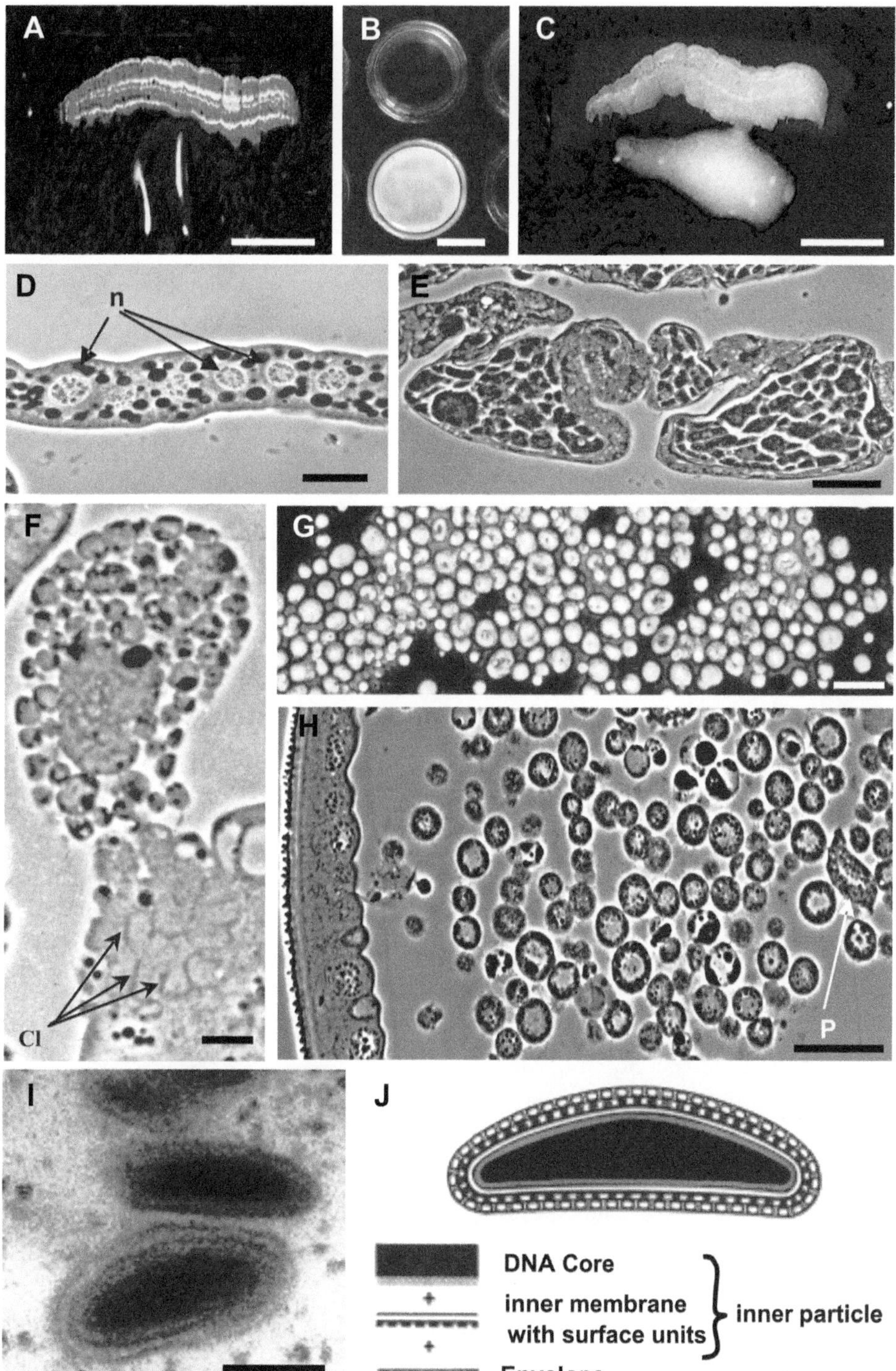


**Figure 2.1** Typical pathology associated with *Ascovirus* infections observed in the cabbage looper, *Trichoplusia ni*. A. Healthy larva. The clear pool of liquid below the larva is normal hemolymph. B. Comparison of the appearance of healthy hemolymph (upper well) with hemolymph from an ascovirus-infected larva (lower well) at 7 days post-infection. C. *Ascovirus*-infected *T. ni* larvae showing typical milky-white appearance of infected hemolymph. White opaque hemolymph in lepidopteran larvae is generally diagnostic of *Ascovirus* infection. The dense opacity is due to the accumulation of virion-containing vesicles in the hemolymph (scale bars in A–C = 1 cm). D and E, respectively. Sections through healthy and infected fat body tissue of a larva. Note the extensive hypertrophy of the *Ascovirus*-infected cells in E (n, nuclei; scale bars in D and E = 10 μm). F. Section through a lobe of fat body in which a greatly hypertrophied cell is cleaving into viral vesicles (arrows), (Cl = cleavage planes throughout a cell, scale bar = 5 μm). G. and H. respectively, wet mount of *Ascovirus*-infected hemolymph shown in (B) and (C), and the appearance of blood and spherical viral vesicles as observed in plastic sections, (P = plasmatocyte, a type of insect blood cell, scale bar of (G) = 5 μm, of (H) = 20 μm). I. Transmission electron micrograph of two ascovirus virions of *T. ni* (scale bar = 100 nm). J. Schematic interpretation of ascovirus virion structure based on the appearance of virions in ultrathin sections and negatively stained preparations. (Courtesy of B. A. Federici, University of California, Riverside, USA.)

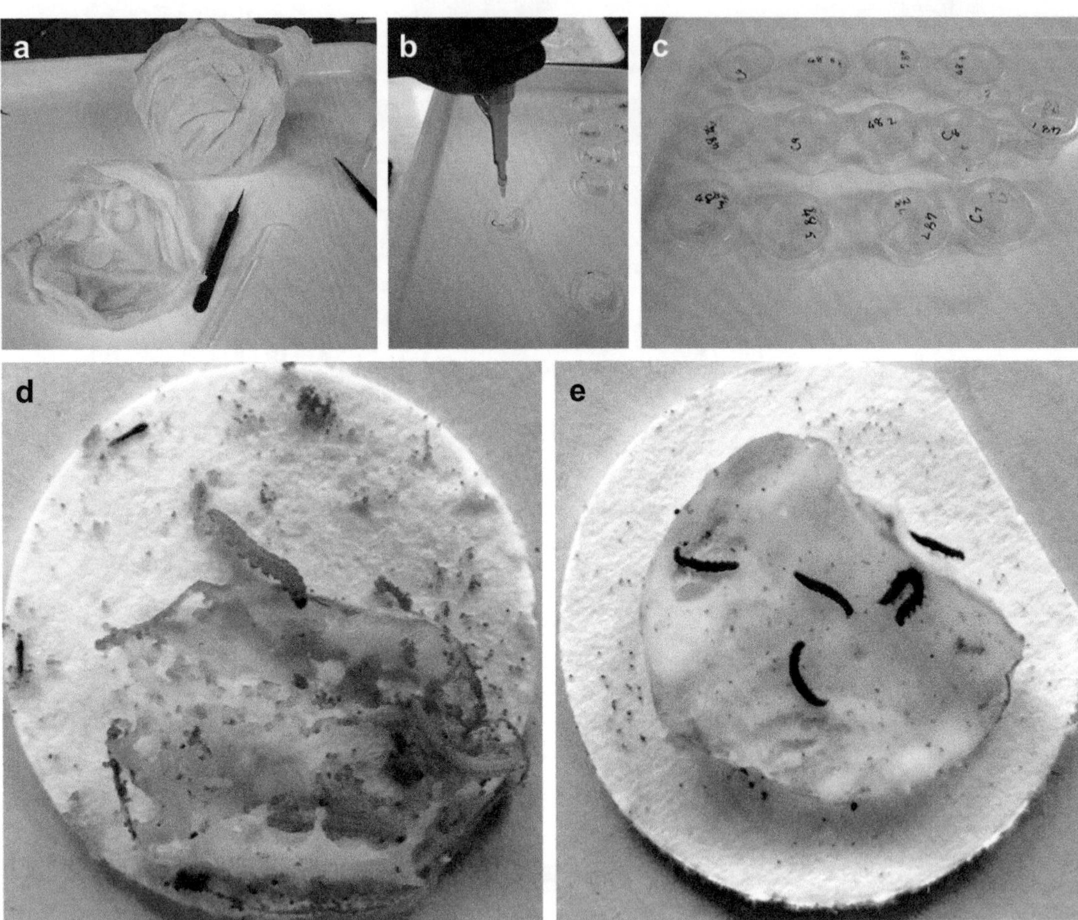


**Figure 4.2** Bioassay of *Yersinia entomophaga* against diamondback moth larvae. a. Discs are cut from cabbage leaves. b. Leaf discs are inoculated with aliquots of known concentration of *Y. entomophaga* toxin or live cells. c. Cups containing six larvae are incubated for 4 days at 22°C. d. Untreated larvae remain healthy and feeding 4 days after treatment. e. Dead larvae 4 days after treatment. (Courtesy of Mark Hurst and Sandra Jones, AgResearch, New Zealand.)

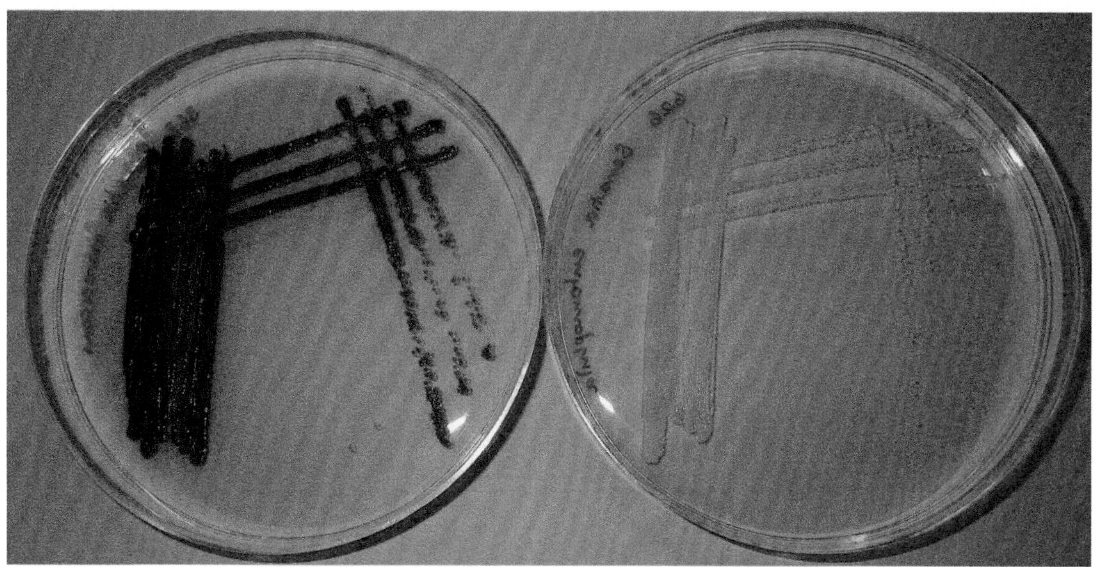

**Figure 5.4** Streak plate cultures of *Serratia marcescens* with red pigmentation (left) and *Serratia entomophila* (right) in 24-h culture on Nutrient Agar.

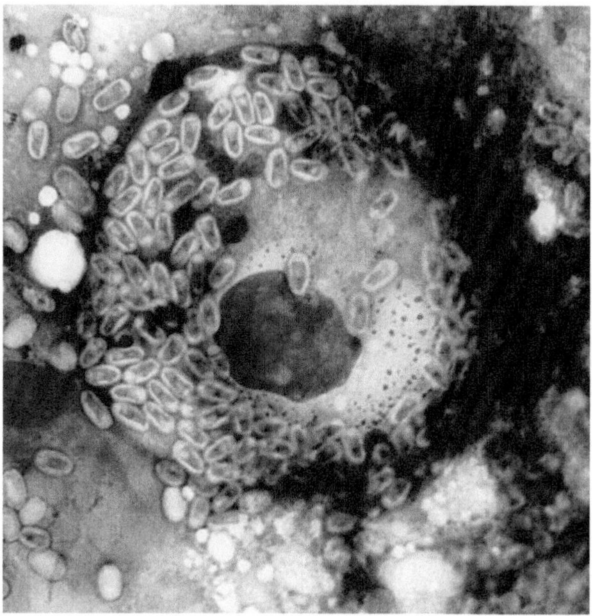

**Figure 11.6** Giemsa-stained environmental spores of a microsporidium infecting a host midgut cell. Note the blue 'thumbprint' or smear that is often diagnostic for mature microsporidian spores. (Photo by LFS.)

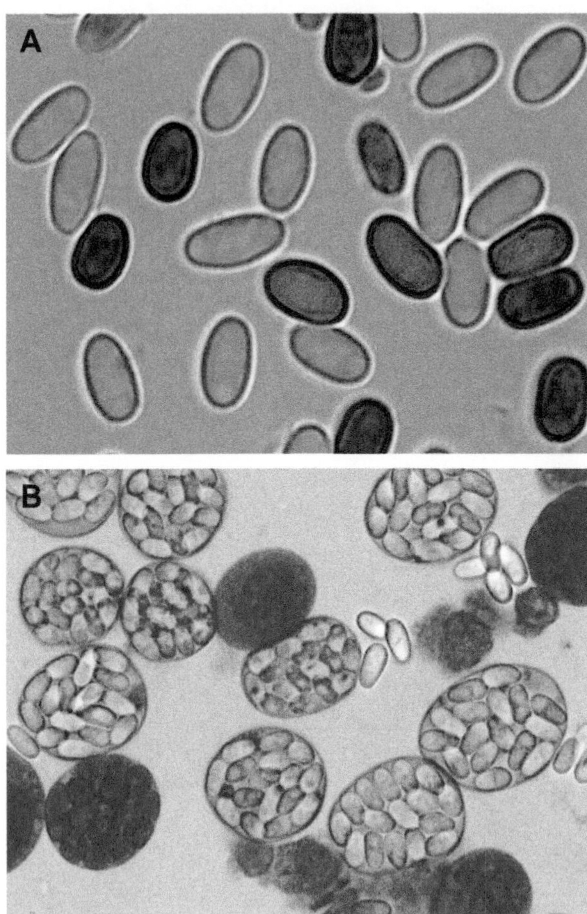


**Figure 11.9** Special stains for microsporidia. A. The spore wall of *Vairimorpha octospores* is thicker than that of the diplokaryotic spores, allowing octospores to take up 1% eosine stain. B. The sporophorous vesicle of *Vavraia culicis* is stained with 1% Congo Red; the spores are not. (Photos by JV.)

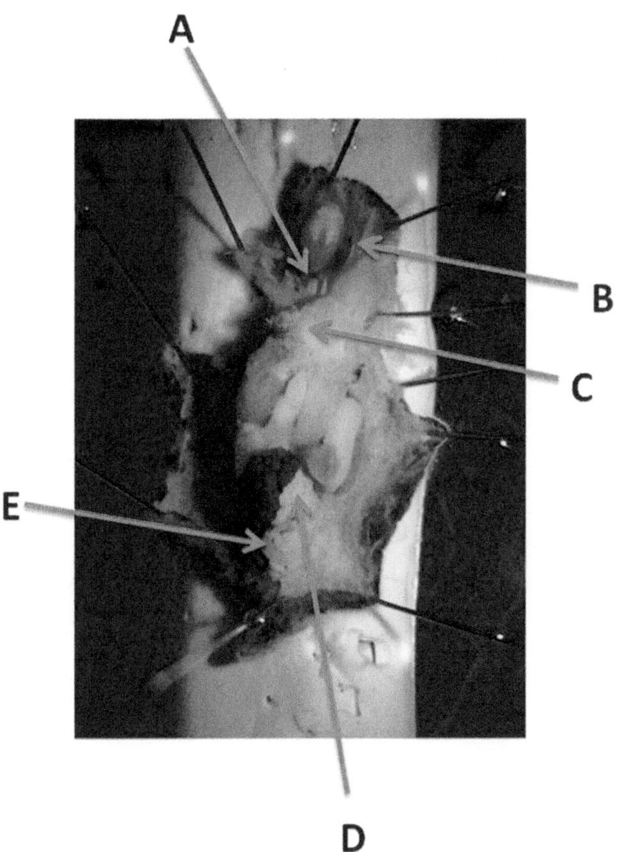


**Figure 13.3** Dissected *Deroceras reticulatum* showing sites of infection for common ciliate and nematode parasites. A. Esophagus, directly behind the buccal mass is where most *Angiostoma* nematodes are found. B. Proximal reproductive structures and salivary glands where *Agfa* nematodes are found. C. Kidney where ciliate protozoan *Tetrahymena rostrata* is found. D. Digestive gland (hepatopancreas) where *Angiostoma glanicola* nematodes and *Tetrahymena limacis* ciliates are found. E. lower reaches of intestinal tract where Alaninematidae nematodes are found.

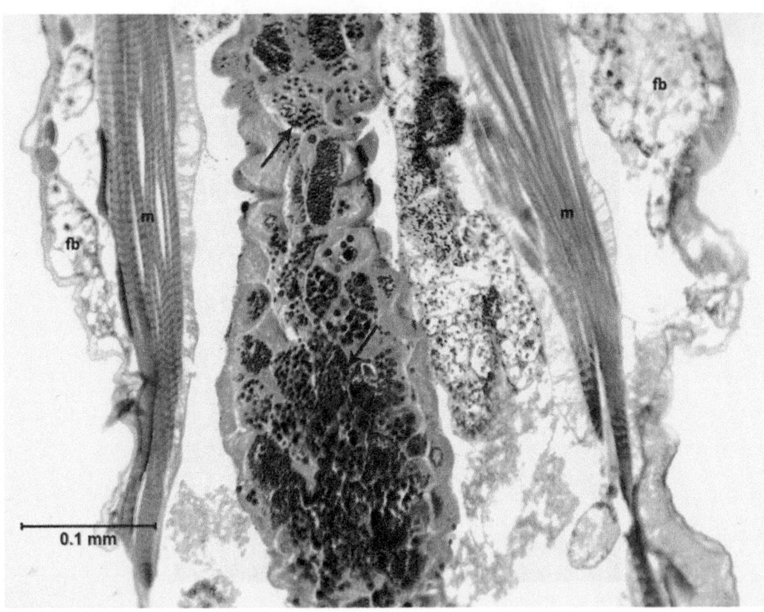


**Figure 15.1** Heidenhain's haematoxylin eosin stained longitudinal section of the larval midgut region of a midge (*Chironomus* sp.) infected with a cypovirus. Note the dense inclusion bodies of the cypovirus (arrows) in the cytoplasm of midgut cells. Muscle (m); fat body (fb).

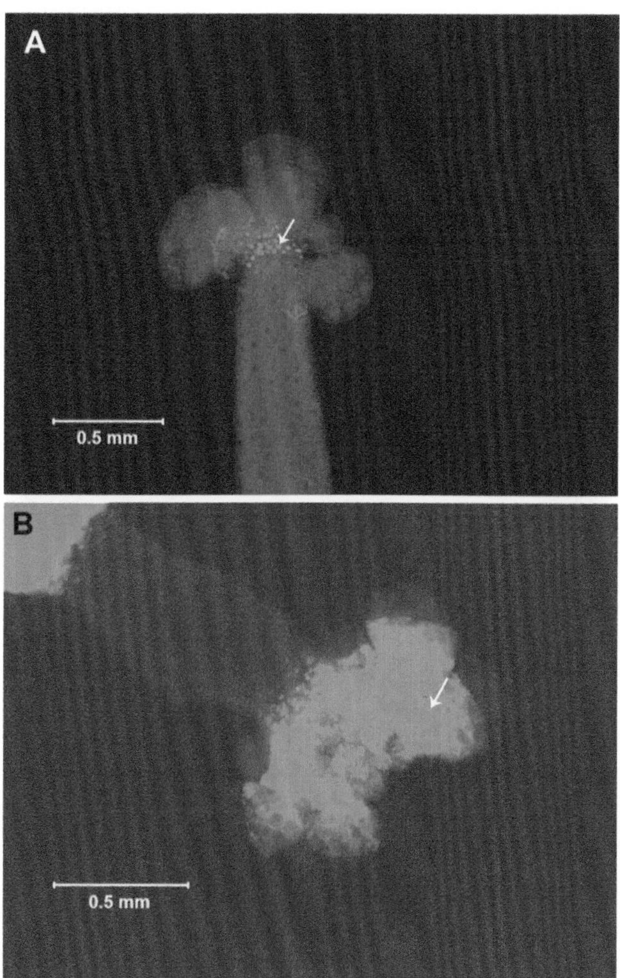


**Figure 15.5** A. Localization of *Culex nigripalpus* baculovirus (CuniNPV) replication 2 h post-infection in gastric caecum of *Culex quinquefasciatus* using a fluorescent labeled probe cocktail against early expressed genes cun16, cun65, cun86, and cun103 (arrow). B. Localization of CuniNPV replication 48 h post-infection in gastric caecum of *Culex quinquefasciatus* using a fluorescent labeled probe cocktail against late expressed genes cun24, cun75, and cun85 (arrow). (Photo credit to Bo Liu and Lei Zhou.)

Printed and bound by CPI Group (UK) Ltd, Croydon, CR0 4YY
08/05/2025
01864917-0001